BARRON'S

AP CALCULUS

14TH EDITION

David Bock, M.S.
Ithaca High School (retired)
Ithaca, New York

Dennis Donovan, M.S.
Xaverian Brothers High School
Westwood, Massachusetts

Shirley O. Hockett, M.A.
Formerly, Professor of Mathematics
Ithaca College
Ithaca, New York

About the Authors

David Bock taught AP Calculus during his 35 years at Ithaca High School, and served for several years as an Exam Reader for the College Board. He also taught mathematics at Tompkins-Cortland Community College, Ithaca College, and Cornell University. A recipient of several local, state, and national teaching awards, Dave has coauthored five textbooks, and now leads workshops for AP teachers.

Dennis Donovan has been a math teacher for 22 years, teaching AP Calculus for the past 19 years (8 years AB, 11 years BC). He has served as an AP Calculus Reader and as one of nine national Question Leaders for the AP Calculus exam. Dennis leads professional development workshops for math teachers as a consultant for the College Board and a T^3 Regional Instructor for Texas Instruments.

IN MEMORIAM
Shirley O. Hockett
(1920-2013)

Shirley Hockett taught mathematics for 45 years, first at Cornell University and later at Ithaca College, where she was named Professor Emerita in 1991. An outstanding teacher, she won numerous awards and authored six mathematics textbooks. Shirley's experiences as an Exam Reader and Table Leader for AP Calculus led her to write the first ever AP Calculus review book, published in 1971. Her knowledge of calculus, attention to detail, pedagogical creativity, and dedication to students continue to shine throughout this book. On behalf of the thousands of AP Calculus students who have benefited from her tireless efforts, we gratefully dedicate this review book to Shirley.

All inquiries should be addressed to:
Barron's Educational Series, Inc.
250 Wireless Boulevard
Hauppauge, New York 11788
www.barronseduc.com

ISBN: 978-1-4380-0859-2 (book)
ISBN: 978-1-4380-7674-4 (book/CD-ROM package)

International Standard Serial No. (book): 1940-3089
International Standard Serial No. (package): 1940-3119

PRINTED IN THE UNITED STATES OF AMERICA
9 8 7 6 5 4 3 2 1

Contents

As you review the content in this book to work toward earning that **5** on your AP CALCULUS AB exam, here are five things that you **MUST** know above everything else:

1

Learn the basic facts:

- derivatives (p. 113) and antiderivatives (p. 211) of common functions;
- the product (p. 113), quotient (p. 113), and chain rules (p. 114) for finding derivatives;
- the midpoint, left and right rectangle, and trapezoid approximations for estimating definite integrals (p. 248);
- finding antiderivatives by substitution (p. 213);
- the important theorems: Rolle's theorem (p. 129), the Mean Value theorem (p. 128), and especially the Fundamental Theorem of Calculus (p. 241).

(Barron's *AP Calculus Flash Cards* are a great way to study these!)

2

Understand that a derivative is an instantaneous rate of change, and be able to apply that concept to:

- using L'Hôpital's rule to find limits of indeterminate forms $\left(\text{only } \frac{0}{0} \text{ and } \frac{\infty}{\infty}\right)$ (p. 130);
- find equations of tangent lines (p. 259);
- determine where a function is increasing/decreasing (p. 160), concave up/down (p. 161), or has maxima, minima, or points of inflection (pp. 161, 167);
- analyze the speed, velocity, and acceleration of an object in motion (p. 177);
- solve related rates problems (p. 185), using implicit differentiation (p. 125) when necessary.

3

Understand that integrals represent accumulation functions based on antiderivatives, and be able to apply those concepts to:

- the average value of a function (p. 261);
- area (p. 281) and volume (p. 288);
- position of object in motion and distance traveled (p. 333);
- total amount when given the rate of accumulation (p. 338);
- differential equations, including solutions and slope fields (p. 352).

4

Be able to apply any of the above calculus concepts to functions defined algebraically, graphically, or in tables.

5

Be able to maximize your score on the exam by:

- answering *all* the multiple-choice questions;
- knowing how and when to use your calculator, and when not to;
- understanding what work you need to show;
- knowing how to explain, interpret, and justify answers when a question requires that. (The free-response solutions in this book model such answers.)

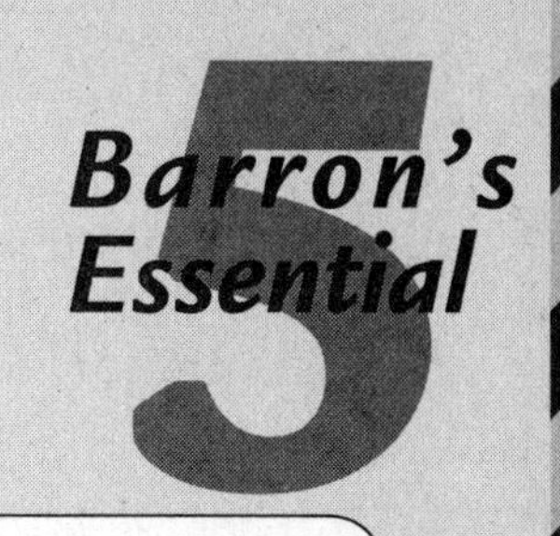

As you review the content in this book to work toward earning that **5** on your AP CALCULUS BC exam, here are five things that you **MUST** know above everything else:

1. **Master the Essential 5 listed for the AB Calculus Exam.** These form the core for questions that determine your AB subscore, and provide the essential knowledge base you'll need for questions related to the additional BC topics.

2. **Understand how to extend AB Calculus concepts to more advanced situations, including:**
   - using limits to analyze improper integrals (p. 297);
   - solving logistic differential equations (p. 371) and estimating solutions using Euler's method (p. 357);
   - finding antiderivatives using integration by parts (p. 220) or partial fractions (p. 218);
   - finding arc lengths (p. 295).

3. **Be able to apply calculus concepts to parametrically defined functions (pp. 81, 123, 335) and polar functions (pp. 78, 187, 286).**

4. **Know how to analyze the position, velocity, speed, acceleration, and distance traveled for an object in motion in two dimensions by applying calculus concepts to vectors (p. 179).**

5. **Understand infinite series.** You must be able to:
   - determine whether a series converges or diverges (p. 391);
   - use Taylor's theorem to represent functions as power series (p. 406);
   - determine the interval of convergence for a power series (p. 402);
   - find bounds on the error for estimates based on series (pp. 401, 413).

Introduction

This book is intended for students who are preparing to take either of the two Advanced Placement Examinations in Mathematics offered by the College Entrance Examination Board, and for their teachers. It is based on the May 2014 course description published by the College Board, and covers the topics listed there for both Calculus AB and Calculus BC.

Candidates who are planning to take the CLEP Examination on Calculus with Elementary Functions are referred to the section of this Introduction on that examination on page 9.

THE COURSES

Calculus AB and BC are both full-year courses in the calculus of functions of a single variable. Both courses emphasize:

(1) student understanding of concepts and applications of calculus over manipulation and memorization;

(2) developing the student's ability to express functions, concepts, problems, and conclusions analytically, graphically, numerically, and verbally, and to understand how these are related; and

(3) using a graphing calculator as a tool for mathematical investigations and for problem-solving.

Both courses are intended for those students who have already studied college-preparatory mathematics: algebra, geometry, trigonometry, analytic geometry, and elementary functions (linear, polynomial, rational, exponential, logarithmic, trigonometric, inverse trigonometric, and piecewise). The AB topical course outline that follows can be covered in a full high-school academic year even if some time is allotted to studying elementary functions. The BC course assumes that students already have a thorough knowledge of all the topics noted above.

TOPICS THAT MAY BE TESTED ON THE CALCULUS AB EXAM

1. Functions and Graphs

Rational, trigonometric, inverse trigonometric, exponential, and logarithmic functions.

2. Limits and Continuity

Intuitive definitions; one-sided limits; functions becoming infinite; asymptotes and graphs; indeterminate limits of the form $\frac{0}{0}$ or $\frac{\infty}{\infty}$ using algebra; $\lim\limits_{\theta \to 0} \frac{\sin \theta}{\theta}$; estimating limits using tables or graphs.

Definition of continuity (in terms of limits); kinds of discontinuities; theorems about continuous functions; Extreme Value and Intermediate Value Theorems.

3. Differentiation

Definition of derivative as the limit of a difference quotient and as instantaneous rate of change; derivatives of power, exponential, logarithmic, trig and inverse trig functions; product, quotient, and chain rules; differentiability and continuity; estimating a derivative numerically and graphically; implicit differentiation; derivative of the inverse of a function; the Mean Value Theorem; recognizing a given limit as a derivative; L'Hôpital's Rule.

4. Applications of Derivatives

Rates of change; slope; critical points; average velocity; tangent line to a curve at a point and local linear approximation; increasing and decreasing functions; using the first and second derivatives for the following: local (relative) max or min, concavity, inflection points, curve sketching, global (absolute) max or min and optimization problems; relating a function and its derivatives graphically; motion along a line; related rates; differential equations and slope fields.

5. The Definite Integral

Definite integral as the limit of a Riemann sum; area; definition of definite integral; properties of the definite integral; use of Riemann sums (left, right and midpoint evaluations) and trapezoidal sums to approximate a definite integral; estimating definite integrals from tables and graphs; comparing approximating sums; average value of a function; Fundamental Theorem of Calculus; graphing a function from its derivative; accumulated change as integral of rate of change.

6. Integration

Antiderivatives and basic formulas; antiderivatives by substitution; applications of antiderivatives; separable differential equations; motion problems.

7. Applications of Integration to Geometry

Area of a region, including between two curves; volume of a solid of known cross section, including a solid of revolution.

8. Further Applications of Integration and Riemann Sums

Velocity and distance problems involving motion along a line; other applications involving the use of integrals of rates as net change or the use of integrals as accumulation functions; average value of a function over an interval.

9. Differential Equations

Basic definitions; geometric interpretations using slope fields; solving first-order separable differential equations analytically; exponential growth and decay.

TOPICS THAT MAY BE TESTED ON THE CALCULUS BC EXAM

BC ONLY

Any of the topics listed above for the Calculus AB exam may be tested on the BC exam. The following additional topics are restricted to the BC exam.

1. Functions and Graphs

Parametrically defined functions; polar functions; vector functions.

BC ONLY

2. Limits and Continuity

No additional topics.

3. Differentiation

Derivatives of polar, vector, and parametrically defined functions; indeterminate forms.

4. Applications of Derivatives

Tangents to parametrically defined curves; slopes of polar curves; analysis of curves defined parametrically or in polar or vector form.

5. The Definite Integral

Integrals involving parametrically defined functions.

6. Integration

By parts; by partial fractions (involving nonrepeating linear factors only); improper integrals.

7. Applications of Integration to Geometry

Area of a region bounded by parametrically defined or polar curves; arc length.

8. Further Applications of Integration and Riemann Sums

Velocity and distance problems involving motion along a planar curve; velocity and acceleration vectors.

9. Differential Equations

Euler's method; applications of differential equations, including logistic growth.

10. Sequences and Series

Definition of series as a sequence of partial sums and of its convergence as the limit of that sequence; harmonic, geometric, and p-series; integral, ratio, comparison and limit comparison tests for convergence; alternating series and error bound; power series, including interval and radius of convergence; absolute and conditional convergence; Taylor polynomials and graphs; finding a power series for a function; Maclaurin and Taylor series; Lagrange error bound for Taylor polynomials; computations using series.

THE EXAMINATIONS

The Calculus AB and BC Examinations and the course descriptions are prepared by committees of teachers from colleges or universities and from secondary schools. The examinations are intended to determine the extent to which a student has mastered the subject matter of the course.

Each examination is 3 hours and 15 minutes long, as follows:

Section I has two parts. Part A has 30 multiple-choice questions for which 60 minutes are allowed. The use of calculators is *not* permitted in Part A.

Part B has 15 multiple-choice questions for which 45 minutes are allowed. Some of the questions in Part B require the use of a graphing calculator.

Section II, the free-response section, has a total of six questions in two parts:

Part A has two questions, of which some parts *require* the use of a graphing calculator. After 30 minutes, however, *you will no longer be permitted to use a calculator.* If you finish Part A early, you will not be permitted to start work on Part B.

Part B has four questions and you are allotted an additional 60 minutes, but *you are not allowed to use a calculator.* You may work further on the Part A questions (without your calculator).

The section that follows gives important information on the use (and misuse!) of the graphing calculator.

THE GRAPHING CALCULATOR: USING YOUR GRAPHING CALCULATOR ON THE AP EXAM

The Four Calculator Procedures

Each student is expected to bring a graphing calculator to the AP Exam. Different models of calculators vary in their features and capabilities; however, there are four procedures you must be able to perform on your calculator:

C1. Produce the graph of a function within an arbitrary viewing window.
C2. Solve an equation numerically.
C3. Compute the derivative of a function numerically.
C4. Compute definite integrals numerically.

Guidelines for Calculator Use

1. On multiple-choice questions in Section I, Part B, *you may use any feature or program on your calculator.* Warning: Don't rely on it too much! Only a few of these questions require the calculator, and in some cases using it may be too time-consuming or otherwise disadvantageous.

2. On the free-response questions of Section II Part A:

(a) You may use the calculator to perform any of the four listed procedures. When you do, you need only write the equation, derivative, or definite integral (called the "setup") that will produce the solution, then write the calculator result to the required degree of accuracy (three places after the decimal point unless otherwise specified). Note especially that a setup must be presented in standard algebraic or calculus notation, not in calculator syntax. For example, you *must* include in your work the setup $\int_0^{\pi} \cos(t)\, dt$ even if you use your calculator to evaluate the integral.

(b) For a solution for which you use a calculator capability other than the four listed above, you must write down the mathematical steps that yield the answer. A correct answer alone will not earn full credit and will likely earn no credit.

(c) You must provide *mathematical reasoning* to support your answer. Calculator results alone will not be sufficient.

The Procedures Explained

Here is more detailed guidance for the four allowed procedures.

C1. "Produce the graph of a function within an arbitrary viewing window." More than likely, you will not have to produce a graph on the exam that will be graded. However, you must be able to graph a wide variety of functions, both simple and complex, and be able to analyze those graphs. Skills you need include, but are not limited to, typing complex functions correctly into your calculator including correct notation, which will ensure that the graph on the screen is what the question writer intended you to see, and finding a window that accurately represents the graph and its features. Note, on rare occasions you may wish to draw a graph in your exam booklet to justify an answer in the free-response section; such a graph must be clearly labeled as to what is being graphed, and there should be an accompanying sentence or two explaining why the graph you produced justifies the answer.

C2. "Solve an equation numerically" is equivalent to "Find the zeros of a function" or "Find the point of intersection of two curves." Remember: you must first show your setup—write the equation out algebraically; then it is sufficient just to write down the calculator solution.

C3. "Compute the derivative of a function numerically." When you seek the value of the derivative of a function at a specific point, you may use your calculator. First, indicate what you are finding—for example, $f'(6)$—then write the numerical answer obtained from your calculator. Note that if you need to find the derivative of the function, rather than its value at a particular point, you must write the derivative symbolically. Note: some calculators are able to perform symbolic operations.

C4. "Compute definite integrals numerically." If, for example, you need to find the area under a curve, you must first show your setup. Write the complete integral, including the integrand in terms of a single variable, with the limits of integration. You may then simply write the calculator answer; you need not compute an antiderivative.

Accuracy

Calculator answers must be correct to three decimal places. To achieve this required accuracy, never type in decimal numbers unless they came from the original question. Do *not* round off numbers at intermediate steps, as this is likely to produce error accumulations resulting in loss of credit. If necessary, store intermediate answers in the calculator's memory. Do *not* copy them down on paper; storing is faster and avoids transcription errors. Round off, or truncate, only after your calculator produces the final answer.

Sample Solutions of Free-Response Questions

The following set of examples illustrates proper use of your calculator on the examination. In all of these examples, the function is

$$f(x) = \frac{10x}{x^2 + 4} \quad \text{for} \quad 0 \leqslant x \leqslant 4.$$

1. Before getting to the questions asked in the problem, it is always a good idea to have a picture of the graph that is being used in the problem. Graph f in $[0,4] \times [0,3]$.

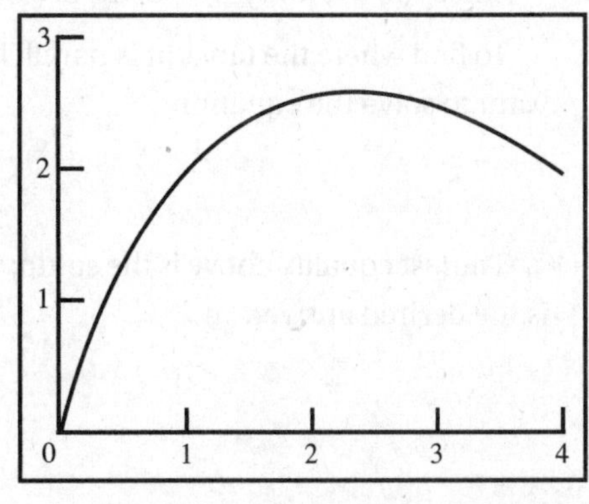

Viewing window $[0,4] \times [0,3]$

2. Write the tangent line for $f(x)$ at $x = 1$.

Note that $f(1) = 2$. Then, using your calculator, evaluate the derivative:

$$f'(1) = 1.2$$

Then write the tangent-line (or local linear) approximation

$$f(x) \simeq f(1) + f'(1)(x - 1)$$
$$\simeq 2 + 1.2(x - 1) = 1.2x + 0.8$$

You need not simplify, as we have, after the last equals sign just above.

3. Find the coordinates of any maxima of f. Justify your answer.

Since finding a maximum is not one of the four allowed procedures, you must use calculus and show your work, writing the derivative algebraically and setting it equal to zero to find any critical numbers:

$$f'(x) = \frac{(x^2 + 4)10 - 10x(2x)}{(x^2 + 4)^2} = \frac{40 - 10x^2}{(x^2 + 4)^2}$$
$$= \frac{10(2 - x)(2 + x)}{(x^2 + 4)^2}$$

Then $f'(x) = 0$ at $x = 2$ and at $x = -2$; but -2 is not in the specified domain.

We analyze the signs of f' (which is easier here than it would be to use the second-derivative test) to assure that $x = 2$ does yield a maximum for f. (Note that the signs analysis alone is *not* sufficient justification.)

		incr		decr	
f	0		2		4
f'		+		−	

Since f' is positive to the left of $x = 2$ and negative to the right of $x = 2$, f does have a maximum at

$$\left(2, \frac{10(2)}{2^2 + 4}\right) = \left(2, \frac{5}{2}\right)$$

—but you may leave $f(2)$ in its unsimplified form, without evaluating to $\frac{5}{2}$.

You may use your calculator's maximum-finder to *verify* the result you obtain analytically, but that would not suffice as a solution or justification.

4. Find the x-coordinate of the point where the line tangent to the curve $y = f(x)$ is parallel to the secant on the interval [0,4].

Since $f(0) = 0$ and $f(4) = 2$, the secant passes through (0,0) and (4,2) and has slope $m = \frac{1}{2}$.

To find where the tangent is parallel to the secant, we find $f'(x)$ as in Example 3. We then want to solve the equation

$$f'(x) = \frac{1}{2} \Rightarrow \frac{40 - 10x^2}{(x^2 + 4)^2} = \frac{1}{2}$$

The last equality above is the setup; we use the calculator to solve the equation: $x = 1.458$ is the desired answer.

5. Estimate the area under the curve $y = f(x)$ using a Trapezoidal Sum with four equal subintervals:

$$\int_0^4 f(x)\,dx \approx \left(\frac{f(0)+f(1)}{2}\right)(1) + \left(\frac{f(1)+f(2)}{2}\right)(1) + \left(\frac{f(2)+f(3)}{2}\right)(1) + \left(\frac{f(3)+f(4)}{2}\right)(1)$$

$$= \left(\frac{0+2}{2}\right)(1) + \left(\frac{2+(5/2)}{2}\right)(1) + \left(\frac{(5/2)+(30/13)}{2}\right)(1) + \left(\frac{(30/13)+2}{2}\right)(1)$$

You may leave the answer in this form or simplify it to 7.808.

6. Find the volume of the solid generated when the curve $y = f(x)$ on [0,4] is rotated about the x-axis.

Using disks, we have

$$\Delta V = \pi R^2 H = \pi y^2\,\Delta x,$$

$$V = \pi \int_0^4 y^2 dx$$

Note that the equation above is not yet the setup: the definite integral must be in terms of x alone:

$$V = \pi \int_0^4 \left(\frac{10x}{x^2+4}\right)^2 dx$$

Now we have shown the setup. Using the calculator we can evaluate V:

$$V = 55.539$$

A Note About Solutions in This Book

Students should be aware that in this book we sometimes do *not* observe the restrictions cited above on the use of the calculator. In providing explanations for solutions to illustrative examples or to exercises we often exploit the capabilities of the calculator to the fullest. Indeed, students are encouraged to do just that on any question of Section I, Part B, of the AP examination for which they use a calculator. However, to avoid losing credit, you must carefully observe the restrictions imposed on when and how the calculator may be used in answering questions in Section II of the examination.

Additional Notes and Reminders

- **Syntax.** Learn the proper syntax for your calculator: the correct way to enter operations, functions, and other commands. Parentheses, commas, variables, or parameters that are missing or entered in the wrong order can produce error messages, waste time, or (worst of all) yield wrong answers.

- **Radians.** Keep your calculator set in radian mode. Almost all questions about angles and trigonometric functions use radians. If you ever need to change to degrees for a specific calculation, return the calculator to radian mode as soon as that calculation is complete.

- **Trigonometric Functions.** Many calculators do not have keys for the secant, cosecant, or cotangent function. To obtain these functions, use their reciprocals.

For example, $\sec\left(\frac{\pi}{8}\right) = 1/\cos\left(\frac{\pi}{8}\right)$.

Evaluate inverse functions such as arcsin, arccos, and arctan on your calculator. Those function keys are usually denoted as $\sin^{-1}$, $\cos^{-1}$, and $\tan^{-1}$.

Don't confuse reciprocal functions with inverse functions. For example:

$$\cos^{-1}\left(\frac{1}{2}\right) = \arccos\left(\frac{1}{2}\right) = \frac{\pi}{3};$$

$$(\cos 2)^{-1} = \frac{1}{\cos 2} = \sec 2 \approx -2.403;$$

$$\cos 2^{-1} = \cos\left(\frac{1}{2}\right) \approx 0.878;$$

$$\cos^{-1}(2) = \arccos 2, \text{ which does not exist.}$$

- **Numerical Derivatives.** You may be misled by your calculator if you ask for the derivative of a function at a point where the function is not differentiable, because the calculator evaluates numerical derivatives using the difference quotient (or the symmetric difference quotient). For example, if $f(x) = |x|$, then $f'(0)$ does not exist. Yet the calculator may find the value of the derivative as

$$\frac{f(x + 0.001) - f(0)}{0.001} = 1 \quad \text{or} \quad \frac{f(x + 0.001) - f(x - 0.001)}{0.002} = 0.$$

Remember: always be sure f is differentiable at a before asking the calculator to evaluate $f'(a)$.

- **Improper Integrals.** Most calculators can compute only definite integrals. Avoid using yours to obtain an improper integral, such as

$$\int_0^{\pi} \frac{1}{x^2}\,dx \quad \text{or} \quad \int_0^2 \frac{dx}{(x-1)^{2/3}}$$

- **Final Answers to Section II Questions.** Although we usually express a final answer in this book in simplest form (often evaluating it on the calculator), this is hardly ever necessary on Section II questions of the AP Examination. According to the directions printed on the exam, "unless otherwise specified" (1) you need not simplify algebraic or numerical answers; (2) answers involving decimals should be correct to three places after the decimal point. However, be aware that if you try to simplify, you must do so correctly or you will lose credit.

- **Use Your Calculator Wisely.** Bear in mind that you will not be allowed to use your calculator at all on Part A of Section I. In Part B of Section I and Part A of Section II *only a few questions* will require one. As repeated often in this section, the questions that require a calculator *will not be identified.* You will have to be sensitive not only to when it is necessary to use the calculator but also to when it is efficient to do so.

The calculator is a marvelous tool, capable of illustrating complicated concepts with detailed pictures and of performing tasks that would otherwise be excessively time-consuming—or even impossible. But the completion of calculations and the displaying of graphs on the calculator can be slow. Sometimes it is faster to find an answer using arithmetic, algebra, and analysis without recourse to the calculator. Before you start pushing buttons, take a few seconds to decide on the best way to attack a problem.

GRADING THE EXAMINATIONS

Each completed AP examination paper receives a grade according to the following five-point scale:

5. Extremely well qualified
4. Well qualified

3. Qualified
2. Possibly qualified
1. No recommendation

Many colleges and universities accept a grade of 3 or better for credit or advanced placement or both; some also consider a grade of 2, while others require a grade of 4. (Students may check AP credit policies at individual colleges' websites.) More than 59 percent of the candidates who took the 2016 Calculus AB Examination earned grades of 3, 4, or 5. Nearly 81 percent of the 2016 BC candidates earned 3 or better. More than 433,000 students altogether took the 2016 AP Calculus examinations.

The multiple-choice questions in Section I are scored by machine. Students should note that *the score will be the number of questions answered correctly*. Since no points can be earned if answers are left blank and there is no deduction for wrong answers, *students should answer every question*. For questions they cannot do, students should try to eliminate as many of the choices as possible and then pick the best remaining answer.

The problems in Section II are graded by college and high-school teachers called "readers." The answers in any one examination booklet are evaluated by different readers, and for each reader all scores given by preceding readers are concealed, as are the student's name and school. Readers are provided sample solutions for each problem, with detailed scoring scales and point distributions that allow partial credit for correct portions of a student's answer. Problems in Section II are all counted equally.

In the determination of the overall grade for each examination, the two sections are given equal weight. The total raw score is then converted into one of the five grades listed above. Students should not think of these raw scores as percents in the usual sense of testing and grading. A student who averages 6 out of 9 points on the Section II questions and performs similarly well on Section I's multiple-choice questions will typically earn a 5. Many colleges offer credit for a score of 3, historically awarded for earning over 40 of 108 possible points.

Students who take the BC examination are given not only a Calculus-BC grade but also a Calculus-AB subscore grade. The latter is based on the part of the BC examination dealing with topics in the AB syllabus.

In general, students will not be expected to answer all the questions correctly in either Section I or II.

Great care is taken by all involved in the scoring and reading of papers to make certain that they are graded consistently and fairly so that a student's overall AP grade reflects as accurately as possible his or her achievement in calculus.

THE CLEP CALCULUS EXAMINATION

Many colleges grant credit to students who perform acceptably on tests offered by the College Level Examination Program (CLEP). The CLEP calculus examination is one such test.

The College Board's *CLEP Official Study Guide: 16th Edition* provides descriptions of all CLEP examinations, test-taking tips, and suggestions on reference and supplementary materials. According to the *Guide*, the calculus examination covers topics usually taught in a one-semester college calculus course. It is assumed that students taking the exam will have studied college-preparatory mathematics (algebra, plane and solid geometry, analytic geometry, and trigonometry).

There are 45 multiple-choice questions on the CLEP calculus exam, for which 90 minutes are allowed. A calculator may *not* be used during the examination.

Approximately 60 percent of the questions are on limits and differential calculus and about 40 percent on integral calculus. The specific topics that may be tested on the CLEP calculus exam are essentially those on pages 1 and 2 under the heading "Topics That May Be Tested on the Calculus AB Exam." (Note that the *only* topics listed as applications of the definite integral for the CLEP calculus test are "average value of a function on an interval" and "area.")

Since any topic that may be tested on the CLEP calculus exam is included in this book on the AP Exam, a candidate who plans to take the CLEP exam will benefit from a review of the AB topics covered here. The multiple-choice questions in Part A of Chapter 11 and in Part A of Section I of each of the four AB Practice Examinations will provide good models for questions on the CLEP calculus test.

A complete description of the knowledge and skills required and of the specific topics that may be tested on the CLEP exam can be downloaded from the College Board's web site at *www.collegeboard.com/clep*.

THIS REVIEW BOOK

This book consists of the following parts:

Diagnostic tests for both AB and BC Calculus are practice AP exams. They are followed by solutions keyed to the corresponding topical review chapter.

Topical Review and Practice includes 10 chapters with notes on the main topics of the Calculus AB and BC syllabi and with numerous carefully worked-out examples. Each chapter concludes with a set of multiple-choice questions, usually divided into calculator and no-calculator sections, followed immediately by answers and solutions.

This review is followed by further practice: (1) Chapter 11, which includes a set of multiple-choice questions on miscellaneous topics and an answer key; (2) Chapter 12, a set of miscellaneous free-response problems that are intended to be similar to those in Section II of the AP examinations. They are followed by solutions.

The next part of the book, titled Practice Examinations: Sections I and II, has three AB and three BC practice exams that simulate the actual AP examinations. Each is followed by answers and explanations.

In this book, review material on topics covered only in Calculus BC is preceded by an asterisk (*), as are both multiple-choice questions and free-response-type problems that are likely to occur only on a BC Examination.

THE TEACHER WHO USES THIS BOOK WITH A CLASS may profitably do so in any of several ways. If the book is used throughout a year's course, the teacher can assign all or part of each set of multiple-choice questions and some miscellaneous exercises after the topic has been covered. These sets can also be used for review purposes shortly before examination time. The Practice Examinations will also be very helpful in reviewing toward the end of the year. Teachers may also assemble examinations by choosing appropriate problems from the sample Miscellaneous Practice Questions in Chapters 11 and 12.

STUDENTS WHO USE THIS BOOK INDEPENDENTLY will improve their performance by studying the illustrative examples carefully and trying to complete practice problems before referring to the solution keys.

Since many FIRST-YEAR MATHEMATICS COURSES IN COLLEGES follow syllabi much like that proposed by the College Board for high-school Advanced Placement courses, college students and teachers may also find the book useful.

FLASH CARDS

Being able to answer AP exam questions quickly and correctly depends in part on knowing many fundamental facts, such as

- common math formulas (e.g., area, volume, trig identities);
- definitions of key terms (e.g., continuous, differentiable, integrable);
- important theorems (e.g., Mean Value Theorem, Fundamental Theorem of Calculus); and
- derivatives and antiderivatives of common functions.

Barron's *AP Calculus Flash Cards* (ISBN 978-1-4380-7400-9) provide a great way to study these facts and more. These 400 cards will help you learn the most important information you'll need to know for the AP Calculus examination.

Diagnostic Tests

ANSWER SHEET
Diagnostic Test Calculus AB

Part A

1. Ⓐ Ⓑ Ⓒ Ⓓ
2. Ⓐ Ⓑ Ⓒ Ⓓ
3. Ⓐ Ⓑ Ⓒ Ⓓ
4. Ⓐ Ⓑ Ⓒ Ⓓ
5. Ⓐ Ⓑ Ⓒ Ⓓ
6. Ⓐ Ⓑ Ⓒ Ⓓ
7. Ⓐ Ⓑ Ⓒ Ⓓ
8. Ⓐ Ⓑ Ⓒ Ⓓ
9. Ⓐ Ⓑ Ⓒ Ⓓ
10. Ⓐ Ⓑ Ⓒ Ⓓ
11. Ⓐ Ⓑ Ⓒ Ⓓ
12. Ⓐ Ⓑ Ⓒ Ⓓ
13. Ⓐ Ⓑ Ⓒ Ⓓ
14. Ⓐ Ⓑ Ⓒ Ⓓ
15. Ⓐ Ⓑ Ⓒ Ⓓ
16. Ⓐ Ⓑ Ⓒ Ⓓ
17. Ⓐ Ⓑ Ⓒ Ⓓ
18. Ⓐ Ⓑ Ⓒ Ⓓ
19. Ⓐ Ⓑ Ⓒ Ⓓ
20. Ⓐ Ⓑ Ⓒ Ⓓ
21. Ⓐ Ⓑ Ⓒ Ⓓ
22. Ⓐ Ⓑ Ⓒ Ⓓ
23. Ⓐ Ⓑ Ⓒ Ⓓ
24. Ⓐ Ⓑ Ⓒ Ⓓ
25. Ⓐ Ⓑ Ⓒ Ⓓ
26. Ⓐ Ⓑ Ⓒ Ⓓ
27. Ⓐ Ⓑ Ⓒ Ⓓ
28. Ⓐ Ⓑ Ⓒ Ⓓ
29. Ⓐ Ⓑ Ⓒ Ⓓ
30. Ⓐ Ⓑ Ⓒ Ⓓ

Part B

31. Ⓐ Ⓑ Ⓒ Ⓓ
32. Ⓐ Ⓑ Ⓒ Ⓓ
33. Ⓐ Ⓑ Ⓒ Ⓓ
34. Ⓐ Ⓑ Ⓒ Ⓓ
35. Ⓐ Ⓑ Ⓒ Ⓓ
36. Ⓐ Ⓑ Ⓒ Ⓓ
37. Ⓐ Ⓑ Ⓒ Ⓓ
38. Ⓐ Ⓑ Ⓒ Ⓓ
39. Ⓐ Ⓑ Ⓒ Ⓓ
40. Ⓐ Ⓑ Ⓒ Ⓓ
41. Ⓐ Ⓑ Ⓒ Ⓓ
42. Ⓐ Ⓑ Ⓒ Ⓓ
43. Ⓐ Ⓑ Ⓒ Ⓓ
44. Ⓐ Ⓑ Ⓒ Ⓓ
45. Ⓐ Ⓑ Ⓒ Ⓓ

Diagnostic Test Calculus AB

SECTION I

Part A

TIME: 60 MINUTES

The use of calculators is not permitted for this part of the examination. There are 30 questions in Part A, for which 60 minutes are allowed.

Directions: Choose the best answer for each question.

1. $\lim\limits_{x\to\infty} \dfrac{3x^2 - 4}{2 - 7x - x^2}$ is

(A) -3 (B) 0 (C) 3 (D) ∞

2. $\lim\limits_{h\to 0} \dfrac{\cos\left(\frac{\pi}{2} + h\right) - \cos\left(\frac{\pi}{2}\right)}{h}$ is

(A) 1 (B) nonexistent (C) 0 (D) -1

3. If, for all x, $f'(x) = (x - 2)^4(x - 1)^3$, it follows that the function f has

(A) a relative minimum at $x = 1$
(B) a relative maximum at $x = 1$
(C) both a relative minimum at $x = 1$ and a relative maximum at $x = 2$
(D) relative minima at $x = 1$ and at $x = 2$

4. Let $F(x) = \int_0^x \frac{10}{1 + e^t}\, dt$. Which of the following statements is (are) true?

I. $F'(0) = 5$
II. $F(2) < F(6)$
III. F is concave upward.

(A) I only (B) II only (C) I and II only (D) I and III only

5. If $f(x) = 10^x$ and $10^{1.04} \simeq 10.96$, which is closest to $f'(1)$?

(A) 0.92 (B) 0.96 (C) 10.5 (D) 24

GO ON TO THE NEXT PAGE

6. If f is differentiable, we can use the line tangent to f at $x = a$ to approximate values of f near $x = a$. Suppose that for a certain function f this method always underestimates the correct values. If so, then in an interval surrounding $x = a$, the graph of f must be

(A) increasing (B) decreasing (C) concave upward (D) concave downward

7. If $f(x) = \cos x \sin 3x$, then $f'\left(\frac{\pi}{6}\right)$ is equal to

(A) $\frac{1}{2}$ (B) $-\frac{\sqrt{3}}{2}$ (C) $\frac{\sqrt{3}}{2}$ (D) $-\frac{1}{2}$

8. $\int_0^1 \frac{x}{x^2+1}\,dx$ is equal to

(A) $\frac{\pi}{4}$ (B) $\ln\sqrt{2}$ (C) $\frac{1}{2}(\ln 2 - 1)$ (D) $\ln 2$

9. The graph of f'' is shown below. If $f'(1) = 0$, then $f'(x) = 0$ at what other value of x on the interval $[0,8]$?

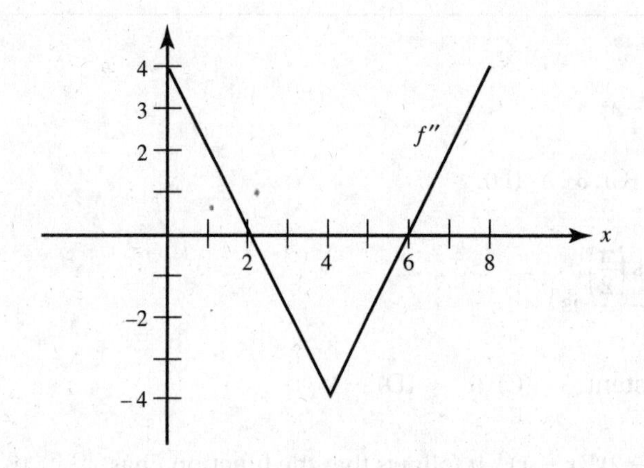

(A) 2 (B) 3 (C) 4 (D) 7

Questions 10 and 11. Use the following table, which shows the values of differentiable functions f and g.

x	f	f'	g	g'
1	2	$\frac{1}{2}$	−3	5
2	3	1	0	4
3	4	2	2	3
4	6	4	3	$\frac{1}{2}$

10. If $P(x) = g^2(x)$, then $P'(3)$ equals

(A) 4 (B) 6 (C) 9 (D) 12

GO ON TO THE NEXT PAGE

11. If $H(x) = f^{-1}(x)$, then $H'(3)$ equals

(A) $-\frac{1}{16}$ (B) $-\frac{1}{8}$ (C) $\frac{1}{2}$ (D) 1

12. The total area of the region bounded by the graph of $y = x\sqrt{1 - x^2}$ and the x-axis is

(A) $\frac{1}{3}$ (B) $\frac{1}{2}$ (C) $\frac{2}{3}$ (D) 1

13. The graph of $y = \frac{1 - x}{x - 3}$ is concave upward when

(A) $x > 3$ (B) $1 < x < 3$ (C) $x < 1$ (D) $x < 3$

14. As an ice block melts, the rate at which its mass, M, decreases is directly proportional to the square root of the mass. Which equation describes this relationship?

(A) $\sqrt{M(t)} = kt$ (B) $\frac{dM}{dt} = k\sqrt{t}$ (C) $\frac{dM}{dt} = k\sqrt{M}$ (D) $\frac{dM}{dt} = \frac{k}{\sqrt{M}}$

15. The average (mean) value of $\tan x$ on the interval from $x = 0$ to $x = \frac{\pi}{3}$ is

(A) $\ln\frac{1}{2}$ (B) $\frac{3}{\pi}\ln 2$ (C) $\frac{\sqrt{3}}{2}$ (D) $\frac{9}{\pi}$

16. If $y = x^2 \ln x$, for $x > 0$, then y'' is equal to

(A) $3 + \ln x$ (B) $3 + 2\ln x$ (C) $3 + 3\ln x$ (D) $2 + x + \ln x$

17. Water is poured at a constant rate into the conical reservoir shown in the figure. If the depth of the water, h, is graphed as a function of time, the graph is

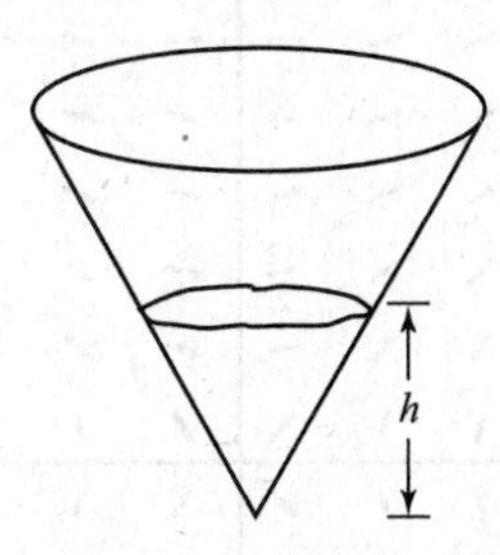

(A) constant (B) linear (C) concave upward (D) concave downward

18. If $f(x) = \begin{cases} x^2 & \text{for } x \le 1 \\ 2x - 1 & \text{for } x > 1 \end{cases}$, then

(A) $f(x)$ is not continuous at $x = 1$
(B) $f(x)$ is continuous at $x = 1$ but $f'(1)$ does not exist
(C) $f'(1) = 2$
(D) $\lim_{x \to 1} f(x)$ does not exist

19. $\lim_{x \to 2} \frac{|x - 2|}{x - 2}$ is

(A) $-\infty$ (B) -1 (C) ∞ (D) nonexistent

GO ON TO THE NEXT PAGE

Questions 20 and 21. The graph below consists of a quarter-circle and two line segments, and represents the velocity of an object during a 6-second interval.

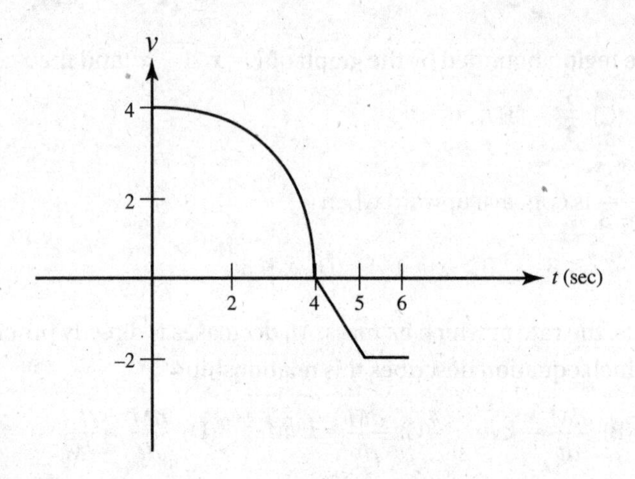

20. The object's average speed (in units/sec) during the 6-second interval is

(A) $\frac{4\pi + 3}{6}$ (B) $\frac{4\pi - 3}{6}$ (C) -1 (D) 1

21. The object's acceleration (in units/sec^2) at $t = 4.5$ is

(A) 0 (B) -1 (C) -2 (D) $-\frac{1}{4}$

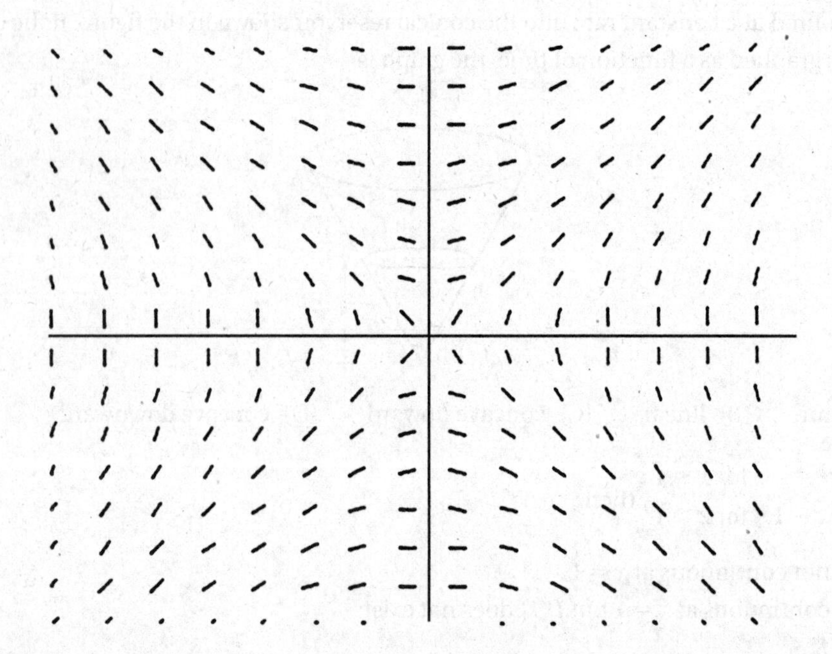

22. The slope field shown above is for which of the following differential equations?

(A) $\frac{dy}{dx} = -\frac{y}{x}$ (B) $\frac{dy}{dx} = \frac{1}{xy}$ (C) $\frac{dy}{dx} = -\frac{x}{y}$ (D) $\frac{dy}{dx} = \frac{x}{y}$

GO ON TO THE NEXT PAGE

23. If y is a differentiable function of x, then the slope of the curve of $xy^2 - 2y + 4y^3 = 6$ at the point where $y = 1$ is

(A) $-\frac{1}{18}$ (B) $-\frac{1}{26}$ (C) $\frac{5}{18}$ (D) $-\frac{11}{18}$

24. In the following, $L(n)$, $R(n)$, $M(n)$, and $T(n)$ denote, respectively, left, right, midpoint, and trapezoidal sums with n equal subdivisions. Which of the following is *not* equal exactly to $\int_{-3}^{3} |x|\, dx$?

(A) $L(2)$ (B) $T(3)$ (C) $M(4)$ (D) $R(6)$

25. The table shows some values of a differentiable function f and its derivative f':

x	0	1	2	3
$f(x)$	3	4	2	8
$f'(x)$	4	−1	1	10

Find $\int_{0}^{3} f'(x)dx$.

(A) 5 (B) 6 (C) 11.5 (D) 14

26. The solution of the differential equation $\frac{dy}{dx} = 2xy^2$ for which $y = -1$ when $x = 1$ is

(A) $y = -\frac{1}{x^2}$ for $x \neq 0$ (B) $y = -\frac{1}{x^2}$ for $x > 0$ (C) $\ln y^2 = x^2 - 1$ for all x (D) $y = -\frac{1}{x}$ for $x > 0$

27. The base of a solid is the region bounded by the parabola $y^2 = 4x$ and the line $x = 2$. Each plane section perpendicular to the x-axis is a square. The volume of the solid is

(A) 8 (B) 16 (C) 32 (D) 64

28. Which of the following could be the graph of $y = \frac{x^2}{e^x}$?

(A)

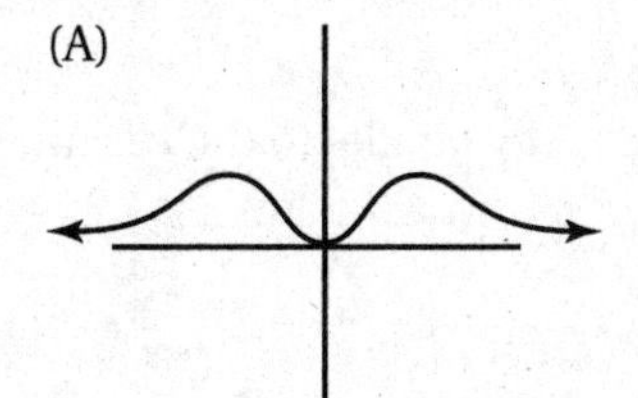

(B)

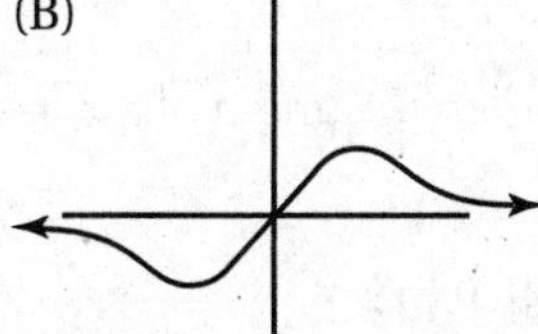

(C) (D)

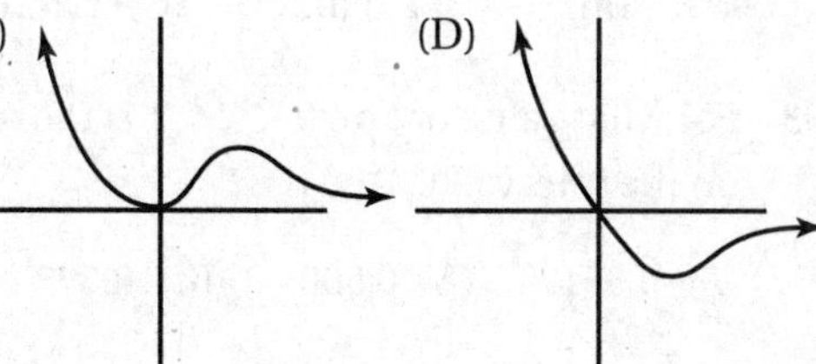

29. If $F(3) = 8$ and $F'(3) = -4$ then $F(3.02)$ is approximately

(A) 7.92 (B) 7.98 (C) 8.02 (D) 8.08

30. If $F(x) = \int_{0}^{2x} \frac{1}{1 - t^3} dt$, then $F'(x) =$

(A) $\frac{1}{1 - x^3}$ (B) $\frac{2}{1 - 2x^3}$ (C) $\frac{1}{1 - 8x^3}$ (D) $\frac{2}{1 - 8x^3}$

STOP

END OF PART A, SECTION I

Part B

TIME: 45 MINUTES

Some questions in this part of the examination require the use of a graphing calculator. There are 15 questions in Part B, for which 45 minutes are allowed.

Directions: Choose the best answer for each question. If the exact numerical value of the correct answer is not listed as a choice, select the choice that is closest to the exact numerical answer.

Questions 31 and 32. Refer to the graph of *f′* below.

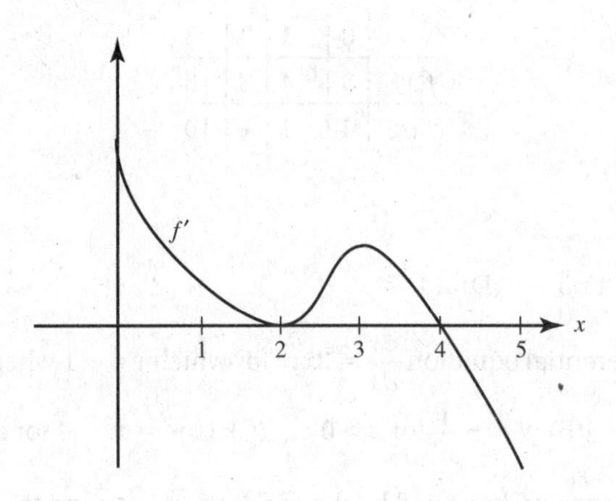

31. f has a local maximum at $x =$

    (A) 3 only (B) 4 only (C) 2 and 4 (D) 3 and 4

32. The graph of f has a point of inflection at $x =$

    (A) 2 only (B) 3 only (C) 2 and 3 only (D) 2 and 4 only

33. For what value of c on $0 < x < 1$ is the tangent to the graph of $f(x) = e^x - x^2$ parallel to the secant line on the interval (0,1)?

    (A) 0.351 (B) 0.500 (C) 0.693 (D) 0.718

34. Find the volume of the solid generated when the region bounded by the y-axis, $y = e^x$, and $y = 2$ is rotated around the y-axis.

    (A) 0.386 (B) 0.592 (C) 1.216 (D) 3.998

35. The table below shows the "hit rate" for an Internet site, measured at various intervals during a day. Use a trapezoid approximation with 6 subintervals to estimate the total number of people who visited that site.

Time	Midnight	6 A.M.	8 A.M.	Noon	5 P.M.	8 P.M.	Midnight
People per minute	5	2	3	8	10	16	5

    (A) 5280 (B) 10,080 (C) 10,440 (D) 10,560

GO ON TO THE NEXT PAGE

36. The acceleration of a particle moving along a straight line is given by $a = 6t$. If, when $t = 0$, its velocity, v, is 1 and its position, s, is 3, then at any time t

(A) $s = t^3 + 3$ (B) $s = t^3 + t + 3$ (C) $s = \frac{t^3}{3} + t + 3$ (D) $s = \frac{t^3}{3} + \frac{t^2}{3} + 3$

37. If $y = f(x^2)$ and $f'(x) = \sqrt{5x - 1}$ then $\frac{dy}{dx}$ is equal to

(A) $2x\sqrt{5x^2 - 1}$ (B) $\sqrt{5x - 1}$ (C) $2x\sqrt{5x - 1}$ (D) $\frac{\sqrt{5x - 1}}{2x}$

38. Find the area of the first quadrant region bounded by $y = x^2$, $y = \cos(x)$, and the y-axis.

(A) 0.292 (B) 0.508 (C) 0.547 (D) 0.921

39. If the substitution $x = 2t + 1$ is used, which of the following is equivalent to $\int_0^3 \sqrt[4]{2t + 1}\, dt$?

(A) $\int_0^3 \sqrt[4]{x}\, dx$ (B) $\frac{1}{2}\int_0^3 \sqrt[4]{x}\, dx$ (C) $\frac{1}{2}\int_{-1/2}^1 \sqrt[4]{x}\, dx$ (D) $\int_1^7 \frac{1}{2}\sqrt[4]{x}\, dx$

40. An object moving along a line has velocity $v(t) = t\cos t - \ln(t + 2)$, where $0 \leq t \leq 10$. How many times does the object reverse direction?

(A) one (B) two (C) three (D) four

41. A 26-foot ladder leans against a building so that its foot moves away from the building at the rate of 3 feet per second. When the foot of the ladder is 10 feet from the building, the top is moving down at the rate of r feet per second, where r is

(A) 0.80 (B) 1.25 (C) 7.20 (D) 12.50

$f(3) = 3$	$g(3) = -3$	$h(3) = 3$
$f'(3) = 2$	$g'(3) = -4$	$h'(3) = 4$
$f''(3) = 1/2$	$g''(3) = 8$	$h''(3) = 2$

42. Given three twice-differentiable functions, $f(x)$, $g(x)$, and $h(x)$. The table above gives values for the functions and their first and second derivatives at $x = 3$. Find $\lim_{x \to 3} \frac{(f(x))^2 - 3h(x)}{g(x) + h(x)}$.

(A) $-\frac{1}{5}$ (B) $\frac{1}{2}$ (C) 1 (D) nonexistent

GO ON TO THE NEXT PAGE

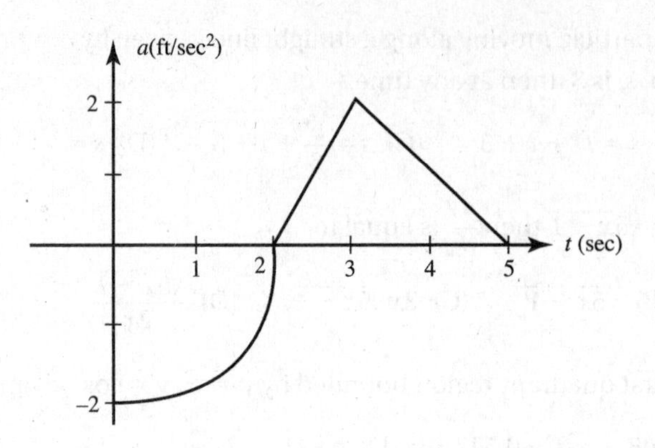

43. The graph above shows an object's acceleration (in ft/sec^2). It consists of a quarter-circle and two line segments. If the object was at rest at $t = 5$ seconds, what was its initial velocity?

(A) -2 ft/sec (B) $3 - \pi$ ft/sec (C) $\pi - 3$ ft/sec (D) $\pi + 3$ ft/sec

44. Water is leaking from a tank at the rate of $R(t) = 5\arctan\left(\frac{t}{5}\right)$ gallons per hour, where t is the number of hours since the leak began. To the nearest gallon, how much water will leak out during the first day?

(A) 7 (B) 12 (C) 24 (D) 124

45. Find the y-intercept of the line tangent to $y = (x^3 - 4x^2 + 8)e^{\cos x^2}$ at $x = 2$.

(A) 0 (B) 2.081 (C) 4.161 (D) 21.746

STOP

END OF SECTION I

SECTION II

Part A

TIME: 30 MINUTES

2 PROBLEMS

A graphing calculator is required for some of these problems. See instructions on page 4.

1. When a faulty seam opened at the bottom of an elevated hopper, grain began leaking out onto the ground. After a while, a worker spotted the growing pile below and began making repairs. The following table shows how fast the grain was leaking (in cubic feet per minute) at various times during the 20 minutes it took to repair the hopper.

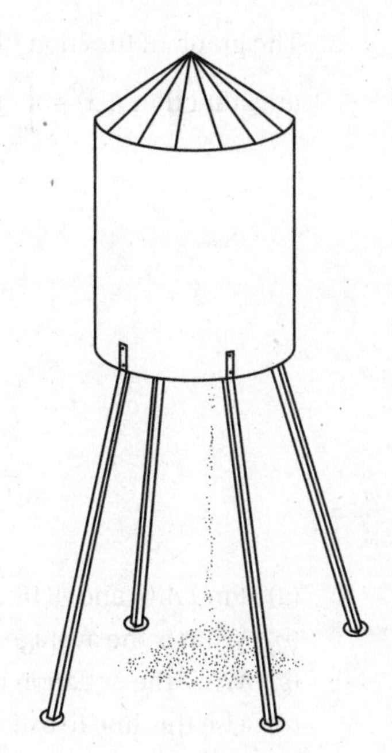

t (min)	0	4	5	7	10	12	18	20
$L(t)$ (ft^3/min)	4	7	9	8	6	5	2	0

(a) Estimate $L'(15)$.

(b) Explain in this context what your answer to part a means.

(c) The falling grain forms a conical pile that the worker estimates to be 5 times as far across as it is deep. The pile was 3 feet deep when the repairs had been half completed. How fast was the depth increasing then?

NOTE: The volume of a cone with height h and radius r is given by: $V = \frac{1}{3}\pi r^2 h$.

(d) Use a trapezoidal sum with seven subintervals as indicated in the table to approximate $\int_0^{20} L(t)\,dt$. Using correct units, explain the meaning of $\int_0^{20} L(t)\,dt$ in the context of the problem.

2. An object in motion along the x-axis has velocity $v(t) = (t + e^t)\sin t^2$ for $1 \le t \le 3$.

(a) At what time, t, is the object moving to the left?

(b) Is the speed of the object increasing or decreasing when $t = 2$? Justify your answer.

(c) At $t = 1$ this object's position was $x = 10$. What is the position of the object at $t = 3$?

STOP

END OF PART A, SECTION II

Part B

TIME: 60 MINUTES

4 PROBLEMS

No calculator is allowed for any of these problems.
If you finish Part B before time has expired, you may return to work on Part A, but you may not use a calculator.

3. The graph of function f consists of the semicircle and line segment shown in the figure. Define the area function $A(x) = \int_0^x f(t)dt$ for $0 \le x \le 18$.

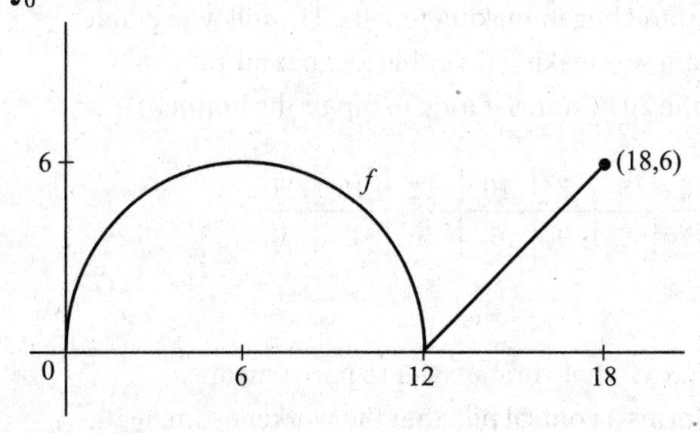

(a) Find $A(6)$ and $A(18)$.
(b) What is the average value of f on the interval $0 \le x \le 18$?
(c) Write the equation of the line tangent to the graph of A at $x = 6$.
(d) Use this line to estimate the area between f and the x-axis on $[0,7]$.
(e) Give the coordinates of any points of inflection on the graph of A. Justify your answer.

4. Consider the curve: $2x^2 - 4xy + 3y^2 = 16$.

(a) Show $\frac{dy}{dx} = \frac{2y - 2x}{3y - 2x}$.
(b) Verify that there exists a point Q where the curve has both an x-coordinate of 4 and a slope of zero. Find the y-coordinate of point Q.
(c) Find $\frac{d^2y}{dx^2}$ at point Q. Classify point Q as a local maximum, local minimum, or neither. Justify your answer.

GO ON TO THE NEXT PAGE

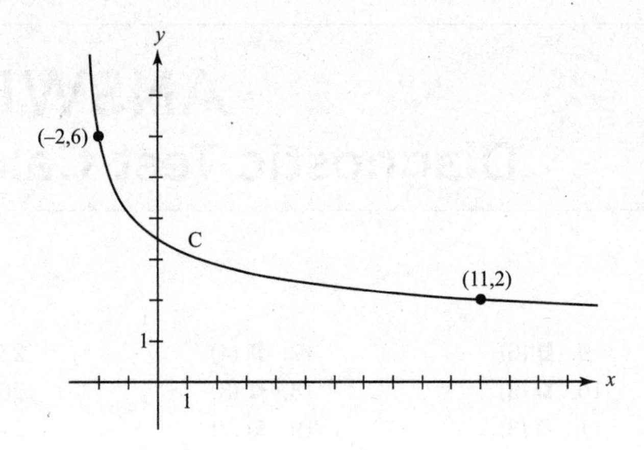

5. The graph above represents the curve C, given by $f(x) = \dfrac{6}{\sqrt[3]{2x+5}}$ for $-2 \le x \le 11$.

(a) Let R represent the region between C and the x-axis. Find the area of R.

(b) Set up, but do not solve, an equation to find the value of k such that the line $x = k$ divides R into two regions of equal area.

(c) Set up but do not evaluate an integral for the volume of the solid generated when R is rotated around the x-axis.

6. Let $y = f(x)$ be the function that has an x-intercept at (2,0) and satisfies the differential equation $x^2 e^y \dfrac{dy}{dx} = 4$.

(a) Solve the differential equation, expressing y as a function of x and specifying the domain of the function.

(b) Find an equation of each horizontal asymptote to the graph of $y = f(x)$.

STOP

END OF TEST

ANSWER KEY
Diagnostic Test Calculus AB

Part A

1. **A** (2)	9. **B** (6)	17. **D** (4)	25. **A** (6)
2. **D** (3)	10. **D** (3)	18. **C** (3)	26. **B** (9)
3. **A** (4)	11. **D** (3)	19. **B** (2)	27. **C** (7)
4. **C** (6)	12. **C** (7)	20. **A** (6)	28. **C** (2)
5. **D** (3)	13. **D** (4)	21. **C** (6)	29. **A** (4)
6. **C** (4)	14. **C** (9)	22. **D** (9)	30. **D** (4)
7. **D** (3)	15. **B** (6)	23. **A** (3)	
8. **B** (5)	16. **B** (3)	24. **B** (6)	

Part B

31. **B** (4)	35. **C** (6)	39. **D** (6)	43. **C** (6)
32. **C** (4)	36. **B** (8)	40. **B** (8)	44. **D** (8)
33. **A** (3)	37. **A** (3)	41. **B** (4)	45. **C** (4)
34. **B** (7)	38. **C** (7)	42. **B** (3)	

NOTE: Chapters that review and offer additional practice for each topic are specified in parentheses.

ANSWERS EXPLAINED

Section I Multiple-Choice

Part A

1. **(A)** Use the Rational Function Theorem (page 96); the ratio of the coefficients of the highest power of x is $\frac{3}{-1} = -3$.

2. **(D)** $\lim_{h \to 0} \frac{f\left(\frac{\pi}{2} + h\right) - f\left(\frac{\pi}{2}\right)}{h} = f'\left(\frac{\pi}{2}\right) \cdot f'(x) = -\sin(x)$, so $f'\left(\frac{\pi}{2}\right) = -\sin\left(\frac{\pi}{2}\right) = -1$.

3. **(A)** Since $f'(1) = 0$ and f' changes from negative to positive there, f reaches a minimum at $x = 1$. Although $f'(2) = 0$ as well, f' does not change sign there, and thus f has neither a maximum nor a minimum at $x = 2$.

4. **(C)** $F'(x) = \frac{10}{1 + e^x} > 0$, and $F''(x) = \frac{-10e^x}{(1 + e^x)^2} < 0$.

5. **(D)** $f'(1) \approx \frac{f(1.04) - f(1)}{1.04 - 1} = \frac{0.96}{0.04}$.

6. **(C)** The graph must look like one of these two:

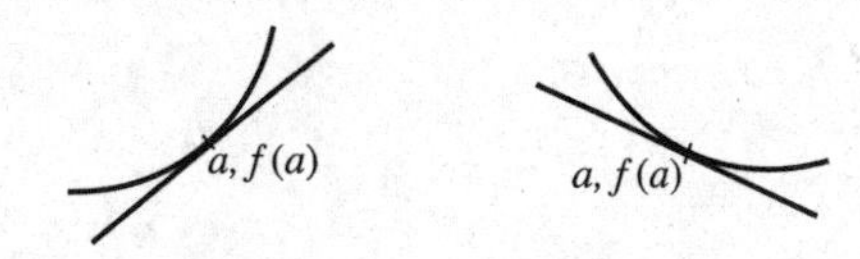

7. **(D)** $F'(x) = 3 \cos x \cos 3x - \sin x \sin 3x$.

$$F'\left(\frac{\pi}{6}\right) = 3\cos\left(\frac{\pi}{6}\right)\cos\left(\frac{\pi}{2}\right) - \sin\left(\frac{\pi}{6}\right)\sin\left(\frac{\pi}{2}\right) = 3 \cdot \frac{\sqrt{3}}{2} \cdot 0 - \frac{1}{2} \cdot 1$$

8. **(B)** $\int_0^1 \frac{x\,dx}{x^2 + 1} = \frac{1}{2} \ln (x^2 + 1)\Big|_0^1$

$$= \frac{1}{2}(\ln 2 - \ln 1)$$

$$= \ln \sqrt{2}$$

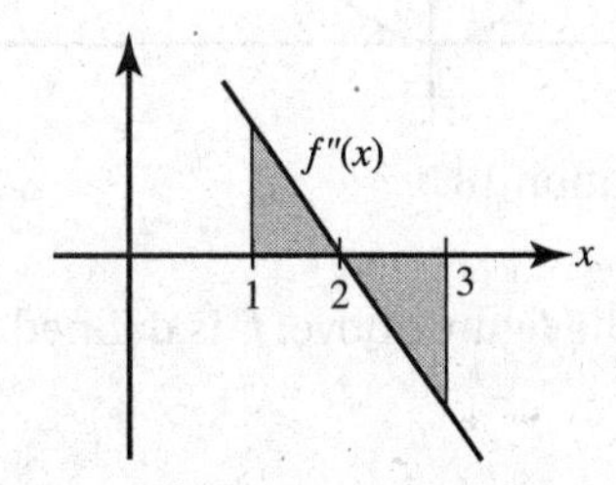

9. **(B)** Let $f'(x) = \int_1^x f''(t)\,dt$. Then f' increases for $1 < x < 2$, then begins to decrease. In the figure above, the area below the x-axis, from 2 to 3, is equal in magnitude to that above the x-axis, hence, $\int_1^3 f''(t)\,dt = 0$.

10. **(D)** $P'(x) = 2g(x) \cdot g'(x)$; $P'(3) = 2g(3) \cdot g'(3) = 2 \cdot 2 \cdot 3 = 12$.

11. **(D)** Note that $H(3) = f^{-1}(3) = 2$. Therefore

$$H'(3) = \frac{1}{f'(H(3))} = \frac{1}{f'(2)} = 1.$$

12. **(C)** Note that the domain of y is all x such that $|x| \leqq 1$ and that the graph is symmetric to the origin. The area is given by

$$2\int_0^1 x\sqrt{1-x^2}\,dx.$$

13. **(D)** Since

$$y' = 2(x-3)^{-2} \text{ and } y'' = -4(x-3)^{-3} = \frac{-4}{(x-3)^3},$$

y'' is positive when $x < 3$.

14. **(C)** $\frac{dM}{dt}$ represents the rate of change of mass with respect to time; y is directly proportional to x if $y = kx$.

15. **(B)** $\frac{1}{\pi/3}\int_0^{\pi/3} \tan x\,dx = \frac{3}{\pi}[-\ln \cos x]\Big|_0^{\pi/3} = \frac{3}{\pi}\left(-\ln\frac{1}{2}\right)$.

16. **(B)** $y' = \left(x^2 \cdot \frac{1}{x}\right) + 2x \ln x = x + 2x \ln x$ and $y'' = 1 + \left(2x \cdot \frac{1}{x} + 2 \ln x\right) = 3 + 2 \ln x$

17. **(D)** As the water gets deeper, the depth increases more slowly. Hence, the rate of change of depth decreases: $\frac{d^2h}{dt^2} < 0$.

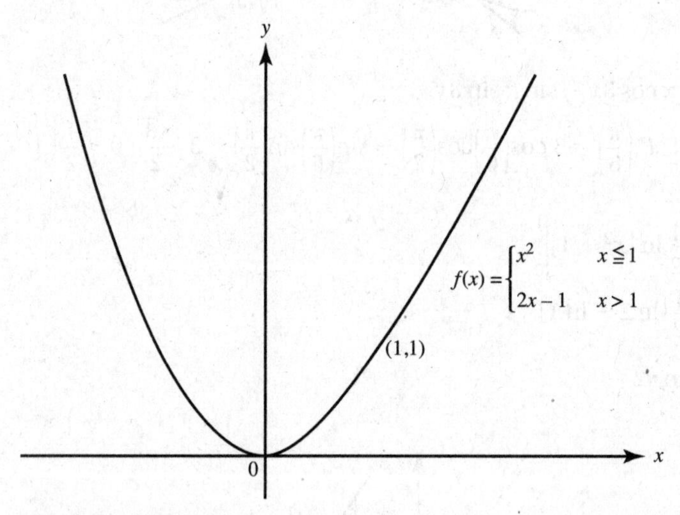

Figure for Solution 18

18. **(C)** The graph of f is shown in the figure above; f is defined and continuous at all x, including $x = 1$. Since

$$\lim_{x\to 1^-} f'(x) = 2 = \lim_{x\to 1^+} f'(x),$$

$f'(1)$ exists and is equal to 2.

19. **(B)** Since $|x-2| = 2-x$ if $x < 2$, the limit as $x \to 2^-$ is $\frac{2-x}{x-2} = -1$.

20. **(A)** Average speed $= \dfrac{\text{distance covered in 6 sec}}{\text{time elapsed}}$

$$= \frac{\frac{1}{4}\pi(4^2) + \frac{1}{2}(1 \cdot 2) + 1 \cdot 2}{6}.$$

Note that the distance covered in 6 seconds is $\int_0^6 |v(t)|\, dt$, the area between the velocity curve and the t-axis.

21. **(C)** Acceleration is the slope of the velocity curve, $\dfrac{-2-0}{5-4}$.

22. **(D)** Slopes are: 1 along $y = x$, -1 along $y = -x$, 0 along $x = 0$, and undefined along $y = 0$.

23. **(A)** Differentiating implicitly yields $2xyy' + y^2 - 2y' + 12y^2y' = 0$. When $y = 1, x = 4$. Substitute to find y'.

24. **(B)**

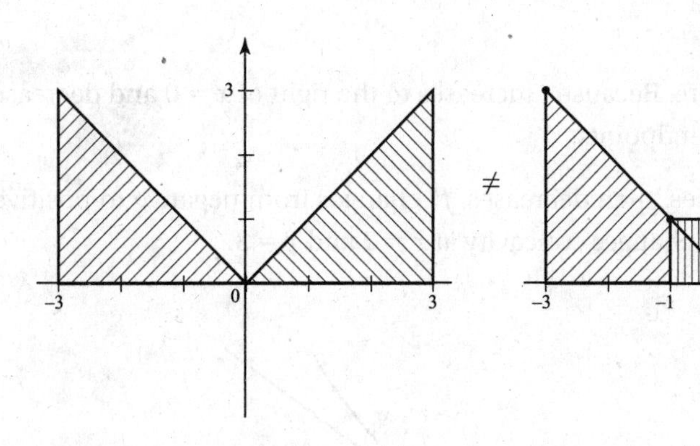

25. **(A)** $\int_0^3 f'(x)dx = f(x)\Big|_0^3 = f(3) - f(0) = 8 - 3 = 5$

26. **(B)** Separate to get $\dfrac{dy}{y^2} = 2x\,dx$, $-\dfrac{1}{y} = x^2 + C$. Since $-(-1) = 1 + C$ implies that $C = 0$, the solution is $-\dfrac{1}{y} = x^2$ or $y = -\dfrac{1}{x^2}$.

This function is discontinuous at $x = 0$. Since the particular solution must be differentiable in an interval containing the initial value $x = 1$, the domain is $x > 0$.

27. **(C)**

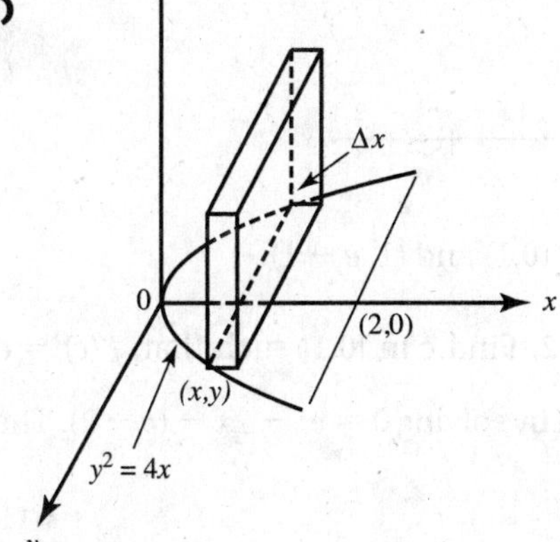

$$\Delta V = (2y)^2 \Delta x;$$

$$V = 4\int_0^2 y^2 dx$$

$$= 4\int_0^2 4x\, dx$$

$$= 32.$$

28. **(C)** Note that $\lim_{x\to\infty} \frac{x^2}{e^x} = 0$, $\lim_{x\to-\infty} \frac{x^2}{e^x} = \infty$, and $\frac{x^2}{e^x} \geqslant 0$ for all x.

29. **(A)** At $x = 3$, the equation of the tangent line is $y - 8 = -4(x - 3)$, so $f(x) \simeq -4(x - 3) + 8$. $f(3.02) \simeq -4(0.02) + 8$.

30. **(D)** Let $u = 2x$ and note that $F'(x) = \frac{1}{1-u^3}$.

    Then $F'(x) = F'(u)u'(x) = 2F'(u) = 2 \cdot \frac{1}{1-(2x)^3}$.

Part B

31. **(B)** The sign diagram shows that f changes from increasing to decreasing at $x = 4$

	0–2	2–4	4–5
f	inc	inc	dec
f'	+	+	−

and thus f has a maximum there. Because f increases to the right of $x = 0$ and decreases to the left of $x = 5$, there are minima at the endpoints.

32. **(C)** Since f' decreases, increases, then decreases, f'' changes from negative to positive, then back to negative. Hence, the graph of f changes concavity at $x = 2$ and $x = 3$.

33. **(A)**

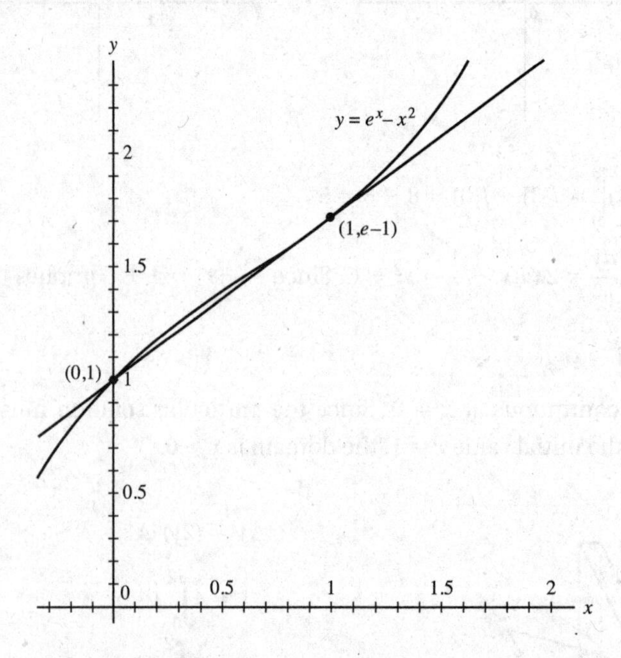

On the curve of $f(x) = e^x - x^2$, the two points labeled are (0,1) and $(1, e - 1)$.

The slope of the secant line is $m = \frac{\Delta y}{\Delta x} = \frac{e-2}{1} = e - 2$. Find c in [0,1] such that, $f'(c) = e - 2$, or $f'(c) - (e - 2) = 0$. Since $f'(x) = e^x - 2x$, c can be calculated by solving $0 = e^x - 2x - (e - 2)$. The answer is 0.351.

34. **(B)**

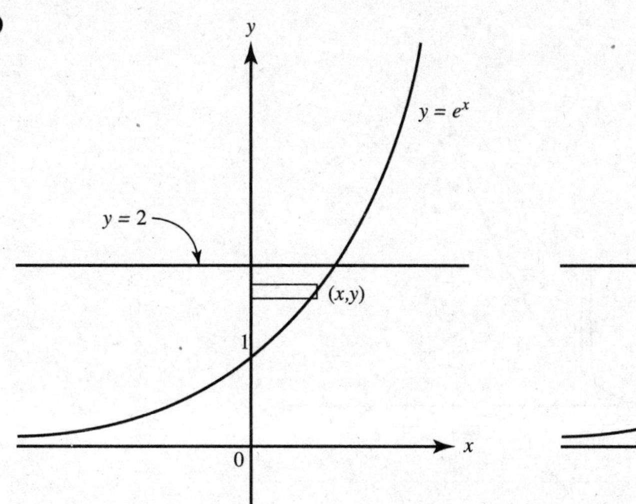

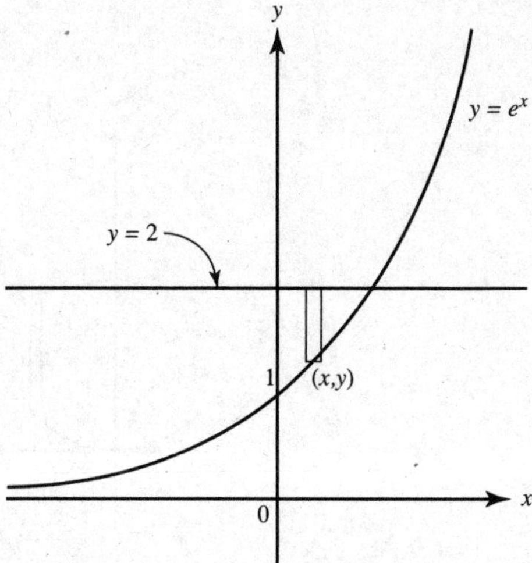

Use disks; then $\Delta V = \pi R^2 H = \pi (\ln y)^2\, \Delta y$. Note that the limits of the definite integral are 1 and 2. Evaluate the integral

$$\pi \int_1^2 (\ln y)^2\, dy = 0.592$$

Alternatively, use shells*; then $\Delta V = 2\pi RHT = 2\pi x(2 - e^x)\, \Delta x$. Here, the upper limit of integration is the value of x for which $e^x = 2$, namely, ln 2. Now evaluate

$$2\pi \int_0^{\ln} x(2 - e^x)\, dx = 0.592$$

35. **(C)** Note that the rate is people *per minute*, so the first interval width from midnight to 6 A.M. is 360 minutes. The total number of people is estimated as the sum of the areas of six trapezoids:

$$T = 360\left(\frac{5+2}{2}\right) + 120\left(\frac{2+3}{2}\right) + 240\left(\frac{3+8}{2}\right) + 300\left(\frac{8+10}{2}\right) +$$

$$180\left(\frac{10+16}{2}\right) + 240\left(\frac{16+5}{2}\right) = 10{,}440$$

36. **(B)** $a = \dfrac{dv}{dt} = 6t$, so $v = 3t^3 + c$.

Since $v = 1$ when $t = 0$, $c = 1$.

Now $v = \dfrac{ds}{dt} = 3t^2 + 1$, so $s = t^3 + t + c$.

Since $s = 3$ when $t = 0$, $c = 3$; then $s = t^3 + t + 3$.

37. **(A)** Let $u = x^2$. Then

$$\frac{dy}{dx} = \frac{dy}{du}\cdot\frac{du}{dx} = \frac{df}{du}\cdot f'(u)\frac{du}{dx} = \sqrt{5u - 1}\cdot 2x = 2x\sqrt{5x^2 - 1}$$

* No question requiring the use of the shells method will appear on the AP exam.

38. **(C)**

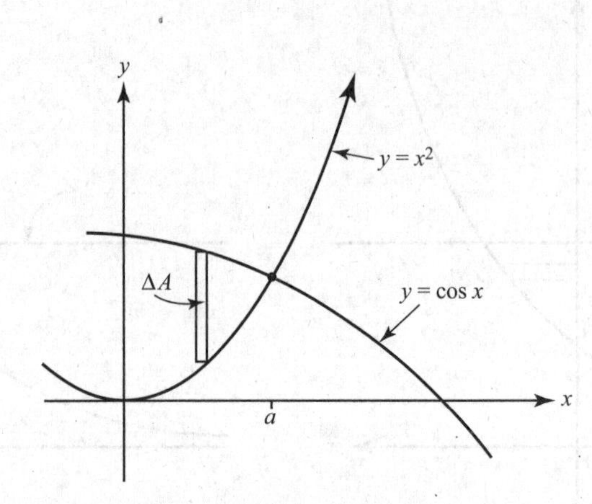

To find a, the point of intersection of $y = x^2$ and $y = \cos(x)$, use your calculator to solve the equation $x^2 - \cos(x) = 0$. (Store the value for later use; $a \approx 0.8241$.)

As shown in the diagram above, $\Delta A = (\cos(x) - x^2)\Delta x$.

Evaluate the area: $A = \int_0^a \left(\cos(x) - x^2\right) dx \approx 0.547$.

39. **(D)** If $x = 2t + 1$, then $t = \frac{x-1}{2}$, so $dt = \frac{1}{2}dx$. When $t = 0$, $x = 1$; when $t = 3$, $x = 7$.

40. **(B)**

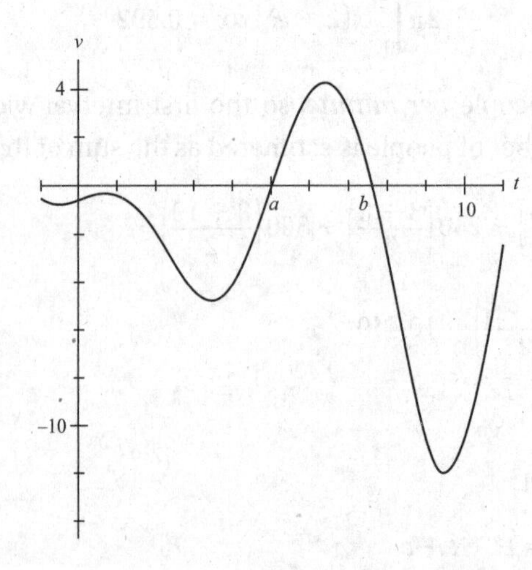

The velocity is graphed in $[-1,11] \times [-15,5]$. The object reverses direction when the velocity changes sign, that is, when the graph crosses the x-axis. There are two such reversals—at $x = a$ and at $x = b$.

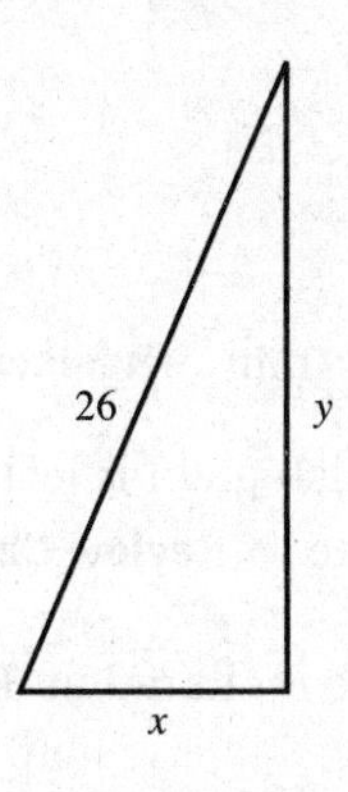

41. **(B)** See the figure above. Since $x^2 + y^2 = 26^2$, it follows that

$$2x\frac{dx}{dt} + 2y\frac{dy}{dt} = 0$$

at any time t. When $x = 10$, then $y = 24$ and it is given that $\frac{dx}{dt} = 3$.

Hence, $2(10)(3) + 2(24)\frac{dy}{dt} = 0$, so $\frac{dy}{dt} = -\frac{5}{4}$.

42. **(B)** The limit $\lim_{x\to 3} \frac{(f(x))^2 - 3h(x)}{g(x) + h(x)}$ by substitution is of the form $\frac{(3)^2 - 3(3)}{-3+3} = \frac{0}{0}$, so apply L'Hôpital's Rule. You get $\lim_{x\to 2} \frac{2f(x)f'(x) - 3h'(x)}{g'(x) + h'(x)}$, which is still of the form $\frac{2(3)(2) - 3(4)}{-4+4} = \frac{0}{0}$; therefore, apply L'Hôpital's Rule again. The new limit is $\lim_{x\to 2} \frac{2f(x)f''(x) + 2f'(x)\cdot f'(x) - 3h''(x)}{g''(x) + h''(x)}$ by substitution is $\frac{2(3)(1/2) + 2(2)(2) - 3(2)}{8+2} = \frac{5}{10} = \frac{1}{2}$.

43. **(C)** $v(5) - v(0) = \int_0^5 a(t)\,dt = -\frac{1}{4}\pi \cdot 2^2 + \frac{1}{2}(3)(2) = -\pi + 3$. Since $v(5) = 0$, $-v(0) = -\pi + 3$; so $v(0) = \pi - 3$.

44. **(D)** $\int_0^{24} 5\arctan\left(\frac{t}{5}\right) dt = 124.102$

45. **(C)** Let $y = (x^3 - 4x^2 + 8)e^{\cos(x^2)}$. The equation of the tangent at point $(2, y(2))$ is $y - y(2) = y'(2)(x - 2)$. Note that $y(2) = 0$. To find the y-intercept, let $x = 0$ and solve for y: $y = -2y'(2)$. A calculator yields $y = 4.161$.

Section II Free-Response

Part A

AB/BC 1. (a) $L'(15) \approx \frac{2-5}{18-12} = -0.5 \text{ ft}^3/\text{min}/\text{min}$ **(Review Chapter 3)**

(b) After 15 minutes the rate at which grain is leaking is slowing down at the rate of one half a cubic foot per minute per minute. **(Review Chapter 3)**

(c) Let h = the height of the cone and r = its radius. The cone's diameter is given to be $5h$, so $r = \frac{5}{2}h$, and the cone's volume,

$$V = \frac{1}{3}\pi r^2 h = \frac{1}{3}\pi\left(\frac{5}{2}h\right)^2 h = \frac{25}{12}\pi h^3.$$

Then

$$\frac{dV}{dt} = \frac{25}{4}\pi h^2 \frac{dh}{dt}.$$

At $t = 10$ the table shows $L = \frac{dV}{dt} = 6$, and it is given that $h = 3$; thus:

$$6 = \frac{25}{4}\pi 3^2 \frac{dh}{dt}, \text{ so } \frac{dh}{dt} = \frac{8}{75\pi} \text{ ft/min.}$$

(Review Chapter 4)

(d) $$\int_0^{20} L(t)dt \approx (4-0)\left(\frac{7+4}{2}\right) + (5-4)\left(\frac{9+7}{2}\right) + (7-5)\left(\frac{8+9}{2}\right) + (10-7)\left(\frac{6+8}{2}\right) +$$

$$(12-10)\left(\frac{5+6}{2}\right) + (18-12)\left(\frac{2+5}{2}\right) + (20-18)\left(\frac{0+2}{2}\right) = 102 \text{ ft}^3$$

The total amount of grain that leaked out of the elevator hopper while the repairs were underway was approximately 102 ft^3.

(Review Chapter 6)

AB 2. Graph $y = (x = e^x)\sin(x^2)$ in $[1,3] \times [-15,20]$. Note that y represents velocity v and x represents time t.

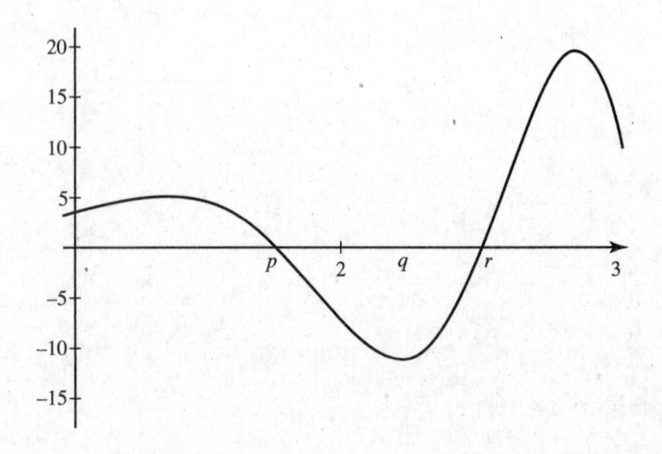

(a) The object moves to the left when the velocity is negative, namely, on the interval $p < t < r$. Use the calculator to solve $(x + e^x)(\sin(x^2)) = 0$; then $p = 1.772$ and $r = 2.507$. The answer is $1.772 < t < 2.507$.

(b) At $t = 2$, the object is speeding up; because $v(2) = -7.106 < 0$ and $a(2) = -30.897 < 0$, the object is accelerating in the direction of motion. (Note: You can use your calculator to find both $v(2)$ and $a(2)$.)

(c) $x(3) = x(1) + \int_1^3 x'(t)dt = x(1) + \int_1^3 v(t)\,dt = 10 + 4.491 = 14.491.$

(Review Chapter 8)

Part B

AB/BC 3. (a) $A(6) = \frac{1}{4}(\pi 6^2) = 9\pi, A(18) = \frac{1}{2}(\pi 6^2) + \frac{1}{2}6 \times 6 = 18\pi + 18$

(b) The average value of $f = \dfrac{\int_0^{18} f(x)dx}{18 - 0} = \dfrac{18\pi + 18}{18} = \pi + 1.$

(c) The line tangent to the graph of A at $x = 6$ passes through point $(6, A(6))$ or $(6, 9\pi)$. Since $A'(x) = f(x)$, the graph of f shows that $A'(6) = f(6) = 6$. Hence, an equation of the line is $y - 9\pi = 6(x - 6)$.

(d) Use the tangent line; then $A(x) = y \approx 6(x - 6) + 9\pi$, so $A(7) \approx 6(7 - 6) + 9\pi = 6 + 9\pi$.

(e) Since f is increasing on $[0,6]$, f' is positive there. Because $f(x) = A'(x)$, $f'(x) = A''(x)$; thus A is concave upward for $[0,6]$. Similarly, the graph of A is concave downward for $[6,12]$, and upward for $[12,18]$. There are points of inflection on the graph of A at $(6,9\pi)$ and $(12,18\pi)$.

AB 4. (a) $4x - 4xy' - 4y + 6yy' = 0$

$(6y - 4x)\,y' = 4x - 4y$

$$\frac{dy}{dx} = \frac{2y - 2x}{3y - 2x}$$

(Review Chapter 3)

(b) $\dfrac{dy}{dx} = 0 \Rightarrow 2y - 2x = 0 \Rightarrow y = x$, when $x = 4$ then $y = 4$.

Verify that (4,4) is on the curve:
$2(4)^2 - 4(4)(4) + 3(4)^2 = 32 - 64 + 48 = 16$
Therefore, the point Q (4,4) is on the curve and the slope there is zero.

(Review Chapter 3)

(c) $$y'' = \frac{(3y - 2x)(2y' - 2) - (2y - 2x)(3y' - 2)}{(3y - 2x)^2}$$

$$y''\Big|_{(4,4)} = \frac{(4)(-2) - (0)(-2)}{(4)^2} = -\frac{1}{2}$$

At $Q = (4,4)$ the curve has a local maximum because $y' = 0$ and $y'' < 0$.

(Review Chapters 3 & 4)

AB/BC 5. (a) Draw elements as shown. Then

$$\Delta A = y\,\Delta x = \frac{6}{\sqrt[3]{2x+5}}\,\Delta x,$$

$$A = \int_{-2}^{11} \frac{6}{\sqrt[3]{2x+5}}\,dx = \frac{6}{2}\int_{-2}^{11} (2x+5)^{-1/3}\,(2\,dx)$$

$$= 3 \cdot \frac{3}{2}(2x+5)^{2/3}\Big|_{-2}^{11} = \frac{9}{2}(27^{2/3} - 1^{2/3}) = 36.$$

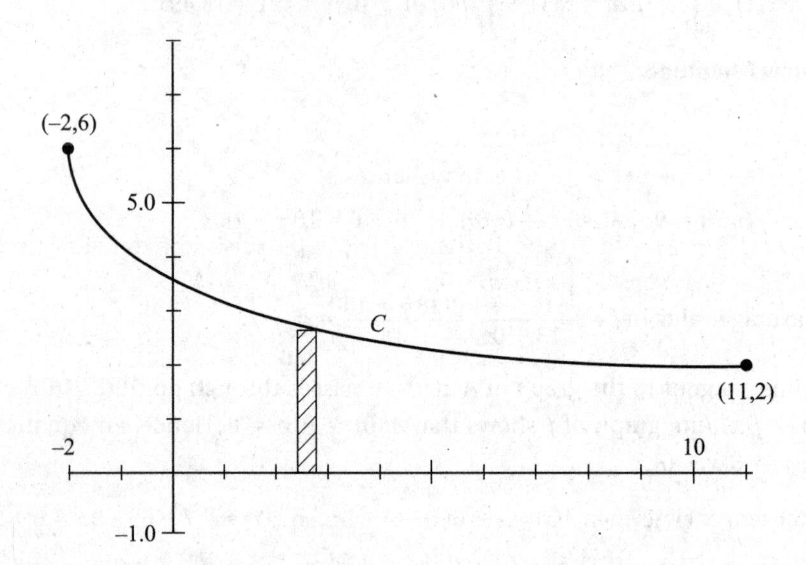

(b) $\displaystyle\int_{-2}^{k} \frac{6}{\sqrt[3]{2x+5}}\,dx = \int_{k}^{11} \frac{6}{\sqrt[3]{2x+5}}\,dx.$

(c) Revolving the element around the x-axis generates disks. Then

$$\Delta V = \pi r^2\,\Delta x = \pi y^2\,\Delta x = \pi\left(\frac{6}{\sqrt[3]{2x+5}}\right)^2 \Delta x, \text{ so}$$

$$V = \pi \int_{-2}^{11} \frac{36}{(2x+5)^{2/3}}\,dx.$$

(Review Chapter 7)

AB 6. (a) The differential equation $x^2 e^y \frac{dy}{dx} = 4$ is separable:

$$\int e^y dy = \int \frac{4}{x^2}\,dx$$

$$e^y = -\frac{4}{x} + c.$$

If $y = 0$ when $x = 2$, then $e^0 = -\frac{4}{2} + c$; thus $c = 3$, and $e^y = -\frac{4}{x} + 3$.

Solving for y gives the solution: $y = \ln\left(3 - \frac{4}{x}\right)$.

Note that $y = \ln\left(3 - \frac{4}{x}\right)$ is defined only if $3 - \frac{4}{x} > 0$.

$\frac{3x - 4}{x} > 0$ only if the numerator and denominator have the same sign.

$$(x > 0 \text{ and } 3x - 4 > 0) \quad \text{OR} \quad (x < 0 \text{ and } 3x - 4 < 0),$$

$$\left(x > 0 \text{ and } x > \frac{4}{3}\right) \quad \text{OR} \quad \left(x < 0 \text{ and } x < \frac{4}{3}\right).$$

$$\left\{x \middle| x < 0 \ \text{ OR } \ x > \frac{4}{3}\right\}.$$

Since the particular solution must be continuous on an interval containing the initial point $x = 2$, the domain is $x > \frac{4}{3}$. **(Review Chapter 9)**

(b) Since $\lim_{x \to \pm\infty} \left(3 - \frac{4}{x}\right) = \ln 3$, the function $y = \ln\left(3 - \frac{4}{x}\right)$ has a horizontal asymptote at $y = \ln 3$. **(Review Chapter 2)**

ANSWER SHEET
Diagnostic Test Calculus BC

Part A

1. Ⓐ Ⓑ Ⓒ Ⓓ	9. Ⓐ Ⓑ Ⓒ Ⓓ	17. Ⓐ Ⓑ Ⓒ Ⓓ	25. Ⓐ Ⓑ Ⓒ Ⓓ
2. Ⓐ Ⓑ Ⓒ Ⓓ	10. Ⓐ Ⓑ Ⓒ Ⓓ	18. Ⓐ Ⓑ Ⓒ Ⓓ	26. Ⓐ Ⓑ Ⓒ Ⓓ
3. Ⓐ Ⓑ Ⓒ Ⓓ	11. Ⓐ Ⓑ Ⓒ Ⓓ	19. Ⓐ Ⓑ Ⓒ Ⓓ	27. Ⓐ Ⓑ Ⓒ Ⓓ
4. Ⓐ Ⓑ Ⓒ Ⓓ	12. Ⓐ Ⓑ Ⓒ Ⓓ	20. Ⓐ Ⓑ Ⓒ Ⓓ	28. Ⓐ Ⓑ Ⓒ Ⓓ
5. Ⓐ Ⓑ Ⓒ Ⓓ	13. Ⓐ Ⓑ Ⓒ Ⓓ	21. Ⓐ Ⓑ Ⓒ Ⓓ	29. Ⓐ Ⓑ Ⓒ Ⓓ
6. Ⓐ Ⓑ Ⓒ Ⓓ	14. Ⓐ Ⓑ Ⓒ Ⓓ	22. Ⓐ Ⓑ Ⓒ Ⓓ	30. Ⓐ Ⓑ Ⓒ Ⓓ
7. Ⓐ Ⓑ Ⓒ Ⓓ	15. Ⓐ Ⓑ Ⓒ Ⓓ	23. Ⓐ Ⓑ Ⓒ Ⓓ	
8. Ⓐ Ⓑ Ⓒ Ⓓ	16. Ⓐ Ⓑ Ⓒ Ⓓ	24. Ⓐ Ⓑ Ⓒ Ⓓ	

Part B

31. Ⓐ Ⓑ Ⓒ Ⓓ	35. Ⓐ Ⓑ Ⓒ Ⓓ	39. Ⓐ Ⓑ Ⓒ Ⓓ	43. Ⓐ Ⓑ Ⓒ Ⓓ
32. Ⓐ Ⓑ Ⓒ Ⓓ	36. Ⓐ Ⓑ Ⓒ Ⓓ	40. Ⓐ Ⓑ Ⓒ Ⓓ	44. Ⓐ Ⓑ Ⓒ Ⓓ
33. Ⓐ Ⓑ Ⓒ Ⓓ	37. Ⓐ Ⓑ Ⓒ Ⓓ	41. Ⓐ Ⓑ Ⓒ Ⓓ	45. Ⓐ Ⓑ Ⓒ Ⓓ
34. Ⓐ Ⓑ Ⓒ Ⓓ	38. Ⓐ Ⓑ Ⓒ Ⓓ	42. Ⓐ Ⓑ Ⓒ Ⓓ	

Diagnostic Test Calculus BC

SECTION I

Part A

TIME: 60 MINUTES

> ***The use of calculators is not permitted for this part of the examination. There are 30 questions in Part A, for which 60 minutes are allowed.***
>
> **Directions:** Choose the best answer for each question.

1. $\lim_{x\to\infty} \frac{3x^2-4}{2-7x-x^2}$ is

   (A) -3 (B) 0 (C) 3 (D) ∞

2. $\lim_{h\to 0} \frac{\cos\left(\frac{\pi}{2}+h\right)-\cos\left(\frac{\pi}{2}\right)}{h}$ is

   (A) 1 (B) nonexistent (C) 0 (D) -1

3. If, for all x, $f'(x) = (x-2)^4(x-1)^3$, it follows that the function f has

   (A) a relative minimum at $x = 1$
   (B) a relative maximum at $x = 1$
   (C) both a relative minimum at $x = 1$ and a relative maximum at $x = 1$
   (D) relative minima at $x = 1$ and at $x = 2$

4. Let $F(x) = \int_0^x \frac{10}{1+e^t}\,dt$. Which of the following statements is (are) true?

   I. $F'(0) = 5$
   II. $F(2) < F(6)$
   III. F is concave upward.

   (A) I only (B) II only (C) I and II only (D) I and III only

5. If $f(x) = 10^x$ and $10^{1.04} \simeq 10.96$, which is closest to $f'(1)$?

   (A) 0.92 (B) 0.96 (C) 10.5 (D) 24

GO ON TO THE NEXT PAGE

6. If f is differentiable, we can use the line tangent to f at $x = a$ to approximate values of f near $x = a$. Suppose that for a certain function f this method always underestimates the correct values. If so, then in an interval surrounding $x = a$, the graph of f must be

(A) increasing (B) decreasing (C) concave upward (D) concave downward

7. The region in the first quadrant bounded by the x-axis, the y-axis, and the curve of $y = e^{-x}$ is rotated about the x-axis. The volume of the solid obtained is equal to

(A) $\frac{1}{2}$ (B) $\frac{\pi}{2}$ (C) π (D) 2π

8. $\int_0^1 \frac{x}{x^2+1}\,dx$ is equal to

(A) $\frac{\pi}{4}$ (B) $\ln\sqrt{2}$ (C) $\frac{1}{2}(\ln 2 - 1)$ (D) $\ln 2$

9. Which series diverges?

(A) $\sum_{n=1}^{\infty} \frac{(-1)^n}{n^5}$ (B) $\sum_{n=1}^{\infty} \frac{(-1)^n}{\sqrt[5]{n}}$ (C) $\sum_{n=1}^{\infty} \frac{(-1)^n}{5n+1}$ (D) $\sum_{n=1}^{\infty} \frac{(-1)^n \cdot n}{5n+1}$

Questions 10 and 11. Use the table below, which shows the values of differentiable functions f and g.

x	f	f'	g	g'
1	2	$\frac{1}{2}$	−3	5
2	3	1	0	4
3	4	2	2	3
4	6	4	3	$\frac{1}{2}$

10. If $P(x) = g^2(x)$, then $P'(3)$ equals

(A) 4 (B) 6 (C) 9 (D) 12

11. If $H(x) = f^{-1}(x)$, then $H'(3)$ equals

(A) $-\frac{1}{16}$ (B) $-\frac{1}{8}$ (C) $\frac{1}{2}$ (D) 1

12. $\int_0^1 xe^x\,dx$ equals

(A) 1 (B) −1 (C) $\frac{e}{2}$ (D) $e - 1$

GO ON TO THE NEXT PAGE

13. The graph of $y = \dfrac{1-x}{x-3}$ is concave upward when

(A) $x > 3$ (B) $1 < x < 3$ (C) $x < 1$ (D) $x < 3$

14. As an ice block melts, the rate at which its mass, M, decreases is directly proportional to the square root of the mass. Which equation describes this relationship?

(A) $\sqrt{M(t)} = kt$ (B) $\dfrac{dM}{dt} = k\sqrt{t}$ (C) $\dfrac{dM}{dt} = k\sqrt{M}$ (D) $\dfrac{dM}{dt} = \dfrac{k}{\sqrt{M}}$

15. The length of the curve $y = 2x^{3/2}$ between $x = 0$ and $x = 1$ is equal to

(A) $\dfrac{2}{27}\left(10^{3/2}\right)$ (B) $\dfrac{2}{27}\left(10^{3/2} - 1\right)$ (C) $\dfrac{2}{3}\left(10^{3/2}\right)$ (D) $\sqrt{5}$

16. If $y = x^2 \ln x$ $(x > 0)$, then y'' is equal to

(A) $3 + \ln x$ (B) $3 + 2\ln x$ (C) $3 + 3\ln x$ (D) $2 + x + \ln x$

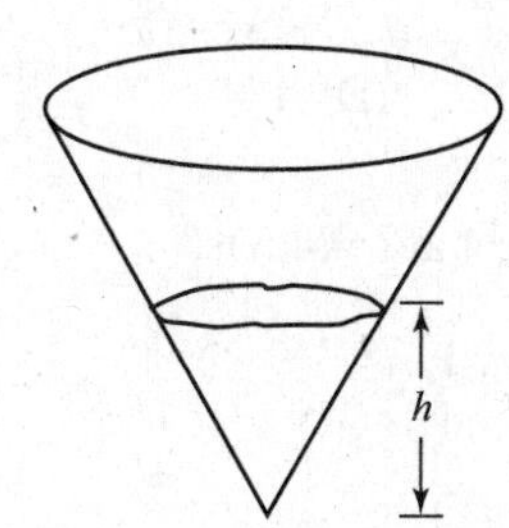

17. Water is poured at a constant rate into the conical reservoir shown above. If the depth of the water, h, is graphed as a function of time, the graph is

(A) constant (B) linear (C) concave upward (D) concave downward

18. A particle moves along the curve given parametrically by $x = \tan t$ and $y = 2\sin t$. At the instant when $t = \frac{\pi}{3}$, the particle's speed equals

(A) $\sqrt{3}$ (B) $\sqrt{5}$ (C) $\sqrt{6}$ (D) $\sqrt{17}$

19. Suppose $\dfrac{dy}{dx} = \dfrac{10x}{x+y}$ and $y = 2$ when $x = 0$. Use Euler's method with two steps to estimate y at $x = 1$.

(A) 1 (B) 2 (C) 3 (D) 5

GO ON TO THE NEXT PAGE

Questions 20 and 21. The graph below consists of a quarter-circle and two line segments, and represents the velocity of an object during a 6-second interval.

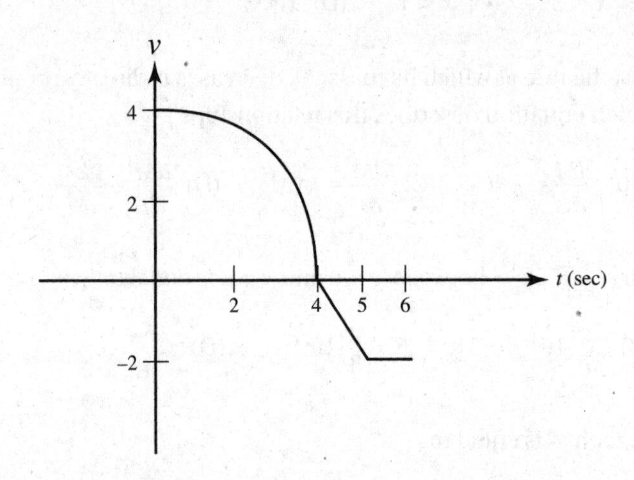

20. The object's average speed (in units/sec) during the 6-second interval is

(A) $\frac{4\pi + 3}{6}$ (B) $\frac{4\pi - 3}{6}$ (C) -1 (D) 1

21. The object's acceleration (in units/sec^2) at $t = 4.5$ is

(A) 0 (B) -1 (C) -2 (D) $-\frac{1}{4}$

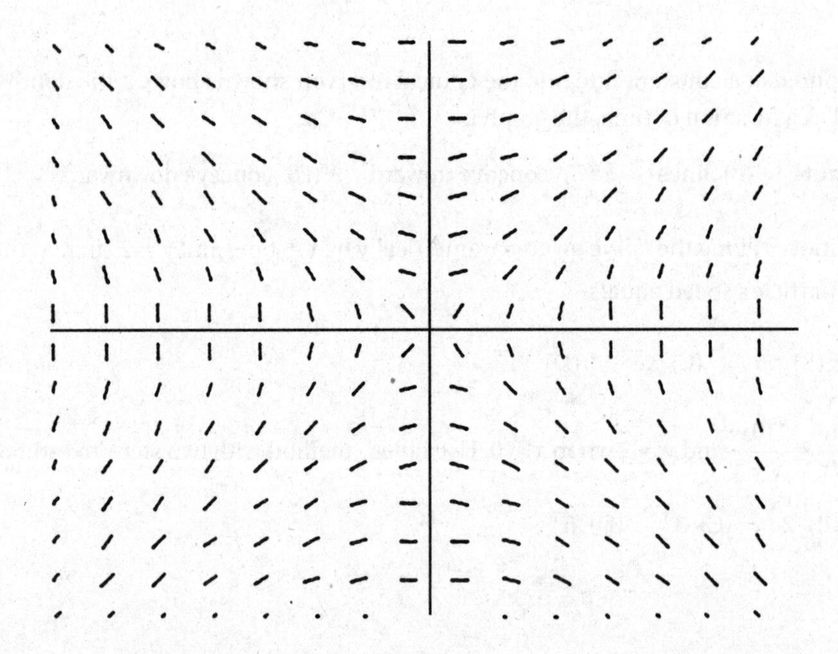

22. The slope field shown above is for which of the following differential equations?

(A) $\frac{dy}{dx} = -\frac{y}{x}$ (B) $\frac{dy}{dx} = \frac{1}{xy}$ (C) $\frac{dy}{dx} = -\frac{x}{y}$ (D) $\frac{dy}{dx} = \frac{x}{y}$

GO ON TO THE NEXT PAGE

23. If y is a differentiable function of x, then the slope of the curve of $xy^2 - 2y + 4y^3 = 6$ at the point where $y = 1$ is

(A) $-\frac{1}{18}$ (B) 0 (C) $\frac{5}{18}$ (D) $\frac{1}{3}$

24. For the function f shown in the graph, which has the smallest value on the interval $2 \le x \le 6$?

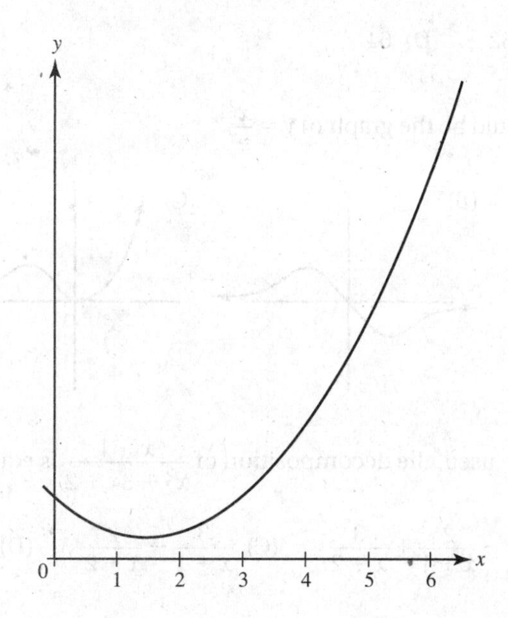

(A) $\int_2^6 f(x)dx$

(B) The left Riemann Sum with 8 equal subintervals.

(C) The midpoint Riemann Sum with 8 equal subintervals.

(D) The trapezoidal approximation with 8 equal subintervals.

25. The table shows some values of a differentiable function f and its derivative f':

x	0	1	2	3
$f(x)$	3	4	2	8
$f'(x)$	4	−1	1	10

Find $\int_0^3 f'(x)dx$.

(A) 5 (B) 6 (C) 11.5 (D) 14

GO ON TO THE NEXT PAGE

DIAGNOSTIC TEST CALCULUS BC

26. The solution of the differential equation $\frac{dy}{dx} = 2xy^2$ for which $y = -1$ when $x = 1$ is

(A) $y = -\frac{1}{x^2}$ for $x \neq 0$ (B) $y = -\frac{1}{x^2}$ for $x > 0$ (C) $\ln y^2 = x^2 - 1$ for all x (D) $y = -\frac{1}{x}$ for $x > 0$

27. The base of a solid is the region bounded by the parabola $y^2 = 4x$ and the line $x = 2$. Each plane section perpendicular to the x-axis is a square. The volume of the solid is

(A) 8 (B) 16 (C) 32 (D) 64

28. Which of the following could be the graph of $y = \frac{x^2}{e^x}$?

(A)
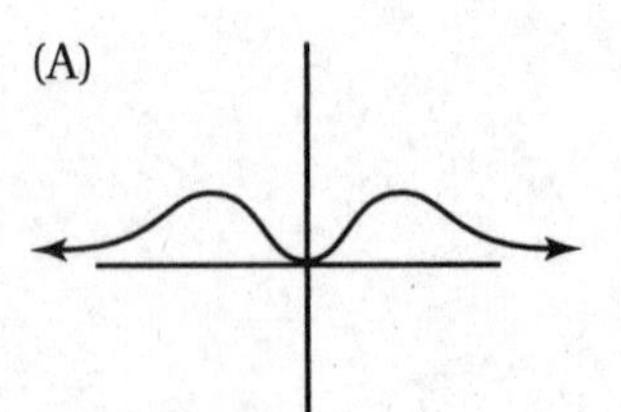

(B)
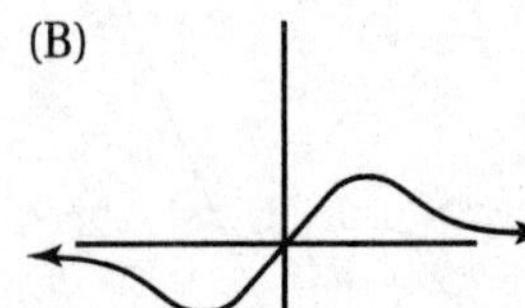

(C)
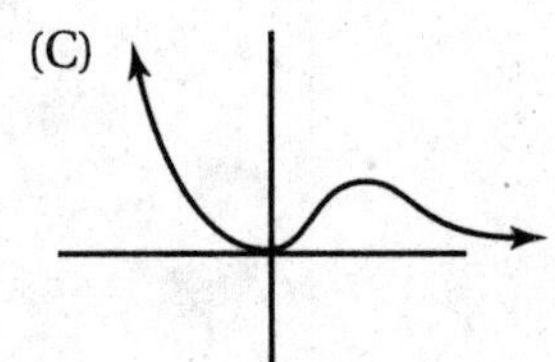

(D)
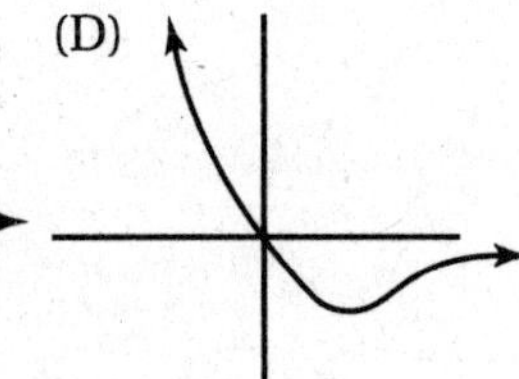

29. When partial fractions are used, the decomposition of $\frac{x-1}{x^2+3x+2}$ is equal to

(A) $\frac{4}{x+1} - \frac{3}{x+2}$ (B) $-\frac{2}{x+1} + \frac{3}{x+2}$ (C) $\frac{3}{x+1} - \frac{2}{x+2}$ (D) $\frac{2}{x+1} - \frac{5}{x+2}$

30. If $F(x) = \int_0^{2x} \frac{1}{1-t^3}\, dt$, then $F'(x) =$

(A) $\frac{1}{1-x^3}$ (B) $\frac{2}{1-2x^3}$ (C) $\frac{1}{1-8x^3}$ (D) $\frac{2}{1-8x^3}$

STOP

END OF PART A, SECTION I

Part B

TIME: 45 MINUTES

> ***Some questions in this part of the examination require the use of a graphing calculator. There are 15 questions in Part B, for which 45 minutes are allowed.***
>
> **Directions:** Choose the best answer for each question. If the exact numerical value of the correct answer is not listed as a choice, select the choice that is closest to the exact numerical answer.

31. The series

$$(x-1)-\frac{(x-1)^2}{2!}+\frac{(x-1)^3}{3!}-\frac{(x-1)^4}{4!}+\cdots$$

converges

(A) for all real x (B) if $0 \leqq x < 2$ (C) if $0 < x \leqq 2$ (D) only if $x = 1$

32. If $f(x)$ is continuous at the point where $x = a$, which of the following statements may be false?

(A) $\lim_{x\to a} f(x) = f(a)$. (B) $f'(a)$ exists. (C) $f(a)$ is defined. (D) $\lim_{x\to a^-} f(x) = \lim_{x\to a^+} f(x)$.

33. A Maclaurin polynomial is to be used to approximate $y = \sin x$ on the interval $-\pi \leqq x \leqq \pi$. What is the least number of terms needed to guarantee no error greater than 0.1?

(A) 3 (B) 4 (C) 5 (D) 6

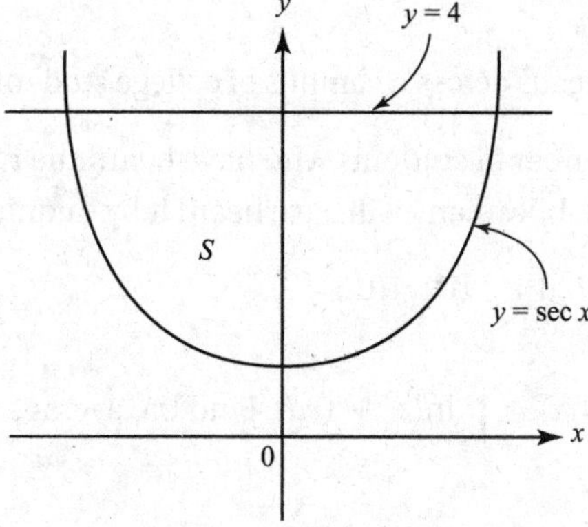

34. The region S in the figure above is bounded by $y = \sec x$ and $y = 4$. What is the volume of the solid formed when S is rotated about the x-axis?

(A) 11.385 (B) 23.781 (C) 53.126 (D) 108.177

DIAGNOSTIC TEST CALCULUS BC

GO ON TO THE NEXT PAGE

x	$f'(x)$
2	1.4
2.5	1.5
3	2.3

35. Values of $f'(x)$ are given in the table above. Using Euler's method with a step size of 0.5, approximate $f(3)$, if $f(2) = 6$.

(A) 7.90 (B) 7.45 (C) 4.55 (D) 4.10

36. If $x = 2t - 1$ and $y = 3 - 4t^2$, then $\frac{dy}{dx}$ is

(A) $4t$ (B) $-4t$ (C) $-\frac{1}{4t}$ (D) $-8t$

37. For the substitution $x = \sin\theta$, which integral is equivalent to $\int_0^1 \frac{\sqrt{1-x^2}}{x}dx$?

(A) $\int_0^1 \cot\theta\, d\theta$ (B) $\int_0^{\pi/2} \cot\theta\, d\theta$ (C) $\int_0^{\pi/2} \frac{\cos^2\theta}{\sin\theta}d\theta$ (D) $\int_0^1 \frac{\cos^2\theta}{\sin\theta}d\theta$

38. The coefficient of x^3 in the Taylor series of $\ln(1 - x)$ about $x = 0$ (the Maclaurin series) is

(A) $-\frac{2}{3}$ (B) $-\frac{1}{3}$ (C) $-\frac{1}{6}$ (D) $\frac{1}{3}$

39. The rate at which a rumor spreads across a campus of college students is given by $\frac{dP}{dt} = 0.16(1200 - P)$, where $P(t)$ represents the number of students who have heard the rumor after t days. If 200 students heard the rumor today ($t = 0$), how many will have heard it by midnight the day after tomorrow ($t = 2$)?

(A) 320 (B) 474 (C) 726 (D) 1015

40. Given function f, defined by $f(x) = \int_2^x \ln(t^2 + 1)\,dt$. Find the average rate of change of f on the interval $[-1,1]$.

(A) -1.433 (B) 0 (C) 0.264 (D) 0.693

41. A 26-foot ladder leans against a building so that its foot moves away from the building at the rate of 3 feet per second. When the foot of the ladder is 10 feet from the building, the top is moving down at the rate of r feet per second, where r is

(A) 0.80 (B) 1.25 (C) 7.20 (D) 12.50

GO ON TO THE NEXT PAGE

$f(3) = 3$	$g(3) = -3$	$h(3) = 3$
$f'(3) = 2$	$g'(3) = -4$	$h'(3) = 4$
$f''(3) = 1/2$	$g''(3) = 8$	$h''(3) = 2$

42. Given three twice-differentiable functions, $f(x)$, $g(x)$, and $h(x)$. The table above gives values for the functions and their first and second derivatives at $x = 3$. Find $\lim\limits_{x \to 3} \dfrac{(f(x))^2 - 3h(x)}{g(x) + h(x)}$.

(A) $-\frac{1}{5}$ (B) $\frac{1}{2}$ (C) 1 (D) nonexistent

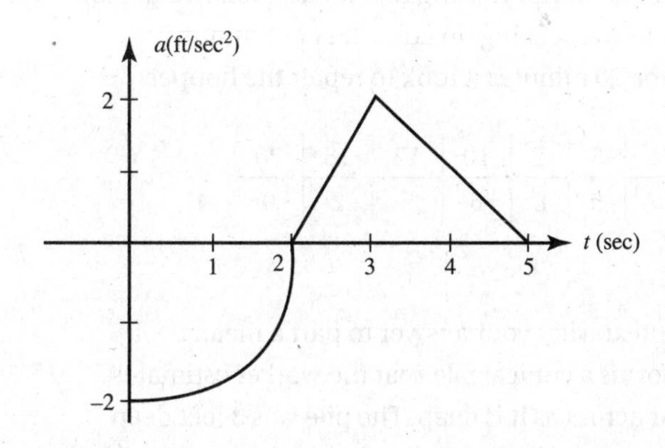

43. The graph above shows an object's acceleration (in ft/sec^2). It consists of a quarter-circle and two line segments. If the object was at rest at $t = 5$ seconds, what was its initial velocity?

(A) -2 ft/sec (B) $3 - \pi$ ft/sec (C) $\pi - 3$ ft/sec (D) $\pi + 3$ ft/sec

44. Water is leaking from a tank at the rate of $R(t) = 5\arctan\left(\frac{t}{5}\right)$ gallons per hour, where t is the number of hours since the leak began. How many gallons will leak out during the first day?

(A) 7 (B) 12 (C) 24 (D) 124

45. The first-quadrant area inside the rose $r = 3\sin 2\theta$ is approximately

(A) 1.5 (B) 1.767 (C) 3 (D) 3.534

STOP

END OF SECTION I

SECTION II

Part A

TIME: 30 MINUTES

2 PROBLEMS

A graphing calculator is required for some of these problems. See instructions on page 4.

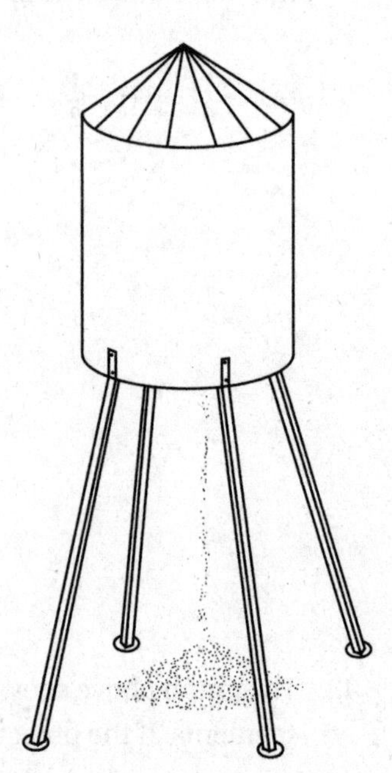

1. When a faulty seam opened at the bottom of an elevated hopper, grain began leaking out onto the ground. After a while, a worker spotted the growing pile below and began making repairs. The following table shows how fast the grain was leaking (in cubic feet per minute) at various times during the 20 minutes it took to repair the hopper.

t (min)	0	4	5	7	10	12	18	20
$L(t)$ (ft^3/min)	4	7	9	8	6	5	2	0

(a) Estimate $L'(15)$.

(b) Explain in this context what your answer to part a means.

(c) The falling grain forms a conical pile that the worker estimates to be 5 times as far across as it is deep. The pile was 3 feet deep when the repairs had been half completed. How fast was the depth increasing then?

NOTE: The volume of a cone with height h and radius r is given by $V=\frac{1}{3}\pi r^2 h$.

(d) Use a trapezoidal sum with seven subintervals as indicated in the table to approximate $\int_0^{20} L(t)\,dt$. Using correct units, explain the meaning of $\int_0^{20} L(t)\,dt$ in the context of the problem.

2. A particle is moving in the plane with position $(x(t), y(t))$ at time t. It is known that $\frac{dx}{dt} = -2t$ and $\frac{dy}{dt} = e^t$. The position at time $t = 0$ is $x(0) = 4$ and $y(0) = 3$.

(a) Find the speed of the particle at time $t = 2$, and find the acceleration vector at time $t = 2$.

(b) Find the slope of the tangent line to the path of the particle at $t = 2$.

(c) Find the position of the particle at $t = 2$.

(d) Find the total distance traveled by the particle on the interval $0 \le t \le 2$.

END OF PART A, SECTION II

Part B

TIME: 60 MINUTES

4 PROBLEMS

No calculator is allowed for any of these problems.
If you finish Part B before time has expired, you may return to work on Part A, but you may not use a calculator.

3. The graph of function f consists of the semicircle and line segment shown in the figure. Define the area function $A(x) = \int_0^x f(t)dt$ for $0 \le x \le 18$.

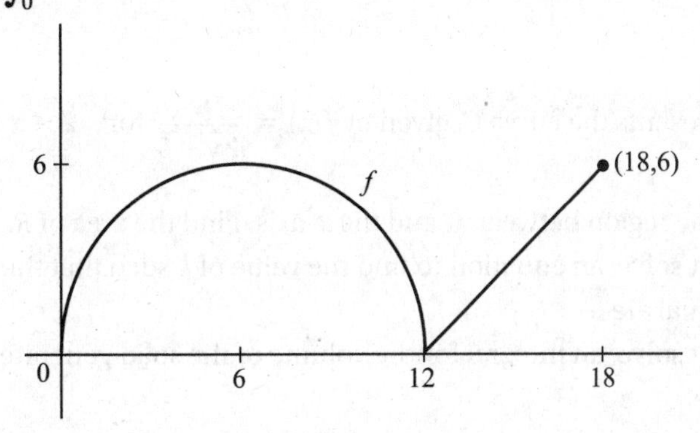

(a) Find $A(6)$ and $A(18)$.
(b) What is the average value of f on the interval $0 \le x \le 18$?
(c) Write the equation of the line tangent to the graph of A at $x = 6$.
(d) Use this line to estimate the area between f and the x-axis on $[0,7]$.
(e) Give the coordinates of any points of inflection on the graph of A. Justify your answer.

4. Let f be the function satisfying the differential equation $\frac{dy}{dx} = 2x(y^2 + 1)$ and passing through $(0, -1)$.

(a) Sketch the slope field for this differential equation at the points shown.

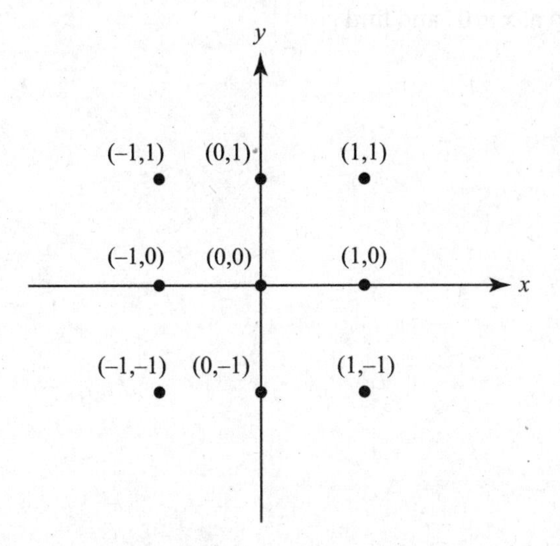

(b) Use Euler's method with a step size of 0.5 to estimate $f(1)$.
(c) Solve the differential equation, expressing f as a function of x.

GO ON TO THE NEXT PAGE

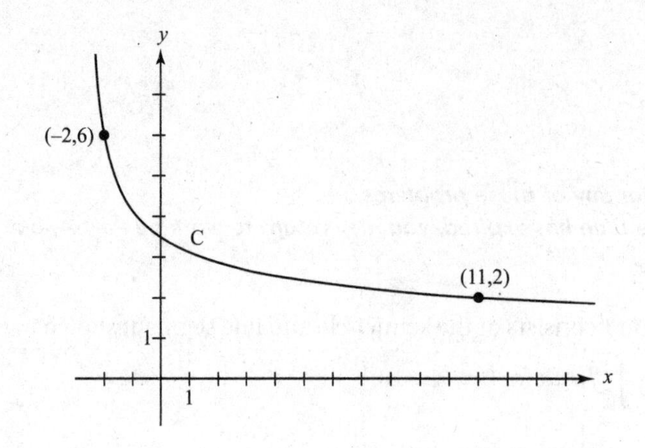

5. The graph above represents the curve C, given by $f(x) = \dfrac{6}{\sqrt[3]{2x+5}}$ for $-2 \le x \le 11$.

(a) Let R represent the region between C and the x-axis. Find the area of R.
(b) Set up, but do not solve, an equation to find the value of k such that the line $x = k$ divides R into two regions of equal area.
(c) Set up, but do not solve, an integral for the volume of the solid generated when R is rotated around the x-axis.

6. The function p is given by the series

$$p(x) = 2 + 2(x-2) + 2(x-2)^2 + \cdots + 2(x-2)^n + \cdots = \sum_{n=0}^{\infty} 2(x-2)^n$$

(a) Find the interval of convergence for p. Justify your answer.
(b) The series that defines p is the Taylor series about $x = 2$. Find the sum of the series for p.
(c) Let $q(x) = \int_2^x p(t)dt$. Find $q\left(\frac{3}{2}\right)$, if it exists, or explain why it cannot be determined.
(d) Let r be defined as $r(x) = p(x^3 + 2)$. Find the first three terms and the general term for the Taylor series for r centered at $x = 0$, and find $r\left(-\frac{1}{2}\right)$.

END OF TEST

DIAGNOSTIC TEST CALCULUS BC

ANSWER KEY
Diagnostic Test Calculus BC

Part A

1. **A** (2)
2. **D** (3)
3. **A** (4)
4. **C** (6)
5. **D** (3)
6. **C** (4)
7. **B** (3)
8. **B** (5)
9. **D** (10)
10. **D** (3)
11. **D** (3)
12. **A** (5)
13. **D** (4)
14. **C** (9)
15. **B** (7)
16. **B** (3)
17. **D** (4)
18. **D** (8)
19. **C** (9)
20. **A** (6)
21. **C** (6)
22. **D** (9)
23. **A** (3)
24. **B** (6)
25. **A** (6)
26. **B** (9)
27. **C** (7)
28. **C** (2)
29. **B** (5)
30. **D** (6)

Part B

31. **A** (10)
32. **B** (2)
33. **B** (10)
34. **D** (7)
35. **B** (9)
36. **B** (3)
37. **C** (6)
38. **B** (10)
39. **B** (9)
40. **C** (4)
41. **B** (4)
42. **B** (3)
43. **C** (6)
44. **D** (8)
45. **D** (7)

NOTE: Chapters that review and offer additional practice for each topic are specified in parentheses.

ANSWERS EXPLAINED

The explanations for questions not given below will be found in the answer section for Calculus AB Diagnostic Test on pages 29–35. Some questions in Section I of Diagnostic Tests AB1 and BC1 are identical.

Section I Multiple-Choice

Part A

7. **(B)** The volume is given by $\pi\int_0^{\infty} e^{-2x}dx$, an improper integral.

$$\lim_{k\to\infty}\pi\int_0^k e^{-2x}dx = \lim_{x\to\infty}\left(-\frac{1}{2}\pi\int_0^k e^{-2x}(-2dx)\right)$$
$$= \lim_{x\to\infty}\left(-\frac{1}{2}\pi e^{-2x}\Big|_0^k\right)$$
$$= \lim_{k\to\infty}\left(-\frac{1}{2}\pi(e^{-2k}-e^0)\right)$$
$$= \frac{\pi}{2}$$

9. **(D)** $\lim_{n\to\infty}\frac{n}{5n+1} = \frac{1}{5} \neq 0$. Series fails the nth Term Test.

12. **(A)** Use the Parts Formula with $u = x$ and $dv = e^x\,dx$. Then $du = dx$ and $v = e^x$, and

$$\left(xe^x - \int e^x dx\right)\Big|_0^1 = (xe^x - e^x)\Big|_0^1 = (e - e) - (0 - 1).$$

15. **(B)** The arc length is given by the integral $\int_0^1 \sqrt{1 + (3x^{1/2})^2}\,dx$ which is

$$\frac{1}{9}\int_0^1 (1 + 9x)^{1/2}(9dx) = \frac{1}{9}\cdot\frac{2}{3}\cdot(1 + 9x)^{3/2}\Big|_0^1 = \frac{2}{27}(10^{3/2} - 1).$$

18. **(D)** $\mathbf{v}(t) = \langle \sec^2 t, \cos t\rangle$, so $\mathbf{v}\left(\frac{\pi}{3}\right) = \left\langle 2^2, 2\cdot\frac{1}{2}\right\rangle$ and $|\mathbf{v}| = \sqrt{4^2 + 1^2}$.

19. **(C)** The initial point given (x_0, y_0) is given to be $(0, 2)$; use this initial point with step size $\Delta x = 0.5$.

Note: $f'(x_n, y_n) = \frac{dy}{dx}\Big|_{(x_n,\, y_n)}$

$x_1 = 0.5;\ y_1 = y_0 + f'(0, 2)(0.5) = 2 + (0)(0.5) = 2$

$x_2 = 1;\ y_2 = y_1 + f'(0.5, 2)(0.5) = 2 + \left(\frac{10\cdot 0.5}{0.5 + 2}\right)(0.5) = 2 + (2)(0.5) = 3$

24. **(B)** Since function f is increasing on the interval [2,6], rectangles based on left endpoints of the subintervals will all lie completely below the curve, and thus have smaller areas than any of the other sums or the definite integral. See pages 253–255.

29. **(B)**

$$\frac{x-1}{x^2+3x+2} = \frac{x-1}{(x+1)(x+2)} = \frac{A}{x+1} + \frac{B}{x+2}.$$
$$x - 1 = A(x + 2) + B(x + 1);$$
$$x = -2 \text{ implies } -3 = -B, \text{ or } B = 3;$$
$$x = -1 \text{ implies } -2 = A, \text{ or } A = -2.$$

Part B

31. **(A)** Use the Ratio Test, page 398:

$$\lim_{n\to\infty}\left|\frac{(x-1)^{n+1}}{(n+1)}\cdot\frac{n!}{(x-1)^n}\right|\lim_{n\to\infty}\frac{1}{n+1}|x-1|,$$

which equals zero if $x \neq 1$. The series also converges if $x = 1$ (each term equals 0).

32. **(B)** The absolute value function $f(x) = |x|$ is continuous at $x = 0$, but $f'(0)$ does not exist.

33. **(B)** The Maclaurin series is

$$\sin x = x - \frac{x^3}{3!} + \frac{x^5}{5!} - \frac{x^7}{7!} + \frac{x^9}{9!} - \cdots$$

When an alternating series satisfies the Alternating Series Test, the sum is approximated by using a finite number of terms, and the error is less than the first term omitted. On the interval $-\pi \leqq x \leqq \pi$, the maximum error (numerically) occurs when $x = \pi$. Since

$$\frac{\pi^7}{7!} < 0.6 \text{ and } \frac{\pi^9}{9!} < 0.09,$$

four terms will suffice to assure no error greater than 0.1.

34. **(D)** S is the region bounded by $y = \sec x$, the y-axis, and $y = 4$.

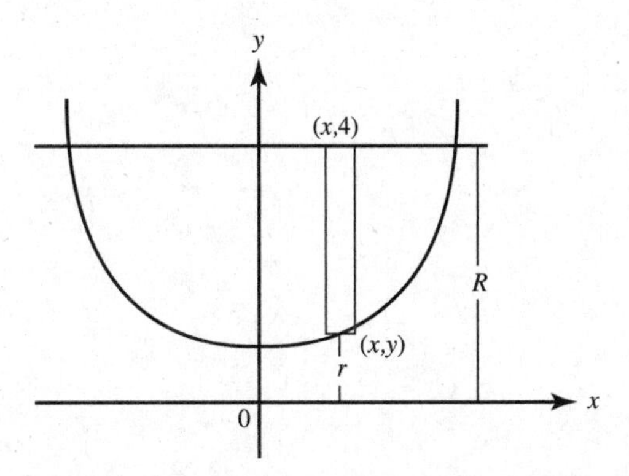

We revolve region S about the x-axis. Using washers, $\Delta V = \pi(R^2 - r^2)\,\Delta x$ (here $R = 4$ and $r = \sec x$). Symmetry allows us to double the volume generated by the first-quadrant portion of S. We find the upper limit of integration by solving $\sec x = 4$ and store the result ($x = 1.31811607$) as A. Then

$$V = 2\pi\int_0^A (16 - \sec^2 x)\,dx = 108.177$$

35. **(B)** $f(2) = 6 \Rightarrow f(2.5) = f(2) + f'(2)(0.5) = 6 + (1.4)(0.5) = 6.7$
$f(2.5) = 6.7 \Rightarrow f(3) = f(2.5) + f'(2.5)(0.5) = 6.7 + (1.5)(0.5) = 7.45$

36. **(B)** $\frac{dx}{dt} = 2$ and $\frac{dy}{dt} = -8t$. Hence $\frac{dy}{dx} = \frac{dy/dt}{dx/dt} = \frac{-8t}{2}$.

37. **(C)** $\sqrt{1-\sin^2\theta} = \cos\theta$, $dx = \cos\theta\, d\theta$, $\sin^{-1} 0 = 0$, $\sin^{-1} 1 = \frac{\pi}{2}$.

38. **(B)** The power series for $\ln(1-x)$, if $x < 1$, is $-x - \frac{x^2}{2} - \frac{x^3}{3} - \cdots$

39. **(B)** Solve by separation of variables; then

$$\frac{dP}{1200 - P} = 0.16\,dt,$$
$$-\ln(1200 - P) = 0.16t + C,$$
$$1200 - P = ce^{-0.16t}.$$

Use $P(0) = 200$; then $c = 1000$, so $P(x) = 1200 - 1000e^{-0.16t}$. Now $P(2) = 473.85$.

40. **(C)** Average Rate of Change:

$$\frac{f(1) - f(-1)}{1 - (-1)} = \frac{\int_2^1 \ln(t^2 + 1)\,dt - \int_2^{-1} \ln(t^2 + 1)\,dt}{2} = \frac{\int_{-1}^1 \ln(t^2 + 1)\,dt}{2} = 0.264$$

45. **(D)** The first quadrant area is $\frac{1}{2}\int_0^{\pi/2} (3 \sin 2\theta)^2\,d\theta \approx 3.534$.

DIAGNOSTIC TEST CALCULUS BC

Section II Free-Response

Part A

BC 1. See solution for AB/BC 1 on page 36.

BC 2. (a) Speed $= \sqrt{(x'(2))^2 + (y'(2))^2} = \sqrt{(-4)^2 + (e^2)^2} = 8.402.$

(b) Slope $= \dfrac{y'(2)}{x'(2)} = \dfrac{e^2}{-4} = -1.847.$

(c) $x(2) = x(0) + \int_0^2 x'(t)dt = 4 + \int_0^2 (-2t)dt = 0.$

$y(2) = y(0) + \int_0^2 y'(t)dt = 3 + \int_0^2 e^t\, dt = 9.389.$

(d) Total Distance $= \int_0^2 \sqrt{(x'(t))^2 + (y'(t))^2}\, dt = 7.566.$

(Review Chapters 4, 6, & 8)

DIAGNOSTIC TEST CALCULUS BC

Part B

BC 3. See solution for AB/BC 3, page 37.

BC 4. (a) Using the differential equation, evaluate the derivative at each point, then sketch a short segment having that slope. For example, at $(-1,-1)$, $\frac{dy}{dx} = 2(-1)\left((-1)^2+1\right) = -4$; draw a segment at $(-1,-1)$ that decreases steeply. Repeat this process at each of the other points. The result is shown below.

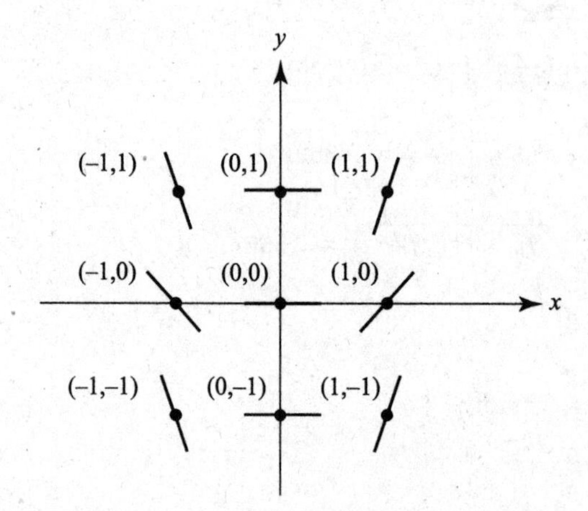

(b) At $(0,-1)$, $\frac{dy}{dx} = 2(0)\left((-1)^2+1\right) = 0$. For $\Delta x = 0.5$ and $\frac{\Delta y}{\Delta x} = 0$, $\Delta y = 0$, so move to $(0+0.5, -1+0) = (0.5, -1)$.

At $(0.5,-1)$, $\frac{dy}{dx} = 2(0.5)\left((-1)^2+1\right) = 2$. Thus, for $\Delta x = 0.5$ and $\frac{\Delta y}{\Delta x} = 2$, $\Delta y = 1$.

Move to $(0.5+0.5, -1+1) = (1, 0)$, then $f(1) \approx 0$.

(c) The differential equation $\frac{dy}{dx} = 2x(y^2+1)$ is separable:

$$\int \frac{dy}{y^2+1} = \int 2x\,dx$$

$$\arctan(y) = x^2 + c$$

$$y = \tan(x^2 + c)$$

It is given that f passes through $(0,-1)$, so $-1 = \tan(0^2 + c)$ and $c = -\frac{\pi}{4}$.

The solution is $f(x) = \tan\left(x^2 - \frac{\pi}{4}\right)$.

BC 5. See solution for AB/BC 5 on page 38.

BC 6. (a) The series is a geometric series with common ratio $(x-2)$. The series converges if $|x-2|<1$, so the interval of convergence is $1<x<3$.

NOTE: Because this is geometric, there is no need to check endpoints.

(b) Geometric series with first term 2 and common ratio $(x-2)$.

$$p(x)=\frac{2}{1-(x-2)}=\frac{2}{3-x} \text{ for } 1<x<3.$$

(c) $$q\left(\frac{3}{2}\right)=\int_{2}^{3/2}\frac{2}{3-x}\,dx=-\int_{3/2}^{2}\frac{2}{3-x}\,dx=\left(2\ln|3-x|\right)\Big|_{3/2}^{2}=-2\ln\left(\frac{3}{2}\right).$$

(d) $r(x)=p(x^3+2)=2+(x^3+2-2)+2(x^3+2-2)^2+\cdots$

$r(x)=2+2x^3+2x^6+\cdots+2x^{2n}+\cdots$

$$r\left(-\frac{1}{2}\right)=p\left(\left(-\frac{1}{2}\right)^3+2\right)=p\left(\frac{15}{8}\right)=\frac{2}{3-\frac{15}{8}}=\frac{16}{9}.$$

(Review Chapter 10)

DIAGNOSTIC TEST CALCULUS BC

Topical Review and Practice

1 Functions

CONCEPTS AND SKILLS

In this chapter you will review precalculus topics. Although these topics are not directly tested on the AP exam, reviewing them will reinforce some basic principles:

- **general properties of functions: domain, range, composition, inverse;**
- **special functions: absolute value, greatest integer; polynomial, rational, trigonometric, exponential, and logarithmic;**

and the BC topics,

- **parametrically defined curves**
- **polar curves**

A. DEFINITIONS

Function

Domain

Range

A1. A *function* f is a correspondence that associates with each element a of a set called the *domain* one and only one element b of a set called the *range.* We write

$$f(a) = b$$

to indicate that b is the value of f at a. The elements in the domain are called *inputs,* and those in the range are called *outputs.*

A function is often represented by an equation, a graph, or a table.

A vertical line cuts the graph of a function in at most one point.

Example 1

The domain of $f(x) = x^2 - 2$ is the set of all real numbers; its range is the set of all reals greater than or equal to -2. Note that

$$\begin{aligned} &f(0) = 0^2 - 2 = -2, \qquad f(-1) = (-1)^2 - 2 = -1, \\ &f(\sqrt{3}) = (\sqrt{3})^2 - 2 = 1, \qquad f(c) = c^2 - 2, \\ &f(x + h) - f(x) = [(x + h)^2 - 2] - [x^2 - 2] \\ &\qquad = x^2 + 2hx + h^2 - 2 - x^2 + 2 = 2hx + h^2. \end{aligned}$$

Example 2

Find the domains of: (a) $f(x) = \dfrac{4}{x-1}$; (b) $g(x) = \dfrac{x}{x^2-9}$; (c) $h(x) = \dfrac{\sqrt{4-x}}{x}$.

SOLUTIONS:

(a) The domain of $f(x) = \dfrac{4}{x-1}$ is the set of all reals except $x = 1$ (which we shorten to "$x \neq 1$").

(b) The domain of $g(x) = \dfrac{x}{x^2 - 9}$ is $x \neq 3, -3$.

(c) The domain of $h(x) = \dfrac{\sqrt{4-x}}{x}$ is $x \leqq 4, x \neq 0$ (which is a short way of writing $\{x \mid x$ is real, $x < 0$ or $0 < x \leqq 4\}$).

A2. Two functions f and g with the same domain may be combined to yield their sum and difference: $f(x) + g(x)$ and $f(x) - g(x)$, also written as $(f + g)(x)$ and $(f - g)(x)$, respectively; or their product and quotient: $f(x)g(x)$ and $f(x)/g(x)$, also written as $(fg)(x)$ and $(f/g)(x)$, respectively. The quotient is defined for all x in the shared domain except those values for which $g(x)$, the denominator, equals zero.

Example 3

If $f(x) = x^2 - 4x$ and $g(x) = x + 1$, then find $\dfrac{f(x)}{g(x)}$ and $\dfrac{g(x)}{f(x)}$.

SOLUTIONS: $\dfrac{f(x)}{g(x)} = \dfrac{x^2 - 4x}{x + 1}$ and has domain $x \neq -1$;

$\dfrac{g(x)}{f(x)} = \dfrac{x+1}{x^2 - 4x} = \dfrac{x+1}{x(x-4)}$ and has domain $x \neq 0, 4$.

Composition

A3. The *composition* (or *composite*) of f with g, written as $f(g(x))$ and read as "f of g of x," is the function obtained by replacing x wherever it occurs in $f(x)$ by $g(x)$. We also write $(f \circ g)(x)$ for $f(g(x))$. The domain of $(f \circ g)(x)$ is the set of all x in the domain of g for which $g(x)$ is in the domain of f.

Example 4A

If $f(x) = 2x - 1$ and $g(x) = x^2$, then does $f(g(x)) = g(f(x))$?

SOLUTION: $f(g(x)) = 2(x^2) - 1 = 2x^2 - 1$

$g(f(x)) = (2x - 1)^2 = 4x^2 - 4x + 1$.

In general, $f(g(x)) \neq g(f(x))$.

Example 4B

If $f(x) = 4x^2 - 1$ and $g(x) = \sqrt{x}$, find $f(g(x))$ and $g(f(x))$.

SOLUTIONS: $f(g(x)) = 4x - 1 \quad (x \geq 0)$; $g(f(x)) = \sqrt{4x^2 - 1} \quad \left(|x| \geq \dfrac{1}{2}\right)$.

Symmetry

A4. A function f is $\begin{matrix} odd \\ even \end{matrix}$ if, for all x in the domain of f, $\begin{matrix} f(-x) = -f(x) \\ f(-x) = f(x) \end{matrix}$.

The graph of an odd function is symmetric about the origin; the graph of an even function is symmetric about the y-axis.

➡ Example 5

The graphs of $f(x) = \frac{1}{2}x^3$ and $g(x) = 3x^2 - 1$ are shown in Figure N1–1; $f(x)$ is odd, $g(x)$ even.

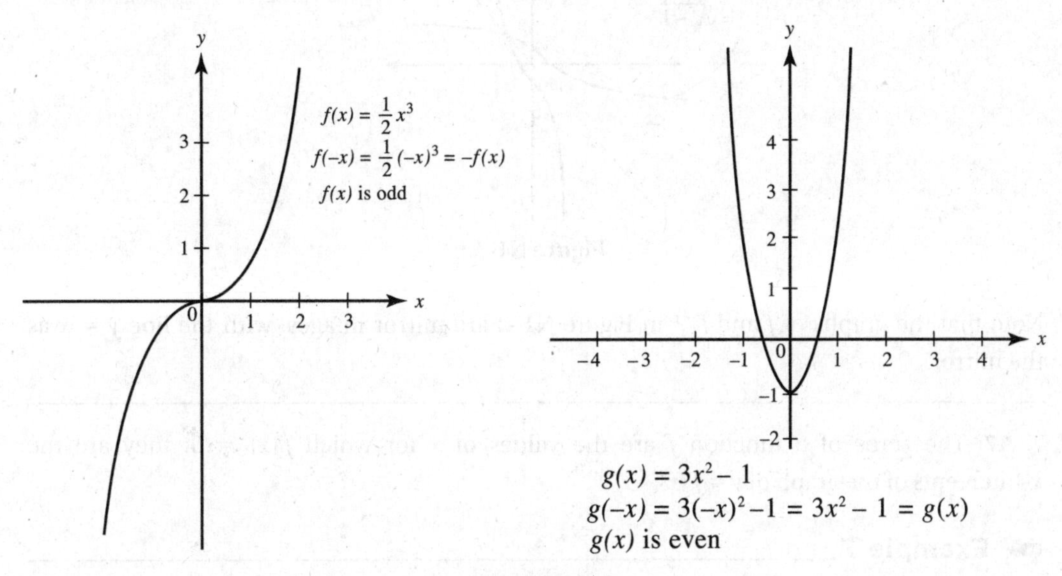

Figure N1–1

A5. If a function f yields a single output for each input and also yields a single input for every output, then f is said to be *one-to-one*. Geometrically, this means that any horizontal line cuts the graph of f in at most one point. The function sketched at the left in Figure N1–1 is one-to-one; the function sketched at the right is not. A function that is increasing (or decreasing) on an interval I is one-to-one on that interval (see pages 160–161 for definitions of increasing and decreasing functions).

A6. If f is one-to-one with domain X and range Y, then there is a function f^{-1}, with domain Y and range X, such that

$$f^{-1}(y_0) = x_0 \qquad \text{if and only if} \qquad f(x_0) = y_0.$$

Inverse

The function f^{-1} is the *inverse* of f. It can be shown that f^{-1} is also one-to-one and that its inverse is f. The graphs of a function and its inverse are symmetric with respect to the line $y = x$.

> **To find the inverse of $y = f(x)$, interchange x and y, then solve for y.**

➡ Example 6

Find the inverse of the one-to-one function $f(x) = x^3 - 1$.

SOLUTION: Interchange x and y: $x = y^3 - 1$

Solve for y: $y = \sqrt[3]{x + 1} = f^{-1}(x)$.

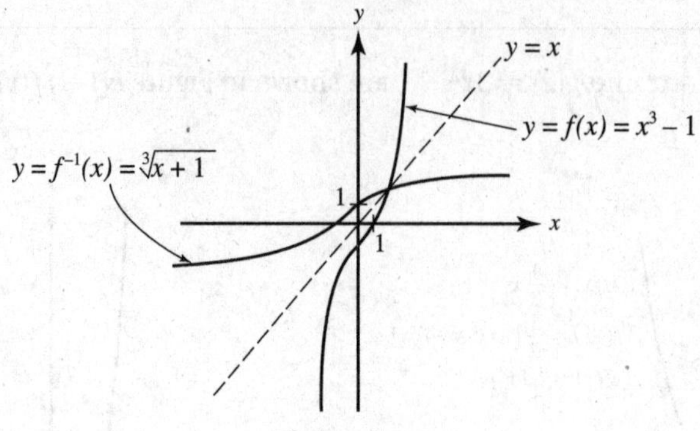

Figure N1–2

Note that the graphs of f and f^{-1} in Figure N1–2 are mirror images, with the line $y = x$ as the mirror.

Zeros

A7. The zeros of a function f are the values of x for which $f(x) = 0$; they are the x-intercepts of the graph of $y = f(x)$.

➡ Example 7

Find zeros of $f(x) = x^4 - 2x^2$.

SOLUTION: The zeros are the x's for which $x^4 - 2x^2 = 0$. The function has three zeros, since $x^4 - 2x^2 = x^2(x^2 - 2)$ equals zero if $x = 0, +\sqrt{2}$, or $-\sqrt{2}$.

B. SPECIAL FUNCTIONS

The *absolute-value* function $f(x) = |x|$ and the *greatest-integer* function $g(x) = [x]$ are sketched in Figure N1–3.

Absolute value

Greatest integer

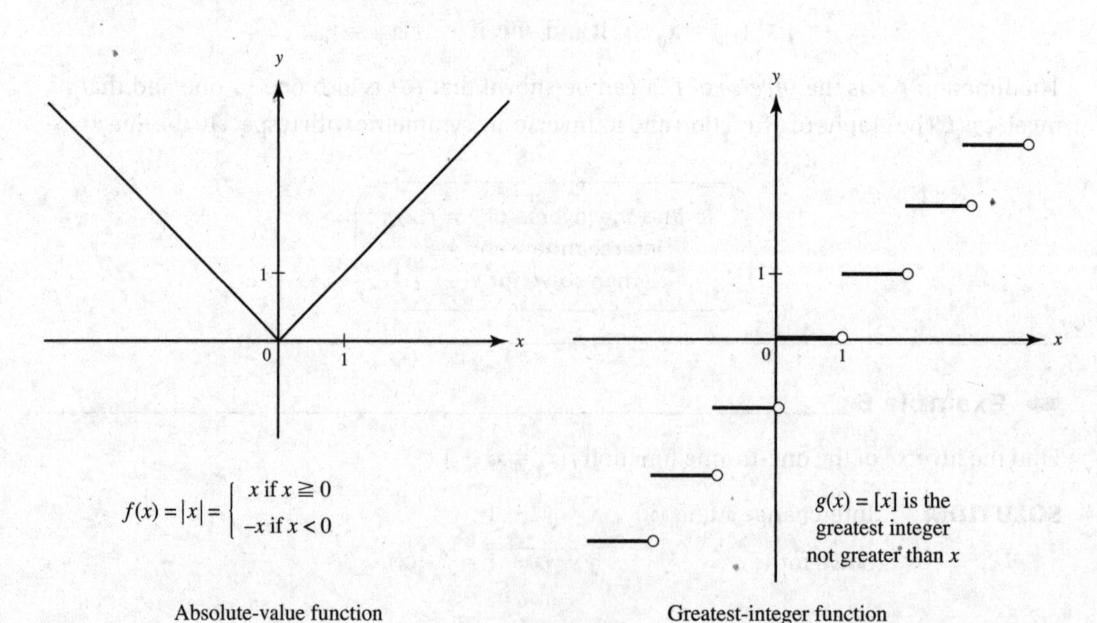

Figure N1–3

➡ Example 8

A function f is defined on the interval $[-2, 2]$ and has the graph shown in Figure N1–4.

(a) Sketch the graph of $y = |f(x)|$.
(b) Sketch the graph of $y = f(|x|)$.
(c) Sketch the graph of $y = -f(x)$.
(d) Sketch the graph of $y = f(-x)$.

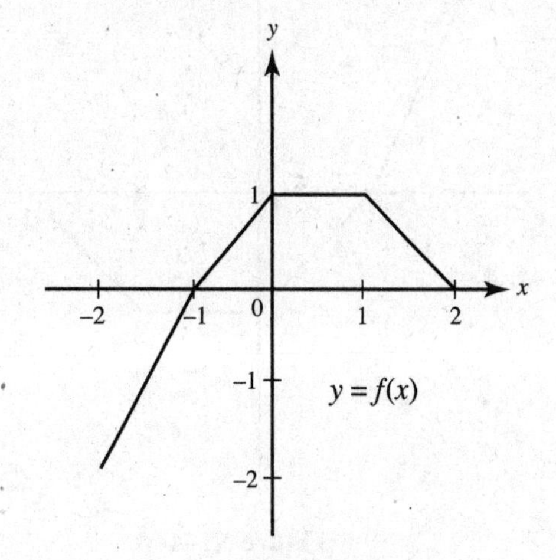

Figure N1–4

SOLUTIONS: The graphs are shown in Figures N1–4a through N1–4d.

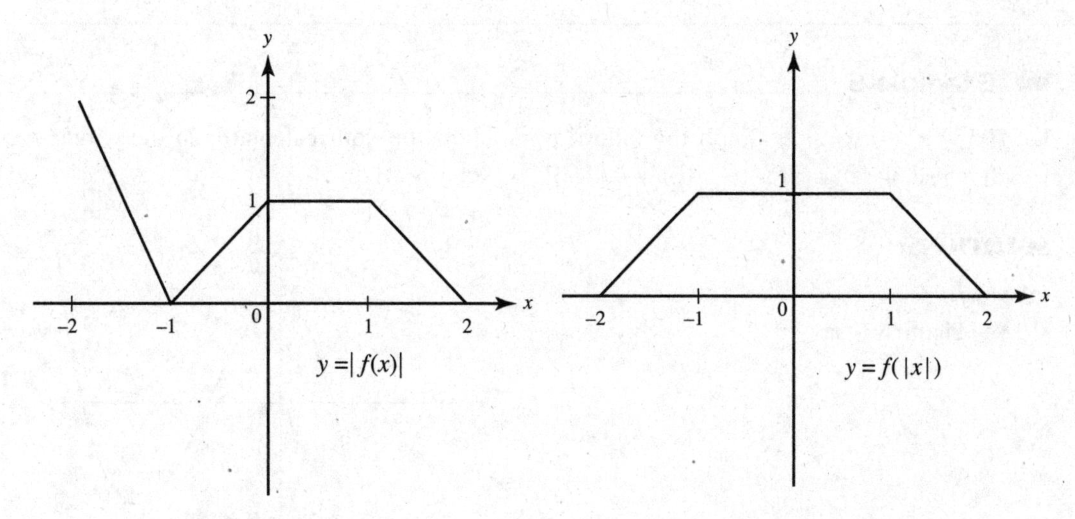

Figure N1–4a **Figure N1–4b**

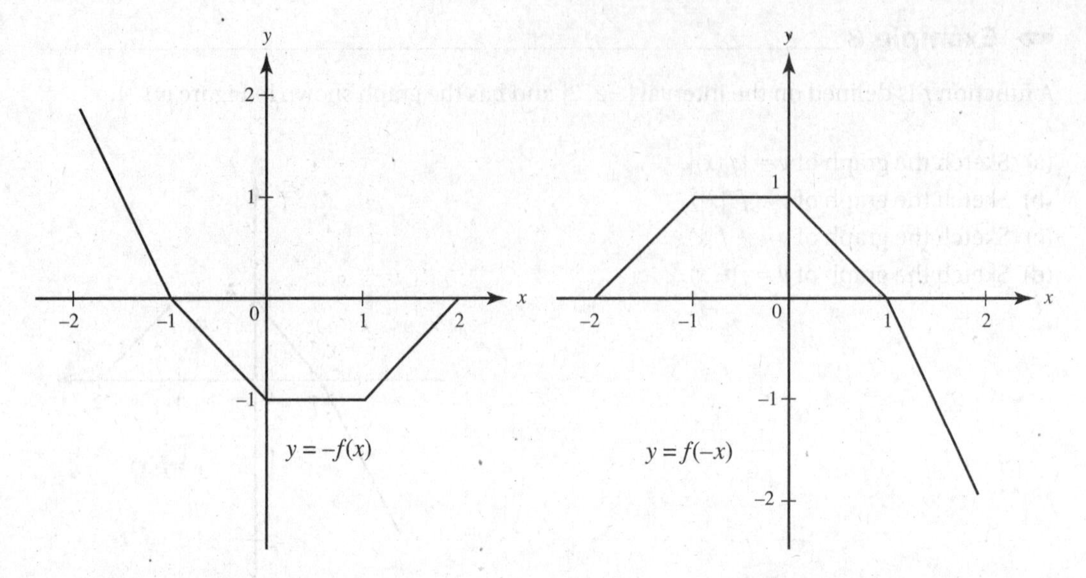

Figure N1–4c **Figure N1–4d**

Note that graph (c) of $y = -f(x)$ is the reflection of $y = f(x)$ in the x-axis, whereas graph (d) of $y = f(-x)$ is the reflection of $y = f(x)$ in the y-axis. How do the graphs of $|f(x)|$ and $f(|x|)$ compare with the graph of $f(x)$?

➡ Example 9

Let $f(x) = x^3 - 3x^2 + 2$. Graph the following functions on your calculator in the window $[-3,3] \times [-3,3]$: (a) $y = f(x)$; (b) $y = |f(x)|$; (c) $y = f(|x|)$.

SOLUTIONS:

(a) $y = f(x)$

See Figure N1–5a.

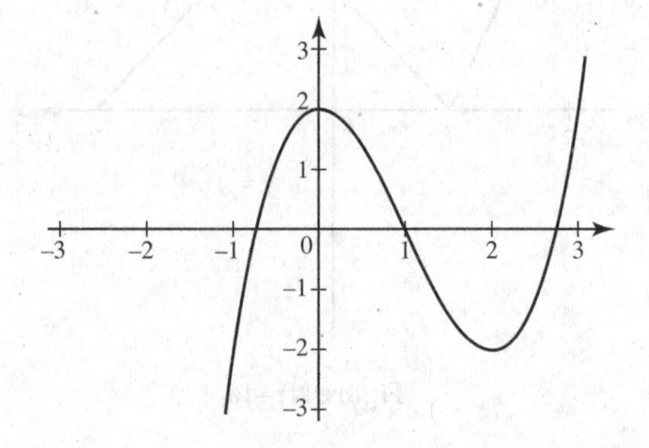

Figure N1–5a

(b) $y = |f(x)|$

See Figure N1–5b.

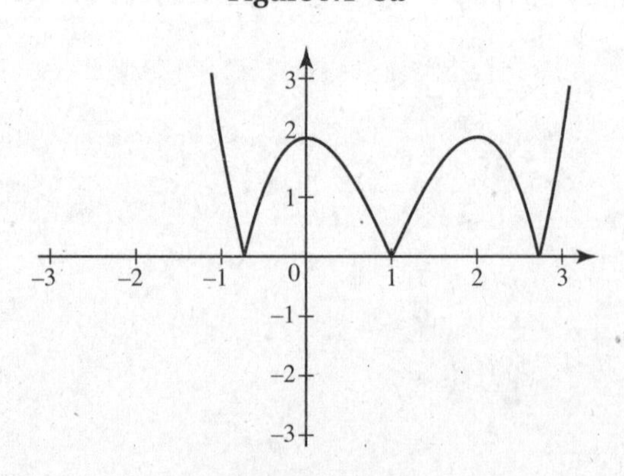

Figure N1–5b

(c) $y = f(|x|)$

See Figure N1–5c.

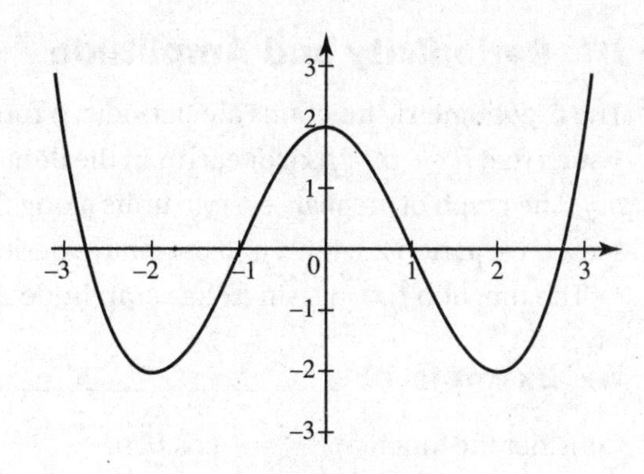

Figure N1–5c

Note how the graphs for (b) and (c) compare with the graph for (a).

C. POLYNOMIAL AND OTHER RATIONAL FUNCTIONS

C1. Polynomial Functions

Polynomial functions

A *polynomial function* is of the form

$$f(x) = a_0x^n + a_1x^{n-1} + a_2x^{n-2} + \cdots + a_{n-1}x + a_n,$$

where n is a positive integer or zero, and the a_k's, the *coefficients,* are constants. If $a_0 \neq 0$, the degree of the polynomial is n.

A *linear* function, $f(x) = mx + b$, is of the first degree; its graph is a straight line with slope m, the constant rate of change of $f(x)$ (or y) with respect to x, and b is the line's y-intercept.

A *quadratic* function, $f(x) = ax^2 + bx + c$, has degree 2; its graph is a parabola that opens up if $a > 0$, down if $a < 0$, and whose axis is the line $x = -\frac{b}{2a}$.

A *cubic*, $f(x) = a_0x^3 + a_1x^2 + a_2x + a_3$, has degree 3; calculus enables us to sketch its graph easily; and so on. The domain of every polynomial is the set of all reals.

C2. Rational Functions

Rational functions

A *rational function* is of the form

$$f(x) = \frac{P(x)}{Q(x)},$$

where $P(x)$ and $Q(x)$ are polynomials. The domain of f is the set of all reals for which $Q(x) \neq 0$.

D. TRIGONOMETRIC FUNCTIONS

The fundamental trigonometric identities, graphs, and reduction formulas are given in the Appendix.

D1. Periodicity and Amplitude

Trigonometric functions

The trigonometric functions are periodic. A function f is *periodic* if there is a positive number p such that $f(x + p) = f(x)$ for each x in the domain of f. The smallest such p is called the *period* of f. The graph of f repeats every p units along the x-axis. The functions $\sin x$, $\cos x$, $\csc x$, and $\sec x$ have period 2π; $\tan x$ and $\cot x$ have period π.

The function $f(x) = A \sin bx$ has amplitude A and period $\frac{2\pi}{b}$; $g(x) = \tan cx$ has period $\frac{\pi}{c}$.

➡ Example 10

Consider the function $f(x) = \frac{1}{k}\cos(kx)$.

(a) For what value of k does f have period 2?

(b) What is the amplitude of f for this k?

SOLUTIONS:

(a) Function f has period $\frac{2\pi}{k}$; since this must equal 2, we solve the equation $\frac{2\pi}{k} = 2$, getting $k = \pi$.

(b) It follows that the amplitude of f that equals $\frac{1}{k}$ has a value of $\frac{1}{\pi}$.

➡ Example 11

Consider the function $f(x) = 3 - \sin\frac{\pi x}{3}$.

Find (a) the period and (b) the maximum value of f.

(c) What is the smallest positive x for which f is a maximum?

(d) Sketch the graph.

SOLUTIONS:

(a) The period of f is $2\pi \div \frac{\pi}{3}$, or 6.

(b) Since the maximum value of $-\sin x$ is $-(-1)$ or $+1$, the maximum value of f is $3 + 1$ or 4.

(c) $-\left(\sin\frac{\pi x}{3}\right)$ equals $+1$ when $\sin\frac{\pi x}{3} = -1$, that is, when $\frac{\pi x}{3} = \frac{3\pi}{2}$. Solving yields $x = \frac{9}{2}$.

(d) We graph $y = 3 - \sin\frac{\pi x}{3}$ in $[-5,8] \times [0,5]$:

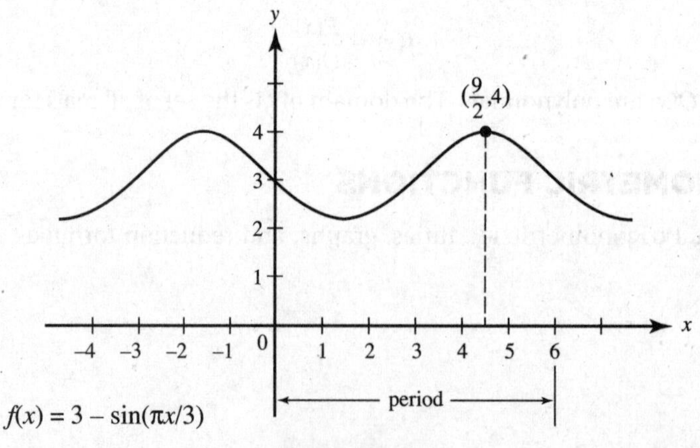

Figure N1–6

D2. Inverses

Inverse trig functions

We obtain *inverses* of the trigonometric functions by limiting the domains of the latter so each trigonometric function is one-to-one over its restricted domain. For example, we restrict

$$\sin x \text{ to } \quad -\frac{\pi}{2} \leqq x \leqq \frac{\pi}{2},$$

$$\cos x \text{ to } \quad 0 \leqq x \leqq \pi,$$

$$\tan x \text{ to } \quad -\frac{\pi}{2} < x < \frac{\pi}{2}.$$

The graphs of $f(x) = \sin x$ on $\left[-\frac{\pi}{2}, \frac{\pi}{2}\right]$ and of its inverse $f^{-1}(x) = \sin^{-1}x$ are shown in Figure N1–7. The inverse trigonometric function $\sin^{-1}x$ is also commonly denoted by arcsin x, which denotes *the* angle whose sine is x. The graph of $\sin^{-1}x$ is, of course, the reflection of the graph of $\sin x$ in the line $y = x$.

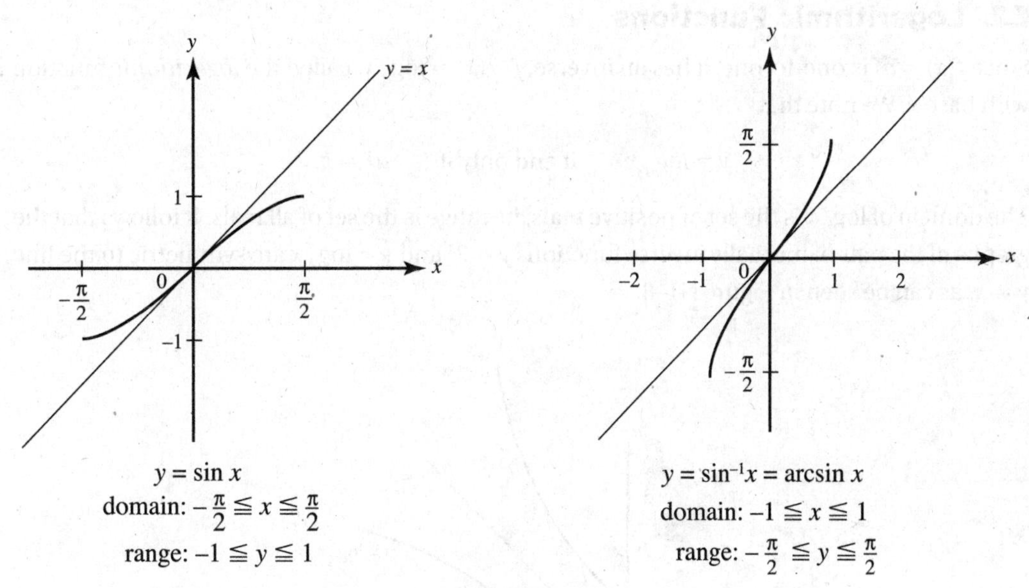

$y = \sin x$
domain: $-\frac{\pi}{2} \leqq x \leqq \frac{\pi}{2}$
range: $-1 \leqq y \leqq 1$

$y = \sin^{-1}x = \arcsin x$
domain: $-1 \leqq x \leqq 1$
range: $-\frac{\pi}{2} \leqq y \leqq \frac{\pi}{2}$

Figure N1–7

Also, for other inverse trigonometric functions,

$y = \cos^{-1} x$ (or arccos x) has domain $-1 \leqq x \leqq 1$ and range $0 \leqq y \leqq \pi$;

$y = \tan^{-1} x$ (or arctan x) has domain the set of reals and range $-\frac{\pi}{2} < y < \frac{\pi}{2}$.

Note also that

$$\sec^{-1}(x) = \cos^{-1}\left(\frac{1}{x}\right), \quad \csc^{-1}(x) = \sin^{-1}\left(\frac{1}{x}\right), \quad \text{and} \quad \cot^{-1}(x) = \frac{\pi}{2} - \tan^{-1}(x).$$

E. EXPONENTIAL AND LOGARITHMIC FUNCTIONS

E1. Exponential Functions

Exponential functions

The following laws of exponents hold for all rational m and n, provided that $a > 0, a \neq 1$:

$$a^0 = 1; \qquad a^1 = a; \qquad a^m \cdot a^n = a^{m+n}; \qquad a^m \div a^n = a^{m-n};$$

$$(a^m)^n = a^{mn}; \qquad a^{-m} = \frac{1}{a^m}.$$

The exponential function $f(x) = a^x$ $(a > 0, a \neq 1)$ is thus defined for all real x; its range is the set of positive reals. The graph of $y = a^x$, when $a = 2$, is shown in Figure N1–8.

Of special interest and importance in calculus is the exponential function $f(x) = e^x$, where e is an irrational number whose decimal approximation to five decimal places is 2.71828. We define e on page 97.

E2. Logarithmic Functions

Log functions

Since $f(x) = a^x$ is one-to-one, it has an inverse, $f^{-1}(x) = \log_a x$, called the *logarithmic* function with base a. We note that

$$y = \log_a x \qquad \text{if and only if} \qquad a^y = x.$$

The domain of $\log_a x$ is the set of positive reals; its range is the set of all reals. It follows that the graphs of the pair of mutually inverse functions $y = 2^x$ and $y = \log_2 x$ are symmetric to the line $y = x$, as can be seen in Figure N1–8.

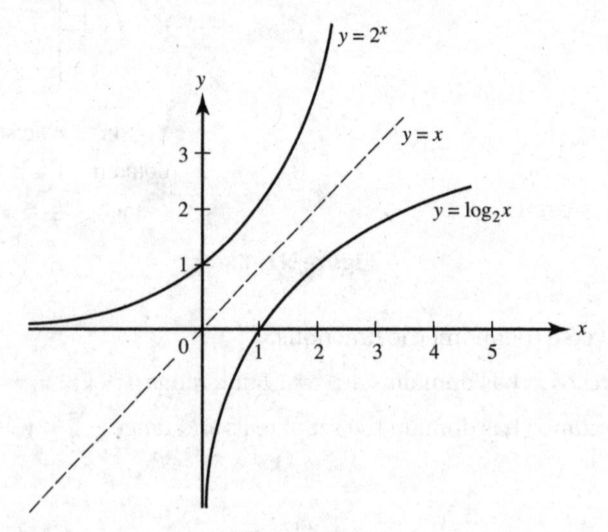

Figure N1–8

The logarithmic function $\log_a x$ $(a > 0, a \neq 1)$ has the following properties:

$$\log_a 1 = 0; \qquad \log_a a = 1; \qquad \log_a mn = \log_a m + \log_a n;$$

$$\log_a \frac{m}{n} = \log_a m - \log_a n; \qquad \log_a x^m = m \log_a x.$$

The logarithmic base e is so important and convenient in calculus that we use a special symbol:

$$\log_e x = \ln x.$$

Logarithms with base e are called *natural* logarithms. The domain of $\ln x$ is the set of positive reals; its range is the set of all reals. The graphs of the mutually inverse functions $\ln x$ and e^x are given in the Appendix.

F. PARAMETRICALLY DEFINED FUNCTIONS

BC ONLY

If the x- and y-coordinates of a point on a graph are given as functions f and g of a third variable, say t, then

$$x = f(t), \qquad y = g(t)$$

Parametric equations

are called *parametric equations* and t is called the *parameter.* When t represents time, as it often does, then we can view the curve as that followed by a moving particle as the time varies.

➡ Example 12

Find the Cartesian equation of, and sketch, the curve defined by the parametric equations

$$x = 4 \sin t, \qquad y = 5 \cos t \qquad (0 \leqq t \leqq 2\pi).$$

SOLUTION: We can eliminate the parameter t as follows:

$$\sin t = \frac{x}{4}, \qquad \cos t = \frac{y}{5}.$$

Since $\sin^2 t + \cos^2 t = 1$, we have

$$\left(\frac{x}{4}\right)^2 + \left(\frac{y}{5}\right)^2 = 1 \text{ or } \frac{x^2}{16} + \frac{y^2}{25} = 1$$

The curve is the ellipse shown in Figure N1–9.

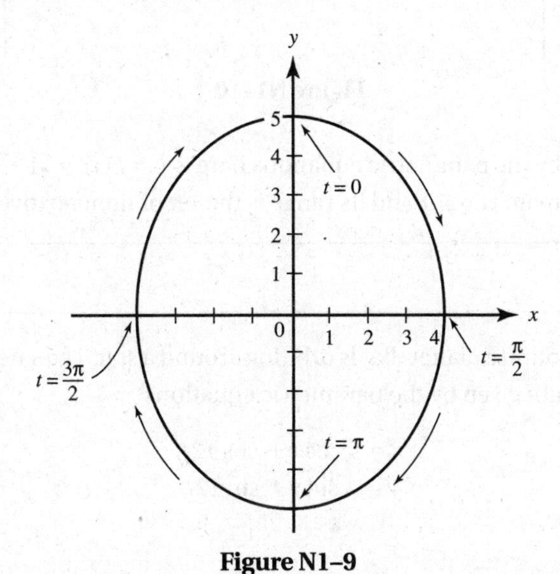

Figure N1–9

Note that, as t increases from 0 to 2π, a particle moving in accordance with the given parametric equations starts at point (0, 5) (when $t = 0$) and travels in a clockwise direction along the ellipse, returning to (0, 5) when $t = 2\pi$.

BC ONLY

➥ Example 13

Given the pair of parametric equations,

$$x = 1 - t, \qquad y = \sqrt{t} \qquad (t \geqq 0),$$

write an equation of the curve in terms of x and y, and sketch the graph.

SOLUTION: We can eliminate t by squaring the second equation and substituting for t in the first; then we have

$$y^2 = t \qquad \text{and} \qquad x = 1 - y^2.$$

We see the graph of the equation $x = 1 - y^2$ on the left in Figure N1–10. At the right we see only the upper part of this graph, the part defined by the parametric equations for which t and y are both restricted to nonnegative numbers.

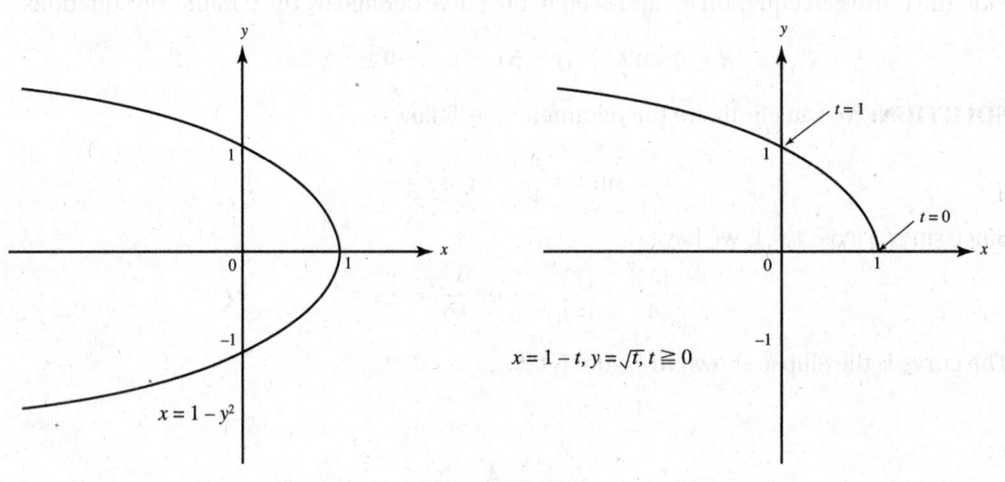

Figure N1–10

The function defined by the parametric equations here is $y = F(x) = \sqrt{1 - x}$, whose graph is at the right above; its domain is $x \leqq 1$ and its range is the set of nonnegative reals.

➥ Example 14

A satellite is in orbit around a planet that is orbiting around a star. The satellite makes 12 orbits each year. Graph its path given by the parametric equations

$$x = 4 \cos t + \cos 12t,$$
$$y = 4 \sin t + \sin 12t.$$

BC ONLY

SOLUTION: Shown below is the graph of the satellite's path using the calculator's parametric mode for $0 \le t \le 2\pi$.

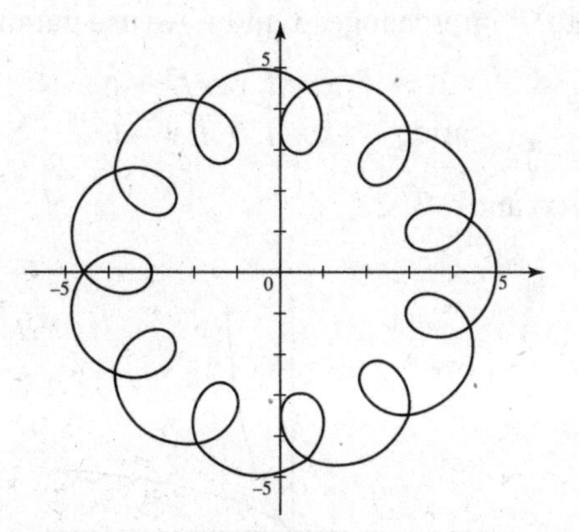

Figure N1–11

➡ Example 15

Graph $x = y^2 - 6y + 8$.

SOLUTION: We encounter a difficulty here. The calculator is constructed to graph y as a function of x: it accomplishes this by scanning horizontally across the window and plotting points in varying vertical positions. Ideally, we want the calculator to scan *down* the window and plot points at appropriate horizontal positions. But it won't do that.

One alternative is to interchange variables, entering x as Y_1 and y as X, thus entering $Y_1 = X^2 - 6X + 8$. But then, during all subsequent processing we must remember that we have made this interchange.

Less risky and more satisfying is to switch to parametric mode: Enter $x = t^2 - 6t + 8$ and $y = t$. Then graph these equations in $[-10,10] \times [-10,10]$, for t in $[-10,10]$. See Figure N1–12.

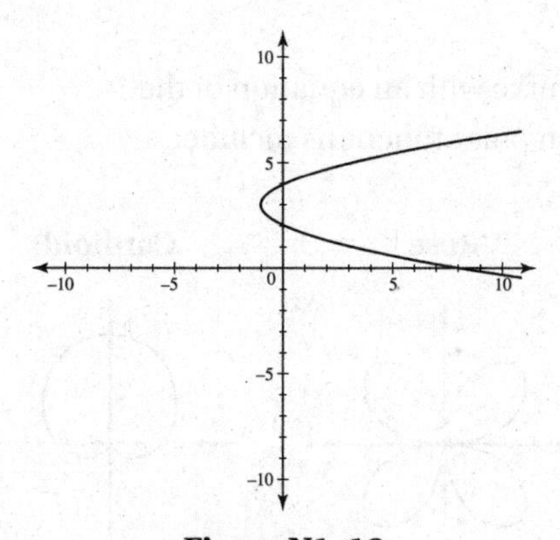

Figure N1–12

BC ONLY

Example 16

Let $f(x) = x^3 + x$; graph $f^{-1}(x)$.

SOLUTION: Recalling that f^{-1} interchanges x and y, we use parametric mode to graph

$$f: x = t, y = t^3 + t$$
$$\text{and } f^{-1}: x = t^3 + t, y = t.$$

Figure N1–13 shows both $f(x)$ and $f^{-1}(x)$.

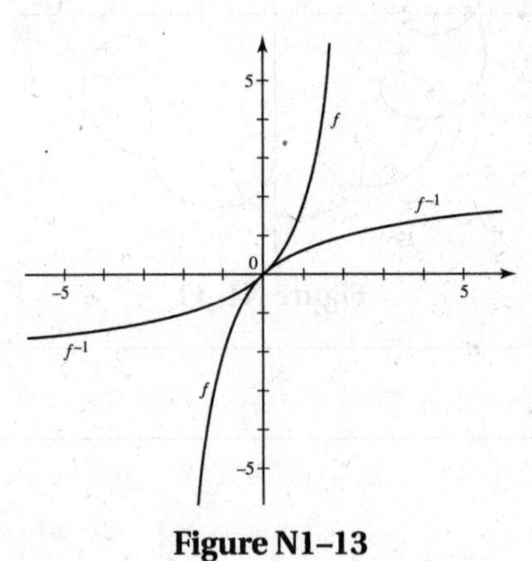

Figure N1–13

Parametric equations give rise to vector functions, which will be discussed in connection with motion along a curve in Chapter 4.

G. POLAR FUNCTIONS

Polar coordinates of the form (r, θ) identify the location of a point by specifying θ, an angle of rotation from the positive x-axis, and r, a distance from the origin, as shown in Figure N1–14.

Figure N1–14

Polar functions

A **polar function** defines a curve with an equation of the form $r = f(\theta)$. Some common polar functions include:

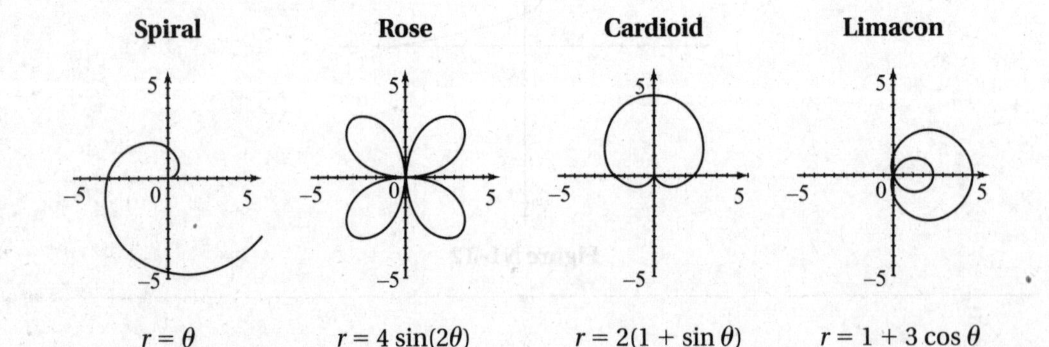

Example 17

Consider the polar function $r = 2 + 4\sin\theta$.

(a) For what values of θ in the interval $[0,2\pi]$ does the curve pass through the origin?

(b) For what value of θ in the interval $[0, \pi/2]$ does the curve intersect the circle $r = 3$?

SOLUTION:

(a) At the origin $r = 0$, so we want $2 + 4\sin\theta = 0$. Solving for θ yields $\sin\theta = -\frac{1}{2}$ which occurs at $\theta = \frac{7\pi}{6}$ and $\theta = \frac{11\pi}{6}$.

(b) The curves $r = 2 + 4\sin\theta$ and $r = 3$ intersect when $2 + 4\sin\theta = 3$, or $\sin\theta = \frac{1}{4}$. From the calculator we find $\theta = \arcsin\frac{1}{4} \approx 0.253$.

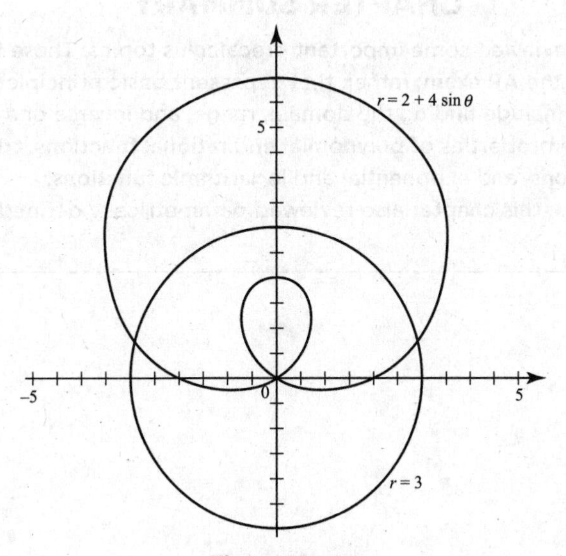

Figure N1–15

A polar function may also be expressed parametrically:

$$x = r\cos\theta, \qquad y = r\sin\theta$$

In this form, the curve $r = 2 + 4\sin\theta$ from Example 17 would be defined by:

$$x(\theta) = (2 + 4\sin\theta)\cos\theta, \qquad y(\theta) = (2 + 4\sin\theta)\sin\theta$$

BC ONLY

➡ Example 18

Find the (x,y) coordinates of the point on $r = 1 + \cos\theta$ where $\theta = \frac{\pi}{3}$.

SOLUTION: At $\theta = \frac{\pi}{3}$, $r = 1 + \cos\frac{\pi}{3} = 1 + \frac{1}{2} = \frac{3}{2}$.

Since $x = r\cos\theta = \frac{3}{2}\cos\frac{\pi}{3} = \frac{3}{2}\cdot\frac{1}{2} = \frac{3}{4}$ and $y = r\sin\theta = \frac{3}{2}\sin\frac{\pi}{3} = \frac{3}{2}\cdot\frac{\sqrt{3}}{2} = \frac{3\sqrt{3}}{4}$, the point is $\left(\frac{3}{4}, \frac{3\sqrt{3}}{4}\right)$.

CHAPTER SUMMARY

This chapter has reviewed some important precalculus topics. These topics are not directly tested on the AP exam; rather, they represent basic principles important in calculus. These include finding the domain, range, and inverse of a function; and understanding the properties of polynomial and rational functions, trigonometric and inverse trig functions, and exponential and logarithmic functions.

For BC students, this chapter also reviewed parametrically defined functions and polar curves.

PRACTICE EXERCISES

Directions: Answer Questions 1–34 *without* using your calculator.

1. If $f(x) = x^3 - 2x - 1$, then $f(-2) =$

   (A) -17 (B) -13 (C) -5 (D) -1 (E) 3

2. The domain of $f(x) = \frac{x-1}{x^2+1}$ is

   (A) all $x \neq 1$ (B) all $x \neq 1, -1$ (C) all $x \neq -1$
   (D) $x \geqq 1$ (E) all reals

3. The domain of $g(x) = \frac{\sqrt{x-2}}{x^2-x}$ is

   (A) all $x \neq 0, 1$ (B) $x \leqq 2, x \neq 0, 1$ (C) $x \leqq 2$
   (D) $x \geqq 2$ (E) $x > 2$

4. If $f(x) = x^3 - 3x^2 - 2x + 5$ and $g(x) = 2$, then $g(f(x)) =$

   (A) $2x^3 - 6x^2 - 2x + 10$ (B) $2x^2 - 6x + 1$ (C) -6
   (D) -3 (E) 2

5. With the functions and choices as in Question 4, which choice is correct for $f(g(x))$?

6. If $f(x) = x^3 + Ax^2 + Bx - 3$ and if $f(1) = 4$ and $f(-1) = -6$, what is the value of $2A + B$?

   (A) 12 (B) 8 (C) 0 (D) -2
   (E) It cannot be determined from the given information.

7. Which of the following equations has a graph that is symmetric with respect to the origin?

   (A) $y = \frac{x-1}{x}$ (B) $y = 2x^4 + 1$ (C) $y = x^3 + 2x$
   (D) $y = x^3 + 2$ (E) $y = \frac{x}{x^3+1}$

8. Let g be a function defined for all reals. Which of the following conditions is not sufficient to guarantee that g has an inverse function?

   (A) $g(x) = ax + b, a \neq 0$. (B) g is strictly decreasing. (C) g is symmetric to the origin.
   (D) g is strictly increasing. (E) g is one-to-one.

9. Let $y = f(x) = \sin(\arctan x)$. Then the range of f is

   (A) $\{y \mid 0 < y \leqq 1\}$ (B) $\{y \mid -1 < y < 1\}$ (C) $\{y \mid -1 \leqq y \leqq 1\}$
   (D) $\left\{y \mid -\frac{\pi}{2} < y < \frac{\pi}{2}\right\}$ (E) $\left\{y \mid -\frac{\pi}{2} \leqq y \leqq \frac{\pi}{2}\right\}$

10. Let $g(x) = |\cos x - 1|$. The maximum value attained by g on the closed interval $[0, 2\pi]$ is for x equal to

(A) -1 (B) 0 (C) $\frac{\pi}{2}$ (D) 2 (E) π

11. Which of the following functions is not odd?

(A) $f(x) = \sin x$ (B) $f(x) = \sin 2x$ (C) $f(x) = x^3 + 1$
(D) $f(x) = \frac{x}{x^2 + 1}$ (E) $f(x) = \sqrt[3]{2x}$

12. The roots of the equation $f(x) = 0$ are 1 and -2. The roots of $f(2x) = 0$ are

(A) 1 and -2 (B) $\frac{1}{2}$ and -1 (C) $-\frac{1}{2}$ and 1
(D) 2 and -4 (E) -2 and 4

13. The set of zeros of $f(x) = x^3 + 4x^2 + 4x$ is

(A) $\{-2\}$ (B) $\{0, -2\}$ (C) $\{0, 2\}$ (D) $\{2\}$ (E) $\{2, -2\}$

14. The values of x for which the graphs of $y = x + 2$ and $y^2 = 4x$ intersect are

(A) -2 and 2 (B) -2 (C) 2 (D) 0 (E) none of these

15. The function whose graph is a reflection in the y-axis of the graph of $f(x) = 1 - 3^x$ is

(A) $g(x) = 1 - 3^{-x}$ (B) $g(x) = 1 + 3^x$ (C) $g(x) = 3^x - 1$
(D) $g(x) = \log_3 (x - 1)$ (E) $g(x) = \log_3 (1 - x)$

16. Let $f(x)$ have an inverse function $g(x)$. Then $f(g(x)) =$

(A) 1 (B) x (C) $\frac{1}{x}$ (D) $f(x) \cdot g(x)$ (E) none of these

17. The function $f(x) = 2x^3 + x - 5$ has exactly one real zero. It is between

(A) -2 and -1 (B) -1 and 0 (C) 0 and 1
(D) 1 and 2 (E) 2 and 3

18. The period of $f(x) = \sin \frac{2\pi}{3} x$ is

(A) $\frac{1}{3}$ (B) $\frac{2}{3}$ (C) $\frac{3}{2}$ (D) 3 (E) 6

19. The range of $y = f(x) = \ln (\cos x)$ is

(A) $\{y \mid -\infty < y \leqq 0\}$ (B) $\{y \mid 0 < y \leqq 1\}$ (C) $\{y \mid -1 < y < 1\}$
(D) $\left\{y \mid -\frac{\pi}{2} < y < \frac{\pi}{2}\right\}$ (E) $\{y \mid 0 \leqq y \leqq 1\}$

20. If $\log_b (3^b) = \frac{b}{2}$, then $b =$

(A) $\frac{1}{9}$ (B) $\frac{1}{3}$ (C) $\frac{1}{2}$ (D) 3 (E) 9

21. Let f^{-1} be the inverse function of $f(x) = x^3 + 2$. Then $f^{-1}(x) =$

(A) $\frac{1}{x^3 - 2}$ (B) $(x + 2)^3$ (C) $(x - 2)^3$
(D) $\sqrt[3]{x + 2}$ (E) $\sqrt[3]{x - 2}$

22. The set of x-intercepts of the graph of $f(x) = x^3 - 2x^2 - x + 2$ is

(A) $\{1\}$ (B) $\{-1, 1\}$ (C) $\{1, 2\}$
(D) $\{-1, 1, 2\}$ (E) $\{-1, -2, 2\}$

23. If the domain of f is restricted to the open interval $\left(-\frac{\pi}{2}, \frac{\pi}{2}\right)$, then the range of $f(x) = e^{\tan x}$ is

(A) the set of all reals (B) the set of positive reals
(C) the set of nonnegative reals (D) $\{y \mid 0 < y \leqq 1\}$ (E) none of these

24. Which of the following is a reflection of the graph of $y = f(x)$ in the x-axis?

(A) $y = -f(x)$ (B) $y = f(-x)$ (C) $y = |f(x)|$
(D) $y = f(|x|)$ (E) $y = -f(-x)$

25. The smallest positive x for which the function $f(x) = \sin\left(\frac{x}{3}\right) - 1$ is a maximum is

(A) $\frac{\pi}{2}$ (B) π (C) $\frac{3\pi}{2}$ (D) 3π (E) 6π

26. $\tan\left(\arccos\left(-\frac{\sqrt{2}}{2}\right)\right) =$

(A) -1 (B) $-\frac{\sqrt{3}}{3}$ (C) $-\frac{1}{2}$ (D) $\frac{\sqrt{3}}{3}$ (E) 1

27. If $f^{-1}(x)$ is the inverse of $f(x) = 2e^{-x}$, then $f^{-1}(x) =$

(A) $\ln\left(\frac{2}{x}\right)$ (B) $\ln\left(\frac{x}{2}\right)$ (C) $\left(\frac{1}{2}\right)\ln x$ (D) $\sqrt{\ln x}$ (E) $\ln(2 - x)$

28. Which of the following functions does not have an inverse function?

(A) $y = \sin x \left(-\frac{\pi}{2} \leqq x \leqq \frac{\pi}{2}\right)$ (B) $y = x^3 + 2$ (C) $y = \frac{x}{x^2 + 1}$
(D) $y = \frac{1}{2}e^x$ (E) $y = \ln(x - 2)$ (where $x > 2$)

29. Suppose that $f(x) = \ln x$ for all positive x and $g(x) = 9 - x^2$ for all real x. The domain of $f(g(x))$ is

(A) $\{x \mid x \leqq 3\}$ (B) $\{x \mid |x| \leqq 3\}$ (C) $\{x \mid |x| > 3\}$
(D) $\{x \mid |x| < 3\}$ (E) $\{x \mid 0 < x < 3\}$

30. Suppose (as in Question 29) that $f(x) = \ln x$ for all positive x and $g(x) = 9 - x^2$ for all real x. The range of $y = f(g(x))$ is

(A) $\{y \mid y > 0\}$ (B) $\{y \mid 0 < y \leqq \ln 9\}$ (C) $\{y \mid y \leqq \ln 9\}$
(D) $\{y \mid y < 0\}$ (E) none of these

BC ONLY

31. The curve defined parametrically by $x(t) = t^2 + 3$ and $y(t) = t^2 + 4$ is part of a(n)

(A) line (B) circle (C) parabola (D) ellipse (E) hyperbola

32. Which equation includes the curve defined parametrically by $x(t) = \cos^2(t)$ and $y(t) = 2\sin(t)$?

(A) $x^2 + y^2 = 4$ (B) $x^2 + y^2 = 1$ (C) $4x^2 + y^2 = 4$
(D) $4x + y^2 = 4$ (E) $x + 4y^2 = 1$

33. Find the smallest value of θ in the interval $[0,2\pi]$ for which the rose $r = 2\cos(5\theta)$ passes through the origin.

(A) 0 (B) $\frac{\pi}{30}$ (C) $\frac{\pi}{20}$ (D) $\frac{\pi}{10}$ (E) $\frac{\pi}{5}$

34. For what value of θ in the interval $[0,\pi]$ do the polar curves $r = 3$ and $r = 2 + 2\cos\theta$ intersect?

(A) $\frac{\pi}{6}$ (B) $\frac{\pi}{4}$ (C) $\frac{\pi}{3}$ (D) $\frac{\pi}{2}$ (E) $\frac{2\pi}{3}$

35. On the interval $[0,2\pi]$ there is one point on the curve $r = \theta - 2\cos\theta$ whose x-coordinate is 2. Find the y-coordinate there.

(A) −4.594 (B) −3.764 (C) 1.979 (D) 4.263 (E) 5.201

Answer Key

1. **C**	10. **E**	19. **A**	28. **C**
2. **E**	11. **C**	20. **E**	29. **D**
3. **D**	12. **B**	21. **E**	30. **C**
4. **E**	13. **B**	22. **D**	31. **A**
5. **D**	14. **E**	23. **B**	32. **D**
6. **B**	15. **A**	24. **A**	33. **D**
7. **C**	16. **B**	25. **C**	34. **C**
8. **C**	17. **D**	26. **A**	35. **B**
9. **B**	18. **D**	27. **A**	

Answers Explained

1. **(C)** $f(-2) = (-2)^3 - 2(-2) - 1 = -5$.
2. **(E)** The denominator, $x^2 + 1$, is never 0.
3. **(D)** Since $x - 2$ may not be negative, $x \geqq 2$. The denominator equals 0 at $x = 0$ and $x = 1$, but these values are not in the interval $x \geqslant 2$.

4. **(E)** Since $g(x) = 2$, g is a constant function. Thus, for all $f(x)$, $g(f(x)) = 2$.
5. **(D)** $f(g(x)) = f(2) = -3$.
6. **(B)** Solve the pair of equations

$$\left\{\begin{array}{rl} 4 = & 1 + A + B - 3 \\ -6 = & -1 + A - B - 3 \end{array}\right\}.$$

Add to get A; substitute in either equation to get B. $A = 2$ and $B = 4$.

7. **(C)** The graph of $f(x)$ is symmetrical to the origin if $f(-x) = -f(x)$. In (C), $f(-x) = (-x)^3 + 2(-x) = -x^3 - 2x = -(x^3 + 2x) = -f(x)$.
8. **(C)** For g to have an inverse function it must be one-to-one. Note, on page 325, that although the graph of $y = xe^{-x^2}$ is symmetric to the origin, it is not one-to-one.
9. **(B)** Note that $-\frac{\pi}{2} < \arctan x < \frac{\pi}{2}$; the sine function varies from -1 to 1 as the argument varies from $-\frac{\pi}{2}$ to $\frac{\pi}{2}$.
10. **(E)** The maximum value of g is 2, attained when $\cos x = -1$. On $[0,2\pi]$, $\cos x = -1$ for $x = \pi$.
11. **(C)** f is odd if $f(-x) = -f(x)$. In (C), $f(-x) = (-x)^3 + 1 = -x^3 + 1 \neq -f(x)$
12. **(B)** Since $f(q) = 0$ if $q = 1$ or $q = -2$, $f(2x) = 0$ if $2x$, a replacement for q, equals 1 or -2.
13. **(B)** $f(x) = x(x^2 + 4x + 4) = x(x + 2)^2$; $f(x) = 0$ for $x = 0$ and $x = -2$.
14. **(E)** Solving simultaneously yields $(x + 2)^2 = 4x$; $x^2 + 4x + 4 = 4x$; $x^2 + 4 = 0$. There are no real solutions.
15. **(A)** The reflection of $y = f(x)$ in the y-axis is $y = f(-x)$.
16. **(B)** If g is the inverse of f, then f is the inverse of g. This implies that the function f assigns to each value $g(x)$ the number x.
17. **(D)** Since f is continuous (see page 97), then, if f is negative at a and positive at b, f must equal 0 at some intermediate point. Since $f(1) = -2$ and $f(2) = 13$, this point is between 1 and 2.
18. **(D)** The function $\sin bx$ has period $\frac{2\pi}{b}$. Then $2\pi \div \frac{2\pi}{3} = 3$.
19. **(A)** Since $\ln q$ is defined only if $q > 0$, the domain of $\ln \cos x$ is the set of x for which $\cos x > 0$, that is, when $0 < \cos x \leqq 1$. Thus $-\infty < \ln \cos x \leqq 0$.
20. **(E)** $\log_b 3^b = \frac{b}{2}$ implies $b \log_b 3 = \frac{b}{2}$. Then $\log_b 3 = \frac{1}{2}$ and $3 = b^{1/2}$. So $3^2 = b$.
21. **(E)** Interchange x and y: $x = y^3 + 2$

Solve for y: $y^3 = x - 2$, so $y = \sqrt[3]{x - 2} = f^{-1}(x)$.

22. **(D)** Since $f(1) = 0$, $x - 1$ is a factor of f. Since $f(x)$ divided by $x - 1$ yields $x^2 - x - 2$, $f(x) = (x - 1)(x + 1)(x - 2)$; the roots are $x = 1, -1$, and 2.
23. **(B)** If $-\frac{\pi}{2} < x < \frac{\pi}{2}$, then $-\infty < \tan x < \infty$ and $0 < e^{\tan x} < \infty$.
24. **(A)** The reflection of $f(x)$ in the x-axis is $-f(x)$.
25. **(C)** $f(x)$ attains its maximum when $\sin\left(\frac{x}{3}\right)$ does. The maximum value of the sine function is 1; the smallest positive occurrence is at $\frac{\pi}{2}$. Set $\frac{x}{3}$ equal to $\frac{\pi}{2}$.
26. **(A)** $\arccos\left(\frac{-\sqrt{2}}{2}\right) = \frac{3\pi}{4}$; $\tan\left(\frac{3\pi}{4}\right) = -1$.
27. **(A)** Interchange x and y: $x = 2e^{-y}$

Solve for y: $e^{-y} = \frac{x}{2}$, so $-y = \ln\frac{x}{2}$ and $y = -\ln\frac{x}{2}$.

Thus $f^{-1}(x) = \ln\frac{2}{x}$.

28. **(C)** The function in (C) is not one-to-one since, for each y between $-\frac{1}{2}$ and $\frac{1}{2}$ (except 0), there are two x's in the domain.

29. **(D)** The domain of the ln function is the set of positive reals. The function $g(x) > 0$ if $x^2 < 9$.

30. **(C)** Since the domain of $f(g)$ is $(-3, 3)$, $\ln(9 - x^2)$ takes on every real value less than or equal to $\ln 9$.

31. **(A)** Substituting $t^2 = x - 3$ in $y(t) = t^2 + 4$ yields $y = x + 1$.

32. **(D)** Using the identity $\cos^2(t) + \sin^2(t) = 1$, $x + \left(\frac{y}{2}\right)^2 = 1$.

33. **(D)** $2 \cos 5\theta = 0$ when $5\theta = \frac{\pi}{2}$.

34. **(C)** If $2 + 2 \cos \theta = 3$, then $\cos \theta = \frac{1}{2}$.

35. **(B)** For polar functions $x = r \cos \theta$. Solving $(\theta - 2 \cos \theta)\cos \theta = 2$ yields $\theta \approx 5.201$, and thus $y = r \sin \theta = (5.201 - 2 \cos 5.201)\sin 5.201$.

Limits and Continuity

2

CONCEPTS AND SKILLS

In this chapter, you will review

- **general properties of limits;**
- **how to find limits using algebraic expressions, tables, and graphs;**
- **horizontal and vertical asymptotes;**
- **continuity;**
- **removable, jump, and infinite discontinuities;**
- **and some important theorems, including the Squeeze Theorem, the Extreme Value Theorem, and the Intermediate Value Theorem.**

A. DEFINITIONS AND EXAMPLES

Limit

The number L is the *limit of the function* $f(x)$ as x approaches c if, as the values of x get arbitrarily close (but not equal) to c, the values of $f(x)$ approach (or equal) L. We write

$$\lim_{x \to c} f(x) = L.$$

In order for $\lim_{x \to c} f(x)$ to exist, the values of f must tend to the same number L as x approaches c from either the left or the right. We write

$$\lim_{x \to c^-} f(x)$$

One-sided limits

for the *left-hand* limit of f at c (as x approaches c through values *less* than c), and

$$\lim_{x \to c^+} f(x)$$

for the *right-hand* limit of f at c (as x approaches c through values *greater* than c).

Example 1

The greatest-integer function $g(x) = [x]$, shown in Figure N2–1, has different left-hand and right-hand limits at *every* integer. For example,

$$\lim_{x \to 1^-} [x] = 0 \quad \text{but} \quad \lim_{x \to 1^+} [x] = 1.$$

This function, therefore, does not have a limit at $x = 1$ or, by the same reasoning, at any other integer.

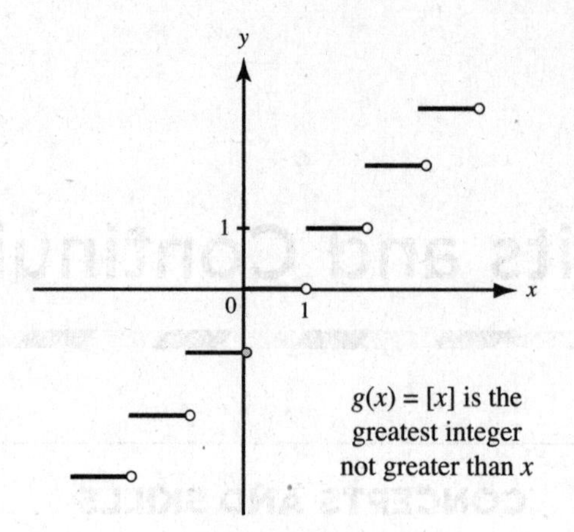

Figure N2–1

However, $[x]$ does have a limit at every nonintegral real number. For example,

$$\lim_{x\to0.6}[x]=0;\quad \lim_{x\to0.99}[x]=0;\quad \lim_{x\to2.01}[x]=2;\quad \lim_{x\to2.95}[x]=2$$

$$\lim_{x\to-3.1}[x]=-4;\quad \lim_{x\to-2.9}[x]=-3;\quad \lim_{x\to-0.9}[x]=-1;\quad \lim_{x\to-0.01}[x]=-1.$$

➡ Example 2

Suppose the function $y = f(x)$, graphed in Figure N2–2, is defined as follows:

$$f(x)=\begin{cases} x+1 & (-2<x<0) \\ 2 & (x=0) \\ -x & (0<x<2) \\ 0 & (x=2) \\ x-4 & (2<x\leqslant 4) \end{cases}$$

Determine whether limits of f, if any, exist at

(a) $x = -2$, (b) $x = 0$,
(c) $x = 2$, (d) $x = 4$.

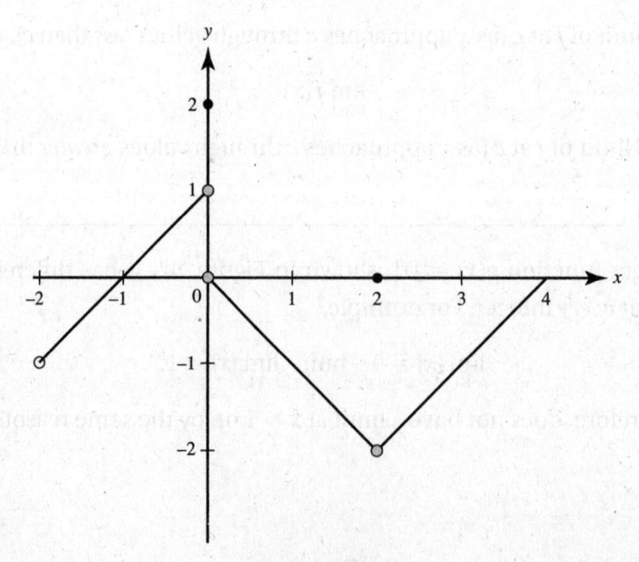

Figure N2–2

SOLUTIONS:

(a) $\lim_{x \to 2^+} f(x) = -1$, so the right-hand limit exists at $x = -2$, even though f is not defined at $x = -2$.

(b) $\lim_{x \to 0} f(x)$ does not exist. Although f is defined at $x = 0$ ($f(0) = 2$), we observe that $\lim_{x \to 0^-} f(x) = 1$ whereas $\lim_{x \to 0^+} f(x) = 0$. For the limit to exist at a point, the left-hand and right-hand limits must be the same.

(c) $\lim_{x \to 2} f(x) = -2$. This limit exists because $\lim_{x \to 2^-} f(x) = \lim_{x \to 2^+} f(x) = -2$. Indeed, the limit exists at $x = 2$ even though it is different from the value of f at 2 ($f(2) = 0$).

(d) $\lim_{x \to 4^-} f(x) = 0$, so the left-hand limit exists at $x = 4$.

Example 3

Prove that $\lim_{x \to 0} |x| = 0$.

SOLUTION: The graph of $|x|$ is shown in Figure N2–3.
We examine both left- and right-hand limits of the absolute-value function as $x \to 0$. Since

$$|x| = \begin{cases} -x \text{ if } x < 0 \\ x \text{ if } x > 0, \end{cases}$$

it follows that $\lim_{x \to 0^-} |x| = \lim_{x \to 0^-} (-x) = 0$ and $\lim_{x \to 0^+} |x| = \lim_{x \to 0^+} x = 0$.
Since the left-hand and right-hand limits both equal 0, $\lim_{x \to 0} |x| = 0$.
Note that $\lim_{x \to c} |x| = c$ if $c > 0$ but equals $-c$ if $c < 0$.

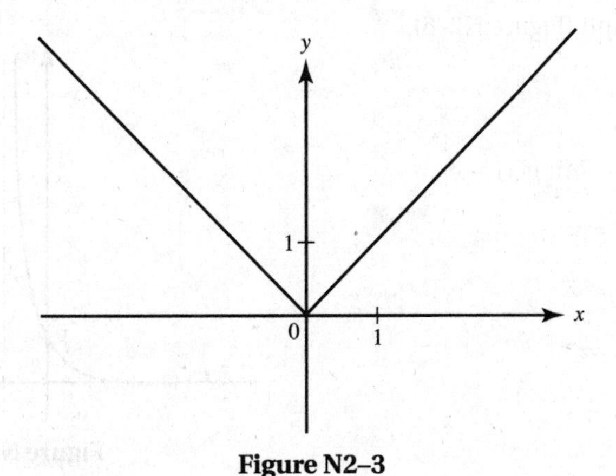

Figure N2–3

Definition

The function $f(x)$ is said to *become infinite* (positively or negatively) as x approaches c if $f(x)$ can be made arbitrarily large (positively or negatively) by taking x sufficiently close to c. We write

$$\lim_{x \to c} f(x) = +\infty \text{ (or } \lim_{x \to c} f(x) = -\infty).$$

Since for the limit to exist it must be a finite number, neither of the preceding limits exists.

This definition can be extended to include x approaching c from the left or from the right. The following examples illustrate these definitions.

Example 4

Describe the behavior of $f(x) = \frac{1}{x}$ near $x = 0$ using limits.

SOLUTION: The graph (Figure N2–4) shows that:

$$\lim_{x \to 0^-} \frac{1}{x} = -\infty,$$

$$\lim_{x \to 0^+} \frac{1}{x} = +\infty.$$

$$\lim_{x \to 0} \frac{1}{x} \text{ does not exist.}$$

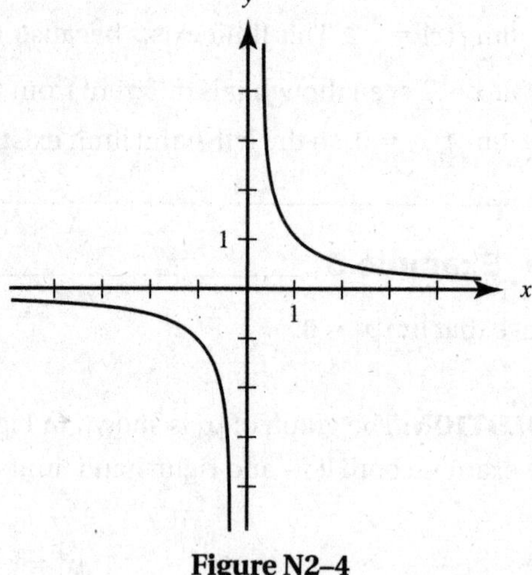

Figure N2–4

Example 5

Describe the behavior of $g(x) = \frac{1}{(x-1)^2}$ near $x = 1$ using limits.

SOLUTION: The graph (Figure N2–5) shows that:

$$\lim_{x \to 1^-} g(x) = \lim_{x \to 1^+} g(x) = \infty$$

$$\lim_{x \to 1^-} g(x) = \infty$$

$$\lim_{x \to 1^+} g(x) = \infty$$

$$\lim_{x \to 1} g(x) = \infty$$

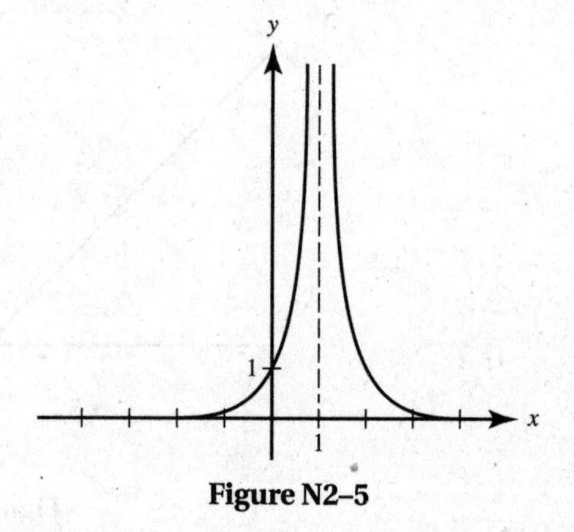

Figure N2–5

NOTE: Using $+\infty$ or $-\infty$ to indicate a limit is describing the behavior of the function and not actually a limit. Remember that none of the limits in Examples 4 and 5 exists!

Definition

We write

$$\lim_{x \to \infty} f(x) = L \text{ (or } \lim_{x \to -\infty} f(x) = L)$$

if the difference between $f(x)$ and L can be made arbitrarily small by making x sufficiently large positively (or negatively).

In Examples 4 and 5, note that $\lim_{x \to \infty} f(x) = \lim_{x \to -\infty} f(x) = 0$ and $\lim_{x \to \infty} g(x) = \lim_{x \to -\infty} g(x) = 0$.

Example 6

From the graph of $h(x) = 1 + \frac{3}{x-2} = \frac{x+1}{x-2}$ (Figure N2–6), describe the behavior of h using limits.

SOLUTION:

$$\lim_{x \to -\infty} h(x) = \lim_{x \to +\infty} h(x) = 1$$

$$\lim_{x \to 2^-} h(x) = -\infty,$$

$$\lim_{x \to 2^+} h(x) = +\infty.$$

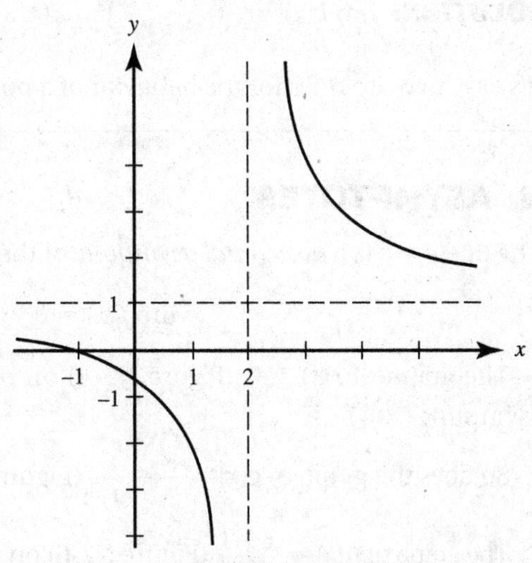

Figure N2–6

Definition

The theorems that follow in §C of this chapter confirm the conjectures made about limits of functions from their graphs.

Finally, if the function $f(x)$ becomes infinite as x becomes infinite, then one or more of the following may hold:

$$\lim_{x \to +\infty} f(x) = +\infty \text{ or } -\infty \quad \text{or} \quad \lim_{x \to -\infty} f(x) = +\infty \text{ or } -\infty.$$

End Behavior of Polynomials

Every polynomial whose degree is greater than or equal to 1 becomes infinite as x does. It becomes positively or negatively infinite, depending only on the sign of the leading coefficient and the degree of the polynomial.

➡ Example 7

For each function given below, describe $\lim_{x\to+\infty}$ and $\lim_{x\to-\infty}$.

(a) $f(x) = x^3 - 3x^2 + 7x + 2$

SOLUTION: $\lim_{x\to+\infty} f(x) = +\infty$, $\lim_{x\to-\infty} f(x) = -\infty$.

(b) $g(x) = -4x^4 + 1{,}000{,}000x^3 + 100$

SOLUTION: $\lim_{x\to+\infty} g(x) = -\infty$, $\lim_{x\to-\infty} g(x) = -\infty$.

(c) $h(x) = -5x^3 + 3x^2 - 4\pi + 8$

SOLUTION: $\lim_{x\to+\infty} h(x) = -\infty$, $\lim_{x\to-\infty} h(x) = +\infty$.

(d) $k(x) = \pi - 0.001x$

SOLUTION: $\lim_{x\to+\infty} k(x) = -\infty$, $\lim_{x\to-\infty} k(x) = +\infty$.

It's easy to write rules for the behavior of a polynomial as x becomes infinite!

B. ASYMPTOTES

Horizontal asymptote

The line $y = b$ is a *horizontal asymptote* of the graph of $y = f(x)$ if

$$\lim_{x\to\infty} f(x) = b \quad \text{or} \quad \lim_{x\to-\infty} f(x) = b.$$

The graph of $f(x) = \frac{1}{x}$ (Figure N2–4) on page 90 has the x-axis ($y = 0$) as the horizontal asymptote.

So does the graph of $g(x) = \frac{1}{(x-1)^2}$ (Figure N2–5) on page 90.

The graph of $h(x) = \frac{x+1}{x-2}$ (Figure N2-6) on page 91, has the line $y = 1$ as the horizontal asymptote.

Note, unlike vertical asymptotes, horizontal asymptotes can be crossed. The graph of $p(x) = \frac{3x^2-8x+7}{x^2-4x+5}$ has a horizontal asymptote at $y = 3$, as shown to the right.

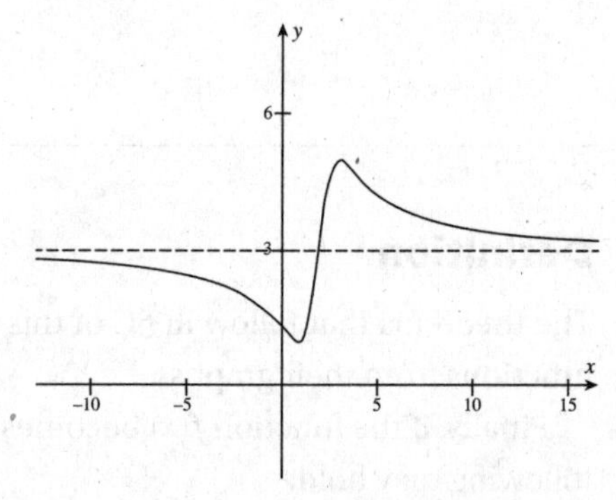

Vertical asymptote

The line $x = a$ is a *vertical asymptote* of the graph of $y = f(x)$ if one or more of the following holds:

$$\lim_{x\to a^-} f(x) = +\infty \quad \text{or} \quad \lim_{x\to a^-} f(x) = -\infty$$

or

$$\lim_{x\to a^+} f(x) = +\infty \quad \text{or} \quad \lim_{x\to a^+} f(x) = -\infty.$$

The graph of $f(x) = \frac{1}{x}$ (Figure N2–4) has $x = 0$ (the y-axis) as the vertical asymptote.

The graph of $g(x) = \frac{1}{(x-1)^2}$ (Figure N2–5) has $x = 1$ as the vertical asymptote.

The graph of $h(x) = \frac{x+1}{x-2}$ (Figure N2–6) has the line $x = 2$ as the vertical asymptote.

Example 8

From the graph of $k(x) = \dfrac{2x-4}{x-3}$ in Figure N2–7, describe the asymptotes of k using limits.

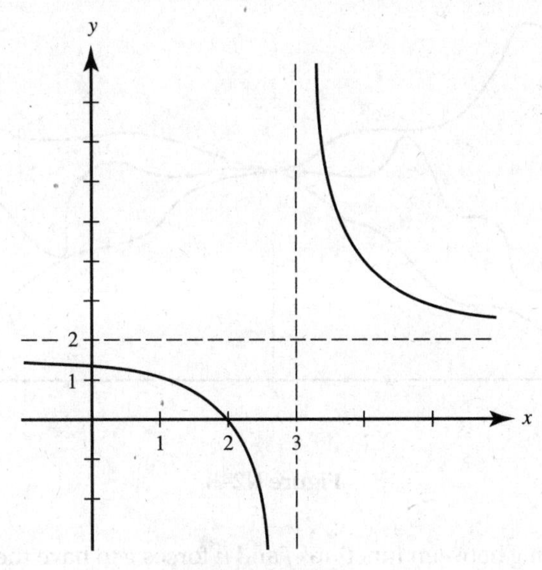

Figure N2–7

SOLUTION: We see that $y = 2$ is a horizontal asymptote, since

$$\lim_{x\to+\infty} k(x) = \lim_{x\to-\infty} k(x) = 2.$$

Also, $x = 3$ is a vertical asymptote; the graph shows that

$$\lim_{x\to3^-} k(x) = -\infty \quad \text{and} \quad \lim_{x\to3^+} k(x) = +\infty.$$

C. THEOREMS ON LIMITS

If $\lim f(x)$ and $\lim g(x)$ are finite numbers, then:

(1) $\lim kf(x) = k \lim f(x)$.
(2) $\lim\left[f(x) + g(x)\right] = \lim f(x) + \lim g(x)$.
(3) $\lim f(x)g(x) = (\lim f(x))(\lim g(x))$.
(4) $\lim \dfrac{f(x)}{g(x)} = \dfrac{\lim f(x)}{\lim g(x)}$ (if $\lim g(x) \neq 0$).
(5) $\lim\limits_{x\to c} k = k$

Sandwich (Squeeze) Theorem

(6) THE SQUEEZE OR SANDWICH THEOREM. If $f(x) \leq g(x) \leq h(x)$ and if $\lim_{x\to c} f(x) = \lim_{x\to c} h(x) = L$, then $\lim_{x\to c} g(x) = L$.

Figure N2–8 illustrates this theorem.

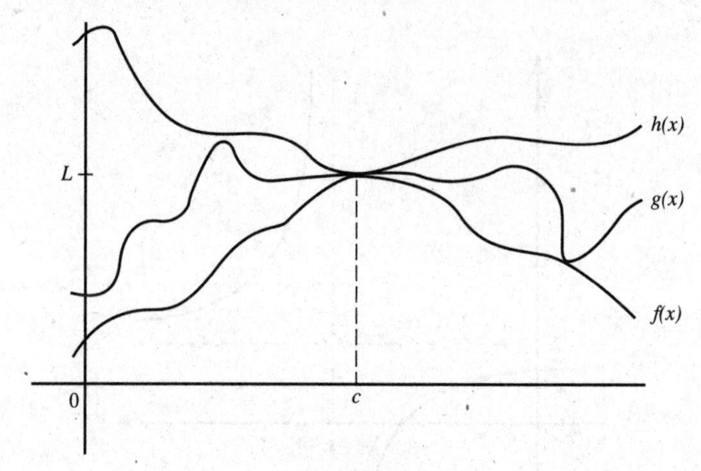

Figure N2–8

Squeezing function g between functions f and h forces g to have the same limit L at $x = c$ as do f and g.

Example 9

$$\begin{aligned}\lim_{x\to 2}\left(5x^2-3x+1\right) &= 5\lim_{x\to 2} x^2 - 3\lim_{x\to 2} x + \lim_{x\to 2} 1\\ &= 5\cdot 4 \quad -3\cdot 2 \quad +1\\ &= 15.\end{aligned}$$

Example 10

$$\begin{aligned}\lim_{x\to 0}(x\cos 2x) &= \lim_{x\to 0} x \cdot \lim_{x\to 0}(\cos 2x)\\ &= 0 \quad \cdot 1\\ &= 0.\end{aligned}$$

Example 11

$$\begin{aligned}\lim_{x\to -1}\frac{3x^2-2x-1}{x^2+1} &= \lim_{x\to -1}\left(3x^2-2x-1\right) \div \lim_{x\to -1}\left(x^2+1\right)\\ &= (3+2-1) \quad \div (1+1)\\ &= 2.\end{aligned}$$

Example 12

$$\lim_{x\to 3}\frac{x^2-9}{x-3}=\lim_{x\to 3}\frac{(x-3)(x+3)}{x-3}=\lim_{x\to 3}(x+3)=6$$

since, by the definition of $\lim_{x\to c} f(x)$ in §A, x must be different from 3 as $x\to 3$, the factor $x-3$ may be removed *before* taking the limit.

Example 13

$$\lim_{x\to -2}\frac{x^3+8}{x^2-4}=\lim_{x\to -2}\frac{(x+2)(x^2-2x+4)}{(x+2)(x-2)}=\lim_{x\to -2}\frac{x^2-2x+4}{x-2}=\frac{4+4+4}{-4}=-3.$$

Example 14

$\lim_{x\to 0}\frac{x}{x^3}=\lim_{x\to 0}\frac{1}{x^2}=\infty$. As $x\to 0$, the numerator approaches 1 while the denominator approaches 0; the limit does *not* exist.

Example 15

$$\lim_{x\to 1}\frac{x^2-1}{x^2-1}=\lim_{x\to 1}1=1.$$

Example 16

$$\lim_{\Delta x\to 0}\frac{(3+\Delta x)^2-3^2}{\Delta x}=\lim_{\Delta x\to 0}\frac{6\Delta x+\Delta x^2}{\Delta x}=\lim_{\Delta x\to 0}6+\Delta x=6.$$

Example 17

$$\lim_{h\to 0}\frac{1}{h}\left(\frac{1}{2+h}-\frac{1}{2}\right)=\lim_{h\to 0}\frac{2-(2+h)}{2h(2+h)}=\lim_{h\to 0}\frac{-h}{2h(2+h)}$$

$$=\lim_{h\to 0}-\frac{1}{2(2+h)}=-\frac{1}{4}.$$

D. LIMIT OF A QUOTIENT OF POLYNOMIALS

To find $\lim_{x\to\infty}\frac{P(x)}{Q(x)}$, where $P(x)$ and $Q(x)$ are polynomials in x, we can divide both numerator and denominator by the highest power of x that occurs and use the fact that $\lim_{x\to\infty}\frac{1}{x}=0$.

Example 18

$$\lim_{x\to\infty}\frac{3-x}{4+x+x^2}=\lim_{x\to\infty}\frac{\frac{3}{x^2}-\frac{1}{x}}{\frac{4}{x^2}+\frac{1}{x}+1}=\frac{0-0}{0+0+1}=0.$$

Example 19

$$\lim_{x\to\infty}\frac{4x^4+5x+1}{37x^3-9}=\lim_{x\to\infty}\frac{4+\frac{5}{x^3}+\frac{1}{x^4}}{\frac{37}{x}-\frac{9}{x^4}}=\infty \text{ (no limit).}$$

Example 20

$$\lim_{x\to\infty}\frac{x^3-4x^2+7}{3-6x-2x^3}=\lim_{x\to\infty}\frac{1-\frac{4}{x}+\frac{7}{x^3}}{\frac{3}{x^3}-\frac{6}{x^2}-2}=\frac{1-0+0}{0-0-2}=-\frac{1}{2}.$$

The Rational Function Theorem

We see from Examples 18, 19, and 20 that: if the degree of $P(x)$ is less than that of $Q(x)$, then $\lim_{x\to\infty}\frac{P(x)}{Q(x)}=0$; if the degree of $P(x)$ is higher than that of $Q(x)$, then $\lim_{x\to\infty}\frac{P(x)}{Q(x)}=\infty$ or $-\infty$ (i.e., does not exist); and if the degrees of $P(x)$ and $Q(x)$ are the same, then $\lim_{x\to\infty}\frac{P(x)}{Q(x)}=\frac{a_n}{b_n}$, where a_n and b_n are the coefficients of the highest powers of x in $P(x)$ and $Q(x)$, respectively.

This theorem holds also when we replace "$x\to\infty$" by "$x\to-\infty$."

Note also that:

(i) when $\lim_{x\to\pm\infty}\frac{P(x)}{Q(x)}=0$, then $y=0$ is a horizontal asymptote of the graph of $y=\frac{P(x)}{Q(x)}$;

(ii) when $\lim_{x\to\pm\infty}\frac{P(x)}{Q(x)}=+\infty$ or $-\infty$, then the graph of $y=\frac{P(x)}{Q(x)}$ has no horizontal asymptotes;

(iii) when $\lim_{x\to\pm\infty}\frac{P(x)}{Q(x)}=\frac{a_n}{b_n}$, then $y=\frac{a_n}{b_n}$ is a horizontal asymptote of the graph of $y=\frac{P(x)}{Q(x)}$.

Example 21

$$\lim_{x\to\infty}\frac{100x^2-19}{x^3+5x^2+2}=0; \qquad \lim_{x\to-\infty}\frac{x^3-5}{1+x^2}=-\infty \text{ (no limit);} \qquad \lim_{x\to\infty}\frac{x-4}{13+5x}=\frac{1}{5};$$

$$\lim_{x\to-\infty}\frac{4+x^2-3x^3}{x+7x^3}=-\frac{3}{7}; \qquad \lim_{x\to\infty}\frac{x^3+1}{2-x^2}=-\infty \text{ (no limit).}$$

E. OTHER BASIC LIMITS

E1. The basic trigonometric limit is:

$$\lim_{\theta\to 0}\frac{\sin\theta}{\theta}=1 \text{ if } \theta \text{ is measured in radians.}$$

Example 22

Prove that $\lim_{x\to\infty}\frac{\sin x}{x}=0$.

SOLUTION: Since, for all x, $-1\le \sin x\le 1$, it follows that, if $x>0$, then $-\frac{1}{x}\le\frac{\sin x}{x}\le\frac{1}{x}$. But as $x\to\infty$, $-\frac{1}{x}$ and $\frac{1}{x}$ both approach 0; therefore by the Squeeze theorem, $\frac{\sin x}{x}$ must also approach 0. To obtain graphical confirmation of this fact, and of the additional fact that $\lim_{x\to-\infty} \frac{\sin x}{x}$ also equals 0, graph

$$y_1 = \frac{\sin x}{x}, y_2 = \frac{1}{x}, \text{ and } y_3 = -\frac{1}{x}$$

in $[-4\pi, 4\pi] \times [-1, 1]$. Observe, as $x \to \pm\infty$, that y_2 and y_3 approach 0 and that y_1 is squeezed between them.

Example 23

Find $\lim_{x\to0} \frac{\sin 3x}{x}$.

SOLUTION: $\lim_{x\to0} \frac{\sin 3x}{x} = \lim_{x\to0} \frac{3\sin 3x}{3x} = 3\lim_{x\to0} \frac{\sin 3x}{3x} = 3 \cdot 1 = 3.$

E2. The number e can be defined as follows:

Limit definition of *e*

$$e = \lim_{n\to\infty} \left(1 + \frac{1}{n}\right)^n.$$

The value of e can be approximated on a graphing calculator to a large number of decimal places by evaluating

$$y_1 = \left(1 + \frac{1}{x}\right)^x$$

for large values of x.

F. CONTINUITY

If a function is continuous over an interval, we can draw its graph without lifting pencil from paper. The graph has no holes, breaks, or jumps on the interval.

Conceptually, if $f(x)$ is continuous at a point $x = c$, then the closer x is to c, the closer $f(x)$ gets to $f(c)$. This is made precise by the following definition:

Definition

Continuous

The function $y = f(x)$ is continuous at $x = c$ if

(1) $f(c)$ exists; (that is, c is in the domain of f);

(2) $\lim_{x\to c} f(x)$ exists;

(3) $\lim_{x\to c} f(x) = f(c)$.

A function is continuous over the closed interval $[a,b]$ if it is continuous at each x such that $a \leq x \leq b$.

A function that is not continuous at $x = c$ is said to be discontinuous at that point. We then call $x = c$ a *point of discontinuity*.

Continuous Functions

Polynomials are continuous everywhere; namely, at every real number.

Rational functions, $\frac{P(x)}{Q(x)}$, are continuous at each point in their domain; that is, except where $Q(x) = 0$. The function $f(x) = \frac{1}{x}$, for example, is continuous except at $x = 0$, where f is not defined.

The absolute value function $f(x) = |x|$ (sketched in Figure N2–3, page 89) is continuous everywhere.

The trigonometric, inverse trigonometric, exponential, and logarithmic functions are continuous at each point in their domains.

Functions of the type $\sqrt[n]{x}$ (where n is a positive integer $\geqslant 2$) are continuous at each x for which $\sqrt[n]{x}$ is defined.

The greatest-integer function $f(x) = [x]$ (Figure N2–1, page 88) is discontinuous at each integer, since it does not have a limit at any integer.

Kinds of Discontinuities

In Example 2, page 88, $y = f(x)$ is defined as follows:

$$f(x) = \begin{cases} x + 1 & (-2 < x < 0) \\ 2 & (x = 0) \\ -x & (0 < x < 2) \\ 0 & (x = 2) \\ x - 4 & (2 < x \leqslant 4) \end{cases}$$

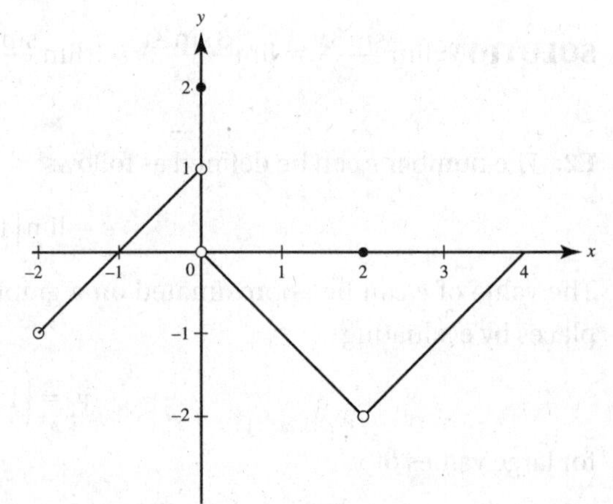

The graph of f is shown at the right.

We observe that f is not continuous at $x = -2$, $x = 0$, or $x = 2$.

At $x = -2$, f is not defined.

At $x = 0$, f is defined; in fact, $f(0) = 2$. However, since $\lim\limits_{x\to 0^-} f(x) = 1$ and $\lim\limits_{x\to 0^+} f(x) = 0$, $\lim\limits_{x\to 0} f(x)$ does not exist. Where the left- and right-hand limits exist, but are different, the function has a *jump discontinuity.* The greatest-integer (or step) function, $y = [x]$, has a jump discontinuity at every integer. (See page 88.)

Jump discontinuity

At $x = 2$, f is defined; in fact, $f(2) = 0$. Also, $\lim\limits_{x\to 2} f(x) = -2$; the limit exists. However, $\lim\limits_{x\to 2} f(x) \neq f(2)$. This discontinuity is called *removable.* If we were to redefine the function at $x = 2$ to be $f(2) = -2$, the new function would no longer have a discontinuity there. We cannot, however, "remove" a jump discontinuity by any redefinition whatsoever.

Removable discontinuity

Whenever the graph of a function $f(x)$ has the line $x = a$ as a vertical asymptote, then $f(x)$ becomes positively or negatively infinite as $x \to a^+$ or as $x \to a^-$. The function is then said to have an *infinite discontinuity.* See, for example, Figure N2–4 (page 90) for $f(x) = \frac{1}{x}$, Figure N2–5 (page 90) for $g(x) = \frac{1}{(x-1)^2}$, or Figure N2–7 (page 93) for $k(x) = \frac{2x-4}{x-3}$. Each of these functions exhibits an infinite discontinuity.

Infinite discontinuity

➡ Example 24

$f(x) = \frac{x-1}{x^2+x} = \frac{x-1}{x(x+1)}$ is not continuous at $x = 0$ or $= -1$, since the function is not defined for either of these numbers. Note also that neither $\lim\limits_{x\to 0} f(x)$ nor $\lim\limits_{x\to -1} f(x)$ exists.

➡ Example 25

Discuss the continuity of f, as graphed in Figure N2–9.

SOLUTION: $f(x)$ is continuous on $[(0,1), (1,3), \text{and } (3,5)]$. The discontinuity at $x = 1$ is removable; the one at $x = 3$ is not. (Note that f is continuous from the right at $x = 0$ and from the left at $x = 5$.)

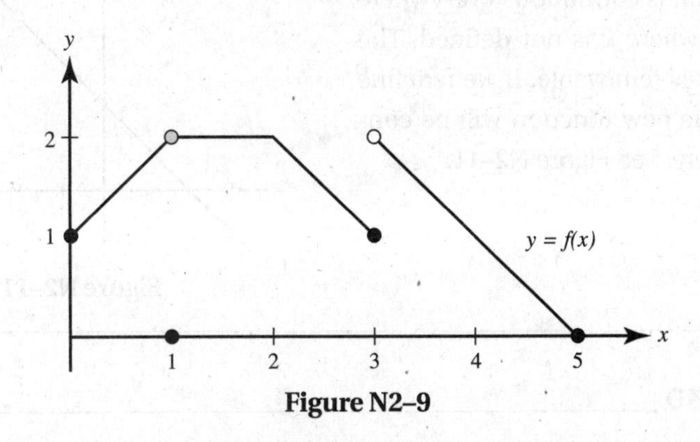

Figure N2–9

In Examples 26 through 31, we determine whether the functions are continuous at the points specified:

➡ Example 26

Is $f(x) = \frac{1}{2}x^4 - \sqrt{3}x^2 + 7$ continuous at $x = -1$?

SOLUTION: Since f is a polynomial, it is continuous everywhere, including, of course, at $x = -1$.

➡ Example 27

Is $g(x) = \frac{1}{x-3}$ continuous (a) at $x = 3$; (b) at $x = 0$?

SOLUTION: This function is continuous except where the denominator equals 0 (where g has an infinite discontinuity). It is not continuous at $x = 3$, but is continuous at $x = 0$.

➡ Example 28

Is $h(x) = \begin{cases} \frac{4}{x-2} & \text{if } x \neq 2 \\ 1 & \text{if } x = 2 \end{cases}$ continuous

(a) at $x = 2$; (b) at $x = 3$?

SOLUTIONS:

(a) $h(x)$ has an infinite discontinuity at $x = 2$; this discontinuity is not removable.

(b) $h(x)$ is continuous at $x = 3$ and at every other point different from 2.

See Figure N2–10.

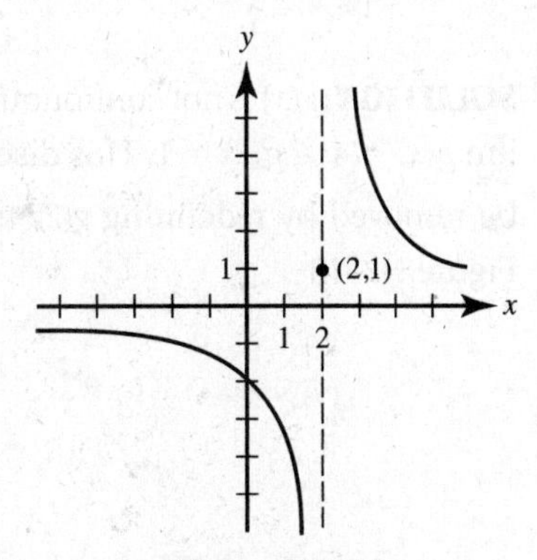

Figure N2–10

➡ Example 29

Is $k(x) = \dfrac{x^2 - 4}{x - 2}$ $(x \neq 2)$ continuous at $x = 2$?

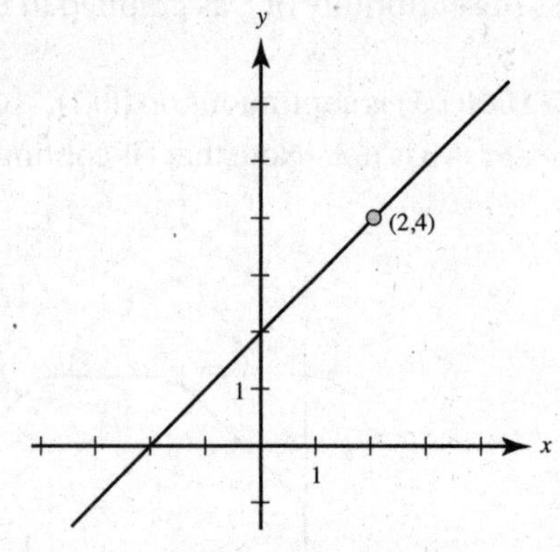

Figure N2–11

SOLUTION: Note that $k(x) = x + 2$ for all $x \neq 2$. The function is continuous everywhere except at $x = 2$, where k is not defined. The discontinuity at 2 is removable. If we redefine $f(2)$ to equal 4, the new function will be continuous everywhere. See Figure N2–11.

➡ Example 30

Is $f(x) = \begin{cases} x^2 + 2 & x \leq 1 \\ 4 & x > 1 \end{cases}$ continuous at $x = 1$?

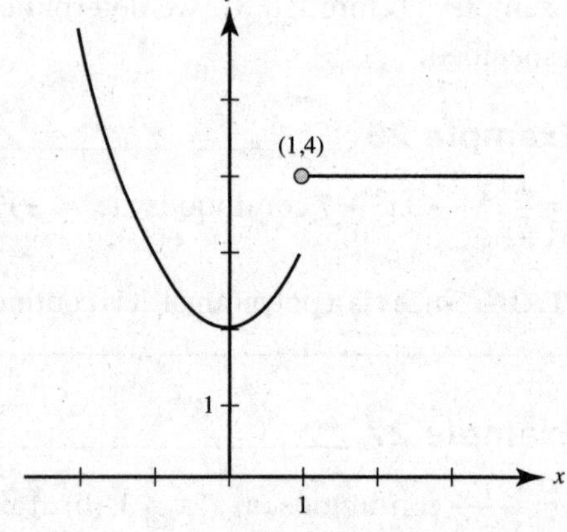

Figure N2–12

SOLUTION: $f(x)$ is not continuous at $x = 1$ since $\lim\limits_{x \to 1^-} f(x) = 3 \neq \lim\limits_{x \to 1^+} f(x) = 4$. This function has a jump discontinuity at $x = 1$ (which cannot be removed). See Figure N2–12.

➡ Example 31

Is $g(x) = \begin{cases} x^2 & x \neq 2 \\ 1 & x = 2 \end{cases}$ continuous at $x = 2$?

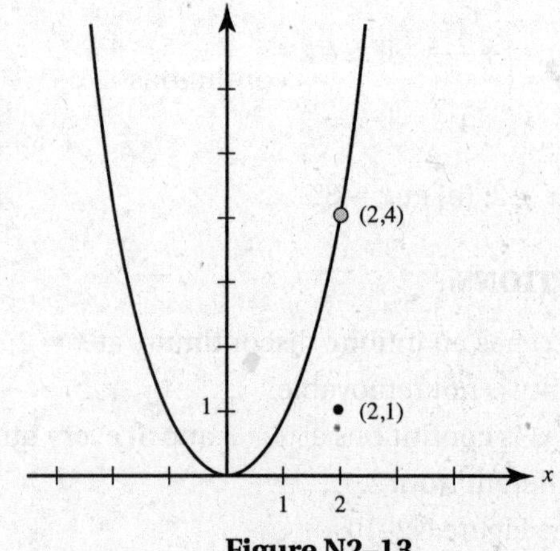

Figure N2–13

SOLUTION: $g(x)$ is not continuous at $x = 2$ since $\lim\limits_{x \to 2} g(x) = 4 \neq g(2) = 1$. This discontinuity can be removed by redefining $g(2)$ to equal 4. See Figure N2–13.

Theorems on Continuous Functions

Extreme Value Theorem

(1) **The Extreme Value Theorem.** If f is continuous on the closed interval $[a,b]$, then f attains a minimum value and a maximum value somewhere in that interval.

Intermediate Value Theorem

(2) **The Intermediate Value Theorem.** If f is continuous on the closed interval $[a,b]$, and M is a number such that $f(a) \leq M \leq f(b)$, then there is at least one number, c, in the interval $[a,b]$, such that $f(c) = M$.

Note an important special case of the Intermediate Value Theorem:

If f is continuous on the closed interval $[a,b]$, and $f(a)$ and $f(b)$ have opposite signs, then f has a zero in that interval (there is a value, c, in $[a,b]$ where $f(c) = 0$).

(3) **The Continuous Functions Theorem.** If functions f and g are both continuous at $x = c$, then so are the following functions:

(a) kf, where k is a constant;

(b) $f \pm g$;

(c) $f \cdot g$;

(d) $\frac{f}{g}$, provided that $g(c) \neq 0$.

Example 32

Show that $f(x) = \frac{x^2 - 5}{x + 1}$ has a root between $x = 2$ and $x = 3$.

SOLUTION: The rational function f is discontinuous only at $x = -1$. $f(2) = -\frac{1}{3}$, and $f(3) = 1$. Since f is continuous on the interval $[2,3]$ and $f(2)$ and $f(3)$ have opposite signs, there is a value, c, in the interval where $f(c) = 0$, by the Intermediate Value Theorem.

CHAPTER SUMMARY

In this chapter, we have reviewed the concept of a limit. We've practiced finding limits using algebraic expressions, graphs, and the Squeeze (Sandwich) Theorem. We have used limits to find horizontal and vertical asymptotes and to assess the continuity of a function. We have reviewed removable, jump, and infinite discontinuities. We have also looked at the very important Extreme Value Theorem and Intermediate Value Theorem.

PRACTICE EXERCISES

Part A. Directions: Answer these questions *without* using your calculator.

1. $\lim_{x\to 2} \frac{x^2-4}{x^2+4}$ is

   (A) 1 (B) 0 (C) $-\frac{1}{2}$ (D) -1 (E) ∞

2. $\lim_{x\to \infty} \frac{4-x^2}{x^2-1}$ is

   (A) 1 (B) 0 (C) -4 (D) -1 (E) ∞

3. $\lim_{x\to 3} \frac{x-3}{x^2-2x-3}$ is

   (A) 0 (B) 1 (C) $\frac{1}{4}$ (D) ∞ (E) none of these

4. $\lim_{x\to 0} \frac{x}{x}$ is

   (A) 1 (B) 0 (C) ∞ (D) -1 (E) nonexistent

5. $\lim_{x\to 2} \frac{x^3-8}{x^2-4}$ is

   (A) 4 (B) 0 (C) 1 (D) 3 (E) ∞

6. $\lim_{x\to \infty} \frac{4-x^2}{4x^2-x-2}$ is

   (A) -2 (B) $-\frac{1}{4}$ (C) 1 (D) 2 (E) nonexistent

7. $\lim_{x\to -\infty} \frac{5x^3+27}{20x^2+10x+9}$ is

   (A) $-\infty$ (B) -1 (C) 0 (D) 3 (E) ∞

8. $\lim_{x\to \infty} \frac{3x^2+27}{x^3-27}$ is

   (A) 3 (B) ∞ (C) 1 (D) -1 (E) 0

9. $\lim_{x\to \infty} \frac{2^{-x}}{2^x}$ is

   (A) -1 (B) 1 (C) 0 (D) ∞ (E) none of these

10. $\lim_{x\to -\infty} \frac{2^{-x}}{2^x}$ is

    (A) -1 (B) 1 (C) 0 (D) ∞ (E) none of these

11. $\lim_{x\to 0} \frac{\sin 5x}{x}$

    (A) $= 0$ (B) $= \frac{1}{5}$ (C) $= 1$ (D) $= 5$ (E) does not exist

12. $\lim_{x\to 0} \frac{\sin 2x}{3x}$

    (A) $= 0$ (B) $= \frac{2}{3}$ (C) $= 1$ (D) $= \frac{3}{2}$ (E) does not exist

13. The graph of $y = \arctan x$ has

(A) vertical asymptotes at $x = 0$ and $x = \pi$

(B) horizontal asymptotes at $y = \pm\frac{\pi}{2}$

(C) horizontal asymptotes at $y = 0$ and $y = \pi$

(D) vertical asymptotes at $x = \pm\frac{\pi}{2}$

(E) no asymptotes

14. The graph of $y = \frac{x^2 - 9}{3x - 9}$ has

(A) a vertical asymptote at $x = 3$

(B) a horizontal asymptote at $y = \frac{1}{3}$

(C) a removable discontinuity at $x = 3$

(D) an infinite discontinuity at $x = 3$

(E) no asymptotes or discontinuities

15. $\lim\limits_{x\to 0} \frac{\sin x}{x^2 + 3x}$ is

(A) 1 (B) $\frac{1}{3}$ (C) 3 (D) ∞ (E) $\frac{1}{4}$

16. $\lim\limits_{x\to 0} \sin\frac{1}{x}$ is

(A) ∞ (B) 1 (C) nonexistent (D) -1 (E) none of these

17. Which statement is true about the curve $y = \frac{2x^2 + 4}{2 + 7x - 4x^2}$?

(A) The line $x = -\frac{1}{4}$ is a vertical asymptote.

(B) The line $x = 1$ is a vertical asymptote.

(C) The line $y = -\frac{1}{4}$ is a horizontal asymptote.

(D) The graph has no vertical or horizontal asymptote.

(E) The line $y = 2$ is a horizontal asymptote.

18. $\lim\limits_{x\to\infty} \frac{2x^2 + 1}{(2 - x)(2 + x)}$ is

(A) -4 (B) -2 (C) 1 (D) 2 (E) nonexistent

19. $\lim\limits_{x\to 0} \frac{|x|}{x}$ is

(A) 0 (B) nonexistent (C) 1 (D) -1 (E) none of these

20. $\lim\limits_{x\to\infty} x\sin\frac{1}{x}$ is

(A) 0 (B) ∞ (C) nonexistent (D) -1 (E) 1

21. $\lim\limits_{x\to\pi} \frac{\sin(\pi - x)}{(\pi - x)}$ is

(A) 1 (B) 0 (C) ∞ (D) π (E) nonexistent

22. Let $f(x) = \begin{cases} \dfrac{x^2 - 1}{x - 1} & \text{if } x \neq 1 \\ 4 & \text{if } x = 1 \end{cases}$.

Which of the following statements is (are) true?

I. $\lim_{x \to 1} f(x)$ exists

II. $f(1)$ exists

III. f is continuous at $x = 1$

(A) I only (B) II only (C) I and II

(D) none of I, II, or III (E) all of I, II, or III

23. If $\begin{cases} f(x) = \dfrac{x^2 - x}{2x} & \text{for } x \neq 0, \\ f(0) = k, \end{cases}$

and if f is continuous at $x = 0$, then $k =$

(A) -1 (B) $-\frac{1}{2}$ (C) 0 (D) $\frac{1}{2}$ (E) 1

24. Suppose $\begin{cases} f(x) = \dfrac{3x(x - 1)}{x^2 - 3x + 2} & \text{for } x \neq 1, 2, \\ f(1) = -3, \\ f(2) = 4. \end{cases}$

Then $f(x)$ is continuous

(A) except at $x = 1$

(B) except at $x = 2$

(C) except at $x = 1$ or 2

(D) except at $x = 0$, 1, or 2

(E) at each real number

25. The graph of $f(x) = \dfrac{4}{x^2 - 1}$ has

(A) one vertical asymptote, at $x = 1$

(B) the y-axis as vertical asymptote

(C) the x-axis as horizontal asymptote and $x = \pm 1$ as vertical asymptotes

(D) two vertical asymptotes, at $x = \pm 1$, but no horizontal asymptote

(E) no asymptote

26. The graph of $y = \dfrac{2x^2 + 2x + 3}{4x^2 - 4x}$ has

(A) a horizontal asymptote at $y = +\frac{1}{2}$ but no vertical asymptote

(B) no horizontal asymptote but two vertical asymptotes, at $x = 0$ and $x = 1$

(C) a horizontal asymptote at $y = \frac{1}{2}$ and two vertical asymptotes, at $x = 0$ and $x = 1$

(D) a horizontal asymptote at $x = 2$ but no vertical asymptote

(E) a horizontal asymptote at $y = \frac{1}{2}$ and two vertical asymptotes, at $x = \pm 1$

27. Let $f(x) = \begin{cases} \dfrac{x^2 + x}{x} & \text{if } x \neq 0 \\ 1 & \text{if } x = 0 \end{cases}$.

Which of the following statements is (are) true?

I. $f(0)$ exists

II. $\lim_{x \to 0} f(x)$ exists

III. f is continuous at $x = 0$

(A) I only (B) II only (C) I and II only

(D) I, II, and III (E) none of I, II, or III

Part B. Directions: Some of the following questions require the use of a graphing calculator.

28. If $[x]$ is the greatest integer not greater than x, then $\lim_{x \to 1/2} [x]$ is

(A) $\frac{1}{2}$ (B) 1 (C) nonexistent (D) 0 (E) none of these

29. (With the same notation) $\lim_{x \to -2} [x]$ is

(A) -3 (B) -2 (C) -1 (D) 0 (E) nonexistent

30. $\lim_{x \to \infty} \sin x$

(A) is -1 **(B)** is infinity **(C)** oscillates between -1 and 1

(D) is zero **(E)** does not exist

31. The function $f(x) = \begin{cases} x^{2/x} & (x \neq 0) \\ 0 & (x = 0) \end{cases}$

(A) is continuous everywhere

(B) is continuous except at $x = 0$

(C) has a removable discontinuity at $x = 0$

(D) has an infinite discontinuity at $x = 0$

(E) has $x = 0$ as a vertical asymptote

Questions 32–36 are based on the function f shown in the graph and defined below:

$$f(x) = \begin{cases} 1 - x & (-1 \leqslant x < 0) \\ 2x^2 - 2 & (0 \leqslant x \leqslant 1) \\ -x + 2 & (1 < x < 2) \\ 1 & (x = 2) \\ 2x - 4 & (2 < x \leqslant 3) \end{cases}$$

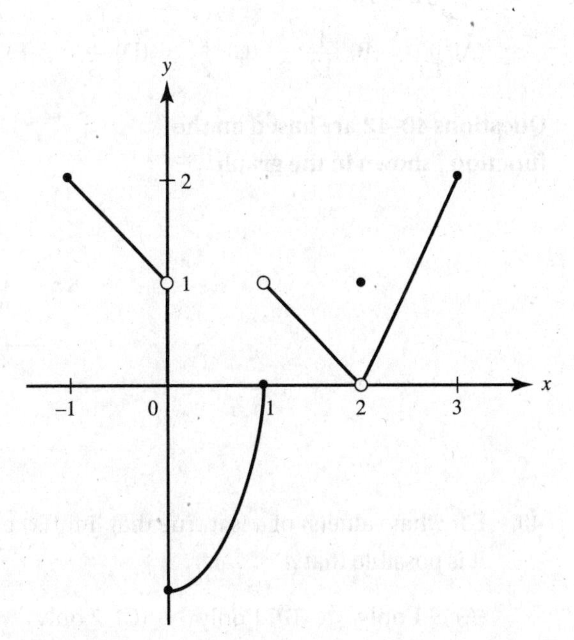

32. $\lim_{x\to2} f(x)$

(A) equals 0 (B) equals 1 (C) equals 2
(D) does not exist (E) none of these

33. The function f is defined on $[-1,3]$

(A) if $x \neq 0$ (B) if $x \neq 1$ (C) if $x \neq 2$ (D) if $x \neq 3$ (E) at each x in $[-1,3]$

34. The function f has a removable discontinuity at

(A) $x = 0$ (B) $x = 1$ (C) $x = 2$ (D) $x = 3$ (E) none of these

35. On which of the following intervals is f continuous?

(A) $-1 \leqslant x \leqslant 0$ (B) $0 < x < 1$ (C) $1 \leqslant x \leqslant 2$
(D) $2 \leqslant x \leqslant 3$ (E) none of these

36. The function f has a jump discontinuity at

(A) $x = -1$ (B) $x = 1$ (C) $x = 2$ (D) $x = 3$ (E) none of these

CHALLENGE 37. $\lim_{x\to0} \sqrt{3 + \arctan\frac{1}{x}}$ is

(A) $-\infty$ (B) $\sqrt{3 - \frac{\pi}{2}}$ (C) $\sqrt{3 + \frac{\pi}{2}}$ (D) ∞ (E) nonexistent

38. Suppose $\lim_{x\to-3^-} f(x) = -1$, $\lim_{x\to-3^+} f(x) = -1$, and $f(-3)$ is not defined. Which of the following statements is (are) true?

I. $\lim_{x\to-3} f(x) = -1$
II. f is continuous everywhere except at $x = -3$.
III. f has a removable discontinuity at $x = -3$.

(A) None of I, II, or III (B) I only (C) III only
(D) I and III only (E) I, II, and III

CHALLENGE 39. If $y = \frac{1}{2 + 10^{1/x}}$, then $\lim_{x\to0} y$ is

(A) 0 (B) $\frac{1}{12}$ (C) $\frac{1}{2}$ (D) $\frac{1}{3}$ (E) nonexistent

Questions 40–42 are based on the function f shown in the graph.

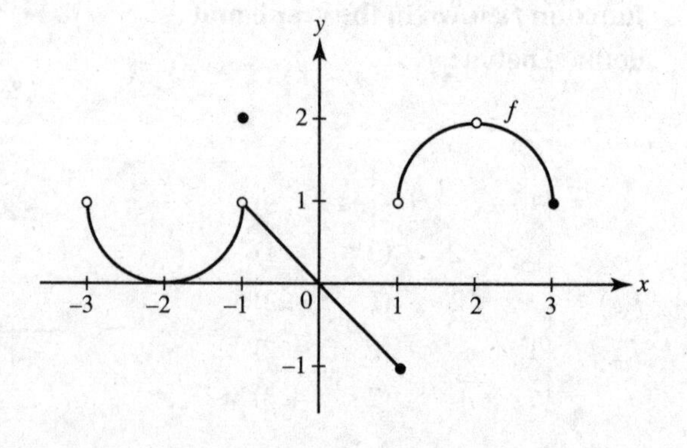

40. For what value(s) of a is it true that $\lim_{x\to a} f(x)$ exists and $f(a)$ exists, but $\lim_{x\to a} f(x) \neq f(a)$? It is possible that $a =$

(A) -1 only (B) 1 only (C) 2 only (D) -1 or 1 only (E) -1 or 2 only

41. $\lim_{x \to a} f(x)$ does not exist for $a =$

(A) -1 only (B) 1 only (C) 2 only (D) 1 and 2 only (E) -1, 1, and 2

42. Which statements about limits at $x = 1$ are true?

I. $\lim_{x \to 1^-} f(x)$ exists.

II. $\lim_{x \to 1^+} f(x)$ exists.

III. $\lim_{x \to 1} f(x)$ exists.

(A) none of I, II, or III (B) I only (C) II only

(D) I and II only (E) I, II, and III

Answer Key

1. **B**	12. **B**	23. **B**	34. **C**
2. **D**	13. **B**	24. **B**	35. **B**
3. **C**	14. **C**	25. **C**	36. **B**
4. **A**	15. **B**	26. **C**	37. **E**
5. **D**	16. **C**	27. **D**	38. **D**
6. **B**	17. **A**	28. **D**	39. **E**
7. **A**	18. **B**	29. **E**	40. **A**
8. **E**	19. **B**	30. **E**	41. **B**
9. **C**	20. **E**	31. **A**	42. **D**
10. **D**	21. **A**	32. **A**	
11. **D**	22. **C**	33. **E**	

Answers Explained

1. **(B)** The limit as $x \to 2$ is $0 \div 8$.

2. **(D)** Use the Rational Function Theorem (page 96). The degrees of $P(x)$ and $Q(x)$ are the same.

3. **(C)** Remove the common factor $x - 3$ from numerator and denominator.

4. **(A)** The fraction equals 1 for all nonzero x.

5. **(D)** Note that $\frac{x^3 - 8}{x^2 - 4} = \frac{(x-2)(x^2 + 2x + 4)}{(x-2)(x+2)}$.

6. **(B)** Use the Rational Function Theorem.

7. **(A)** Use the Rational Function Theorem.

8. **(E)** Use the Rational Function Theorem.

9. **(C)** The fraction is equivalent to $\frac{1}{2^{2x}}$; the denominator approaches ∞.

10. **(D)** Since $\frac{2^{-x}}{2^x} = 2^{-2x}$, therefore, as $x \to -\infty$, the fraction $\to +\infty$.

11. **(D)** $\lim_{x \to 0} \frac{\sin 5x}{x} = \lim_{x \to 0} \frac{\sin 5x}{x} \cdot \frac{5}{5} = 5 \lim_{x \to 0} \frac{\sin 5x}{5x} = 5$

12. **(B)** $\lim_{x \to 0} \frac{\sin 2x}{3x} = \frac{1}{3} \lim_{x \to 0} \frac{\sin 2x}{x} \cdot \frac{2}{2} = \frac{2}{3} \lim_{x \to 0} \frac{\sin 2x}{2x} = \frac{2}{3}$

13. **(B)** Because the graph of $y = \tan x$ has vertical asymptotes at $x = \pm\frac{\pi}{2}$, the graph of the inverse function $y = \arctan x$ has horizontal asymptotes at $y = \pm\frac{\pi}{2}$.

14. **(C)** Since $\frac{x^2 - 9}{3x - 9} = \frac{(x-3)(x+3)}{3(x-3)} = \frac{x+3}{3}$ (provided $x \neq 3$), y can be defined to be equal to 2 at $x = 3$, removing the discontinuity at that point.

15. **(B)** Note that $\frac{\sin x}{x^2 + 3x} = \frac{\sin x}{x(x+3)} = \frac{\sin x}{x} \cdot \frac{1}{x+3} \rightarrow 1 \cdot \frac{1}{3}$.

16. **(C)** As $x \rightarrow 0$, $\frac{1}{x}$ takes on varying finite values as it increases. Since the sine function repeats, $\sin \frac{1}{x}$ oscillates, taking on, infinitely many times, each value between -1 and 1. The calculator graph of $Y_1 = \sin(1/X)$ exhibits this oscillating discontinuity at $x = 0$.

17. **(A)** Note that, since $y = \frac{2x^2 + 4}{(2-x)(1+4x)}$, both $x = 2$ and $x = -\frac{1}{4}$ are vertical asymptotes. Also, $y = -\frac{1}{2}$ is a horizontal asymptote.

18. **(B)** $\frac{2x^2 + 1}{(2-x)(2+x)} = \frac{2x^2 + 1}{4 - x^2}$. Use the Rational Function Theorem (page 96).

19. **(B)** Since $|x| = x$ if $x > 0$ but equals $-x$ if $x < 0$, $\lim\limits_{x \to 0^+} \frac{|x|}{x} = \lim\limits_{x \to 0^+} \frac{x}{x} = 1$ while $\lim\limits_{x \to 0^-} \frac{|x|}{x} = \lim\limits_{x \to 0^-} \frac{-x}{x} = -1$.

20. **(E)** Note that $x \sin \frac{1}{x}$ can be rewritten as $\frac{\sin \frac{1}{x}}{\frac{1}{x}}$ and that, as $x \rightarrow -\infty$, $\frac{1}{x} \rightarrow 0$.

21. **(A)** As $x \rightarrow \pi$, $(\pi - x) \rightarrow 0$.

22. **(C)** Since $f(x) = x + 1$ if $x \neq 1$, $\lim\limits_{x \to 1} f(x)$ exists (and is equal to 2).

23. **(B)** $f(x) = \frac{x(x-1)}{2x} = \frac{(x-1)}{2}$, for all $x \neq 0$. For f to be continuous at $x = 0$, $\lim\limits_{x \to 0} f(x)$ must equal $f(0)$. $\lim\limits_{x \to 0} f(x) = -\frac{1}{2}$.

24. **(B)** Only $x = 1$ and $x = 2$ need be checked. Since $f(x) = \frac{3x}{x-2}$ for $x \neq 1, 2$, and $\lim\limits_{x \to 1} f(x) = -3 = f(1)$, f is continuous at $x = 1$. Since $\lim\limits_{x \to 2} f(x)$ does not exist, f is not continuous at $x = 2$.

25. **(C)** As $x \rightarrow \pm\infty$, $y = f(x) \rightarrow 0$, so the x-axis is a horizontal asymptote. Also, as $x \rightarrow \pm 1$, $y \rightarrow \infty$, so $x = \pm 1$ are vertical asymptotes.

26. **(C)** As $x \rightarrow \infty$, $y \rightarrow \frac{1}{2}$; the denominator (but not the numerator) of y equals 0 at $x = 0$ and at $x = 1$.

27. **(D)** The function is defined at 0 to be 1, which is also $\lim\limits_{x \to 0} \frac{x^2 + x}{x} = \lim\limits_{x \to 0} (x + 1)$.

28. **(D)** See Figure N2–1 on page 88.

29. **(E)** Note, from Figure N2–1, that $\lim\limits_{x \to -2^-} [x] = -3$ but $\lim\limits_{x \to -2^+} [x] = -2$.

30. **(E)** As $x \rightarrow \infty$, the function $\sin x$ oscillates between -1 and 1; hence the limit does not exist.

31. **(A)** Note that $\frac{x^2}{x} = x$ if $x \neq 0$ and that $\lim\limits_{x \to 0} f = 0$.

32. **(A)** $\lim\limits_{x \to 2^-} f(x) = \lim\limits_{x \to 2^+} f(x) = 0$.

33. **(E)** Verify that f is defined at $x = 0, 1, 2$, and 3 (as well as at all other points in $[-1,3]$).

34. **(C)** Note that $\lim\limits_{x \to 2^-} f(x) = \lim\limits_{x \to 2^+} f(x) = 0$. However, $f(2) = 1$. Redefining $f(2)$ as 0 removes the discontinuity.

35. **(B)** The function is not continuous at $x = 0, 1$, or 2.

36. **(B)** $\lim\limits_{x \to 1^-} f(x) = 0 \neq \lim\limits_{x \to 1^+} f(x) = 1$.

37. **(E)** As $x \rightarrow 0^-$, $\arctan \frac{1}{x} \rightarrow -\frac{\pi}{2}$, so $y \rightarrow \sqrt{3 - \frac{\pi}{2}}$. As $x \rightarrow 0^+$, $y \rightarrow \sqrt{3 + \frac{\pi}{2}}$. The graph has a jump discontinuity at $x = 0$. (Verify with a calculator.)

38. **(D)** No information is given about the domain of f except in the neighborhood of $x = -3$.

39. **(E)** As $x \to 0^+$, $10^{1/x} \to \infty$ and therefore $y \to 0$. As $x \to 0^-$, $\frac{1}{x} \to -\infty$, so $10^{1/x} \to 0$ and therefore $y \to \frac{1}{2}$. Because the two one-sided limits are not equal, the limit does not exist. (Verify with a calculator.)

40. **(A)** $\lim_{x\to -1} f(x) = 1$, but $f(-1) = 2$. The limit does not exist at $a = 1$ and $f(2)$ does not exist.

41. **(B)** $\lim_{x\to -1} f(x) = 1$ and $\lim_{x\to -2} f(x) = 2$.

42. **(D)** $\lim_{x\to 1^-} f(x) = -1$ and $\lim_{x\to 1^+} f(x) = 1$, but since these two limits are not the same, $\lim_{x\to 1} f(x)$ does not exist.

Differentiation

3

CONCEPTS AND SKILLS

In this chapter, you will review

- **derivatives as instantaneous rates of change;**
- **estimating derivatives using graphs and tables;**
- **derivatives of basic functions;**
- **the product, quotient, and chain rules;**
- **implicit differentiation;**
- **derivatives of inverse functions;**
- **Rolle's Theorem and the Mean Value Theorem;**
- **L'Hôpital's Rule for evaluating limits of indeterminate forms** $\left(\text{only } \frac{0}{0} \text{ and } \frac{\infty}{\infty}\right)$.

In addition, BC Calculus students will review

- **derivatives of parametrically defined functions.**

A. DEFINITION OF DERIVATIVE

Derivative

At any x in the domain of the function $y = f(x)$, the *derivative* is defined as

$$\lim_{\Delta x \to 0} \frac{f(x + \Delta x) - f(x)}{\Delta x} \text{ or } \lim_{\Delta x \to 0} \frac{\Delta y}{\Delta x}. \tag{1}$$

Differentiable

The function is said to be *differentiable* at every x for which this limit exists, and its derivative may be denoted by $f'(x)$, y', $\frac{dy}{dx}$, or $D_x y$. Frequently Δx is replaced by h or some other symbol.

The derivative of $y = f(x)$ at $x = a$, denoted by $f'(a)$ or $y'(a)$, may be defined as follows:

$$f'(a) = \lim_{h \to 0} \frac{f(a + h) - f(a)}{h}. \tag{2}$$

Difference quotient

Average rate of change

Instantaneous rate of change

Slope of a curve

The fraction $\frac{f(a + h) - f(a)}{h}$ is called the *difference quotient for f at a* and represents the *average rate of change of f from a to a + h*. Geometrically, it is the slope of the secant PQ to the curve $y = f(x)$ through the points $P(a,f(a))$ and $Q(a + h,f(a + h))$. The limit, $f'(a)$, of the difference quotient is the (*instantaneous*) *rate of change of f* at point a. Geometrically (see Figure N3-1a), the derivative $f'(a)$ is the limit of the slope of secant PQ as Q approaches P; that is, as h approaches zero. This limit is the *slope of the curve at P*. The *tangent to the curve at P* is the line through P with this slope.

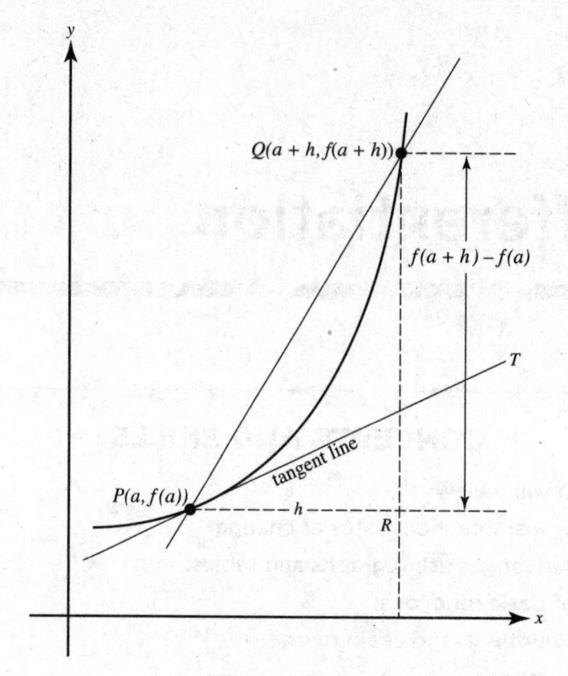

Figure N3–1a

In Figure N3–1a, PQ is the secant line through $(a, f(a))$ and $(a + h, f(a + h))$. The average rate of change from a to $a + h$ equals $\frac{RQ}{PR}$, which is the slope of secant PQ.

PT is the tangent to the curve at P. As h approaches zero, point Q approaches point P along the curve, PQ approaches PT, and the slope of PQ approaches the slope of PT, which equals $f'(a)$.

If we replace $(a + h)$ by x, in (2) above, so that $h = x - a$, we get the equivalent expression

$$f'(a) = \lim_{x \to a} \frac{f(x) - f(a)}{x - a}. \quad (3)$$

See Figure N3–1b.

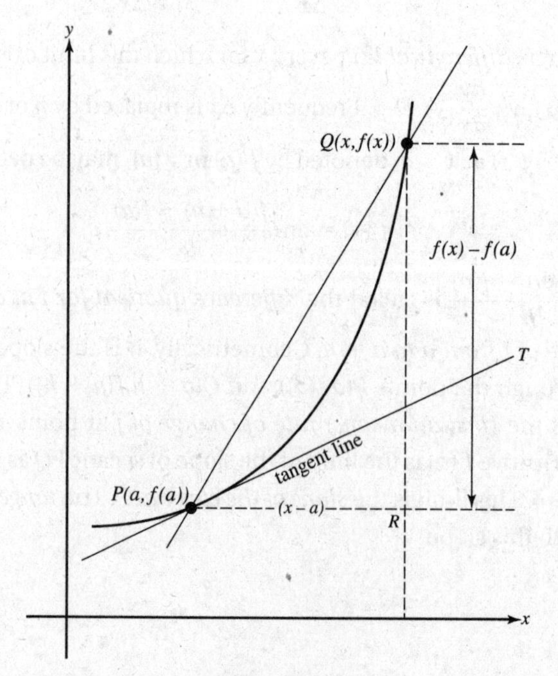

Figure N3–1b

The second derivative, denoted by $f''(x)$ or $\frac{d^2y}{dx^2}$ or y'', is the (first) derivative of $f'(x)$. Also, $f''(a)$ is the second derivative of $f(x)$ at $x = a$.

B. FORMULAS

The formulas in this section for finding derivatives are so important that familiarity with them is essential. If a and n are constants and u and v are differentiable functions of x, then:

$$\frac{da}{dx} = 0 \quad (1)$$

$$\frac{d}{dx}au = a\frac{du}{dx} \quad (2)$$

$$\frac{d}{dx}u^a = au^{a-1}\frac{du}{dx} \quad \text{(the Power Rule)}; \frac{d}{dx}x^n = nx^{n-1} \quad (3)$$

$$\frac{d}{dx}(u+v) = \frac{d}{dx}u + \frac{d}{dx}v; \ \frac{d}{dx}(u-v) = \frac{d}{dx}u - \frac{d}{dx}v \quad (4)$$

$$\frac{d}{dx}(uv) = u\frac{dv}{dx} + v\frac{du}{dx} \quad \text{(the Product Rule)} \quad (5)$$

Product rule

$$\frac{d}{dx}\left(\frac{u}{v}\right) = \frac{v\frac{du}{dx} - u\frac{dv}{dx}}{v^2} \quad (v \neq 0) \quad \text{(the Quotient Rule)} \quad (6)$$

Quotient rule

$$\frac{d}{dx}\sin u = \cos u\frac{du}{dx} \quad (7)$$

$$\frac{d}{dx}\cos u = -\sin u\frac{du}{dx} \quad (8)$$

$$\frac{d}{dx}\tan u = \sec^2 u\frac{du}{dx} \quad (9)$$

$$\frac{d}{dx}\cot u = -\csc^2 u\frac{du}{dx} \quad (10)$$

$$\frac{d}{dx}\sec u = \sec u \tan u\frac{du}{dx} \quad (11)$$

$$\frac{d}{dx}\csc u = -\csc u \cot u\frac{du}{dx} \quad (12)$$

$$\frac{d}{dx}\ln u = \frac{1}{u}\frac{du}{dx} \quad (13)$$

$$\frac{d}{dx}e^u = e^u\frac{du}{dx} \quad (14)$$

$$\frac{d}{dx}a^u = a^u \ln a\frac{du}{dx} \quad (15)$$

$$\frac{d}{dx}\sin^{-1} u = \frac{d}{dx}\arcsin u = \frac{1}{\sqrt{1-u^2}}\frac{du}{dx} \quad (-1 < u < 1) \quad (16)$$

$$\frac{d}{dx}\cos^{-1} u = \frac{d}{dx}\arccos u = -\frac{1}{\sqrt{1-u^2}}\frac{du}{dx} \quad (-1 < u < 1) \quad (17)$$

$$\frac{d}{dx}\tan^{-1} u = \frac{d}{dx} \text{ arctan } u = \frac{1}{1+u^2}\frac{du}{dx} \tag{18}$$

$$\frac{d}{dx}\cot^{-1} u = \frac{d}{dx} \text{ arccot } u = -\frac{1}{1+u^2}\frac{du}{dx} \tag{19}$$

$$\frac{d}{dx}\sec^{-1} u = \frac{d}{dx} \text{ arcsec } u = \frac{1}{|u|\sqrt{u^2-1}}\frac{du}{dx} \quad (|u|>1) \tag{20}$$

$$\frac{d}{dx}\csc^{-1} u = \frac{d}{dx} \text{ arccsc } u = -\frac{1}{|u|\sqrt{u^2-1}}\frac{du}{dx} \quad (|u|>1) \tag{21}$$

C. THE CHAIN RULE; THE DERIVATIVE OF A COMPOSITE FUNCTION

Formula (3) on page 113 says that

$$\frac{d}{dx}u^a = au^{a-1}\frac{du}{dx}.$$

This formula is an application of the *Chain Rule*. For example, if we use formula (3) to find the derivative of $(x^2 - x + 2)^4$, we get

$$\frac{d}{dx}(x^2 - x + 2)^4 = 4(x^2 - x + 2)^3 \cdot (2x - 1).$$

In this last equation, if we let $y = (x^2 - x + 2)^4$ and let $u = x^2 - x + 2$, then $y = u^4$. The preceding derivative now suggests one form of the Chain Rule:

$$\frac{dy}{dx} = \frac{dy}{du} \cdot \frac{du}{dx} = 4u^3 \cdot \frac{du}{dx} = 4(x^2 - x + 2)^3 \cdot (2x - 1)$$

as before. Formula (3) on page 113 gives the general case where $y = u^n$ and u is a differentiable function of x.

Now suppose we think of y as the composite function $f(g(x))$, where $y = f(u)$ and $u = g(x)$ are differentiable functions. Then

Chain rule

$$\begin{aligned}(f(g(x)))' &= f'(g(x)) \cdot g'(x) \\ &= f'(u) \cdot g'(x) \\ &= \frac{dy}{du} \cdot \frac{du}{dx},\end{aligned}$$

as we obtained above. The Chain Rule tells us how to differentiate the composite function: "Find the derivative of the 'outside' function first, then multiply by the derivative of the 'inside' one."

For example:

$$\frac{d}{dx}(x^3 + 1)^{10} = 10(x^3 + 1)^9 \cdot 3x^2 = 30x^2(x^3 + 1)^9,$$

$$\frac{d}{dx}\sqrt{7x-2} = \frac{d}{dx}(7x-2)^{1/2} = \frac{1}{2}(7x-2)^{-1/2} \cdot 7,$$

$$\frac{d}{dx}\frac{3}{(2-4x^2)^4} = \frac{d}{dx}3(2-4x^2)^{-4} = 3 \cdot (-4)(2-4x^2)^{-5} \cdot (-8x),$$

$$\frac{d}{dx}\sin\left(\frac{\pi}{2}-x\right)=\cos\left(\frac{\pi}{2}-x\right)\cdot(-1),$$

$$\frac{d}{dx}\cos^3 2x=\frac{d}{dx}(\cos 2x)^3=3(\cos 2x)^2\cdot(-\sin 2x\cdot 2).$$

Many of the formulas listed above in §B and most of the illustrative examples that follow use the Chain Rule. Often the chain rule is used more than once in finding a derivative.

Note that the algebraic simplifications that follow are included only for completeness.

Example 1

If $y = 4x^3 - 5x + 7$, find $y'(1)$ and $y''(1)$.

SOLUTION: $y' = \frac{dy}{dx} = 12x^2 - 5$ and $y'' = \frac{d^2y}{dx^2} = 24x$.

Then $y'(1) = 12\cdot 1^2 - 5 = 7$ and $y''(1) = 24\cdot 1 = 24$.

Example 2

If $f(x) = (3x + 2)^5$, find $f'(x)$.

SOLUTION: $f'(x) = 5(3x + 2)^4 \cdot 3 = 15(3x + 2)^4$.

Example 3

If $y = \sqrt{3 - x - x^2}$, find $\frac{dy}{dx}$.

SOLUTION: $y = (3 - x - x^2)^{1/2}$ so, $\frac{dy}{dx} = \frac{1}{2}(3 - x - x^2)^{-1/2}(-1 - 2x)$

$$= -\frac{1 + 2x}{2\sqrt{3 - x - x^2}}.$$

Example 4

If $y = \frac{5}{\sqrt{(1 - x^2)^3}}$, find $\frac{dy}{dx}$.

SOLUTION: $y = 5(1 - x^2)^{-3/2}$ so $\frac{dy}{dx} = \frac{-15}{2}(1 - x^2)^{-5/2}(-2x)$

$$= \frac{15x}{(1 - x^2)^{5/2}}.$$

Example 5

If $s(t) = (t^2 + 1)(1 - t)^2$, find $s'(t)$.

SOLUTION: $s'(t) = (t^2 + 1)\cdot 2(1 - t)(-1) + (1 - t)^2 \cdot 2t$ (Product Rule)

$$= 2(1 - t)(-1 + t - 2t^2).$$

➥ Example 6

If $f(t) = e^{2t} \sin 3t$, find $f'(0)$.

SOLUTION: $f'(t) = e^{2t}(\cos 3t \cdot 3) + \sin 3t(e^{2t} \cdot 2)$ (Product Rule)

$= e^{2t}(3 \cos 3t + 2 \sin 3t)$

Then, $f'(0) = 1(3 \cdot 1 + 2 \cdot 0) = 3$.

➥ Example 7

If $f(v) = \frac{2v}{1 - 2v^2}$, find $f'(v)$.

SOLUTION: $f'(v) = \frac{(1 - 2v^2) \cdot 2 - 2v(-4v)}{(1 - 2v^2)^2} = \frac{2 + 4v^2}{(1 - 2v^2)^2}$. (Quotient Rule)

Note that neither $f(v)$ nor $f'(v)$ exists where the denominator equals zero, namely, where $1 - 2v^2 = 0$ or where v equals $\pm\frac{\sqrt{2}}{2}$.

➥ Example 8

If $f(x) = \frac{\sin x}{x^2}$, $x \neq 0$, find $f'(x)$.

SOLUTION: $f'(x) = \frac{x^2 \cos x - \sin x \cdot 2x}{x^4} = \frac{x \cos x - 2 \sin x}{x^3}$.

➥ Example 9

If $y = \tan\left(2x^2 + 1\right)$, find y'.

SOLUTION: $y' = 4x \sec^2\left(2x^2 + 1\right)$.

➥ Example 10

If $x = \cos^3 (1 - 3\theta)$, find $\frac{dx}{d\theta}$.

SOLUTION: $\frac{dx}{d\theta} = -3 \cos^2 (1 - 3\theta) \sin (1 - 3\theta)(-3)$

$= 9 \cos^2 (1 - 3\theta) \sin (1 - 3\theta)$.

➥ Example 11

If $y = e^{(\sin x) + 1}$, find $\frac{dy}{dx}$.

SOLUTION: $\frac{dy}{dx} = \cos x \cdot e^{(\sin x) + 1}$.

➡ Example 12

If $y = (x + 1)\ln^2(x + 1)$, find $\frac{dy}{dx}$.

SOLUTION: $\frac{dy}{dx} = (x + 1)\frac{2\ln(x + 1)}{x + 1} + \ln^2(x + 1)$ (Product and Chain Rules)

$$= 2\ln(x + 1) + \ln^2(x + 1).$$

➡ Example 13

If $g(x) = \left(1 + \sin^2 3x\right)^4$, find $g'\left(\frac{\pi}{2}\right)$.

SOLUTION: $g'(x) = 4\left(1 + \sin^2 3x\right)^3(2\sin 3x\cos 3x)\cdot(3)$

$$= 24\left(1 + \sin^2 3x\right)^3(\sin 3x\cos 3x).$$

Then $g'\left(\frac{\pi}{2}\right) = 24\left(1 + (-1)^2\right)^3(-1\cdot 0) = 24\cdot 8\cdot 0 = 0.$

➡ Example 14

If $y = \sin^{-1}x + x\sqrt{1 - x^2}$, find y'.

SOLUTION: $y' = \frac{1}{\sqrt{1 - x^2}} + \frac{x(-2x)}{2\sqrt{1 - x^2}} + \sqrt{1 - x^2}$

$$= \frac{1 - x^2 + 1 - x^2}{\sqrt{1 - x^2}} = 2\sqrt{1 - x^2}.$$

➡ Example 15

If $u = \ln\sqrt{v^2 + 2v - 1}$, find $\frac{du}{dv}$.

SOLUTION: $u = \frac{1}{2}\ln\left(v^2 + 2v - 1\right)$ so

$$\frac{du}{dv} = \frac{1}{2}\frac{2v + 2}{v^2 + 2v - 1} = \frac{v + 1}{v^2 + 2v - 1}.$$

➡ Example 16

If $s = e^{-t}(\sin t - \cos t)$, find s'.

SOLUTION: $s' = e^{-t}(\cos t + \sin t) + (\sin t - \cos t)(-e^{-t})$

$$= e^{-t}(2\cos t) = 2e^{-t}\cos t.$$

➡ Example 17

Let $y = 2u^3 - 4u^2 + 5u - 3$ and $u = x^2 - x$. Find $\frac{dy}{dx}$.

SOLUTION:
$$\frac{dy}{dx} = (6u^2 - 8u + 5)(2x - 1)$$
$$= \left[6(x^2 - x)^2 - 8(x^2 - x) + 5\right](2x - 1).$$

➡ Example 18

If $y = \sin(ax + b)$, with a and b constants, find $\frac{dy}{dx}$.

SOLUTION: $\frac{dy}{dx} = [\cos(ax + b)] \cdot a = a\cos(ax + b).$

➡ Example 19

If $f(x) = ae^{kx}$ (with a and k constants), find f' and f''.

SOLUTION: $f'(x) = kae^{kx}$ and $f'' = k^2ae^{kx}$.

➡ Example 20

If $y = \ln(kx)$, where k is a constant, find $\frac{dy}{dx}$.

SOLUTION: We can use both formula (13), page 113, and the Chain Rule to get

$$\frac{dy}{dx} = \frac{1}{kx} \cdot k = \frac{1}{x}.$$

Alternatively, we can rewrite the given function using a property of logarithms: $\ln(kx) = \ln k + \ln x$. Then,

$$\frac{dy}{dx} = 0 + \frac{1}{x} = \frac{1}{x}.$$

➡ Example 21

Given $f(u) = u^2 - u$ and $u = g(x) = x^3 - 5$ and $F(x) = f(g(x))$, evaluate $F'(2)$.

SOLUTION: $F'(2) = f'(g(2))g'(2) = f'(3) \cdot (12) = 5 \cdot 12 = 60.$

Now, since $g'(x) = 3x^2$, $g'(2) = 12$, and since $f'(u) = 2u - 1$, $f'(3) = 5$.
Of course, we get exactly the same answer as follows.
Since $F(x) = (x^3 - 5)^2 - (x^3 - 5)$,

$$F'(x) = 2(x^3 - 5) \cdot 3x^2 - 3x^2,$$
$$F'(2) = 2 \cdot (3) \cdot 12 - 12 = 60.$$

D. DIFFERENTIABILITY AND CONTINUITY

If a function f has a derivative at $x = c$, then f is continuous at $x = c$.

This statement is an immediate consequence of the definition of the derivative of $f'(c)$ in the form

$$f'(c) = \lim_{x \to c} \frac{f(x) - f(c)}{x - c}.$$

If $f'(c)$ exists, then it follows that $\lim_{x \to c} f(x) = f(c)$, which guarantees that f is continuous at $x = c$.

If f is differentiable at c, its graph cannot have a hole or jump at c, nor can $x = c$ be a vertical asymptote of the graph. The tangent to the graph of f cannot be vertical at $x = c$; there cannot be a corner or cusp at $x = c$.

Each of the "prohibitions" in the preceding paragraph (each "cannot") tells how a function may fail to have a derivative at c. These cases are illustrated in Figures N3–2 (a) through (f).

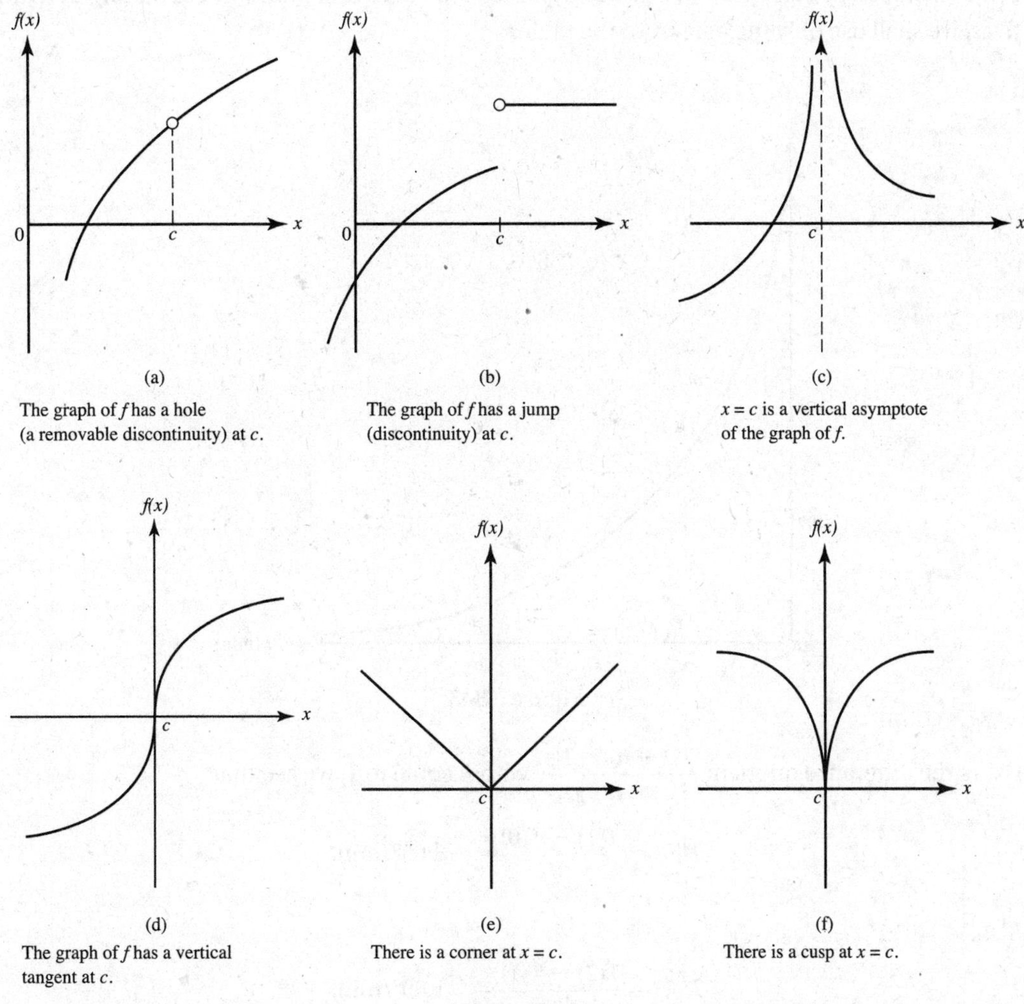

Figure N3–2

The graph in (e) is for the absolute function, $f(x) = |x|$. Since $f'(x) = -1$ for all negative x but $f'(x) = +1$ for all positive x, $f'(0)$ does not exist.

We may conclude from the preceding discussion that, although differentiability implies continuity, the converse is false. The functions in (d), (e), and (f) in Figure N3–2 are all continuous at $x = 0$, but not one of them is differentiable at the origin.

E. ESTIMATING A DERIVATIVE

E1. Numerically

➡ Example 22

The table shown gives the temperatures of a polar bear on a very cold arctic day (t = minutes; T = degrees Fahrenheit):

t	0	1	2	3	4	5	6	7	8
T	98	94.95	93.06	91.90	91.17	90.73	90.45	90.28	90.17

Our task is to estimate the derivative of T numerically at various times. One possible graph of $T(t)$ is sketched in Figure N3–3, but this assumes the curve is smooth. In estimating derivatives, we shall use only the data from the table.

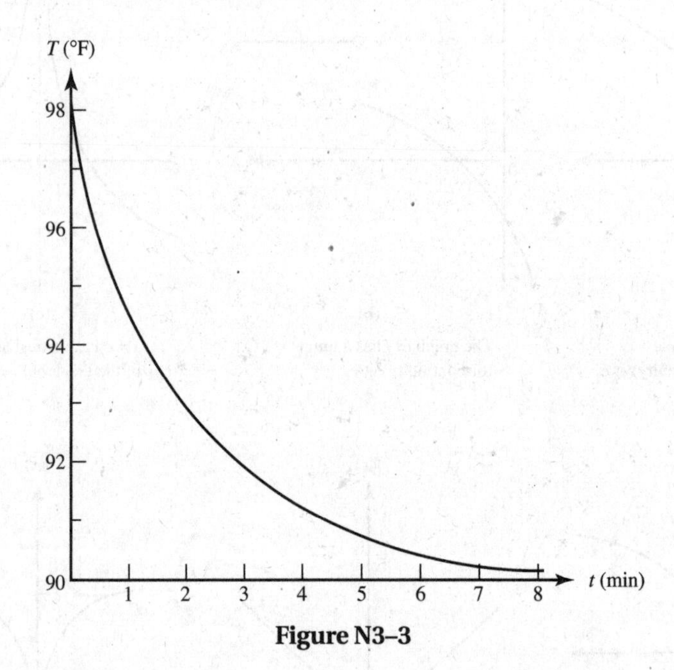

Figure N3–3

Using the difference quotient $\frac{T(t+h)-T(t)}{h}$ with h equal to 1, we see that

$$T'(0) \simeq \frac{T(1)-T(0)}{1} = -3.05°/\text{min}.$$

Also,

$$T'(1) \simeq \frac{T(2)-T(1)}{1} = -1.89°/\text{min},$$

$$T'(2) \simeq \frac{T(3)-T(2)}{1} = -1.16°/\text{min},$$

$$T'(3) \simeq \frac{T(4)-T(3)}{1} = -0.73°/\text{min},$$

and so on.

The following table shows the *approximate* values of $T'(t)$ obtained from the difference quotients above:

t	0	1	2	3	4	5	6	7
$T'(t)$	−3.05	−1.89	−1.16	−0.73	−0.47	−0.28	−0.17	−0.11

Note that the entries for $T'(t)$ also represent the approximate slopes of the T curve at times 0.5, 1.5, 2.5, and so on.

From a Symmetric Difference Quotient

Symmetric difference quotient

In Example 22 we approximated a derivative numerically from a table of values. We can also estimate $f'(a)$ numerically using the *symmetric difference quotient,* which is defined as follows:

$$f'(a) \simeq \frac{f(a+h) - f(a-h)}{2h}.$$

Note that the symmetric difference quotient is equal to

$$\frac{1}{2}\left[\frac{f(a+h) - f(a)}{h} + \frac{f(a) - f(a-h)}{h}\right].$$

We see that it is just the average of two difference quotients. Many calculators use the symmetric difference quotient in finding derivatives.

Example 23

For the function $f(x) = x^4$, approximate $f'(1)$ using the symmetric difference quotient with $h = 0.01$.

SOLUTION:

$$f'(1) \simeq \frac{(1.01)^4 - (0.99)^4}{2(0.01)} = 4.0004.$$

The exact value of $f'(1)$, of course, is 4.

The use of the symmetric difference quotient is particularly convenient when, as is often the case, obtaining a derivative precisely (with formulas) is cumbersome and an approximation is all that is needed for practical purposes.

A word of caution is in order. Sometimes a wrong result is obtained using the symmetric difference quotient. On page 119 we noted that $f(x) = |x|$ does not have a derivative at $x = 0$, since $f'(x) = -1$ for all $x < 0$ but $f'(x) = 1$ for all $x > 0$. Our calculator (which uses the symmetric difference quotient) tells us (incorrectly!) that $f'(0) = 0$. Note that, if $f(x) = |x|$, the symmetric difference quotient gives 0 for $f'(0)$ for every $h \neq 0$. If, for example, $h = 0.01$, then we get

$$f'(0) \simeq \frac{|0.01| - |-0.01|}{0.02} = \frac{0}{0.02} = 0,$$

which, as previously noted, is incorrect. The graph of the derivative of $f(x) = |x|$, which we see in Figure N3–4, shows that $f'(0)$ does not exist.

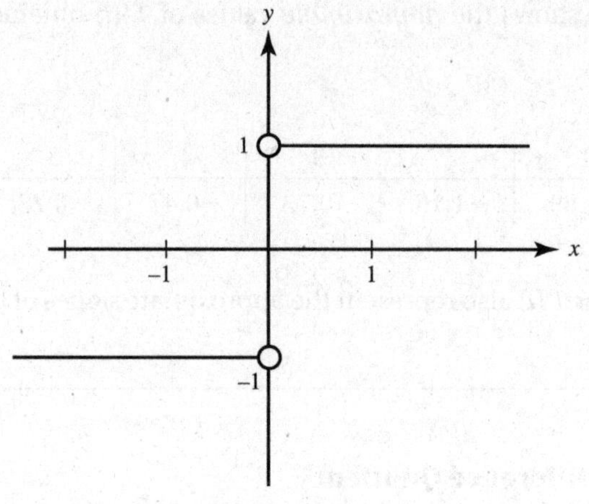

Figure N3–4

E2. Graphically

If we have the graph of a function $f(x)$, we can use it to graph $f'(x)$. We accomplish this by estimating the slope of the graph of $f(x)$ at enough points to assure a smooth curve for $f'(x)$. In Figure N3–5 we see the graph of $y = f(x)$. Below it is a table of the approximate slopes estimated from the graph.

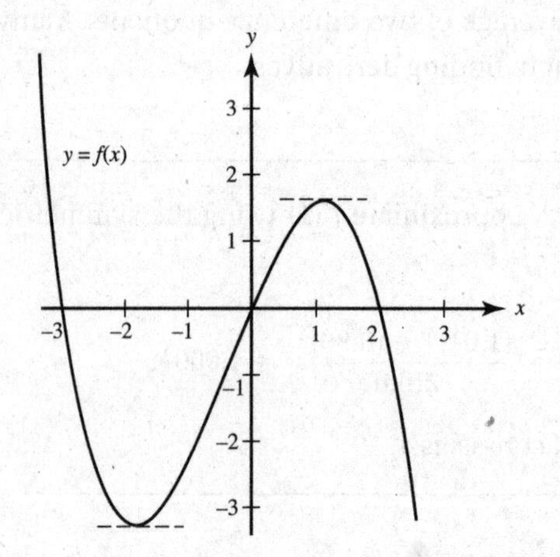

Figure N3–5

x	−3	−2.5	−2	−1.5	−1	0	0.5	1	1.5	2	2.5
$f'(x)$	−6	−3	−0.5	1	2	2	1.5	0.5	−2	−4	−7

Figure N3–6 was obtained by plotting the points from the table of slopes above and drawing a smooth curve through these points. The result is the graph of $y = f'(x)$.

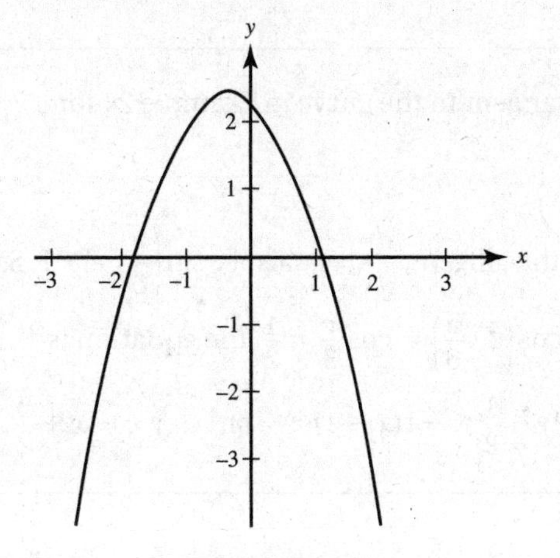

Figure N3–6

From the graphs above we can make the following observations:

(1) At the points where the slope of f (in Figure N3–5) equals 0, the graph of f' (Figure N3–6) has x-intercepts: approximately $x = -1.8$ and $x = 1.1$. We've drawn horizontal broken lines at these points on the curve in Figure N3-5.

(2) On intervals where f $\begin{matrix}\text{decreases}\\ \text{increases}\end{matrix}$, the derivative is $\begin{matrix}\text{negative}\\ \text{positive}\end{matrix}$. We see here that f decreases for $x < -1.8$ (approximately) and for $x > 1.1$ (approximately), and that f increases for $-1.8 < x < 1.1$ (approximately). In Chapter 4 we discuss other behaviors of f that are reflected in the graph of f'.

F. DERIVATIVES OF PARAMETRICALLY DEFINED FUNCTIONS

BC ONLY

Parametric differentiation

Parametric equations were defined on page 75.

If $x = f(t)$ and $y = g(t)$ are differentiable functions of t, then

$$\frac{dy}{dx} = \frac{\frac{dy}{dt}}{\frac{dx}{dt}} \quad \text{and} \quad \frac{d^2y}{dx^2} = \frac{d}{dx}\left(\frac{dy}{dx}\right) = \frac{\frac{d}{dt}\left(\frac{dy}{dx}\right)}{\frac{dx}{dt}}.$$

➡ Example 24

If $x = 2 \sin \theta$ and $y = \cos 2\theta$, find $\frac{dy}{dx}$ and $\frac{d^2y}{dx^2}$.

SOLUTION: $$\frac{dy}{dx} = \frac{\frac{dy}{d\theta}}{\frac{dx}{d\theta}} = \frac{-2 \sin 2\theta}{2 \cos \theta} = -\frac{2 \sin \theta \cos \theta}{\cos \theta} = -2 \sin \theta.$$

Also,

$$\frac{d^2y}{dx^2} = \frac{\frac{d}{d\theta}\left(\frac{dy}{dx}\right)}{\frac{dx}{d\theta}} = \frac{-2 \cos \theta}{2 \cos \theta} = -1.$$

BC ONLY

➡ Example 25

Find the equation of the tangent to the curve in Example 24 for $\theta = \frac{\pi}{6}$.

SOLUTION:

When $\theta = \frac{\pi}{6}$, the slope of the tangent, $\frac{dy}{dx}$, equals $-2 \sin\left(\frac{\pi}{6}\right) = -1$. Since $x = 2 \sin\left(\frac{\pi}{6}\right) = 1$ and $y = \cos\left(2 \cdot \frac{\pi}{6}\right) = \cos\frac{\pi}{3} = \frac{1}{2}$, the equation is

$$y - \frac{1}{2} = -1(x - 1) \quad \text{or} \quad y = -x + \frac{3}{2}.$$

➡ Example 26

Suppose two objects are moving in a plane during the time interval $0 \leqslant t \leqslant 4$. Their positions at time t are described by the parametric equations

$$x_1 = 2t, \quad y_1 = 4t - t^2 \quad \text{and} \quad x_2 = t + 1, \quad y_2 = 4 - t.$$

(a) Find all collision points. Justify your answer.

(b) Use a calculator to help you sketch the paths of the objects, indicating the direction in which each object travels.

SOLUTION:

(a) Equating x_1 and x_2 yields $t = 1$. When $t = 1$, both y_1 and y_2 equal 3. So $t = 1$ yields a *true* collision point (not just an intersection point) at (2,3). (An *intersection point* is any point that is on both curves, but not necessarily at the same time.)

(b) Using parametric mode, we graph both curves with t in [0,4], in the window $[0,8] \times [0,4]$ as shown in Figure N3–7.

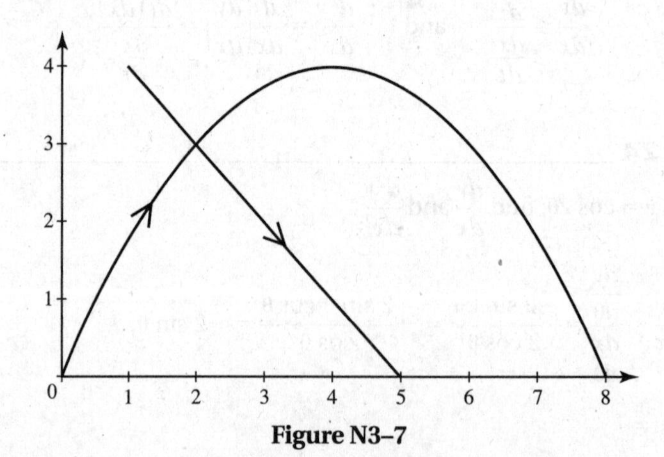

Figure N3–7

We've inserted arrows to indicate the direction of motion.

Note that if your calculator can draw the curves in simultaneous graphing mode, you can watch the objects as they move, seeing that they do indeed pass through the intersection point at the same time.

G. IMPLICIT DIFFERENTIATION

When a functional relationship between x and y is defined by an equation of the form $F(x,y) = 0$, we say that the equation defines y *implicitly* as a function of x. Some examples are $x^2 + y^2 - 9 = 0$, $y^2 - 4x = 0$, and $\cos(xy) = y^2 - 5$ (which can be written as $\cos(xy) - y^2 + 5 = 0$). Sometimes two (or more) explicit functions are defined by $F(x,y) = 0$. For example, $x^2 + y^2 - 9 = 0$ defines the two functions $y_1 = +\sqrt{9 - x^2}$ and $y_2 = -\sqrt{9 - x^2}$, the upper and lower halves, respectively, of the circle centered at the origin with radius 3. Each function is differentiable except at the points where $x = 3$ and $x = -3$.

Implicit differentiation is the technique we use to find a derivative when y is not defined explicitly in terms of x but is differentiable.

In the following examples, we differentiate both sides with respect to x, using appropriate formulas, and then solve for $\frac{dy}{dx}$.

Example 27

If $x^2 + y^2 - 9 = 0$, then

$$2x + 2y\frac{dy}{dx} = 0 \quad \text{and} \quad \frac{dy}{dx} = -\frac{x}{y}.$$

Note that the derivative above holds for every point on the circle, and exists for all y different from 0 (where the tangents to the circle are vertical).

Example 28

If $x^2 - 2xy + 3y^2 = 2$, find $\frac{dy}{dx}$.

SOLUTION: $2x - 2\left(x\frac{dy}{dx} + y \cdot 1\right) + 6y\frac{dy}{dx} = 0$

$$\frac{dy}{dx}(6y - 2x) = 2y - 2x, \text{ so } \frac{dy}{dx} = \frac{y - x}{3y - x}.$$

Example 29

If $x \sin y = \cos(x + y)$, find $\frac{dy}{dx}$.

SOLUTION: $x \cos y\frac{dy}{dx} + \sin y = -\sin(x + y)\left(1 + \frac{dy}{dx}\right),$

$$\frac{dy}{dx} = -\frac{\sin y + \sin(x + y)}{x \cos y + \sin(x + y)}.$$

Example 30

Find $\frac{dy}{dx}$ and $\frac{d^2y}{dx^2}$ using implicit differentiation on the equation $x^2 + y^2 = 1$.

SOLUTION: $$2x + 2y\frac{dy}{dx} = 0 \quad \rightarrow \quad \frac{dy}{dx} = -\frac{x}{y}. \tag{1}$$

Then

$$\frac{d^2y}{dx^2} = -\frac{y \cdot 1 - x\left(\frac{dy}{dx}\right)}{y^2} = -\frac{y - x\left(-\frac{x}{y}\right)}{y^2} \tag{2}$$

$$= -\frac{y^2 + x^2}{y^3} = -\frac{1}{y^3}, \tag{3}$$

where we substituted for $\frac{dy}{dx}$ from (1) in (2), then used the given equation to simplify in (3).

Example 31

Using implicit differentiation, verify the formula for the derivative of the inverse sine function, $y = \sin^{-1} x = \arcsin x$, with domain $[-1,1]$ and range $\left[-\frac{\pi}{2}, \frac{\pi}{2}\right]$.

SOLUTION: $$y = \sin^{-1} x \quad \leftrightarrow \quad x = \sin y.$$

Now we differentiate with respect to x:

$$1 = \cos y \frac{dy}{dx},$$

$$\frac{dy}{dx} = \frac{1}{\cos y} = \frac{1}{+\sqrt{1 - \sin^2 y}} = \frac{1}{\sqrt{1 - x^2}},$$

where we chose the positive sign for $\cos y$ since $\cos y$ is nonnegative if $-\frac{\pi}{2} < y < \frac{\pi}{2}$. Note that this derivative exists only if $-1 < x < 1$.

Example 32

Where is the tangent to the curve $4x^2 + 9y^2 = 36$ vertical?

SOLUTION: We differentiate the equation implicitly to get $\frac{dy}{dx}$: $8x + 18y\frac{dy}{dx} = 0$, so $\frac{dy}{dx} = -\frac{4x}{9y}$. Since the tangent line to a curve is vertical when $\frac{dx}{dy} = 0$, we conclude that $-\frac{9y}{4x}$ must equal zero; that is, y must equal zero with $x \neq 0$. When we substitute $y = 0$ in the original equation, we get $x = \pm 3$. The points $(\pm 3,0)$ are the ends of the major axis of the ellipse, where the tangents are indeed vertical.

H. DERIVATIVE OF THE INVERSE OF A FUNCTION

Suppose f and g are inverse functions. What is the relationship between their derivatives? Recall that the graphs of inverse functions are the reflections of each other in the line $y = x$, and that at corresponding points their x- and y-coordinates are interchanged.

Figure N3–8 shows a function f passing through point (a,b) and the line tangent to f at that point. The slope of the curve there, $f'(a)$, is represented by the ratio of the legs of the triangle, $\frac{dy}{dx}$. When this figure is reflected across the line $y = x$, we obtain the graph of f^{-1}, passing through point (b,a), with the horizontal and vertical sides of the slope triangle interchanged. Note that the slope of the line tangent to the graph of f^{-1} at $x = b$ is represented by $\frac{dx}{dy}$, the reciprocal of the slope of f at $x = a$. We have, therefore,

Derivative of inverse function

$$\left(f^{-1}\right)'(b) = \frac{1}{f'(a)} \quad \text{or} \quad \left(f^{-1}\right)'(x) = \frac{1}{f'\left(f^{-1}(x)\right)}.$$

Simply put, the derivative of the inverse of a function at a point is the *reciprocal* of the derivative of the function *at the corresponding point.*

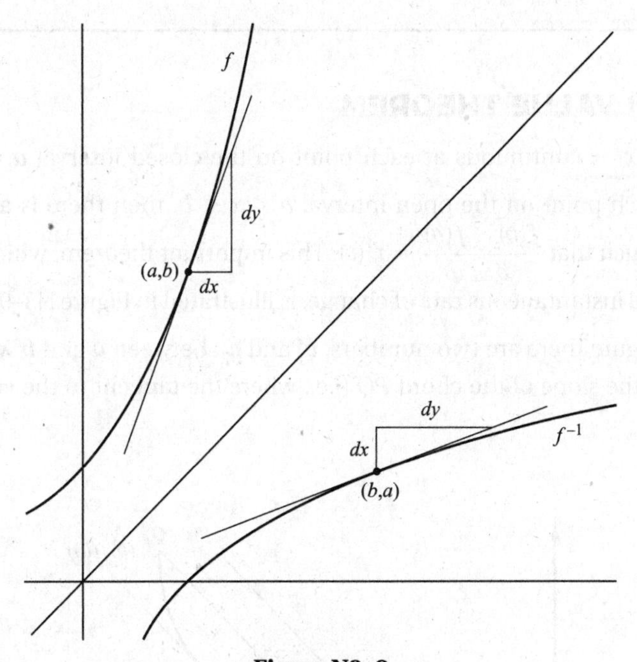

Figure N3–8

➡ Example 33

If $f(3) = 8$ and $f'(3) = 5$, what do we know about f^{-1}?

SOLUTION: Since f passes through the point (3,8), f^{-1} must pass through the point (8,3). Furthermore, since the graph of f has slope 5 at (3,8), the graph of f^{-1} must have slope $\frac{1}{5}$ at (8,3).

➡ Example 34

x	$f(x)$	$f'(x)$
2	6	$\frac{1}{3}$
6	8	$\frac{3}{2}$

A function f and its derivative take on the values shown in the table. If g is the inverse of f, find $g'(6)$.

SOLUTION: To find the slope of g at the point where $x = 6$, we must look at the point on f where $y = 6$, namely, (2,6). Since $f'(2) = \frac{1}{3}$, $g'(6) = 3$.

➡ Example 35

Let $y = f(x) = x^3 + x - 2$, and let g be the inverse function. Evaluate $g'(0)$.

SOLUTION: Since $f'(x) = 3x^2 + 1$, $g'(y) = \dfrac{1}{3x^2 + 1}$. To find x when $y = 0$, we must solve the equation $x^3 + x - 2 = 0$. Note by inspection that $x = 1$, so

$$g'(0) = \frac{1}{3(1)^2 + 1} = \frac{1}{4}.$$

I. THE MEAN VALUE THEOREM

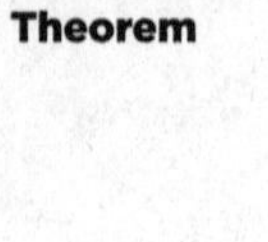

Mean Value Theorem

If the function $f(x)$ is continuous at each point on the closed interval $a \leqslant x \leqslant b$ and has a derivative at each point on the open interval $a < x < b$, then there is at least one number c, $a < c < b$, such that $\dfrac{f(b) - f(a)}{b - a} = f'(c)$. This important theorem, which relates average rate of change and instantaneous rate of change, is illustrated in Figure N3–9. For the function sketched in the figure there are two numbers, c_1 and c_2, between a and b where the slope of the curve equals the slope of the chord PQ (i.e., where the tangent to the curve is parallel to the secant line).

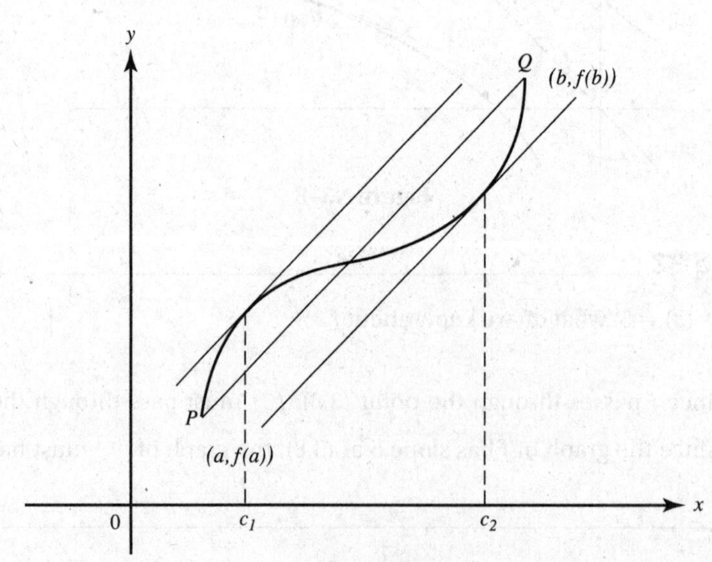

Figure N3–9

We will often refer to the Mean Value Theorem by its initials, MVT.

If, in addition to the hypotheses of the MVT, it is given that $f(a) = f(b) = k$, then there is a number, c, between a and b such that $f'(c) = 0$. This special case of the MVT is called Rolle's Theorem, as seen in Figure N3–10 for $k = 0$.

Rolle's Theorem

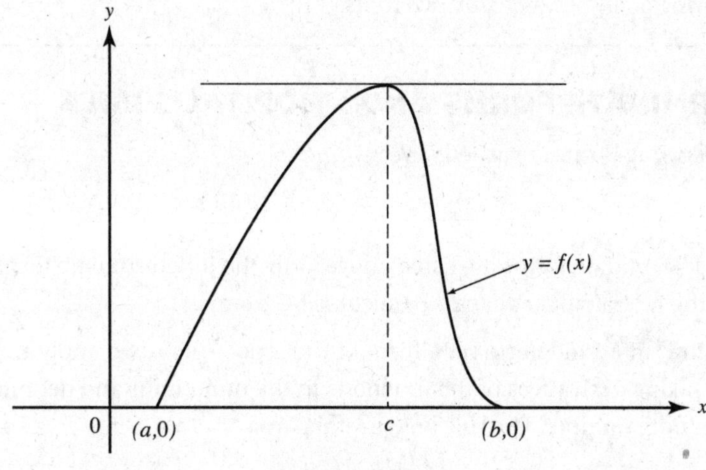

Figure N3–10

The Mean Value Theorem is one of the most useful laws when properly applied.

➡ Example 36

You left home one morning and drove to a cousin's house 300 miles away, arriving 6 hours later. What does the Mean Value Theorem say about your speed along the way?

SOLUTION: Your journey was continuous, with an average speed (the average rate of change of distance traveled) given by

$$\frac{\Delta\text{distance}}{\Delta\text{time}} = \frac{300\text{ miles}}{6\text{ hours}} = 50\text{ mph}.$$

Furthermore, the derivative (your instantaneous speed) existed everywhere along your trip. The MVT, then, guarantees that at least at one point your instantaneous speed was equal to your average speed for the entire 6-hour interval. Hence, your car's speedometer must have read exactly 50 mph at least once on your way to your cousin's house.

➡ Example 37

Demonstrate Rolle's Theorem using $f(x) = x \sin x$ on the interval $[0,\pi]$.

SOLUTION: First, we check that the conditions of Rolle's Theorem are met:

(1) $f(x) = x \sin x$ is continuous on $(0,\pi)$ and exists for all x in $[0,\pi]$.

(2) $f'(x) = x \cos x + \sin x$ exists for all x in $(0,\pi)$.

(3) $f(0) = 0 \sin 0 = 0$ and $f(\pi) = \pi \sin \pi = 0$.

Hence there must be a point, $x = c$, in the interval $0 < x < \pi$ where $f'(c) = 0$. Using the calculator to solve $x \cos x + \sin x = 0$, we find $c = 2.029$ (to three decimal places). As predicted by Rolle's Theorem, $0 \le c \le \pi$.

Note that this result indicates that at $x = c$ the line tangent to f is horizontal. The MVT (here as Rolle's Theorem) tells us that any function that is continuous and differentiable must have at least one turning point between any two roots.

J. INDETERMINATE FORMS AND L'HÔPITAL'S RULE

Limits of the following forms are called *indeterminate*:

$$\frac{0}{0} \text{ or } \frac{\infty}{\infty},\ 0 \cdot \infty,\ \infty - \infty,\ 0^0,\ 1^\infty,\ \infty^0$$

Although all indeterminate forms are listed above, only the indeterminate forms $\frac{0}{0}$ and $\frac{\infty}{\infty}$ are tested on both the AP Calculus AB and AP Calculus BC exams.

To find the limit of an indeterminate form of the type $\frac{0}{0}$ or $\frac{\infty}{\infty}$, we apply L'Hôpital's Rule, which involves taking derivatives of the functions in the numerator and denominator. In the following, a is a finite number. The rule has several parts:

(a) If $\lim_{x\to a} f(x) = \lim_{x\to a} g(x) = 0$ and if $\lim_{x\to a} \frac{f'(x)}{g'(x)}$ exists*, then

$$\lim_{x\to a} \frac{f(x)}{g(x)} = \lim_{x\to a} \frac{f'(x)}{g'(x)};$$

if $\lim_{x\to a} \frac{f'(x)}{g'(x)}$ does not exist, then L'Hôpital's Rule cannot be applied.

(b) If $\lim_{x\to a} f(x) = \lim_{x\to a} g(x) = \infty$, the same consequences follow as in case (a). The rules in (a) and (b) both hold for one-sided limits.

(c) If $\lim_{x\to\infty} f(x) = \lim_{x\to\infty} g(x) = 0$ and if $\lim_{x\to\infty} \frac{f'(x)}{g'(x)}$ exists, then

$$\lim_{x\to\infty} \frac{f(x)}{g(x)} = \lim_{x\to\infty} \frac{f'(x)}{g'(x)};$$

if $\lim_{x\to\infty} \frac{f'(x)}{g'(x)}$ does not exist, then L'Hôpital's Rule cannot be applied. (Here the notation "$x \to \infty$" represents either "$x \to +\infty$" or "$x \to -\infty$.")

(d) If $\lim_{x\to\infty} f(x) = \lim_{x\to\infty} g(x) = \infty$, the same consequences follow as in case (c).

In applying any of the above rules, if we obtain $\frac{0}{0}$ or $\frac{\infty}{\infty}$ again, we can apply the rule once more, repeating the process until the form we obtain is no longer indeterminate.

➡ Example 38

$\lim_{x\to 3} \frac{x^2 - 9}{x - 3}$ is of type $\frac{0}{0}$ and thus equals $\lim_{x\to 3} \frac{2x}{1} = 6$.

(Compare with Example 12, page 95.)

---

*The limit can be finite or infinite ($+\infty$ or $-\infty$).

➡ Example 39

$\lim_{x\to 0} \frac{\tan x}{x}$ is of type $\frac{0}{0}$ and therefore equals $\lim_{x\to 0} \frac{\sec^2 x}{1} = 1$.

➡ Example 40

$\lim_{x\to -2} \frac{x^3+8}{x^2-4}$ (Example 13, page 95) is of type $\frac{0}{0}$ and thus equals $\lim_{x\to -2} \frac{3x^2}{2x} = -3$, as before. Note that $\lim_{x\to -2} \frac{3x^2}{2x}$ is *not* the limit of an indeterminate form!

➡ Example 41

$\lim_{h\to 0} \frac{e^h-1}{h}$ is of type $\frac{0}{0}$ and therefore equals $\lim_{h\to 0} \frac{e^h}{1} = 1$.

➡ Example 42

$\lim_{x\to\infty} \frac{x^3-4x^2+7}{3-6x-2x^3}$ (Example 20, page 96) is of type $\frac{\infty}{\infty}$, so that it equals $\lim_{x\to\infty} \frac{3x^2-8x}{-6-6x^2}$, which is again of type $\frac{\infty}{\infty}$. Apply L'Hôpital's Rule twice more:

$$\lim_{x\to\infty} \frac{6x-8}{-12x} = \lim_{x\to\infty} \frac{6}{-12} = -\frac{1}{2}.$$

For this problem, it is easier and faster to apply the Rational Function Theorem!

➡ Example 43

Find $\lim_{x\to\infty} \frac{\ln x}{x}$.

SOLUTION: $\lim_{x\to\infty} \frac{\ln x}{x}$ is of type $\frac{\infty}{\infty}$ and equals $\lim_{x\to\infty} \frac{1/x}{1} = 0$.

➡ Example 44

Find $\lim_{x\to 2} \frac{x^3+8}{x^2+4}$.

SOLUTION: $\lim_{x\to 2} \frac{x^3+8}{x^2+4} = \frac{16}{8} = 2$.

BEWARE: L'Hôpital's Rule applies only to indeterminate forms $\frac{0}{0}$ and $\frac{\infty}{\infty}$. Trying to use it in other situations leads to incorrect results, like this:

$$\lim_{x\to 2} \frac{x^3+8}{x^2+4} = \lim_{x\to 2} \frac{3x^2}{2x} = 3 \text{ (WRONG!)}$$

For more practice, redo the Practice Exercises on pages 102–107, applying L'Hôpital's Rule wherever possible.

NOTE: Below is a description of how to determine the limit for other indeterminate forms by transforming them into $\frac{0}{0}$ or $\frac{\infty}{\infty}$, but only limits originally of the form $\frac{0}{0}$ and $\frac{\infty}{\infty}$ will be tested on the AP Calculus exam as given in Examples 38–44. Examples 45, 46, and 47 are presented to complete the discussion of indeterminate forms and L'Hôpital's Rule, but questions similar to those in Examples 45–47 will not appear on the AP Calculus exam.

L'Hôpital's Rule can be applied also to indeterminate forms of the types $0 \cdot \infty$ and $\infty - \infty$, if the forms can be transformed to either $\frac{0}{0}$ or $\frac{\infty}{\infty}$.

Example 45

Find $\lim_{x \to \infty} x \sin \frac{1}{x}$.

SOLUTION: $\lim_{x \to \infty} x \sin \frac{1}{x}$ is of the type $\infty \cdot 0$. Since $x \sin \frac{1}{x} = \frac{\sin 1/x}{1/x}$ and, as $x \to \infty$, the latter is the indeterminate form $\frac{0}{0}$, we see that

$$\lim_{x \to \infty} x \sin \frac{1}{x} = \lim_{x \to \infty} \frac{-\frac{1}{x^2} \cos \frac{1}{x}}{-\frac{1}{x^2}} = \lim_{x \to \infty} \cos \frac{1}{x} = 1.$$

(Note the easier solution $\lim_{x \to \infty} x \sin \frac{1}{x} = \lim_{x \to \infty} \frac{\sin \frac{1}{x}}{\frac{1}{x}} = 1$.)

Other indeterminate forms, such as 0^0, 1^∞, and ∞^0, may be resolved by taking the natural logarithm and then applying L'Hôpital's Rule.

Example 46

Find $\lim_{x \to 0} (1 + x)^{1/x}$.

SOLUTION: $\lim_{x \to 0} (1 + x)^{1/x}$ is of type 1^∞. Let $y = (1 + x)^{1/x}$, so that $\ln y = \frac{1}{x} \ln (1 + x)$. Then $\lim_{x \to 0} \ln y = \lim_{x \to 0} \frac{\ln(1 + x)}{x}$, which is of type $\frac{0}{0}$. Thus,

$$\lim_{x \to 0} \ln y = \lim_{x \to 0} \frac{\frac{1}{1 + x}}{1} = \frac{1}{1} = 1,$$

and since $\lim_{x \to 0} \ln y = 1$, $\lim_{x \to 0} y = e^1 = e$.

Example 47

Find $\lim_{x \to \infty} x^{1/x}$.

SOLUTION: $\lim_{x \to \infty} x^{1/x}$ is of type ∞^0. Let $y = x^{1/x}$, so that $\ln y = \frac{1}{x} \ln x = \frac{\ln x}{x}$ (which, as $x \to \infty$, is of type $\frac{\infty}{\infty}$). Then $\lim_{x \to \infty} \ln y = \frac{1/x}{1} = 0$, and $\lim_{x \to \infty} y = e^0 = 1$.

K. RECOGNIZING A GIVEN LIMIT AS A DERIVATIVE

It is often extremely useful to evaluate a limit by recognizing that it is merely an expression for the definition of the derivative of a specific function (often at a specific point). The relevant definition is the limit of the difference quotient:

$$f'(c) = \lim_{h\to 0} \frac{f(c+h) - f(c)}{h}.$$

Example 48

Find $\lim_{h\to 0} \frac{(2+h)^4 - 2^4}{h}$.

SOLUTION: $\lim_{h\to 0} \frac{(2+h)^4 - 2^4}{h}$ is the derivative of $f(x) = x^4$ at the point where $x = 2$. Since $f'(x) = 4x^3$, the value of the given limit is $f'(2) = 4(2^3) = 32$.

Example 49

Find $\lim_{h\to 0} \frac{\sqrt{9+h} - 3}{h}$.

SOLUTION: $\lim_{h\to 0} \frac{\sqrt{9+h} - 3}{h} = f'(9)$, where $f(x) = \sqrt{x}$. The value of the limit is $\frac{1}{2}x^{-1/2}$ when $x = 9$, or $\frac{1}{6}$.

Example 50

Find $\lim_{h\to 0} \frac{1}{h}\left(\frac{1}{2+h} - \frac{1}{2}\right)$.

SOLUTION: $\lim_{h\to 0} \frac{1}{h}\left(\frac{1}{2+h} - \frac{1}{2}\right) = f'(2)$, where $f(x) = \frac{1}{x}$.

Verify that $f'(2) = -\frac{1}{4}$ and compare with Example 17, page 95.

Example 51

Find $\lim_{h\to 0} \frac{e^h - 1}{h}$.

SOLUTION: $\lim_{h\to 0} \frac{e^h - 1}{h} = f'(0)$, where $f(x) = e^x$. The limit has value e^0 or 1 (see also Example 41, on page 131).

➡ Example 52

Find $\lim_{x\to 0} \frac{\sin x}{x}$.

SOLUTION: $\lim_{x\to 0} \frac{\sin x}{x}$ is $f'(0)$, where $f(x) = \sin x$, because we can write

$$f'(0) = \lim_{x\to 0} \frac{\sin(0+x) - \sin 0}{x} = \lim_{x\to 0} \frac{\sin x}{x}.$$

The answer is 1, since $f'(x) = \cos x$ and $f'(0) = \cos 0 = 1$. Of course, we already know that the given limit is the basic trigonometric limit with value 1. Also, L'Hôpital's Rule yields 1 as the answer immediately.

CHAPTER SUMMARY

In this chapter we have reviewed differentiation. We've defined the derivative as the instantaneous rate of change of a function, and looked at estimating derivatives using tables and graphs. We've reviewed the formulas for derivatives of basic functions, as well as the product, quotient, and chain rules. We've looked at derivatives of implicitly defined functions and inverse functions, and reviewed two important theorems: Rolle's Theorem and the Mean Value Theorem.

For BC Calculus students, we've reviewed derivatives of parametrically defined functions and the use of L'Hopital's Rule for evaluating limits of indeterminate forms.

PRACTICE EXERCISES

Part A. Directions: Answer these questions *without* using your calculator.

In each of Questions 1–20 a function is given. Choose the alternative that is the derivative, $\frac{dy}{dx}$, of the function.

1. $y = x^5 \tan x$

(A) $5x^4 \tan x$ (B) $x^5 \sec^2 x$ (C) $5x^4 \sec^2 x$

(D) $5x^4 + \sec^2 x$ (E) $5x^4 \tan x + x^5 \sec^2 x$

2. $y = \frac{2-x}{3x+1}$

(A) $-\frac{7}{(3x+1)^2}$ (B) $\frac{6x-5}{(3x+1)^2}$ (C) $-\frac{9}{(3x+1)^2}$ (D) $\frac{7}{(3x+1)^2}$ (E) $\frac{7-6x}{(3x+1)^2}$

3. $y = \sqrt{3-2x}$

(A) $\frac{1}{2\sqrt{3-2x}}$ (B) $-\frac{1}{\sqrt{3-2x}}$ (C) $-\frac{(3-2x)^{3/2}}{3}$

(D) $-\frac{1}{3-2x}$ (E) $\frac{2}{3}(3-2x)^{3/2}$

4. $y = \frac{2}{(5x+1)^3}$

(A) $-\frac{30}{(5x+1)^2}$ (B) $-30(5x+1)^{-4}$ (C) $\frac{-6}{(5x+1)^4}$

(D) $-\frac{10}{3}(5x+1)^{-4/3}$ (E) $\frac{30}{(5x+1)^4}$

5. $y = 3x^{2/3} - 4x^{1/2} - 2$

(A) $2x^{1/3} - 2x^{-1/2}$ (B) $3x^{-1/3} - 2x^{-1/2}$ (C) $\frac{9}{5}x^{5/3} - 8x^{3/2}$

(D) $\frac{2}{x^{1/3}} - \frac{2}{x^{1/2}} - 2$ (E) $2x^{-1/3} - 2x^{-1/2}$

6. $y = 2\sqrt{x} - \frac{1}{2\sqrt{x}}$

(A) $x + \frac{1}{x\sqrt{x}}$ (B) $x^{-1/2} + x^{-3/2}$ (C) $\frac{4x-1}{4x\sqrt{x}}$ (D) $\frac{1}{\sqrt{x}} + \frac{1}{4x\sqrt{x}}$ (E) $\frac{4}{\sqrt{x}} + \frac{1}{x\sqrt{x}}$

7. $y = \sqrt{x^2 + 2x - 1}$

(A) $\frac{x+1}{y}$ (B) $4y(x+1)$ (C) $\frac{1}{2\sqrt{x^2+2x-1}}$

(D) $-\frac{x+1}{(x^2+2x-1)^{3/2}}$ (E) none of these

8. $y = \frac{x^2}{\cos x}$

(A) $\frac{2x}{\sin x}$ (B) $-\frac{2x}{\sin x}$ (C) $\frac{2x \cos x - x^2 \sin x}{\cos^2 x}$

(D) $\frac{2x \cos x + x^2 \sin x}{\cos^2 x}$ (E) $\frac{2x \cos x + x^2 \sin x}{\sin^2 x}$

9. $y = \ln \dfrac{e^x}{e^x - 1}$

(A) $x - \dfrac{e^x}{e^x - 1}$ (B) $\dfrac{1}{e^x - 1}$ (C) $-\dfrac{1}{e^x - 1}$ (D) 0 (E) $\dfrac{e^x - 2}{e^x - 1}$

10. $y = \tan^{-1} \dfrac{x}{2}$

(A) $\dfrac{4}{4 + x^2}$ (B) $\dfrac{1}{2\sqrt{4 - x^2}}$ (C) $\dfrac{2}{\sqrt{4 - x^2}}$ (D) $\dfrac{1}{2 + x^2}$ (E) $\dfrac{2}{x^2 + 4}$

11. $y = \ln(\sec x + \tan x)$

(A) $\sec x$ (B) $\dfrac{1}{\sec x}$ (C) $\tan x + \dfrac{\sec^2 x}{\tan x}$

(D) $\dfrac{1}{\sec x + \tan x}$ (E) $-\dfrac{1}{\sec x + \tan x}$

12. $y = \dfrac{e^x - e^{-x}}{e^x + e^{-x}}$

(A) 0 (B) 1 (C) $\dfrac{2}{(e^x + e^{-x})^2}$ (D) $\dfrac{4}{(e^x + e^{-x})^2}$ (E) $\dfrac{1}{e^{2x} + e^{-2x}}$

13. $y = \ln\left(\sqrt{x^2 + 1}\right)$

(A) $\dfrac{1}{\sqrt{x^2 + 1}}$ (B) $\dfrac{2x}{\sqrt{x^2 + 1}}$ (C) $\dfrac{1}{2(x^2 + 1)}$ (D) $\dfrac{x}{x^2 + 1}$ (E) $\dfrac{2x}{x^2 + 1}$

14. $y = \sin\left(\dfrac{1}{x}\right)$

(A) $\cos\left(\dfrac{1}{x}\right)$ (B) $\cos\left(-\dfrac{1}{x^2}\right)$ (C) $-\dfrac{1}{x^2}\cos\left(\dfrac{1}{x}\right)$

(D) $-\dfrac{1}{x^2}\sin\left(\dfrac{1}{x}\right) + \dfrac{1}{x}\cos\left(\dfrac{1}{x}\right)$ (E) $\cos(\ln x)$

15. $y = \dfrac{1}{2\sin 2x}$

(A) $-\csc 2x \cot 2x$ (B) $\dfrac{1}{4\cos 2x}$ (C) $-4\csc 2x \cot 2x$

(D) $\dfrac{\cos 2x}{2\sqrt{\sin 2x}}$ (E) $-\csc^2 2x$

16. $y = e^{-x}\cos 2x$

(A) $-e^{-x}(\cos 2x + 2\sin 2x)$
(B) $e^{-x}(\sin 2x - \cos 2x)$
(C) $2e^{-x}\sin 2x$
(D) $-e^{-x}(\cos 2x + \sin 2x)$
(E) $-e^{-x}\sin 2x$

17. $y = \sec^2(x)$

(A) $2\sec x$ (B) $2\sec x \tan x$ (C) $2\sec^2 x \tan x$
(D) $\sec^2 x \tan^2 x$ (E) $\tan x$

18. $y = x\ln^3 x$

(A) $\dfrac{3\ln^2 x}{x}$ (B) $3\ln^2 x$ (C) $3x\ln^2 x + \ln^3 x$

(D) $3(\ln x + 1)$ (E) none of these

19. $y = \frac{1+x^2}{1-x^2}$

(A) $-\frac{4x}{(1-x^2)^2}$ (B) $\frac{4x}{(1-x^2)^2}$ (C) $-\frac{4x^3}{(1-x^2)^2}$ (D) $\frac{2x}{1-x^2}$ (E) $\frac{4}{1-x^2}$

20. $y = \sin^{-1} x - \sqrt{1-x^2}$

(A) $\frac{1}{2\sqrt{1-x^2}}$ (B) $\frac{2}{\sqrt{1-x^2}}$ (C) $\frac{1+x}{\sqrt{1-x^2}}$ (D) $\frac{x^2}{\sqrt{1-x^2}}$ (E) $\frac{1}{\sqrt{1+x}}$

In each of Questions 21–24, y is a differentiable function of x. Choose the alternative that is the derivative $\frac{dy}{dx}$.

21. $x^3 - y^3 = 1$

(A) x (B) $3x^2$ (C) $\sqrt[3]{3x^2}$ (D) $\frac{x^2}{y^2}$ (E) $\frac{3x^2-1}{y^2}$

22. $x + \cos(x+y) = 0$

(A) $\csc(x+y) - 1$ (B) $\csc(x+y)$ (C) $\frac{x}{\sin(x+y)}$

(D) $\frac{1}{\sqrt{1-x^2}}$ (E) $\frac{1-\sin x}{\sin y}$

23. $\sin x - \cos y - 2 = 0$

(A) $-\cot x$ (B) $-\cot y$ (C) $\frac{\cos x}{\sin y}$

(D) $-\csc y \cos x$ (E) $\frac{2-\cos x}{\sin y}$

24. $3x^2 - 2xy + 5y^2 = 1$

(A) $\frac{3x+y}{x-5y}$ (B) $\frac{y-3x}{5y-x}$ (C) $3x + 5y$ (D) $\frac{3x+4y}{x}$ (E) none of these

BC ONLY

25. If $x = t^2 + 1$ and $y = 2t^3$, then $\frac{dy}{dx} =$

(A) $3t$ (B) $6t^2$ (C) $\frac{6t^2}{t^2+1}$ (D) $\frac{6t^2}{(t^2+1)^2}$ (E) $\frac{2t^4+6t^2}{(t^2+1)^2}$

26. If $f(x) = x^4 - 4x^3 + 4x^2 - 1$, then the set of values of x for which the derivative equals zero is

(A) $\{1, 2\}$ (B) $\{0, -1, -2\}$ (C) $\{-1, +2\}$ (D) $\{0\}$ (E) $\{0, 1, 2\}$

27. If $f(x) = 16\sqrt{x}$, then $f'''(4)$ is equal to

(A) -32 (B) -16 (C) -4 (D) -2 (E) $-\frac{1}{2}$

28. If $f(x) = \ln x^3$, then $f''(3)$ is

(A) $-\frac{1}{3}$ (B) -1 (C) -3 (D) 1 (E) 3

29. If a point moves on the curve $x^2 + y^2 = 25$, then, at $(0, 5)$, $\frac{d^2y}{dx^2}$ is

(A) 0 (B) $\frac{1}{5}$ (C) -5 (D) $-\frac{1}{5}$ (E) nonexistent

BC ONLY

30. If $x = t^2 - 1$ and $y = t^4 - 2t^3$, then, when $t = 1$, $\dfrac{d^2y}{dx^2}$ is

(A) 1 (B) -1 (C) 0 (D) 3 (E) $\dfrac{1}{2}$

31. If $f(x) = 5^x$ and $5^{1.002} \simeq 5.016$, which is closest to $f'(1)$?

(A) 0.016 (B) 1.0 (C) 5.0 (D) 8.0 (E) 32.0

32. If $y = e^x(x - 1)$, then $y''(0)$ equals

(A) -2 (B) -1 (C) 0 (D) 1 (E) e

BC ONLY

33. If $x = e^\theta \cos \theta$ and $y = e^\theta \sin \theta$, then, when $\theta = \dfrac{\pi}{2}$, $\dfrac{dy}{dx}$ is

(A) 1 (B) 0 (C) $e^{\pi/2}$ (D) nonexistent (E) -1

34. If $x = \cos t$ and $y = \cos 2t$, then $\dfrac{d^2y}{dx^2}$ $(\sin t \neq 0)$ is

(A) $4 \cos t$ (B) 4 (C) $\dfrac{4y}{x}$ (D) -4 (E) $-4 \cot t$

35. $\lim\limits_{h \to 0} \dfrac{(1 + h)^6 - 1}{h}$ is

(A) 0 (B) 1 (C) 6 (D) ∞ (E) nonexistent

36. $\lim\limits_{h \to 0} \dfrac{\sqrt[3]{8 + h} - 2}{h}$ is

(A) 0 (B) $\dfrac{1}{12}$ (C) 1 (D) 192 (E) ∞

37. $\lim\limits_{h \to 0} \dfrac{\ln(e + h) - 1}{h}$ is

(A) 0 (B) $\dfrac{1}{e}$ (C) 1 (D) e (E) nonexistent

38. $\lim\limits_{x \to 0} \dfrac{\cos x - 1}{x}$ is

(A) $-\infty$ (B) -1 (C) 0 (D) 1 (E) ∞

39. If $f(x) = \begin{cases} \dfrac{4x^2 - 4}{x - 1}, & x \neq 1 \\ 4, & x = 1 \end{cases}$, which of these statements are true?

I. $\lim\limits_{x \to 1} f(x)$ exists.
II. f is continuous at $x = 1$.
III. f is differentiable at $x = 1$.

(A) none (B) I only (C) I and II only (D) I and III only (E) I, II, and III

40. If $g(x) = \begin{cases} x^2, & x \leq 3 \\ 6x - 9, & x > 3 \end{cases}$, which of these statements are true?

I. $\lim\limits_{x \to 3} g(x)$ exists.
II. g is continuous at $x = 3$.
III. g is differentiable at $x = 3$.

(A) I only (B) II only (C) III only (D) I and II only (E) I, II, and III

41. The function $f(x) = x^{2/3}$ on $[-8, 8]$ does not satisfy the conditions of the Mean Value Theorem because

(A) $f(0)$ is not defined
(B) $f(x)$ is not continuous on $[-8, 8]$
(C) $f'(-1)$ does not exist
(D) $f(x)$ is not defined for $x < 0$
(E) $f'(0)$ does not exist

42. If $f(x) = 2x^3 - 6x$, at what point on the interval $0 \le x \le \sqrt{3}$, if any, is the tangent to the curve parallel to the secant line on that interval?

(A) 1 (B) -1 (C) $\sqrt{2}$ (D) 0 (E) nowhere

43. If h is the inverse function of f and if $f(x) = \frac{1}{x}$, then $h'(3) =$

(A) -9 (B) $-\frac{1}{9}$ (C) $\frac{1}{9}$ (D) 3 (E) 9

44. $\lim_{x\to\infty} \frac{e^x}{x^{50}}$ equals

(A) 0 (B) 1 (C) $\frac{1}{50!}$ (D) $\frac{e}{50!}$ (E) ∞

45. If $\sin(xy) = x$, then $\frac{dy}{dx} =$

(A) $\sec(xy)$ (B) $\frac{\sec(xy)}{x}$ (C) $\frac{\sec(xy) - y}{x}$
(D) $\frac{1 + \sec(xy)}{x}$ (E) $\sec(xy) - 1$

46. $\lim_{x\to 0} \frac{\sin 2x}{x}$ is

(A) 1 (B) 2 (C) $\frac{1}{2}$ (D) 0 (E) ∞

47. $\lim_{x\to 0} \frac{\sin 3x}{\sin 4x}$ is

(A) 1 (B) $\frac{4}{3}$ (C) $\frac{3}{4}$ (D) 0 (E) nonexistent

48. $\lim_{x\to 0} \frac{1 - \cos x}{x}$ is

(A) $-\infty$ (B) 0 (C) 1 (D) 2 (E) ∞

49. $\lim_{x\to 0} \frac{\tan \pi x}{x}$ is

(A) $\frac{1}{\pi}$ (B) 0 (C) 1 (D) π (E) ∞

50. $\lim_{x\to\infty} x^2 \sin \frac{1}{x}$

(A) is 1 (B) is 0 (C) is ∞
(D) oscillates between -1 and 1 (E) is none of these

51. The graph in the xy-plane represented by $x = 3 + 2 \sin t$ and $y = 2 \cos t - 1$, for $-\pi \le t \le \pi$, is

BC ONLY

(A) a semicircle (B) a circle (C) an ellipse
(D) half of an ellipse (E) a hyperbola

52. $\lim_{x \to 0} \frac{\sec x - \cos x}{x^2}$

(A) $= 0$ (B) $= \frac{1}{2}$ (C) $= 1$ (D) $= 2$ (E) does not exist

BC ONLY

In each of Questions 53–56 a pair of equations that represent a curve parametrically is given. Choose the alternative that is the derivative $\frac{dy}{dx}$.

53. $x = t - \sin t$ and $y = 1 - \cos t$

(A) $\frac{\sin t}{1 - \cos t}$ (B) $\frac{1 - \cos t}{\sin t}$ (C) $\frac{\sin t}{\cos t - 1}$

(D) $\frac{1 - x}{y}$ (E) $\frac{1 - \cos t}{t - \sin t}$

BC ONLY

54. $x = \cos^3 \theta$ and $y = \sin^3 \theta$

(A) $\tan^3 \theta$ (B) $-\cot \theta$ (C) $\cot \theta$ (D) $-\tan \theta$ (E) $-\tan^2 \theta$

55. $x = 1 - e^{-t}$ and $y = t + e^{-t}$

(A) $\frac{e^{-t}}{1 - e^{-t}}$ (B) $e^{-t} - 1$ (C) $e^t + 1$ (D) $e^t - e^{-2t}$ (E) $e^t - 1$

56. $x = \frac{1}{1 - t}$ and $y = 1 - \ln(1 - t)$ $(t < 1)$

(A) $\frac{1}{1 - t}$ (B) $t - 1$ (C) $\frac{1}{x}$ (D) $\frac{(1 - t)^2}{t}$ (E) $1 + \ln x$

Part B. Directions: Some of the following questions require the use of a graphing calculator.

In Questions 57–64, differentiable functions f and g have the values shown in the table.

x	f	f'	g	g'
0	2	1	5	−4
1	3	2	3	−3
2	5	3	1	−2
3	10	4	0	−1

57. If $A = f + 2g$, then $A'(3) =$

(A) −2 (B) 2 (C) 7 (D) 8 (E) 10

58. If $B = f \cdot g$, then $B'(2) =$

(A) −20 (B) −7 (C) −6 (D) −1 (E) 13

59. If $D = \frac{1}{g}$, then $D'(1) =$

(A) $-\frac{1}{2}$ (B) $-\frac{1}{3}$ (C) $-\frac{1}{9}$ (D) $\frac{1}{9}$ (E) $\frac{1}{3}$

60. If $H(x) = \sqrt{f(x)}$, then $H'(3) =$

(A) $\frac{1}{4}$ (B) $\frac{1}{2\sqrt{10}}$ (C) 2 (D) $\frac{2}{\sqrt{10}}$ (E) $4\sqrt{10}$

61. If $K(x) = \left(\frac{f}{g}\right)(x)$, then $K'(0) =$

(A) $\frac{-13}{25}$ (B) $-\frac{1}{4}$ (C) $\frac{13}{25}$ (D) $\frac{13}{16}$ (E) $\frac{22}{25}$

62. If $M(x) = f(g(x))$, then $M'(1) =$

(A) -12 (B) -6 (C) 4 (D) 6 (E) 12

63. If $P(x) = f(x^3)$, then $P'(1) =$

(A) 2 (B) 6 (C) 8 (D) 12 (E) 54

64. If $S(x) = f^{-1}(x)$, then $S'(3) =$

(A) -2 (B) $-\frac{1}{25}$ (C) $\frac{1}{4}$ (D) $\frac{1}{2}$ (E) 2

65. The graph of g' is shown here. Which of the following statements is (are) true of g?

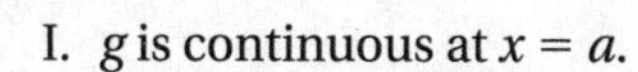

I. g is continuous at $x = a$.
II. g is differentiable at $x = a$.
III. g is increasing in an interval containing $x = a$.

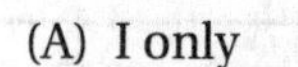

(A) I only (B) III only (C) I and III only
(D) II and III only (E) I, II, and III

66. A function f has the derivative shown. Which of the following statements must be false?

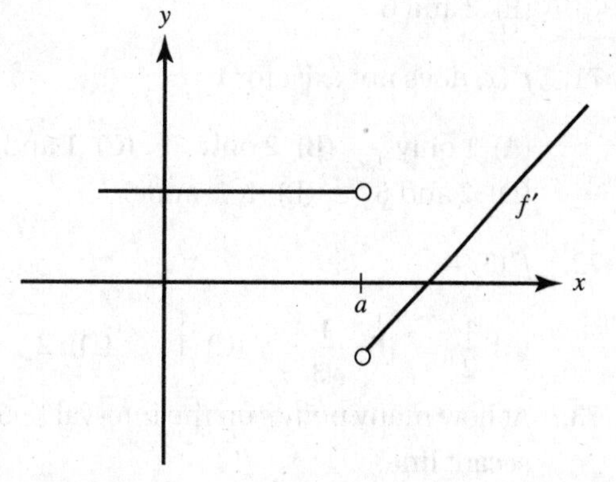

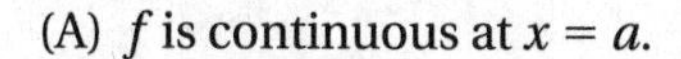

(A) f is continuous at $x = a$.
(B) $f(a) = 0$.
(C) f has a vertical asymptote at $x = a$.
(D) f has a jump discontinuity at $x = a$.
(E) f has a removable discontinuity at $x = a$.

67. The function f whose graph is shown has $f' = 0$ at $x =$

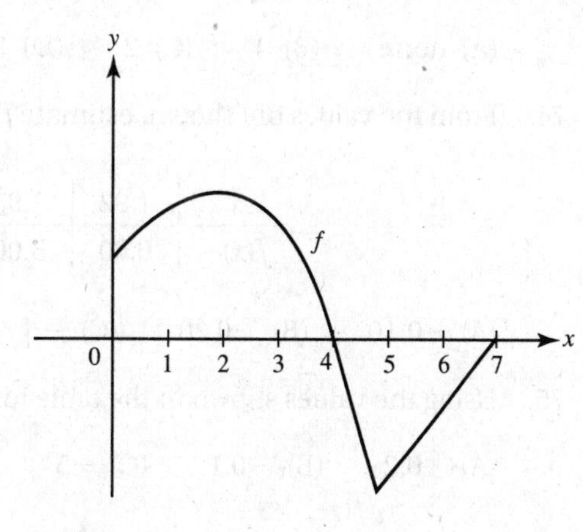

(A) 2 only
(B) 2 and 5
(C) 4 and 7
(D) 2, 4, and 7
(E) 2, 4, 5, and 7

68. A differentiable function f has the values shown. Estimate $f'(1.5)$.

x	1.0	1.3	1.4	1.6
$f(x)$	8	10	14	22

(A) 8 (B) 12 (C) 18 (D) 40 (E) 80

69. Water is poured into a conical reservoir at a constant rate. If $h(t)$ is the rate of change of the depth of the water, then h is

(A) constant
(B) linear and increasing
(C) linear and decreasing
(D) nonlinear and increasing
(E) nonlinear and decreasing

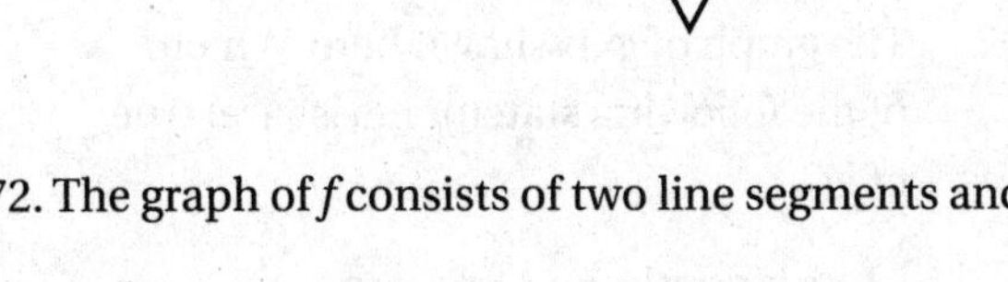

Use the figure to answer Questions 70–72. The graph of f consists of two line segments and a semicircle.

70. $f'(x) = 0$ for $x =$

(A) 1 only
(B) 2 only
(C) 4 only
(D) 1 and 4
(E) 2 and 6

71. $f'(x)$ does not exist for $x =$

(A) 1 only (B) 2 only (C) 1 and 2
(D) 2 and 6 (E) 1, 2, and 6

72. $f'(5) =$

(A) $\frac{1}{2}$ (B) $\frac{1}{\sqrt{3}}$ (C) 1 (D) 2 (E) $\sqrt{3}$

73. At how many points on the interval $[-5,5]$ is a tangent to $y = x + \cos x$ parallel to the secant line?

(A) none (B) 1 (C) 2 (D) 3 (E) more than 3

74. From the values of f shown, estimate $f'(2)$.

x	1.92	1.93	1.95	1.98	2.00
$f(x)$	6.00	5.00	4.40	4.10	4.00

(A) -0.10 (B) -0.20 (C) -5 (D) -10 (E) -25

75. Using the values shown in the table for Question 74, estimate $(f^{-1})'(4)$.

(A) -0.2 (B) -0.1 (C) -5 (D) -10 (E) -25

76. The "left half" of the parabola defined by $y = x^2 - 8x + 10$ for $x \leq 4$ is a one-to-one function; therefore its inverse is also a function. Call that inverse g. Find $g'(3)$.

(A) $-\frac{1}{2}$ (B) $-\frac{1}{6}$ (C) $\frac{1}{6}$ (D) $\frac{1}{2}$ (E) $\frac{11}{2}$

77. The table below shows some points on a function f that is both continuous and differentiable on the closed interval [2,10].

x	2	4	6	8	10
$f(x)$	30	25	20	25	30

Which must be true?

(A) $f(x) > 0$ for $2 < x < 10$
(B) $f'(6) = 0$
(C) $f'(8) > 0$
(D) The maximum value of f on the interval [2, 10] is 30.
(E) For some value of x on the interval [2, 10] $f'(x) = 0$.

78. If f is differentiable and difference quotients overestimate the slope of f at $x = a$ for all $h > 0$, which must be true?

(A) $f'(x) \geq 0$ on $[a, h]$ (B) $f'(x) \leq 0$ on $[a, h]$ (C) $f''(x) \geq 0$ on $[a, h]$
(D) $f''(x) \leq 0$ on $[a, h]$ (E) none of these

79. If $f(u) = \sin u$ and $u = g(x) = x^2 - 9$, then $(f \circ g)'(3)$ equals

(A) 0 (B) 1 (C) 3 (D) 6 (E) 9

80. If $f(x) = \dfrac{x}{(x-1)^2}$, then the set of x's for which $f'(x)$ exists is

(A) all reals
(B) all reals except $x = 1$ and $x = -1$
(C) all reals except $x = -1$
(D) all reals except $x = \frac{1}{3}$ and $x = -1$
(E) all reals except $x = 1$

81. If $y = \sqrt{x^2 + 1}$, then the derivative of y^2 with respect to x^2 is

BC ONLY

(A) 1 (B) $\dfrac{x^2+1}{2x}$ (C) $\dfrac{x}{2(x^2+1)}$ (D) $\dfrac{2}{x}$ (E) $\dfrac{x^2}{x^2+1}$

82. If $y = x^2 + x$, then the derivative of y with respect to $\dfrac{1}{1-x}$ is

BC ONLY

(A) $(2x + 1)(x - 1)^2$ (B) $\dfrac{2x+1}{(1-x)^2}$ (C) $2x + 1$

(D) $\dfrac{3-x}{(1-x)^3}$ (E) none of these

83. If $f(x) = \dfrac{1}{x^2+1}$ and $g(x) = \sqrt{x}$, then the derivative of $f(g(x))$ is

(A) $\dfrac{-\sqrt{x}}{(x^2+1)^2}$ (B) $-(x + 1)^{-2}$ (C) $\dfrac{-2x}{(x^2+1)^2}$

(D) $\dfrac{1}{(x+1)^2}$ (E) $\dfrac{1}{2\sqrt{x}\,(x+1)}$

84. If $f(a) = f(b) = 0$ and $f(x)$ is continuous on $[a, b]$, then

(A) $f(x)$ must be identically zero
(B) $f'(x)$ may be different from zero for all x on $[a, b]$
(C) there exists at least one number c, $a < c < b$, such that $f'(c) = 0$
(D) $f'(x)$ must exist for every x on (a, b)
(E) none of the preceding is true

85. Suppose $y = f(x) = 2x^3 - 3x$. If $h(x)$ is the inverse function of f, then $h'(-1) =$

(A) -1 (B) $\frac{1}{5}$ (C) $\frac{1}{3}$ (D) 1 (E) 3

86. Suppose $f(1) = 2$, $f'(1) = 3$, and $f'(2) = 4$. Then $(f^{-1})'(2)$

(A) equals $-\frac{1}{3}$ (B) equals $-\frac{1}{4}$ (C) equals $\frac{1}{4}$

(D) equals $\frac{1}{3}$ (E) cannot be determined

87. If $f(x) = x^3 - 3x^2 + 8x + 5$ and $g(x) = f^{-1}(x)$, then $g'(5) =$

(A) 8 (B) $\frac{1}{8}$ (C) 1 (D) $\frac{1}{53}$ (E) 53

88. Suppose $\lim_{x \to 0} \frac{g(x) - g(0)}{x} = 1$. It follows necessarily that

(A) g is not defined at $x = 0$
(B) g is not continuous at $x = 0$
(C) the limit of $g(x)$ as x approaches 0 equals 1
(D) $g'(0) = 1$
(E) $g'(1) = 0$

Use this graph of $y = f(x)$ for Questions 89 and 90.

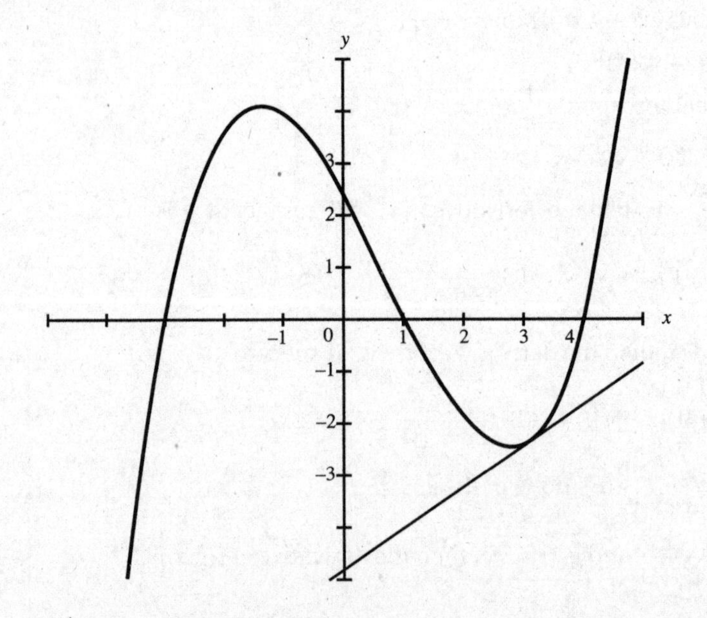

89. $f'(3)$ is most closely approximated by

(A) 0.3 (B) 0.8 (C) 1.5 (D) 1.8 (E) 2

90. The rate of change of $f(x)$ is least at $x \approx$

(A) -3 (B) -1.3 (C) 0 (D) 0.7 (E) 2.7

Use the following definition of the *symmetric difference quotient for* $f'(x_0)$ for Questions 91–93: For small values of h,

$$f'(x_0) = \frac{f(x_0 + h) - f(x_0 - h)}{2h}.$$

91. For $f(x) = 5^x$, what is the estimate of $f'(2)$ obtained by using the symmetric difference quotient with $h = 0.03$?

(A) 25.029 (B) 40.236 (C) 40.252 (D) 41.223 (E) 80.503

92. To how many places is the symmetric difference quotient accurate when it is used to approximate $f'(0)$ for $f(x) = 4^x$ and $h = 0.08$?

(A) 1 (B) 2 (C) 3 (D) 4 (E) more than 4

93. To how many places is $f'(x_0)$ accurate when it is used to approximate $f'(0)$ for $f(x) = 4^x$ and $h = 0.001$?

(A) 1 (B) 2 (C) 3 (D) 4 (E) more than 4

94. The value of $f'(0)$ obtained using the symmetric difference quotient with $f(x) = |x|$ and $h = 0.001$ is

(A) -1 (B) 0 (C) ± 1 (D) 1 (E) indeterminate

95. If $\frac{d}{dx}f(x) = g(x)$ and $h(x) = \sin x$, then $\frac{d}{dx}f(h(x))$ equals

(A) $g(\sin x)$ (B) $\cos x \cdot g(x)$ (C) $g'(x)$
(D) $\cos x \cdot g(\sin x)$ (E) $\sin x \cdot g(\sin x)$

96. Let $f(x) = 3^x - x^3$. The tangent to the curve is parallel to the secant through (0,1) and (3,0) for $x =$

(A) 0.984 only (B) 1.244 only (C) 2.727 only
(D) 0.984 and 2.804 only (E) 1.244 and 2.727 only

Questions 97–101 are based on the following graph of $f(x)$, sketched on $-6 \leq x \leq 7$. Assume the horizontal and vertical grid lines are equally spaced at unit intervals.

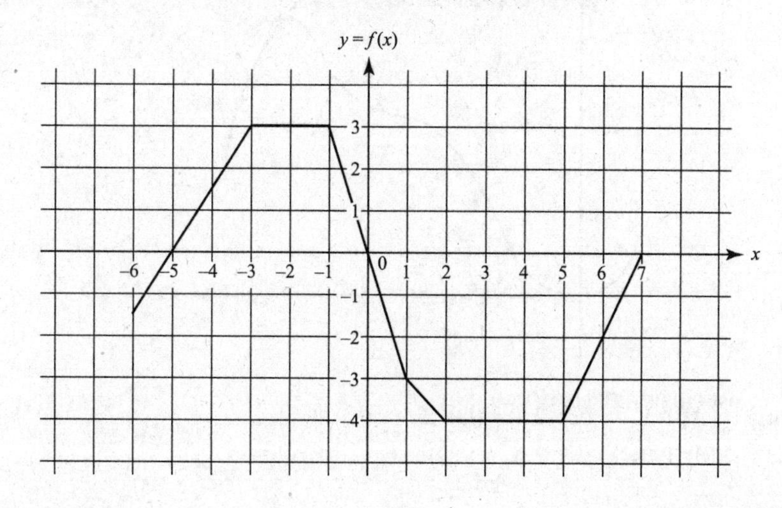

97. On the interval $1 < x < 2$, $f(x)$ equals

(A) $-x - 2$ (B) $-x - 3$ (C) $-x - 4$ (D) $-x + 2$ (E) $x - 2$

98. Over which of the following intervals does $f'(x)$ equal zero?

I. (–6,–3) II. (–3,–1) III. (2,5)

(A) I only (B) II only (C) I and II only
(D) I and III only (E) II and III only

99. How many points of discontinuity does $f'(x)$ have on the interval $-6 < x < 7$?

(A) none (B) 2 (C) 3 (D) 4 (E) 5

100. For $-6 < x < -3$, $f'(x)$ equals

(A) $-\frac{3}{2}$ (B) -1 (C) 1 (D) $\frac{3}{2}$ (E) 2

101. Which of the following statements about the graph of $f'(x)$ is false?

(A) It consists of six horizontal segments.
(B) It has four jump discontinuities.
(C) $f'(x)$ is discontinuous at each x in the set $\{-3,-1,1,2,5\}$.
(D) $f'(x)$ ranges from -3 to 2.
(E) On the interval $-1 < x < 1$, $f'(x) = -3$.

102. The table gives the values of a function f that is differentiable on the interval [0,1]:

x	0.10	0.20	0.26	0.40	0.52	0.60
$f(x)$	0.171	0.288	0.357	0.384	0.375	0.336

According to this table, the best approximation of $f'(0.10)$ is

(A) 0.12 (B) 1.08 (C) 1.17 (D) 1.77 (E) 2.88

103. At how many points on the interval $[a,b]$ does the function graphed satisfy the Mean Value Theorem?

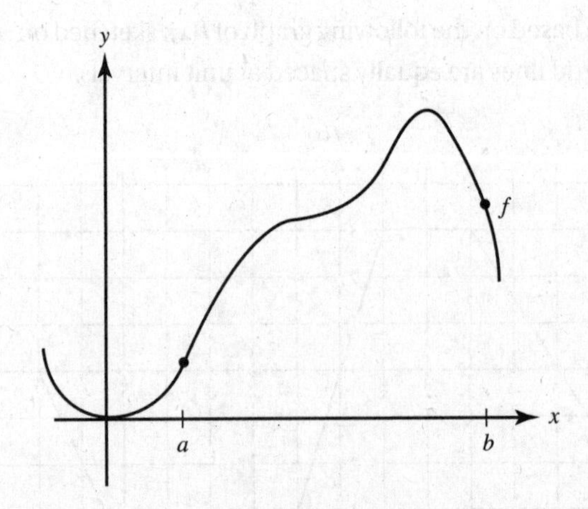

(A) none (B) 1 (C) 2 (D) 3 (E) 4

Answer Key

1. **E**	27. **E**	53. **A**	79. **D**
2. **A**	28. **A**	54. **D**	80. **E**
3. **B**	29. **D**	55. **E**	81. **A**
4. **B**	30. **E**	56. **C**	82. **A**
5. **E**	31. **D**	57. **B**	83. **B**
6. **D**	32. **D**	58. **B**	84. **B**
7. **A**	33. **E**	59. **E**	85. **C**
8. **D**	34. **B**	60. **D**	86. **D**
9. **C**	35. **C**	61. **C**	87. **B**
10. **E**	36. **B**	62. **A**	88. **D**
11. **A**	37. **B**	63. **B**	89. **B**
12. **D**	38. **C**	64. **D**	90. **D**
13. **D**	39. **B**	65. **E**	91. **C**
14. **C**	40. **E**	66. **C**	92. **B**
15. **A**	41. **E**	67. **A**	93. **E**
16. **A**	42. **A**	68. **D**	94. **B**
17. **C**	43. **B**	69. **E**	95. **D**
18. **E**	44. **E**	70. **C**	96. **E**
19. **B**	45. **C**	71. **E**	97. **A**
20. **C**	46. **B**	72. **B**	98. **E**
21. **D**	47. **C**	73. **D**	99. **E**
22. **A**	48. **B**	74. **C**	100. **D**
23. **D**	49. **D**	75. **A**	101. **B**
24. **B**	50. **C**	76. **B**	102. **C**
25. **A**	51. **B**	77. **E**	103. **D**
26. **E**	52. **C**	78. **C**	

Answers Explained

Many of the explanations provided include intermediate steps that would normally be reached on the way to a final algebraically simplified result. You may not need to reach the final answer.

NOTE: The formulas or rules cited in parentheses in the explanations are given on pages 113 and 114.

1. **(E)** By the Product Rule, (5),

$$y' = x^5(\tan x)' + (x^5)'(\tan x).$$

2. **(A)** By the Quotient Rule, (6),

$$y' = \frac{(3x+1)(-1) - (2-x)(3)}{(3x+1)^2} = -\frac{7}{(3x+1)^2}.$$

3. **(B)** Since $y = (3 - 2x)^{1/2}$, by the Power Rule, (3),

$$y' = \frac{1}{2}(3 - 3x)^{-1/2} \cdot (-2) = -\frac{1}{\sqrt{3-2x}}.$$

4. **(B)** Since $y = 2(5x+1)^{-3}$, $y' = -6(5x+1)^{-4}(5)$.

5. **(E)** $y' = 3\left(\frac{2}{3}\right)x^{-1/3} - 4\left(\frac{1}{2}\right)x^{-1/2}$

6. **(D)** Rewrite: $y = 2x^{1/2} - \frac{1}{2}x^{-1/2}$; so $y' = x^{-1/2} + \frac{1}{4}x^{-3/2}$.

7. **(A)** Rewrite: $y = (x^2 + 2x - 1)^{1/2}$; then $y = \frac{1}{2}(x^2 + 2x - 1)^{-1/2}(2x + 2)$.

8. **(D)** Use the Quotient Rule:

$$y' = \frac{2x\cos x - x^2(-\sin x)}{\cos^2 x}.$$

9. **(C)** Since

$$\begin{aligned} y &= \ln e^x - \ln(e^x - 1) \\ &= x - \ln(e^x - 1), \end{aligned}$$

then

$$y' = 1 - \frac{e^x}{e^x - 1} = \frac{e^x - 1 - e^x}{e^x - 1} = -\frac{1}{e^x - 1}.$$

10. **(E)** Use formula (18): $y' = \dfrac{\frac{1}{2}}{1 + \frac{x^2}{4}}$.

11. **(A)** Use formulas (13), (11), and (9):

$$y' = \frac{\sec x \tan x + \sec^2 x}{\sec x + \tan x} = \frac{\sec x(\tan x + \sec x)}{\sec x + \tan x}.$$

12. **(D)** By the Quotient Rule,

$$\begin{aligned} y' &= \frac{(e^x + e^{-x})(e^x + e^{-x}) - (e^x - e^{-x})(e^x - e^{-x})}{(e^x + e^{-x})^2} \\ &= \frac{(e^{2x} + 2 + e^{-2x}) - (e^{2x} - 2 + e^{-2x})}{(e^x + e^{-x})^2} = \frac{4}{(e^x + e^{-x})^2}. \end{aligned}$$

13. **(D)** Since $y = \frac{1}{2}\ln(x^2 + 1)$,

$$y' = \frac{1}{2} \cdot \frac{2x}{x^2 + 1}.$$

14. **(C)** $y' = \sin'\left(\frac{1}{x}\right) \cdot \left(\frac{1}{x}\right)' = \cos\left(\frac{1}{x}\right) \cdot \left(-\frac{1}{x^2}\right)$.

15. **(A)** Since $y = \frac{1}{2}\csc 2x$, $y' = \frac{1}{2}(-\csc 2x \cot 2x \cdot 2)$.

16. **(A)** $y' = e^{-x}(-2\sin 2x) + \cos 2x(-e^{-x})$.

17. **(C)** $y' = (2\sec x)(\sec x \tan x)$.

18. **(E)** $y' = \dfrac{x(3\ln^2 x)}{x} + \ln^3 x$. The correct answer is $3\ln^2 x + \ln^3 x$.

19. **(B)** $y' = \dfrac{(1 - x^2)(2x) - (1 + x^2)(-2x)}{(1 - x^2)^2}$.

20. **(C)** $y' = \dfrac{1}{\sqrt{1 - x^2}} - \dfrac{1 \cdot (-2x)}{2\sqrt{1 - x^2}}$.

21. **(D)** Let y' be $\frac{dy}{dx}$; then $3x^2 - 3y^2y' = 0$; $y' = \frac{-3x^2}{-3y^2}$.

22. **(A)** $1 - \sin(x + y)(1 + y') = 0$; $\frac{1 - \sin(x + y)}{\sin(x + y)} = y'$.

23. **(D)** $\cos x + \sin y \cdot y' = 0$; $y' = -\frac{\cos x}{\sin y}$.

24. **(B)** $6x - 2(xy' + y) + 10yy' = 0$; $y'(10y - 2x) = 2y - 6x$.

25. **(A)** $\frac{dy}{dx} = \frac{dy / dt}{dx / dt} = \frac{6t^2}{2t}$.

26. **(E)** $f'(x) = 4x^3 - 12x^2 + 8x = 4x(x - 1)(x - 2)$.

27. **(E)** $f'(x) = 8x^{-1/2}$; $f''(x) = -4x^{-3/2} = -\frac{4}{x^{3/2}}$; $f''(4) = -\frac{4}{8}$.

28. **(A)** $f(x) = 3 \ln x$; $f'(x) = \frac{3}{x}$; $f''(x) = \frac{-3}{x^2}$. Replace x by 3.

29. **(D)** $2x + 2yy' = 0$; $y' = -\frac{x}{y}$; $y'' = -\frac{y - xy'}{y^2}$. At (0,5), $y'' = -\frac{5 - 0}{25}$.

30. **(E)** $\frac{dy}{dx} = \frac{4t^3 - 6t^2}{2t} = 2t^2 - 3t\,(t \neq 0)$; $\frac{d^2y}{dx^2} = \frac{4t - 3}{2t}$. Replace t by 1.

31. **(D)** $f'(1) \simeq \frac{5^{1.002} - 5^1}{0.002} = \frac{5.016 - 5}{0.002}$.

32. **(D)** $y' = e^x \cdot 1 + e^x(x - 1) = xe^x$;
$y'' = xe^x + e^x$ and $y''(0) = 0 \cdot 1 + 1 = 1$.

33. **(E)** When simplified, $\frac{dy}{dx} = \frac{\cos\theta + \sin\theta}{\cos\theta - \sin\theta}$.

34. **(B)** Since (if $\sin t \neq 0$)

$$\frac{dy}{dt} = -2\sin 2t = -4\sin t \cos t \text{ and } \frac{dx}{dt} = -\sin t,$$

then $\frac{dy}{dx} = 4\cos t$. Thus:

$$\frac{d^2y}{dx^2} = -\frac{4\sin t}{-\sin t}.$$

NOTE: **Since each of the limits in Questions 35–39 yields an indeterminate form of the type $\frac{0}{0}$, we can apply L'Hôpital's Rule in each case, getting identical answers.**

35. **(C)** **The given limit is the derivative of $f(x) = x^6$ at $x = 1$.**

36. **(B)** **The given limit is the definition for $f'(8)$, where $f(x) = \sqrt[3]{x}$;**

$$f'(x) = \frac{1}{3x^{2/3}}.$$

37. **(B)** **The given limit is $f'(e)$, where $f(x) = \ln x$.**

38. **(C)** **The given limit is the derivative of $f(x) = \cos x$ at $x = 0$; $f'(x) = -\sin x$.**

39. **(B)** $\lim\limits_{x\to1}\dfrac{4x^2-4}{x-1}=\lim\limits_{x\to1}\dfrac{4(x+1)(x-1)}{(x-1)}=\lim\limits_{x\to1}4(x+1)=8$, but $f(1)=4$.

Thus f is discontinuous at $x=1$, so it cannot be differentiable.

40. **(E)** $\lim\limits_{x\to3^-}x^2=\lim\limits_{x\to3^+}(6x-9)=9$, so the limit exists. Because $g(3)=9$, g is continuous at $x=3$.

Since $g'(x)=\begin{cases}2x, x<3\\6, x>3\end{cases}$, so $g'(3)=6$.

41. **(E)** Since $f'(x)=\dfrac{2}{3x^{1/3}}$, $f'(0)$ is not defined; $f'(x)$ must be defined on $(-8,8)$.

42. **(A)** Note that $f(0)=f(\sqrt{3})=0$ and that $f'(x)$ exists on the given interval. By the MVT, there is a number, c, in the interval such that $f'(c)=0$. If $c=1$, then $6c^2-6=0$. (-1 is not in the interval.)

43. **(B)** Since the inverse, h, of $f(x)=\dfrac{1}{x}$ is $h(x)=\dfrac{1}{x}$, then $h'(x)=-\dfrac{1}{x^2}$. Replace x by 3.

44. **(E)** After 50(!) applications of L'Hôpital's Rule we get $\lim\limits_{x\to\infty}\dfrac{e^x}{50!}$, which "equals" ∞. A perfunctory examination of the limit, however, shows immediately that the answer is ∞. In fact, $\lim\limits_{x\to\infty}\dfrac{e^x}{x^n}$ for any positive integer n, no matter how large, is ∞.

45. **(C)** $\cos(xy)(xy'+y)=1$; $x\cos(xy)y'=1-y\cos(xy)$;

$$y'=\frac{1-y\cos(xy)}{x\cos(xy)}.$$

NOTE: In Questions 46–50 the limits are all indeterminate forms of the type $\frac{0}{0}$. We have therefore applied L'Hôpital's Rule in each one. The indeterminacy can also be resolved by introducing $\dfrac{\sin a}{a}$, which approaches 1 as a approaches 0. The latter technique is presented in square brackets.

46. **(B)** $\lim\limits_{x\to0}\dfrac{\sin 2x}{x}=\lim\limits_{x\to0}\dfrac{2\cos 2x}{1}=\dfrac{2\cdot1}{1}=2.$

[Using $\sin 2x=2\sin x\cos x$ yields $\lim\limits_{x\to0}2\left(\dfrac{\sin x}{x}\right)\cos x=2\cdot1\cdot1=2.$]

47. **(C)** $\lim\limits_{x\to0}\dfrac{\sin 3x}{\sin 4x}=\lim\limits_{x\to0}\dfrac{3\cos 3x}{4\cos 4x}=\dfrac{3\cdot1}{4\cdot1}=\dfrac{3}{4}.$

[We rewrite $\dfrac{\sin 3x}{\sin 4x}$ as $\dfrac{\sin 3x}{3x}\cdot\dfrac{4x}{\sin 4x}\cdot\dfrac{3}{4}$. As $x\to0$, so do $3x$ and $4x$; the fraction approaches $1\cdot1\cdot\dfrac{3}{4}$.]

48. **(B)** $\lim\limits_{x\to0}\dfrac{1-\cos x}{x}=\lim\limits_{x\to0}\dfrac{\sin x}{1}=0.$

[We can replace $1-\cos x$ by $2\sin^2\dfrac{x}{2}$, getting

$$\lim_{x\to0}\frac{2\sin^2\frac{x}{2}}{x}=\lim_{x\to0}\frac{\sin^2\frac{x}{2}}{\frac{x}{2}}=\lim_{x\to0}\sin\frac{x}{2}\left(\frac{\sin\frac{x}{2}}{\frac{x}{2}}\right)=0\cdot1.]$$

49. **(D)** $\lim\limits_{x\to0}\dfrac{\tan\pi x}{x}=\lim\limits_{x\to0}\dfrac{(\sec^2\pi x)\cdot\pi}{1}=1\cdot\pi=\pi.$

[$\frac{\tan \pi x}{x} = \frac{\sin \pi x}{x \cos \pi x} = \pi \cdot \frac{\sin \pi x}{\pi x} \cdot \frac{1}{\cos x}$; as x (or πx) approaches 0, the original fraction approaches $\pi \cdot 1 \cdot \frac{1}{1} = \pi$.]

50. **(C)** The limit is easiest to obtain here if we rewrite:

$$\lim_{x\to\infty} x^2 \sin\frac{1}{x} = \lim_{x\to\infty} x \frac{\sin(1/x)}{(1/x)} = \infty \cdot 1 = \infty.$$

51. **(B)** Since $x - 3 = 2 \sin t$ and $y + 1 = 2 \cos t$,

$$(x-3)^2 + (y+1)^2 = 4.$$

This is the equation of a circle with center at $(3,-1)$ and radius 2. In the domain given, $-\pi \leqslant t \leqslant \pi$, the entire circle is traced by a particle moving counterclockwise, starting from and returning to $(3, -3)$.

52. **(C)** Use L'Hôpital's Rule; then

$$\lim_{x\to 0} \frac{\sec x - \cos x}{x^2} = \lim_{x\to 0} \frac{\sec x \tan x + \sin x}{2x}$$

$$= \lim_{x\to 0} \frac{\sec^3 x + \sec x \tan^2 x + \cos x}{2} = \frac{1 + 1 \cdot 0 + 1}{2} = 1.$$

53. **(A)** $\frac{dy}{dx} = \frac{\frac{dy}{dt}}{\frac{dx}{dt}} = \frac{\sin t}{1 - \cos t}.$

54. **(D)** $\frac{dy}{dx} = \frac{\frac{dy}{d\theta}}{\frac{dx}{d\theta}} = \frac{3 \sin^2 \theta \cos \theta}{-3 \cos^2 \theta \sin \theta}.$

55. **(E)** $\frac{dy}{dx} = \frac{\frac{dy}{dt}}{\frac{dx}{dt}} = \frac{1 - e^{-t}}{e^{-t}} = e^t - 1.$

56. **(C)** Since $\frac{dy}{dt} = \frac{1}{1-t}$ and $\frac{dx}{dt} = \frac{1}{(1-t)^2}$, then

$$\frac{dy}{dx} = 1 - t = \frac{1}{x}.$$

57. **(B)** $(f + 2g)'(3) = f'(3) + 2g'(3) = 4 + 2(-1)$

58. **(B)** $(f \cdot g)'(2) = f(2) \cdot g'(2) + g(2) \cdot f'(2) = 5(-2) + 1(3)$

59. **(E)** $\left(\frac{1}{g}\right)'(1) = -1 \cdot \frac{1}{[g(1)]^2} \cdot g'(1) = -1 \cdot \frac{1}{3^2}(-3).$

60. **(D)** $\left(\sqrt{f}\right)'(3) = \frac{1}{2}[f(3)]^{-\frac{1}{2}} \cdot f'(3) = \frac{1}{2}\left(10^{-\frac{1}{2}}\right) \cdot 4.$

61. **(C)** $\left(\frac{f}{g}\right)'(0) = \frac{g(0) \cdot f'(0) - f(0) \cdot g'(0)}{[g(0)]^2} = \frac{5(1) - 2(-4)}{5^2}.$

62. **(A)** $M'(1) = f'(g(1)) \cdot g'(1) = f'(3)g'(1) = 4(-3).$

63. **(B)** $[f(x^3)]' = f'(x^3) \cdot 3x^2$, so $P'(1) = f'(1^3) \cdot 3 \cdot 1^2 = 2 \cdot 3.$

64. **(D)** $f(S(x)) = x$ implies that $f'(S(x)) \cdot S'(x) = 1$, so

$$S'(3) = \frac{1}{f'\left(S(3)\right)} = \frac{1}{f'\left(f^{-1}(3)\right)} = \frac{1}{f'(1)}.$$

65. **(E)** Since $g'(a)$ exists, g is differentiable and thus continuous; $g'(a) > 0$.

66. **(C)** Near a vertical asymptote the slopes must approach $\pm\infty$.

67. **(A)** There is only one horizontal tangent.

68. **(D)** Use the symmetric difference quotient; then

$$f'(1.5) \approx \frac{f(1.6) - f(1.4)}{1.6 - 1.4} = \frac{8}{0.2}.$$

69. **(E)** Since the water level rises more slowly as the cone fills, the rate of depth change is decreasing, as in (C) and (E). However, at every instant the portion of the cone containing water is similar to the entire cone; the volume is proportional to the cube of the depth of the water. The rate of change of depth (the derivative) is therefore not linear, as in (C).

70. **(C)** The only horizontal tangent is at $x = 4$. Note that $f'(1)$ does not exist.

71. **(E)** The graph has corners at $x = 1$ and $x = 2$; the tangent line is vertical at $x = 6$.

72. **(B)** Consider triangle ABC: $AB = 1$; radius $AC = 2$; thus, $BC = \sqrt{3}$ and AC has $m = -\sqrt{3}$. The tangent line is perpendicular to the radius.

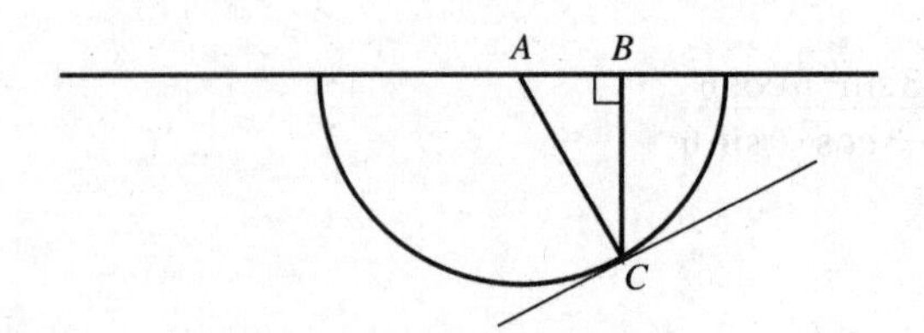

73. **(D)** The graph of $y = x + \cos x$ is shown in window $[-5,5] \times [-6,6]$. The average rate of change is represented by the slope of secant segment $\overline{AB}$. There appear to be 3 points at which tangent lines are parallel to $\overline{AB}$.

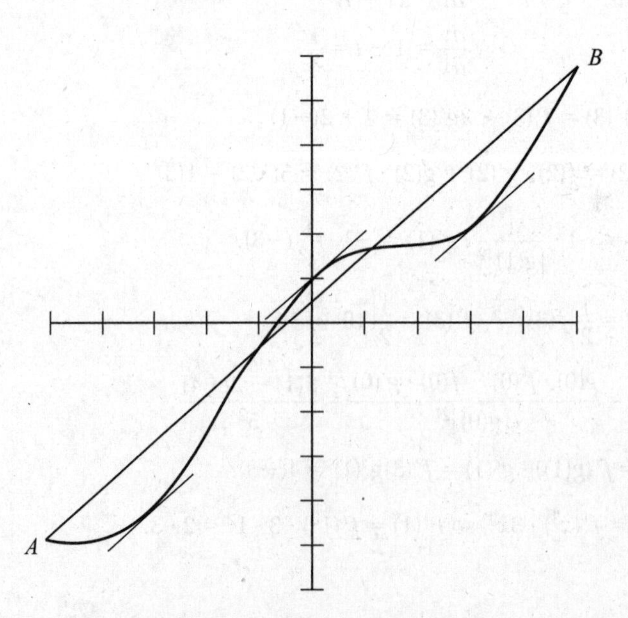

74. **(C)** $f'(2) \approx \dfrac{f(2) - f(1.98)}{2 - 1.98} = \dfrac{4.00 - 4.10}{0.02}$

75. **(A)** Since an estimate of the answer for Question 74 is $f'(2) \approx -5$, then

$(f^{-1})'(4) = \dfrac{1}{f'(2)} \approx \dfrac{1}{-5} = -0.2.$

76. **(B)** When $x = 3$ on g^{-1}, $y = 3$ on the original half-parabola. $3 = x^2 - 8x + 10$ at $x = 1$ (and at $x = 7$, but that value is not in the given domain).

$g'(3) = \dfrac{1}{y'(1)} = \dfrac{1}{2x - 8}\Big|_{x=1} = -\dfrac{1}{6}.$

77. **(E)** f satisfies Rolle's Theorem on [2,10].

78. **(C)** The diagrams show secant lines (whose slope is the difference quotient) with greater slopes than the tangent line. In both cases, f is concave upward.

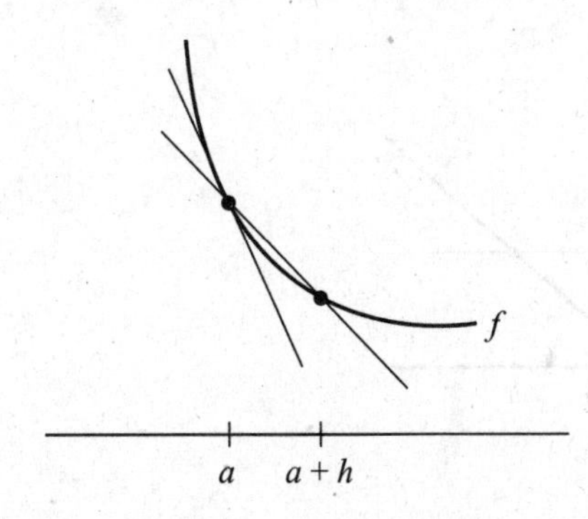

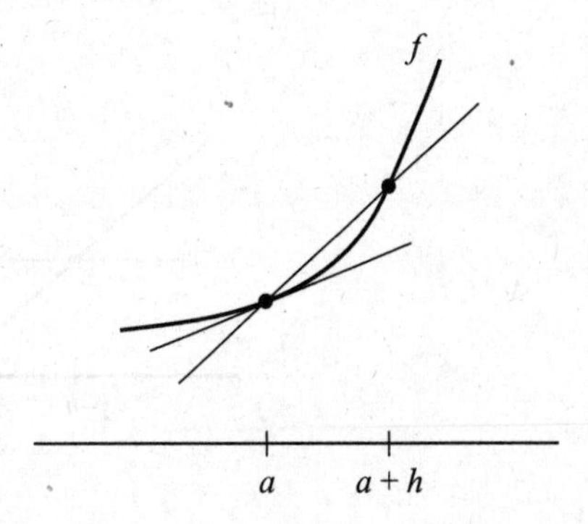

79. **(D)** $(f \circ g)'$ at $x = 3$ equals $f'(g(3)) \cdot g'(3)$ equals $\cos u$ (at $u = 0$) times $2x$ (at $x = 3$) $= 1 \cdot 6 = 6$.

80. **(E)** Here $f'(x)$ equals $\dfrac{-x - 1}{(x - 1)^3}$.

81. **(A)** $\dfrac{dy^2}{dx^2} = \dfrac{\dfrac{dy^2}{dx}}{\dfrac{dx^2}{dx}}$. Since $y^2 = x^2 + 1$, $\dfrac{dy^2}{dx^2} = \dfrac{2x}{2x}$.

82. **(A)** $\dfrac{dy}{d\left(\dfrac{1}{1-x}\right)} = \dfrac{\dfrac{dy}{dx}}{\dfrac{d\left(\dfrac{1}{1-x}\right)}{dx}} = \dfrac{2x + 1}{\dfrac{1}{(1-x)^2}}$.

83. **(B)** Note that $f(g(x)) = \dfrac{1}{x+1}$.

84. **(B)** Sketch the graph of $f(x) = 1 - |x|$; note that $f(-1) = f(1) = 0$ and that f is continuous on $[-1,1]$. Only (B) holds.

85. **(C)** Since $f'(x) = 6x^2 - 3$, therefore $h'(x) = \dfrac{1}{6x^2 - 3}$; also, $f(x)$, or $2x^3 - 3x$, equals -1, by observation, for $x = 1$. So $h'(-1)$ or $\dfrac{1}{6x^2 - 3}$ (when $x = 1$) equals $\dfrac{1}{6 - 3} = \dfrac{1}{3}$.

86. **(D)** $(f^{-1})'(2) = \dfrac{1}{f'(1)} = \dfrac{1}{3}$.

87. **(B)** Since $f(0) = 5$, $g'(5) = \frac{1}{f'(0)} = \frac{1}{3x^2 - 6x + 8}\bigg|_{x=0} = \frac{1}{8}$.

88. **(D)** The given limit is the derivative of $g(x)$ at $x = 0$.

89. **(B)** The tangent line appears to contain $(3,-2.6)$ and $(4,-1.8)$.

90. **(D)** $f'(x)$ is least at the point of inflection of the curve, at about 0.7.

91. **(C)** $\frac{5^{2.03} - 5^{1.97}}{2.03 - 1.97} \approx 40.25158$

92. **(B)** By calculator, $f'(0) = 1.386294805$ and $\frac{4^{0.08} - 4^{-0.08}}{0.16} = 1.3891\ldots$.

93. **(E)** Now $\frac{4^{0.001} - 4^{-0.001}}{0.002} = 1.386294805$.

94. **(B)** Note that any line determined by two points equidistant from the origin will necessarily be horizontal.

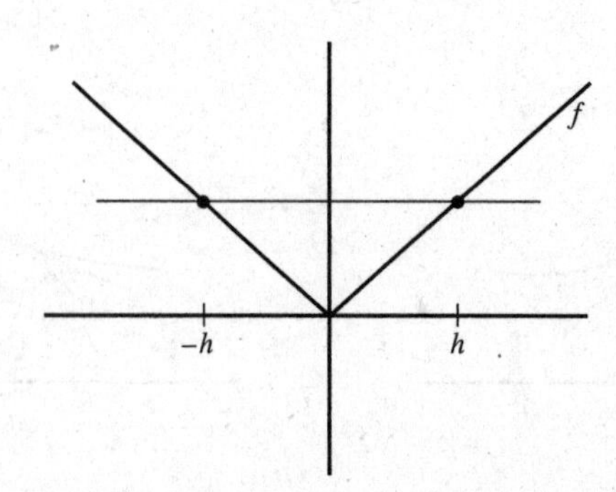

Therefore, the symmetric difference quotient yields:

$$\frac{f(0 + 0.001) - f(0 - 0.001)}{2(0.001)} = \frac{0.001 - 0.001}{0.002} = 0$$

95. **(D)** Note that $\frac{d}{dx}f(h(x)) = f'(h(x)) \cdot h'(x) = g(h(x)) \cdot h'(x) = g(\sin x) \cdot \cos x$.

96. **(E)** Since $f(x) = 3^x - x^3$, then $f'(x) = 3^x \ln 3 - 3x^2$. Furthermore, f is continuous on $[0,3]$ and f' is differentiable on $(0,3)$, so the MVT applies. We therefore seek c such that $f'(c) = \frac{f(3) - f(0)}{3} = -\frac{1}{3}$. Solving $3^x \ln 3 - 3x^2 = -\frac{1}{3}$ with a calculator, we find that c may be either 1.244 or 2.727. These values are the x-coordinates of points on the graph of $f(x)$ at which the tangents are parallel to the secant through points $(0,1)$ and $(3,0)$ on the curve.

97. **(A)** The line segment passes through $(1,-3)$ and $(2,-4)$.

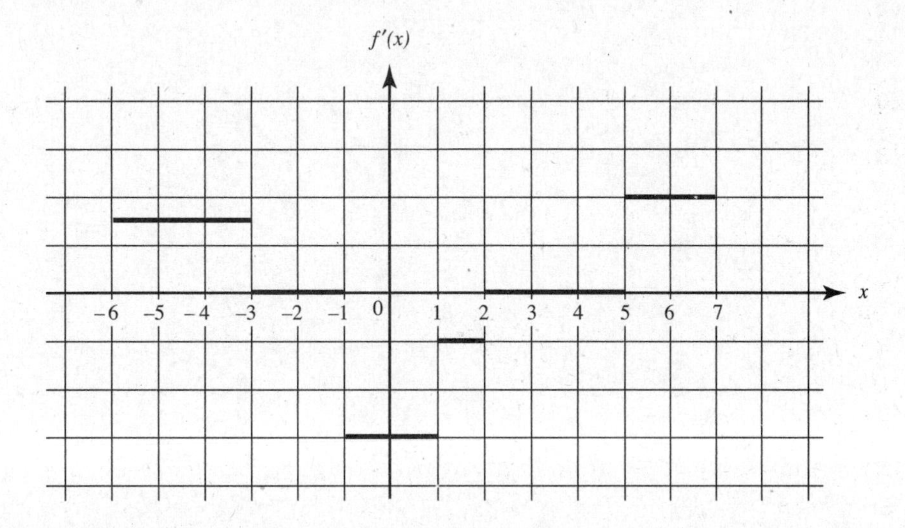

Use the graph of $f'(x)$, shown above, for Questions 98–101.

98. **(E)** $f'(x) = 0$ when the slope of $f(x)$ is 0; that is, when the graph of f is a horizontal segment.

99. **(E)** The graph of $f'(x)$ jumps at each corner of the graph of $f(x)$, namely, at x equal to $-3, -1, 1, 2$, and 5.

100. **(D)** On the interval $(-6,-3)$, $f(x) = \frac{3}{2}(x + 5)$.

101. **(B)** Verify that all choices but (B) are true. The graph of $f'(x)$ has five (not four) jump discontinuities.

102. **(C)** The best approximation to $f'(0.10)$ is $\frac{f(0.20) - f(0.10)}{0.20 - 0.10}$.

103. **(D)**

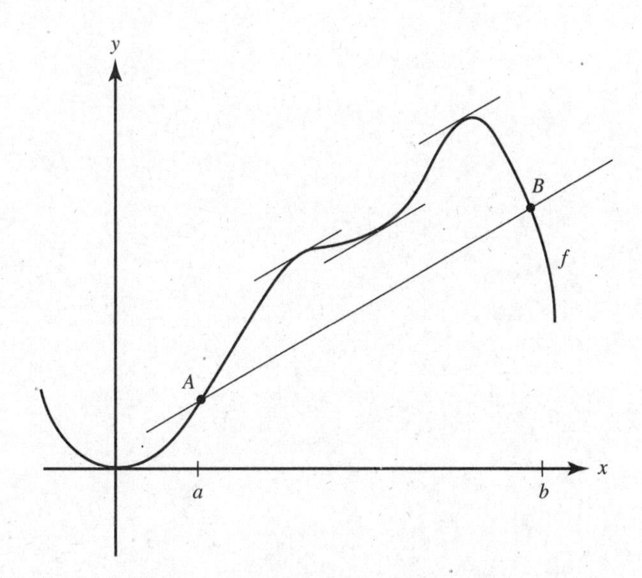

The average rate of change is represented by the slope of secant segment $\overline{AB}$. There appear to be 3 points at which the tangent lines are parallel to $\overline{AB}$.

Applications of Differential Calculus

4

CONCEPTS AND SKILLS

In this chapter, we review how to use derivatives to

- find slopes of curves and equations of tangent lines;
- find a function's maxima, minima, and points of inflections;
- describe where the graph of a function is increasing, decreasing, concave upward, and concave downward;
- analyze motion along a line;
- create local linear approximations;
- and work with related rates.

For BC Calculus students, we also review how to

- find the slope of parametric and polar curves
- and use vectors to analyze motion along parametrically defined curves.

A. SLOPE; CRITICAL POINTS

Slope of a curve

If the derivative of $y = f(x)$ exists at $P(x_1, y_1)$, then the *slope* of the curve at P (which is defined to be the slope of the tangent to the curve at P) is $f'(x_1)$, the derivative of $f(x)$ at $x = x_1$.

Critical point

Any c in the domain of f such that either $f'(c) = 0$ or $f'(c)$ is undefined is called a *critical point* or *critical value* of f. If f has a derivative everywhere, we find the critical points by solving the equation $f'(x) = 0$.

Example 1

For $f(x) = 4x^3 - 6x^2 - 8$, what are the critical points?

SOLUTION: $f'(x) = 12x^2 - 12x = 12x(x - 1)$,

which equals zero if x is 0 or 1. Thus, 0 and 1 are critical points.

Example 2

Find any critical points of $f(x) = 3x^3 + 2x$.

SOLUTION: $f'(x) = 9x^2 + 2$.

Since $f'(x)$ never equals zero (indeed, it is always positive), f has no critical values.

Example 3

Find any critical points of $f(x) = (x - 1)^{1/3}$.

SOLUTION: $f'(x) = \frac{1}{3(x-1)^{2/3}}$.

Although f' is never zero, $x = 1$ is a critical value of f because f' does not exist at $x = 1$.

Average and Instantaneous Rates of Change

Both average and instantaneous rates of change were defined in Chapter 3. If as x varies from a to $a + h$, the function f varies from $f(a)$ to $f(a + h)$, then we know that the difference quotient

$$\frac{f(a+h) - f(a)}{h}$$

is the average rate of change of f over the interval from a to $a + h$.

Thus, the *average velocity* of a moving object over some time interval is the change in distance divided by the change in time, the average rate of growth of a colony of fruit flies over some interval of time is the change in size of the colony divided by the time elapsed, the average rate of change in the profit of a company on some gadget with respect to production is the change in profit divided by the change in the number of gadgets produced.

The (instantaneous) rate of change of f at a, or the derivative of f at a, is the limit of the average rate of change as $h \to 0$:

$$f'(a) = \lim_{h \to 0} \frac{f(a+h) - f(a)}{h}.$$

On the graph of $y = f(x)$, the rate at which the y-coordinate changes with respect to the x-coordinate is $f'(x)$, the slope of the curve. The rate at which $s(t)$, the distance traveled by a particle in t seconds, changes with respect to time is $s'(t)$, the velocity of the particle; the rate at which a manufacturer's profit $P(x)$ changes relative to the production level x is $P'(x)$.

Example 4

Let $G = 400(15 - t)^2$ be the number of gallons of water in a cistern t minutes after an outlet pipe is opened. Find the average rate of drainage during the first 5 minutes and the rate at which the water is running out at the end of 5 minutes.

SOLUTION: The average rate of change during the first 5 min equals

$$\frac{G(5) - G(0)}{5} = \frac{400 \cdot 100 - 400 \cdot 225}{5} = -10{,}000 \text{ gal/min.}$$

The average rate of drainage during the first 5 min is 10,000 gal/min.

The instantaneous rate of change at $t = 5$ is $G'(5)$. Since

$$G'(t) = -800(15 - t),$$

$G'(5) = -800(10) = -8000$ gal/min. Thus the rate of drainage at the end of 5 min is 8000 gal/min.

B. TANGENTS TO A CURVE

The *equation of the tangent* to the curve $y = f(x)$ at point $P(x_1, y_1)$ is

$$y - y_1 = f'(x_1)(x - x_1).$$

Tangent to a curve

If the tangent to a curve is horizontal at a point, then the derivative at the point is 0. If the tangent is vertical at a point, then the derivative does not exist at the point.

Tangents to Parametrically Defined Curves

BC ONLY

If the curve is defined parametrically, say in terms of t (as in Chapter 1, page 75), then we obtain the slope at any point from the parametric equations. We then evaluate the slope and the x- and y-coordinates by replacing t by the value specified in the question (see Example 9, page 160).

Example 5

Find the equation of the tangent to the curve of $f(x) = x^3 - 3x^2$ at the point $(1, -2)$.

SOLUTION: Since $f'(x) = 3x^2 - 6x$ and $f'(1) = -3$, the equation of the tangent is

$$y + 2 = -3(x - 1) \quad \text{or} \quad y + 3x = 1.$$

Example 6

Find the equation of the tangent to $x^2y - x = y^3 - 8$ at the point where $x = 0$.

SOLUTION: Here we differentiate implicitly to get $\frac{dy}{dx} = \frac{1 - 2xy}{x^2 - 3y^2}$.

Since $y = 2$ when $x = 0$ and the slope at this point is $\frac{1 - 0}{0 - 12} = -\frac{1}{12}$, the equation of the tangent is

$$y - 2 = -\frac{1}{12}x \quad \text{or} \quad 12y + x = 24.$$

Example 7

Find the coordinates of any point on the curve of $y^2 - 4xy = x^2 + 5$ for which the tangent is horizontal.

SOLUTION: Since $\frac{dy}{dx} = \frac{x + 2y}{y - 2x}$ and the tangent is horizontal when $\frac{dy}{dx} = 0$, then $x = -2y$. If we substitute this in the equation of the curve, we get

$$y^2 - 4y(-2y) = 4y^2 + 5$$
$$5y^2 = 5.$$

Thus $y = \pm 1$ and $x = \pm 2$. The points, then, are $(2, -1)$ and $(-2, 1)$.

Example 8

Find the x-coordinate of any point on the curve of $y = \sin^2(x + 1)$ for which the tangent is parallel to the line $3x - 3y - 5 = 0$.

SOLUTION: Since $\frac{dy}{dx} = 2\sin(x + 1)\cos(x + 1) = \sin 2(x + 1)$ and since the given line has slope 1, we seek x such that $\sin 2(x + 1) = 1$. Then

$$2(x + 1) = \frac{\pi}{2} + 2n\pi \qquad (n \text{ an integer})$$

or

$$x + 1 = \frac{\pi}{4} + n\pi \qquad \text{and} \qquad x = \frac{\pi}{4} + n\pi - 1.$$

BC ONLY

Example 9

Find the equation of the tangent to $F(t) = (\cos t, 2\sin^2 t)$ at the point where $t = \frac{\pi}{3}$.

SOLUTION: Since $\frac{dx}{dt} = -\sin t$ and $\frac{dy}{dt} = 4\sin t\cos t$, we see that

$$\frac{dy}{dx} = \frac{4\sin t\cos t}{-\sin t} = -4\cos t.$$

At $t = \frac{\pi}{3}$, $x = \frac{1}{2}$, $y = 2\left(\frac{\sqrt{3}}{2}\right)^2 = \frac{3}{2}$, and $\frac{dy}{dx} = -2$. The equation of the tangent is

$$y - \frac{3}{2} = -2\left(x - \frac{1}{2}\right) \qquad \text{or} \qquad 4x + 2y = 5.$$

C. INCREASING AND DECREASING FUNCTIONS

Case I. Functions with Continuous Derivatives

Increasing Decreasing

A function $y = f(x)$ is said to be $\begin{matrix} \textit{increasing} \\ \textit{decreasing} \end{matrix}$ on an interval if for all a and b in the interval such that $a < b$, $\begin{matrix} f(b) \ge f(a) \\ f(b) \le f(a) \end{matrix}$. To find intervals over which $f(x)$ $\begin{matrix} \text{increases} \\ \text{decreases} \end{matrix}$, that is, over which the curve $\begin{matrix} \text{rises} \\ \text{falls} \end{matrix}$, analyze the signs of the derivative to determine where $\begin{matrix} f'(x) \ge 0 \\ f'(x) \le 0 \end{matrix}$.

Example 10

For what values of x is $f(x) = x^4 - 4x^3$, increasing? decreasing?

SOLUTION: $f'(x) = 4x^3 - 12x^2 = 4x^2(x - 3)$.

With critical values at $x = 0$ and $x = 3$, we analyze the signs of f' in three intervals:

		0		3	
f'	$-$		$-$		$+$

The derivative changes sign only at $x = 3$. Thus,

$$\text{if } x < 3 \qquad f'(x) \le 0 \text{ and } f \text{ is decreasing;}$$
$$\text{if } x > 3 \qquad f'(x) > 0 \text{ and } f \text{ is increasing.}$$

Note that f is decreasing at $x = 0$ even though $f'(0) = 0$. (See Figure N4–5 on page 166.)

Case II. Functions Whose Derivatives Have Discontinuities

Here we proceed as in Case I, but also consider intervals bounded by any points of discontinuity of f or f'.

➡ Example 11

For what values of x is $f(x) = \frac{1}{x+1}$ increasing? decreasing?

SOLUTION: $f'(x) = -\frac{1}{(x+1)^2}$.

We note that neither f nor f' is defined at $x = -1$; furthermore, $f'(x)$ never equals zero. We need therefore examine only the signs of $f'(x)$ when $x < -1$ and when $x > -1$.

When $x < -1$, $f'(x) < 0$; when $x > -1$, $f'(x) < 0$. Therefore, f decreases on both intervals. The curve is a hyperbola whose center is at the point $(-1, 0)$.

D. MAXIMUM, MINIMUM, CONCAVITY, AND INFLECTION POINTS: DEFINITIONS

Local (relative) max/min

The curve of $y = f(x)$ has a *local* (or *relative*) $\begin{matrix}\textit{maximum}\\ \textit{minimum}\end{matrix}$ at a point where $x = c$ if $\begin{matrix} f(c) \geqslant f(x) \\ f(c) \leqslant f(x)\end{matrix}$ for all x in the immediate neighborhood of c. If a curve has a local $\begin{matrix}\text{maximum}\\ \text{minimum}\end{matrix}$ at $x = c$, then the curve changes from $\begin{matrix}\text{rising}\\ \text{falling}\end{matrix}$ to $\begin{matrix}\text{falling}\\ \text{rising}\end{matrix}$ as x increases through c. If a function is differentiable on the closed interval $[a, b]$ and has a local maximum or minimum at $x = c$ $(a < c < b)$, then $f'(c) = 0$. The converse of this statement is not true.

If $f(c)$ is either a local maximum or a local minimum, then $f(c)$ is called a *local extreme value* or *local extremum*. (The plural of *extremum* is *extrema*.)

Global (absolute) max/min

The *global* or *absolute* $\begin{matrix}\textit{maximum}\\ \textit{minimum}\end{matrix}$ of a function on $[a, b]$ occurs at $x = c$ if $\begin{matrix} f(c) \geqslant f(x) \\ f(c) \leqslant f(x)\end{matrix}$ for all x on $[a, b]$.

Concavity

A curve is said to be *concave* $\begin{matrix}\textit{upward}\\ \textit{downward}\end{matrix}$ on an interval (a, b) if the curve lies $\begin{matrix}\text{above}\\ \text{below}\end{matrix}$ the tangent lines at each point in the interval (a, b). If $\begin{matrix} y'' > 0 \\ y'' < 0\end{matrix}$ at every point in an interval (a, b), the curve is concave $\begin{matrix}\text{up}\\ \text{down}\end{matrix}$. In Figure N4–1, the curves sketched in (a) and (b) are concave downward, while in (c) and (d) they are concave upward.

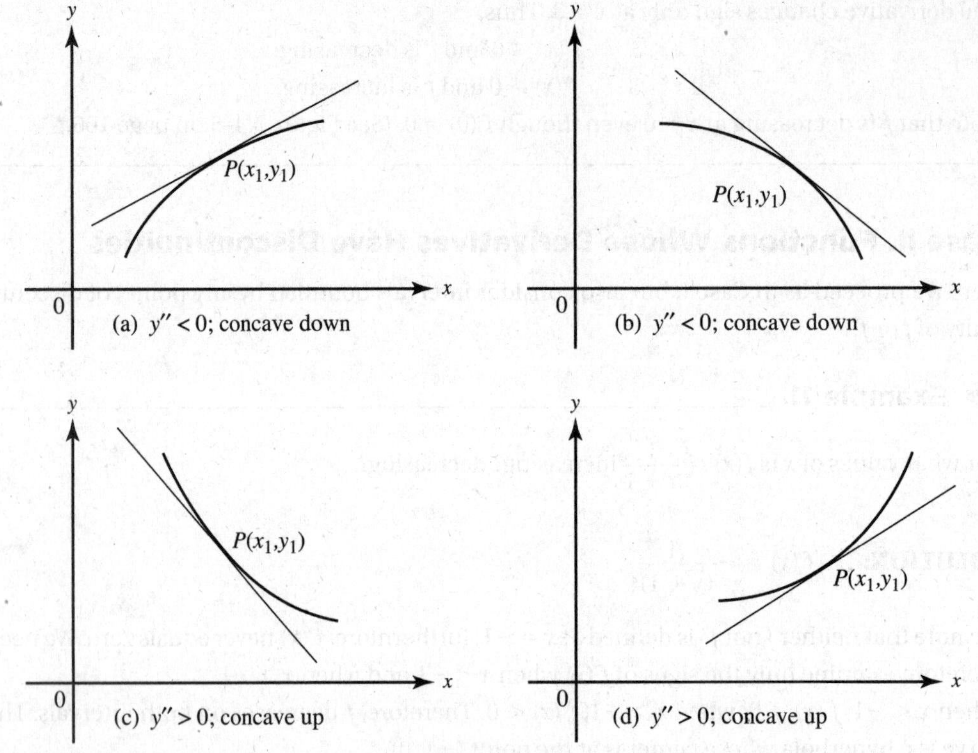

Figure N4–1

Point of inflection

A *point of inflection* is a point where the curve changes its concavity from upward to downward or from downward to upward. See §I, page 174, for a table relating a function and its derivatives. It tells how to graph the derivatives of f, given the graph of f. On pages 176 and 253 we graph f, given the graph of f'.

E. MAXIMUM, MINIMUM, AND INFLECTION POINTS: CURVE SKETCHING

Case I. Functions That Are Everywhere Differentiable

The following procedure is suggested to determine any maximum, minimum, or inflection point of a curve and to sketch the curve.

(1) Find y' and y''.

(2) Find all critical points of y, that is, all x for which $y' = 0$. At each of these x's the tangent to the curve is horizontal.

Second Derivative Test

(3) Let c be a number for which y' is 0; investigate the sign of y'' at c. If $y''(c) > 0$, then c yields a local minimum; if $y''(c) < 0$, then c yields a local maximum. This procedure is known as the *Second Derivative Test* (for extrema). See Figure N4–2. If $y''(c) = 0$, the Second Derivative Test fails and we must use the test in step (4) below.

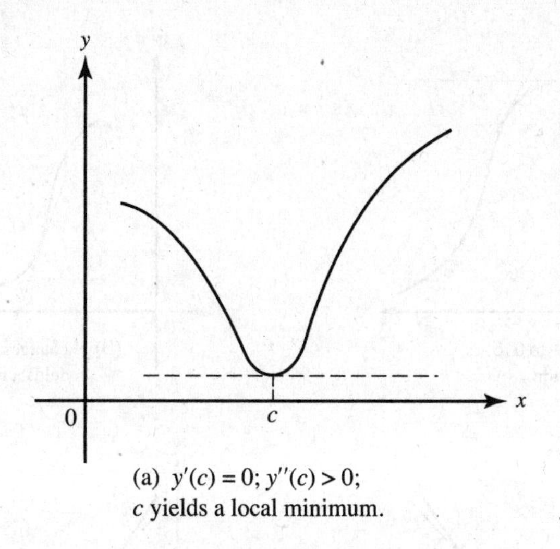

(a) $y'(c) = 0$; $y''(c) > 0$;
c yields a local minimum.

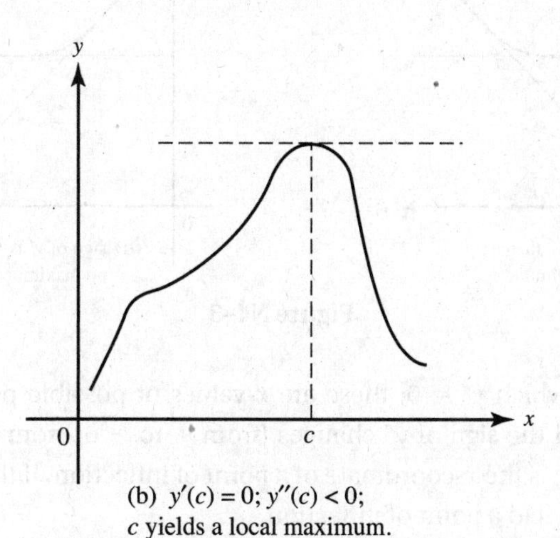

(b) $y'(c) = 0$; $y''(c) < 0$;
c yields a local maximum.

Figure N4–2

(4) If $y'(c) = 0$ and $y''(c) = 0$, investigate the signs of y' as x increases through c. If $y'(x) > 0$ for x's (just) less than c but $y'(x) < 0$ for x's (just) greater than c, then the situation is that indicated in Figure N4–3a, where the tangent lines have been sketched as x increases through c; here c yields a local maximum. If the situation is reversed and the sign of y' changes from $-$ to $+$ as x increases through c, then c yields a local minimum. Figure N4–3b shows this case. The schematic sign pattern of y', $+ \; 0 \; -$ or $- \; 0 \; +$, describes each situation completely. If y' does not change sign as x increases through c, then c yields neither a local maximum nor a local minimum. Two examples of this appear in Figures N4–3c and N4–3d.

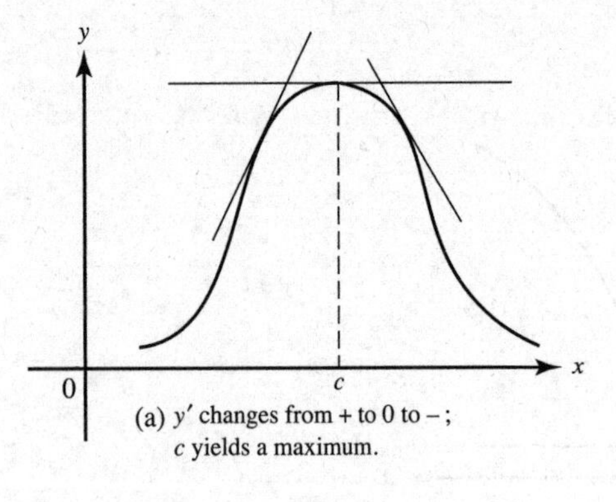

(a) y' changes from + to 0 to −; c yields a maximum.

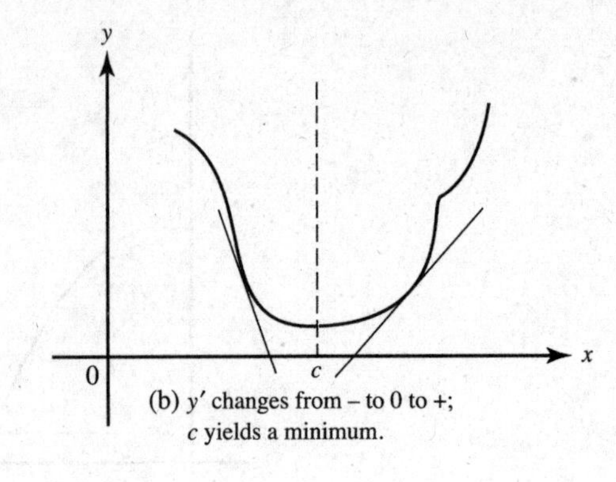

(b) y' changes from − to 0 to +; c yields a minimum.

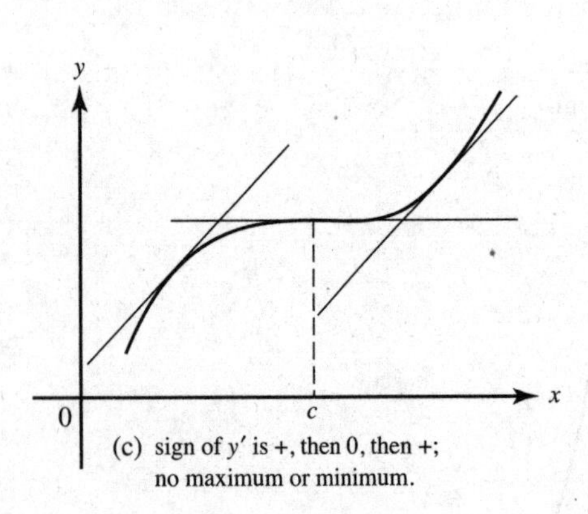

(c) sign of y' is +, then 0, then +; no maximum or minimum.

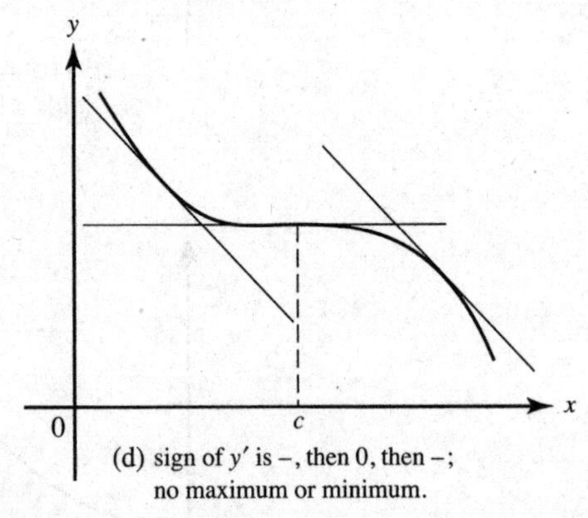

(d) sign of y' is −, then 0, then −; no maximum or minimum.

Figure N4–3

(5) Find all x's for which $y'' = 0$; these are x-values of possible points of inflection. If c is such an x and the sign of y'' changes (from + to − or from − to +) as x increases through c, then c is the x-coordinate of a point of inflection. If the signs do not change, then c does not yield a point of inflection.

The crucial points found as indicated in (1) through (5) above should be plotted along with the intercepts. Care should be exercised to ensure that the tangent to the curve is horizontal whenever $\frac{dy}{dx} = 0$ and that the curve has the proper concavity.

➡ Example 12

Find any maximum, minimum, or inflection points on the graph of $f(x) = x^3 - 5x^2 + 3x + 6$, and sketch the curve.

SOLUTION: For the steps listed above:

(1) Here $f'(x) = 3x^2 - 10x + 3$ and $f''(x) = 6x - 10$.

(2) $f'(x) = (3x - 1)(x - 3)$, which is zero when $x = \frac{1}{3}$ or 3.

(3) Since $f'\left(\frac{1}{3}\right) = 0$ and $f''\left(\frac{1}{3}\right) < 0$, we know that the point $\left(\frac{1}{3}, f\left(\frac{1}{3}\right)\right)$ is a local maximum; since $f'(3) = 0$ and $f''(3) > 0$, the point $(3, f(3))$ is a local minimum. Thus, $\left(\frac{1}{3}, \frac{175}{27}\right)$ is a local maximum and $(3, -3)$ a local minimum.

(4) is unnecessary for this problem.

(5) $f''(x) = 0$ when $x = \frac{5}{3}$, and f'' changes from negative to positive as x increases through $\frac{5}{3}$, so the graph of f has an inflection point. See Figure N4–4.

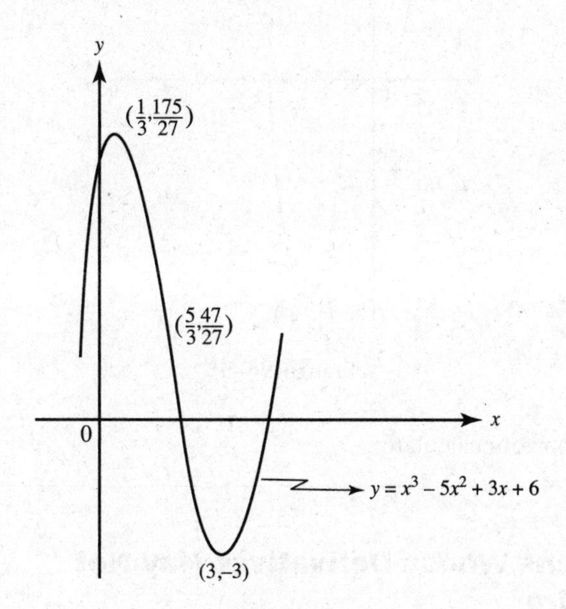

Figure N4–4

Verify the graph and information obtained above on your graphing calculator.

Example 13

Sketch the graph of $f(x) = x^4 - 4x^3$.

SOLUTION:

(1) $f'(x) = 4x^3 - 12x^2$ and $f''(x) = 12x^2 - 24x$.

(2) $f'(x) = 4x^2(x - 3)$, which is zero when $x = 0$ or $x = 3$.

(3) Since $f''(x) = 12x(x - 2)$ and $f''(3) > 0$ with $f'(3) = 0$, the point $(3, -27)$ is a local minimum. Since $f''(0) = 0$, the second-derivative test fails to tell us whether $x = 0$ yields a local maximum or a local minimum.

(4) Since $f'(x)$ does not change sign as x increases through 0, the point (0, 0) yields neither a local maximum nor a local minimum.

(5) $f''(x) = 0$ when x is 0 or 2; f'' changes signs as x increases through 0 (+ to −), and also as x increases through 2 (− to +). Thus both (0, 0) and (2, −16) are inflection points of the curve.

The curve is sketched in Figure N4–5 on page 166.

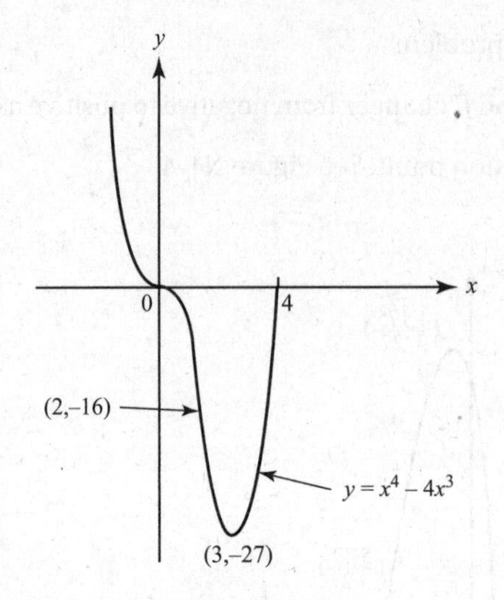

Figure N4–5

Verify the preceding on your calculator.

Case II. Functions Whose Derivatives May Not Exist Everywhere

If there are values of x for which a first or second derivative does not exist, we consider those values separately, recalling that a local maximum or minimum point is one of transition between intervals of rise and fall and that an inflection point is one of transition between intervals of upward and downward concavity.

➥ Example 14

Sketch the graph of $y = x^{2/3}$.

SOLUTION: $\frac{dy}{dx} = \frac{2}{3x^{1/3}}$ and $\frac{d^2y}{dx^2} = -\frac{2}{9x^{4/3}}$.

Neither derivative is zero anywhere; both derivatives fail to exist when $x = 0$. As x increases through 0, $\frac{dy}{dx}$ changes from – to +; (0, 0) is therefore a minimum. Note that the tangent is vertical at the origin, and that since $\frac{d^2y}{dx^2}$ is negative everywhere except at 0, the curve is everywhere concave down. See Figure N4–6.

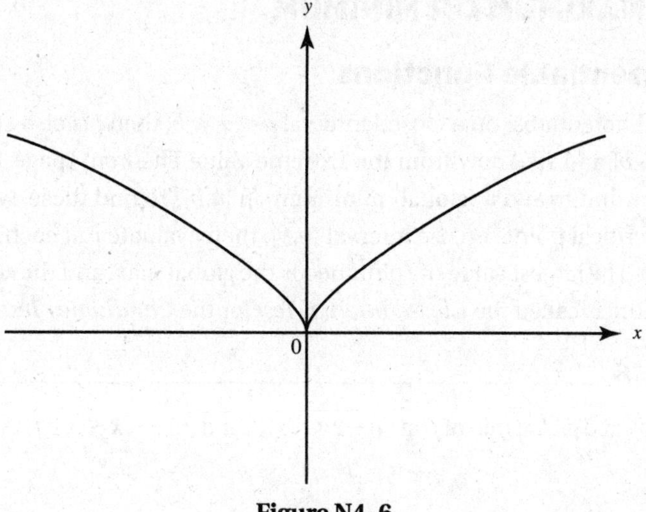

Figure N4–6

➥ Example 15

Sketch the graph of $y = x^{1/3}$.

SOLUTION: $\frac{dy}{dx} = \frac{1}{3x^{2/3}}$ and $\frac{d^2y}{dx^2} = -\frac{2}{9x^{5/3}}$.

As in Example 14, neither derivative ever equals zero and both fail to exist when $x = 0$. Here, however, as x increases through 0, $\frac{dy}{dx}$ does not change sign.

Since $\frac{dy}{dx}$ is positive for all x except 0, the curve rises for all x and can have neither maximum nor minimum points. The tangent is again vertical at the origin. Note here that $\frac{d^2y}{dx^2}$ does change sign (from + to −) as x increases through 0, so that (0, 0) is a point of inflection of the curve. See Figure N4–7.

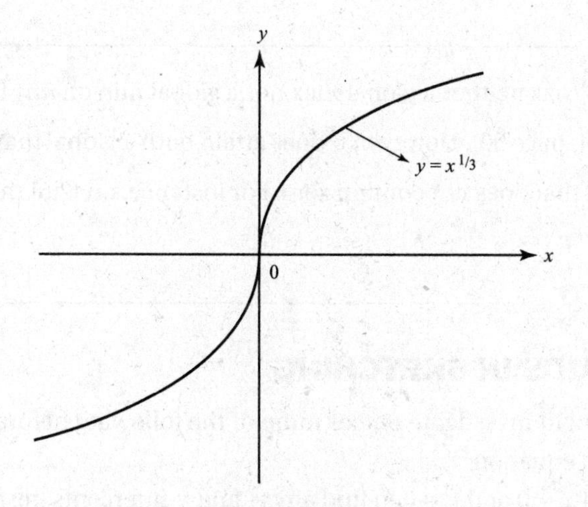

Figure N4–7

Verify the graph on your calculator.

F. GLOBAL MAXIMUM OR MINIMUM

Case I. Differentiable Functions

Closed Interval Test (Candidates Test)

If a function f is differentiable on a closed interval $a \leq x \leq b$, then f is also continuous on the closed interval $[a,b]$ and we know from the Extreme Value Theorem (page 101) that f attains both a (global) maximum and a (global) minimum on $[a,b]$. To find these, we solve the equation $f'(x) = 0$ for critical points on the interval $[a,b]$, then evaluate f at each of those and also at $x = a$ and $x = b$. The largest value of f obtained is the global max, and the smallest the global min. This procedure is called the *Closed Interval Test*, or the *Candidates Test.*

Example 16

Find the global max and global min of f on (a) $-2 \leq x \leq 3$, and (b) $0 \leq x \leq 3$, if $f(x) = 2x^3 - 3x^2 - 12x$.

SOLUTION:

(a) $f'(x) = 6x^2 - 6x - 12 = 6(x + 1)(x - 2)$, which equals zero if $x = -1$ or 2. Since $f(-2) = -4$, $f(-1) = 7$, $f(2) = -20$, and $f(3) = -9$, the global max of f occurs at $x = -1$ and equals 7, and the global min of f occurs at $x = 2$ and equals -20.

(b) Only the critical value 2 lies in $[0,3]$. We now evaluate f at 0, 2, and 3. Since $f(0) = 0$, $f(2) = -20$, and $f(3) = -9$, the global max of f equals 0 and the global min equals -20.

Case II. Functions That Are Not Everywhere Differentiable

We proceed as for Case I but now evaluate f also at each point in a given interval for which f is defined but for which f' does not exist.

Example 17

The absolute-value function $f(x) = |x|$ is defined for all real x, but $f'(x)$ does not exist at $x = 0$. Since $f'(x) = -1$ if $x < 0$, but $f'(x) = 1$ if $x > 0$, we see that f has a global min at $x = 0$.

Example 18

The function $f(x) = \frac{1}{x}$ has neither a global max nor a global min on *any* interval that contains zero (see Figure N2–4, page 90). However, it does attain both a global max and a global min on every closed interval that does not contain zero. For instance, on $[2,5]$ the global max of f is $\frac{1}{2}$, the global min $\frac{1}{5}$.

G. FURTHER AIDS IN SKETCHING

It is often very helpful to investigate one or more of the following before sketching the graph of a function or of an equation:

(1) Intercepts. Set $x = 0$ and $y = 0$ to find any y- and x-intercepts, respectively.

(2) Symmetry. Let the point (x, y) satisfy an equation. Then its graph is symmetric about

the x-axis if $(x, -y)$ also satisfies the equation;
the y-axis if $(-x, y)$ also satisfies the equation;
the origin if $(-x, -y)$ also satisfies the equation.

(3) Asymptotes. The line $y = b$ is a horizontal asymptote of the graph of a function f if either $\lim_{x\to\infty} f(x) = b$ or $\lim_{x\to-\infty} f(x) = b$. If $f(x) = \frac{P(x)}{Q(x)}$, inspect the degrees of $P(x)$ and $Q(x)$, then use the Rational Function Theorem, page 96. The line $x = c$ is a vertical asymptote of the rational function $\frac{P(x)}{Q(x)}$ if $Q(c) = 0$ but $P(c) \neq 0$.

(4) Points of discontinuity. Identify points not in the domain of a function, particularly where the denominator equals zero.

➡ Example 19

Sketch the graph of $y = \frac{2x+1}{x-1}$.

SOLUTION: If $x = 0$, then $y = -1$. Also, $y = 0$ when the numerator equals zero, which is when $x = -\frac{1}{2}$. A check shows that the graph does not possess any of the symmetries described above. Since $y \to 2$ as $x \to \pm\infty$, $y = 2$ is a horizontal asymptote; also, $x = 1$ is a vertical asymptote. The function is defined for all reals except $x = 1$; the latter is the only point of discontinuity.

We find derivatives: $y' = -\frac{3}{(x-1)^2}$ and $y'' = \frac{6}{(x-1)^3}$.

From y' we see that the function decreases everywhere (except at $x = 1$), and from y'' that the curve is concave down if $x < 1$, up if $x > 1$. See Figure N4–8.

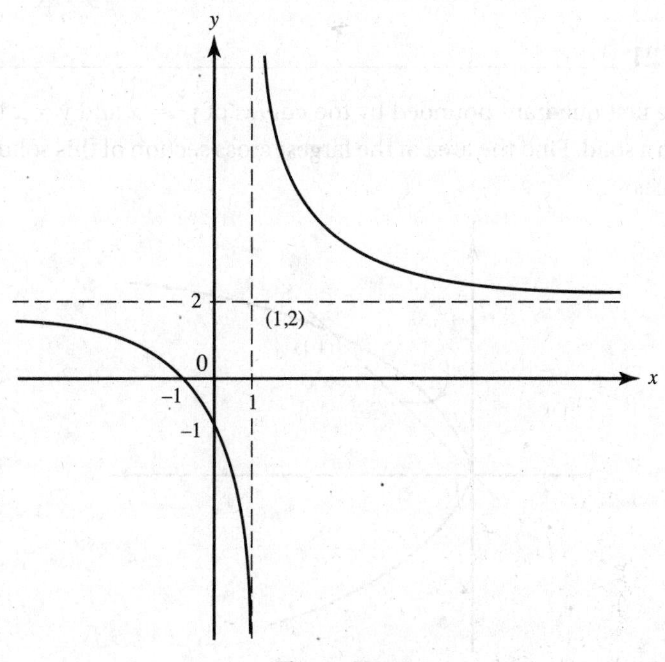

Figure N4–8

Verify the preceding on your calculator, using [−4,4] × [−4,8].

➡ Example 20

Describe any symmetries of the graphs of

(a) $3y^2 + x = 2$; (b) $y = x + \frac{1}{x}$; (c) $x^2 - 3y^2 = 27$.

SOLUTIONS:

(a) Suppose point (x, y) is on this graph. Then so is point $(x, -y)$, since $3(-y)^2 + x = 2$ is equivalent to $3y^2 + x = 2$. Then (a) is symmetric about the x-axis.

(b) Note that point $(-x, -y)$ satisfies the equation if point (x, y) does:.

$$(-y) = (-x) + \frac{1}{(-x)} \leftrightarrow y = x + \frac{1}{x}.$$

Therefore the graph of this function is symmetric about the origin.

(c) This graph is symmetric about the x-axis, the y-axis, and the origin. It is easy to see that, if point (x, y) satisfies the equation, so do points $(x, -y)$, $(-x, y)$, and $(-x, -y)$.

H. OPTIMIZATION: PROBLEMS INVOLVING MAXIMA AND MINIMA

The techniques described above can be applied to problems in which a function is to be maximized (or minimized). Often it helps to draw a figure. If y, the quantity to be maximized (or minimized), can be expressed explicitly in terms of x, then the procedure outlined above can be used. If the domain of y is restricted to some closed interval, one should always check the endpoints of this interval so as not to overlook possible extrema. Often, implicit differentiation, sometimes of two or more equations, is indicated.

➡ Example 21

The region in the first quadrant bounded by the curves of $y^2 = x$ and $y = x$ is rotated about the y-axis to form a solid. Find the area of the largest cross section of this solid that is perpendicular to the y-axis.

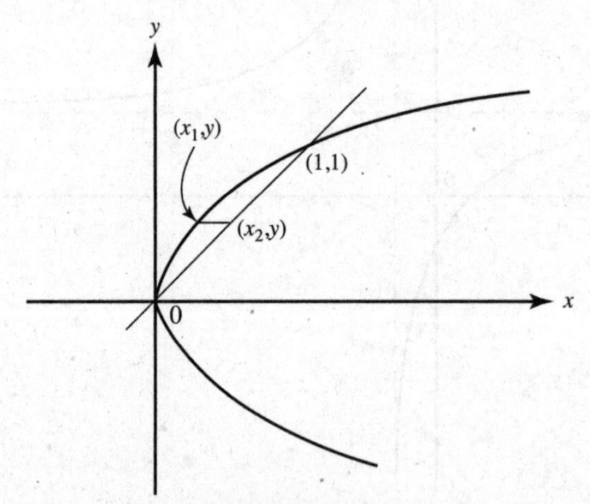

Figure N4–9

SOLUTION: See Figure N4–9. The curves intersect at the origin and at (1,1), so $0 < y < 1$. A cross section of the solid is a ring whose area A is the difference between the areas of two circles, one with radius x_2, the other with radius x_1. Thus

$$A = \pi x_2^2 - \pi x_1^2 = \pi(y^2 - y^4); \quad \frac{dA}{dy} = \pi(2y - 4y^3) = 2\pi y(1 - 2y^2).$$

The only relevant zero of the first derivative is $y = \frac{1}{\sqrt{2}}$. There the area A is

$$A = \pi\left(\frac{1}{2} - \frac{1}{4}\right) = \frac{\pi}{4}.$$

Note that $\frac{d^2A}{dy^2} = \pi\left(2 - 12y^2\right)$ and that this is negative when $y = \frac{1}{\sqrt{2}}$, assuring a maximum there. Note further that A equals zero at each endpoint of the interval [0,1] so that $\frac{\pi}{4}$ is the global maximum area.

➥ Example 22

The volume of a cylinder equals V cubic inches, where V is a constant. Find the proportions of the cylinder that minimize the total surface area.

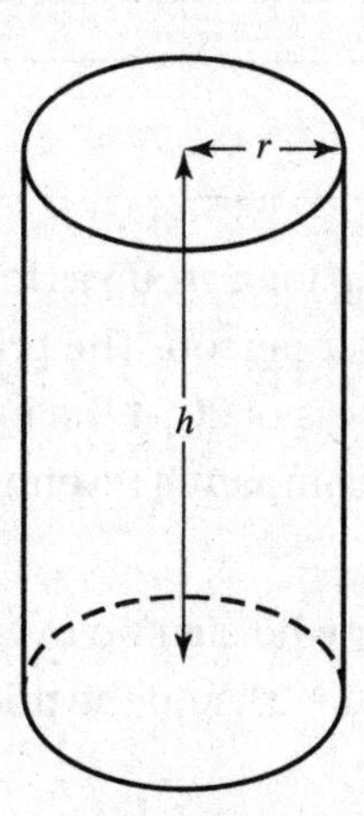

Figure N4–10

SOLUTION: We know that the volume is

$$V = \pi r^2 h \tag{1}$$

where r is the radius and h the height. We seek to minimize S, the total surface area, where

$$S = 2\pi r^2 + 2\pi rh \tag{2}$$

Solving (1) for h, we have $h = \frac{V}{\pi r^2}$, which we substitute in (2):

$$S = 2\pi r^2 + 2\pi r\frac{V}{\pi r^2} = 2\pi r^2 + \frac{2V}{r}. \tag{3}$$

Differentiating (3) with respect to r yields

$$\frac{dS}{dr} = 4\pi r - \frac{2V}{r^2}.$$

Now we set $\frac{dS}{dr}$ equal to zero to determine the conditions that make S a minimum:

$$4\pi r - \frac{2V}{r^2} = 0$$
$$4\pi r = \frac{2V}{r^2}$$
$$4\pi r = \frac{2(\pi r^2 h)}{r^2}$$
$$2r = h.$$

The total surface area of a cylinder of fixed volume is thus a minimum when its height equals its diameter.

(Note that we need not concern ourselves with the possibility that the value of r that renders $\frac{dS}{dr}$ equal to zero will produce a maximum surface area rather than a minimum one. With V fixed, we can choose r and h so as to make S as large as we like.)

Example 23

A charter bus company advertises a trip for a group as follows: At least 20 people must sign up. The cost when 20 participate is \$80 per person. The price will drop by \$2 per ticket for each member of the traveling group in excess of 20. If the bus can accommodate 28 people, how many participants will maximize the company's revenue?

SOLUTION: Let x denote the number who sign up in excess of 20. Then $0 \leqq x \leqq 8$. The total number who agree to participate is $(20 + x)$, and the price per ticket is $(80 - 2x)$ dollars. Then the revenue R, in dollars, is

$$R = (20 + x)(80 - 2x),$$
$$R'(x) = (20 + x)(-2) + (80 - 2x) \cdot 1$$
$$= 40 - 4x.$$

$R'(x)$ is zero if $x = 10$. Although $x = 10$ yields maximum R—note that $R''(x) = -4$ and is always negative—this value of x is not within the restricted interval. We therefore evaluate R at the endpoints 0 and 8: $R(0) = 1600$ and $R(8) = 28 \cdot 64 = 1792$; 28 participants will maximize revenue.

Example 24

A utilities company wants to deliver gas from a source S to a plant P located across a straight river 3 miles wide, then downstream 5 miles, as shown in Figure N4–11. It costs \$4 per foot to lay the pipe in the river but only \$2 per foot to lay it on land.

(a) Express the cost of laying the pipe in terms of u.
(b) How can the pipe be laid most economically?

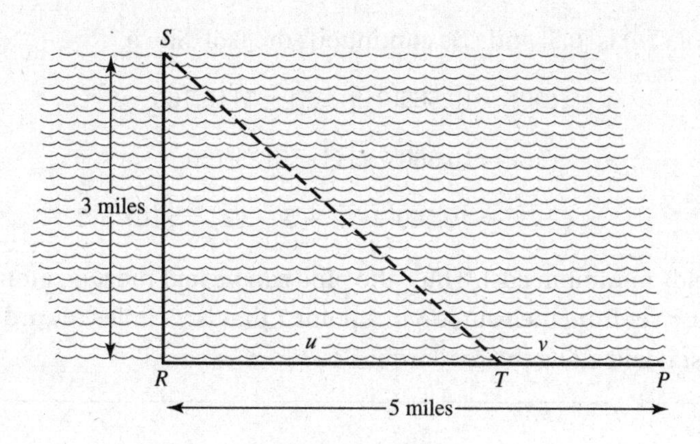

Figure N4–11

SOLUTIONS:

(a) Note that the problem "allows" us to (1) lay all of the pipe in the river, along a line from S to P; (2) lay pipe along SR, in the river, then along RP on land; or (3) lay some pipe in the river, say, along ST, and lay the rest on land along TP. When T coincides with P, we have case (1), with $v = 0$; when T coincides with R, we have case (2), with $u = 0$. Case (3) includes both (1) and (2).

In any event, we need to find the lengths of pipe needed (that is, the distances involved); then we must figure out the cost.

In terms of u:

	In the River	On Land
Distances:		
miles	$ST = \sqrt{9 + u^2}$	$TP = v = 5 - u$
feet	$ST = 5280\sqrt{9 + u^2}$	$TP = 5280(5 - u)$
Costs (dollars):	$4(5280)\sqrt{9 + u^2}$	$2[5280(5 - u)]$

If $C(u)$ is the total cost,

$$\begin{aligned} C(u) &= 21{,}120\sqrt{9 + u^2} + 10{,}560(5 - u) \\ &= 10{,}560\left(2\sqrt{9 + u^2} + 5 - u\right). \end{aligned}$$

(b) We now minimize $C(u)$:

$$C'(u) = 10{,}560\left(2 \cdot \frac{1}{2}\frac{2u}{\sqrt{9 + u^2}} - 1\right) = 10{,}560\left(\frac{2u}{\sqrt{9 + u^2}} - 1\right).$$

We now set $C'(u)$ equal to zero and solve for u:

$$\frac{2u}{\sqrt{9 + u^2}} - 1 = 0 \quad \rightarrow \quad \frac{2u}{\sqrt{9 + u^2}} = 1 \quad \rightarrow \quad \frac{4u^2}{\sqrt{9 + u^2}} = 1,$$

where, in the last step, we squared both sides; then

$$4u^2 = 9 + u^2, \qquad 3u^2 = 9, \qquad u^2 = 3, \qquad u = \sqrt{3},$$

where we discard $u = -\sqrt{3}$ as meaningless for this problem.

The domain of $C(u)$ is $[0,5]$ and C is continuous on $[0,5]$. Since

$$C(0) = 10{,}560(2\sqrt{9} + 5) = \$116{,}160,$$

$$C(5) = 10{,}560(2\sqrt{34}) \approx \$123{,}150,$$

$$C(\sqrt{3}) = 10{,}560(2\sqrt{12} + 5 - \sqrt{3}) = \$107{,}671.$$

So $u = \sqrt{3}$ yields minimum cost. Thus, the pipe can be laid most economically if some of it is laid in the river from the source S to a point T that is $\sqrt{3}$ miles toward the plant P from R, and the rest is laid along the road from T to P.

I. RELATING A FUNCTION AND ITS DERIVATIVES GRAPHICALLY

The following table shows the characteristics of a function f and their implications for f's derivatives. These are crucial in obtaining one graph from another. The table can be used reading from left to right or from right to left.

Note that the slope at $x = c$ of any graph of a function is equal to the ordinate at c of the derivative of the function.

	f	f'	f''
ON AN INTERVAL	increasing	≥ 0	
	decreasing	≤ 0	
AT c	local maximum	$x<c \quad x=c \quad x>c$ $+ \quad 0 \quad -$ (f' is decreasing)	$f''(c) < 0$
	local minimum	$- \quad 0 \quad +$ (f' is increasing)	$f''(c) > 0$
	neither local maximum nor local minimum	$+ \quad 0 \quad +$ $- \quad 0 \quad -$ (f' does not change sign)	
AT c	point of inflection	$f'(c)$ is a minimum; f' changes from decreasing to increasing	$x<c \quad x=c \quad x>c$ $- \quad 0 \quad +$
		$f'(c)$ is a maximum; f' changes from increasing to decreasing	$+ \quad 0 \quad -$
ON AN INTERVAL	concave up	f' is increasing	$f'' \geq 0$
	concave down	f' is decreasing	$f'' \leq 0$

If $f'(c)$ does not exist, check the signs of f' as x increases through c: plus-to-minus yields a local maximum; minus-to-plus yields a local minimum; no sign change means no maximum or minimum, but check the possibility of a point of inflection.

AN IMPORTANT NOTE:

Tables and number lines showing sign changes of the function and its derivatives can be very helpful in organizing all of this information. *Note, however, that the AP Exam requires that students write sentences that describe the behavior of the function based on the sign of its derivative.*

➡ Example 25A

Given the graph of $f(x)$ shown in Figure N4–12, sketch $f'(x)$.

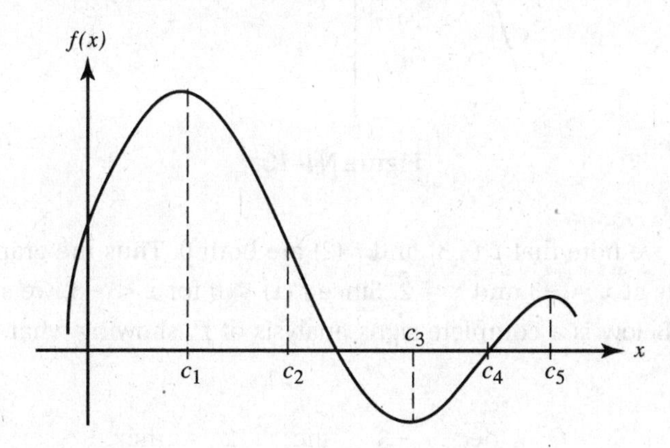

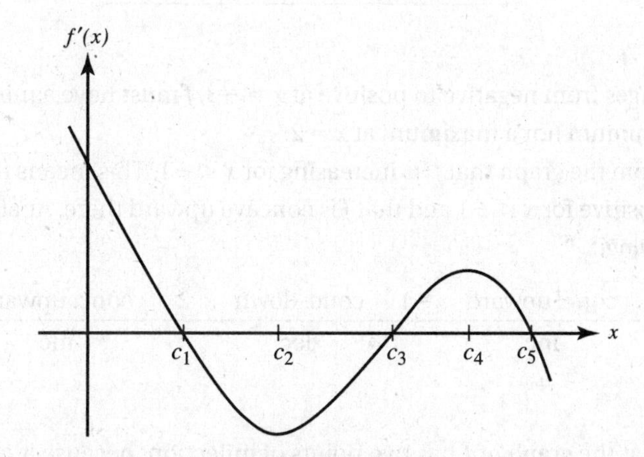

Figure N4–12

Point $x =$	Behavior of f	Behavior of f'
c_1	$f(c_1)$ is a local max	$f'(c_1) = 0$; f' changes sign from + to −
c_2	c_2 is an inflection point of f; the graph of f changes concavity from down to up	f' changes from decreasing to increasing; $f'(c_2)$ is a local minimum
c_3	$f(c_3)$ is a local minimum	$f'(c_3) = 0$; f' changes sign from − to +
c_4	c_4 is an inflection point of f; the graph of f changes concavity from up to down	f' changes from increasing to decreasing; $f'(c_4)$ is a local maximum
c_5	$f(c_5)$ is a local maximum	$f'(c_5) = 0$; f' changes sign from + to −

➡ Example 25B

Given the graph of $f'(x)$ shown in Figure N4–13, sketch a possible graph of f.

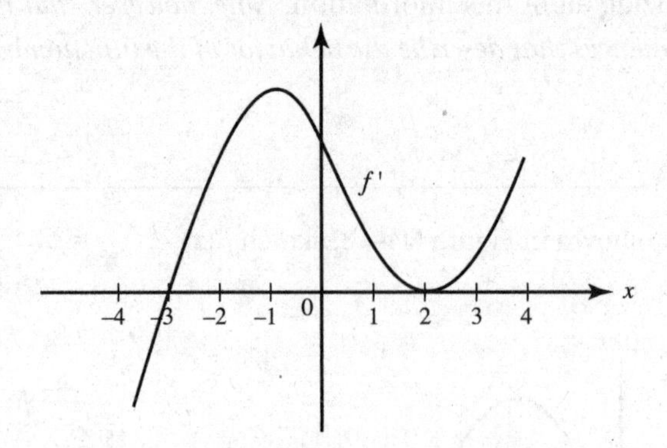

Figure N4–13

SOLUTION: First, we note that $f'(-3)$ and $f'(2)$ are both 0. Thus the graph of f must have horizontal tangents at $x = -3$ and $x = 2$. Since $f'(x) < 0$ for $x < -3$, we see that f must be decreasing there. Below is a complete signs analysis of f', showing what it implies for the behavior of f.

f	dec	-3	inc	2	inc
f'	$-$		$+$		$+$

Because f' changes from negative to positive at $x = -3$, f must have a minimum there, but f has neither a minimum nor a maximum at $x = 2$.

We note next from the graph that f' is increasing for $x < -1$. This means that the derivative of f', f'', must be positive for $x < -1$ and that f is concave upward there. Analyzing the signs of f'' yields the following:

f	conc. upward	-1	conc. down	2	conc. upward
f'	inc		dec		inc
f''	$+$		$-$		$+$

We conclude that the graph of f has two points of inflection, because it changes concavity from upward to downward at $x = -1$ and back to upward at $x = 2$. We use the information obtained to sketch a possible graph of f, shown in Figure N4–14. Note that other graphs are possible; in fact, any vertical translation of this f will do!

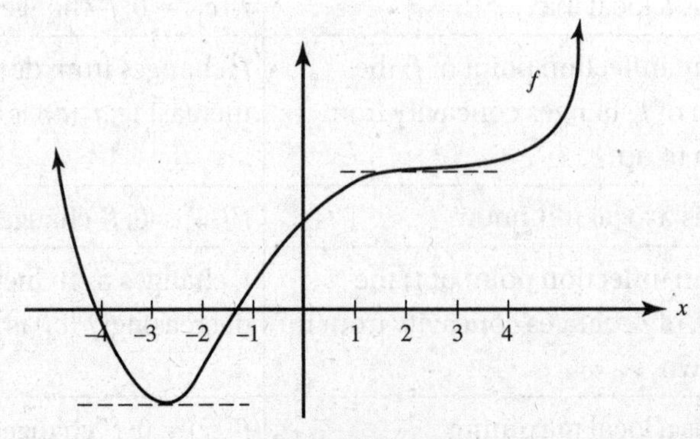

Figure N4–14

J. MOTION ALONG A LINE

If a particle moves along a line according to the law $s = f(t)$, where s represents the position of the particle P on the line at time t, then the velocity v of P at time t is given by $\frac{ds}{dt}$ and its acceleration a by $\frac{dv}{dt}$ or by $\frac{d^2s}{dt^2}$. The speed of the particle is $|v|$, the magnitude of v. If the line of motion is directed positively to the right, then the motion of the particle P is subject to the following: At any instant,

Velocity

Acceleration

Speed

(1) if $v > 0$, then P is moving to the right (its position s is increasing); if $v < 0$, then P is moving to the left (its position s is decreasing);
(2) if $a > 0$, then v is increasing; if $a < 0$, then v is decreasing;
(3) if a and v are both positive or both negative, then (1) and (2) imply that the speed of P is increasing or that P is accelerating; if a and v have opposite signs, then the speed of P is decreasing or P is decelerating;
(4) if s is a continuous function of t, then P reverses direction whenever v is zero and a is different from zero; note that zero velocity does not necessarily imply a reversal in direction.

➡ Example 26

A particle moves along a horizontal line such that its position $s = 2t^3 - 9t^2 + 12t - 4$, for $t \geqq 0$.
(a) Find all t for which the particle is moving to the right.
(b) Find all t for which the velocity is increasing.
(c) Find all t for which the speed of the particle is increasing.
(d) Find the speed when $t = \frac{3}{2}$.
(e) Find the total distance traveled between $t = 0$ and $t = 4$.

SOLUTION: $v = \frac{ds}{dt} = 6t^2 - 18t + 12 = 6(t^2 - 3t + 2) = 6(t - 2)(t - 1)$

and $$a = \frac{dv}{dt} = \frac{d^2s}{dt^2} = 12t - 18 = 12\left(t - \frac{3}{2}\right).$$

Velocity $v = 0$ at $t = 1$ and $t = 2$, and:

$$\begin{array}{llll} \text{if} & t < 1, & \text{then} & v > 0, \\ & 1 < t < 2, & & v < 0, \\ & t > 2, & & v > 0. \end{array}$$

Acceleration $a = 0$ at $t = \frac{3}{2}$, and:

$$\begin{array}{llll} \text{if} & t < \frac{3}{2}, & \text{then} & a < 0, \\ & t > \frac{3}{2}, & & a > 0. \end{array}$$

These signs of v and a immediately yield the answers, as follows:
(a) The particle moves to the right when $t < 1$ or $t > 2$.
(b) v increases when $t > \frac{3}{2}$.
(c) The speed $|v|$ is increasing when v and a are both positive, that is, for $t > 2$, and when v and a are both negative, that is, for $1 < t < \frac{3}{2}$.

(d) The speed when $t = \frac{3}{2}$ equals $|v| = \left|-\frac{3}{2}\right| = \frac{3}{2}$.

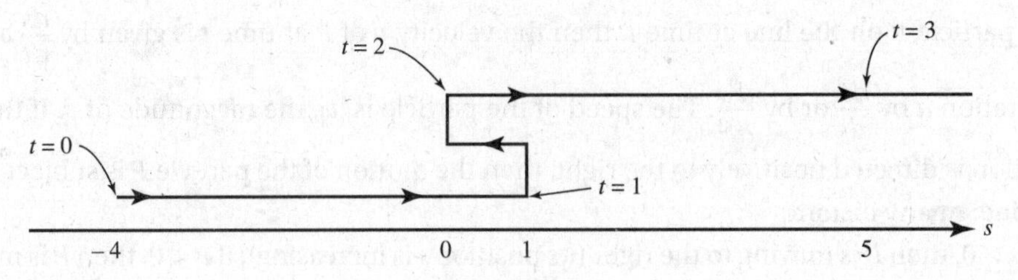

Figure N4–15

(e) *P*'s motion can be indicated as shown in Figure N4–15. *P* moves to the right if $t < 1$, reverses its direction at $t = 1$, moves to the left when $1 < t < 2$, reverses again at $t = 2$, and continues to the right for all $t > 2$. The position of *P* at certain times *t* are shown in the following table:

t:	0	1	2	4
s:	−4	1	0	28

Thus *P* travels a total of 34 units between times $t = 0$ and $t = 4$.

Example 27

Answer the questions of Example 26 if the law of motion is

$$s = t^4 - 4t^3.$$

SOLUTION: Since $v = 4t^3 - 12t^2 = 4t^2(t - 3)$ and $a = 12t^2 - 24t = 12t(t - 2)$, the signs of v and a are as follows:

$$\begin{array}{llll} \text{if} & t < 3, & \text{then} & v < 0 \\ & 3 < t, & & v > 0; \\ \text{if} & t < 0, & \text{then} & a > 0 \\ & 0 < t < 2, & & a < 0 \\ & 2 < t, & & a > 0. \end{array}$$

Thus

(a) s increases if $t > 3$.

(b) v increases if $t < 0$ or $t > 2$.

(c) Since v and a have the same sign if $0 < t < 2$ or if $t > 3$, the speed increases on these intervals.

(d) The speed when $t = \frac{3}{2}$ equals $|v| = \left|-\frac{27}{2}\right| = \frac{27}{2}$.

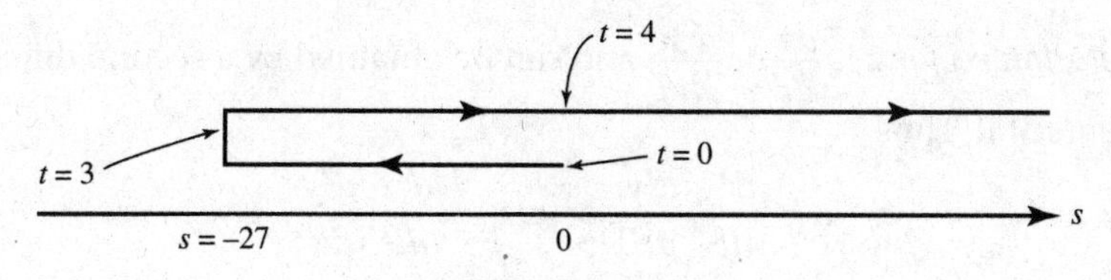

Figure N4–16

(e) The motion is shown in Figure N4–16. The particle moves to the left if $t < 3$ and to the right if $t > 3$, stopping instantaneously when $t = 0$ and $t = 3$, but reversing direction only when $t = 3$. Thus:

t:	0	3	4
s:	0	−27	0

The particle travels a total of 54 units between $t = 0$ and $t = 4$.

(Compare with Example 13, page 165, where the function $f(x) = x^4 - 4x^3$ is investigated for maximum and minimum values; also see the accompanying Figure N4–5 on page 166.)

K. MOTION ALONG A CURVE: VELOCITY AND ACCELERATION VECTORS

BC ONLY

Vector

If a point P moves along a curve defined parametrically by $P(t) = (x(t), y(t))$, where t represents time, then the vector from the origin to P is called the *position vector,* with x as its *horizontal component* and y as its *vertical component.* The set of position vectors for all values of t in the domain common to $x(t)$ and $y(t)$ is called *vector function.*.

Components

A vector may be symbolized either by a boldface letter (**R**) or an italic letter with an arrow written over it ($\vec{R}$). The position vector, then, may be written as $\vec{R}(t) = \langle x, y\rangle$ or as $\mathbf{R} = \langle x, y\rangle$. In print the boldface notation is clearer, and will be used in this book; when writing by hand, the arrow notation is simpler.

The *velocity vector* is the derivative of the vector function (the position vector):

Velocity vector

$$\mathbf{v} = \frac{d\mathbf{R}}{dt} = \left\langle \frac{dx}{dt}, \frac{dy}{dt} \right\rangle \quad \text{or} \quad \vec{v}(t) = \left\langle \frac{dx}{dy}, \frac{dy}{dt} \right\rangle.$$

Alternative notations for $\frac{dx}{dt}$ and $\frac{dy}{dt}$ are v_x and v_y, respectively; these are the components of **v** in the horizontal and vertical directions, respectively. The slope of **v** is

$$\frac{\frac{dy}{dt}}{\frac{dx}{dt}} = \frac{dy}{dx},$$

which is the slope of the curve; the *magnitude* of **v** is the vector's length:

Magnitude

$$|\mathbf{v}| = \sqrt{\left(\frac{dx}{dt}\right)^2 + \left(\frac{dy}{dt}\right)^2} = \sqrt{v_x^{\,2} + v_y^{\,2}}.$$

Thus, if the vector **v** is drawn initiating at P, it will be tangent to the curve at P and its magnitude will be the *speed* of the particle at P.

BC ONLY

Acceleration vector

The *acceleration vector* **a** is $\frac{d\mathbf{v}}{dt}$ or $\frac{d^2\mathbf{R}}{dt^2}$, and can be obtained by a second differentiation of the components of **R**. Thus

$$\mathbf{a} = \left\langle \frac{d^2x}{dt^2}, \frac{d^2y}{dt^2} \right\rangle, \text{ or } \vec{a}(t) = \left\langle \frac{d^2x}{dt^2}, \frac{d^2y}{dt^2} \right\rangle,$$

and its magnitude is the vector's length:

$$|\mathbf{a}| = \sqrt{\left(\frac{d^2x}{dt^2}\right)^2 + \left(\frac{d^2y}{dt^2}\right)^2} = \sqrt{a_x^{\,2} + a_y^{\,2}},$$

where we have used a_x and a_y for $\frac{d^2x}{dt^2}$ and $\frac{d^2y}{dt^2}$, respectively.

➡ Example 28

A particle moves according to the equations $x = 3 \cos t, y = 2 \sin t$.

(a) Find a single equation in x and y for the path of the particle and sketch the curve.

(b) Find the velocity and acceleration vectors at any time t, and show that $\mathbf{a} = -\mathbf{R}$ at all times.

(c) Find **R**, **v**, and **a** when (1) $t_1 = \frac{\pi}{6}$, (2) $t_2 = \pi$, and draw them on the sketch.

(d) Find the speed of the particle and the magnitude of its acceleration at each instant in (c).

(e) When is the speed a maximum? A minimum?

SOLUTIONS:

(a) Since $\frac{x^2}{9} = \cos^2 t$ and $\frac{y^2}{4} = \sin^2 t$, therefore

$$\frac{x^2}{9} + \frac{y^2}{4} = 1$$

and the particle moves in a counterclockwise direction along an ellipse, starting, when $t = 0$, at (3, 0) and returning to this point when $t = 2\pi$.

(b) We have

$$\mathbf{R} = \langle 3 \cos t, 2 \sin t \rangle$$
$$\mathbf{v} = \langle -3 \sin t, 2 \cos t \rangle$$
$$\mathbf{a} = \langle -3 \cos t, -2 \sin t \rangle = -\mathbf{R}.$$

The acceleration, then, is always directed toward the center of the ellipse.

(c) At $t_1 = \frac{\pi}{6}$,

$$\mathbf{R}_1 = \left\langle \frac{3\sqrt{3}}{2}, 1 \right\rangle$$
$$\mathbf{v}_1 = \left\langle -\frac{3}{2}, \sqrt{3} \right\rangle$$
$$\mathbf{a}_1 = \left\langle -\frac{3\sqrt{3}}{2}, -1 \right\rangle.$$

At $t_2 = \pi$,

$$\mathbf{R}_2 = \langle -3, 0 \rangle$$
$$\mathbf{v}_2 = \langle 0, -2 \rangle$$
$$\mathbf{a}_2 = \langle 3, 0 \rangle$$

BC ONLY

The curve, and **v** and **a** at t_1 and t_2, are sketched in Figure N4–18, below.

(d) At $t_1 = \frac{\pi}{6}$,

$$|\mathbf{v}_1| = \sqrt{\frac{9}{4} + 3} = \frac{\sqrt{21}}{2},$$

$$|\mathbf{a}_1| = \sqrt{\frac{27}{4} + 1} = \frac{\sqrt{31}}{2},$$

At $t_2 = \pi$,

$$|\mathbf{v}_2| = \sqrt{0 + 4} = 2,$$

$$|\mathbf{a}_2| = \sqrt{9 + 0} = 3.$$

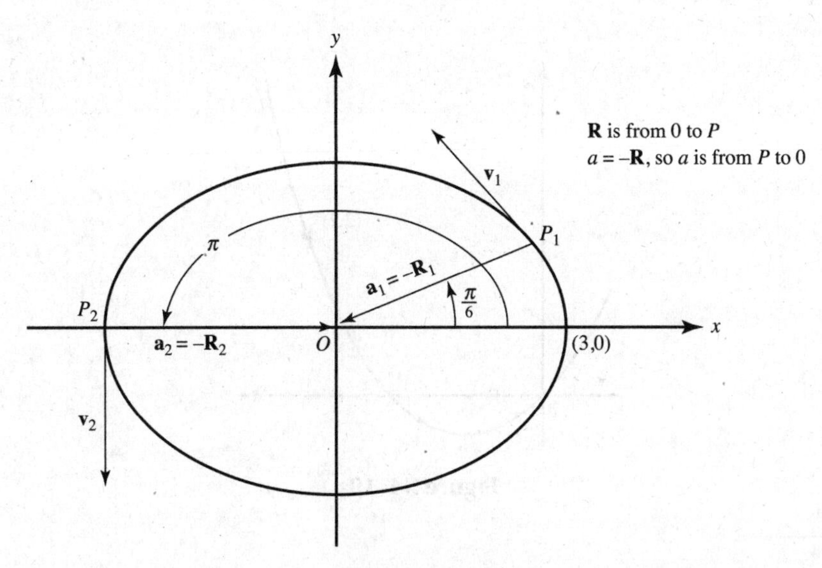

Figure N4–18

(e) For the speed $|\mathbf{v}|$ at any time t

$$\begin{aligned} \mathbf{v} &= \sqrt{9\sin^2 t + 4\cos^2 t} \\ &= \sqrt{4\sin^2 t + 4\cos^2 t + 5\sin^2 t} \\ &= \sqrt{4 + 5\sin^2 t}. \end{aligned}$$

We see immediately that the speed is a maximum when $t = \frac{\pi}{2}$ or $\frac{3\pi}{2}$, and a minimum when $t = 0$ or π. The particle goes fastest at the ends of the minor axis and most slowly at the ends of the major axis. Generally one can determine maximum or minimum speed by finding $\frac{d}{dt}|\mathbf{v}|$, setting it equal to zero, and applying the usual tests to sort out values of t that yield maximum or minimum speeds.

Example 29

A particle moves along the parabola $y = x^2 - x$ with constant speed $\sqrt{10}$. Find **v** at (2, 2).

SOLUTION: Since

$$v_y = \frac{dy}{dt} = (2x - 1)\frac{dx}{dt} = (2x - 1)v_x \qquad (1)$$

and

BC ONLY

$$v_x^2 + v_y^2 = 10, \qquad (2)$$

$$v_x^2 + (2x-1)^2 v_x^2 = 10. \qquad (3)$$

Relation (3) holds at all times; specifically, at (2, 2), $v_x^2 + 9v_x^2 = 10$ so that $v_x = \pm 1$. From (1), then, we see that $v_y = \pm 3$. Therefore **v** at (2, 2) is either $\langle 1,3 \rangle$ or $\langle -1,-3 \rangle$. The former corresponds to counterclockwise motion along the parabola, as shown in Figure N4–19a; the latter to clockwise motion, indicated in Figure N4–19b.

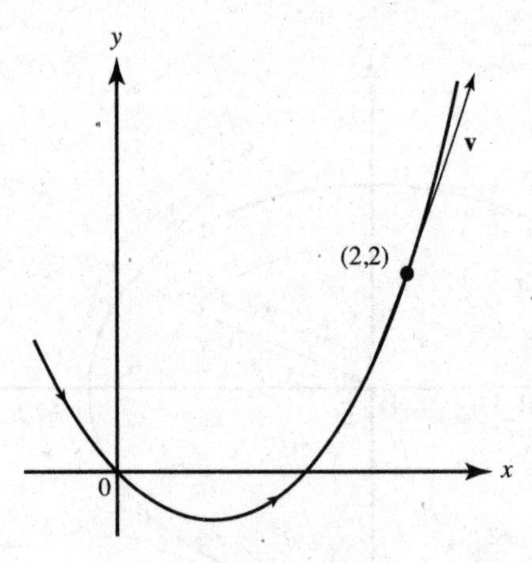

Figure N4–19a

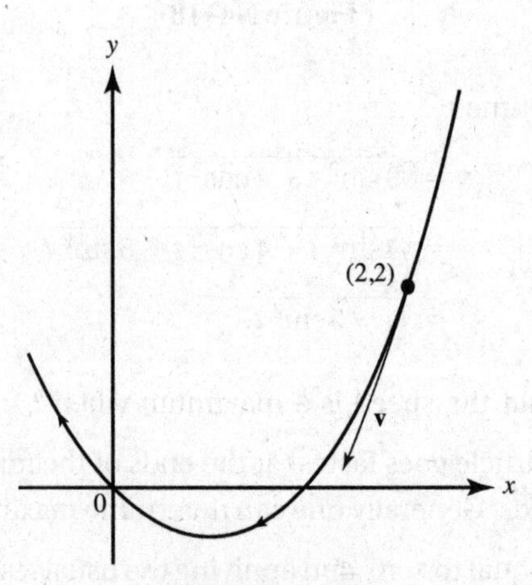

Figure N4–19b

L. TANGENT-LINE APPROXIMATIONS

Local linear approximation

If $f'(a)$ exists, then the *local linear approximation* of $f(x)$ at a is

$$f(a) + f'(a)(x - a).$$

Since the equation of the tangent line to $y = f(x)$ at $x = a$ is

$$y - f(a) = f'(a)(x - a),$$

we see that the y value on the tangent line is an approximation for the actual or true value of f. Local linear approximation is therefore also called *tangent-line approximation.*[†] For values of x close to a, we have

Tangent-line approximation

$$f(x) \simeq f(a) + f'(a)(x - a), \tag{1}$$

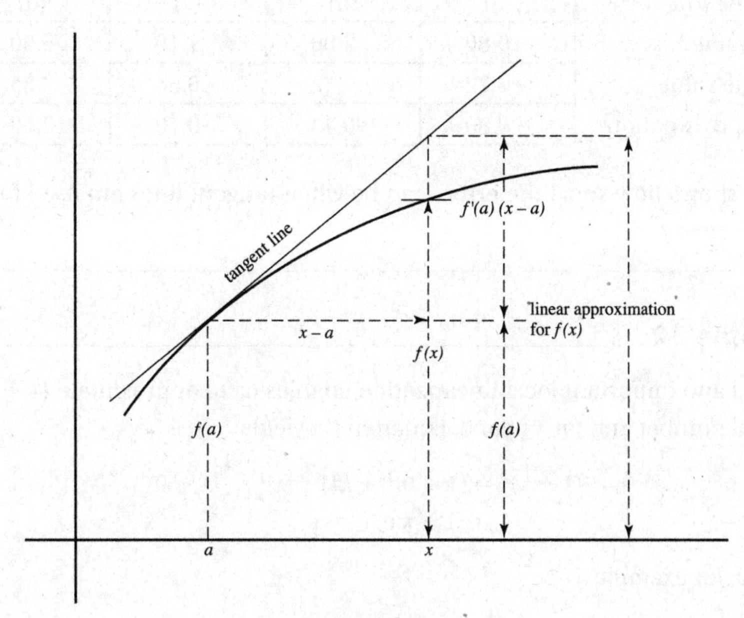

Figure N4–20

where $f(a) + f'(a)(x - a)$ is the linear or tangent-line approximation for $f(x)$, and $f'(a)(x - a)$ is the approximate change in f as we move along the curve from a to x. See Figure N4–20.

In general, the closer x is to a, the better the approximation is to $f(x)$.

Example 30

Find tangent-line approximations for each of the following functions at the values indicated:

(a) $\sin x$ at $a = 0$ (b) $\cos x$ at $a = \frac{\pi}{2}$

(c) $2x^3 - 3x$ at $a = 1$ (d) $\sqrt{1 + x}$ at $a = 8$

SOLUTIONS:

(a) At $a = 0$, $\sin x \simeq \sin(0) + \cos(0)(x - 0) \simeq 0 + 1 \cdot x \simeq x$

(b) At $a = \frac{\pi}{2}$, $\cos x \simeq \cos\frac{\pi}{2} - \sin\left(\frac{\pi}{2}\right)\left(x - \frac{\pi}{2}\right) \simeq -x + \frac{\pi}{2}$

(c) At $a = 1$, $2x^3 - 3x \simeq -1 + 3(x - 1) \simeq 3x - 4$

(d) At $a = 8$, $\sqrt{1 + x} = \sqrt{1 + 8} + \frac{1}{2\sqrt{1 + 8}}(x - 8) = 3 + \frac{1}{6}(x - 8)$

[†]Local linear approximation is also referred to as "local linearization" or even "best linear approximation" (the latter because it is better than any other linear approximation).

➡ Example 31

Using the tangent lines obtained in Example 30 and a calculator, we evaluate each function, then its linear approximation, at the indicated x-values:

Function	(a)	(b)	(c)	(d)
x-value	−0.80	2.00	1.10	5.50
True value	−0.72	−0.42	−0.64	2.55
Approximation	−0.80	−0.43	−0.70	2.58

Example 31 shows how small the errors can be when tangent lines are used for approximations and x is near a.

➡ Example 32

A very useful and important local linearization enables us to approximate $(1 + x)^k$ by $1 + kx$ for k any real number and for x near 0. Equation (1) yields

$$(1 + x)^k \simeq (1 + 0)^k + k(1 + x)^{k-1}_{\text{at } 0} \cdot (x - 0)$$
$$\simeq 1 + kx. \qquad (2)$$

Then, near 0, for example,

$$\sqrt{1 + x} \simeq 1 + \frac{1}{2}x \quad \text{and} \quad (1 + x)^3 \simeq 1 + 3x.$$

➡ Example 33

Estimate the value of $\dfrac{3}{(1 - x)^2}$ at $x = 0.05$.

SOLUTION: Use the line tangent to $f(x) = \dfrac{3}{(1 - x)^2}$ at $x = 0$; $f(0) = 3$.

$f'(x) = \dfrac{6}{(1 - x)^3}$, so $f'(0) = 6$; hence, the line is $y = 6x + 3$.

Our tangent-line approximation, then, is $\dfrac{3}{(1 - x)^2} \approx 6x + 3$.

At $x = 0.05$, we have $f(0.05) \approx 6(0.05) + 3 = 3.3$.

To determine if this is an overestimate or an underestimate, we examine the second derivative.

$f''(x) = \dfrac{18}{(1 - x)^4} > 0$ for $x \neq 1$; therefore, $f(x)$ is always concave up.

Therefore the tangent line at $x = 0$ lies below the graph of $f(x)$ on the interval [0, 0.05]. The estimate is on the tangent line, so it must be an underestimate. We can verify this by evaluating $f(0.05)$ or by graphing $f(x)$ and its tangent line at $x = 0$ on a calculator.

M. RELATED RATES

If several variables that are functions of time t are related by an equation, we can obtain a relation involving their (time) rates of change by differentiating with respect to t.

Example 34

If one leg AB of a right triangle increases at the rate of 2 inches per second, while the other leg AC decreases at 3 inches per second, find how fast the hypotenuse is changing when $AB = 6$ feet and $AC = 8$ feet.

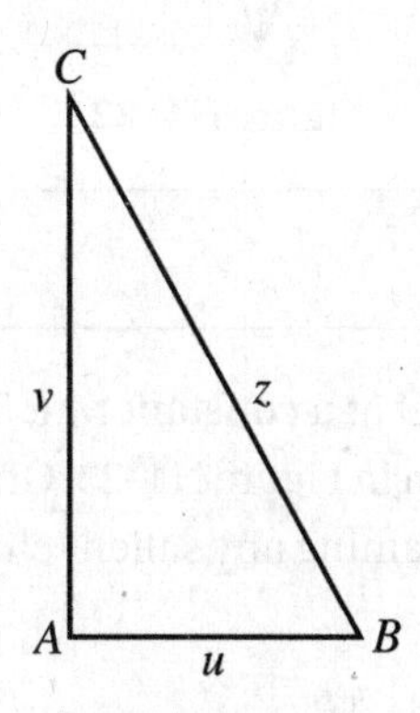

Figure N4–21

SOLUTION: See Figure N4–21. Let u, v, and z denote the lengths respectively of AB, AC, and BC. We know that $\frac{du}{dt} = \frac{1}{6}$ (ft/sec) and $\frac{dv}{dt} = -\frac{1}{4}$. Since (at any time) $z^2 = u^2 + v^2$, then

$$2z\frac{dz}{dt} = 2u\frac{du}{dt} + 2v\frac{dv}{dt} \quad \text{and} \quad \frac{dz}{dt} = \frac{u\frac{du}{dt} + v\frac{dv}{dt}}{z}.$$

At the instant in question, $u = 6$, $v = 8$, and $z = 10$, so

$$\frac{dz}{dt} = \frac{6\left(\frac{1}{6}\right) + 8\left(-\frac{1}{4}\right)}{10} = -\frac{1}{10} \text{ ft/sec.}$$

Example 35

The diameter and height of a paper cup in the shape of a cone are both 4 inches, and water is leaking out at the rate of $\frac{1}{2}$ cubic inch per second. Find the rate at which the water level is dropping when the diameter of the surface is 2 inches.

SOLUTION: See Figure N4–22. We know that $\frac{dV}{dt} = -\frac{1}{2}$ and that $h = 2r$.

Here, $V = \frac{1}{3}\pi r^2 h = \frac{\pi h^3}{12}$, so

$$\frac{dV}{dt} = \frac{\pi h^2}{4}\frac{dh}{dt} \quad \text{and} \quad \frac{dh}{dt} = -\frac{1}{2}\cdot\frac{4}{\pi h^2} = -\frac{2}{\pi h^2} \text{ at any time.}$$

When the diameter is 2 in., so is the height, and $\frac{dh}{dt} = -\frac{1}{2\pi}$. The water level is thus dropping at the rate of $\frac{1}{2\pi}$ in./sec.

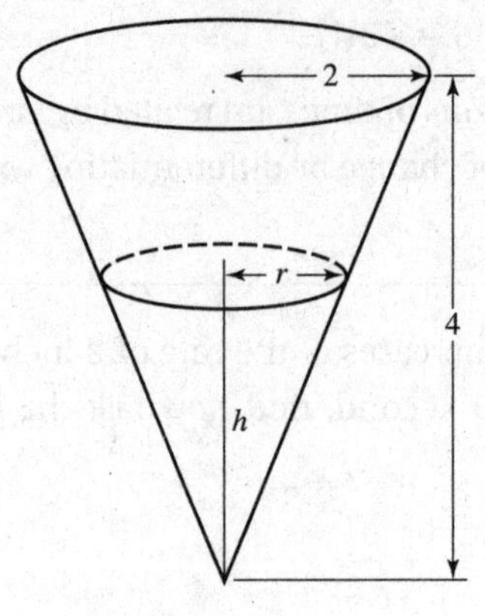

Figure N4–22

➥ Example 36

Suppose liquid is flowing into a vessel at a constant rate. The vessel has the shape of a hemisphere capped by a cylinder, as shown in Figure N4–23. Graph $y = h(t)$, the height (= depth) of the liquid at time t, labeling and explaining any salient characteristics of the graph.

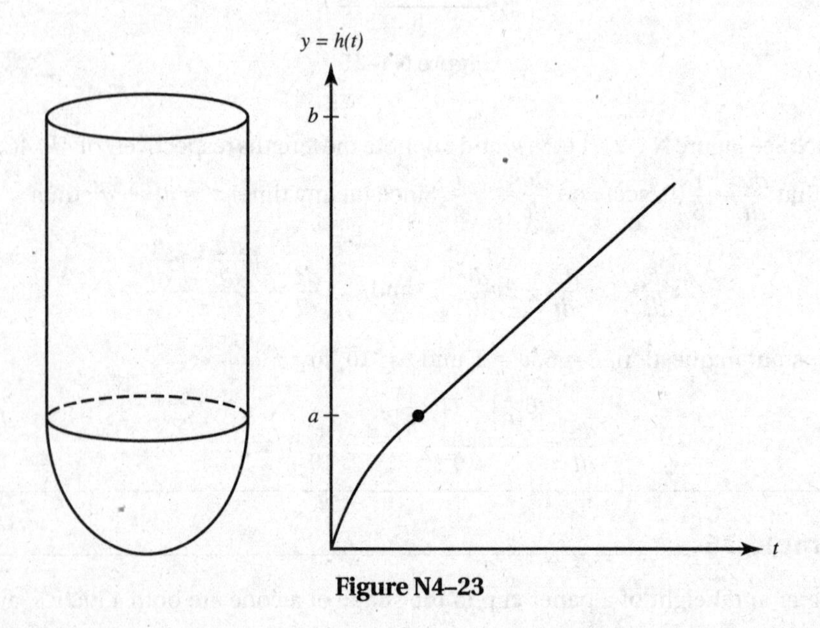

Figure N4–23

SOLUTION: Liquid flowing in at a constant rate means the change in volume is constant per unit of time. Obviously, the depth of the liquid increases as t does, so $h'(t)$ is positive throughout. To maintain the constant increase in volume per unit of time, when the radius grows, $h'(t)$ must decrease. Thus, the rate of increase of h decreases as h increases from 0 to a (where the cross-sectional area of the vessel is largest). Therefore, since $h'(t)$ decreases, $h''(t) < 0$ from 0 to a and the curve is concave down.

As h increases from a to b, the radius of the vessel (here cylindrical) remains constant, as do the cross-sectional areas. Therefore $h'(t)$ is also constant, implying that $h(t)$ is linear from a to b.

Note that the inflection point at depth a does not exist, since $h''(t) < 0$ for all values less than a but is equal to 0 for all depths greater than or equal to a.

N. SLOPE OF A POLAR CURVE

BC ONLY

We know that, if a smooth curve is given by the parametric equations

$$x = f(t) \quad \text{and} \quad y = g(t),$$

then

$$\frac{dy}{dx} = \frac{g'(t)}{f'(t)} \quad \text{provided that } f'(t) \neq 0.$$

To find the slope of a polar curve $r = f(\theta)$, we must first express the curve in parametric form. Since

$$x = r\cos\theta \quad \text{and} \quad y = r\sin\theta,$$

therefore,

$$x = f(\theta)\cos\theta \quad \text{and} \quad y = f(\theta)\sin\theta.$$

If $f(\theta)$ is differentiable, so are x and y; then

$$\frac{dx}{d\theta} = f'(\theta)\cos\theta - f(\theta)\sin\theta,$$

$$\frac{dy}{d\theta} = f'(\theta)\sin\theta + f(\theta)\cos\theta.$$

Also, if $\frac{dx}{d\theta} \neq 0$, then

$$\frac{dy}{dx} = \frac{dy/d\theta}{dx/d\theta} = \frac{f'(\theta)\sin\theta + f(\theta)\cos\theta}{f'(\theta)\cos\theta - f(\theta)\sin\theta} = \frac{r'\sin\theta + r\cos\theta}{r'\cos\theta - r\sin\theta}.$$

In doing an exercise, it is far easier simply to express the polar equation parametrically, then find dy/dx, rather than to memorize the formula.

Example 37

(a) Find the slope of the cardioid $r = 2(1 + \cos\theta)$ at $\theta = \frac{\pi}{6}$. See Figure N4–24.

(b) Where is the tangent to the curve horizontal?

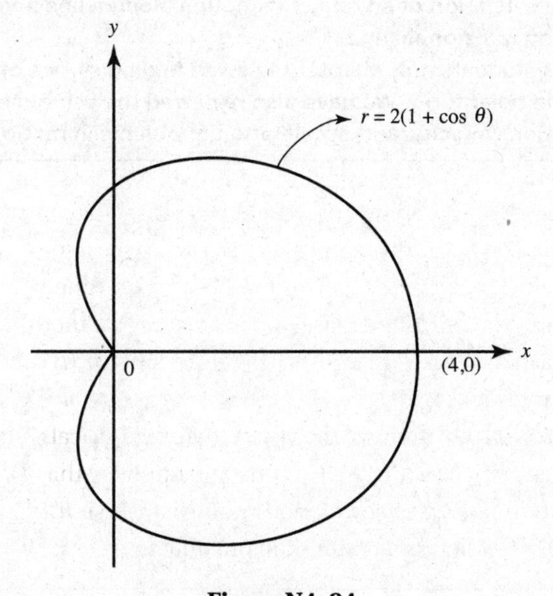

Figure N4–24

BC ONLY

SOLUTIONS:

(a) Use $r = 2(1 + \cos\theta)$, $x = r\cos\theta$, $y = r\sin\theta$, and $r' = -2\sin\theta$; then

$$\frac{dy}{dx} = \frac{dy/d\theta}{dx/d\theta} = \frac{(-2\sin\theta)\sin\theta + 2(1+\cos\theta)(\cos\theta)}{(-2\sin\theta)\cos\theta - 2(1+\cos\theta)(\sin\theta)}.$$

At $\theta = \frac{\pi}{6}$, $\frac{dy}{dx} = -1$.

(b) Since the cardioid is symmetric to $\theta = 0$ we need consider only the upper half of the curve for part (b). The tangent is horizontal where $\frac{dy}{d\theta} = 0$ (provided $\frac{dx}{d\theta} \neq 0$). Since $\frac{dy}{d\theta}$ factors into $2(2\cos\theta - 1)(\cos\theta + 1)$, which equals 0 for $\cos\theta = \frac{1}{2}$ or -1, $\theta = \frac{\pi}{3}$ or π. From part (a), $\frac{dx}{d\theta} \neq 0$ at $\frac{\pi}{3}$, but $\frac{dx}{d\theta}$ does equal 0 at π. Therefore, the tangent is horizontal only at $\frac{\pi}{3}$ (and, by symmetry, at $\frac{5\pi}{3}$).

It is obvious from Figure N4–24 that $r'(\theta)$ does *not* give the slope of the cardioid. As θ varies from 0 to $\frac{2\pi}{3}$, the slope varies from $-\infty$ to 0 to $+\infty$ (with the tangent rotating counterclockwise), taking on every real value. However, $r'(\theta)$ equals $-2\sin\theta$, which takes on values only between -2 and 2!

CHAPTER SUMMARY

In this chapter we reviewed many applications of derivatives. We've seen how to find slopes of curves and used that skill to write equations of lines tangent to a curve. Those lines often provide very good approximations for values of functions. We have looked at ways derivatives can help us understand the behavior of a function. The first derivative can tell us whether a function is increasing or decreasing and locate maximum and minimum points. The second derivative can tell us whether the graph of the function is concave upward or concave downward and locate points of inflection. We've reviewed how to use derivatives to determine the velocity and acceleration of an object in motion along a line and to describe relationships among rates of change.

For BC Calculus students, this chapter reviewed finding slopes of curves defined parametrically or in polar form. We have also reviewed the use of vectors to describe the position, velocity, and acceleration of objects in motion along curves.

PRACTICE EXERCISES

Part A. Directions: Answer these questions *without* using your calculator.

1. The slope of the curve $y^3 - xy^2 = 4$ at the point where $y = 2$ is

   (A) -2 (B) $\frac{1}{4}$ (C) $-\frac{1}{2}$ (D) $\frac{1}{2}$ (E) 2

2. The slope of the curve $y^2 - xy - 3x = 1$ at the point $(0, -1)$ is

   (A) -1 (B) -2 (C) $+1$ (D) 2 (E) -3

3. The equation of the tangent to the curve $y = x \sin x$ at the point $\left(\frac{\pi}{2}, \frac{\pi}{2}\right)$ is

   (A) $y = x - \pi$ (B) $y = \frac{\pi}{2}$ (C) $y = \pi - x$

   (D) $y = x + \frac{\pi}{2}$ (E) $y = x$

4. The tangent to the curve of $y = xe^{-x}$ is horizontal when x is equal to

   (A) 0 (B) 1 (C) -1 (D) $\frac{1}{e}$ (E) e

5. The minimum value of the slope of the curve $y = x^5 + x^3 - 2x$ is

   (A) 0 (B) 2 (C) 6 (D) -2 (E) -6

6. The equation of the tangent to the hyperbola $x^2 - y^2 = 12$ at the point (4, 2) on the curve is

   (A) $x - 2y + 6 = 0$ (B) $y = 2x$ (C) $y = 2x - 6$

   (D) $y = \frac{x}{2}$ (E) $x + 2y = 6$

7. The tangent to the curve $y^2 - xy + 9 = 0$ is vertical when

   (A) $y = 0$ (B) $y = \pm\sqrt{3}$ (C) $y = \frac{1}{2}$

   (D) $y = \pm 3$ (E) none of these

8. The best approximation, in cubic inches, to the increase in volume of a sphere when the radius is increased from 3 to 3.1 in. is

   (A) $\frac{0.04\pi}{3}$ (B) 0.04π (C) 1.2π (D) 3.6π (E) 36π

9. When $x = 3$, the equation $2x^2 - y^3 = 10$ has the solution $y = 2$. When $x = 3.04$, $y \simeq$

   (A) 1.6 (B) 1.96 (C) 2.04 (D) 2.14 (E) 2.4

10. If the side e of a square is increased by 1%, then the area is increased approximately

    (A) $0.02e$ (B) $0.02e^2$ (C) $0.01e^2$ (D) 1% (E) $0.01e$

11. The edge of a cube has length 10 in., with a possible error of 1%. The possible error, in cubic inches, in the volume of the cube is

    (A) 1 (B) 3 (C) 10 (D) 30 (E) 100

12. The function $f(x) = x^4 - 4x^2$ has

(A) one local minimum and two local maxima
(B) one local minimum and one local maximum
(C) two local maxima and no local minimum
(D) two local minima and no local maximum
(E) two local minima and one local maximum

13. The number of inflection points of the curve in Question 12 is

(A) 0 (B) 1 (C) 2 (D) 3 (E) 4

14. The maximum value of the function $y = -4\sqrt{2-x}$ is

(A) 0 (B) -4 (C) 4 (D) -2 (E) 2

15. The total number of local maximum and minimum points of the function whose derivative, for all x, is given by $f'(x) = x(x-3)^2(x+1)^4$ is

(A) 0 (B) 1 (C) 2 (D) 3 (E) 6

16. For which curve shown below are both f' and f'' negative?

(A)

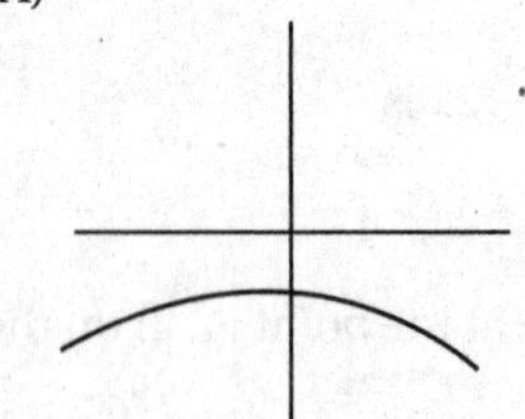

(B)

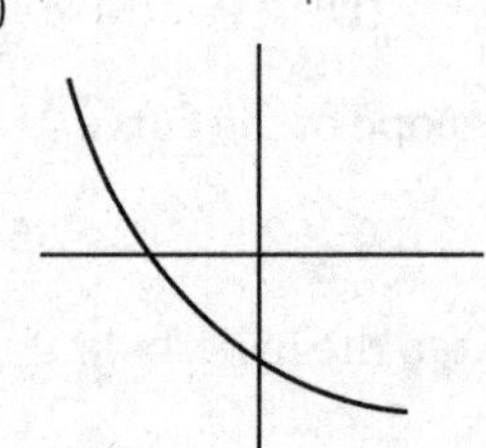

(C)

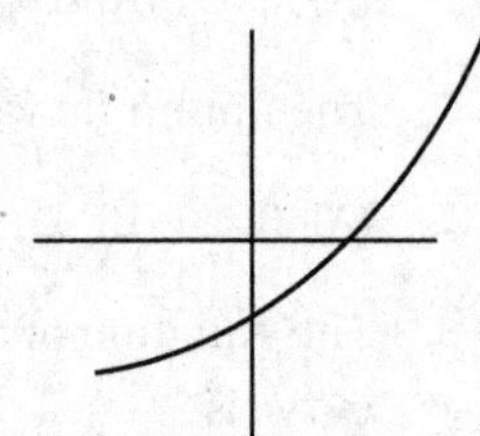

(D)

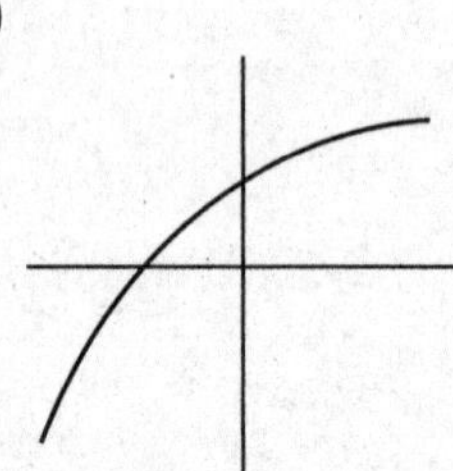

(E)

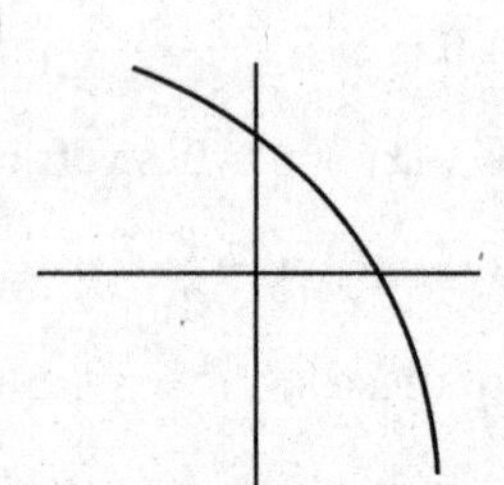

17. For which curve shown in question 16 is f'' positive but f' negative?

In Questions 18–21, the position of a particle moving along a horizontal line is given by $s = t^3 - 6t^2 + 12t - 8$.

18. The object is moving to the right for

(A) $t < 2$ (B) all t except $t = 2$ (C) $1 < t < 3$
(D) $t < 1$ or $t > 3$ (E) $t > 2$

19. The minimum value of the speed is

(A) 0 (B) 1 (C) 2 (D) 3 (E) 8

20. The acceleration is positive

(A) when $t > 2$ (B) for all t, $t \neq 2$ (C) when $t < 2$
(D) for $1 < t < 3$ (E) for $1 < t < 2$

21. The speed of the particle is decreasing for

(A) $t < 2$ (B) $t < 3$ (C) all t
(D) $t < 1$ or $t > 2$ (E) $t > 2$

In Questions 22–24, a particle moves along a horizontal line and its position at time t is $s = t^4 - 6t^3 + 12t^2 + 3$.

22. The particle is at rest when t is equal to

(A) 1 or 2 (B) 0 (C) $\frac{9}{4}$ (D) 0, 1, or 2 (E) 0, 2, or 3

23. The velocity, v, is increasing when

(A) $t > 1$ (B) $1 < t < 2$ (C) $t < 2$
(D) $t < 1$ or $t > 2$ (E) $t > 0$

24. The speed of the particle is increasing for

(A) $0 < t < 1$ or $t > 2$ (B) $1 < t < 2$ (C) $t < 2$
(D) $t < 0$ or $t > 2$ (E) $t < 0$

25. The displacement from the origin of a particle moving on a line is given by $s = t^4 - 4t^3$. The maximum displacement during the time interval $-2 \leqq t \leqq 4$ is

(A) 27 (B) 3 (C) $12\sqrt{3} + 3$
(D) 48 (E) 16

26. If a particle moves along a line according to the law $s = t^5 + 5t^4$, then the number of times it reverses direction is

(A) 0 (B) 1 (C) 2 (D) 3 (E) 4

BC ONLY

In Questions 27–30, $\mathbf{R} = \left\langle 3 \cos \frac{\pi}{3}t, 2 \sin \frac{\pi}{3}t \right\rangle$ is the (position) vector $\langle x, y \rangle$ from the origin to a moving point $P(x, y)$ at time t.

27. A single equation in x and y for the path of the point is

(A) $x^2 + y^2 = 13$ (B) $9x^2 + 4y^2 = 36$ (C) $2x^2 + 3y^2 = 13$
(D) $4x^2 + 9y^2 = 1$ (E) $4x^2 + 9y^2 = 36$

28. When $t = 3$, the speed of the particle is

(A) $\frac{2\pi}{3}$ (B) 2 (C) 3 (D) π (E) $\frac{\sqrt{13}}{3}\pi$

29. The magnitude of the acceleration when $t = 3$ is

(A) 2 (B) $\frac{\pi^2}{3}$ (C) 3 (D) $\frac{2\pi^2}{9}$ (E) π

30. At the point where $t = \frac{1}{2}$, the slope of the curve along which the particle moves is

(A) $-\frac{2\sqrt{3}}{9}$ (B) $-\frac{\sqrt{3}}{2}$ (C) $\frac{2}{\sqrt{3}}$ (D) $-\frac{2\sqrt{3}}{3}$ (E) 2

31. A balloon is being filled with helium at the rate of 4 ft^3/min. The rate, in square feet per minute, at which the surface area is increasing when the volume is $\frac{32\pi}{3}$ ft^3 is

(A) 4π (B) 2 (C) 4 (D) 1 (E) 2π

32. A circular conical reservoir, vertex down, has depth 20 ft and radius at the top 10 ft. Water is leaking out so that the surface is falling at the rate of $\frac{1}{2}$ ft/hr. The rate, in cubic feet per hour, at which the water is leaving the reservoir when the water is 8 ft deep is

(A) 4π (B) 8π (C) 16π (D) $\frac{1}{4\pi}$ (E) $\frac{1}{8\pi}$

33. A local minimum value of the function $y = \frac{e^x}{x}$ is

(A) $\frac{1}{e}$ (B) 1 (C) -1 (D) e (E) 0

CHALLENGE

34. The area of the largest rectangle that can be drawn with one side along the x-axis and two vertices on the curve of $y = e^{-x^2}$ is

(A) $\sqrt{\frac{2}{e}}$ (B) $\sqrt{2e}$ (C) $\frac{2}{e}$ (D) $\frac{1}{\sqrt{2e}}$ (E) $\frac{2}{e^2}$

CHALLENGE

35. A line is drawn through the point (1, 2) forming a right triangle with the positive x- and y-axes. The slope of the line forming the triangle of least area is

(A) -1 (B) -2 (C) -4 (D) $-\frac{1}{2}$ (E) -3

36. The point(s) on the curve $x^2 - y^2 = 4$ closest to the point (6, 0) is (are)

(A) (2, 0) (B) $(\sqrt{5}, \pm 1)$ (C) $(3, \pm\sqrt{5})$

(D) $(\sqrt{13}, \pm\sqrt{3})$ (E) none of these

37. The sum of the squares of two positive numbers is 200; their minimum product is

(A) 100 (B) $25\sqrt{7}$ (C) 28 (D) $24\sqrt{14}$ (E) none of these

38. The first-quadrant point on the curve $y^2x = 18$ that is closest to the point (2, 0) is

(A) (2, 3) (B) $(6, \sqrt{3})$ (C) $(3, \sqrt{6})$ (D) $(1, 3\sqrt{2})$ (E) none of these

39. If h is a small negative number, then the local linear approximation for $\sqrt[3]{27 + h}$ is

(A) $3 + \frac{h}{27}$ (B) $3 - \frac{h}{27}$ (C) $\frac{h}{27}$ (D) $-\frac{h}{27}$ (E) $3 - \frac{h}{9}$

40. If $f(x) = xe^{-x}$, then at $x = 0$

(A) f is increasing (B) f is decreasing (C) f has a relative maximum
(D) f has a relative minimum (E) f' does not exist

41. A function f has a derivative for each x such that $|x| < 2$ and has a local minimum at $(2, -5)$. Which statement below must be true?

(A) $f'(2) = 0$.
(B) f' exists at $x = 2$.
(C) The graph is concave up at $x = 2$.
(D) $f'(x) < 0$ if $x < 2$, $f'(x) > 0$ if $x > 2$.
(E) None of the preceding is necessarily true.

42. The height of a rectangular box is 10 in. Its length increases at the rate of 2 in./sec; its width decreases at the rate of 4 in./sec. When the length is 8 in. and the width is 6 in., the rate, in cubic inches per second, at which the volume of the box is changing is

(A) 200 (B) 80 (C) -80 (D) -200 (E) -20

43. The tangent to the curve $x^3 + x^2y + 4y = 1$ at the point $(3, -2)$ has slope

(A) -3 (B) $-\frac{23}{9}$ (C) $-\frac{27}{13}$ (D) $-\frac{11}{9}$ (E) $-\frac{15}{13}$

44. If $f(x) = ax^4 + bx^2$ and $ab > 0$, then

(A) the curve has no horizontal tangents
(B) the curve is concave up for all x
(C) the curve is concave down for all x
(D) the curve has no inflection point
(E) none of the preceding is necessarily true

45. A function f is continuous and differentiable on the interval [0,4], where f' is positive but f'' is negative. Which table could represent points on f?

(A)

x	0	1	2	3	4
y	10	12	14	16	18

(B)

x	0	1	2	3	4
y	10	12	15	19	24

(C)

x	0	1	2	3	4
y	10	18	24	28	30

(D)

x	0	1	2	3	4
y	30	28	24	18	10

(E)

x	0	1	2	3	4
y	24	19	15	12	10

46. The equation of the tangent to the curve with parametric equations $x = 2t + 1, y = 3 - t^3$ at the point where $t = 1$ is

BC ONLY

(A) $2x + 3y = 12$ (B) $3x + 2y = 13$ (C) $6x + y = 20$
(D) $3x - 2y = 5$ (E) none of these

47. Approximately how much less than 4 is $\sqrt[3]{63}$?

(A) $\frac{1}{48}$ (B) $\frac{1}{16}$ (C) $\frac{1}{3}$ (D) $\frac{2}{3}$ (E) 1

48. The best linear approximation for $f(x) = \tan x$ near $x = \frac{\pi}{4}$ is

(A) $1 + \frac{1}{2}\left(x - \frac{\pi}{4}\right)$ (B) $1 + \left(x - \frac{\pi}{4}\right)$ (C) $1 + \sqrt{2}\left(x - \frac{\pi}{4}\right)$

(D) $1 + 2\left(x - \frac{\pi}{4}\right)$ (E) $2 + 2\left(x - \frac{\pi}{4}\right)$

49. When h is near zero, e^{kh}, using the tangent-line approximation, is approximately

(A) k (B) kh (C) 1 (D) $1 + k$ (E) $1 + kh$

50. If $f(x) = cx^2 + dx + e$ for the function shown in the graph, then

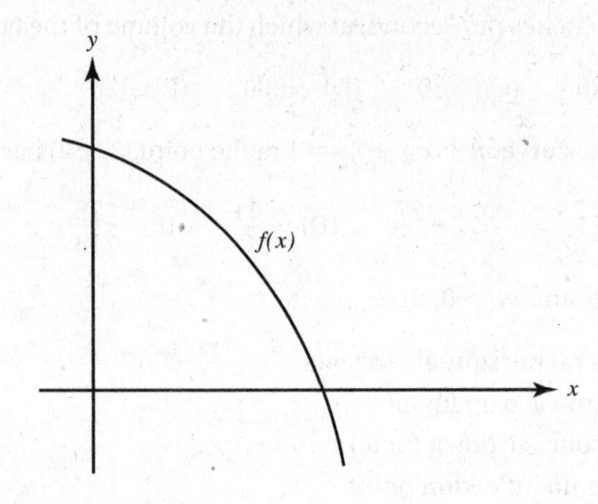

(A) c, d, and e are all positive
(B) $c > 0, d < 0, e < 0$
(C) $c > 0, d < 0, e > 0$
(D) $c < 0, d > 0, e > 0$
(E) $c < 0, d < 0, e > 0$

Part B. Directions: Some of the following questions require the use of a graphing calculator.

51. The point on the curve $y = \sqrt{2x + 1}$ at which the tangent is parallel to the line $x - 3y = 6$ is

(A) $(4, 3)$ (B) $(0, 1)$ (C) $(1, \sqrt{3})$
(D) $(4, -3)$ (E) $(2, \sqrt{5})$

52. The equation of the tangent to the curve $x^2 = 4y$ at the point on the curve where $x = -2$ is

(A) $x + y - 3 = 0$ (B) $y - 1 = 2x(x + 2)$ (C) $x - y + 3 = 0$
(D) $x + y - 1 = 0$ (E) $x + y + 1 = 0$

53. The table shows the velocity at time t of an object moving along a line. Estimate the acceleration (in ft/sec^2) at $t = 6$ sec.

t (sec)	0	4	8	10
vel.	18	16	10	0

(A) -6 (B) -1.8 (C) -1.5 (D) 1.5 (E) 6

Use the graph shown, sketched on [0, 7], for Questions 54–56.

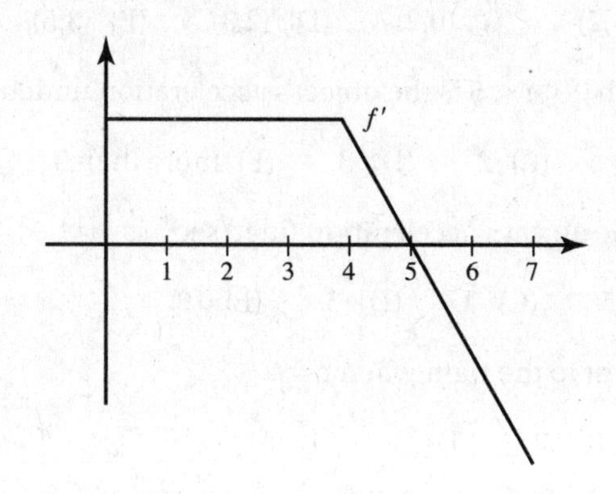

54. From the graph it follows that

(A) f is discontinuous at $x = 4$
(B) f is decreasing for $4 < x < 7$
(C) f is constant for $0 < x < 4$
(D) $f(5) < f(0)$
(E) $f(2) < f(3)$

55. Which statement best describes f at $x = 5$?

(A) f has a root. (B) f has a maximum. (C) f has a minimum.
(D) The graph of f has a point of inflection. (E) f is discontinuous.

56. For which interval is the graph of f concave downward?

(A) (0,4) (B) (4,5) (C) (5,7) (D) (4,7) (E) none of these

Use the graph shown for Questions 57–63. It shows the velocity of an object moving along a straight line during the time interval $0 \leq t \leq 5$.

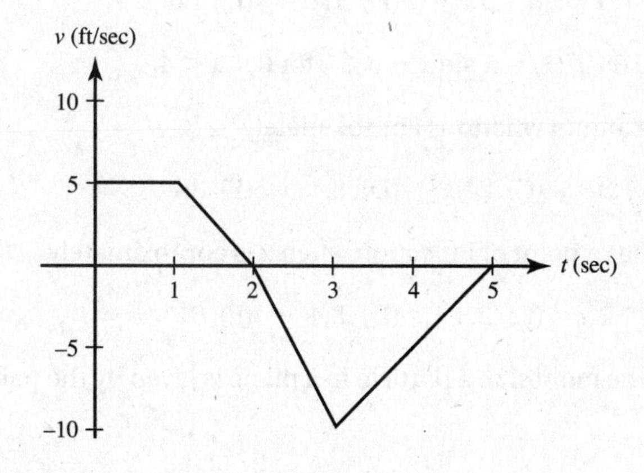

57. The object attains its maximum speed when $t =$

(A) 0 (B) 1 (C) 2 (D) 3 (E) 5

58. The speed of the object is increasing during the time interval

(A) (0,1) (B) (1,2) (C) (0,2) (D) (2,3) (E) (3,5)

59. The acceleration of the object is positive during the time interval

(A) (0,1) (B) (1,2) (C) (0,2) (D) (2,3) (E) (3,5)

60. How many times on $0 < t < 5$ is the object's acceleration undefined?

(A) none (B) 1 (C) 2 (D) 3 (E) more than 3

61. During $2 < t < 3$ the object's acceleration (in ft/sec^2) is

(A) -10 (B) -5 (C) 0 (D) 5 (E) 10

62. The object is farthest to the right when $t =$

(A) 0 (B) 1 (C) 2 (D) 3 (E) 5

63. The object's average acceleration (in ft/sec^2) for the interval $0 \leqslant t \leqslant 3$ is

(A) -15 (B) -5 (C) -3 (D) -1 (E) none of these

64. The line $y = 3x + k$ is tangent to the curve $y = x^3$ when k is equal to

(A) 1 or -1 (B) 0 (C) 3 or -3 (D) 4 or -4 (E) 2 or -2

65. The two tangents that can be drawn from the point (3,5) to the parabola $y = x^2$ have slopes

(A) 1 and 5 (B) 0 and 4 (C) 2 and 10

(D) 2 and $-\frac{1}{2}$ (E) 2 and 4

66. The table shows the velocity at various times of an object moving along a line. An estimate of its acceleration (in ft/sec^2) at $t = 1$ is

t (sec)	1.0	1.5	2.2	2.5
v (ft/sec)	12.2	13.0	13.4	13.7

(A) 0.8 (B) 1.0 (C) 1.2 (D) 1.4 (E) 1.6

For Questions 67 and 68, $f'(x) = x \sin x - \cos x$ for $0 < x < 4$.

67. f has a local maximum when x is approximately

(A) 0.9 (B) 1.2 (C) 2.3 (D) 3.4 (E) 3.7

68. The graph of f has a point of inflection when x is approximately

(A) 0.9 (B) 1.2 (C) 2.3 (D) 3.4 (E) 3.7

BC ONLY

In Questions 69–72, the motion of a particle in a plane is given by the pair of equations $x = 2t$ and $y = 4t - t^2$.

69. The particle moves along

(A) an ellipse (B) a circle (C) a hyperbola (D) a line (E) a parabola

70. The speed of the particle at any time t is

(A) $\sqrt{6 - 2t}$ (B) $2\sqrt{t^2 - 4t + 5}$ (C) $2\sqrt{t^2 - 2t + 5}$

(D) $\sqrt{8}(|t - 2|)$ (E) $2(|3 - t|)$

BC ONLY

71. The minimum speed of the particle is

(A) 2 (B) $2\sqrt{2}$ (C) 0 (D) 1 (E) 4

72. The acceleration of the particle

(A) depends on t
(B) is always directed upward
(C) is constant both in magnitude and in direction
(D) never exceeds 1 in magnitude
(E) is none of these

73. If a particle moves along a curve with constant speed, then

(A) the magnitude of its acceleration must equal zero
(B) the direction of acceleration must be constant
(C) the curve along which the particle moves must be a straight line
(D) its velocity and acceleration vectors must be perpendicular
(E) the curve along which the particle moves must be a circle

74. A particle is moving on the curve of $y = 2x - \ln x$ so that $\frac{dx}{dt} = -2$ at all times t. At the point (1,2), $\frac{dy}{dt}$ is

(A) 4 (B) 2 (C) -4 (D) 1 (E) -2

In Questions 75–76, a particle is in motion along the polar curve $r = 6 \cos 2\theta$ such that $\frac{d\theta}{dt} = \frac{1}{3}$ radian/sec when $\theta = \frac{\pi}{6}$.

75. At that point, find the rate of change (in units per second) of the particle's distance from the origin.

(A) $-6\sqrt{3}$ (B) $-2\sqrt{3}$ (C) $-\sqrt{3}$ (D) $2\sqrt{3}$ (E) $6\sqrt{3}$

76. At that point, what is the horizontal component of the particle's velocity?

(A) $-\frac{21}{2}$ (B) $-\frac{7}{2}$ (C) -2 (D) $\frac{1}{2}$ (E) $\frac{3}{2}$

Use the graph of f' on [0,5], shown below, for Questions 77 and 78.

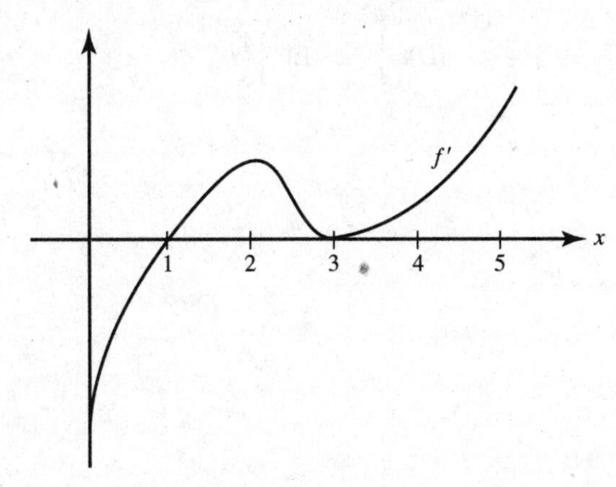

77. f has a local minimum at $x =$

(A) 0 (B) 1 (C) 2 (D) 3 (E) 5

78. The graph of f has a point of inflection at $x =$

(A) 1 only (B) 2 only (C) 3 only
(D) 2 and 3 only (E) 1, 2, and 3

79. It follows from the graph of f', shown at the right, that

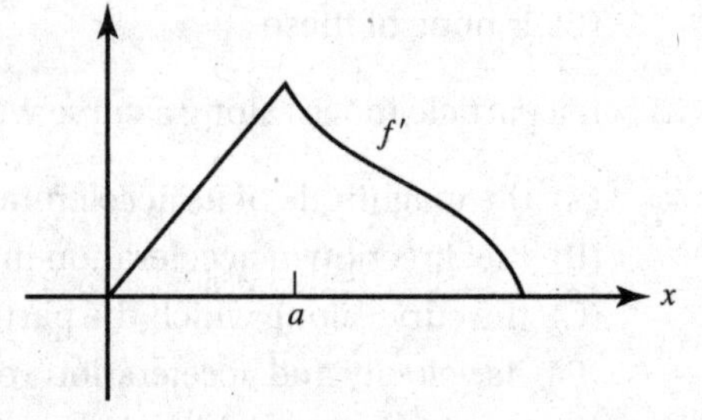

(A) f is not continuous at $x = a$
(B) f is continuous but not differentiable at $x = a$
(C) f has a relative maximum at $x = a$
(D) The graph of f has a point of inflection at $x = a$
(E) f is always concave upward

80. A vertical circular cylinder has radius r ft and height h ft. If the height and radius both increase at the constant rate of 2 ft/sec, then the rate, in square feet per second, at which the lateral surface area increases is

(A) $4\pi r$ (B) $2\pi(r + h)$ (C) $4\pi(r + h)$ (D) $4\pi rh$ (E) $4\pi h$

81. A tangent drawn to the parabola $y = 4 - x^2$ at the point (1, 3) forms a right triangle with the coordinate axes. The area of the triangle is

(A) $\frac{5}{4}$ (B) $\frac{5}{2}$ (C) $\frac{25}{2}$ (D) 1 (E) $\frac{25}{4}$

82. Two cars are traveling along perpendicular roads, car A at 40 mph, car B at 60 mph. At noon, when car A reaches the intersection, car B is 90 mi away, and moving toward it. At 1 P.M. the rate, in miles per hour, at which the distance between the cars is changing is

(A) -40 (B) 68 (C) 4 (D) -4 (E) 40

83. For Question 82, if t is the number of hours of travel after noon, then the cars are closest together when t is

(A) 0 (B) $\frac{27}{26}$ (C) $\frac{9}{5}$ (D) $\frac{3}{2}$ (E) $\frac{14}{13}$

The graph for Questions 84 and 85 shows the velocity of an object moving along a straight line during the time interval $0 \leqslant t \leqslant 12$.

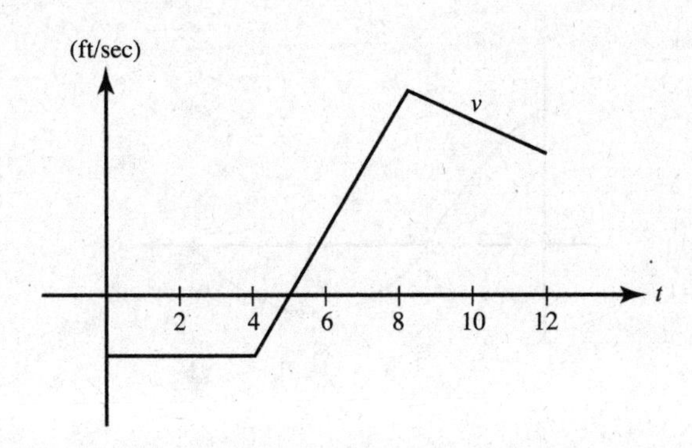

84. For what t does this object attain its maximum acceleration?

    (A) $0 < t < 4$ (B) $4 < t < 8$ (C) $t = 5$ (D) $t = 8$ (E) $t = 12$

85. The object reverses direction at $t =$

    (A) 4 only (B) 5 only (C) 8 only (D) 5 and 8 (E) none of these

86. The graph of f' is shown below. If we know that $f(2) = 10$, then the local linearization of f at $x = 2$ is $f(x) \simeq$

    (A) $\frac{x}{2} + 2$ (B) $\frac{x}{2} + 9$ (C) $3x - 3$ (D) $3x + 4$ (E) $10x - 17$

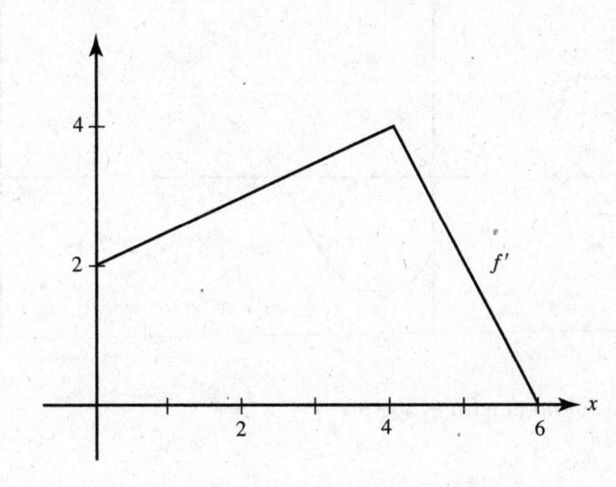

87. Given f' as graphed, which could be the graph of f?

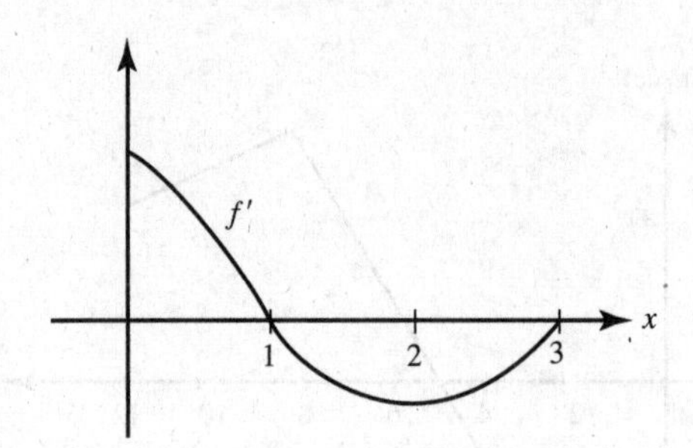

(A)

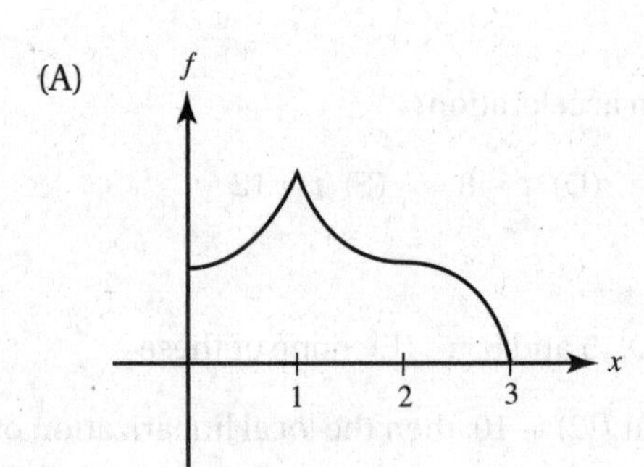

(B)

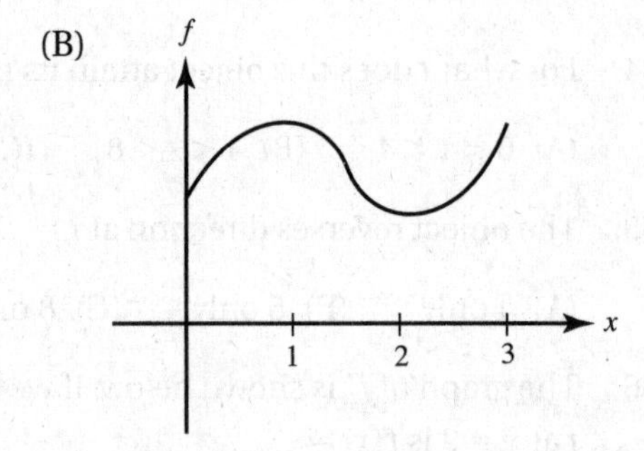

(C) (D)

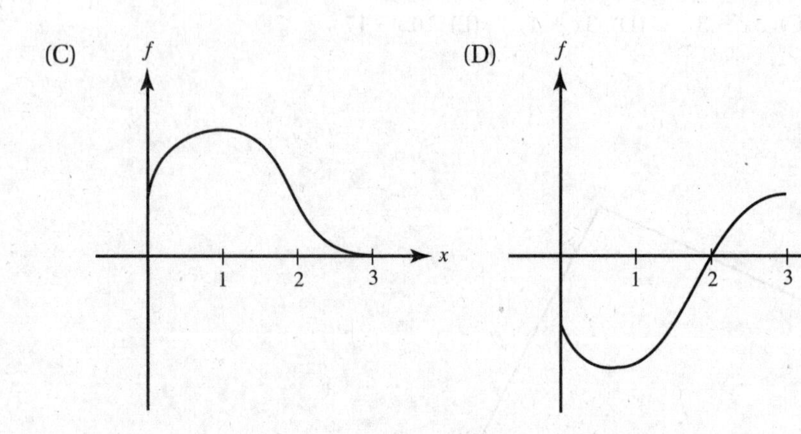

(E)

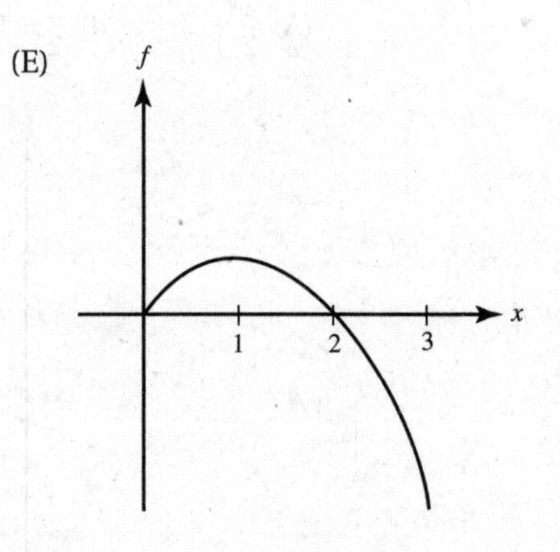

Use the following graph for Questions 88–90.

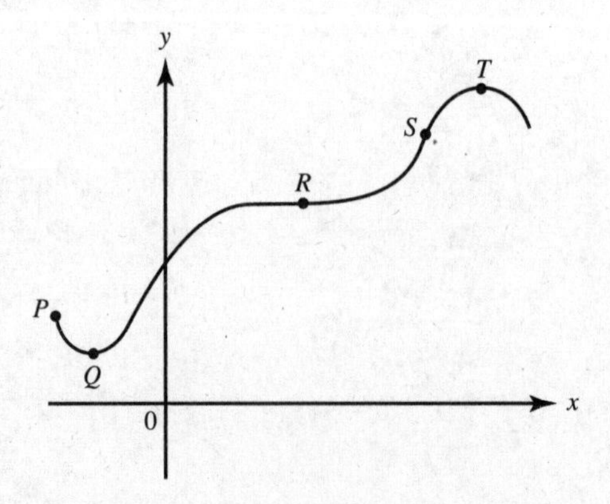

88. At which labeled point do both $\frac{dy}{dx}$ and $\frac{d^2y}{dx^2}$ equal zero?

(A) P (B) Q (C) R (D) S (E) T

89. At which labeled point is $\frac{dy}{dx}$ positive and $\frac{d^2y}{dx^2}$ equal to zero?

(A) P (B) Q (C) R (D) S (E) T

90. At which labeled point is $\frac{dy}{dx}$ equal to zero and $\frac{d^2y}{dx^2}$ negative?

(A) P (B) Q (C) R (D) S (E) T

91. If $f(6) = 30$ and $f'(x) = \frac{x^2}{x+3}$, estimate $f(6.02)$ using the line tangent to f at $x = 6$.

(A) 29.92 (B) 30.02 (C) 30.08
(D) 30.16 (E) 34.00

92. The local linear approximation for $f(x) = \sqrt{x^2 + 16}$ near $x = -3$ is

(A) $5 - \frac{3}{5}(x - 3)$ (B) $5 + \frac{3}{5}(x - 3)$ (C) $5 - \frac{3}{5}(x + 3)$

(D) $3 - \frac{5}{3}(x - 3)$ (E) $3 + \frac{3}{5}(x + 3)$

Answer Key

1. **D**	20. **A**	39. **A**	58. **D**	77. **B**
2. **A**	21. **A**	40. **A**	59. **E**	78. **D**
3. **E**	22. **B**	41. **E**	60. **D**	79. **D**
4. **B**	23. **D**	42. **D**	61. **A**	80. **C**
5. **D**	24. **A**	43. **E**	62. **C**	81. **E**
6. **C**	25. **D**	44. **D**	63. **B**	82. **D**
7. **D**	26. **C**	45. **C**	64. **E**	83. **B**
8. **D**	27. **E**	46. **B**	65. **C**	84. **B**
9. **C**	28. **A**	47. **A**	66. **E**	85. **B**
10. **B**	29. **B**	48. **D**	67. **D**	86. **D**
11. **D**	30. **D**	49. **E**	68. **C**	87. **C**
12. **E**	31. **C**	50. **E**	69. **E**	88. **C**
13. **C**	32. **B**	51. **A**	70. **B**	89. **D**
14. **A**	33. **D**	52. **E**	71. **A**	90. **E**
15. **B**	34. **A**	53. **C**	72. **C**	91. **C**
16. **E**	35. **B**	54. **E**	73. **D**	92. **C**
17. **B**	36. **C**	55. **B**	74. **E**	
18. **B**	37. **E**	56. **D**	75. **B**	
19. **E**	38. **C**	57. **D**	76. **B**	

ANSWERS EXPLAINED

1. **(D)** Substituting $y = 2$ yields $x = 1$. We find y' implicitly.

$$3y^2y' - (2xyy' + y^2) = 0; \quad (3y^2 - 2xy)y' - y^2 = 0.$$

Replace x by 1 and y by 2; solve for y'.

2. **(A)** $2yy' - (xy' + y) - 3 = 0$. Replace x by 0 and y by -1; solve for y'.

3. **(E)** Find the slope of the curve at $x = \frac{\pi}{2}$: $y' = x\cos x + \sin x$. At $x = \frac{\pi}{2}$, $y' = \frac{\pi}{2} \cdot 0 + 1$. The equation is $y - \frac{\pi}{2} = 1\left(x - \frac{\pi}{2}\right)$.

4. **(B)** Since $y' = e^{-x}(1 - x)$ and $e^{-x} > 0$ for all x, $y' = 0$ when $x = 1$.

5. **(D)** The slope $y' = 5x^4 + 3x^2 - 2$. Let $g = y'$. Since $g'(x) = 20x^3 + 6x = 2x(10x^2 + 3)$, $g'(x) = 0$ only if $x = 0$. Since $g''(x) = 60x^2 + 6$, g'' is always positive, assuring that $x = 0$ yields the minimum slope. Find y' when $x = 0$.

6. **(C)** Since $2x - 2yy' = 0$, $y' = \frac{x}{y}$. At (4, 2), $y' = 2$. The equation of the tangent at (4, 2) is $y - 2 = 2(x - 4)$.

7. **(D)** Since $y' = \frac{y}{2y - x}$, the tangent is vertical for $x = 2y$. Substitute in the given equation and solve for y.

8. **(D)** Since $V = \frac{4}{3}\pi r^3$, therefore, $dV = 4\pi r^2 dr$. The approximate increase in volume is $dV \approx 4\pi(3^2)(0.1)$ in^3.

9. **(C)** Differentiating implicitly yields $4x - 3y^2y' = 0$. So $y' = \frac{4x}{3y^2}$. The linear approximation for the true value of y when x changes from 3 to 3.04 is

$$y_{\text{at } x = 3} + y'_{\text{at point }(3,2)} \cdot (3.04 - 3).$$

Since it is given that, when $x = 3$, $y = 2$, the approximate value of y is

$$2 + \frac{4x}{3y^2}_{\text{at}(3,2)} \cdot (0.04)$$

or

$$2 + \frac{12}{12} \cdot (0.04) = 2.04.$$

10. **(B)** We want to approximate the change in area of the square when a side of length e increases by $0.01e$. The answer is

$$A'(e)(0.01e) \quad \text{or} \quad 2e(0.01e).$$

11. **(D)** Since $V = e^3$, $V' = 3e^2$. Therefore at $e = 10$, the slope of the tangent line is 300. The change in volume is approximately $300(\pm 0.1) = 30$ in.^{3}

12. **(E)** $f'(x) = 4x^3 - 8x = 4x(x^2 - 2)$. $f' = 0$ if $x = 0$ or $\pm\sqrt{2}$.

$f''(x) = 12x^2 - 8$; f'' is positive if $x = \pm\sqrt{2}$, negative if $x = 0$.

13. **(C)** Since $f''(x) = 4(3x^2 - 2)$, it equals 0 if $x = \pm\sqrt{\frac{2}{3}}$. Since f'' changes sign from positive to negative at $x = -\sqrt{\frac{2}{3}}$ and from negative to positive at $x = +\sqrt{\frac{2}{3}}$ both locate inflection points.

14. **(A)** The domain of y is $\{x \mid x \leq 2\}$. Note that y is negative for each x in the domain except 2, where $y = 0$.

15. **(B)** $f'(x)$ changes sign (from negative to positive) as x passes through zero only.

16. **(E)** The graph must be decreasing and concave downward.

17. **(B)** The graph must be concave upward but decreasing.

18. **(B)** The object moves to the right when v is positive. Since $v = \frac{ds}{dt} = 3(t-2)^2$, $v > 0$ for all $t \neq 2$.

19. **(E)** The speed $= |v|$. From Question 18, $|v| = v$. The least value of v is 0.

20. **(A)** The acceleration $a = \frac{dv}{dt}$. From Question 18, $a = 6(t-2)$.

21. **(A)** The speed is decreasing when v and a have opposite signs. The answer is $t < 2$, since for all such t the velocity is positive while the acceleration is negative. For $t > 2$, both v and a are positive.

22. **(B)** The particle is at rest when $v = 0$; $v = 2t(2t^2 - 9t + 12) = 0$ only if $t = 0$. Note that the discriminant of the quadratic factor $(b^2 - 4ac)$ is negative.

23. **(D)** Since $a = 12(t-1)(t-2)$, we check the signs of a in the intervals $t < 1$, $1 < t < 2$, and $t > 2$. We choose those where $a > 0$.

24. **(A)** From Questions 22 and 23 we see that $v > 0$ if $t > 0$ and that $a > 0$ if $t < 1$ or $t > 2$. So both v and a are positive if $0 < t < 1$ or $t > 2$. There are no values of t for which both v and a are negative.

25. **(D)** See the figure, which shows the motion of the particle during the time interval $-2 \leq t \leq 4$. The particle is at rest when $t = 0$ or 3, but reverses direction only at 3. The endpoints need to be checked here, of course. Indeed, the maximum displacement occurs at one of those, namely, when $t = -2$.

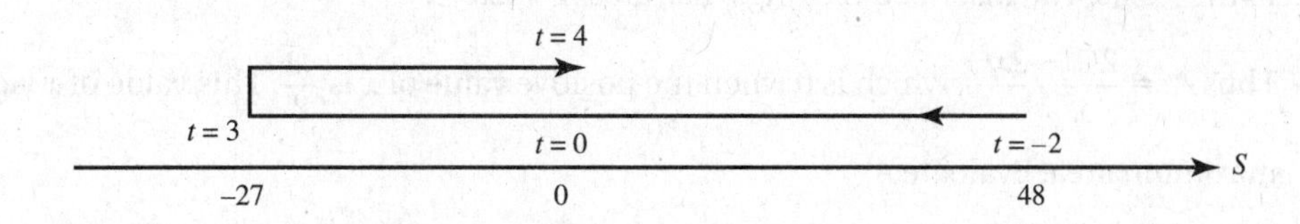

26. **(C)** Since $v = 5t^3(t+4)$, $v = 0$ when $t = -4$ or 0. Note that v does change sign at each of these times.

27. **(E)** Since $x = 3\cos\frac{\pi}{3}t$ and $y = 2\sin\frac{\pi}{3}t$, we note that $\left(\frac{x}{3}\right)^2 + \left(\frac{y}{2}\right)^2 = 1$.

28. **(A)** Note that $\mathbf{v} = \left\langle -\pi \sin\frac{\pi}{3}t, \frac{2\pi}{3}\cos\frac{\pi}{3}t \right\rangle$. At $t = 3$,

$$|\mathbf{v}| = \sqrt{(-\pi \cdot 0)^2 + \left(\frac{2\pi}{3} \cdot -1\right)^2}.$$

29. **(B)** $\mathbf{a} = \left\langle -\frac{\pi^2}{3}\cos\frac{\pi}{3}t, \frac{2\pi^2}{9}\sin\frac{\pi}{3}t \right\rangle$. At $t = 3$,

$$|\mathbf{a}| = \sqrt{\left(\frac{-\pi^2}{3} \cdot -1\right)^2 + \left(\frac{-2\pi^2}{9} \cdot 0\right)^2}.$$

30. **(D)** The slope of the curve is the slope of **v**, namely, $\frac{\frac{dy}{dt}}{\frac{dx}{dt}}$. At $t = \frac{1}{2}$, the slope is equal to

$$\frac{\frac{2\pi}{3} \cdot \cos\frac{\pi}{6}}{-\pi \cdot \sin\frac{\pi}{6}} = -\frac{2}{3}\cot\frac{\pi}{6}.$$

31. **(C)** Since $V = \frac{4}{3}\pi r^3, \frac{dV}{dt} = 4\pi r^2 \frac{dr}{dt}$. Since $\frac{dV}{dt} = 4, \frac{dr}{dt} = \frac{1}{\pi r^2}$. When $V = \frac{32\pi}{3}$, $r = 2$ and $\frac{dr}{dt} = \frac{1}{4\pi}$.

$$S = 4\pi r^2; \frac{dS}{dt} = 8\pi r\frac{dr}{dt};$$

when $r = 2, \frac{ds}{dt} = 8\pi(2)\left(\frac{1}{4\pi}\right) = 4.$

32. **(B)** See Figure N4–22 on page 186. Replace the printed measurements of the radius and height by 10 and 20, respectively. We are given here that $r = \frac{h}{2}$ and that $\frac{dh}{dt} = -\frac{1}{2}$. Since $V = \frac{1}{3}\pi r^2 h$, we have $V = \frac{\pi}{3}\frac{h^3}{4}$, so

$$\frac{dV}{dt} = \frac{\pi h^2}{4}\frac{dh}{dt} = \frac{-\pi h^2}{8}.$$

Replace h by 8.

33. **(D)** $y' = \frac{e^x(x-1)}{x^2}$ $(x \neq 0)$. Since $y' = 0$ if $x = 1$ and changes from negative to positive as x increases through 1, $x = 1$ yields a minimum. Evaluate y at $x = 1$.

34. **(A)** The domain of y is $-\infty < x < \infty$. The graph of y, which is nonnegative, is symmetric to the y-axis. The inscribed rectangle has area $A = 2xe^{-x^2}$.

Thus $A' = \frac{2(1-2x^2)}{e^{x^2}}$, which is 0 when the positive value of x is $\frac{\sqrt{2}}{2}$. This value of x yields maximum area. Evaluate A.

35. **(B)** See the figure. If we let m be the slope of the line, then its equation is $y - 2 = m(x - 1)$ with intercepts as indicated in the figure.

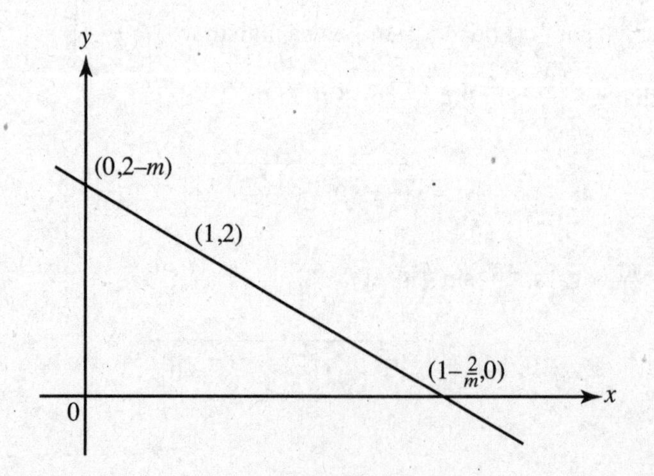

The area A of the triangle is given by

$$A = \frac{1}{2}(2 - m)\left(1 - \frac{2}{m}\right) = \frac{1}{2}\left(4 - \frac{4}{m} - m\right).$$

Then $\frac{dA}{dm} = \frac{1}{2}\left(\frac{4}{m^2} - 1\right)$ and equals 0 when $m = \pm 2$; m must be negative.

36. **(C)** Let $q = (x - 6)^2 + y^2$ be the quantity to be minimized. Then

$$q = (x - 6)^2 + (x^2 - 4);$$

$q' = 0$ when $x = 3$. Note that it suffices to minimize the square of the distance.

37. **(E)** Minimize, if possible, xy, where $x^2 + y^2 = 200$ $(x, y > 0)$. The derivative of the product is $\frac{2(100 - x^2)}{\sqrt{200 - x^2}}$, which equals 0 for $x = 10$. The derivative is positive to the left of that point and negative to the right, showing that $x = 10$ yields a maximum product. No minimum exists.

38. **(C)** Minimize $q = (x - 2)^2 + \frac{18}{x}$. Since

$$q' = 2(x - 2) - \frac{18}{x^2} = \frac{2(x^3 - 2x^2 - 9)}{x^2},$$

$q' = 0$ if $x = 3$. Since q' is negative to the left of $x = 3$ and positive to the right, the minimum value of q occurs at $x = 3$.

39. **(A)** The best approximation for $\sqrt[3]{27 + h}$ when h is small is the local linear (or tangent line) approximation. If we let $f(h) = \sqrt[3]{27 + h}$, then $f'(h) = \frac{1}{3(27 + h)^{2/3}}$ and $f'(0) = \frac{1}{3 \cdot 9}$. The approximation for $f(h)$ is $f(0) + f'(0) \cdot h$, which equals $3 + \frac{1}{27}h$.

40. **(A)** Since $f'(x) = e^{-x}(1 - x)$, $f'(0) > 0$.

41. **(E)** The graph shown serves as a counterexample for A–D.

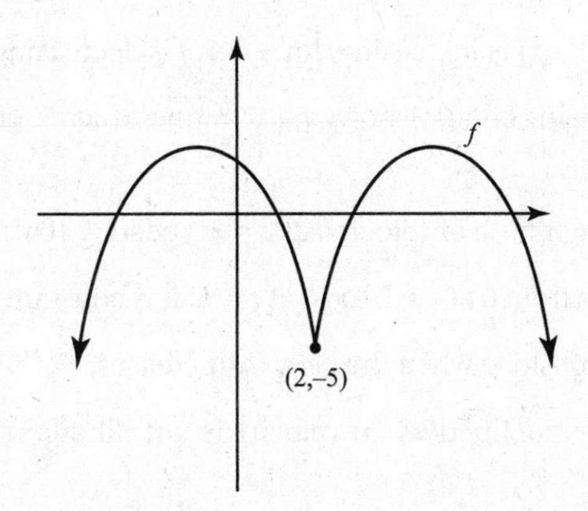

42. **(D)** Since $V = 10\ell w$, $V' = 10\left(\ell\frac{dw}{dt} + w\frac{d\ell}{dt}\right) = 10(8 \cdot -4 + 6 \cdot 2)$.

43. **(E)** We differentiate implicitly: $3x^2 + x^2y' + 2xy + 4y' = 0$. Then $y' = -\frac{3x^2 + 2xy}{x^2 + 4}$. At $(3, -2)$, $y' = -\frac{27 - 12}{9 + 4} = -\frac{15}{13}$.

44. **(D)** Since $ab > 0$, a and b have the same sign; therefore $f''(x) = 12ax^2 + 2b$ never equals 0. The curve has one horizontal tangent at $x = 0$.

45. **(C)** Since the first derivative is positive, the function must be increasing. However, the negative second derivative indicates that the rate of increase is slowing down, as seen in table C.

46. **(B)** Since $\frac{dy}{dx} = -\frac{3t^2}{2}$, therefore, at $t = 1$, $\frac{dy}{dx} = -\frac{3}{2}$. Also, $x = 3$ and $y = 2$.

47. **(A)** Let $f(x) = x^{1/3}$, and find the slope of the tangent line at (64, 4). Since $f'(x) = \frac{1}{3}x^{-2/3}$, $f'(64) = \frac{1}{48}$. If we move one unit to the left of 64, the tangent line will drop approximately $\frac{1}{48}$ unit.

48. **(D)** $\tan x \simeq \tan\left(\frac{\pi}{4}\right) + \sec^2\left(\frac{\pi}{4}\right)\left(x - \frac{\pi}{4}\right) = 1 + 2\left(x - \frac{\pi}{4}\right)$

49. **(E)** $e^{kh} \simeq e^{k \cdot 0} + ke^{k \cdot 0}(h - 0) = 1 + kh$

50. **(E)** Since the curve has a positive y-intercept, $e > 0$. Note that $f'(x) = 2cx + d$ and $f''(x) = 2c$. Since the curve is concave down, $f''(x) < 0$, implying that $c < 0$. Since the curve is decreasing at $x = 0$, $f'(0)$ must be negative, implying, since $f'(0) = d$, that $d < 0$. Therefore $c < 0$, $d < 0$, and $e > 0$.

51. **(A)** $y' = \frac{1}{\sqrt{2x+1}}$. Solving the equation of the line for y yields $y = \frac{1}{3}x - 2$, so to be parallel the slope of the tangent must also be $\frac{1}{3}$. If $\frac{1}{\sqrt{2x+1}} = \frac{1}{3}$, then $2x + 1 = 9$.

52. **(E)** The slope $y' = \frac{2x}{4}$; at the given point $y' = -\frac{4}{4} = -1$ and $y = 1$. The equation is therefore

$$y - 1 = -1(x + 2) \qquad \text{or} \qquad x + y + 1 = 0.$$

53. **(C)** $a \simeq \frac{\Delta v}{\Delta t} = \frac{v(8) - v(4)}{8 - 4} = \frac{10 - 16}{4}$ ft/sec^2.

54. **(E)** Because $f'(x) > 0$ on the interval $2 < x < 3$, the value of f increases.

55. **(B)** For $x < 5$, $f' > 0$, so f is increasing; for $x > 5$, f is decreasing.

56. **(D)** The graph of f being concave downward implies that $f'' < 0$, which implies that f' is decreasing.

57. **(D)** Speed is the magnitude of velocity; at $t = 3$, speed $= 10$ ft/sec.

58. **(D)** Speed increases from 0 at $t = 2$ to 10 at $t = 3$; it is constant or decreasing elsewhere.

59. **(E)** Acceleration is positive when the *velocity* increases.

60. **(D)** Acceleration is undefined when velocity is not differentiable. Here that occurs at $t = 1, 2, 3$.

61. **(A)** Acceleration is the derivative of velocity. Since the velocity is linear, its derivative is its slope.

62. **(C)** Positive velocity implies motion to the right ($t < 2$); negative velocity ($t > 2$) implies motion to the left.

63. **(B)** The average rate of change of velocity is $\frac{v(3) - v(0)}{3 - 0} = \frac{-10 - 5}{-3}$ ft/sec^2.

64. **(E)** The slope of $y = x^3$ is $3x^2$. It is equal to 3 when $x = \pm 1$. At $x = 1$, the equation of the tangent is

$$y - 1 = 3(x - 1) \quad \text{or} \quad y = 3x + 2.$$

At $x = -1$, the equation is

$$y + 1 = 3(x + 1) \quad \text{or} \quad y = 3x + 2.$$

65. **(C)** Let the tangent to the parabola from (3, 5) meet the curve at (x_1, y_1). Its equation is $y - 5 = 2x_1(x - 3)$. Since the point (x_1, y_1) is on both the tangent and the parabola, we solve simultaneously:

$$y_1 - 5 = 2x_1(x_1 - 3) \quad \text{and} \quad y_1 = x_1^2$$

The points of tangency are (5, 25) and (1, 1). The slopes, which equal $2x_1$, are 10 and 2.

66. **(E)** $a \simeq \dfrac{\Delta v}{\Delta t} = \dfrac{v(1.5) - v(1.0)}{0.5} = \dfrac{13.2 - 12.2}{0.5}$ ft/sec^2.

67. **(D)** The graph of $f'(x) = x \sin x - \cos x$ is drawn here in the window $[0,4] \times [-3,3]$:

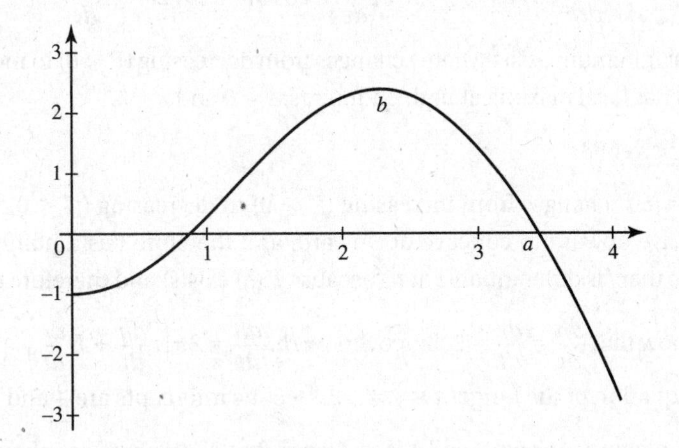

A local maximum exists at $x = 0$, where f' changes from positive to negative; use your calculator to approximate a.

68. **(C)** f'' changes sign when f' changes from increasing to decreasing (or vice versa). Again, use your calculator to approximate the x-coordinate at b.

69. **(E)** Eliminating t yields the equation $y = -\dfrac{1}{4}x^2 + 2x$.

70. **(B)** $|\mathbf{v}| = \sqrt{\left(\dfrac{dx}{dt}\right)^2 + \left(\dfrac{dy}{dt}\right)^2} = \sqrt{2^2 + (4 - 2t)^2}$.

71. **(A)** Since $|\mathbf{v}| = 2\sqrt{t^2 - 4t + 5}$, $\dfrac{d|\mathbf{v}|}{dt} = \dfrac{2t - 4}{\sqrt{t^2 - 4t + 5}} = 0$ if $t = 2$. We note that, as t increases through 2, the sign of $|\mathbf{v}|'$ changes from negative to positive, assuring a minimum of $|\mathbf{v}|$ at $t = 2$. Evaluate $|\mathbf{v}|$ at $t = 2$.

72. **(C)** The direction of **a** is $\tan^{-1}\dfrac{\frac{d^2y}{dt^2}}{\frac{d^2x}{dt^2}}$. Since $\dfrac{d^2x}{dt^2} = 0$ and $\dfrac{d^2y}{dt^2} = -2$, the acceleration is always directed downward. Its magnitude, $\sqrt{0^2 + (-2)^2}$, is 2 for all t.

73. **(D)** Using the notations v_x, v_y, a_x, and a_y, we are given that $|\mathbf{v}| = \sqrt{v_x^2 + v_y^2} = k$, where k is a constant. Then

$$\frac{d|\mathbf{v}|}{dt} = \frac{v_x a_x + v_y a_y}{|\mathbf{v}|} = 0 \quad \text{or} \quad \frac{v_x}{v_y} = -\frac{a_y}{a_x}.$$

74. **(E)** $\dfrac{dy}{dt} = \left(2 - \dfrac{1}{x}\right)\dfrac{dx}{dt} = \left(2 - \dfrac{1}{x}\right)(-2).$

75. **(B)** The rate of change of the distance from the origin with respect to time is given by $\dfrac{dr}{dt} = -6 \sin 2\theta \cdot 2 \cdot \dfrac{d\theta}{dt}$.

76. **(B)** In parametric form, $x = r\cos\theta = 6\cos 2\theta \cos\theta$; hence:

$$\frac{dx}{dt} = 6\cos 2\theta(-\sin\theta)\frac{d\theta}{dt} + \cos\theta(-6\sin 2\theta \cdot 2)\frac{d\theta}{dt}$$

77. **(B)** A local minimum exists where f changes from decreasing ($f' < 0$) to increasing ($f' > 0$). Note that f has local maxima at both endpoints, $x = 0$ and $x = 5$.

78. **(D)** See Answer 68.

79. **(D)** At $x = a$, f' changes from increasing ($f'' > 0$) to decreasing ($f'' < 0$). Thus f changes from concave upward to concave downward, and therefore has a point of inflection at $x = a$. Note that f is differentiable at a (because $f'(a)$ exists) and therefore continuous at a.

80. **(C)** We know that $\dfrac{dh}{dt} = \dfrac{dr}{dt} = 2$. Since $S = 2\pi rh$, $\dfrac{dS}{dt} = 2\pi\left(r\dfrac{dh}{dt} + h\dfrac{dr}{dt}\right)$.

81. **(E)** The equation of the tangent is $y = -2x + 5$. Its intercepts are $\dfrac{5}{2}$ and 5.

82. **(D)** See the figure. At noon, car A is at O, car B at N; the cars are shown t hours after noon. We know that $\dfrac{dx}{dt} = -60$ and that $\dfrac{dy}{dt} = 40$.

Using $s^2 = x^2 + y^2$, we get

$$\frac{ds}{dt} = \frac{x\frac{dx}{dt} + y\frac{dy}{dt}}{s} = \frac{-60x + 40y}{s}.$$

At 1 P.M., $x = 30$, $y = 40$, and $s = 50$.

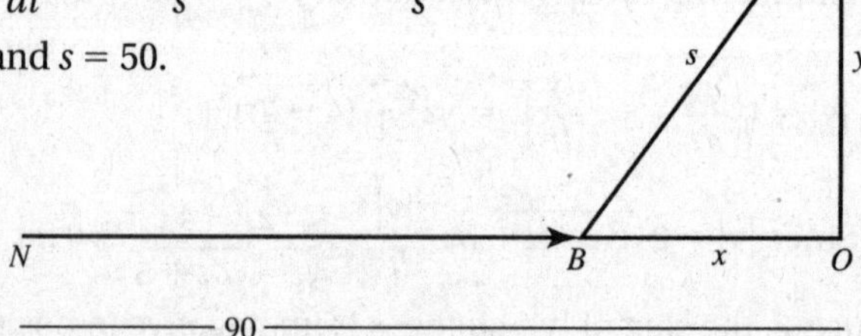

83. **(B)** $\dfrac{ds}{dt}$ (from Question 82) is zero when $y = \dfrac{3}{2}x$. Note that $x = 90 - 60t$ and $y = 40t$.

84. **(B)** Maximum acceleration occurs when the derivative (slope) of velocity is greatest.

85. **(B)** The object changes direction only when velocity changes sign. Velocity changes sign from negative to positive at $t = 5$.

86. **(D)** From the graph, $f'(2) = 3$, and we are told the line passes through (2,10). We therefore have $f(x) \simeq 10 + 3(x - 2) = 3x + 4$.

87. **(C)** At $x = 1$ and 3, $f'(x) = 0$; therefore f has horizontal tangents.

    For $x < 1$, $f' > 0$; therefore f is increasing.
    For $x > 1$, $f' < 0$, so f is decreasing.
    For $x < 2$, f' is decreasing, so $f'' < 0$ and the graph of f is concave downward.
    For $x > 2$, f' is increasing, so $f'' > 0$ and the graph of f is concave upward.

88. **(C)** Note that $\frac{dy}{dx} = 0$ at Q, R, and T. At Q, $\frac{d^2y}{dx^2} > 0$; at T, $\frac{d^2y}{dx^2} < 0$.

89. **(D)** Only at S does the graph both rise and change concavity.

90. **(E)** Only at T is the tangent horizontal and the curve concave down.

91. **(C)** Since $f'(6) = 4$, the equation of the tangent at (6, 30) is $y - 30 = 4(x - 6)$. Therefore $f(x) \simeq 4x + 6$ and $f(6.02) \simeq 30.08$.

92. **(C)** $\sqrt{x^2 + 16} \simeq \sqrt{9 + 16} + \frac{(-3)}{\sqrt{(-3)^2 + 16}}(x + 3) = 5 - \frac{3}{5}(x + 3)$

Antidifferentiation

5

CONCEPTS AND SKILLS

In this chapter, we review

- **indefinite integrals,**
- **formulas for antiderivatives of basic functions,**
- **and techniques for finding antiderivatives (including substitution).**

For BC Calculus students, we review two important techniques of integration:

- **integration by parts,**
- **and integration by partial fractions.**

A. ANTIDERIVATIVES

Antiderivative

The *antiderivative* or *indefinite integral* of a function $f(x)$ is a function $F(x)$ whose derivative is $f(x)$. Since the derivative of a constant equals zero, the antiderivative of $f(x)$ is not unique; that is, if $F(x)$ is an integral of $f(x)$, then so is $F(x) + C$, where C is any constant. The arbitrary constant C is called the *constant of integration.* The indefinite integral of $f(x)$ is written as $\int f(x)\,dx$; thus

Indefinite integral

$$\int f(x)\,dx = F(x) + C \quad \text{if} \quad \frac{dF(x)}{dx} = f(x).$$

The function $f(x)$ is called the *integrand.* The Mean Value Theorem can be used to show that, if two functions have the same derivative on an interval, then they differ at most by a constant; that is, if $\frac{dF(x)}{dx} = \frac{dG(x)}{dx}$, then

$$F(x) - G(x) = C \quad (C \text{ a constant}).$$

B. BASIC FORMULAS

Familiarity with the following fundamental integration formulas is essential.

$$\int kf(x)\,dx = k\int f(x)\,dx \qquad (k \neq 0) \tag{1}$$

$$\int [f(x) + g(x)]\,dx = \int f(x)\,dx + \int g(x)\,dx \tag{2}$$

$$\int u^n\,du = \frac{u^{n+1}}{n+1} + C \qquad (n \neq -1) \tag{3}$$

$$\int \frac{du}{u} = \ln|u| + C \tag{4}$$

$$\int \cos u \, du = \sin u + C \tag{5}$$

$$\int \sin u \, du = -\cos u + C \tag{6}$$

$$\int \tan u \, du = \ln\ |\sec u| + C \tag{7}$$

$$\text{or } -\ln\ |\cos u| + C$$

$$\int \cot u \, du = \ln\ |\sin u| + C \tag{8}$$

$$\text{or } -\ln\ |\csc u| + C$$

$$\int \sec^2 u \, du = \tan u + C \tag{9}$$

$$\int \csc^2 u \, du = -\cot u + C \tag{10}$$

$$\int \sec u \tan u \, du = \sec u + C \tag{11}$$

$$\int \csc u \cot u \, du = -\csc u + C \tag{12}$$

$$\int \sec u \, du = \ln\ |\sec u + \tan u| + C \tag{13}$$

$$\int \csc u \, du = \ln\ |\csc u - \cot u| + C \tag{14}$$

$$\int e^u \, du = e^u + C \tag{15}$$

$$\int a^u \, du = \frac{a^u}{\ln a} + C \qquad (a > 0, a \neq 1) \tag{16}$$

$$\int \frac{du}{\sqrt{1 - u^2}} = \sin^{-1} u + C \tag{17}$$

$$\text{or } \arcsin u + C$$

$$\int \frac{du}{1 + u^2} = \tan^{-1} u + C \tag{18}$$

$$\text{or } \arctan u + C$$

$$\int \frac{du}{u\sqrt{u^2 - 1}} = \sec^{-1} |u| + C \tag{19}$$

$$\text{or } \operatorname{arcsec} |u| + C$$

All the references in the following set of examples are to the preceding basic formulas. In all of these, whenever u is a function of x, we define du to be $u'(x)\, dx$; when u is a function of t, we define du to be $u'(t)\, dt$; and so on.

➡ Example 1

$$\int 5x\,dx = 5\int x\,dx \text{ by (1),}$$

$$= 5\left(\frac{x^2}{2}\right) + C \text{ by (3).}$$

➡ Example 2

$$\int \left(x^4 + \sqrt[3]{x^2} - \frac{2}{x^2} - \frac{1}{3\sqrt[3]{x}}\right)dx = \int \left(x^4 + x^{2/3} - 2x^{-2} - \frac{1}{3}x^{-1/3}\right) dx$$

$$= \int x^4\,dx + \int x^{2/3}\,dx - 2\int x^{-2}\,dx - \frac{1}{3}\int x^{-1/3}\,dx \text{ by (1) and (2)}$$

$$= \frac{x^5}{5} + \frac{x^{5/3}}{\frac{5}{3}} - \frac{2x^{-1}}{-1} - \frac{1}{3}\frac{x^{2/3}}{\frac{2}{3}} + C \text{ by (3)}$$

$$= \frac{x^5}{5} + \frac{3}{5}x^{5/3} + \frac{2}{x} - \frac{1}{2}x^{2/3} + C.$$

➡ Example 3

Similarly, $\int (3 - 4x + 2x^3)\,dx = \int 3\,dx - 4\int x\,dx + 2\int x^3\,dx$

$$= 3x - \frac{4x^2}{2} + \frac{2x^4}{4} + C$$

$$= 3x - 2x^2 + \frac{x^4}{2} + C.$$

➡ Example 4

$\int 2(1 - 3x)^2 dx$ is integrated most efficiently by using formula (3) with $u = 1 - 3x$ and $du = u'(x)\,dx = -3\,dx$.

$$2\int (1 - 3x)^2\,dx = \frac{2}{-3}\int (1 - 3x)^2(-3\,dx)$$

$$= -\frac{2}{9}u^3 + C \text{ by (3)}$$

$$= -\frac{2}{9}(1 - 3x)^3 + C.$$

➡ Example 5

$\int (2x^3 - 1)^5 \cdot x^2\,dx = \frac{1}{6}\int (2x^3 - 1)^5 \cdot (6x^2\,dx) = \frac{1}{6}\int u^5\,du$, where $u = 2x^3 - 1$ and $du = u'(x)\,dx = 6x^2\,dx$; this, by formula (3), equals

$$\frac{1}{6} \cdot \frac{u^6}{6} + C = \frac{1}{36}(2x^3 - 1)^6 + C.$$

➡ Example 6

$\int \sqrt[3]{1 - x}\,dx = \int (1 - x)^{1/3}\,dx = -\int (1 - x)^{1/3}(-1\,dx) = -\int u^{1/3}\,du$, where $u = 1 - x$ and $du = -1\,dx$; this, by formula (3) yields $-\frac{u^{4/3}}{\frac{4}{3}} + C = -\frac{3}{4}(1 - x)^{4/3} + C.$

Example 7

$$\int \frac{x}{\sqrt{3-4x^2}}\,dx = \int (3-4x^2)^{-1/2} \cdot (x\,dx) = -\frac{1}{8}\int (3-4x^2)^{-1/2}\,(-8x\,dx) = -\frac{1}{8}\int u^{-1/2}\,du$$

$$= -\frac{1}{8}\frac{u^{1/2}}{\frac{1}{2}} + C \text{ by (3)} = -\frac{1}{4}\sqrt{3-4x^2} + C.$$

Example 8

$$\int \frac{4x^2}{(x^3-1)^3}\,dx = 4\int (x^3-1)^{-3} \cdot x^2\,dx = \frac{4}{3}\int (x^3-1)^{-3}\,(3x^2\,dx)$$

$$= \frac{4}{3}\frac{(x^3-1)^{-2}}{-2} + C = -\frac{2}{3}\frac{1}{(x^3-1)^2} + C.$$

Example 9

$\int \frac{(1+\sqrt{x})^4}{\sqrt{x}}\,dx = \int (1+x^{1/2})^4 \cdot \frac{1}{x^{1/2}}\,dx$. Now let $u = 1 + x^{1/2}$, and note that $du = \frac{1}{2}x^{1/2}\,dx$; this gives $2\int (1+x^{1/2})^4\left(\frac{1}{2x^{1/2}}\,dx\right) = \frac{2}{5}(1+\sqrt{x})^5 + C.$

Example 10

$$\int (2-y)^2 \cdot \sqrt{y}\,dy = \int \left(4-4y+y^2\right) \cdot y^{1/2}\,dy = \int \left(4y^{1/2} - 4y^{3/2} + y^{5/2}\right)dy$$

$$= 4 \cdot \frac{2}{3}y^{3/2} - 4 \cdot \frac{2}{5} \cdot y^{5/2} + \frac{2}{7} \cdot y^{7/2} + C \text{ by (2)}$$

$$= \frac{8}{3}y^{3/2} + -\frac{8}{5}y^{5/2} + \frac{2}{7}y^{7/2} + C.$$

Example 11

$$\int \frac{x^3-x-4}{2x^2}\,dx = \frac{1}{2}\int \left(x - \frac{1}{x} - \frac{4}{x^2}\right)dx = \frac{1}{2}\left(\frac{x^2}{2} - \ln|x| + \frac{4}{x}\right) + C.$$

Example 12

$$\int \frac{3x-1}{\sqrt[3]{1-2x+3x^2}}\,dx = \int (1-2x+3x^2)^{-1/3}(3x-1)\,dx$$

$$= \frac{1}{2}\int (1-2x+3x^2)^{-1/3}(6x-2)\,dx \text{ or } \frac{1}{2} \cdot \frac{3}{2}(1-2x+3x^2)^{2/3} + C \text{ by (3)}$$

$$= \frac{3}{4}(1-2x+3x^2)^{2/3} + C.$$

Example 13

$$\int \frac{2x^2 - 4x + 3}{(x-1)^2}\,dx = \int \frac{2x^2 - 4x + 3}{x^2 - 2x + 1}\,dx = \int \left(2 + \frac{1}{(x-1)^2}\right) dx = \int 2\,dx + \frac{dx}{(x-1)^2} = 2x - \frac{1}{x-1} + C.$$

This example illustrates the following principle:

If the degree of the numerator of a rational function is not less than that of the denominator, divide until a remainder of lower degree is obtained.

Example 14

$\int \frac{du}{u-3} = \ln|u-3| + C$ by (4).

Example 15

$\int \frac{z\,dz}{1-4z^2} = -\frac{1}{8}\ln|1-4z^2| + C$ by (4).

Example 16

$\int \frac{\cos x}{5 + 2\sin x}\,dx = \frac{1}{2}\int \frac{2\cos x\,dx}{5 + 2\sin x} = \frac{1}{2}\ln(5 + 2\sin x) + C$ by (4) with $u = 5 + 2\sin x$. The absolute-value sign is not necessary here since $(5 + 2\sin x) > 0$ for all x.

Example 17

$\int \frac{e^x}{1-2e^x}\,dx = -\frac{1}{2}\int \frac{-2e^x}{1-2e^x}\,dx = -\frac{1}{2}\ln|1-2e^x| + C$ by (4).

Example 18

$\int \frac{x}{1-x}\,dx = \int \left(-1 + \frac{1}{1-x}\right) dx$ (by long division) $= -x - \ln|1-x| + C.$

Example 19

$\int \sin(1-2y)\,dy = -\frac{1}{2}\int \sin(1-2y)(-2\,dy)$

$= -\frac{1}{2}[-\cos(1-2y)] + C$ by (6) $= \frac{1}{2}\cos(1-2y) + C.$

Example 20

$\int \sin^2\frac{x}{2}\cos\frac{x}{2}\,dx = 2\int \left(\sin\frac{x}{2}\right)^2 \cos\frac{x}{2}\,\frac{dx}{2} = \frac{2}{3}\sin^3\frac{x}{2} + C$ by (3).

Example 21

$\int \frac{\sin x}{1 + 3\cos x}\,dx = -\frac{1}{3}\int \frac{-3\sin x\,dx}{1 + 3\cos x} = \frac{1}{3}\ln|1 + 3\cos x| + C.$

➡ Example 22

$\int e^{\tan y} \sec^2 y\, dy = e^{\tan y} + C$ by (15) with $u = \tan y$.

➡ Example 23

$\int e^x \tan e^x\, dx = \ln |\sec e^x| + C$ by (7) with $u = e^x$.

➡ Example 24

$\int \frac{\cos z}{\sin^2 z}\, dz = \int \csc z \cot z\, dz = -\csc z + C$ by (12).

➡ Example 25

$\int \tan t \sec^2 t\, dt = \frac{\tan^2 t}{2} + C$ by (3) with $u = \tan t$ and $du = u'(t)\, dt = \sec^2 t\, dt$.

➡ Example 26

(a) $\int \frac{dz}{\sqrt{9 - z^2}} = \frac{1}{3} \int \frac{dz}{\sqrt{1 - \left(\frac{z}{3}\right)^2}} = 3 \cdot \frac{1}{3} \int \frac{\frac{1}{3} dz}{\sqrt{1 - \left(\frac{z}{3}\right)^2}} = \sin^{-1} \frac{z}{3} + C$ by (17) with $u = \frac{z}{3}$.

(b) $\int \frac{z\, dz}{\sqrt{9 - z^2}} = -\frac{1}{2} \int (9 - z^2)^{-1/2}(-2z\, dz) = -\frac{1}{2} \frac{(9 - z^2)^{1/2}}{1/2} + C$ by (3) $= -\sqrt{9 - z^2} + C$

with $u = 9 - z^2, n = -\frac{1}{2}$

(c) $\int \frac{z dz}{9 - z^2} = -\frac{1}{2} \int \frac{(-2z\, dz)}{9 - z^2} = -\frac{1}{2} \ln |9 - z^2| + C$ by (4) with $u = 9 - z^2$.

(d) $\int \frac{z dz}{(9 - z^2)^2} = -\frac{1}{2} \int (9 - z^2)^{-2}(-2z\, dz) = \frac{1}{2(9 - z^2)} + C$ by (3).

(e) $\int \frac{dz}{9 + z^2} = \frac{1}{9} \int \frac{dz}{1 + \left(\frac{z}{3}\right)^2} = 3 \cdot \frac{1}{9} \int \frac{\frac{1}{3} dz}{1 + \left(\frac{z}{3}\right)^2} = \frac{1}{3} \tan^{-1} \frac{z}{3} + C$ by (18) with $u = \frac{z}{3}$.

➡ Example 27

$\int \frac{dx}{\sqrt{x}(1 + 2\sqrt{x})} = \int \frac{x^{-1/2}\, dx}{1 + 2\sqrt{x}} = \ln (1 + 2\sqrt{x}) + C$ by (4) with $u = 1 + 2\sqrt{x}$ and $du = \frac{dx}{\sqrt{x}}$.

➡ Example 28

$\int \sin x \cos x\, dx = \frac{1}{2} \sin^2 x + C$ by (3) with $u = \sin x$; OR $= -\frac{1}{2} \cos^2 x + C$ by (3) with $u = \cos x$;

OR $= -\frac{1}{4} \cos 2x + C$ by (6), where we use the trigonometric identity $\sin 2x = 2 \sin x \cos x$.

Example 29

$\int \frac{\cos\sqrt{x}}{\sqrt{x}}\,dx = 2\int \cos\left(x^{1/2}\right)\left(\frac{1}{2}x^{-1/2}\,dx\right) = 2\sin\sqrt{x} + C$ by (5) with $u = \sqrt{x}$.

Example 30

$\int \sin^2 y\,dy = \int \left(\frac{1}{2} - \frac{\cos 2y}{2}\right)dy = \frac{y}{2} - \frac{\sin 2y}{4} + C$, using the trigonometric identity $\sin^2\theta = \frac{1}{2}(1 - \cos 2\theta)$.

Example 31

$\int \frac{x\,dx}{x^4+1} = \frac{1}{2}\int \frac{2x\,dx}{1+\left(x^2\right)^2} = \frac{1}{2}\tan^{-1}x^2 + C$ by (18) with $u = x^2$.

Example 32

$$\int \frac{dx}{x^2+4x+13} = \int \frac{dx}{9+(x+2)^2} = \frac{1}{9}\int \frac{dx}{1+\left(\frac{x+2}{3}\right)^2} = 3\cdot\frac{1}{9}\int \frac{\frac{1}{3}dx}{1+\left(\frac{x+2}{3}\right)^2}$$

$$= \frac{1}{3}\tan^{-1}\frac{x+2}{3} + C \text{ by (18) with } u = \frac{x+2}{3}.$$

Example 33

$$\int \frac{dy}{\sqrt{6y-y^2}} = \int \frac{dy}{\sqrt{9-\left(y^2-6y+9\right)}} = \frac{1}{3}\int \frac{dy}{\sqrt{1-\left(\frac{y-3}{3}\right)^2}} = 3\cdot\frac{1}{3}\int \frac{\frac{1}{3}dy}{\sqrt{1-\left(\frac{y-3}{3}\right)^2}}$$

$$= \sin^{-1}\frac{y-3}{3} + C \text{ by (17) with } u = \frac{y-3}{3}.$$

Example 34

$$\int \frac{e^x}{9+e^{2x}}\,dx = \frac{1}{9}\int \frac{e^x\,dx}{1+\left(\frac{e^x}{3}\right)^2} = 3\cdot\frac{1}{9}\int \frac{\frac{e^x}{3}dx}{1+\left(\frac{e^x}{3}\right)^2} = \frac{1}{3}\tan^{-1}\frac{e^x}{3} + C \text{ by (18) with } u = \frac{e^x}{3}.$$

Example 35

$\int \frac{e^x - e^{-x}}{e^x + e^{-x}}\,dx = \ln\left(e^x + e^{-x}\right) + C$ by (4) with $u = e^x + e^{-x}$.

Example 36

$\int \frac{x+1}{x^2+1}\,dx = \int \frac{x\,dx}{x^2+1} + \int \frac{dx}{x^2+1} = \frac{1}{2}\ln\left(x^2+1\right) + \tan^{-1}x + C$ by (4) and (18).

Example 37

$$\int \frac{dt}{\sin^2 2t} = \int \csc^2 2t\,dt = \frac{1}{2}\int \csc^2 2t(2\,dt) = -\frac{1}{2}\cot 2t + C \text{ by (10).}$$

Example 38

$$\int \cos^2 4z\,dz = \int \left(\frac{1}{2} + \frac{\cos 8z}{2}\right) dz = \frac{z}{2} + \frac{\sin 8z}{16} + C, \text{ using the trig identity } \cos^2\theta = \frac{1}{2}(1 + \cos 2\theta).$$

Example 39

$$\int \frac{\sin 2x}{1 + \sin^2 x}\,dx = \ln(1 + \sin^2 x) + C \text{ by (4).}$$

Example 40

$$\int x^2 e^{-x^3} dx = -\frac{1}{3}\int e^{-x^3}(-3x^2\,dx) = -\frac{1}{3}e^{-x^3} + C \text{ by (15) with } u = -x^3.$$

Example 41

$$\int \frac{dy}{y\sqrt{1 + \ln y}} = \int (1 + \ln y)^{-1/2}\left(\frac{1}{y}\,dy\right) = 2\sqrt{1 + \ln y} + C \text{ by (3).}$$

BC ONLY

†C. INTEGRATION BY PARTIAL FRACTIONS

The method of partial fractions makes it possible to express a rational function $\frac{f(x)}{g(x)}$ as a sum of simpler fractions. Here $f(x)$ and $g(x)$ are real polynomials in x and it is assumed that $\frac{f(x)}{g(x)}$ is a proper fraction; that is, that $f(x)$ is of lower degree than $g(x)$. If not, we divide $f(x)$ by $g(x)$ to express the given rational function as the sum of a polynomial and a proper rational function. Thus,

$$\frac{x^3 - x^2 - 2}{x(x-1)} = x - \frac{2}{x(x-1)},$$

where the fraction on the right is proper.

Theoretically, every real polynomial can be expressed as a product of (powers of) real linear factors and (powers of) real quadratic factors.†

In the following, the capital letters denote constants to be determined. We consider only nonrepeating linear factors. For each distinct linear factor $(x - a)$ of $g(x)$ we set up one partial fraction of the type $\frac{A}{x-a}$. The techniques for determining the unknown constants are illustrated in the following examples.

†In the Topical Outline for Calculus BC, integration by partial fractions is restricted to "simple partial fractions (nonrepeating linear factors only)."

BC ONLY

Example 42

Find $\int \frac{x^2 - x + 4}{x^3 - 3x^2 + 2x}\,dx$.

SOLUTION: We factor the denominator and then set

$$\frac{x^2 - x + 4}{x(x-1)(x-2)} = \frac{A}{x} + \frac{B}{x-1} + \frac{C}{x-2}, \qquad (1)$$

where the constants A, B, and C are to be determined. It follows that

$$x^2 - x + 4 = A(x-1)(x-2) + Bx(x-2) + Cx(x-1). \qquad (2)$$

Since the polynomial on the right in (2) is to be identical to the one on the left, we can find the constants by either of the following methods:

METHOD ONE. We expand and combine the terms on the right in (2), getting

$$x^2 - x + 4 = (A + B + C)x^2 - (3A + 2B + C)x + 2A.$$

We then *equate coefficients of like powers in x* and solve simultaneously. Thus

using the coefficients of x^2, we get	$1 = A + B + C$;
using the coefficients of x, we get	$-1 = -(3A + 2B + C)$;
using the constant coefficient,	$4 = 2A$

These equations yield $A = 2$, $B = -4$, $C = 3$.

METHOD TWO. Although equation (1) above is meaningless for $x = 0$, $x = 1$, or $x = 2$, it is still true that equation (2) must hold even for these special values. We see, in (2), that

if $x = 0$,	then $4 = 2A$ and $A = 2$;
if $x = 1$,	then $4 = -B$ and $B = -4$;
if $x = 2$,	then $6 = 2C$ and $C = 3$.

The second method is shorter than the first and more convenient when the denominator of the given fraction can be decomposed into nonrepeating linear factors.
Finally, then, the original integral equals

$$\int \left(\frac{2}{x} - \frac{4}{x-1} + \frac{3}{x-2}\right) dx = 2\ln|x| - 4\ln|x-1| + 3\ln|x-2| + C'$$
$$= \ln\frac{x^2|x-2|^3}{(x-1)^4} + C'.$$

[The symbol "C'" appears here for the constant of integration because C was used in simplifying the original rational function.]

BC ONLY

D. INTEGRATION BY PARTS

Parts Formula

The Integration by Parts Formula stems from the equation for the derivative of a product:

$$\frac{d}{dx}(uv) = u\frac{dv}{dx} + v\frac{du}{dx} \quad \text{or, or more conveniently } d(uv) = u\,dv + v\,du.$$

Hence, $u\,dv = d(uv) - v\,du$ and integrating gives us $\int u\,dv = \int d(uv) - \int v\,du$, or

$$\int u\,dv = uv - \int v\,du,$$

the Integration by Parts Formula. Success in using this important technique depends on being able to separate a given integral into parts u and dv so that (a) dv can be integrated, and (b) $\int v\,du$ is no more difficult to calculate than the original integral.

The following acronym may help to determine which function to designate as u when using Integration by Parts. The acronym is LIPET. Choose u in this order: (1) a **L**ogarithmic function, (2) an **I**nverse trigonometric function, (3) a **P**olynomial function, (4) an **E**xponential function, and (5) A **T**rigonometric function. *NOTE*: Some may advise to swap the last two and always choose Exponential last.

➡ Example 43

Find $\int x \cos x\,dx$.

SOLUTION: We let $u = x$ and $dv = \cos x\,dx$. Then $du = dx$ and $v = \sin x$. Thus, the Parts Formula yields

$$\int x\cos x\,dx = x\sin x - \int \sin x\,dx = x\sin x + \cos x + C.$$

➡ Example 44

Find $\int x^2 e^x\,dx$.

SOLUTION: We let $u = x^2$ and $dv = e^x\,dx$. Then $du = 2x\,dx$ and $v = e^x$,

so $\int x^2 e^x\,dx = x^2 e^x - \int 2x\,e^x\,dx$. We use the Parts Formula again, this time letting $u = x$ and $dv = e^x dx$ so that $du = dx$ and $v = e^x$. Thus,

$$\int x^2 e^x\,dx = x^2 e^x - 2(xe^x - \int e^x\,dx) = x^2 e^x - 2xe^x + 2e^x + C.$$

➡ Example 45

Find $I = \int e^x \cos x\,dx$.

SOLUTION: To integrate, we can let $u = e^x$ and $dv = \cos x\,dx$; then $du = e^x dx$, $v = \sin x$. Thus,

$$I = e^x \sin x - \int e^x \sin x\,dx.$$

BC ONLY

To evaluate the integral on the right, again we let $u = e^x$, $dv = \sin x\,dx$, so that $du = e^x dx$ and $v = -\cos x$. Then,

$$\begin{aligned} I &= e^x \sin x - \left(-e^x \cos x + \int e^x \cos x\,dx\right) \\ &= e^x \sin x + e^x \cos x - I. \\ 2I &= e^x(\sin x + \cos x), \\ I &= \frac{1}{2}e^x(\sin x + \cos x) + C. \end{aligned}$$

Example 46

Find $\int x^4 \ln x\,dx$.

SOLUTION: We let $u = \ln x$ and $dv = x^4\,dx$. Then, $du = \frac{1}{x}dx$ and $v = \frac{x^5}{5}$. Thus,

$$\int x^4 \ln x\,dx = \frac{x^5}{5}\ln x - \frac{1}{5}\int x^4\,dx = \frac{x^5}{5}\ln x - \frac{x^5}{25} + C.$$

The Tic-Tac-Toe Method[1]

This method of integrating is extremely useful when repeated integration by parts is necessary. To integrate $\int u(x)v(x)\,dx$, we construct a table as follows:

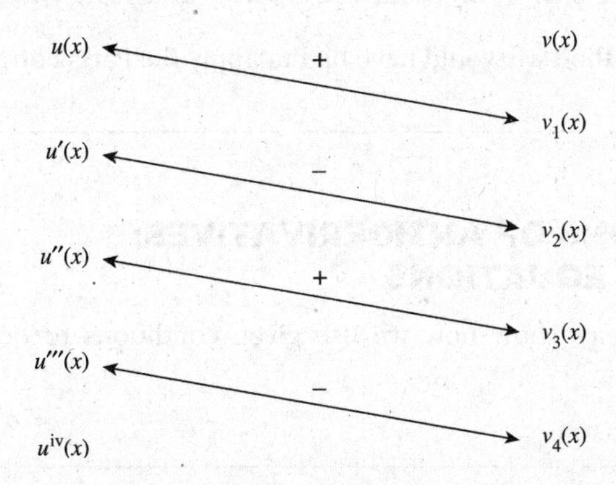

Here the column at the left contains the successive derivatives of $u(x)$. The column at the right contains the successive antiderivatives of $v(x)$ (always with $C = 0$); that is, $v_1(x)$ is the antiderivative of $v(x)$, $v_2(x)$ is the antiderivative of $v_1(x)$, and so on. The diagonal arrows join the pairs of factors whose products form the successive terms of the desired integral; above each arrow is the sign of that term. By the tic-tac-toe method,

$$\int u(x)v(x)dx = u(x)v_1(x) - u'(x)v_2(x) + u''(x)v_3(x) - u'''(x)v_4(x) + \ldots.$$

[1]This method was described by K. W. Folley in Vol. 54 (1947) of the *American Mathematical Monthly* and was referred to in the movie *Stand and Deliver*.

BC ONLY

Example 47

To integrate $\int x^4 \cos x\, dx$ by the tic-tac-toe method, we let $u(x) = x^4$ and $v(x) = \cos x$, and get the following table:

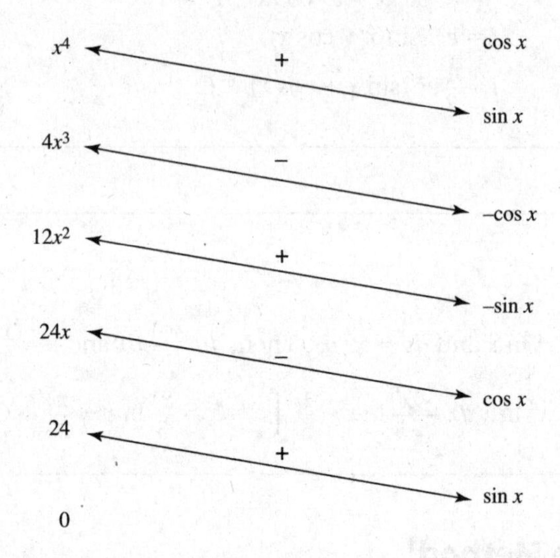

The method yields

$$\int x^4 \cos x\, dx = x^4 \sin x - (-4x^3 \cos x) + (-12x^2 \sin x) - 24x \cos x + 24 \sin x + C$$
$$= x^4 \sin x + 4x^3 \cos x - 12x^2 \sin x - 24x \cos x + 24 \sin x + C.$$

With the ordinary method we would have had to apply the Parts Formula four times to perform this integration.

E. APPLICATIONS OF ANTIDERIVATIVES; DIFFERENTIAL EQUATIONS

The following examples show how we use given conditions to determine constants of integration.

Example 48

Find $f(x)$ if $f'(x) = 3x^2$ and $f(1) = 6$.

$$f(x) = \int 3x^2\, dx = x^3 + C.$$

SOLUTION:

Since $f(1) = 6$, $1^3 + C$ must equal 6; so C must equal $6 - 1$ or 5, and $f(x) = x^3 + 5$.

Example 49

Find a curve whose slope at each point (x, y) equals the reciprocal of the x-value if the curve contains the point $(e, -3)$.

SOLUTION: We are given that $\frac{dy}{dx} = \frac{1}{x}$ and that $y = -3$ when $x = e$. This equation is also solved by integration. Since $\frac{dy}{dx} = \frac{1}{x}$, $dy = \frac{1}{x}\,dx$.

Thus, $y = \ln x + C$. We now use the given condition, by substituting the point $(e, -3)$, to determine C. Since $-3 = \ln e + C$, we have $-3 = 1 + C$, and $C = -4$. Then, the solution of the given equation subject to the given condition is

$$y = \ln x - 4.$$

Differential Equations: Motion Problems

Differential equation

An equation involving a derivative is called a *differential equation*. In Examples 48 and 49, we solved two simple differential equations. In each one we were given the derivative of a function and the value of the function at a particular point. The problem of finding the function is called an *initial-value problem* and the given condition is called the *initial condition*.

In Examples 50 and 51, we use the velocity (or the acceleration) of a particle moving on a line to find the position of the particle. Note especially how the initial conditions are used to evaluate constants of integration.

Example 50

The velocity of a particle moving along a line is given by $v(t) = 4t^3 - 3t^2$ at time t. If the particle is initially at $x = 3$ on the line, find its position when $t = 2$.

SOLUTION: Since

$$v(t) = \frac{dx}{dt} = 4t^3 - 3t^2,$$

$$x = \int (4t^3 - 3t^2)\,dt = t^4 - t^3 + C.$$

Since $x(0) = 0^4 - 0^3 + C = 3$, we see that $C = 3$, and that the position function is $x(t) = t^4 - t^3 + 3$. When $t = 2$, we see that

$$x(2) = 2^4 - 2^3 + 3 = 16 - 8 + 3 = 11.$$

Example 51

Suppose that $a(t)$, the acceleration of a particle at time t, is given by $a(t) = 4t - 3$, that $v(1) = 6$, and that $f(2) = 5$, where $f(t)$ is the position function.

(a) Find $v(t)$ and $f(t)$.

(b) Find the position of the particle when $t = 1$.

SOLUTIONS:

(a)

$$a(t) = v'(t) = \frac{dv}{dt} = 4t - 3,$$

$$v = \int (4t - 3)dt = 2t^2 - 3t + C_1.$$

Using $v(1) = 6$, we get $6 = 2(1)^2 - 3(1) + C_1$, and $C_1 = 7$, from which it follows that $v(t) = 2t^2 - 3t + 7$. Since

$$v(t) = f'(t) = \frac{df}{dt},$$

$$f(t) = \int (2t^2 - 3t + 7)\, dt = \frac{2t^3}{3} - \frac{3t^2}{2} + 7t + C_2.$$

Using $f(2) = 5$, we get $5 = \frac{2}{3}(2)^3 - \frac{3}{2}(2)^2 + 7(2) + C_2$, $5 = \frac{16}{3} - 6 + 14 + C_2$, so $C_2 = -\frac{25}{3}$. Thus,

$$f(t) = \frac{2}{3}t^3 - \frac{3}{2}t^2 + 7t - \frac{25}{3}.$$

(b) When $t = 1$, $f(1) = \frac{2}{3} - \frac{3}{2} + 7 - \frac{25}{3} = -\frac{13}{6}$.

For more examples of motion along a line, see Chapter 8, Further Applications of Integration, and Chapter 9, Differential Equations.

CHAPTER SUMMARY

In this chapter, we have reviewed basic skills for finding indefinite integrals. We've looked at the antiderivative formulas for all of the basic functions and reviewed techniques for finding antiderivatives of other functions.

We've also reviewed the more advanced techniques of integration by partial fractions and integration by parts, both topics only for the BC Calculus course.

PRACTICE EXERCISES

Directions: Answer these questions *without* using your calculator.

1. $\int(3x^2 - 2x + 3)\,dx =$

   (A) $x^3 - x^2 + C$ (B) $3x^3 - x^2 + 3x + C$ (C) $x^3 - x^2 + 3x + C$

   (D) $\frac{1}{2}(3x^2 - 2x + 3)^2 + C$ (E) $3x^3 - x^2 + C$

2. $\int\left(x - \frac{1}{2x}\right)^2 dx =$

   (A) $\frac{1}{3}\left(x - \frac{1}{2x}\right)^3 + C$ (B) $x^2 - 1 + \frac{1}{4x^2} + C$ (C) $\frac{x^3}{3} - 2x - \frac{1}{4x} + C$

   (D) $\frac{x^3}{3} - x - \frac{4}{x} + C$ (E) $\frac{x^3}{3} - x - \frac{1}{4x} + C$

3. $\int\sqrt{4 - 2t}\,dt =$

   (A) $-\frac{1}{3}(4 - 2t)^{3/2} + C$ (B) $\frac{2}{3}(4 - 2t)^{3/2} + C$ (C) $-\frac{1}{6}(4 - 2t)^3 + C$

   (D) $+\frac{1}{2}(4 - 2t)^2 + C$ (E) $\frac{4}{3}(4 - 2t)^{3/2} + C$

4. $\int(2 - 3x)^5\,dx =$

   (A) $\frac{1}{6}(2 - 3x)^6 + C$ (B) $-\frac{1}{2}(2 - 3x)^6 + C$ (C) $\frac{1}{2}(2 - 3x)^6 + C$

   (D) $-\frac{1}{18}(2 - 3x)^6 + C$ (E) $\left(2x - \frac{3}{2}x^2\right)^5 + C$

5. $\int\frac{1 - 3y}{\sqrt{2y - 3y^2}}\,dy =$

   (A) $4\sqrt{2y - 3y^2} + C$ (B) $\frac{1}{4}(2y - 3y^2)^2 + C$ (C) $\frac{1}{2}\ln\sqrt{2y - 3y^2} + C$

   (D) $\frac{1}{4}(2y - 3y^2)^{1/2} + C$ (E) $\sqrt{2y - 3y^2} + C$

6. $\int\frac{dx}{3(2x - 1)^2} =$

   (A) $\frac{-3}{2x - 1} + C$ (B) $\frac{1}{6 - 12x} + C$ (C) $+\frac{6}{2x - 1} + C$

   (D) $\frac{2}{3\sqrt{2x - 1}} + C$ (E) $\frac{1}{3}\ln|2x - 1| + C$

7. $\int\frac{2\,du}{1 + 3u} =$

   (A) $\frac{2}{3}\ln|1 + 3u| + C$ (B) $-\frac{1}{3(1 + 3u)^2} + C$ (C) $2\ln|1 + 3u| + C$

   (D) $\frac{3}{(1 + 3u)^2} + C$ (E) $\frac{4}{2 + 3u} + C$

8. $\int \frac{t}{\sqrt{2t^2 - 1}}\,dt =$

(A) $\frac{1}{2}\ln\sqrt{2t^2 - 1} + C$ (B) $4\ln\sqrt{2t^2 - 1} + C$ (C) $8\sqrt{2t^2 - 1} + C$

(D) $-\frac{1}{4(2t^2 - 1)} + C$ (E) $\frac{1}{2}\sqrt{2t^2 - 1} + C$

9. $\int \cos 3x\,dx =$

(A) $3\sin 3x + C$ (B) $-\sin 3x + C$ (C) $-\frac{1}{3}\sin 3x + C$

(D) $\frac{1}{3}\sin 3x + C$ (E) $\frac{1}{2}\cos^2 3x + C$

10. $\int \frac{x\,dx}{1 + 4x^2} =$

(A) $\frac{1}{8}\ln(1 + 4x^2) + C$ (B) $\frac{1}{8(1 + 4x^2)^2} + C$ (C) $\frac{1}{4}\sqrt{1 + 4x^2} + C$

(D) $\frac{1}{2}\ln|1 + 4x^2| + C$ (E) $\frac{1}{2}\tan^{-1} 2x + C$

11. $\int \frac{dx}{1 + 4x^2} =$

(A) $\tan^{-1}(2x) + C$ (B) $\frac{1}{8}\ln(1 + 4x^2) + C$ (C) $\frac{1}{8(1 + 4x^2)^2} + C$

(D) $\frac{1}{2}\tan^{-1}(2x) + C$ (E) $\frac{1}{8x}\ln|1 + 4x^2| + C$

12. $\int \frac{x}{(1 + 4x^2)^2}\,dx =$

(A) $\frac{1}{8}\ln(1 + 4x^2)^2 + C$ (B) $\frac{1}{4}\sqrt{1 + 4x^2} + C$ (C) $-\frac{1}{8(1 + 4x^2)} + C$

(D) $-\frac{1}{3(1 + 4x^2)^3} + C$ (E) $-\frac{1}{(1 + 4x^2)} + C$

13. $\int \frac{x\,dx}{\sqrt{1 + 4x^2}} =$

(A) $\frac{1}{8}\sqrt{1 + 4x^2} + C$ (B) $\frac{\sqrt{1 + 4x^2}}{4} + C$ (C) $\frac{1}{2}\sin^{-1} 2x + C$

(D) $\frac{1}{2}\tan^{-1} 2x + C$ (E) $\frac{1}{8}\ln\sqrt{1 + 4x^2} + C$

14. $\int \frac{dy}{\sqrt{4 - y^2}} =$

(A) $\frac{1}{2}\sin^{-1}\frac{y}{2} + C$ (B) $-\sqrt{4 - y^2} + C$ (C) $\sin^{-1}\frac{y}{2} + C$

(D) $-\frac{1}{2}\ln\sqrt{4 - y^2} + C$ (E) $-\frac{1}{3(4 - y^2)^{3/2}} + C$

15. $\int \frac{y\,dy}{\sqrt{4 - y^2}} =$

(A) $\frac{1}{2}\sin^{-1}\frac{y}{2} + C$ (B) $-\sqrt{4 - y^2} + C$ (C) $\sin^{-1}\frac{y}{2} + C$

(D) $-\frac{1}{2}\ln\sqrt{4 - y^2} + C$ (E) $2\sqrt{4 - y^2} + C$

16. $\int \frac{2x+1}{2x}\,dx =$

(A) $x + \frac{1}{2}\ln|x| + C$ (B) $1 + \frac{1}{2}x^{-1} + C$ (C) $x + 2\ln|x| + C$

(D) $x + \ln|2x| + C$ (E) $\frac{1}{2}\left(2x - \frac{1}{x^2}\right) + C$

17. $\int \frac{(x-2)^3}{x^2}\,dx =$

(A) $\frac{(x-2)^4}{4x^2} + C$ (B) $\frac{x^2}{2} - 6x + 6\ln|x| - \frac{8}{x} + C$ (C) $\frac{x^2}{2} - 3x + 6\ln|x| + \frac{4}{x} + C$

(D) $-\frac{(x-2)^4}{4x} + C$ (E) $\frac{x^2}{2} - 6x + 12\ln|x| + \frac{8}{x} + C$

18. $\int \left(\sqrt{t} - \frac{1}{\sqrt{t}}\right)^2 dt =$

(A) $t - 2 + \frac{1}{t} + C$ (B) $\frac{t^3}{3} - 2t - \frac{1}{t} + C$ (C) $\frac{t^2}{2} + \ln|t| + C$

(D) $\frac{t^2}{2} - 2t + \ln|t| + C$ (E) $\frac{t^2}{2} - t - \frac{1}{t^2} + C$

19. $\int (4x^{1/3} - 5x^{3/2} - x^{-1/2})\,dx =$

(A) $3x^{4/3} - 2x^{5/2} - 2x^{1/2} + C$

(B) $3x^{4/3} - 2x^{5/2} + 2x^{1/2} + C$

(C) $6x^{2/3} - 2x^{5/2} - \frac{1}{2}x^2 + C$

(D) $\frac{4}{3}x^{-2/3} - \frac{15}{2}x^{1/2} + \frac{1}{2}x^{-3/2} + C$

(E) $\frac{16}{9}x^{4/3} - \frac{25}{2}x^{5/2} - \frac{1}{2}x^{1/2} + C$

20. $\int \frac{x^3 - x - 1}{x^2}\,dx =$

(A) $\frac{\frac{1}{4}x^4 - \frac{1}{2}x^2 - x}{\frac{1}{3}x^3} + C$

(B) $1 + \frac{1}{x^2} + \frac{2}{x^3} + C$

(C) $\frac{x^2}{2} - \ln|x| - \frac{1}{x} + C$

(D) $\frac{x^2}{2} - \ln|x| + \frac{1}{x} + C$

(E) $\frac{x^2}{2} - \ln x + \frac{2}{x^3} + C$

21. $\int \frac{dy}{\sqrt{y}(1-\sqrt{y})} =$

(A) $4\sqrt{1-\sqrt{y}} + C$ (B) $\frac{1}{2}\ln\left|1-\sqrt{y}\right| + C$ (C) $2\ln\left(1-\sqrt{y}\right) + C$

(D) $2\sqrt{y} - \ln|y| + C$ (E) $-2\ln\left|1-\sqrt{y}\right| + C$

22. $\displaystyle\int \frac{u\,du}{\sqrt{4-9u^2}} =$

(A) $\frac{1}{3}\sin^{-1}\frac{3u}{2} + C$ (B) $-\frac{1}{18}\ln\sqrt{4-9u^2} + C$ (C) $2\sqrt{4-9u^2} + C$

(D) $\frac{1}{6}\sin^{-1}\frac{3}{2}u + C$ (E) $-\frac{1}{9}\sqrt{4-9u^2} + C$

23. $\displaystyle\int \sin\theta\cos\theta\,d\theta =$

(A) $-\frac{\sin^2\theta}{2} + C$ (B) $-\frac{1}{4}\cos 2\theta + C$ (C) $\frac{\cos^2\theta}{2} + C$

(D) $\frac{1}{2}\sin 2\theta + C$ (E) $\cos 2\theta + C$

24. $\displaystyle\int \frac{\sin\sqrt{x}}{\sqrt{x}}\,dx =$

(A) $-2\cos^{1/2}x + C$ (B) $-\cos\sqrt{x} + C$ (C) $-2\cos\sqrt{x} + C$

(D) $\frac{3}{2}\sin^{3/2}x + C$ (E) $\frac{1}{2}\cos\sqrt{x} + C$

25. $\displaystyle\int t\cos(2t)^2\,dt =$

(A) $\frac{1}{8}\sin\left(4t^2\right) + C$ (B) $\frac{1}{2}\cos^2(2t) + C$ (C) $-\frac{1}{8}\sin\left(4t^2\right) + C$

(D) $\frac{1}{4}\sin(2t)^2 + C$ (E) $-\frac{1}{4}\sin(2t)^2 + C$

26. $\displaystyle\int \cos^2 2x\,dx =$

(A) $\frac{x}{2} + \frac{\sin 4x}{8} + C$ (B) $\frac{x}{2} - \frac{\sin 4x}{8} + C$ (C) $\frac{x}{4} + \frac{\sin 4x}{4} + C$

(D) $\frac{x}{4} + \frac{\sin 4x}{16} + C$ (E) $\frac{1}{4}(x + \sin 4x) + C$

27. $\displaystyle\int \sin 2\theta\,d\theta =$

(A) $\frac{1}{2}\cos 2\theta + C$ (B) $-2\cos 2\theta + C$ (C) $-\sin^2\theta + C$

(D) $\cos^2\theta + C$ (E) $-\frac{1}{2}\cos 2\theta + C$

BC ONLY

28. $\displaystyle\int x\cos x\,dx =$

(A) $x\sin x + C$ (B) $x\sin x + \cos x + C$ (C) $x\sin x - \cos x + C$

(D) $\cos x - x\sin x + C$ (E) $\frac{x^2}{2}\sin x + C$

29. $\displaystyle\int \frac{du}{\cos^2 3u} =$

(A) $-\frac{\sec 3u}{3} + C$ (B) $\tan 3u + C$ (C) $u + \frac{\sec 3u}{3} + C$

(D) $\frac{1}{3}\tan 3u + C$ (E) $\frac{1}{3\cos 3u} + C$

30. $\int \frac{\cos x\, dx}{\sqrt{1 + \sin x}} =$

(A) $-\frac{1}{2}(1 + \sin x)^{1/2} + C$

(B) $\ln \sqrt{1 + \sin x} + C$

(C) $2\sqrt{1 + \sin x} + C$

(D) $\ln |1 + \sin x| + C$

(E) $\frac{2}{3(1 + \sin x)^{3/2}} + C$

31. $\int \frac{\cos(\theta - 1)\, d\theta}{\sin^2(\theta - 1)} =$

(A) $2 \ln \sin |\theta - 1| + C$ (B) $-\csc(\theta - 1) + C$ (C) $-\frac{1}{3}\sin^{-3}(\theta - 1) + C$

(D) $-\cot(\theta - 1) + C$ (E) $\csc(\theta - 1) + C$

32. $\int \sec \frac{t}{2}\, dt =$

(A) $\ln \left|\sec \frac{t}{2} + \tan \frac{t}{2}\right| + C$ (B) $2\tan^2 \frac{t}{2} + C$ (C) $2 \ln \cos \frac{t}{2} + C$

(D) $\ln |\sec t + \tan t| + C$ (E) $2 \ln \left|\sec \frac{t}{2} + \tan \frac{t}{2}\right| + C$

33. $\int \frac{\sin 2x\, dx}{\sqrt{1 + \cos^2 x}} =$

(A) $-2\sqrt{1 + \cos^2 x} + C$ (B) $\frac{1}{2}\ln(1 + \cos^2 x) + C$ (C) $\sqrt{1 + \cos^2 x} + C$

(D) $-\ln \sqrt{1 + \cos^2 x} + C$ (E) $2 \ln |\sin x| + C$

34. $\int \sec^{3/2} x \tan x\, dx =$

(A) $\frac{2}{5}\sec^{5/2} x + C$ (B) $-\frac{2}{3}\cos^{-3/2} x + C$ (C) $\sec^{3/2} x + C$

(D) $\frac{2}{3}\sec^{3/2} x + C$ (E) $\frac{2}{3}\sec^{5/2} x \tan x + C$

35. $\int \tan \theta\, d\theta =$

(A) $-\ln |\sec \theta| + C$ (B) $\sec^2 \theta + C$ (C) $\ln |\sin \theta| + C$

(D) $\sec \theta + C$ (E) $-\ln |\cos \theta| + C$

36. $\int \frac{dx}{\sin^2 2x} =$

(A) $\frac{1}{2}\csc 2x \cot 2x + C$ (B) $-\frac{2}{\sin 2x} + C$ (C) $-\frac{1}{2}\cot 2x + C$

(D) $-\cot x + C$ (E) $-\csc 2x + C$

37. $\int \frac{\tan^{-1} y}{1 + y^2}\, dy =$

(A) $\sec^{-1} y + C$ (B) $(\tan^{-1} y)^2 + C$ (C) $\ln(1 + y^2) + C$

(D) $\ln(\tan^{-1} y) + C$ (E) $\frac{1}{2}(\tan^{-1} y)^2 + C$

38. $\int \sin 2\theta \cos \theta \, d\theta =$

(A) $-\frac{2}{3}\cos^3 \theta + C$ (B) $\frac{2}{3}\cos^3 \theta + C$ (C) $\sin^2 \theta \cos \theta + C$

(D) $\cos^3 \theta + C$ (E) $\frac{1}{8}\sin^2 (2\theta) + C$

39. $\int \frac{\sin 2t}{1 - \cos 2t} dt =$

(A) $\frac{2}{(1 - \cos 2t)^2} + C$ (B) $-\ln|1 - \cos 2t| + C$ (C) $\ln\sqrt{|1 - \cos 2t|} + C$

(D) $\sqrt{1 - \cos 2t} + C$ (E) $2\ln|1 - \cos 2t| + C$

40. $\int \cot 2u \, du =$

(A) $\ln|\sin u| + C$ (B) $\frac{1}{2}\ln|\sin 2u| + C$ (C) $-\frac{1}{2}\csc^2 2u + C$

(D) $-\sec 2u + C$ (E) $2\ln|\sin 2u| + C$

41. $\int \frac{e^x}{e^x - 1} dx =$

(A) $x + \ln|e^x - 1| + C$ (B) $x - e^x + C$ (C) $x - \frac{1}{(e^x - 1)^2} + C$

(D) $1 + \frac{1}{e^x - 1} + C$ (E) $\ln|e^x - 1| + C$

BC ONLY

42. $\int \frac{x - 1}{x(x - 2)} dx =$

(A) $\frac{1}{2}\ln|x| + \ln|x - 2| + C$ (B) $\frac{1}{2}\ln\left|\frac{x - 2}{x}\right| + C$ (C) $\ln|x - 2| + \ln|x| + C$

(D) $\frac{1}{2}\ln|x(x - 2)| + C$ (E) $\ln\left|\frac{x}{x - 2}\right| + C$

43. $\int xe^{x^2} dx =$

(A) $\frac{1}{2}e^{x^2} + C$ (B) $e^{x^2}(2x^2 + 1) + C$ (C) $2e^{x^2} + C$

(D) $e^{x^2} + C$ (E) $\frac{1}{2}e^{x^2 + 1} + C$

44. $\int \cos \theta \, e^{\sin \theta} d\theta =$

(A) $e^{\sin \theta + 1} + C$ (B) $e^{\sin \theta} + C$ (C) $-e^{\sin \theta} + C$

(D) $e^{\cos \theta} + C$ (E) $e^{\sin \theta}(\cos \theta - \sin \theta) + C$

45. $\int e^{2\theta} \sin e^{2\theta} d\theta =$

(A) $\cos e^{2\theta} + C$ (B) $2e^{4\theta}(\cos e^{2\theta} + \sin e^{2\theta}) + C$ (C) $-\frac{1}{2}\cos e^{2\theta} + C$

(D) $-2\cos e^{2\theta} + C$ (E) $-\cos\left(\frac{e^{2\theta}}{2}\right) + C$

46. $\int \frac{e^{\sqrt{x}}}{\sqrt{x}} dx =$

(A) $2\sqrt{x}(e^{\sqrt{x}} - 1) + C$ (B) $2e^{\sqrt{x}} + C$ (C) $\frac{e^{\sqrt{x}}}{2}\left(\frac{1}{x} + \frac{1}{x\sqrt{x}}\right) + C$

(D) $\frac{1}{2}e^{\sqrt{x}} + C$ (E) $\frac{\sqrt{x}}{2}e^{\sqrt{x}} + C$

BC ONLY

47. $\int xe^{-x}\,dx =$

(A) $e^{-x}(1 - x) + C$ (B) $\dfrac{e^{1-x}}{1-x} + C$ (C) $-e^{-x}(x + 1) + C$

(D) $-\dfrac{x^2}{2}e^{-x} + C$ (E) $e^{-x}(x + 1) + C$

48. $\int x^2e^x\,dx =$

(A) $e^x(x^2 + 2x) + C$ (B) $e^x(x^2 - 2x - 2) = + C$ (C) $e^x(x^2 - 2x + 2) + C$

(D) $e^x(x - 1)^2 + C$ (E) $e^x(x + 1)^2 + C$

49. $\int \dfrac{e^x + e^{-x}}{e^x - e^{-x}}\,dx =$

(A) $x - \ln\left|e^x - e^{-x}\right| + C$ (B) $x + 2\ln\left|e^x - e^{-x}\right| + C$ (C) $-\dfrac{1}{2}(e^x - e^{-x})^{-2} + C$

(D) $\ln\left|e^x - e^{-x}\right| + C$ (E) $\ln(e^x + e^{-x}) + C$

50. $\int \dfrac{e^x}{1 + e^{2x}}\,dx =$

(A) $\tan^{-1}e^x + C$ (B) $\dfrac{1}{2}\ln(1 + e^{2x}) + C$ (C) $\ln(1 + e^{2x}) + C$

(D) $\dfrac{1}{2}\tan^{-1}e^x + C$ (E) $2\tan^{-1}e^x + C$

51. $\int \dfrac{\ln v\,dv}{v} =$

(A) $\ln\left|\ln v\right| + C$ (B) $\ln\dfrac{v^2}{2} + C$ (C) $\dfrac{1}{2}(\ln v)^2 + C$

(D) $2\ln v + C$ (E) $\dfrac{1}{2}\ln v^2 + C$

52. $\int \dfrac{\ln\sqrt{x}}{x}\,dx =$

(A) $\dfrac{\ln^2\sqrt{x}}{\sqrt{x}} + C$ (B) $\ln^2 x + C$ (C) $\dfrac{1}{2}\ln\left|\ln x\right| + C$

(D) $\dfrac{(\ln\sqrt{x})^2}{2} + C$ (E) $\dfrac{1}{4}\ln^2 x + C$

BC ONLY

53. $\int x^3\ln x\,dx =$

(A) $x^2(3\ln x + 1) + C$ (B) $\dfrac{x^4}{16}(4\ln x - 1) + C$ (C) $\dfrac{x^4}{4}(\ln x - 1) + C$

(D) $3x^2\left(\ln x - \dfrac{1}{2}\right) + C$ (E) $\dfrac{1}{4}\left(x^4\ln x - \dfrac{x^5}{5}\right) + C$

54. $\int \ln\eta\,d\eta =$

(A) $\dfrac{1}{2}\ln^2\eta + C$ (B) $\eta(\ln\eta - 1) + C$ (C) $\dfrac{1}{2}\ln\eta^2 + C$

(D) $\ln\eta(\eta - 1) + C$ (E) $\eta\ln\eta + \eta + C$

55. $\int \ln x^3\,dx =$

(A) $\dfrac{3}{2}\ln^2 x + C$ (B) $3x(\ln x - 1) + C$ (C) $3\ln x(x - 1) + C$

(D) $\dfrac{3x\ln^2 x}{2} + C$ (E) $\dfrac{1}{4}\ln x^4 + C$

BC ONLY

56. $\int \frac{\ln y}{y^2}\, dy =$

(A) $\frac{1}{y}(1 - \ln y) + C$ (B) $\frac{1}{2y}\ln^2 y + C$ (C) $-\frac{1}{3y^3}(4 \ln y + 1) + C$

(D) $-\frac{1}{y}(\ln y + 1) + C$ (E) $\frac{\ln y}{y} - \frac{1}{y} + C$

57. $\int \frac{dv}{v \ln v} =$

(A) $\frac{1}{\ln v^2} + C$ (B) $-\frac{1}{\ln^2 v} + C$ (C) $-\ln |\ln v| + C$

(D) $\ln \frac{1}{v} + C$ (E) $\ln |\ln v| + C$

58. $\int \frac{y-1}{y+1}\, dy =$

(A) $y - 2 \ln |y + 1| + C$ (B) $1 - \frac{2}{y+1} + C$ (C) $\ln \frac{|y|}{(y+1)^2} + C$

(D) $1 - 2 \ln |y + 1| + C$ (E) $\ln \left| \frac{e^y}{y+1} \right| + C$

59. $\int \frac{dx}{x^2 + 2x + 2} =$

(A) $\ln(x^2 + 2x + 2) + C$ (B) $\ln |x + 1| + C$ (C) $\arctan(x + 1) + C$

(D) $\frac{1}{\frac{1}{3}x^3 + x^2 + 2x} + C$ (E) $-\frac{1}{x} + \frac{1}{2}\ln |x| + \frac{x}{2} + C$

60. $\int \sqrt{x}(\sqrt{x} - 1)dx =$

(A) $2(x^{3/2} - x) + C$ (B) $\frac{x^2}{2} - x + C$ (C) $\frac{1}{2}(\sqrt{x} - 1)^2 + C$

(D) $\frac{1}{2}x^2 - \frac{2}{3}x^{3/2} + C$ (E) $x - 2\sqrt{x} + C$

BC ONLY

61. $\int e^\theta \cos \theta \, d\theta =$

(A) $e^\theta(\cos \theta - \sin \theta) + C$

(B) $e^\theta \sin \theta + C$

(C) $\frac{1}{2}e^\theta (\sin \theta + \cos \theta) + C$

(D) $2e^\theta(\sin \theta + \cos \theta) + C$

(E) $\frac{1}{2}e^\theta (\sin \theta - \cos \theta) + C$

62. $\int \frac{(1 - \ln t)^2}{t}\, dt =$

(A) $\frac{1}{3}(1 - \ln t)^3 + C$ (B) $\ln t - 2 \ln^2 t + \ln^3 t + C$ (C) $-2(1 - \ln t) + C$

(D) $\ln t - \ln^2 t + \frac{\ln t^3}{3} + C$ (E) $-\frac{(1 - \ln t)^3}{3} + C$

BC ONLY

63. $\int u \sec^2 u \, du =$

(A) $u \tan u + \ln |\cos u| + C$ (B) $\frac{u^2}{2}\tan u + C$ (C) $\frac{1}{2}\sec u \tan u + C$

(D) $u \tan u - \ln |\sin u| + C$ (E) $u \sec u - \ln |\sec u + \tan u| + C$

64. $\int \frac{2x+1}{4+x^2}\,dx =$

(A) $\ln(x^2+4)+C$ (B) $\ln(x^2+4)+\tan^{-1}\frac{x}{2}+C$ (C) $\frac{1}{2}\tan^{-1}\frac{x}{2}+C$

CHALLENGE

(D) $\ln(x^2+4)+\frac{1}{2}\tan^{-1}\frac{x}{2}+C$ (E) $\ln(x^2+4)+\frac{1}{4}\tan^{-1}x+C$

65. $\int \frac{1-x}{\sqrt{1-x^2}}\,dx =$

CHALLENGE

(A) $\sqrt{1-x^2}+C$ (B) $\sin^{-1}x+C$ (C) $\frac{1}{2}\ln\sqrt{1-x^2}+C$

(D) $\sin^{-1}x+\sqrt{1-x^2}+C$ (E) $\sin^{-1}x+\frac{1}{2}\ln\sqrt{1-x^2}+C$

66. $\int \frac{2x-1}{\sqrt{4x-4x^2}}\,dx =$

(A) $4\ln\sqrt{4x-4x^2}+C$ (B) $\sin^{-1}(1-2x)+C$ (C) $\frac{1}{2}\sqrt{4x-4x^2}+C$

CHALLENGE

(D) $-\frac{1}{4}\ln(4x-4x^2)+C$ (E) $-\frac{1}{2}\sqrt{4x-4x^2}+C$

67. $\int \frac{e^{2x}}{1+e^x}\,dx =$

(A) $\tan^{-1}e^x+C$ (B) $e^x-\ln(1+e^x)+C$ (C) $e^x-x+\ln|1+e^x|+C$

CHALLENGE

(D) $e^x+\frac{1}{(e^x+1)^2}+C$ (E) $e^x\ln(1+e^x)+C$

68. $\int \frac{\cos\theta}{1+\sin^2\theta}\,d\theta =$

(A) $\sec\theta\tan\theta+C$ (B) $\sin\theta-\csc\theta+C$ (C) $\ln(1+\sin^2\theta)+C$

(D) $\tan^{-1}(\sin\theta)+C$ (E) $-\frac{1}{(1+\sin^2\theta)^2}+C$

69. $\int \arctan x\,dx =$

BC ONLY

(A) $\arctan x+C$

(B) $x\arctan x-\ln(1+x^2)+C$

(C) $x\arctan x+\ln(1+x^2)+C$

(D) $x\arctan x+\frac{1}{2}\ln(1+x^2)+C$

(E) $x\arctan x-\frac{1}{2}\ln(1+x^2)+C$

70. $\int \frac{dx}{1-e^x} =$

(A) $-\ln|1-e^x|+C$ (B) $x-\ln|1-e^x|+C$ (C) $\frac{1}{(1-e^x)^2}+C$

CHALLENGE

(D) $e^{-x}\ln|1+e^x|+C$ (E) $x+\ln|1-e^x|+C$

71. $\int \frac{(2-y)^2}{4\sqrt{y}}\,dy =$

(A) $\frac{1}{6}(2-y)^3\sqrt{y}+C$ (B) $2\sqrt{y}-\frac{2}{3}y^{3/2}+\frac{8}{5}y^{5/2}+C$ (C) $\ln|y|-y+2y^2+C$

(D) $2y^{1/2}-\frac{2}{3}y^{3/2}+\frac{1}{10}y^{5/2}+C$ (E) none of these

72. $\int e^{2\ln u}\,du =$

(A) $\frac{1}{3}e^{u^3} + C$ (B) $e^{u^3/3} + C$ (C) $\frac{1}{3}u^3 + C$

(D) $\frac{2}{u}e^{2\ln u} + C$ (E) $e^{1+2\ln u} + C$

73. $\int \frac{dy}{y\left(1 + \ln y^2\right)} =$

(A) $\frac{1}{2}\ln\left|1 + \ln y^2\right| + C$ (B) $-\frac{1}{\left(1 + \ln y^2\right)^2} + C$ (C) $\ln|y| + \frac{1}{2}\ln|\ln y| + C$

(D) $\tan^{-1}\left(\ln|y|\right) + C$ (E) $\frac{6}{3 + \ln y^3} + C$

74. $\int (\tan\theta - 1)^2\,d\theta =$

CHALLENGE

(A) $\sec\theta + \theta + 2\ln|\cos\theta| + C$ (B) $\tan\theta + 2\ln|\cos\theta| + C$
(C) $\tan\theta - 2\sec^2\theta + C$ (D) $\sec\theta + \theta - \tan^2\theta + C$
(E) $\tan\theta - 2\ln|\cos\theta| + C$

75. $\int \frac{d\theta}{1 + \sin\theta} =$

CHALLENGE

(A) $\sec\theta - \tan\theta + C$ (B) $\ln(1 + \sin\theta) + C$ (C) $\ln|\sec\theta + \tan\theta| + C$
(D) $\theta + \ln|\csc\theta - \cot\theta| + C$ (E) $\tan\theta - \sec\theta + C$

76. A particle starting at rest at $t = 0$ moves along a line so that its acceleration at time t is 12t ft/sec^2. How much distance does the particle cover during the first 3 sec?

(A) 16 ft (B) 32 ft (C) 48 ft (D) 54 ft (E) 108 ft

77. The equation of the curve whose slope at point (x, y) is $x^2 - 2$ and which contains the point $(1, -3)$ is

(A) $y = \frac{1}{3}x^3 - 2x$ (B) $y = 2x - 1$ (C) $y = \frac{1}{3}x^3 - \frac{10}{3}$

(D) $y = \frac{1}{3}x^3 - 2x - \frac{4}{3}$ (E) $3y = x^3 - 10$

78. A particle moves along a line with acceleration $2 + 6t$ at time t. When $t = 0$, its velocity equals 3 and it is at position $s = 2$. When $t = 1$, it is at position $s =$

(A) 2 (B) 5 (C) 6 (D) 7 (E) 8

79. Find the acceleration (in ft/sec^2) needed to bring a particle moving with a velocity of 75 ft/sec to a stop in 5 sec.

(A) -3 (B) -6 (C) -15 (D) -25 (E) -30

BC ONLY

80. $\int \frac{x^2}{x^2 - 1}\,dx =$

(A) $x + \frac{1}{2}\ln\left|\frac{x-1}{x+1}\right| + C$ (B) $\ln|x^2 - 1| + C$ (C) $x + \tan^{-1}x + C$

CHALLENGE

(D) $x + \frac{1}{2}\ln\left|\frac{x+1}{x-1}\right| + C$ (E) $1 + \frac{1}{2}\ln\left|\frac{x+1}{x-1}\right| + C$

Answer Key

1. **C**
2. **E**
3. **A**
4. **D**
5. **E**
6. **B**
7. **A**
8. **E**
9. **D**
10. **A**
11. **D**
12. **C**
13. **B**
14. **C**
15. **B**
16. **A**
17. **E**
18. **D**
19. **A**
20. **D**
21. **E**
22. **E**
23. **B**
24. **C**
25. **A**
26. **A**
27. **E**
28. **B**
29. **D**
30. **C**
31. **B**
32. **E**
33. **A**
34. **D**
35. **E**
36. **C**
37. **E**
38. **A**
39. **C**
40. **B**
41. **E**
42. **D**
43. **A**
44. **B**
45. **C**
46. **B**
47. **C**
48. **C**
49. **D**
50. **A**
51. **C**
52. **E**
53. **B**
54. **B**
55. **B**
56. **D**
57. **E**
58. **A**
59. **C**
60. **D**
61. **C**
62. **E**
63. **A**
64. **D**
65. **D**
66. **E**
67. **B**
68. **D**
69. **E**
70. **B**
71. **D**
72. **C**
73. **A**
74. **B**
75. **E**
76. **D**
77. **D**
78. **D**
79. **C**
80. **A**

Answers Explained

All the references in parentheses below are to the basic integration formulas on pages 211 and 212. In general, if u is a function of x, then $du = u'(x)\,dx$.

1. **(C)** Use, first, formula (2), then (3), replacing u by x.

2. **(E)** Hint: Expand. $\int \left(x^2 - 1 + \frac{1}{4x^2}\right) dx = \frac{x^3}{3} - x - \frac{1}{4x} + C.$

3. **(A)** By formula (3), with $u = 4 - 2t$ and $n = \frac{1}{2}$,

$$\int \sqrt{4 - 2t}\, dx = -\frac{1}{2}\int \sqrt{4 - 2t} \cdot (-2dt) = -\frac{1}{2}\frac{(4 - 2dt)^{3/2}}{3/2} + C.$$

4. **(D)** Rewrite: $-\frac{1}{3}\int (2 - 3x)^5(-3\,dx)$

5. **(E)** Rewrite:

$$\int \left(2y - 3y^2\right)^{-1/2}(1 - 3y)\,dy = \frac{1}{2}\int \left(2y - 3y^2\right)^{-1/2}(2 - 6y)\,dy.$$

Use (3).

6. **(B)** Rewrite:

$$\frac{1}{3}\int (2x - 1)^{-2}\,dx = \frac{1}{3}\cdot\frac{1}{2}\int (2x - 1)^{-2} \cdot 2\,dx.$$

Using (3) yields $-\frac{1}{6(2x - 1)} + C.$

7. **(A)** This is equivalent to $\frac{1}{3} \cdot 2 \int \frac{3du}{1 + 3u}$. Use (4).

8. **(E)** Rewrite as $\frac{1}{4}\int \left(2t^2 - 1\right)^{-1/2} \cdot 4t\,dt$. Use (3).

9. **(D)** Use (5) with $u = 3x$; $du = 3\,dx$: $\frac{1}{3}\int \cos(3x)(3\,dx)$

10. **(A)** Use (4). If $u = 1 + 4x^2$, $du = 8x\,dx$: $\frac{1}{8}\int \frac{8x\,dx}{1+4x^2}$

11. **(D)** Use (18). Let $u = 2x$; then $du = 2\,dx$: $\frac{1}{2}\int \frac{2\,dx}{1+(2x)^2}$

12. **(C)** Rewrite as $\frac{1}{8}\int (1+4x^3)^{-2} \cdot (8x\,dx)$. Use (3) with $n = -2$.

13. **(B)** Rewrite as $\frac{1}{8}\int (1+4x^2)^{-1/2} \cdot (8x\,dx)$. Use (3) with $n = -\frac{1}{2}$.

Note carefully the differences in the integrands in Questions 10–13.

14. **(C)** Use (17); rewrite as $\int \frac{\frac{1}{2}\,dy}{\sqrt{1-\left(\frac{y}{2}\right)^2}}$.

15. **(B)** Rewrite as $-\frac{1}{2}\int (4-y^2)^{-1/2} \cdot (-2y\,dy)$. Use (3).

Compare the integrands in Questions 14 and 15, noting the difference.

16. **(A)** Divide to obtain $\int \left(1 + \frac{1}{2} \cdot \frac{1}{x}\right) dx$. Use (2), (3), and (4). Remember that $\int k\,dx = kx + C$ whenever $k \neq 0$.

17. **(E)** $\int \frac{(x-2)^3}{x^2}\,dx = \int \left(x - 6 + \frac{12}{x} - \frac{8}{x^2}\right) dx = \frac{x^2}{x} - 6x + 12 \ln|x| + \frac{8}{x} + C.$
(Note the Binomial Theorem on page 633 with $n = 3$ to expand $(x-2)^3$.)

18. **(D)** The integral is equivalent to $\int \left(t - 2 + \frac{1}{t}\right) dt$. Integrate term by term.

19. **(A)** Integrate term by term.

20. **(D)** Division yields

$$\int \left(x - \frac{1}{x} - \frac{1}{x^2}\right) dx = \int x\,dx - \int \frac{1}{x}\,dx - \int \frac{1}{x^2}\,dx.$$

21. **(E)** Use formula (4) with $u = 1 - \sqrt{y} = 1 - y^{1/2}$. Then $du = -\frac{1}{2\sqrt{y}}\,dy$. Note that the integral can be written as $-2\int \frac{1}{(1-\sqrt{y})}\left(-\frac{1}{2\sqrt{y}}\right) dy$.

22. **(E)** Rewrite as $-\frac{1}{18}\int (4-9u^2)^{-1/2}(-18u\,du)$ and use formula (3).

23. **(B)** The integral is equal to $\frac{1}{2}\int \sin 2\theta\,d\theta$. Use formula (6) with $u = 2\theta$; $du = 2\,d\theta$.

24. **(C)** Use formula (6) with $u = \sqrt{x}$; $du = \frac{1}{2\sqrt{x}}\,dx$; $2\int \sin(\sqrt{x})\left(\frac{1}{2\sqrt{x}}\,dx\right)$

25. **(A)** Use formula (5) with $u = 4t^2$; $du = 8t\,dt$; $\frac{1}{8}\int \cos(4t^2)(8t\,dt)$

26. **(A)** Using the Half-Angle Formula (23) on page 634 with $\alpha = 2x$ yields $\int \left(\frac{1}{2} + \frac{1}{2}\cos 4x\right) dx$.

27. **(E)** Use formula (6): $\frac{1}{2}\int \sin 2\theta\,(2\,d\theta)$. (See Answer 23.)

28. **(B)** Integrate by parts (page 220). Let $u = x$ and $dv = \cos x\,dx$. Then $du = dx$ and $v = \sin x$. The given integral equals $x \sin x - \int \sin x\,dx$.

29. **(D)** Replace $\frac{1}{\cos^2 3u}$ by $\sec^2 3u$; then use formula (9): $\frac{1}{3}\int \sec^2 3u\,(3\,du)$

30. **(C)** Rewrite using $u = 1 + \sin x$ and $du = \cos x\,dx$ as $\int (1 + \sin x)^{-1/2}(\cos x\,dx)$.

Use formula (3).

31. **(B)** The integral is equivalent to $\int \csc(\theta - 1)\cot(\theta - 1)\,d\theta$. Use formula (12).

32. **(E)** Use formula (13) with $u = \frac{t}{2}$; $du = \frac{1}{2}\,dt$: $2\int \sec\frac{t}{2}\left(\frac{1}{2}\,dt\right)$

33. **(A)** Replace $\sin 2x$ by $2\sin x\cos x$; then the integral is equivalent to

$$-\int \frac{-2\sin x\cos x}{\sqrt{1+\cos^2 x}}\,dx = -\int u^{-1/2}\,du$$

where $u = 1 + \cos^2 x$ and $du = -2\sin x\cos x\,dx$. Use formula (3).

34. **(D)** Rewriting in terms of sines and cosines yields

$$\int \frac{\sin x}{\cos^{5/2} x}\,dx = -\int \cos^{-5/2} x(-\sin x)\,dx = -\left(-\frac{2}{3}\right)\cos^{-3/2} x + C.$$

35. **(E)** Use formula (7).

36. **(C)** Replace $\frac{1}{\sin^2 2x}$ by $\csc^2 2x$ and use formula (10): $\frac{1}{2}\int \csc^2 2x\,(2\,dx)$

37. **(E)** Let $u = \tan^{-1} y$; then integrate $\int u\,du$.

38. **(A)** Replacing $\sin 2\theta$ by 2 sin

$$\int 2\sin\theta\cos^2\theta\,d\theta = -2\int (\cos\theta)^2(-\sin\theta\,d\theta) = -\frac{2}{3}\cos^3\theta + C.$$

39. **(C)** $\frac{1}{2}\int \frac{2\sin 2t\,dt}{1-\cos 2t} = \frac{1}{2}\ln|1-\cos 2t| + C.$

40. **(B)** Rewrite as $\frac{1}{2}\int \cot 2u\,(2\,du)$ and use formula (8).

41. **(E)** Use formula (4) with $u = e^x - 1$; $du = e^x\,dx$.

42. **(D)** Use partial fractions; find A and B such that

$$\frac{x-1}{x(x-2)} = \frac{A}{x} + \frac{B}{x-2}.$$

Then $x - 1 = A(x-2) + Bx$.

Set $x = 0$: $-1 = -2A$ and $A = \frac{1}{2}$.

Set $x = 2$: $1 = 2B$ and $B = \frac{1}{2}$.

So the given integral equals

$$\int\left(\frac{1}{2x} + \frac{1}{2(x-2)}\right)dx = \frac{1}{2}\ln|x| + \frac{1}{2}\ln|x-2| + C$$

$$= \frac{1}{2}\ln|x(x-2)| + C.$$

43. **(A)** Use formula (15) with $u = x^2$; $du = 2x\,dx$; $\frac{1}{2}\int e^{x^2}(2x\,dx)$.

44. **(B)** Use formula (15) with $u = \sin\theta$; $du = \cos\theta\,d\theta$.

45. **(C)** Use formula (6) with $u = e^{2\theta}$; $du = 2e^{2\theta}\,d\theta$; $\frac{1}{2}\int \sin e^{2\theta}\left(2e^{2\theta}\,d\theta\right)$.

46. **(B)** Use formula (15) with $u = \sqrt{x} = x^{1/2}$; $du = \frac{1}{2\sqrt{x}}\,dx$.

47. **(C)** Use the Parts Formula. Let $u = x$, $dv = e^{-x}dx$; $du = dx$, $v = -e^{-x}$. Then,

$$-xe^{-x} + \int e^{-x}\,dx = -xe^{-x} - e^{-x} + C.$$

48. **(C)** See Example 44, page 220.

49. **(D)** The integral is of the form $\int \frac{du}{u}$; use (4).

50. **(A)** The integral has the form $\int \frac{du}{1+u^2}$. Use formula (18), with $u = e^x$, $du = e^x\,dx$.

51. **(C)** Let $u = \ln v$; then $du = \frac{dv}{v}$. Use formula (3) for $\int \ln v\left(\frac{1}{v}\,dv\right)$.

52. **(E)** Hint: $\ln\sqrt{x} = \frac{1}{2}\ln x$; the integral is $\frac{1}{2}\int (\ln x)(\frac{1}{x}\,dx)$

53. **(B)** Use parts, letting $u = \ln x$ and $dv = x^3\,dx$. Then $du = \frac{1}{x}\,dx$ and $v = \frac{x^4}{4}$. The integral equals $\frac{x^4}{4}\ln x - \frac{1}{4}\int x^3\,dx$.

54. **(B)** Use parts, letting $u = \ln \eta$ and $dv = dx$. Then $du = \frac{1}{\eta}\,d\eta$ and $v = \eta$. The integral equals $\eta \ln \eta - \int d\eta$.

55. **(B)** Rewrite $\ln x^3$ as $3 \ln x$, and use the method of Answer 54.

56. **(D)** Use parts, letting $u = \ln y$ and $dv = y^{-2}\,dy$. Then $du = \frac{1}{y}\,dy$ and $v = -\frac{1}{y}$. The Parts Formula yields $\frac{-\ln y}{y} + \int \frac{1}{y^2}\,dy$.

57. **(E)** The integral has the form $\int \frac{du}{u}$, where $u = \ln v$: $\int \frac{\left(\frac{1}{v}\,dv\right)}{\ln v}$

58. **(A)** By long division, the integrand is equivalent to $1 - \frac{2}{y+1}$.

59. **(C)** $\int \frac{dx}{(x+1)^2+1} = \int \frac{dx}{1+(x+1)^2}$; use formula (18) with $u = x + 1$.

60. **(D)** Multiply to get $\int (x - \sqrt{x})dx$.

61. **(C)** See Example 45, page 220. Replace x by θ.

62. **(E)** The integral equals $-\int (1-\ln t)^2\left(-\frac{1}{t}\,dt\right)$; it is equivalent to $-\int u^2\,du$, where $u = 1 - \ln t$.

63. **(A)** Replace u by x in the given integral to avoid confusion in applying the Parts Formula. To integrate $\int x\sec^2 x\,dx$, let the variable u in the Parts Formula be x, and let dv be $\sec^2 x\,dx$. Then $du = dx$ and $v = \tan x$, so

$$\int x\sec^2 x\,dx = x\tan x - \int \tan x\,dx$$
$$= x\tan x + \ln|\cos x| + C.$$

64. **(D)** The integral is equivalent to $\int \frac{2x}{4+x^2}\,dx + \int \frac{1}{4+x^2}\,dx$. Use formula (4) on the first integral and (18) on the second.

65. **(D)** The integral is equivalent to $\int \frac{1}{\sqrt{1-x^2}}\,dx - \int \frac{x}{\sqrt{1-x^2}}\,dx$.

Use formula (17) on the first integral. Rewrite the second integral as

$-\frac{1}{2}\int (1-x^2)^{-\frac{1}{2}}(-2x)\,dx$, and use (3).

66. **(E)** Rewrite: $-\frac{1}{4}\int (4x-4x^2)^{-\frac{1}{2}}(4-8x)\,dx$.

67. **(B)** Hint: Divide, getting $\int \left[e^x - \frac{e^x}{1+e^x}\right] dx$.

68. **(D)** Letting $u = \sin\theta$ yields the integral $\int \frac{du}{1+u^2}$. Use formula (18).

69. **(E)** Use integration by parts, letting $u = \arctan x$ and $dv = dx$. Then

$$du = \frac{dx}{1+x^2} \quad \text{and} \quad v = x.$$

The Parts Formula yields

$$x\arctan x - \int \frac{x\,dx}{1+x^2} \qquad \text{or} \qquad x\arctan x - \frac{1}{2}\ln\left(1+x^2\right) + C.$$

70. **(B)** Hint: Note that

$$\frac{1}{1-e^x} = \frac{1-e^x+e^x}{1-e^x} = 1 + \frac{e^x}{1-e^x}.$$

Or multiply the integrand by $\frac{e^{-x}}{e^{-x}}$, recognizing that the correct answer is equivalent to $-\ln\left|e^{-x}-1\right|$.

71. **(D)** Hint: Expand the numerator and divide. Then integrate term by term.

72. **(C)** Hint: Observe that $e^{2\ln u} = u^2$.

73. **(A)** Let $u = 1 + \ln y^2 = 1 + 2\ln|y|$; integrate $\frac{1}{2}\int \frac{\frac{2}{y}\,dy}{1+2\ln|y|}$.

74. **(B)** Hint: Expand and note that

$$\int \left(\tan^2\theta - 2\tan\theta + 1\right) d\theta = \int \sec^2\theta\,d\theta - 2\int \tan\theta\,d\theta.$$

Use formulas (9) and (7).

75. **(E)** $\int \frac{1}{1+\sin\theta}\cdot\frac{1-\sin\theta}{1-\sin\theta}\,d\theta = \int \frac{1-\sin\theta}{\cos^2\theta}\,d\theta = \int \left(\sec^2\theta - \sec\theta\tan\theta\right)$

76. **(D)** Note the initial conditions: when $t = 0$, $v = 0$ and $s = 0$. Integrate twice: $v = 6t^2$ and $s = 2t^3$. Let $t = 3$.

77. **(D)** Since $y' = x^2 - 2$, $y = \frac{1}{3}x^3 - 2x + C$. Replacing x by 1 and y by -3 yields $C = -\frac{4}{3}$.

78. **(D)** When $t = 0$, $v = 3$ and $s = 2$, so

$$v = 2t + 3t^2 + 3 \qquad \text{and} \qquad s = t^2 + t^3 + 3t + 2.$$

Let $t = 1$.

79. **(C)** Let $\frac{dv}{dt} = a$; then

$$v = at + C. \qquad (*)$$

Since $v = 75$ when $t = 0$, therefore $C = 75$. Then (*) becomes

$$v = at + 75$$

so

$$0 = at + 75 \qquad \text{and} \qquad a = -15.$$

80. **(A)** Divide to obtain $\int \left(1 + \frac{1}{x^2 - 1}\right) dx$. Use partial fractions to get

$$\frac{1}{x^2 - 1} = \frac{1}{2(x - 1)} - \frac{1}{2(x + 1)}.$$

Definite Integrals

6

CONCEPTS AND SKILLS

In this chapter, we will review what definite integrals mean and how to evaluate them. We'll look at

- **the all-important Fundamental Theorem of Calculus;**
- **other important properties of definite integrals, including the Mean Value Theorem for Integrals;**
- **analytic methods for evaluating definite integrals;**
- **evaluating definite integrals using tables and graphs;**
- **Riemann Sums;**
- **numerical methods for approximating definite integrals, including left and right rectangular sums, the midpoint rule, and the trapezoid rule;**
- **and the average value of a function.**

For BC students, we'll also review how to work with integrals based on parametrically defined functions.

A. FUNDAMENTAL THEOREM OF CALCULUS (FTC); EVALUATION OF DEFINITE INTEGRAL

If f is continuous on the closed interval $[a,b]$ and $F' = f$, then, according to the Fundamental Theorem of Calculus,

Fundamental Theorem of Calculus

$$\int_a^b f(x)\,dx = F(b) - F(a).$$

Here $\int_a^b f(x)$ is the *definite integral of f from a to b*; $f(x)$ is called the *integrand*; and a and b are called, respectively, the *lower* and *upper limits of integration*.

Definite integral

This important theorem says that if f is the derivative of F then the definite integral of f gives the net change in F as x varies from a to b. It also says that we can evaluate any definite integral for which we can find an antiderivative of a continuous function.

By extension, a definite integral can be evaluated for any function that is bounded and piecewise continuous. Such functions are said to be *integrable*.

Integrable

B. PROPERTIES OF DEFINITE INTEGRALS

The following theorems about definite integrals are important.

Fundamental Theorem of Calculus

THE FUNDAMENTAL THEOREM OF CALCULUS: $\frac{d}{dx}\int_a^x f(t)\,dt = f(x)$ (1)

$$\int_a^b kf(x)\,dx = k\int_a^b f(x)\,dx \qquad (k \text{ a constant}) \tag{2}$$

$$\int_a^a f(x)\,dx = 0 \tag{3}$$

$$\int_a^b f(x)\,dx = -\int_b^a f(x)\,dx \tag{4}$$

$$\int_a^c f(x)\,dx + \int_c^b f(x)\,dx = \int_a^b f(x)\,dx \qquad (a < c < b) \tag{5}$$

If f and g are both integrable functions of x on $[a,b]$, then

$$\int_a^b [f(x) \pm g(x)]\,dx = \int_a^b f(x)\,dx \pm \int_a^b g(x)\,dx \tag{6}$$

Mean Value Theorem for Integrals

THE MEAN VALUE THEOREM FOR INTEGRALS: If f is continuous on $[a,b]$ there exists at least one number c, $a < c < b$, such that

$$\int_a^b f(x)\,dx = f(c)(b-a) \tag{7}$$

By the comparison property, if f and g are integrable on $[a,b]$ and if $f(x) \leq g(x)$ for all x in $[a,b]$, then

$$\int_a^b f(x)\,dx \leq \int_a^b g(x)\,dx \tag{8}$$

The evaluation of a definite integral is illustrated in the following examples. A calculator will be helpful for some numerical calculations.

Example 1

$$\int_{-1}^{2}\left(3x^2 - 2x\right)dx = x^3 - x^2\Big|_{-1}^{2} = (8-4) - (-1-1) = 6.$$

Example 2

$$\int_1^2 \frac{x^2 + x - 2}{2x^2}\,dx = \frac{1}{2}\int_1^2\left(1 + \frac{1}{x} - \frac{2}{x^2}\right)dx = \frac{1}{2}\left(x + \ln x + \frac{2}{x}\right)\Big|_1^2$$

$$= \frac{1}{2}[(2 + \ln 2 + 1) - (1 + 2)] = \frac{1}{2}\ln 2, \text{ or } \ln\sqrt{2}.$$

Example 3

$$\int_5^8 \frac{dy}{\sqrt{9-y}} = -\int_5^8 (9-y)^{-1/2}(-dy) = -2\sqrt{9-y}\,\Big|_5^8 = -2(1-2) = 2.$$

Example 4

$$\int_0^1 \frac{x\,dx}{(2-x^2)^3} = -\frac{1}{2}\int_0^1 (2-x^2)^{-3}(-2x\,dx) = -\frac{1}{2}\frac{(2-x^2)^{-2}}{-2}\Bigg|_0^1 = \frac{1}{4}\left(1-\frac{1}{4}\right) = \frac{3}{16}.$$

Example 5

$$\int_0^3 \frac{dt}{9+t^2} = \frac{1}{3}\tan^{-1}\frac{t}{3}\Bigg|_0^3 = \frac{1}{3}(\tan^{-1}1 - \tan^{-1}0)$$
$$= \frac{1}{3}\left(\frac{\pi}{4}-0\right) = \frac{\pi}{12}.$$

Example 6

$$\int_0^1 (3x-2)^3\,dx = \frac{1}{3}\int_0^1 (3x-2)^3(3\,dx)$$
$$= \frac{(3x-2)^4}{12}\Bigg|_0^1 = \frac{1}{12}(1-16) = -\frac{5}{4}.$$

Example 7

$$\int_0^1 xe^{-x^2}\,dx = -\frac{1}{2}e^{-x^2}\Bigg|_0^1 = -\frac{1}{2}\left(\frac{1}{e}-1\right) = \frac{e-1}{2e}.$$

Example 8

$$\int_{-\pi/4}^{\pi/4} \cos 2x\,dx = \frac{1}{2}\sin 2x\Bigg|_{-\pi/4}^{\pi/4} = \frac{1}{2}(1+1) = 1.$$

Example 9 BC ONLY

$$\int_{-1}^1 xe^x\,dx = (xe^x - e^x)\Bigg|_{-1}^1 = e - e - \left(-\frac{1}{e}-\frac{1}{e}\right) = \frac{2}{e} \text{ (by Parts).}$$

Example 10

$$\int_0^{1/2} \frac{dx}{\sqrt{1-x^2}} = \sin^{-1}x\Bigg|_0^{1/2} = \frac{\pi}{6}.$$

Example 11 BC ONLY

$$\int_0^{e-1} \ln(x+1)\,dx = \left[(x+1)\ln(x+1) - x\right]\Bigg|_0^{e-1} \text{ (Parts Formula)}$$
$$= e\ln e - (e-1) - 0 = 1.$$

BC ONLY

Example 12

Evaluate $\int_{-1}^{1} \frac{dy}{y^2 - 4}$.

SOLUTION: We use the method of partial fractions and set

$$\frac{1}{y^2 - 4} = \frac{A}{y+2} + \frac{B}{y-2}.$$

Solving for A and B yields $A = -\frac{1}{4}$, $B = \frac{1}{4}$. Thus,

$$\int_{-1}^{1} \frac{dy}{y^2 - 4} = \frac{1}{4} \ln \left| \frac{y-2}{y+2} \right| \Bigg|_{-1}^{1} = \frac{1}{4}\left(\ln \frac{1}{3} - \ln 3\right) = -\frac{1}{2} \ln 3.$$

Example 13

$$\int_{\pi/3}^{\pi/2} \tan \frac{\theta}{2} \sec^2 \frac{\theta}{2}\, d\theta = \frac{2}{2} \tan^2 \frac{\theta}{2} \Bigg|_{\pi/3}^{\pi/2} = 1 - \frac{1}{3} = \frac{2}{3}.$$

Example 14

$$\int_0^{\pi/2} \sin^2 \frac{1}{2} x\, dx = \int_0^{\pi/2} \left(\frac{1}{2} - \frac{\cos x}{2}\right) dx = \frac{x}{2} - \frac{\sin x}{2} \Bigg|_0^{\pi/2} = \frac{\pi}{4} - \frac{1}{2}.$$

Example 15

$$\frac{d}{dx} \int_{-1}^{x} \sqrt{1 + \sin^2 t}\, dt = \sqrt{1 + \sin^2 x} \text{ by theorem (1), page 242.}$$

Example 16

$$\frac{d}{dx} \int_x^1 e^{-t^2}\, dt = \frac{d}{dx}\left(-\int_1^x e^{-t^2}\, dt\right) \text{ by theorem (4), page 242,}$$

$$= -\frac{d}{dx} \int_1^x e^{-t^2}\, dt = -e^{-x^2} \text{ by theorem (1).}$$

Example 17

Given $F(x) = \int_1^{x^2} \frac{dt}{3+t}$, find $F'(x)$.

SOLUTION:
$$F'(x) = \frac{d}{dx} \int_1^{x^2} \frac{dt}{3+t}$$

$$= \frac{d}{dx} \int_1^{u} \frac{dt}{3+t} \qquad (\text{where } u = x^2)$$

$$= \frac{d}{du}\left(\int_1^{u} \frac{dt}{3+t}\right) \cdot \frac{du}{dx} \qquad \text{by the Chain Rule}$$

$$= \left(\frac{1}{3+u}\right)(2x) = \frac{2x}{3+x^2}.$$

➡ Example 18

If $F(x) = \int_0^{\cos x} \sqrt{1 - t^3}\, dt$, find $F'(x)$.

SOLUTION: We let $u = \cos x$. Thus

$$\frac{dF}{dx} = \frac{dF}{du} \cdot \frac{du}{dx} = \sqrt{1 - u^3}\,(-\sin x) = -\sin x \sqrt{1 - \cos^3 x}.$$

➡ Example 19

Find $\lim_{h \to 0} \frac{1}{h} \int_x^{x+h} \sqrt{e^t - 1}\, dt$.

SOLUTION: $\lim_{h \to 0} \frac{1}{h} \int_x^{x+h} \sqrt{e^t - 1}\, dt = \sqrt{e^x - 1}$.

Here we have let $f(t) = \sqrt{e^t - 1}$ and noted that

$$\lim_{h \to 0} \frac{1}{h} \int_x^{x+h} f(t) dt = \lim_{h \to 0} \frac{F(x+h) - F(x)}{h} \qquad (*)$$

where

$$\frac{dF(x)}{dx} = f(x) = \sqrt{e^x - 1}.$$

The limit on the right in the starred equation is, by definition, the derivative of $F(x)$, that is, $f(x)$.

➡ Example 20

Reexpress $\int_3^6 x\sqrt{x - 2}\, dx$, in terms of u if $u = \sqrt{x - 2}$.

SOLUTION: When $u = \sqrt{x - 2}$, $u^2 = x - 2$, and $2u\, du = dx$. The limits of the given integral are values of x. When we write the new integral in terms of the variable u, then the limits, if written, must be the values of u that correspond to the given limits. Thus, when $x = 3$, $u = 1$, and when $x = 6$, $u = 2$. Then

$$\int_3^6 x\sqrt{x - 2}\, dx = 2, \int_1^2 (u^2 + 2)u^2 du = 2\int_1^2 (u^4 + 2u^2) du.$$

➡ Example 21

If g' is continuous, find $\lim_{h \to 0} \frac{1}{h} \int_c^{c+h} g'(x)\, dx$.

SOLUTION:

$$\lim_{h \to 0} \frac{1}{h} \int_c^{c+h} g'(x)\, dx = \lim_{h \to 0} \frac{g(c+h) - g(c)}{h} = g'(c).$$

Note that the expanded limit is, by definition, the derivative of $g(x)$ at c.

C. DEFINITION OF DEFINITE INTEGRAL AS THE LIMIT OF A RIEMANN SUM

Most applications of integration are based on the FTC. This theorem provides the tool for evaluating an infinite sum by means of a definite integral. Suppose that a function $f(x)$ is continuous on the closed interval $[a,b]$. Divide the interval into n subintervals of lengths Δx_k (it is not necessary that the widths be of equal length, but the formulation is generally simpler if they are). Choose numbers, one in each subinterval, as follows: x_1 in the first, x_2 in the second, . . . , x_k in the kth, . . . , x_n in the nth. Then $\lim_{n\to\infty} \sum_{k=1}^{n} f(x_k) \cdot \Delta x_k = \int_a^b f(x)dx$, where

Riemann Sum

$\sum_{k=1}^{n} f(x_k) \cdot \Delta x_k$ is called a *Riemann Sum.*

If we choose equal width subintervals on the interval $[a,b]$, then Δx_k is the same for all values of k, and we use $\Delta x = \frac{b-a}{n}$ for each width. Given that the set $\{x_0, x_1, x_2, \ldots, x_n\}$ partitions the interval $[a,b]$ into n equal subintervals where $x_0 = a$, $x_n = b$ and each of $x_1, x_2, \ldots, x_n$ are the x value for the right edge of each subinterval, then $x_k = a + k \cdot \Delta x$ for each k. We can now write the limit as a definite integral: $\lim_{n\to\infty} \sum_{k=1}^{n} f(a + k \cdot \Delta x) \cdot \Delta x = \int_a^b f(x)dx.$

Example 22

Write $\lim_{n\to\infty} \sum_{k=1}^{n} \left(2 + \frac{5k}{n}\right)^2 \cdot \frac{5}{n}$ as a definite integral.

SOLUTION #1: Recall that this limit is of the form $\lim_{n\to\infty} \sum_{k=1}^{n} f(a + k \cdot \Delta x) \cdot \Delta x$. If we assume that $x_k = 2 + \frac{5k}{n}$, then $a = 2$ and $\Delta x = \frac{5}{n} = \frac{b-2}{n}$, so $b = 7$; x_k replaced x in $f(x)$ so $f(x) = x^2$. This gives us the definite integral $\int_2^7 x^2 dx$.

SOLUTION #2: We could also assume that $x_k = \frac{5k}{n}$, then $a = 0$ and $\Delta x = \frac{5}{n} = \frac{b-0}{n}$, so $b = 5$; x_k replaced x in $f(x)$ so $f(x) = (2 + x)^2$. This gives us the definite integral $\int_0^5 (2 + x)^2\, dx$. Indeed, both definite integrals yield the same value: $\frac{335}{3}$.

Example 23

Write $\int_1^4 (x^3 - 2)\, dx$ as the limit of a Riemann Sum.

SOLUTION: The endpoints indicate that $a = 1$ and $b = 4$, so $\Delta x = \frac{4-1}{n} = \frac{3}{n}$ and $x_k = 1 + k \cdot \frac{3}{n} = 1 + \frac{3k}{n}$. Therefore, we can write the definite integral as $\lim_{n\to\infty} \sum_{k=1}^{n} \left(f\left(1 + \frac{3k}{n}\right)\right) \cdot \frac{3}{n} = \lim_{n\to\infty} \sum_{k=1}^{n} \left(\left(1 + \frac{3k}{n}\right)^3 - 2\right) \cdot \frac{3}{n}$.

D. THE FUNDAMENTAL THEOREM AGAIN

Area

If $f(x)$ is nonnegative on $[a,b]$, we see (Figure N6–1) that $f(x_k)\,\Delta x_k$ can be regarded as the area of a typical approximating rectangle. As the number of rectangles increases, or, equivalently, as the width Δx of the rectangles approaches zero, the rectangles become an increasingly better fit to the curve. The sum of their areas gets closer and closer to the exact area under the curve. Finally, the area bounded by the x-axis, the curve, and the vertical lines $x = a$ and $x = b$ is given exactly by

$$\lim_{n\to\infty} \sum_{k=1}^{n} f(x_k)\Delta x \qquad \text{and hence by} \qquad \int_a^b f(x)\,dx.$$

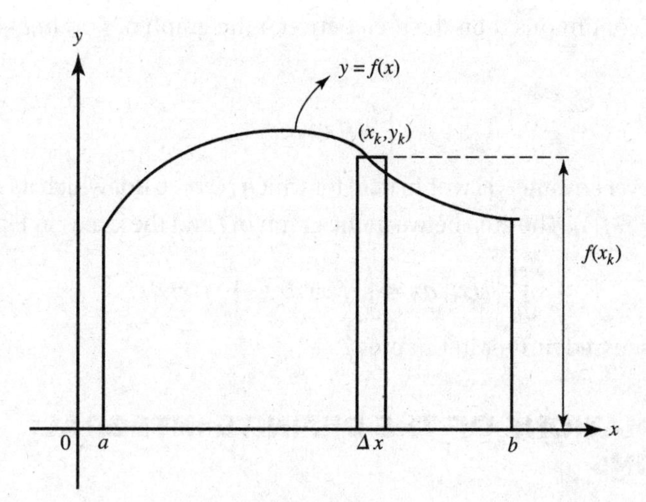

Figure N6–1

What if $f(x)$ is negative? Then *any area above the graph and below the x-axis is counted as negative* (Figure N6–2).

Geometrically, area is always positive, so the shaded area above the curve and below the x-axis equals

$$-\int_a^b f(x)dx,$$

where the integral yields a negative number. Note that every product $f(x_k)\,\Delta x$ in the shaded region is negative, since $f(x_k)$ is negative for all x between a and b.

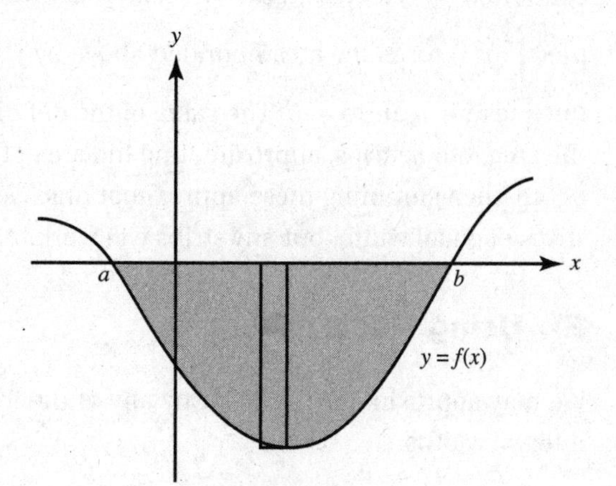

Figure N6–2

We see from Figure N6–3 that the graph of f crosses the x-axis at c, that area A_1 lies above the x-axis, and that area A_2 lies below the x-axis. Since, by property (5) on page 242,

$$\int_a^b f(x)\,dx = \int_a^c f(x)\,dx + \int_c^b f(x)\,dx,$$

therefore

$$\int_a^b f(x)\,dx = A_1 - A_2.$$

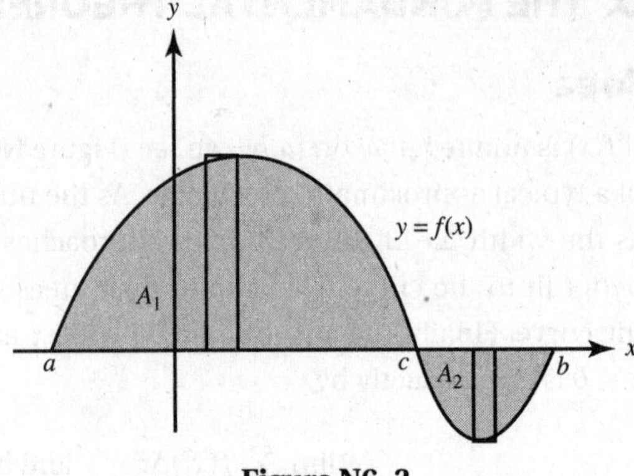

Figure N6–3

Note that if f is continuous then the area between the graph of f on $[a,b]$ and the x-axis is given by

$$\int_a^b |f(x)|\,dx.$$

This implies that, over any interval within $[a,b]$ for which $f(x) < 0$ (for which its graph dips below the x-axis), $|f(x)| = -f(x)$. The area between the graph of f and the x-axis in Figure N6–3 equals

$$\int_a^b |f(x)|\,dx = \int_a^c f(x)\,dx - \int_c^b f(x)\,dx.$$

This topic is discussed further in Chapter 7.

E. APPROXIMATIONS OF THE DEFINITE INTEGRAL; RIEMANN SUMS

It is always possible to approximate the value of a definite integral, even when an integrand cannot be expressed in terms of elementary functions. If f is nonnegative on $[a,b]$, we interpret $\int_a^b f(x)\,dx$ as the area bounded above by $y = f(x)$, below by the x-axis, and vertically by the lines $x = a$ and $x = b$. The value of the definite integral is then approximated by dividing the area into n strips, approximating the area of each strip by a rectangle or other geometric figure, then summing these approximations. We often divide the interval from a to b into n strips of equal width, but any strips will work.

E1. Using Rectangles

We may approximate $\int_a^b f(x)\,dx$ by any of the following sums, where Δx represents the subinterval widths:

Left sum

(1) Left sum: $f(x_0)\,\Delta x_1 + f(x_1)\,\Delta x_2 + \cdots + f(x_{n-1})\,\Delta x_n$, using the value of f at the left endpoint of each subinterval.

Right sum

(2) Right sum: $f(x_1)\,\Delta x_1 + f(x_2)\,\Delta x_2 + \cdots + f(x_n)\,\Delta x_n$, using the value of f at the right end of each subinterval.

Midpoint sum

(3) Midpoint sum: $f\left(\frac{x_0 + x_1}{2}\right)\Delta x_1 + f\left(\frac{x_1 + x_2}{2}\right)\Delta x_2 + \cdots + f\left(\frac{x_{n-1} + x_n}{2}\right)\Delta x_n$, using the value of f at the midpoint of each subinterval.

These approximations are illustrated in Figures N6–4 and N6–5, which accompany Example 24.

Example 24

Approximate $\int_0^2 x^3\,dx$ by using four subintervals of equal width and calculating:

(a) the left sum,

(b) the right sum,

(c) the midpoint sum, and (d) the integral.

SOLUTIONS: Here $\Delta x = \frac{2-0}{4} = \frac{1}{2}$.

(a) For a left sum we use the left-hand altitudes at $x = 0, \frac{1}{2}, 1$, and $\frac{3}{2}$. The approximating sum is

$$(0)^3 \cdot \frac{1}{2} + \left(\frac{1}{2}\right)^3 \cdot \frac{1}{2} + (1)^3 \cdot \frac{1}{2} + \left(\frac{3}{2}\right)^3 \cdot \frac{1}{2} = \frac{9}{4}.$$

The dashed lines in Figure N6–4 show the inscribed rectangles used.

(b) For the right sum we use right-hand altitudes at $x = \frac{1}{2}, 1, \frac{3}{2}$, and 2. The approximating sum is

$$\left(\frac{1}{2}\right)^3 \cdot \frac{1}{2} + (1)^3 \cdot \frac{1}{2} + \left(\frac{3}{2}\right)^3 \cdot \frac{1}{2} + (2)^3 \cdot \frac{1}{2} = \frac{25}{4}.$$

This sum uses the circumscribed rectangles shown in Figure N6–4.

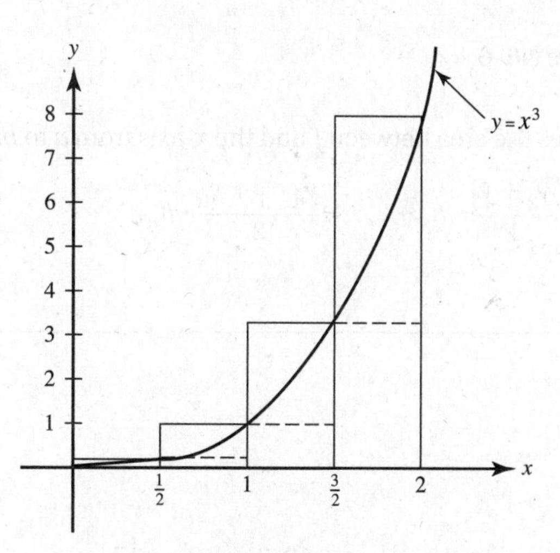

Figure N6–4

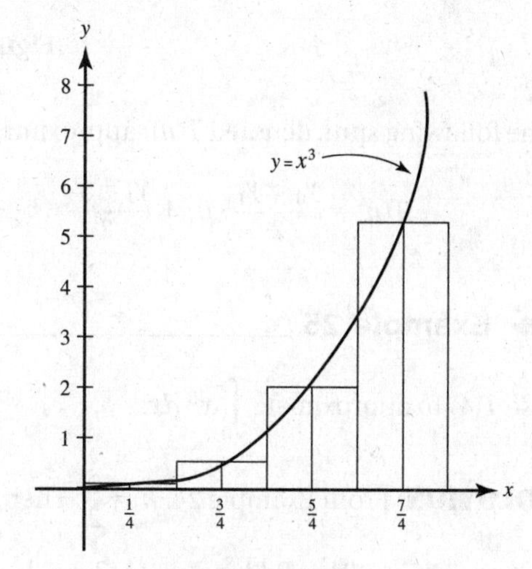

Figure N6–5

(c) The midpoint sum uses the heights at the midpoints of the subintervals, as shown in Figure N6–5. The approximating sum is

$$\left(\frac{1}{4}\right)^3 \cdot \frac{1}{2} + \left(\frac{3}{4}\right)^3 \cdot \frac{1}{2} + \left(\frac{5}{4}\right)^3 \cdot \frac{1}{2} + \left(\frac{7}{4}\right)^3 \cdot \frac{1}{2} = \frac{31}{8} \text{ or } \frac{15.5}{4}.$$

(d) Since the exact value of $\int_0^2 x^3\,dx$ is $\left.\frac{x^4}{4}\right|_0^2$ or 4, the midpoint sum is the best of the three approximations. This is usually the case.

We will denote the three Riemann Sums, with n subintervals, by $L(n)$, $R(n)$, and $M(n)$. (These sums are also sometimes called "rules.")

E2. Using Trapezoids

We now find the areas of the strips in Figure N6–6 by using trapezoids. We denote the bases of the trapezoids by $y_0, y_1, y_2, \ldots, y_n$ and the heights by $\Delta x = h_1, h_2, \ldots, h_n$.

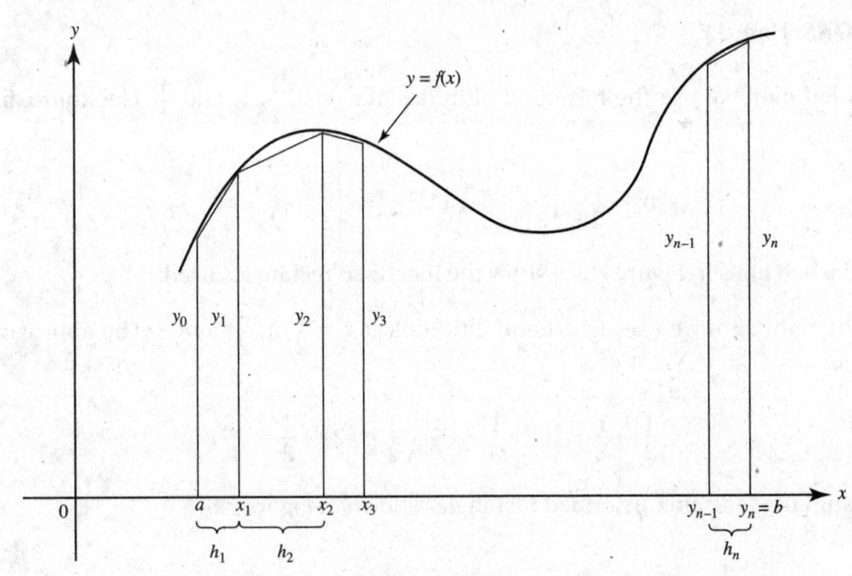

Figure N6–6

The following sum, denoted $T(n)$, approximates the area between f and the x-axis from a to b:

Trapezoid Rule

$$T(n) = \frac{y_0 + y_1}{2} \cdot h_1 + \frac{y_1 + y_2}{2} \cdot h_2 + \frac{y_2 + y_3}{2} \cdot h_3 + \ldots + \frac{y_{n-1} + y_n}{2} \cdot h_n.$$

➡ Example 25

Use $T(4)$ to approximate $\int_0^2 x^3\, dx$.

SOLUTION: From Example 24, $h = \frac{1}{2}$. Then,

$$T(4) = \frac{0^3 + (1/2)^3}{2} \cdot \frac{1}{2} + \frac{(1/2)^3 + 1^3}{2} \cdot \frac{1}{2} + \frac{1^3 + (3/2)^3}{2} \cdot \frac{1}{2} + \frac{(3/2)^3 + 2^3}{2} = \frac{17}{4}.$$

This is better than either $L(4)$ or $R(4)$, but $M(4)$ is the best approximation here.

➡ Example 26

A function f passes through the five points shown. Estimate the area $A = \int_2^{12} f(x)\,dx$ using (a) a left rectangular approximation and (b) a trapezoidal approximation.

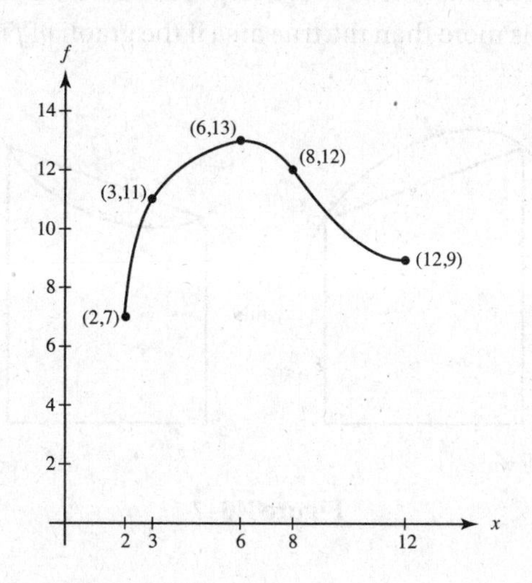

SOLUTION: Note that the subinterval widths are not equal.

(a) In each subinterval, we sketch the rectangle with height determined by the point on f at the left endpoint. Our estimate is the sum of the areas of these rectangles:

$$A \approx 1(7) + 3(11) + 2(13) + 4(12) \approx 114$$

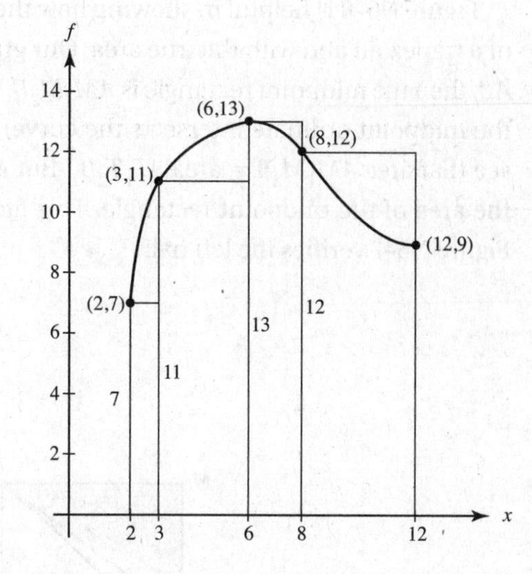

(b) In each subinterval, we sketch trapezoids by drawing segments connecting the points on f. Our estimate is the sum of the areas of these trapezoids:

$$A \approx \left(\frac{7+11}{2}\right)\cdot 1 + \left(\frac{11+13}{2}\right)\cdot 3 + \left(\frac{13+12}{2}\right)\cdot 2 + \left(\frac{12+9}{2}\right)\cdot 4 \approx 112$$

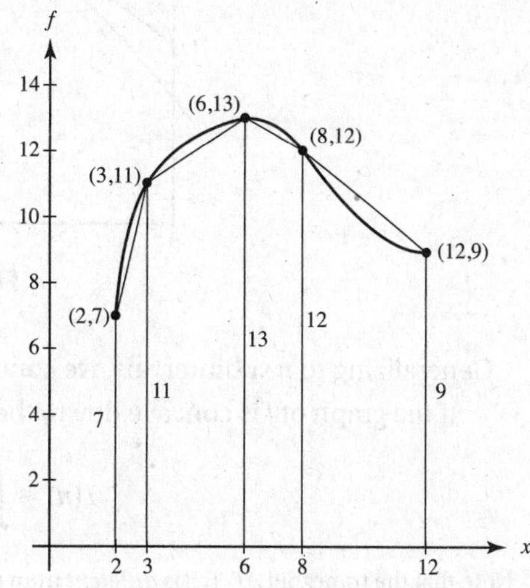

E3. Comparing Approximating Sums

If f is an increasing function on $[a,b]$, then $L(n) \leq \int_a^b f(x)\,dx \leq R(n)$, while if f is decreasing, then $R(n) \leq \int_a^b f(x)\,dx \leq L(n)$.

From Figure N6–7 we infer that the area of a trapezoid is less than the true area if the graph of f is concave down, but is more than the true area if the graph of f is concave up.

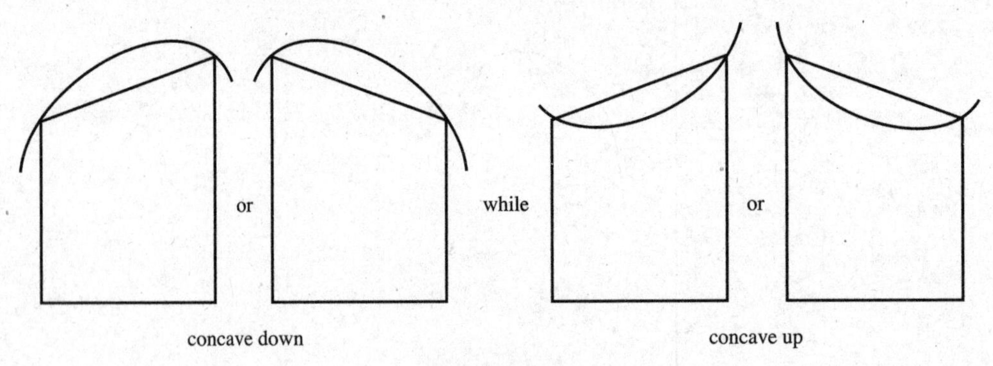

Figure N6–7

Figure N6–8 is helpful in showing how the area of a midpoint rectangle compares with that of a trapezoid and with the true area. Our graph here is concave down. If M is the midpoint of AB, then the midpoint rectangle is AM_1M_2B. We've drawn T_1T_2 tangent to the curve at T (where the midpoint ordinate intersects the curve). Since the shaded triangles have equal areas, we see that area AM_1M_2B = area AT_1T_2B.† But area AT_1T_2B clearly exceeds the true area, as does the area of the midpoint rectangle. This fact justifies the right half of the inequality below; Figure N6–7 verifies the left half.

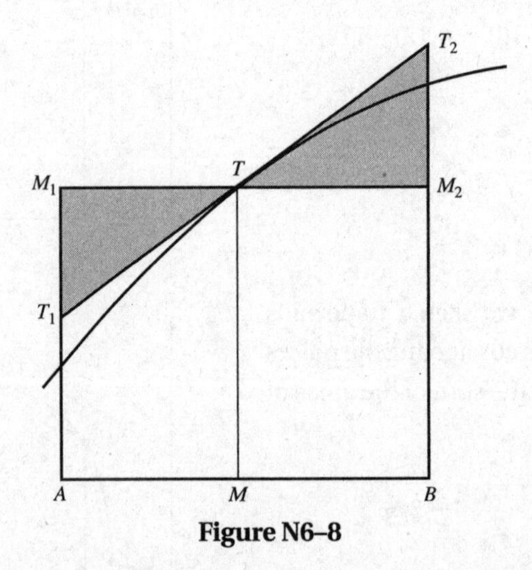

Figure N6–8

Generalizing to n subintervals, we conclude:

If the graph of f is concave down, then

$$T(n) \leq \int_a^b f(x)\,dx \leq M(n).$$

† Note that the trapezoid AT_1T_2B is *different* from the trapezoids in Figure N6–7, which are like the ones we use in applying the trapezoid rule.

If the graph of f is concave up, then

$$M(n) \leq \int_a^b f(x)\,dx \leq T(n).$$

➡ Example 27

Write an inequality including $L(n)$, $R(n)$, $M(n)$, $T(n)$, and $\int_a^b f(t)\,dt$ for the graph of f shown in Figure N6–9.

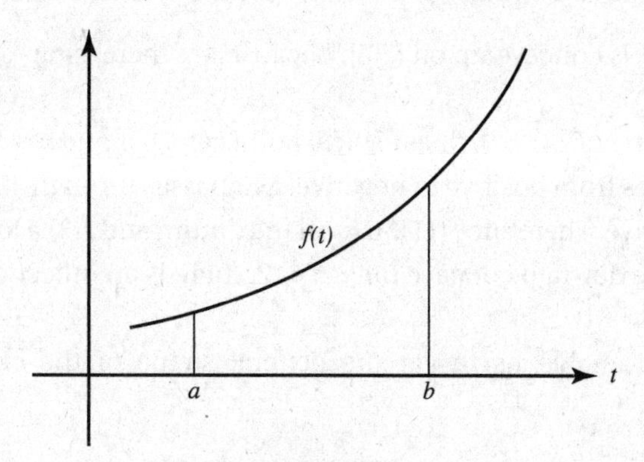

Figure N6–9

SOLUTION: Since f increases on $[a,b]$ and is concave up, the inequality is

$$L(n) \leq M(n) \leq \int_a^b f(x)\,dx \leq T(n) \leq R(n).$$

F. GRAPHING A FUNCTION FROM ITS DERIVATIVE; ANOTHER LOOK

➡ Example 28

Figure N6–10 is the graph of function $f'(x)$; it consists of two line segments and a semicircle. If $f(0) = 1$, sketch the graph of $f(x)$. Identify any critical or inflection points of f and give their coordinates.

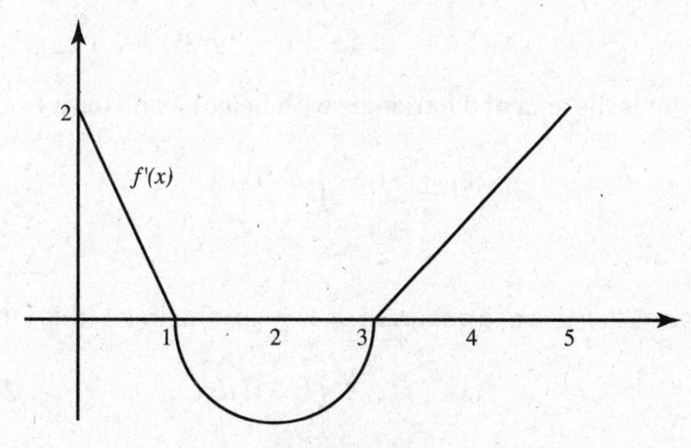

Figure N6–10

SOLUTION: We know that if $f' > 0$ on an interval then f increases on the interval, while if $f' < 0$ then f decreases; also, if f' is increasing on an interval then the graph of f is concave up on the interval, while if f' is decreasing then the graph of f is concave down. These statements lead to the following conclusions:

	f increases on [0,1] and [3,5], because $f' > 0$ there;
but	f decreases on [1,3], because $f' < 0$ there;
also	the graph of f is concave down on [0,2], because f' is decreasing;
but	the graph of f is concave up on [2,5], because f' is increasing.

Additionally, since $f'(1) = f'(3) = 0$, f has critical points at $x = 1$ and $x = 3$. As x passes through 1, the sign of f' changes from positive to negative; as x passes through 3, the sign of f' changes from negative to positive. Therefore $f(1)$ is a local maximum and $f(3)$ a local minimum. Since f changes from concave down to concave up at $x = 2$, there is an inflection point on the graph of f there.

These conclusions enable us to get the general shape of the curve, as displayed in Figure N6–11a.

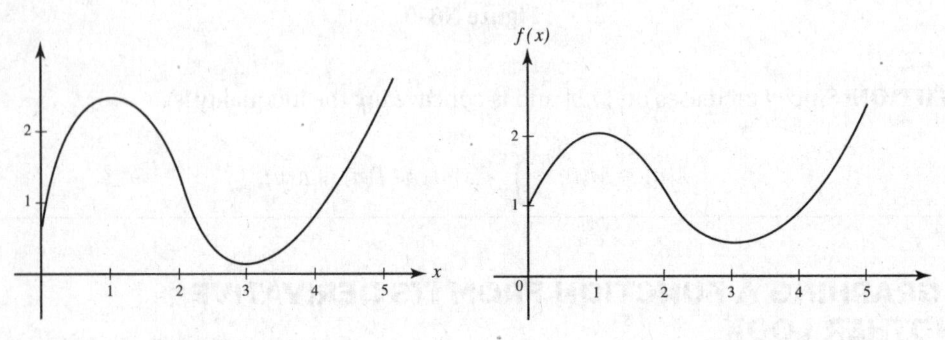

Figure N6–11a **Figure N6–11b**

All that remains is to evaluate $f(x)$ at $x = 1$, 2, and 3. We use the Fundamental Theorem of Calculus to accomplish this, finding f also at $x = 4$ and 5 for completeness.
We are given that $f(0) = 1$. Then

$$f(1) = f(0) + \int_0^1 f(x)\,dx$$
$$= 1 + 1 = 2,$$

where the integral yields the area of the triangle with height 2 and base 1;

$$f(2) = f(1) + \int_1^2 f'(x)\,dx$$
$$= 2 - \frac{\pi}{4} \simeq 1.2,$$

where the integral gives the area of a quadrant of a circle of radius 1 (this integral is negative!);

$$f(3) = f(2) + \int_2^3 f'(x)\,dx$$
$$= 1.2 - \frac{\pi}{4} \quad \text{(why?)} \quad \simeq 0.4,$$

$$f(4) = f(3) + \int_3^4 f'(x)\,dx$$

$$\simeq 0.4 + \quad \frac{1}{2} \quad \simeq 0.9,$$

where the integral is the area of the triangle with height 1 and base 1;

$$f(5) = f(4) + \int_4^5 f'(x)\,dx$$

$$\simeq 0.9 + \quad 1.5 \quad \text{(why?)} \quad \simeq 2.4.$$

So the function $f(x)$ has a local maximum at (1,2), a point of inflection at (2,1.2), and a local minimum at (3,0.4) where we have rounded to one decimal place when necessary.

In Figure N6–11b, the graph of f is shown again, but now it incorporates the information just obtained using the FTC.

➡ Example 29

Readings from a car's speedometer at 10-minute intervals during a 1-hour period are given in the table; t = time in minutes, v = speed in miles per hour:

t	0	10	20	30	40	50	60
v	26	40	55	10	60	32	45

(a) Draw a graph that could represent the car's speed during the hour.
(b) Approximate the distance traveled, using $L(6)$, $R(6)$, and $T(6)$.
(c) Draw a graph that could represent the distance traveled during the hour.

SOLUTIONS:

(a) Any number of curves will do. The graph has only to pass through the points given in the table of speeds, as does the graph in Figure N6–12a.

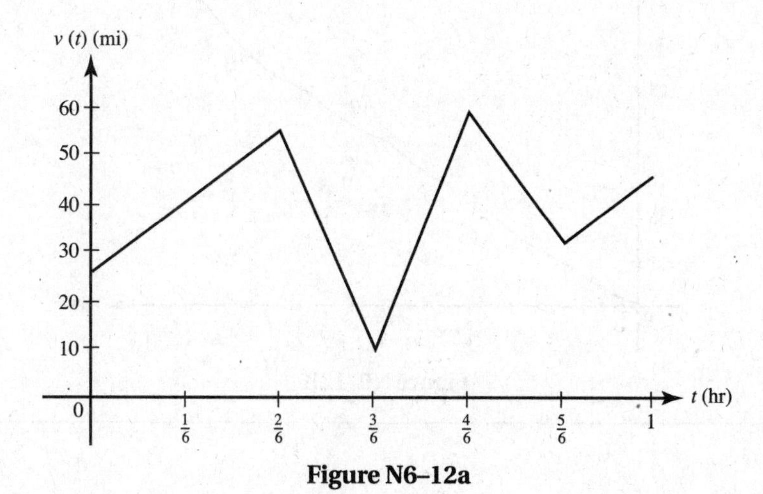

Figure N6–12a

(b) $L(6) = 26 \cdot \frac{1}{6} + 40 \cdot \frac{1}{6} + 55 \cdot \frac{1}{6} + 10 \cdot \frac{1}{6} + 60 \cdot \frac{1}{6} + 32 \cdot \frac{1}{6} = 37\frac{1}{6}$ mi;

$R(6) = 40 \cdot \frac{1}{6} + 55 \cdot \frac{1}{6} + 10 \cdot \frac{1}{6} + 60 \cdot \frac{1}{6} + 32 \cdot \frac{1}{6} + 45 \cdot \frac{1}{6} = 40\frac{1}{3}$ mi;

$$T(6) = \frac{26+40}{2} \cdot \frac{1}{6} + \frac{40+55}{2} \cdot \frac{1}{6} + \frac{55+10}{2} \cdot \frac{1}{6} +$$

$$\frac{10+60}{2} + \frac{60+32}{2} \cdot \frac{1}{6} + \frac{32+45}{2} \cdot \frac{1}{6} = 38\frac{3}{4} \text{ mi.}$$

(c) To calculate the distance traveled during the hour, we use the methods demonstrated in Example 28. We know that, since $v(t) > 0$, $s = \int_a^b v(t)\,dt$ is the distance covered from time a to time b, where $v(t)$ is the speed or velocity. Thus,

$$s(0) = 0,$$

$$s\left(\frac{1}{6}\right) = 0 + \int_0^{1/6} v(t)\,dt = 0 + (26 + 40) \cdot \frac{1}{12} = \frac{66}{12},$$

$$s\left(\frac{2}{6}\right) = \frac{66}{12} + \int_{1/6}^{2/6} v(t)\,dt = \frac{66}{12} + (40 + 55) \cdot \frac{1}{12} = \frac{161}{12},$$

$$\vdots \quad \vdots$$

$$s\left(\frac{6}{6}\right) = \frac{388}{12} + \int_{5/6}^{6/6} v(t)\,dt = \frac{388}{12} + (32 + 45) \cdot \frac{1}{12} = \frac{465}{12}.$$

It is left to the student to complete the missing steps above and to verify the distances in the following table (t = time in minutes, s = distance in miles):

t	0	10	20	30	40	50	60
s	0	5.5	13.4	18.8	24.7	32.3	38.8

Figure N6–12b is one possible graph for the distance covered during the hour.

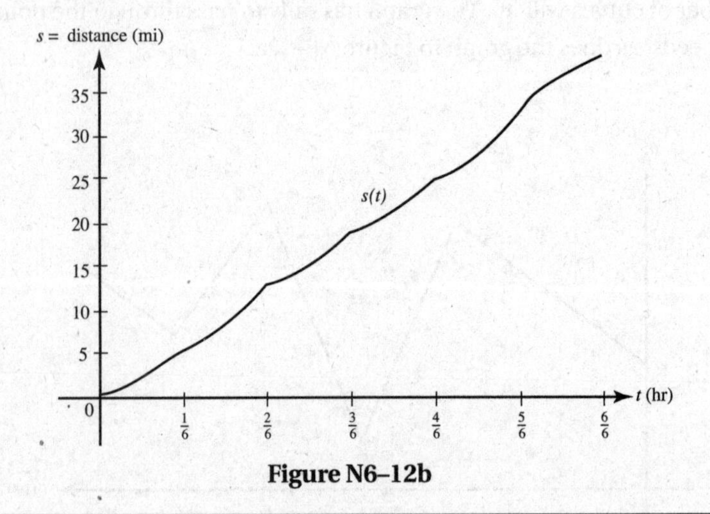

Figure N6–12b

➡ Example 30

The graph of $f(t)$ is given in Figure N6–13. If $F(x) = \int_0^x f(t)\,dt$, fill in the values for $F(x)$ in the table:

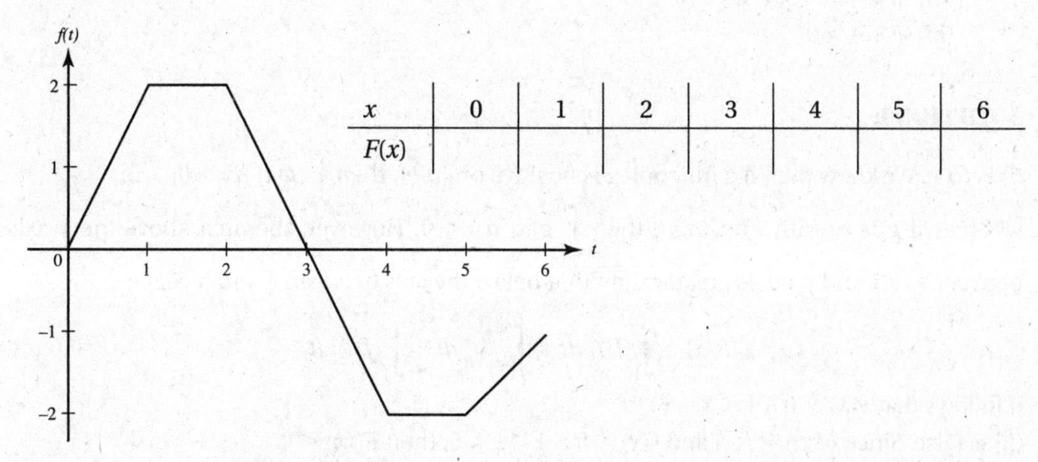

x	0	1	2	3	4	5	6
$F(x)$							

Figure N6–13

SOLUTION: We evaluate $F(x)$ by finding areas of appropriate regions.

$$F(0) = \int_0^1 f(t)dt = 0;$$

$$F(1) = \int_0^1 f(t)dt = \frac{1}{2}(1)(2) = 1 \text{ (the area of a triangle);}$$

$$F(2) = \int_0^2 f(t)dt = \frac{1}{2}(1)(2) + (1)(2) = 3 \text{ (a triangle plus a rectangle);}$$

$$\text{and } F(4) = \int_0^3 f(t)dt = \frac{1}{2}(3+1)(2) - \frac{1}{2}(1)(2) = 3 \text{ (a trapezoid minus a triangle).}$$

Here is the completed table:

x	0	1	2	3	4	5	6
$F(x)$	0	1	3	4	3	1	−0.5

➡ Example 31

The graph of the function $f(t)$ is shown in Figure N6–14.

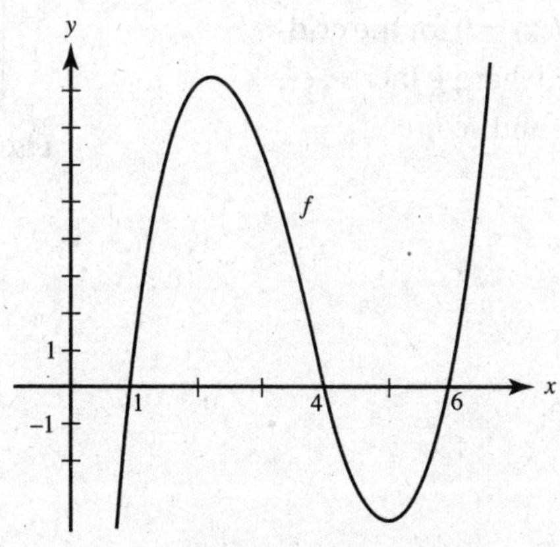

Figure N6–14

Let $F(x) = \int_1^x f(t)\, dt$. Decide whether each statement is true or false; justify your answers:

(i) If $4 < x < 6$, $F(x) > 0$.

(ii) If $4 < x < 5$, $F'(x) > 0$.

(iii) $F''(6) > 0$.

SOLUTIONS:

(i) is true. We know that, if a function g is positive on (a,b), then $\int_a^b g(x)\, dx > 0$, whereas if g is negative on (a,b), then $\int_a^b g(x)\, dx < 0$. However, the area above the x-axis between $x = 1$ and $x = 4$ is greater than that below the axis between 4 and 6. Since

$$F(x) = \int_1^x f(t)\, dt = \int_1^4 f(t)\, dt + \int_4^x f(t)\, dt,$$

it follows that $F(x) > 0$ if $4 < x < 6$.

(ii) is false. Since $F'(x) = f(x)$ and $f(x) < 0$ if $4 < x < 5$, then $F'(x) < 0$.

(iii) is false. Since $F'(x) = f(x)$, $F''(x) = f'(x)$. At $x = 6$, $f'(x) > 0$ (because f is increasing). Therefore, $F''(6) > 0$.

➡ Example 32

Graphs of functions $f(x)$, $g(x)$, and $h(x)$ are given in Figures N6–15a, N6–15b, and N6–15c. Consider the following statements:

(I) $f(x) = g'(x)$ (II) $h(x) = f'(x)$ (III) $g(x) = \int_{-2.5}^x f(t)\, dt$

Which of these statements is (are) true?

(A) I only
(B) II only
(C) III only
(D) all three
(E) none of them

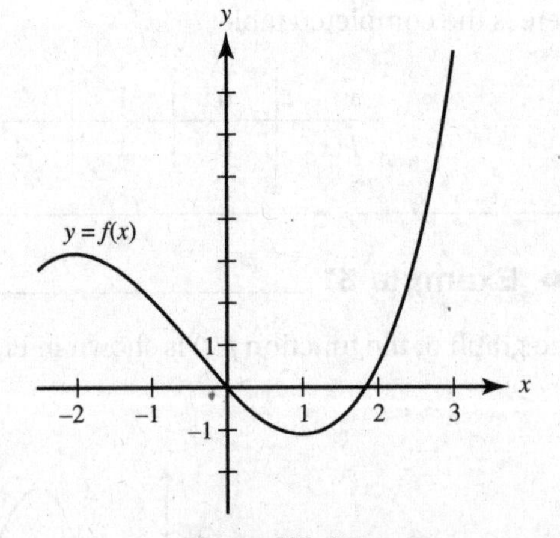

Figure N6–15a

SOLUTION:

The correct answer is D.

I is true since, for example, $f(x) = 0$ for the critical values of g: f is positive where g increases, negative where g decreases, and so on.

II is true for similar reasons.

III is also true. Verify that the value of the integral $g(x)$ increases on the interval $-2.5 < x < 0$ (where $f > 0$), decreases between the zeros of f (where $f < 0$), then increases again when f becomes positive.

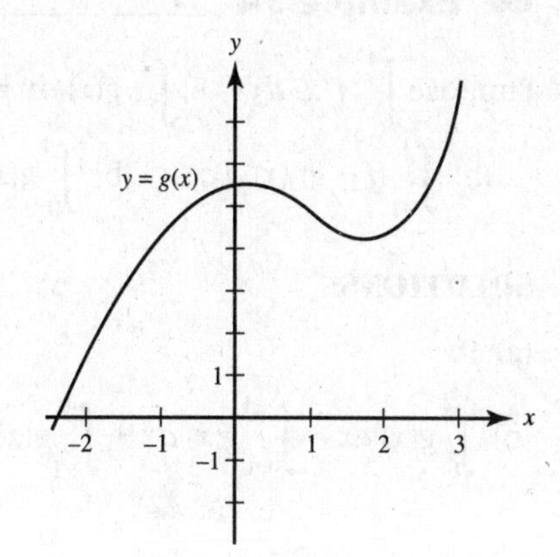

Figure N6–15b

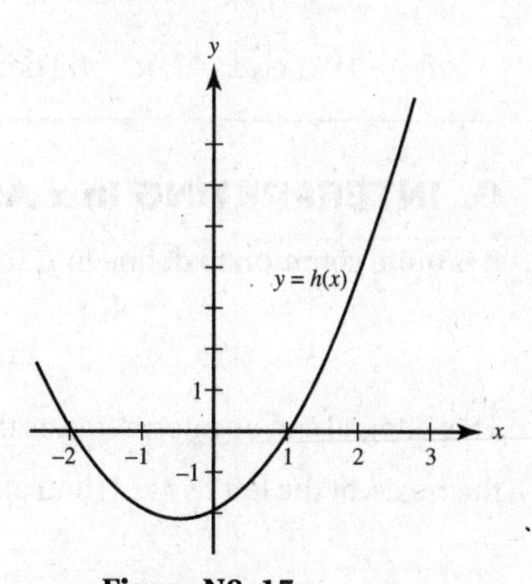

Figure N6–15c

➡ Example 33

Assume the world use of copper has been increasing at a rate given by $f(t) = 15e^{0.015t}$, where t is measured in years, with $t = 0$ the beginning of 2000, and $f(t)$ is measured in millions of tons per year.

(a) What definite integral gives the total amount of copper that was used for the 5-year period from $t = 0$ to the beginning of the year 2005?

(b) Write out the terms in the left sum $L(5)$ for the integral in (a). What do the individual terms of $L(5)$ mean in terms of the world use of copper?

(c) How good an approximation is $L(5)$ for the definite integral in (a)?

SOLUTIONS:

(a) $\int_0^5 15e^{0.015t}\, dt$.

(b) $L(5) = 15e^{0.015 \cdot 0} + 15e^{0.015 \cdot 1} + 15e^{0.015 \cdot 2} + 15e^{0.015 \cdot 3} + 15e^{0.015 \cdot 4}$. The five terms on the right represent the world's use of copper for the 5 years from 2000 until 2005.

(c) The answer to (a), using our calculator, is 77.884 million tons. $L(5) = 77.301$ million tons, so $L(5)$ underestimates the projected world use of copper during the 5-year period by approximately 583,000 tons.

Example 33 is an excellent instance of the FTC: if $f = F'$, then $\int_a^b f(x)\, dx$ gives the total change in F as x varies from a to b.

Example 34

Suppose $\int_{-1}^{4} f(x)\,dx = 6$, $\int_{-1}^{4} g(x)\,dx = -3$, and $\int_{-1}^{0} g(x)\,dx = -1$. Evaluate

(a) $\int_{-1}^{4} (f - g)(x)\,dx$; (b) $\int_{0}^{4} g(x)\,dx$; (c) $\int_{2}^{7} f(x - 3)\,dx$.

SOLUTIONS:

(a) 9.

(b) $\int_{0}^{4} g(x)\,dx = \int_{0}^{-1} g(x)\,dx + \int_{-1}^{4} g(x)\,dx = -\int_{-1}^{0} g(x)\,dx + \int_{-1}^{4} g(x)\,dx$

$= +1 + (-3) = -2.$

(c) To evaluate $\int_{2}^{7} f(x - 3)\,dx$, let $u = x - 3$. Then $du = dx$ and, when $x = 2$, $u = -1$; when $x = 7$, $u = 4$. Therefore $\int_{2}^{7} f(x - 3)\,dx = \int_{-1}^{4} f(u)\,du = 6$.

G. INTERPRETING ln *x* AS AN AREA

It is quite common to define $\ln x$, the natural logarithm of x, as a definite integral, as follows:

$$\ln x = \int_{1}^{x} \frac{1}{t}\,dt \qquad (x > 0).$$

This integral can be interpreted as the area bounded above by the curve $y = \frac{1}{t}$ $(t > 0)$, below by the t-axis, at the left by $t = 1$, and at the right by $t = x$ $(x > 1)$. See Figure N6–16.

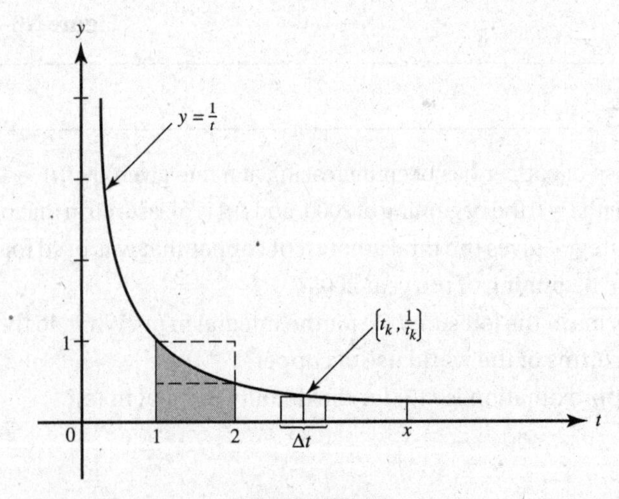

Figure N6–16

Note that if $x = 1$ the above definition yields $\ln 1 = 0$, and if $0 < x < 1$ we can rewrite as follows:

$$\ln x = -\int_{x}^{1} \frac{1}{t}\,dt,$$

showing that $\ln x < 0$ if $0 < x < 1$.

With this definition of $\ln x$ we can approximate $\ln x$ using rectangles or trapezoids.

Example 35

Show that $\frac{1}{2} < \ln 2 < 1$.

SOLUTION: Using the definition of $\ln x$ above yields $\ln 2 = \int_1^2 \frac{1}{t}\,dt$, which we interpret as the area under $y = \frac{1}{t}$, above the t-axis, and bounded at the left by $t = 1$ and at the right by $t = 2$ (the shaded region in Figure N6–16). Since $y = \frac{1}{t}$ is strictly decreasing, the area of the inscribed rectangle (height $\frac{1}{2}$, width 1) is less than ln 2, which, in turn, is less than the area of the circumscribed rectangle (height 1, width 1). Thus

$$\frac{1}{2} \cdot 1 < \ln 2 < 1 \cdot 1 \text{ or } \frac{1}{2} < \ln 2 < 1.$$

Example 36

Find $L(5)$, $R(5)$, and $T(5)$ for $\int_1^6 \frac{120}{x}\,dx$.

SOLUTION: Noting that for $n = 5$ subintervals on the interval [1,6] we have $\Delta x = 1$, we make a table of values for $f(x) = \frac{120}{x}$:

x	1	2	3	4	5	6
$f(x)$	120	60	40	30	24	20

Then:

$$L(5) = 120 \cdot 1 + 60 \cdot 1 + 40 \cdot 1 + 30 \cdot 1 + 24 \cdot 1 = 274;$$

$$R(5) = 60 \cdot 1 + 40 \cdot 1 + 30 \cdot 1 + 24 \cdot 1 + 20 \cdot 1 = 174;$$

$$T(5) = \frac{120+60}{2} \cdot 1 + \frac{60+40}{2} \cdot 1 + \frac{40+30}{2} \cdot 1 + \frac{30+24}{2} \cdot 1 + \frac{24+20}{2} \cdot 1 = 224.$$

NOTE: The calculator finds that $\int_1^6 \frac{120}{x}\,dx$ is approximately 215.011.

H. AVERAGE VALUE

Average value of a function

If the function $y = f(x)$ is integrable on the interval $a \leqslant x \leqslant b$, then we define the *average value* of f from a to b to be

$$\frac{1}{b-a}\int_a^b f(x)\,dx. \qquad (1)$$

Note that (1) is equivalent to

$$(\text{average value of } f) \cdot (b-a) = \int_a^b f(x)\,dx. \qquad (2)$$

If $f(x) \geqslant 0$ for all x on $[a,b]$, we can interpret (2) in terms of areas as follows: The right-hand expression represents the area under the curve of $y = f(x)$, above the x-axis, and bounded by the vertical lines $x = a$ and $x = b$. The left-hand expression of (2) represents the area of a rectangle with the same base $(b - a)$ and with the average value of f as its height. See Figure N6–17.

CAUTION: The average value of a function is not the same as the *average rate of change* (see page 111). Before answering any question about either of these, be sure to reread the question carefully to be absolutely certain which is called for.

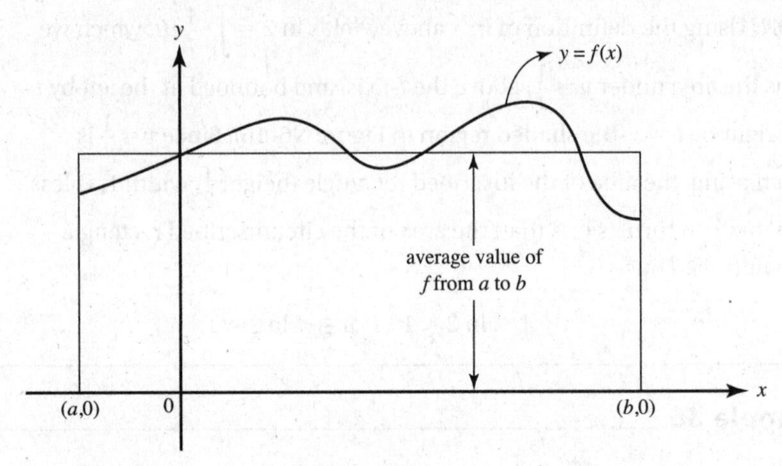

Figure N6–17

➡ Example 37

Find the average value of $f(x) = \ln x$ on the interval [1,4].

SOLUTION: $\frac{1}{4-1}\int_1^4 \ln x\,dx = \frac{1}{3}(x\ln x - x)\Big|_1^4 = \frac{4\ln 4 - 3}{3}.$

➡ Example 38

Find the average value of y for the semicircle $y = \sqrt{4-x^2}$ on [–2,2].

SOLUTION: $\frac{1}{2-(-2)}\int_{-2}^{2}\sqrt{4-x^2}\,dx = \frac{1}{4}\frac{\pi(2^2)}{2} = \frac{\pi}{2}.$

NOTE: We have used the fact that the definite integral equals exactly the area of a semicircle of radius 2.

➡ Example 39

The graphs (a) through (e) in Figure N6–18 show the velocities of five cars moving along an east-west road (the x-axis) at time t, where $0 \leqslant t \leqslant 6$. In each graph the scales on the two axes are the same.

Which graph shows the car

(1) with constant acceleration?
(2) with the greatest initial acceleration?
(3) back at its starting point when $t = 6$?
(4) that is farthest from its starting point at $t = 6$?
(5) with the greatest average velocity?
(6) with the least average velocity?
(7) farthest to the left of its starting point when $t = 6$?

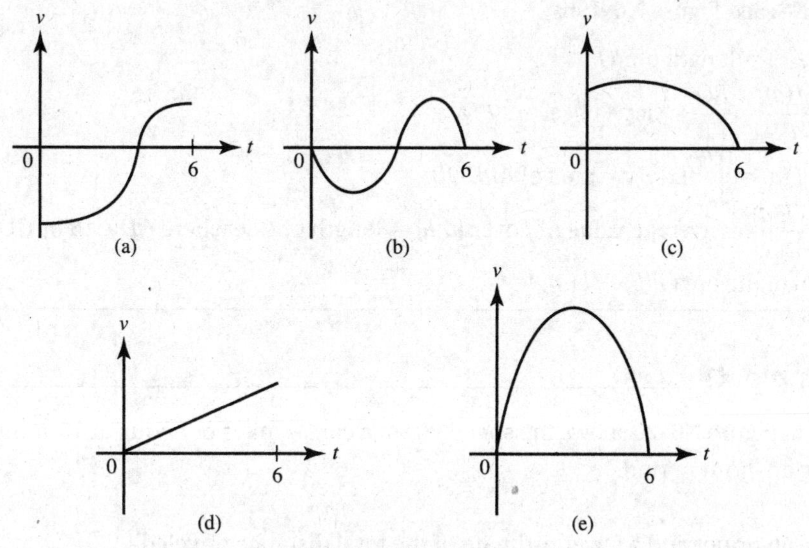

Figure N6–18

SOLUTIONS:

(1) (d), since acceleration is the derivative of velocity and in (d) v', the slope, is constant.
(2) (e), when $t = 0$ the slope of this v-curve (which equals acceleration) is greatest.
(3) (b), since for this car the net distance traveled (given by the net area) equals zero.
(4) (e), since the area under the v-curve is greatest, this car is farthest east.
(5) (e), the average velocity equals the total distance divided by 6, which is the net area divided by 6 (see (4)).
(6) (a), since only for this car is the net area negative.
(7) (a) again, since net area is negative only for this car.

➡ Example 40

Identify each of the following quantities for the function $f(x)$, whose graph is shown in Figure N6–19a (note: $F'(x) = f(x)$):

(a) $f(b) - f(a)$ (b) $\dfrac{f(b) - f(a)}{b - a}$

(c) $F(b) - F(a)$ (d) $\dfrac{F(a) - F(b)}{b - a}$

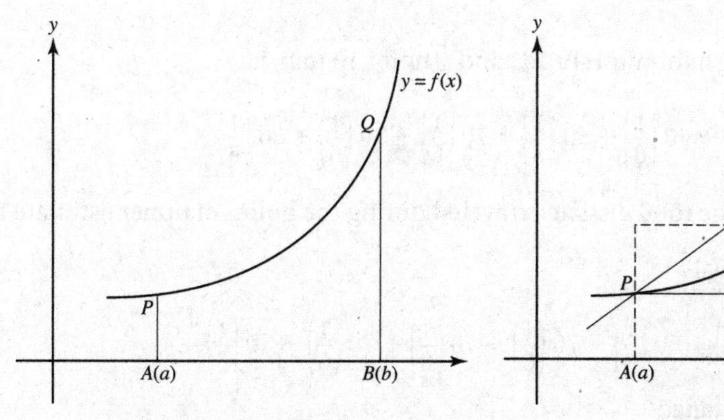

Figure N6–19a **Figure N6–19b**

SOLUTIONS: See Figure N6–19b.

(a) $f(b) - f(a) =$ length of RQ.

(b) $\dfrac{f(b) - f(a)}{b - a} = \dfrac{RQ}{PR} =$ slope of secant PQ.

(c) $F(b) - F(a) = \int_a^b f(x)\,dx =$ area of $APDQB$.

(d) $\dfrac{F(b) - F(a)}{b - a} =$ average value of f over $[a,b]$ = length of CD, where $CD \cdot AB$ or $CD \cdot (b - a)$ is equal to the area $F(b) - F(a)$.

Example 41

The graph in Figure N6–20 shows the speed $v(t)$ of a car, in miles per hour, at 10-minute intervals during a 1-hour period.

(a) Give an upper and a lower estimate of the total distance traveled.
(b) When does the acceleration appear greatest?
(c) Estimate the acceleration when $t = 20$.
(d) Estimate the average speed of the car during the interval $30 \leq t \leq 50$.

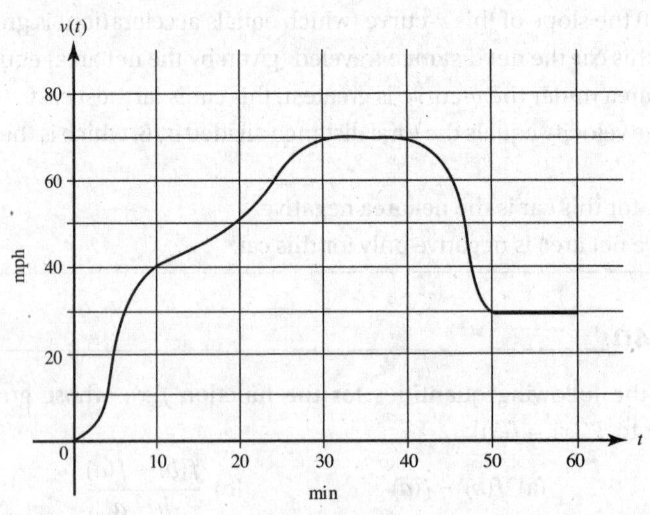

Figure N6–20

SOLUTIONS:

(a) A lower estimate, using minimum speeds and $\frac{1}{6}$ hr for 10 min, is

$$0\left(\frac{1}{6}\right) + 40\left(\frac{1}{6}\right) + 50\left(\frac{1}{6}\right) + 70\left(\frac{1}{6}\right) + 30\left(\frac{1}{6}\right) + 30\left(\frac{1}{6}\right).$$

This yields $36\frac{2}{3}$ mi for the total distance traveled during the hour. An upper estimate uses maximum speeds; it equals

$$40\left(\frac{1}{6}\right) + 50\left(\frac{1}{6}\right) + 70\left(\frac{1}{6}\right) + 70\left(\frac{1}{6}\right) + 70\left(\frac{1}{6}\right) + 30\left(\frac{1}{6}\right)$$

or 55 mi for the total distance.

(b) The acceleration, which is the slope of $v(t)$, appears greatest at $t = 5$ min, when the curve is steepest.

(c) To estimate the acceleration $v'(t)$ at $t = 20$, we approximate the slope of the curve at $t = 20$. The slope of the tangent at $t = 20$ appears to be equal to (10 mph)/(10 min) = (10 mph)/$\left(\frac{1}{6}\text{ hr}\right)$ = 60 mi/hr^2.

(d) The average speed equals the distance traveled divided by the time. We can approximate the distance from $t = 30$ to $t = 50$ by the area under the curve, or, roughly, by the sum of the areas of a rectangle and a trapezoid:

$$70\left(\frac{1}{6}\right) + \frac{70 + 30}{2}\left(\frac{1}{6}\right) = 20 \text{ mi}.$$

Thus the average speed from $t = 30$ to $t = 50$ is

$$\frac{20 \text{ mi}}{20 \text{ min}} = 1 \text{ mi/min} = 60 \text{ mph}.$$

Example 42

Given the graph of $G(x)$ in Figure N6–21a, identify the following if $G'(x) = g(x)$:

(a) $g(b)$ (b) $\int_a^b G(x)\,dx$ (c) $\int_a^b g(x)\,dx$ (d) $\dfrac{\int_a^b g(x)\,dx}{b-a}$.

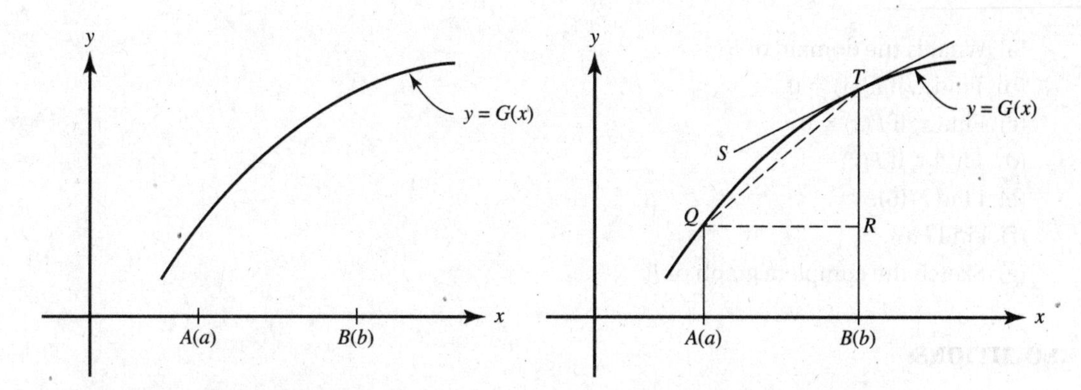

Figure N6–21a **Figure N6–21b**

SOLUTIONS: See Figure N6–21b.

(a) $g(b)$ is the slope of $G(x)$ at b, the slope of line ST.

(b) $\int_a^b G(x)\,dx$ is equal to the area under $G(x)$ from a to b.

(c) $\int_a^b g(x)\,dx = G(b) - G(a) =$ length of BT – length of BR = length of RT.

(d) $\dfrac{\int_a^b g(x)\,dx}{b-a} = \dfrac{\text{length of } RT}{\text{length of } QR} =$ slope of QT.

➡ Example 43

The function $f(t)$ is graphed in Figure N6–22a. Let

$$F(x) = \int_4^{x/2} f(t)\, dt.$$

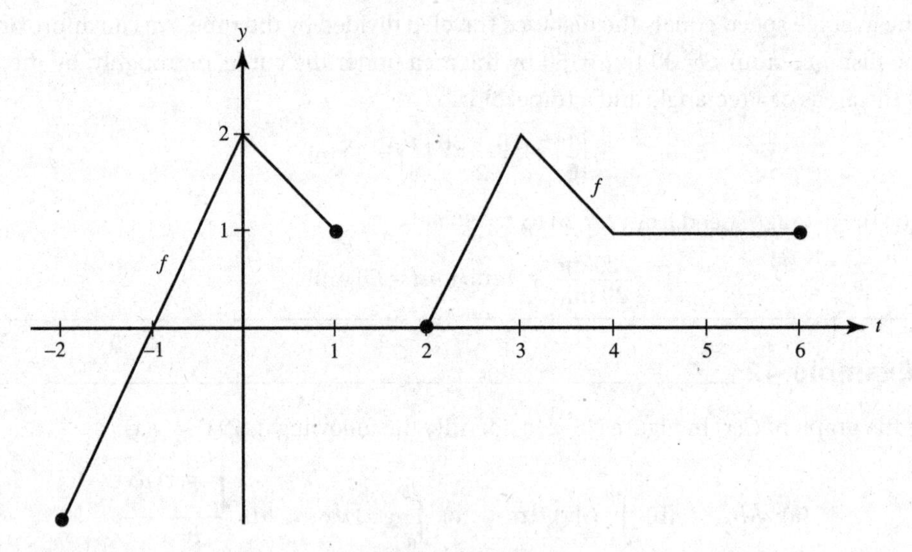

Figure N6–22a

(a) What is the domain of F?
(b) Find x, if $F'(x) = 0$.
(c) Find x, if $F(x) = 0$.
(d) Find x, if $F(x) = 1$.
(e) Find $F'(6)$.
(f) Find $F(6)$.
(g) Sketch the complete graph of F.

SOLUTIONS:

(a) The domain of f is $[-2,1]$ and $[2,6]$. We choose the portion of this domain that contains the lower limit of integration, 4. Thus the domain of F is $2 \leqslant \frac{x}{2} \leqslant 6$, or $4 \leqslant x \leqslant 12$.

(b) Since $F'(x) = f\left(\frac{x}{2}\right) \cdot \frac{1}{2}$, $F'(x) = 0$ if $f\left(\frac{x}{2}\right) = 0$. Then $\frac{x}{2} = 2$ and $x = 4$.

(c) $F(x) = 0$ when $\frac{x}{2} = 4$ or $x = 8$. $F(8) = \int_4^{8/2} f(t)\, dt = 0$.

(d) For $F(x)$ to equal 1, we need a region under f whose left endpoint is 4 with area equal to 1. The region from 4 to 5 works nicely; so $\frac{x}{2} = 5$ and $x = 10$.

(e) $F'(6) = f\left(\frac{6}{2}\right) \cdot \frac{1}{2} = f(3) \cdot \left(\frac{1}{2}\right) = 2 \cdot \left(\frac{1}{2}\right) = 1.$

(f) $F(6) = \int_4^3 f(t)\, dt = -(\text{area of trapezoid}) = -\frac{2+1}{2} \cdot 1 = -\frac{3}{2}.$

(g) In Figure N6–22b we evaluate the areas in the original graph.

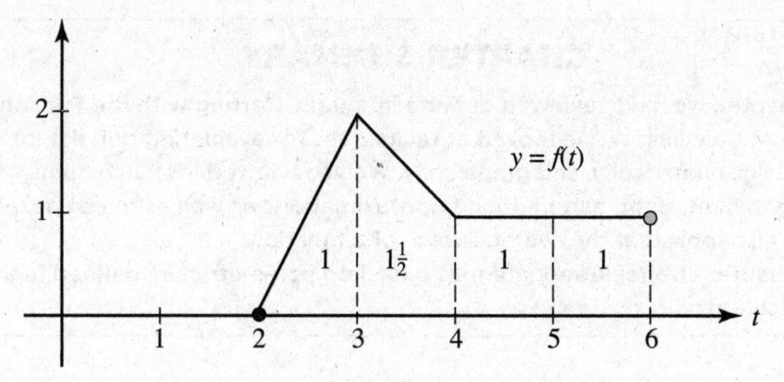

Figure N6–22b

Measured from the lower limit of integration, 4, we have (with "f" as an abbreviation for "$f(t)\,dt$"):

$$F(4) = \int_4^2 f = -2\frac{1}{2}, \qquad F(6) = \int_4^3 f = -1\frac{1}{2},$$

$$F(8) = \int_4^4 f = 0, \qquad F(10) = \int_4^5 f = 1,$$

$$F(12) = \int_4^6 f = 2.$$

We note that, since $F'(= f)$ is linear on (2,4), F is quadratic on (4,8); also, since F' is positive and increasing on (2,3), the graph of F is increasing and concave up on (4,6), while since F' is positive and decreasing on (3,4), the graph of F is increasing but concave down on (6,8). Finally, since F' is constant on (4,6), F is linear on (8,12). (See Figure N6–22c.)

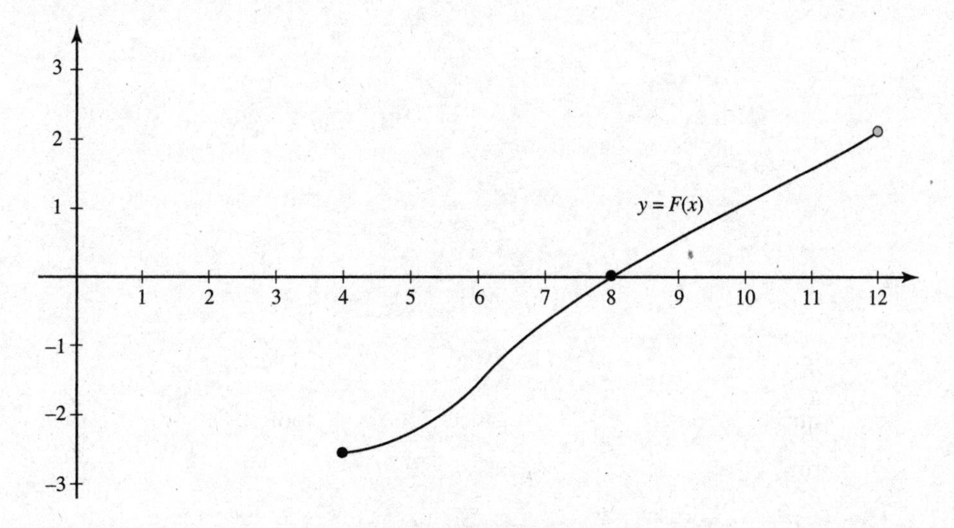

Figure N6–22c

CHAPTER SUMMARY

In this chapter, we have reviewed definite integrals, starting with the Fundamental Theorem of Calculus. We've looked at techniques for evaluating definite integrals algebraically, numerically, and graphically. We've reviewed Riemann Sums, including the left, right, and midpoint approximations as well as the trapezoid rule. We have also looked at the average value of a function.

This chapter also reviewed integrals based on parametrically defined functions, a BC Calculus topic.

PRACTICE EXERCISES

Part A. Directions: Answer these questions *without* using your calculator.

1. $\int_{-1}^{1} x^2 - x - 1\,dx =$

(A) $\frac{2}{3}$ (B) 0 (C) $-\frac{4}{3}$ (D) -2 (E) -1

2. $\int_{1}^{2} \frac{3x-1}{3x}\,dx =$

(A) $\frac{3}{4}$ (B) $1 - \frac{1}{3}\ln 2$ (C) $1 - \ln 2$ (D) $-\frac{1}{3}\ln 2$ (E) 1

3. $\int_{0}^{3} \frac{dt}{\sqrt{4-t}} =$

(A) 1 (B) -2 (C) 4 (D) -1 (E) 2

4. $\int_{-1}^{0} \sqrt{3u+4}\,du =$

(A) 2 (B) $\frac{14}{9}$ (C) $\frac{14}{3}$ (D) 6 (E) $\frac{7}{2}$

5. $\int_{2}^{3} \frac{dy}{2y-3} =$

(A) $\ln 3$ (B) $\frac{1}{2}\ln\frac{3}{2}$ (C) $\frac{16}{9}$ (D) $\ln\sqrt{3}$ (E) $\sqrt{3} - 1$

6. $\int_{0}^{\sqrt{3}} \frac{x}{\sqrt{4-x^2}}\,dx =$

(A) 1 (B) $\frac{\pi}{6}$ (C) $\frac{\pi}{3}$ (D) -1 (E) 2

7. $\int_{0}^{1} (2t-1)^3\,dt =$

(A) $\frac{1}{4}$ (B) 6 (C) $\frac{1}{2}$ (D) 0 (E) 4

8. $\int_{4}^{9} \frac{2+x}{2\sqrt{x}}\,dx =$

(A) $\frac{25}{3}$ (B) $\frac{41}{3}$ (C) $\frac{100}{3}$ (D) $\frac{5}{3}$ (E) $\frac{1}{3}$

9. $\int_{-3}^{3} \frac{dx}{9+x^2} =$

(A) $\frac{\pi}{2}$ (B) 0 (C) $\frac{\pi}{6}$ (D) $-\frac{\pi}{2}$ (E) $\frac{\pi}{3}$

10. $\int_{0}^{1} e^{-x}\,dx =$

(A) $\frac{1}{e} - 1$ (B) $1 - e$ (C) $-\frac{1}{e}$ (D) $1 - \frac{1}{e}$ (E) $\frac{1}{e}$

11. $\int_{0}^{1} xe^{x^2}\,dx =$

(A) $e - 1$ (B) $\frac{1}{2}(e-1)$ (C) $2(e-1)$ (D) $\frac{e}{2}$ (E) $\frac{e}{2} - 1$

12. $\int_0^{\pi/4} \sin 2\theta \, d\theta =$

(A) 2 (B) $\frac{1}{2}$ (C) -1 (D) $-\frac{1}{2}$ (E) -2

13. $\int_1^2 \frac{dz}{3-z} =$

(A) $-\ln 2$ (B) $\frac{3}{4}$ (C) $2(\sqrt{2}-1)$ (D) $\frac{1}{2}\ln 2$ (E) $\ln 2$

14. If we let $x = 2\sin\theta$, then $\int_1^2 \frac{\sqrt{4-x^2}}{x}\,dx$ is equivalent to

(A) $2\int_1^2 \frac{\cos^2\theta}{\sin\theta}\,d\theta$ (B) $\int_{\pi/6}^{\pi/2} \frac{\cos\theta}{\sin\theta}\,d\theta$ (C) $2\int_{\pi/6}^{\pi/2} \frac{\cos^2\theta}{\sin\theta}\,d\theta$

(D) $\int_1^2 \frac{\cos\theta}{\sin\theta}\,d\theta$ (E) $\int_{\pi/3}^{0} \frac{\cos\theta}{\sin\theta}\,d\theta$

15. $\int_0^{\pi} \cos^2\theta \sin\theta \, d\theta =$

(A) $-\frac{2}{3}$ (B) $\frac{1}{3}$ (C) 1 (D) $\frac{2}{3}$ (E) 0

16. $\int_1^e \frac{\ln x}{x}\,dx =$

(A) $\frac{1}{2}$ (B) $\frac{1}{2}(e^2-1)$ (C) 0 (D) 1 (E) $e-1$

BC ONLY

17. $\int_0^1 xe^x\,dx =$

(A) -1 (B) $e+1$ (C) 1 (D) $e-1$ (E) $\frac{1}{2}(e-1)$

18. $\int_0^{\pi/6} \frac{\cos\theta}{1+2\sin\theta}\,d\theta =$

(A) $\ln 2$ (B) $\frac{3}{8}$ (C) $-\frac{1}{2}\ln 2$ (D) $\frac{3}{2}$ (E) $\ln\sqrt{2}$

19. $\int_{\sqrt{2}}^2 \frac{u}{u^2-1}\,du =$

(A) $\ln\sqrt{3}$ (B) $\frac{8}{9}$ (C) $\ln\frac{3}{2}$ (D) $\ln 3$ (E) $1-\sqrt{3}$

20. $\int_{\sqrt{2}}^2 \frac{u\,du}{(u^2-1)^2} =$

(A) $-\frac{1}{3}$ (B) $-\frac{2}{3}$ (C) $\frac{2}{3}$ (D) -1 (E) $\frac{1}{3}$

21. $\int_{\pi/12}^{\pi/4} \frac{\cos 2x\,dx}{\sin^2 2x} =$

(A) $-\frac{1}{4}$ (B) 1 (C) $\frac{1}{2}$ (D) $-\frac{1}{2}$ (E) -1

22. $\int_0^1 \frac{e^{-x}+1}{e^{-x}}\,dx =$

(A) e (B) $2+e$ (C) $\frac{1}{e}$ (D) $1+e$ (E) $e-1$

23. $\int_0^1 \frac{e^x}{e^x+1}\,dx =$

(A) $\ln 2$ (B) e (C) $1+e$ (D) $-\ln 2$ (E) $\ln\frac{e+1}{2}$

24. If we let $x = \tan\theta$, then $\int_1^{\sqrt{3}} \sqrt{1+x^2}\, dx$ is equivalent to

(A) $\int_{\pi/4}^{\pi/3} \sec\theta\, d\theta$ (B) $\int_1^{\sqrt{3}} \sec^3\theta\, d\theta$ (C) $\int_{\pi/4}^{\pi/3} \sec^3\theta\, d\theta$

(D) $\int_{\pi/4}^{\pi/3} \sec^2\theta \tan\theta\, d\theta$ (E) $\int_1^{\sqrt{3}} \sec\theta\, d\theta$

25. If the substitution $u = \sqrt{x+1}$ is used, then $\int_0^3 \frac{dx}{x\sqrt{x+1}}$ is equivalent to

(A) $\int_1^2 \frac{du}{u^2-1}$ (B) $\int_1^2 \frac{2\,du}{u^2-1}$ (C) $2\int_0^3 \frac{du}{(u-1)(u+1)}$

(D) $2\int_1^2 \frac{du}{u(u^2-1)}$ (E) $2\int_0^3 \frac{du}{u(u-1)}$

26. The table shows some values of continuous function f and its first derivative. Evaluate $\int_8^0 f'(x)\, dx$.

x	$f(x)$	$f'(x)$
0	11	3
2	15	2
4	16	−1
6	12	−3
8	7	0

(A) −1/2 (B) −3/8 (C) 3
(D) 4 (E) −1

27. Using $M(3)$, we find that the approximate area of the shaded region below is

(A) 9 (B) 19 (C) 36 (D) 38 (E) 54

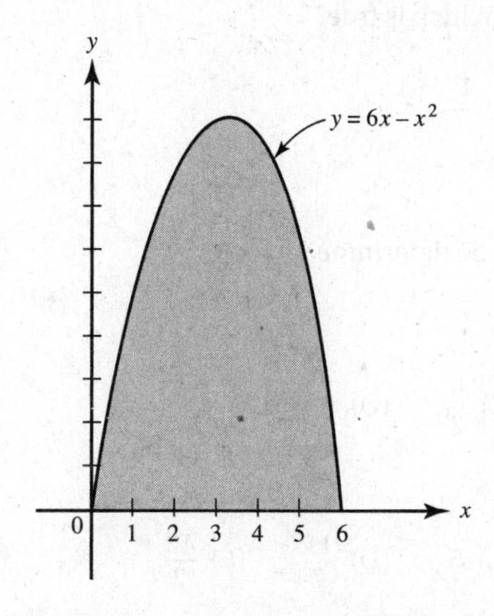

28. The graph of a continuous function f passes through the points (4,2), (6,6), (7,5), and (10,8). Using trapezoids, we estimate that $\int_4^{10} f(x)dx \approx$

(A) 25 (B) 30 (C) 32 (D) 33 (E) 41

29. The area of the shaded region in the figure is equal exactly to ln 3. If we approximate ln 3 using $L(2)$ and $R(2)$, which inequality follows?

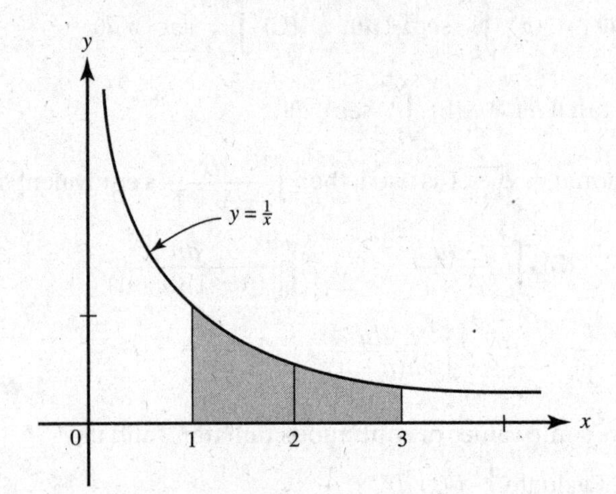

(A) $\frac{1}{2} < \int_1^2 \frac{1}{x}dx < 1$ (B) $\frac{1}{3} < \int_1^3 \frac{1}{x}dx < 2$ (C) $\frac{1}{2} < \int_0^2 \frac{1}{x}dx < 2$

(D) $\frac{1}{3} < \int_2^3 \frac{1}{x}dx < \frac{1}{2}$ (E) $\frac{5}{6} < \int_1^3 \frac{1}{x}dx < \frac{3}{2}$

30. Let $A = \int_0^1 \cos x\, dx$. We estimate A using the L, R, and T approximations with $n = 100$ subintervals. Which is true?

(A) $L < A < T < R$
(B) $L < T < A < R$
(C) $R < A < T < L$
(D) $R < T < A < L$
(E) The order cannot be determined.

31. $\int_{-1}^{3} |x|\, dx =$

(A) $\frac{7}{2}$ (B) 4 (C) $\frac{9}{2}$ (D) 5 (E) $\frac{11}{2}$

32. $\int_{-3}^{2} |x + 1|\, dx =$

(A) $\frac{5}{2}$ (B) $\frac{7}{2}$ (C) 5 (D) $\frac{11}{2}$ (E) $\frac{13}{2}$

33. The average value of $y = \sqrt{64 - x^2}$ on its domain is

(A) 2 (B) 4 (C) 2π (D) 4π (E) undefined

34. The average value of $\cos x$ over the interval $\frac{\pi}{3} \leq x \leq \frac{\pi}{2}$ is

(A) $\frac{3}{\pi}$ (B) $\frac{1}{2}$ (C) $\frac{3(2 - \sqrt{3})}{\pi}$ (D) $\frac{3}{2\pi}$ (E) $\frac{2}{3\pi}$

35. The average value of $\csc^2 x$ over the interval from $x = \frac{\pi}{6}$ to $x = \frac{\pi}{4}$ is

(A) $\frac{3\sqrt{3}}{\pi}$ (B) $\frac{\sqrt{3}}{\pi}$ (C) $\frac{12}{\pi}(\sqrt{3} - 1)$ (D) $3\sqrt{3}$ (E) $3(\sqrt{3} - 1)$

36. Choose the Riemann Sum whose limit is the integral $\int_1^5 x^3 dx$.

(A) $\lim_{n\to\infty} \sum_{k=1}^{n} \left(\left(1 + \frac{k}{n}\right)^3 \cdot \left(\frac{4}{n}\right)\right)$ (B) $\lim_{n\to\infty} \sum_{k=1}^{n} \left(\left(1 + \frac{4k}{n}\right)^3 \cdot \left(\frac{1}{n}\right)\right)$

(C) $\lim_{n\to\infty} \sum_{k=1}^{n} \left(\left(1 + \frac{4k}{n}\right)^3 \cdot \left(\frac{4}{n}\right)\right)$ (D) $\lim_{n\to\infty} \sum_{k=1}^{n} \left(\left(1 + \frac{k}{n}\right)^3 \cdot \left(\frac{1}{n}\right)\right)$

37. Choose the Riemann Sum whose limit is the integral $\int_0^{\pi} \sin(3x)\, dx$.

(A) $\lim_{n\to\infty} \sum_{k=1}^{n} \left(\sin\left(\frac{3k}{n}\right) \cdot \left(\frac{1}{n}\right)\right)$ (B) $\lim_{n\to\infty} \sum_{k=1}^{n} \left(\sin\left(\frac{3\pi k}{n}\right) \cdot \left(\frac{1}{n}\right)\right)$

(C) $\lim_{n\to\infty} \sum_{k=1}^{n} \left(\sin\left(\frac{3k}{n}\right) \cdot \left(\frac{\pi}{n}\right)\right)$ (D) $\lim_{n\to\infty} \sum_{k=1}^{n} \left(\sin\left(\frac{3\pi k}{n}\right) \cdot \left(\frac{\pi}{n}\right)\right)$

38. Choose the integral that is the limit of the Riemann Sum $\lim_{n\to\infty} \sum_{k=1}^{n} \left(\sin\left(\frac{6k}{n} + 2\right) \cdot \left(\frac{3}{n}\right)\right)$.

(A) $\int_2^5 \sin(6x)\, dx$ (B) $\int_0^3 \sin(2x + 2)\, dx$ (C) $\int_2^5 \sin(2x)\, dx$ (D) $\int_0^3 \sin(2x)\, dx$

Part B. Directions: Some of the following questions require the use of a graphing calculator.

39. Find the average value of function f, as shown in the graph below, on the interval [0,5].

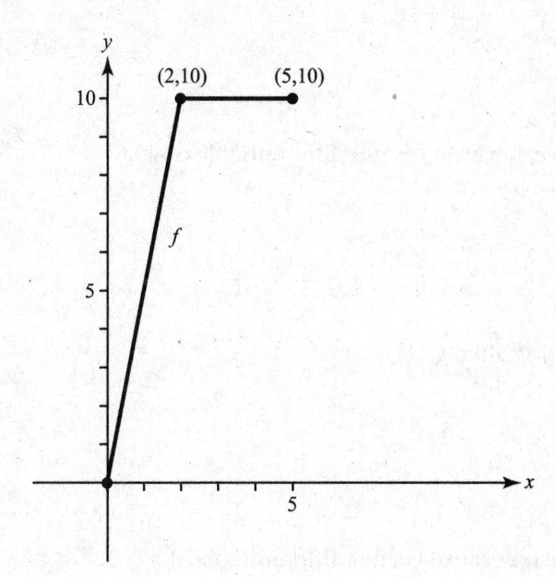

(A) 2 (B) 4 (C) 5 (D) 7 (E) 8

40. The integral $\int_{-4}^{4} \sqrt{16 - x^2}\, dx$ gives the area of

(A) a circle of radius 4
(B) a semicircle of radius 4
(C) a quadrant of a circle of radius 4
(D) an ellipse
(E) half of an ellipse

41. $\int_0^{\pi/4} \sqrt{1 - \cos 2\alpha}\, d\alpha =$

(A) 0.25 (B) 0.414 (C) 1.000 (D) 1.414 (E) 2.000

Use the graph of function *f*, shown below, for Questions 42–45.

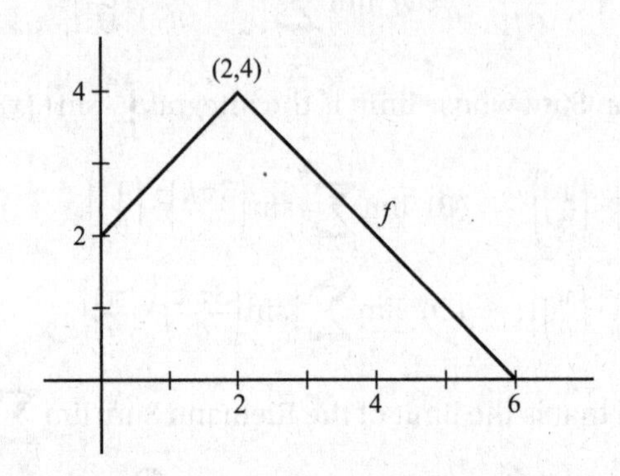

42. In which of these intervals is there a value *c* for which *f*(*c*) is the average value of *f* over the interval [0,6]?

I. [0,2]
II. [2,4]
III. [4,6]

(A) I only
(B) II only
(C) III only
(D) I and II only
(E) none of these, because *f* is not differentiable on [0,6]

43. $\int_0^2 f'(3x)dx =$

(A) -2 (B) $-\frac{2}{3}$ (C) 0 (D) $\frac{2}{3}$ (E) 2

44. Let $g(x) = \int_0^2 xf(t)dt$; then $g'(1)$

(A) = 3.
(B) = 4.
(C) = 6.
(D) = 8.
(E) does not exist, because *f* is not differentiable at $x = 2$.

45. Let $h(x) = x^2 - f(x)$. Find $\int_0^6 h(x)dx$.

(A) 22 (B) 38 (C) 58 (D) 70 (E) 74

46. If *f*(*x*) is continuous on the closed interval [*a*,*b*], then there exists at least one number *c*, $a < c < b$, such that $\int_a^b f(x)\, dx$ is equal to

(A) $\frac{f(c)}{b-a}$ (B) $f'(c)(b-a)$ (C) $f(c)(b-a)$

(D) $\frac{f'(c)}{b-a}$ (E) $f(c)[f(b) - f(a)]$

47. If $f(x)$ is continuous on the closed interval $[a,b]$ and k is a constant, then $\int_a^b kf(x)\,dx$ is equal to

(A) $k(b-a)$ (B) $k[f(b)-f(a)]$ (C) $kF(b-a)$, where $\frac{dF(x)}{dx}=f(x)$

(D) $k\int_a^b f(x)dx$ (B) $\left.\frac{[kf(x)]^2}{2}\right|_a^b$

48. $\frac{d}{dt}\int_0^t \sqrt{x^3+1}\,dx =$

(A) $\sqrt{t^3+1}$ (B) $\frac{\sqrt{t^3+1}}{3t^2}$ (C) $\frac{2}{3}(t^3+1)\left(\sqrt{t^3+1}-1\right)$

(D) $3x^2\sqrt{x^3+1}$ (E) $\sqrt{t^3+1}-1$

49. If $F(u)=\int_1^u (2-x^2)^3\,dx$, then $F'(u)$ is equal to

(A) $-6u(2-u^2)^2$ (B) $\frac{(2-u^2)^4}{4}-\frac{1}{4}$ (C) $(2-u^2)^3-1$

(D) $(2-u^2)^3$ (E) $-2u(2-u^2)^3$

50. $\frac{d}{dx}\int_{\pi/2}^{x^2} \sqrt{\sin t}\,dt =$

(A) $\sqrt{\sin t^2}$ (B) $2x\sqrt{\sin x^2}-1$ (C) $\frac{2}{3}(\sin^{3/2}x^2-1)$

(D) $\sqrt{\sin x^2}-1$ (E) $2x\sqrt{\sin x^2}$

51. If $x=4\cos\theta$ and $y=3\sin\theta$, then $\int_2^4 xy\,dx$ is equivalent to

(A) $48\int_{\pi/3}^{0} \sin\theta\cos^2\theta\,d\theta$ (B) $48\int_2^4 \sin^2\theta\cos\theta\,d\theta$

(C) $36\int_2^4 \sin\theta\cos^2\theta\,d\theta$ (D) $-48\int_0^{\pi/3} \sin\theta\cos^2\theta\,d\theta$

(E) $48\int_0^{\pi/3} \sin^2\theta\cos\theta\,d\theta$

52. A continuous function f takes on the values shown in the table below. Estimate $\int_0^{30} f(x)dx$ using a left rectangular approximation with five subintervals.

x	0	5	12	20	24	30
fx	9	11	7	3	1	2

(A) 144 (B) 165 (C) 170 (D) 186 (E) 196

53. Find the value of x at which the function $y=x^2$ reaches its average value on the interval $[0,10]$.

(A) 4.642 (B) 5 (C) 5.313 (D) 5.774 (E) 7.071

54. The average value of $f(x)=\begin{cases}x^3, & x<2\\ 4x, & x\ge 2\end{cases}$ on the interval $0\le x\le 5$ is

(A) 8 (B) 9.2 (C) 16 (D) 23
(E) undefined because f is not differentiable on this interval

Answer Key

1. **C**	12. **B**	23. **E**	34. **C**	45. **C**
2. **B**	13. **E**	24. **C**	35. **C**	46. **C**
3. **E**	14. **C**	25. **B**	36. **C**	47. **D**
4. **B**	15. **D**	26. **D**	37. **D**	48. **A**
5. **D**	16. **A**	27. **D**	38. **B**	49. **D**
6. **A**	17. **C**	28. **D**	39. **E**	50. **E**
7. **D**	18. **E**	29. **E**	40. **B**	51. **E**
8. **A**	19. **A**	30. **D**	41. **B**	52. **E**
9. **C**	20. **E**	31. **D**	42. **D**	53. **D**
10. **D**	21. **C**	32. **E**	43. **B**	54. **B**
11. **B**	22. **A**	33. **C**	44. **D**	

Answers Explained

1. **(C)** The integral is equal to

$$\left(\frac{1}{3}x^3 - \frac{1}{2}x^2 - x\right)\Bigg|_{-1}^{1} = -\frac{7}{6} - \frac{1}{6}.$$

2. **(B)** Rewrite as $\int_1^2 \left(1 - \frac{1}{3}\cdot\frac{1}{x}\right) dx$. This equals

$$\left(x - \frac{1}{3}\ln x\right)\Bigg|_1^2 = 2 - \frac{1}{3}\ln 2 - 1.$$

3. **(E)** Rewrite as

$$-\int_0^3 (4-t)^{-1/2}(-1\,dt) = -2\sqrt{4-t}\,\Bigg|_0^3 = -2(1-2).$$

4. **(B)** This integral equals

$$\frac{1}{3}\int_{-1}^{0} (3u+4)^{1/2}\cdot 3\,du = \frac{1}{3}\cdot\frac{2}{3}(3u+4)^{3/2}\Bigg|_{-1}^{0}$$
$$= \frac{2}{9}\left(4^{3/2} - 1^{3/2}\right).$$

5. **(D)** $\frac{1}{2}\int_2^3 \frac{2dy}{2y-3} = \frac{1}{2}\ln(2y-3)\Bigg|_2^3 = \frac{1}{2}(\ln 3 - \ln 1)$

6. **(A)** Rewrite as

$$-\frac{1}{2}\int_0^{\sqrt{3}} (4-x^2)^{-1/2}(-2x dx) = -\frac{1}{2}\cdot 2\sqrt{4-x^2}\,\Bigg|_0^{\sqrt{3}} = -(1-2).$$

7. **(D)** $\frac{1}{2}\int_0^1 (2t-1)^3(2dt) = \frac{1}{2}\cdot\frac{(2t-1)^4}{4}\Bigg|_0^1 = \frac{1}{2}\left(\frac{(2\cdot 1 - 1)^4}{4} - \frac{(2\cdot 0 - 1)^4}{4}\right)$

8. **(A)** Divide: $\int_4^9 \left(x^{-1/2} + \frac{1}{2}x^{1/2}\right)dx = \left(2x^{1/2} + \frac{1}{2}\cdot\frac{2}{3}x^{3/2}\right)\Bigg|_4^9$

$$= \left(2\cdot 3 + \frac{1}{2}\cdot 27\right) - \left(2\cdot 2 + \frac{1}{3}\cdot 8\right).$$

9. **(C)** $\frac{1}{9}\int_{-3}^{3} \frac{dx}{1+\frac{x^2}{9}} = 3\cdot\frac{1}{9}\int_{-3}^{3}\frac{\frac{1}{3}dx}{1+\left(\frac{x}{3}\right)^2} = \frac{1}{3}\tan^{-1}\frac{x}{3}\Bigg|_{-3}^{3} = \frac{1}{3}\left(\frac{\pi}{4} - \left(-\frac{\pi}{4}\right)\right).$

10. **(D)** You get $-e^{-x}\Big|_0^1 = -(e^{-1} - 1)$.

11. **(B)** $\frac{1}{2}\int_0^1 e^{x^2}(2x\,dx) = \frac{1}{2}e^{x^2}\Big|_0^1 = \frac{1}{2}(e - 1)$.

12. **(B)** Evaluate $-\frac{1}{2}\cos 2\theta\Big|_0^{\pi/4}$, which equals $-\frac{1}{2}(0 - 1)$.

13. **(E)** $-\int_1^2 \frac{-dx}{3 - z} = -\ln(3 - z)\Big|_1^2 = -(\ln 1 - \ln 2)$.

14. **(C)** If $x = 2\sin\theta$, $\sqrt{4 - x^2} = 2\cos\theta$, $dx = 2\cos\theta\,d\theta$. When $x = 1$, $\theta = \frac{\pi}{6}$; when $x = 2$, $\theta = \frac{\pi}{2}$.

The integral is equivalent to $\int_{\pi/6}^{\pi/2} \frac{(2\cos\theta)(2\cos\theta)\,d\theta}{2\sin\theta}$.

15. **(D)** Evaluate $-\int_0^{\pi} \cos^2\theta(-\sin\theta\,d\theta)$. This equals $-\frac{1}{3}\cos^3\theta\Big|_0^{\pi} = -\frac{1}{3}(-1 - 1)$.

16. **(A)** $\int_1^e (\ln x)\left(\frac{1}{x}dx\right) = \frac{1}{2}\ln^2 x\Big|_1^e = \frac{1}{2}(1 - 0)$.

17. **(C)** Use the Parts Formula with $u = x$ and $dv = e^x\,dx$. Then $du = dx$ and $v = e^x$.

The result is $\left(xe^x - \int e^x\,dx\right)\Big|_0^1 = (xe^x - e^x)\Big|_0^1 = (e - e) - (0 - 1)$.

18. **(E)** $\frac{1}{2}\int_0^{\pi/6} \frac{2\cos\theta\,d\theta}{1 + 2\sin\theta} = \frac{1}{2}\ln(1 + 2\sin\theta)\Big|_0^{\pi/6}$ and get $\frac{1}{2}(\ln(1 + 1) - \ln 1)$.

19. **(A)** $\frac{1}{2}\int_{\sqrt{2}}^2 \frac{2u}{u^2 - 1}du = \frac{1}{2}\ln(u^2 - 1)\Big|_{\sqrt{2}}^2 = \frac{1}{2}(\ln 3 - \ln 1)$.

20. **(E)** $\frac{1}{2}\int_{\sqrt{2}}^2 (u^2 - 1)^{-2}\cdot 2u\,du = -\frac{1}{(2u^2 - 1)}\Big|_{\sqrt{2}}^2 = -\frac{1}{2}\left(\frac{1}{3} - \frac{1}{1}\right)$.

21. **(C)** $\frac{1}{2}\int_{\pi/12}^{\pi/4} \sin^{-2}2x\cos 2x\,(2\,dx) = -\frac{1}{2}\cdot\frac{1}{\sin 2x}\Big|_{\pi/12}^{\pi/4} = -\frac{1}{2}\left(\frac{1}{1} - \frac{1}{1/2}\right)$.

22. **(A)** $\int_0^1 (1 + e^x)\,dx = (x + e^x)$ _0^1 $= (1 + e) - 1$.

23. **(E)** Evaluate $\ln(e^x + 1)\Big|_0^1$, getting $\ln(e + 1) - \ln 2$.

24. **(C)** Note that $dx = \sec^2\theta\,d\theta$ and that $\sqrt{1 + \tan^2\theta} = \sec\theta$. Be sure to express the limits as values of θ: $1 = \tan\theta$ yields $\theta = \frac{\pi}{4}$; $\sqrt{3} = \tan\theta$ yields $\theta = \frac{\pi}{3}$.

25. **(B)** If $u = \sqrt{x + 1}$, then $u^2 = x + 1$, and $2u\,du = dx$. When you substitute for the limits, you get $2\int_1^2 \frac{u\,du}{u(u^2 - 1)}$. Since $u \neq 0$ on its interval of integration, you may divide numerator and denominator by it.

26. **(D)** $\int_8^0 f'(x)\,dx = f(0) - f(8) = 11 - 7 = 4$

27. **(D)** On [0,6] with $n = 3$, $\Delta x = 2$. Heights of rectangles at $x = 1, 3,$ and 5 are 5, 9, and 5, respectively; $M(3) = (5 + 9 + 5)(2)$.

28. **(D)** $\int_4^{10} f(x)\,dx \approx \left(\frac{2+6}{2}\right)\cdot 2 + \left(\frac{6+5}{2}\right)\cdot 1 + \left(\frac{5+8}{2}\right)\cdot 3 \approx 33$

29. **(E)** For $L(2)$ use the circumscribed rectangles: $1\cdot 1 + \frac{1}{2}\cdot 1 = \frac{3}{2}$;

for $R(2)$ use the inscribed rectangles: $\frac{1}{2}\cdot 1 + \frac{1}{3}\cdot 1 = \frac{5}{6}$.

30. **(D)** On [0,1] $f(x) = \cos x$ is decreasing, so $R < L$. Furthermore, f is concave downward, so $T < A$.

31. **(D)** On the interval $[-1,3]$ the area under the graph of $y = |x|$ is the sum of the areas of two triangles: $\frac{1}{2}(1)(1) + \frac{1}{2}(3)(3) = 5$.

32. **(E)** Note that the graph $y = |x + 1|$ is the graph of $y = |x|$ translated one unit to the left. The area under the graph $y = |x + 1|$ on the interval $[-3,2]$ is the sum of the areas of two triangles: $\frac{1}{2}(2)(2) + \frac{1}{2}(3)(3) = \frac{13}{2}$.

33. **(C)** Because $y = \sqrt{64 - x^2}$ is a semicircle of radius 8, its area is 32π. The domain is $[-8,8]$, or 16 units wide. Hence the average height of the function is $\frac{32\pi}{16}$.

34. **(C)** The average value is equal to $\frac{1}{\pi/2 - \pi/3}\int_{\pi/3}^{\pi/2} \cos x\,dx$.

35. **(C)** The average value is equal to $\frac{1}{\pi/4 - \pi/6}\int_{\pi/6}^{\pi/4} \csc^2 x\,dx$.

36. **(C)** From the integral, we get $a = 1, b = 5$, so $\Delta x = \frac{5-1}{n} = \frac{4}{n}$ and $x_k = a + k\cdot\Delta x = 1 + \frac{4k}{n} = 1 + \frac{4k}{n}$. Replace x with x_k and replace dx with Δx in the integrand to get the general term in the summation.

37. **(D)** From the integral, we get $a = 0, b = \pi$, so $\Delta x = \frac{\pi - 0}{n} = \frac{\pi}{n}$ and $x_k = a + k\cdot\Delta x = 0 + \frac{\pi k}{n} = \frac{\pi k}{n}$. Replace x with x_k and replace dx with Δx in the integrand to get the general term in the summation.

38. **(B)** From the Riemann Sum, we see $\Delta x = \frac{3}{n}$, then $k\cdot\Delta x = \frac{3k}{n}$. Notice that the term involving k in the Riemann Sum is not equal to $\frac{3k}{n}$ but $2\left(\frac{3k}{n}\right)$. Thus, we choose $x_k = \frac{3k}{n}$, so $a = 0$ and $\Delta x = \frac{b-0}{n} = \frac{3}{n}$, so $b = 3$. Since x_k replaces x, $f(x) = \sin(2x + 2)$ giving the integral $\int_0^3 \sin(2x + 2)\,dx$.

39. **(E)** The average value is $\frac{1}{5-0}\int_0^5 f(x)dx$. The integral represents the area of a trapezoid: $\frac{1}{2}(5 + 3)\cdot 10 = 40$. The average value is $\frac{1}{5}(40)$.

40. **(B)** Since $x^2 + y^2 = 16$ is a circle, the given integral equals the area of a semicircle of radius 4.

41. **(B)** Use a graphing calculator.

42. **(D)** A vertical line at $x = 2$ divides the area under f into a trapezoid and a triangle; hence, $\int_0^6 f(x)\,dx = \frac{1}{2}(2 + 4)(2) + \frac{1}{2}(4)(4) = 14$. Thus, the average value of f on [0,6] is $\frac{14}{6} = 2\frac{1}{3}$. There are points on f with y-values of $2\frac{1}{3}$ in the intervals [0,2] and [2,4].

43. **(B)** $\int_0^2 f'(3x)dx = \frac{1}{3}\int_0^2 f'(3x)(3dx) = \frac{1}{3}f(3x)\Big|_0^2 = \frac{1}{3}\big(f(6) - f(0)\big) = \frac{1}{3}(0 - 2)$

44. **(D)** $g'(x) = f(2x) \cdot 2$; thus $g'(1) = f(2) \cdot 2$

45. **(C)** $\int_0^6 \big(x^2 - f(x)\big)dx = \int_0^6 x^2 dx - \int_0^6 f(x)dx = \frac{x^3}{3}\Big|_0^6 - 14$. (Why 14? See the solution for Question 42.)

46. **(C)** This is the Mean Value Theorem for Integrals (page 242).

47. **(D)** This is theorem (2) on page 242. Prove by counterexamples that (A), (B), (C), and (D) are false.

48. **(A)** This is a restatement of the Fundamental Theorem. In theorem (1) on page 242, interchange t and x.

49. **(D)** Apply theorem (1) on page 242, noting that

$$F'(u) = \frac{d}{du}\int_a^u f(x)dx = f(u).$$

50. **(E)** Let $y = \int_{\pi/2}^{x^2} \sqrt{\sin t}\,dt$ and $u = x^2$; then

$$y = \int_{\pi/2}^{u} \sqrt{\sin t}\,dt$$

By the Chain Rule, $\frac{dy}{dx} = \frac{dy}{du} \cdot \frac{du}{dx} = \sqrt{\sin u} \cdot 2x$, where theorem (1) on page 242 is used to find $\frac{dy}{du}$. Replace u by x^2.

51. **(E)** Since $dx = -4 \sin \theta\, d\theta$, you get the new integral $-48\int_{\pi/3}^{0} \sin^2 \theta \cos \theta\, d\theta$. Use theorem (4) on page 242 to get the correct answer.

52. **(E)** $L(5) = 9 \cdot 5 + 11 \cdot 7 + 7 \cdot 8 + 3 \cdot 4 + 1 \cdot 6$.

53. **(D)** The average value is $\frac{1}{10 - 0}\int_0^{10} x^2 dx = \frac{1}{10} \cdot \frac{x^3}{3}\Big|_0^{10} = \frac{100}{3}$. Solve $x^2 = \frac{100}{3}$.

54. **(B)** $\frac{1}{5 - 0}\int_0^5 f(x)dx = \frac{1}{5}\left(\int_0^2 x^3 dx + \int_2^5 4x dx\right) = \frac{1}{5}\left(\frac{x^4}{4}\Big|_0^2 + 2x^2\Big|_2^5\right) = \frac{1}{5}(4 + 42)$

Applications of Integration to Geometry

7

CONCEPTS AND SKILLS

In this chapter, we will review using definite integrals to find areas and volumes; specifically

- **area under a curve,**
- **area between two curves,**
- **volumes of solids with known cross sections,**
- **and volumes of solids of revolution (using disks and washers).**

We'll also review related BC topics, including

- **arc length;**
- **arc lengths, areas, and volumes involving parametrically defined functions;**
- **and area and arc length for polar curves.**

Also for BC Calculus students, we'll review the topic of improper integrals, including

- **recognizing when an integral is improper**
- **and techniques for determining whether an improper integral converges or diverges.**

A. AREA

To find an area, we

(1) draw a sketch of the given region and of a typical element;
(2) write the expression for the area of a typical rectangle; and
(3) set up the definite integral that is the limit of the Riemann Sum of n areas as $n \to \infty$.

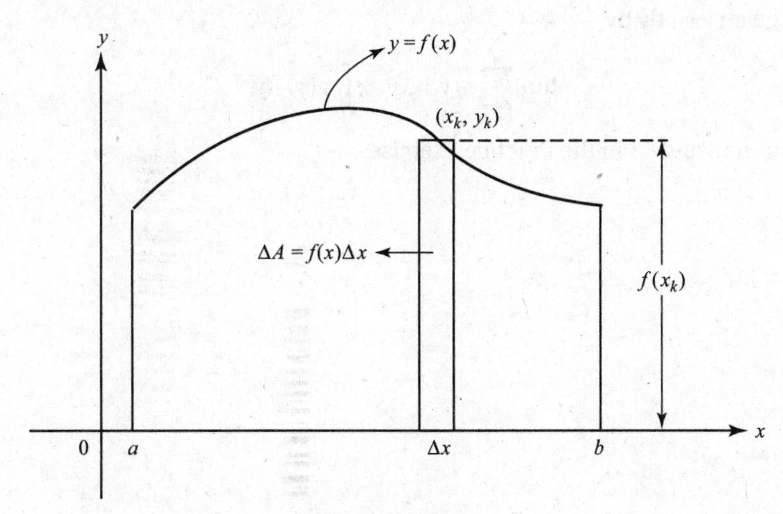

Figure N7–1

If $f(x)$ is nonnegative on $[a,b]$, as in Figure N7–1, then $f(x_k)\ \Delta x$ can be regarded as the area of a typical approximating rectangle, and the area bounded by the x-axis, the curve, and the vertical lines $x = a$ and $x = b$ is given exactly by

$$\lim_{n\to\infty} \sum_{k=1}^{n} f(x_k)\ \Delta x \qquad \text{and hence by} \qquad \int_a^b f(x)\ dx.$$

See Questions 1, 5, and 10 in the Practice Exercises at the end of this chapter.

If $f(x)$ changes sign on the interval (Figure N7–2), we find the values of x for which $f(x) = 0$ and note where the function is positive, where it is negative. The total area bounded by the x-axis, the curve, $x = a$, and $x = b$ is here given exactly by

$$\int_a^c f(x)\ dx - \int_c^d f(x)\ dx + \int_d^b f(x)\ dx,$$

where we have taken into account that $f(x_k)\ \Delta x$ is a negative number if $c < x < d$.

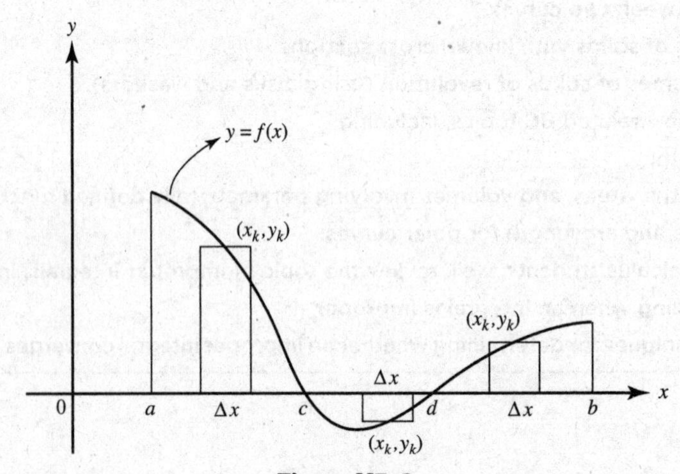

Figure N7–2

See Question 11 in the Practice Exercises.

If x is given as a function of y, say $x = g(y)$, then (Figure N7–3) the subdivisions are made along the y-axis, and the area bounded by the y-axis, the curve, and the horizontal lines $y = a$ and $y = b$ is given exactly by

$$\lim_{n\to\infty} \sum_{k=1}^{n} g(y_k)\ \Delta y = \int_a^b g(y)\ dy.$$

See Questions 3 and 13 in the Practice Exercises.

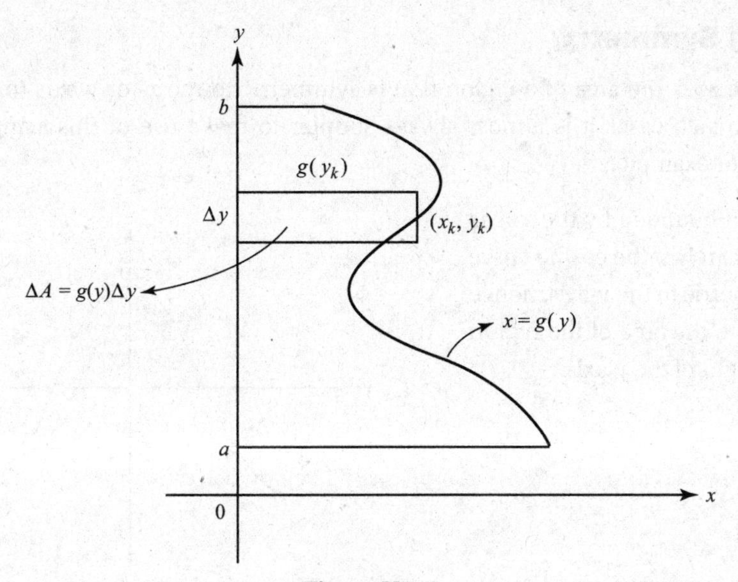

Figure N7–3

A1. Area Between Curves

To find the area between curves (Figure N7–4), we first find where they intersect and then write the area of a typical element for each region between the points of intersection. For the total area bounded by the curves $y = f(x)$ and $y = g(x)$ between $x = a$ and $x = e$, we see that, if they intersect at $[c,d]$, the total area is given exactly by

$$\int_a^c [f(x) - g(x)]\, dx + \int_c^e [g(x) - f(x)]\, dx.$$

See Questions 4, 6, 7, and 9 in the Practice Exercises.

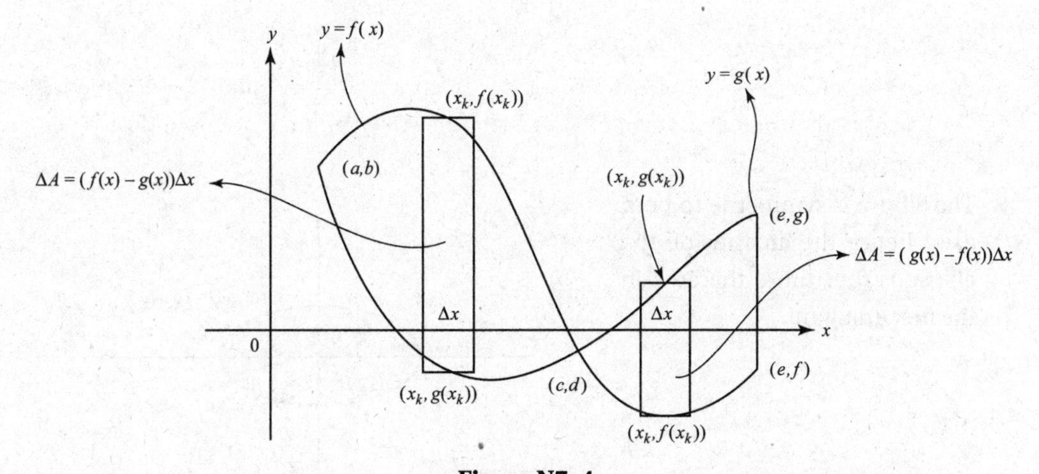

Figure N7–4

A2. Using Symmetry

Frequently we seek the area of a region that is symmetric to the x- or y-axis (or both) or to the origin. In such cases it is almost always simpler to make use of this symmetry when integrating. For example:

- The area bounded by the x-axis and this arch of the cosine curve is symmetric to the y-axis; hence it is twice the area of the region to the right of the y-axis.

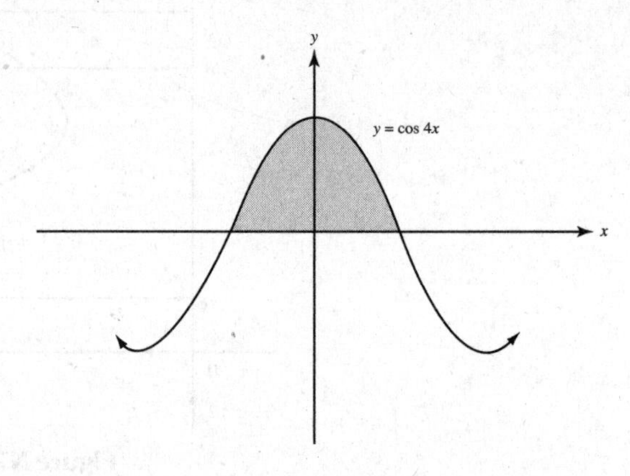

- The area bounded by the parabola and the line is symmetric to the x-axis; hence it is twice the area of the region above the x-axis.

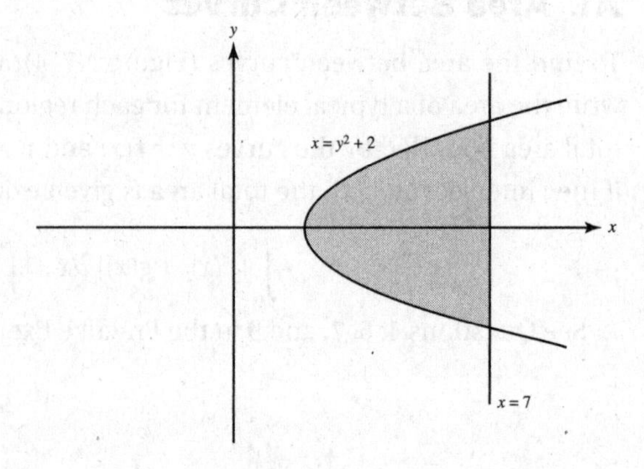

- The ellipse is symmetric to both axes; hence the area inside the ellipse is four times the area in the first quadrant.

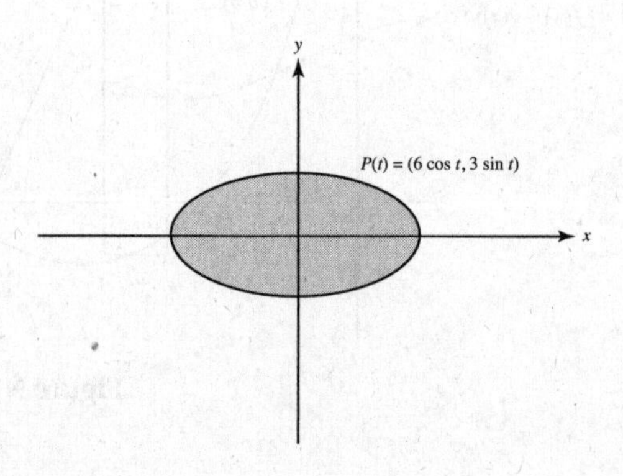

Evaluating $\int_a^b f(x)\,dx$ Using a Graphing Calculator

The calculator is especially useful in evaluating definite integrals when the x-intercepts are not easily determined otherwise or when an explicit antiderivative of f is not obvious (or does not exist).

➡ Example 1

Evaluate $\int_0^1 e^{-x^2}\,dx$.

SOLUTION: The integrand $f(x) = e^{-x^2}$ has no easy antiderivative. The calculator estimates the value of the integral to be 0.747 to three decimal places.

➡ Example 2

In Figure N7–5, find the area under $f(x) = -x^4 + x^2 + x + 10$ and above the x-axis.

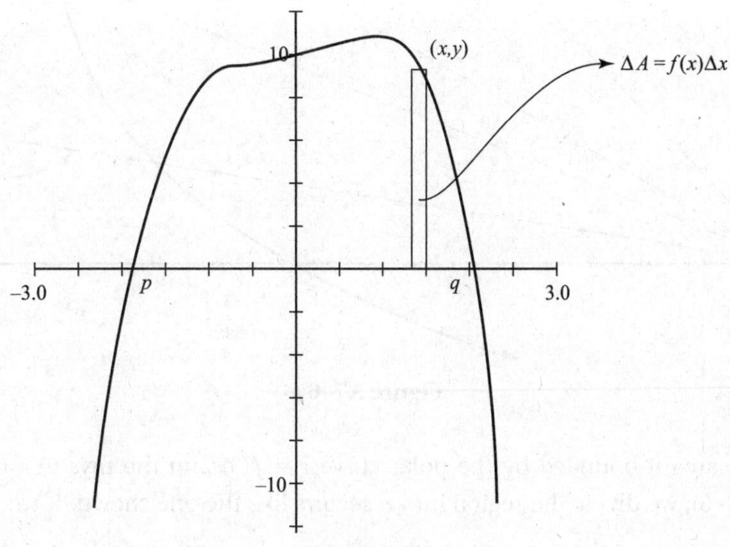

Figure N7–5

SOLUTION: To get an accurate answer for the area $\int_p^q f(x)\,dx$, use the calculator to find the two intercepts, storing them as P and Q, and then evaluate the integral:

$$\int_P^Q \left(-x^4 + x^2 + x + 10\right) dx = 32.832,$$

which is accurate to three decimal places.

BC ONLY

Region Bounded by a Parametric Curve

If x and y are given parametrically, say by $x = f(\theta)$, $y = g(\theta)$, then to evaluate $\int_a^b y\,dx$, we express y, dx, and the limits a and b in terms of θ and $d\theta$, then integrate. Remember that we define dx to be $x'(\theta)\,d\theta$, or $f'(\theta)\,d\theta$.

See Questions 14, 15, and 44 in the Practice Exercises.

BC ONLY

Region Bounded by Polar Curve

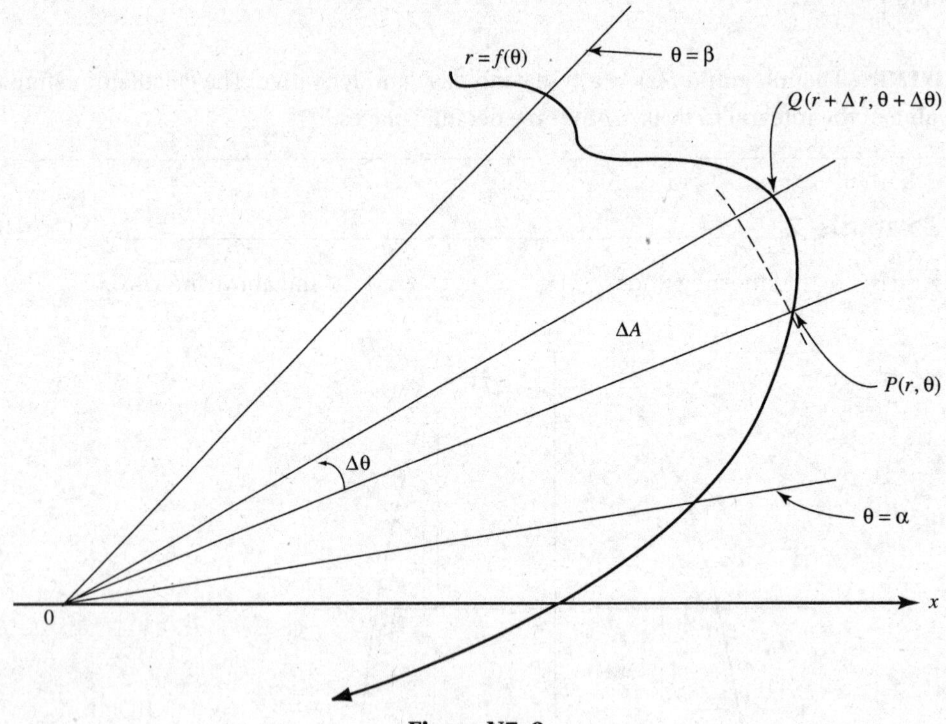

Figure N7–6

To find the area A bounded by the polar curve $r = f(\theta)$ and the rays $\theta = \alpha$ and $\theta = \beta$ (see Figure N7–6), we divide the region into n sectors like the one shown. If we think of that element of area, ΔA, as a circular sector with radius r and central angle $\Delta\theta$, its area is given by $\Delta A = \frac{1}{2}r^2\,\Delta\theta$.

Summing the areas of all such sectors yields the area of the entire region:

$$A = \lim_{n\to\infty} \sum_{k=1}^{n} \frac{1}{2} r_k^{\,2}\,\Delta\theta_k.$$

The expression above is a Riemann Sum, equivalent to this definite integral:

$$A = \int_{\alpha}^{\beta} \frac{1}{2} r^2 d\theta\,.$$

We have assumed above that $f(\theta) \geqq 0$ on $[\alpha, \beta]$. We must be careful in determining the limits α and β in (2); often it helps to think of the required area as that "swept out" (or generated) as the radius vector (from the pole) rotates from $\theta = \alpha$ to $\theta = \beta$. It is also useful to exploit symmetry of the curve wherever possible.

The relations between rectangular and polar coordinates, some common polar equations, and graphs of polar curves are given in the Appendix, starting on page 633.

BC ONLY

Example 3

Find the area inside both the circle $r = 3 \sin \theta$ and the cardioid $r = 1 + \sin \theta$.

SOLUTION: Choosing an appropriate window, graph the curves on your calculator. See Figure N7–7, where one half of the required area is shaded. Since $3 \sin \theta = 1 + \sin \theta$ when $\theta = \frac{\pi}{6}$ or $\frac{5\pi}{6}$, we see that the desired area is twice the sum of two parts: the area of the circle swept out by θ as it varies from 0 to $\frac{\pi}{6}$ plus the area of the cardioid swept out by a radius vector as θ varies from $\frac{\pi}{6}$ to $\frac{\pi}{2}$. Consequently

$$A = 2\left[\int_0^{\pi/6} \frac{9}{2} \sin^2 \theta \, d\theta + \int_{\pi/6}^{\pi/2} \frac{1}{2}(1 + \sin \theta)^2 \, d\theta\right] = \frac{5\pi}{4}.$$

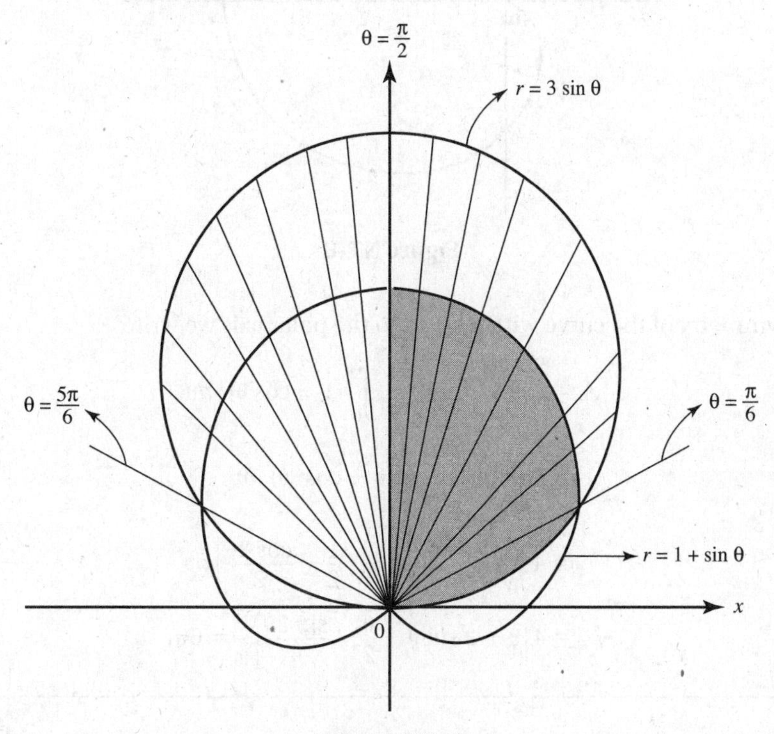

Figure N7–7

See also Questions 46 and 47 in the Practice Exercises.

BC ONLY

Example 4

Find the area enclosed by the cardioid $r = 2(1 + \cos\theta)$.

SOLUTION: We graphed the cardioid on our calculator, using polar mode, in the window $[-2,5] \times [-3,3]$ with θ in $[0,2\pi]$.

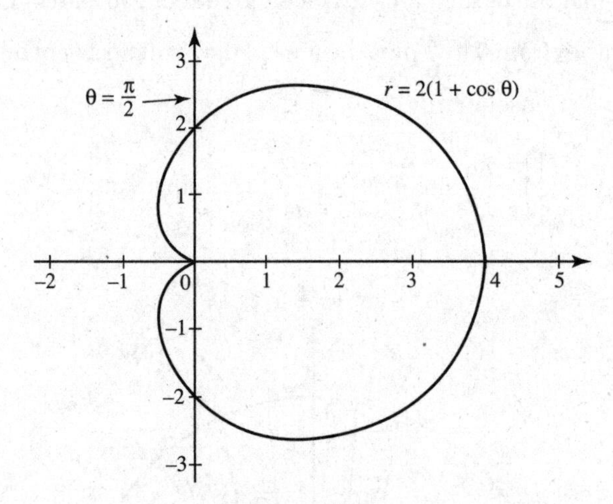

Figure N7–8

Using the symmetry of the curve with respect to the polar axis we write

$$A = 2 \cdot \frac{1}{2}\int_0^{\pi} r^2\, d\theta = 4\int_0^{\pi} (1 + \cos\theta)^2\, d\theta$$

$$= 4\int_0^{\pi} (1 + 2\cos\theta + \cos^2\theta)\, d\theta$$

$$= 4\int_0^{\pi} \left(1 + 2\cos\theta + \frac{1}{2} + \frac{\cos 2\theta}{2}\right) d\theta$$

$$= 4\left[\theta + 2\sin\theta + \frac{\theta}{2} + \frac{\sin 2\theta}{4}\right]\Bigg|_0^{\pi} = 6\pi.$$

B. VOLUME

B1. Solids with Known Cross Sections

If the *area of a cross section* of a solid *is known* and can be expressed in terms of x, then the volume of a typical slice, ΔV, can be determined. The volume of the solid is obtained, as usual, by letting the number of slices increase indefinitely. In Figure N7–9, the slices are taken perpendicular to the x-axis so that $\Delta V = A(x)\,\Delta x$, where $A(x)$ is the area of a cross section and Δx is the thickness of the slice.

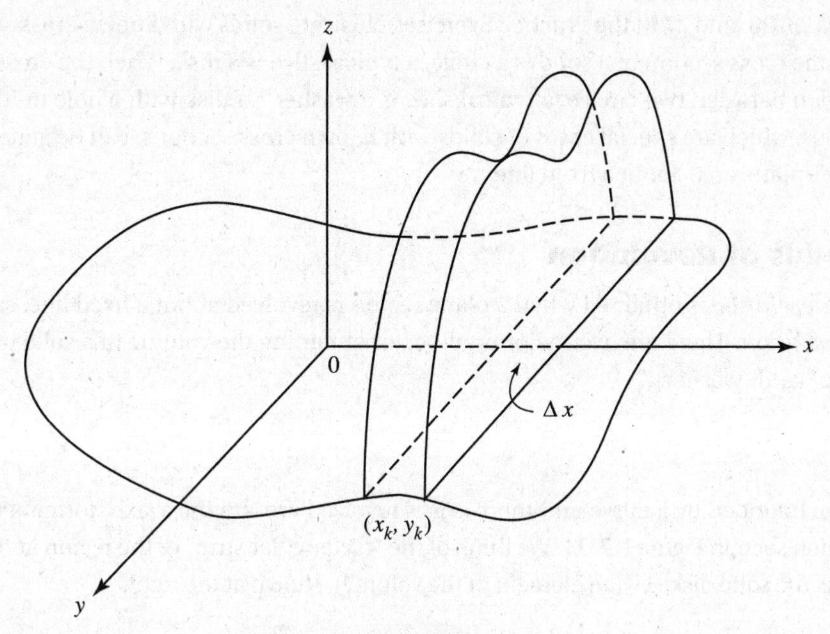

Figure N7–9

➥ Example 5

A solid has as its base the circle $x^2 + y^2 = 9$, and all cross sections parallel to the y-axis are squares. Find the volume of the solid.

SOLUTION:

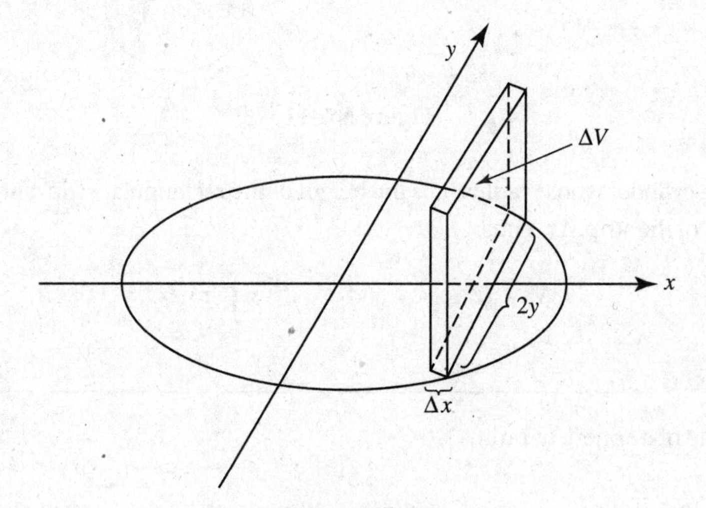

Figure N7–10

In Figure N7–10 the element of volume is a square prism with sides of length $2y$ and thickness Δx, so

$$\Delta V = (2y)^2\,\Delta x = 4y^2\,\Delta x = 4(9 - x^2)\,\Delta x.$$

Now, using symmetry across the y-axis, we find the volume of the solid:

$$V = 2\int_0^3 4\left(9 - x^2\right) dx = 8\int_0^3 \left(9 - x^2\right) dx = 8\left(9x - \frac{x^3}{3}\right)\Bigg|_0^3 = 144.$$

Questions 25, 26, and 27 in the Practice Exercises illustrate solids with known cross sections.

When the cross section of a solid is a circle, a typical slice is a disk. When the cross section is the region between two circles, a typical slice is a washer—a disk with a hole in it. Both of these solids, which are special cases of solids with known cross sections, can be generated by revolving a plane area about a fixed line.

B2. Solids of Revolution

A *solid of revolution* is obtained when a plane region is revolved about a fixed line, called the *axis of revolution*. There are two major methods of obtaining the volume of a solid of revolution "disks" and "washers."

Disks

DISKS

The region bounded by a curve and the x-axis is revolved around the x-axis, forming the solid of revolution seen in Figure N7–11. We think of the "rectangular strip" of the region at the left as generating the solid disk, ΔV (an element of the volume), shown at the right.

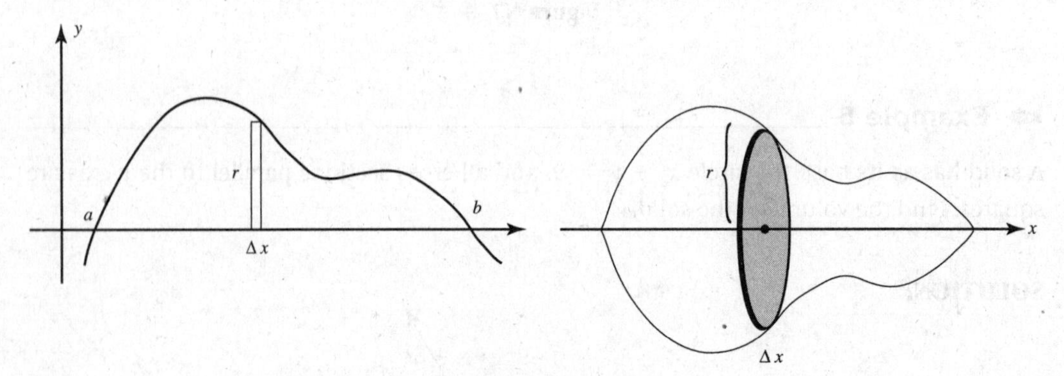

Figure N7–11

This disk is a cylinder whose radius, r, is the height of the rectangular strip, and whose height is the thickness of the strip, Δx. Thus

$$\Delta V = \pi r^2 \Delta x \qquad \text{and} \qquad V = \pi \int_a^b r^2\, dx.$$

Example 6

Find the volume of a sphere of radius r.

SOLUTION: If the region bounded by a semicircle (with center O and radius r) and its diameter is revolved about the x-axis, the solid of revolution obtained is a sphere of radius r, as seen in Figure N7–12.

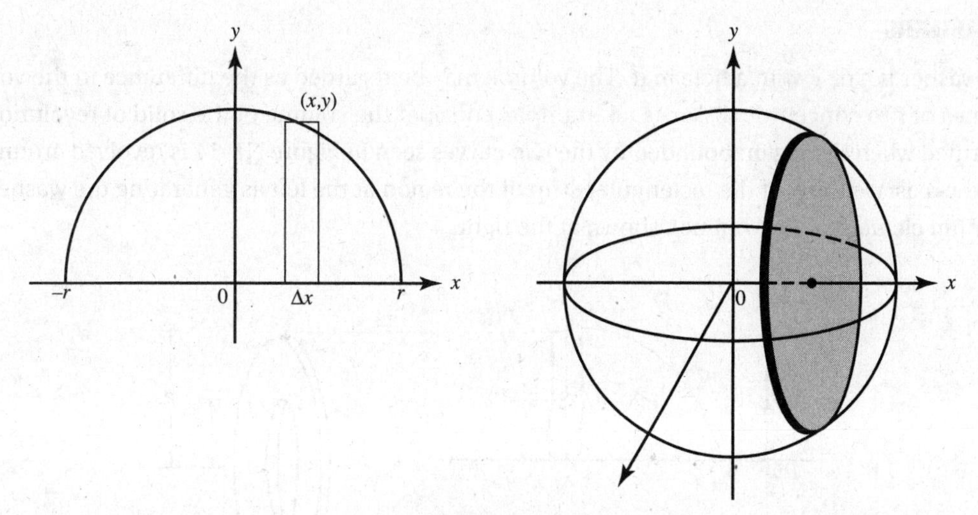

Figure N7–12

The volume ΔV of a typical disk is given by $\Delta V = \pi y^2\,\Delta x$. The equation of the circle is $x^2 + y^2 = r^2$. To find the volume of the sphere, we form a Riemann Sum whose limit as n becomes infinite is a definite integral. Then,

$$V = \int_{-r}^{r} \pi y^2\,dx = \pi \int_{-r}^{r} (r^2 - x^2)\,dx = \pi \left[r^2 x - \frac{x^3}{3} \right]\Bigg|_{-r}^{r} = \frac{4}{3}\pi r^3.$$

➥ Example 7

Find the volume of the solid generated when the region bounded by $y = x^2$, $x = 2$, and $y = 0$ is rotated about the line $x = 2$ as shown in Figure N7–13.

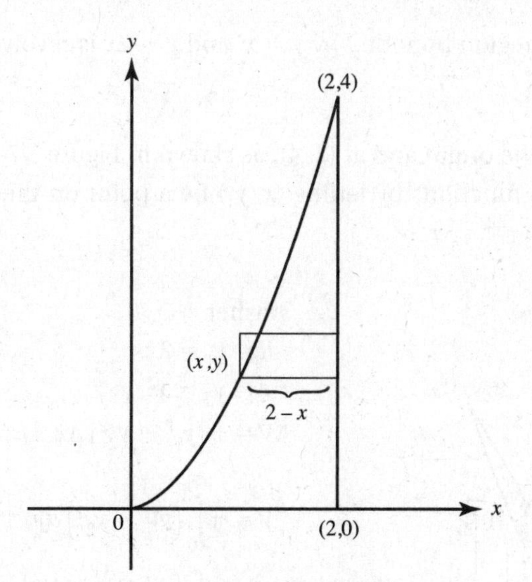

Figure N7–13

SOLUTION:

Disk.

$r = 2 - x.$

$\Delta V = \pi(2 - x)^2\,\Delta y.$

$$V = \pi \int_0^4 (2 - x)^2\,dy$$

$$= \pi \int_0^4 (2 - \sqrt{y})^2\,dy$$

$$= \frac{8\pi}{3}.$$

See Questions 18, 49, 51, 52, and 53 in the Practice Exercises for examples of finding volumes by disks.

WASHERS

A washer is a disk with a hole in it. The volume may be regarded as the difference in the volumes of two concentric disks. As an example, consider the volume of the solid of revolution formed when the region bounded by the two curves seen in Figure N7–14 is revolved around the x-axis. We think of the rectangular strip of the region at the left as generating the washer, ΔV (an element of the volume), shown at the right.

Washers

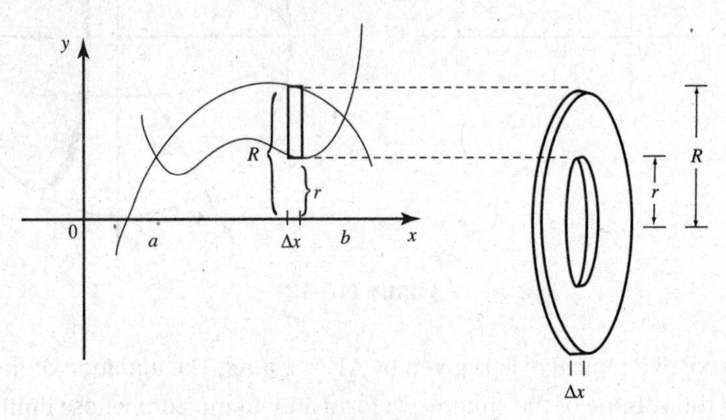

Figure N7–14

This washer's height is the thickness of the rectangular strip, Δx. The washer is a disk whose outer radius, R, is the distance to the top of the rectangular strip, with the disk of inner radius r (the distance to the bottom of the strip) removed. Thus:

$$\Delta V = \pi R^2 \Delta x - \pi r^2 \Delta x = \pi\left(R^2 - r^2\right)\Delta x \quad \text{and} \quad V = \pi\int_a^b \left(R^2 - r^2\right) dx.$$

Example 8

Find the volume obtained when the region bounded by $y = x^2$ and $y = 2x$ is revolved about the x-axis.

SOLUTION: The curves intersect at the origin and at (2, 4), as shown in Figure N7–15. Note that we distinguish between the two functions by letting (x, y_1) be a point on the line and (x, y_2) be a point on the parabola.

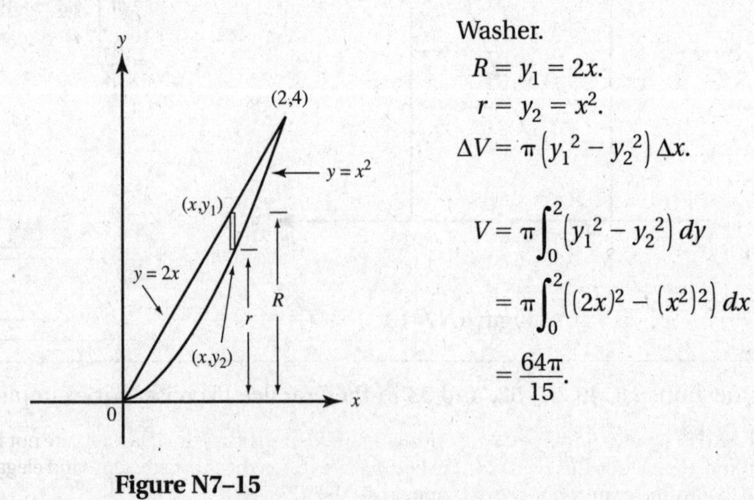

Washer.

$R = y_1 = 2x.$

$r = y_2 = x^2.$

$\Delta V = \pi\left(y_1^{\,2} - y_2^{\,2}\right)\Delta x.$

$$V = \pi\int_0^2 \left(y_1^{\,2} - y_2^{\,2}\right) dy$$

$$= \pi\int_0^2 \left((2x)^2 - (x^2)^2\right) dx$$

$$= \frac{64\pi}{15}.$$

Figure N7–15

➡ Example 9

Find the volume of the solid generated when the region bounded by $y = x^2$, $x = 2$, and $y = 0$ is rotated about the y-axis, as shown in Figure N7–16.

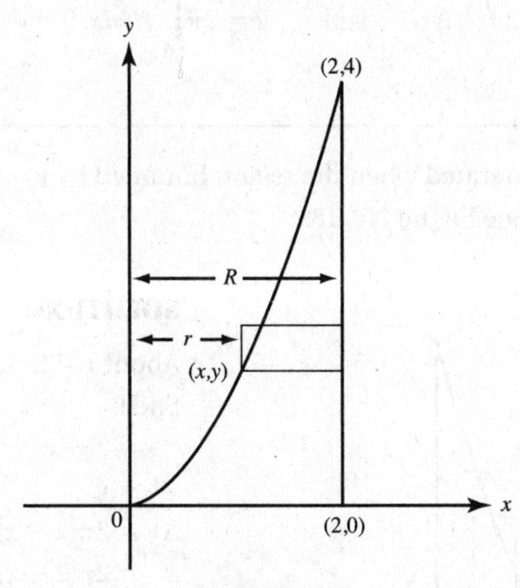

Figure N7–16

SOLUTION:

Washer.

$R = 2.$

$r = x.$

$$\Delta V = \pi\left(2^2 - x^2\right)\Delta y$$

$$V = \pi\int_0^4 \left(2^2 - x^2\right) dy$$

$$= \pi\int_0^4 (4 - y)\, dy$$

$$= 8\pi.$$

See Questions 19, 21, 48, 50, and 54 in the Practice Exercises for examples in which washers are regarded as the differences of two disks.

Occasionally when more than one method is satisfactory we try to use the most efficient. In the answers to each question in the Practice Exercises, a sketch is shown and the type and volume of a typical element are given. The required volume is then found by letting the number of elements become infinite and applying the Fundamental Theorem.

SHELLS‡

A cylindrical shell may be regarded as the outer skin of a cylinder. Its volume is the volume of the rectangular solid formed when this skin is peeled from the cylinder and flattened out. As an example, consider the volume of the solid of revolution formed when the region bounded by the two curves seen in Figure N7–17 is revolved around the y-axis. We think of the rectangular strip of the region at the left as generating the shell, ΔV (an element of the volume), shown at the right.

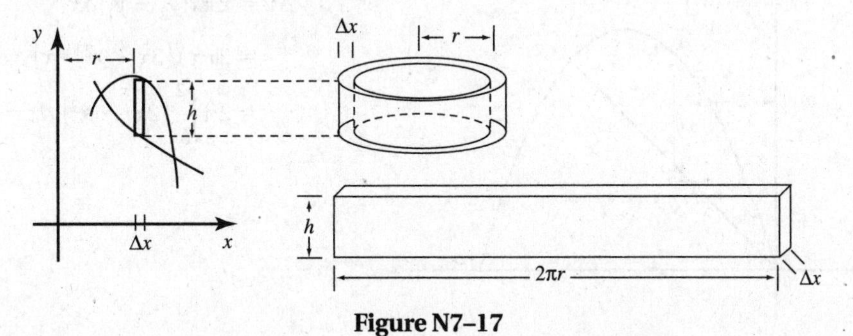

Figure N7–17

‡Examples 10–12 involve finding volumes by the method of shells. Although shells are not included in the Topic Outline, we include this method here because it is often the most efficient (and elegant) way to find a volume. No question *requiring* shells will appear on the AP exam.

This shell's radius, r, is the distance from the axis to the rectangular strip, and its height is the height of the rectangular strip, h. When the shell is unwound and flattened to form a rectangular solid, the length of the solid is the circumference of the cylinder, $2\pi r$, its height is the height of the cylinder, h, and its thickness is the thickness of the rectangular strip, Δx. Thus:

$$\Delta V = 2\pi rh\,\Delta x \quad \text{and} \quad V = 2\pi\int_a^b rh\,dx.$$

Example 10

Find the volume of the solid generated when the region bounded by $y = x^2$, $x = 2$, and $y = 0$ is rotated about the line $x = 2$. See Figure N7–18.

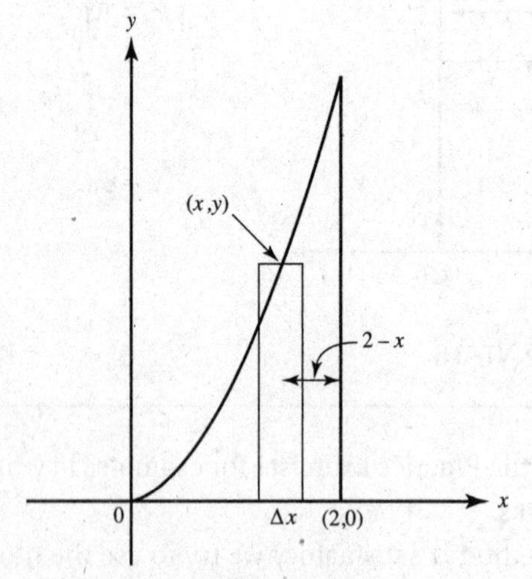

Figure N7–18

SOLUTION:

About $x = 2$.

Shell.

$r = 2 - x.$

$h = y.$

$\Delta V = 2\pi(2 - x)y\,\Delta x$

$= 2\pi(2 - x)x^2\,\Delta x.$

$2\pi\int_0^2 (2 - x)x^2\,dx = \frac{8\pi}{3}.$

(Note that we obtained the same result using disks in Example 7.)

Example 11

The region bounded by $y = 3x - x^2$ and $y = x$ is rotated about the y-axis. Find the volume of the solid obtained. See Figure N7–19.

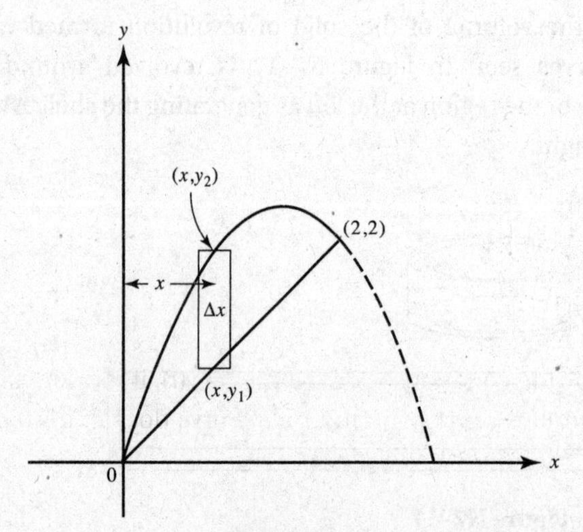

Figure N7–19

SOLUTION:

About the y-axis.

Shell.

$\Delta V = 2\pi x(y_2 - y_1)\Delta x$

$= 2\pi x\left[(3x - x^2) - x\right]\Delta x.$

$= 2\pi\int_0^2 (2x^2 - x^3)\,dx = \frac{8\pi}{3}.$

➡ Example 12

Find the volume obtained when the region bounded by $y = x^2$ and $y = 2x$ is revolved about the x-axis.

SOLUTION: The curves intersect at the origin and at (2,4), as shown in Figure N7–20. Note that we distinguish between the two functions by letting (x_1, y) be a point on the line and (x_2, y) be a point on the parabola.

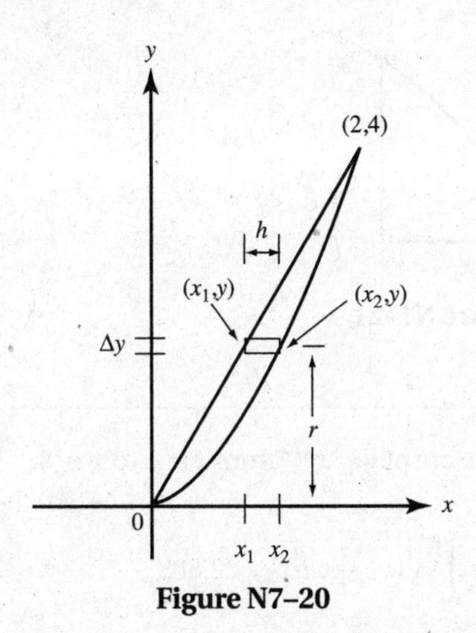

Figure N7–20

Shell.

$$r = y.$$
$$h = x_2 - x_1.$$
$$\Delta V = 2\pi y(x_2 - x_1)\Delta y.$$

$$V = 2\pi \int_0^4 y(x_2 - x_1)dy$$
$$= \pi \int_0^2 y\left(\sqrt{y} - \frac{y}{2}\right) dy$$
$$= \frac{64\pi}{15}.$$

(Note that we obtained the same result using washers in Example 8.)

NOTE: On pages 305 and 306 in Examples 32 and 33 we consider finding the volumes of solids using shells that lead to improper integrals.

C. ARC LENGTH

BC ONLY

If the derivative of a function $y = f(x)$ is continuous on the interval $a \leqq x \leqq b$, then the length s of the arc of the curve of $y = f(x)$ from the point where $x = a$ to the point where $x = b$ is given by

$$s = \int_a^b \sqrt{1 + \left(\frac{dy}{dx}\right)^2}\, dx. \qquad (1)$$

Here a small piece of the curve is equal approximately to $\sqrt{1 + (f'(x))^2}\, \Delta x$. As $\Delta x \to 0$, the sum of these pieces approaches the definite integral above.

If the derivative of the function $x = g(y)$ is continuous on the interval $c \leqslant y \leqslant d$, then the length s of the arc from $y = c$ to $y = d$ is given by

$$s = \int_c^d \sqrt{1 + \left(\frac{dx}{dy}\right)^2}\, dy. \qquad (2)$$

If a curve is defined parametrically by the equations $x = x(t)$ and $y = y(t)$, if the derivatives of the functions $x(t)$ and $y(t)$ are continuous on $[t_a, t_b]$ (and if the curve does not intersect itself), then the length of the arc from $t = t_a$ to $t = t_b$ is given by

$$s = \int_{t_a}^{t_b} \sqrt{\left(\frac{dx}{dy}\right)^2 + \left(\frac{dy}{dx}\right)^2}\, dt. \qquad (3)$$

BC ONLY

The parenthetical clause above is equivalent to the requirement that the curve is traced out just once as t varies from t_a to t_b.

As indicated in Equation (4), formulas (1), (2), and (3) can all be derived easily from the very simple relation

$$ds^2 = dx^2 + dy^2 \qquad (4)$$

and can be remembered by visualizing Figure N7–21.

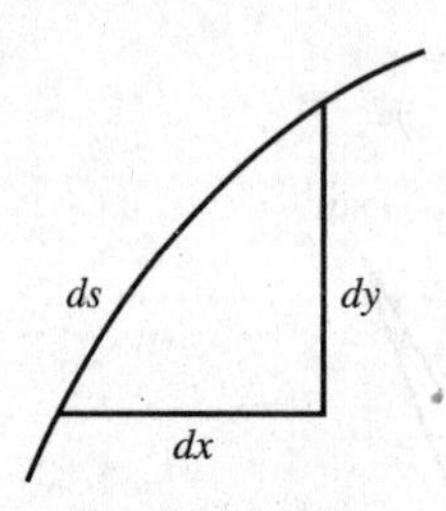

Figure N7–21

Example 13

Find the length, to three decimal places, of the arc of $y = x^{3/2}$ from $x = 1$ to $x = 8$.

SOLUTION: Here $\frac{dy}{dx} = \frac{3}{2}x^{1/2}$, so, by (1), $s = \int_1^8 \sqrt{1 + \frac{9}{4}x}\, dx \approx 22.803$.

Example 14

Find the length, to three decimal places, of the curve $(x - 2)^2 = 4y^3$ from $y = 0$ to $y = 1$.

SOLUTION: Since $x - 2 = 2y^{3/2}$ and $\frac{dx}{dy} = 3y^{1/2}$, Equation (2) above yields

$$s = \int_0^1 \sqrt{1 + 9y}\, dy \approx 2.268.$$

Example 15

The position (x, y) of a particle at time t is given parametrically by $x = t^2$ and $y = \frac{t^3}{3} - t$. Find the distance the particle travels between $t = 1$ and $t = 2$.

SOLUTION: We can use (4): $ds^2 = dx^2 + dy^2$, where $dx = 2t\, dt$ and $dy = (t^2 - 1)\, dt$. Thus,

$$ds = \sqrt{4t^2 + t^4 - 2t^2 + 1}\, dt, \text{ so}$$

$$s = \int_1^2 \sqrt{(t^2 + 1)^2}\, dt = \int_1^2 (t^2 + 1)dt$$

$$= \frac{t^3}{3} + t\Big|_1^2 = \frac{10}{3}.$$

Example 16

BC ONLY

Find the length of the arc of $y = \ln \sec x$ from $x = 0$ to $x = \frac{\pi}{3}$.

SOLUTION: $\frac{dy}{dx} = \frac{\sec x \tan x}{\sec x}$, so

$$s = \int_0^{\pi/3} \sqrt{1 + \tan^2 x}\, dx = \int_0^{\pi/3} \sec x\, dx$$

$$= \ln(\sec x + \tan x)\Big|_0^{\pi/3} = \ln(2 + \sqrt{3}).$$

D. IMPROPER INTEGRALS

Improper integral

There are two classes of improper integrals:

(1) those in which at least one of the limits of integration is infinite (the *interval* is not bounded); and

(2) those of the type $\int_a^b f(x)\, dx$, where $f(x)$ has a point of discontinuity (becoming infinite) at $x = c$, $a \leqq c \leqq b$ (the *function* is not bounded).

Illustrations of improper integrals of class (1) are:

$$\int_0^\infty \frac{dx}{\sqrt[3]{x+1}}; \quad \int_1^\infty \frac{dx}{x}; \quad \int_{-\infty}^\infty \frac{dx}{a^2 + x^2}; \quad \int_{-\infty}^0 e^{-x}\, dx;$$

$$\int_{-\infty}^{-1} \frac{dx}{x^n} \quad (n \text{ a real number}); \quad \int_{-\infty}^0 \frac{dx}{e^x + e^{-x}};$$

$$\int_0^\infty \frac{dx}{(4+x)^2}; \quad \int_{-\infty}^0 e^{-x^2} dx; \quad \int_1^\infty \frac{e^{-x^2}}{x^2}\, dx.$$

The following improper integrals are of class (2):

$$\int_0^1 \frac{dx}{x}; \quad \int_1^2 \frac{dx}{(x-1)^n} \quad (n \text{ a real number}); \quad \int_{-1}^1 \frac{dx}{1 - x^2};$$

$$\int_0^2 \frac{x}{\sqrt{4 - x^2}}\, dx; \quad \int_\pi^{2\pi} \frac{dx}{1 + \sin x}; \quad \int_{-1}^2 \frac{dx}{x(x-1)^2};$$

$$\int_0^{2\pi} \frac{\sin x\, dx}{\cos x + 1}; \quad \int_a^b \frac{dx}{(x-c)^n} \quad (n \text{ real}; a \leqq c \leqq b);$$

$$\int_0^1 \frac{dx}{\sqrt{x + x^4}}; \quad \int_{-2}^2 \sqrt{\frac{2+x}{2-x}}\, dx.$$

Sometimes an improper integral belongs to both classes. Consider, for example,

$$\int_0^\infty \frac{dx}{x}; \quad \int_0^\infty \frac{dx}{\sqrt{x + x^4}}; \quad \int_{-\infty}^1 \frac{dx}{\sqrt{1 - x}}.$$

In each case, the interval is not bounded and the integrand fails to exist at some point on the interval of integration.

BC ONLY

Note, however, that each integral of the following set is *proper*:

$$\int_{-1}^{3} \frac{dx}{\sqrt{x+2}}; \quad \int_{-2}^{2} \frac{dx}{x^2+4}; \quad \int_{0}^{\pi/6} \frac{dx}{\cos x};$$

$$\int_{0}^{e} \ln(x+1)\, dx; \quad \int_{-3}^{3} \frac{dx}{e^x+1}.$$

The integrand, in every example above, is defined at each number on the interval of integration.

Improper integrals of class (1), where the interval is not bounded, are handled as limits:

$$\int_{a}^{\infty} f(x)\, dx = \lim_{b\to\infty} \int_{a}^{b} f(x)\, dx,$$

where f is continuous on $[a,b]$. If the limit on the right exists, the improper integral on the left is said to *converge* to this limit; if the limit on the right fails to exist, we say that the improper integral *diverges* (or is *meaningless*).

Converge

Diverge

The evaluation of improper integrals of class (1) is illustrated in Examples 17–23.

Example 17

Find $\int_{1}^{\infty} \frac{dx}{x^2}$.

SOLUTION: $\int_{1}^{\infty} \frac{dx}{x^2} = \lim_{b\to\infty} \int_{1}^{b} x^{-2}\, dx = \lim_{b\to\infty} -\frac{1}{x}\Big|_{1}^{b} = \lim_{b\to\infty} -\left(\frac{1}{b} - 1\right) = 1$. The given integral thus converges to 1. In Figure N7–22 we interpret $\int_{1}^{\infty} \frac{dx}{x^2}$ as the area above the x-axis, under the curve of $y = 3$, and bounded at the left by the vertical line $x = 1$.

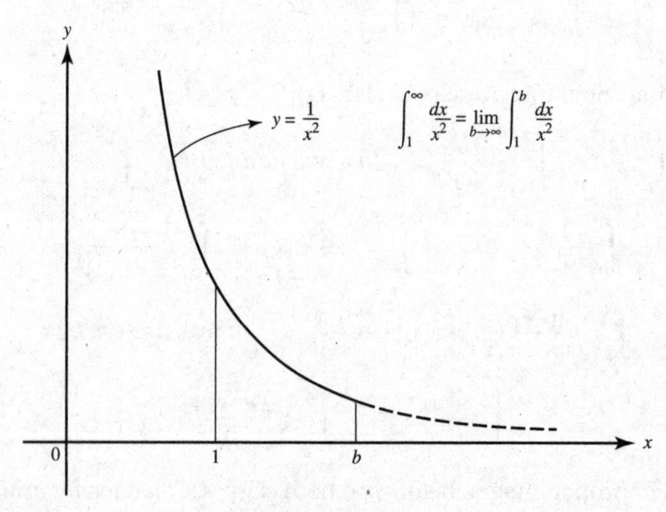

Figure N7–22

BC ONLY

Example 18

$$\int_1^{\infty} \frac{dx}{\sqrt{x}} = \lim_{b\to\infty} \int_1^b x^{-1/2}\,dx = \lim_{b\to\infty} 2\sqrt{x}\,\Big|_1^b = \lim_{b\to\infty} 2\left(\sqrt{b}-1\right) = +\infty.$$

Then $\int_1^{\infty} \frac{dx}{\sqrt{x}}$ diverges.

It can be proved that $\int_1^{\infty} \frac{dx}{x^p}$ converges if $p > 1$ but diverges if $p \leqq 1$. Figure N7–23 gives a geometric interpretation in terms of area of $\int_1^{\infty} \frac{dx}{x^p}$ for $p = \frac{1}{2}, 1, 2$. Only the first-quadrant area under $y = \frac{1}{x^2}$ bounded at the left by $x = 1$ exists. Note that

$$\int_1^{\infty} \frac{dx}{x} = \lim_{b\to\infty} \ln x\,\Big|_1^b = +\infty.$$

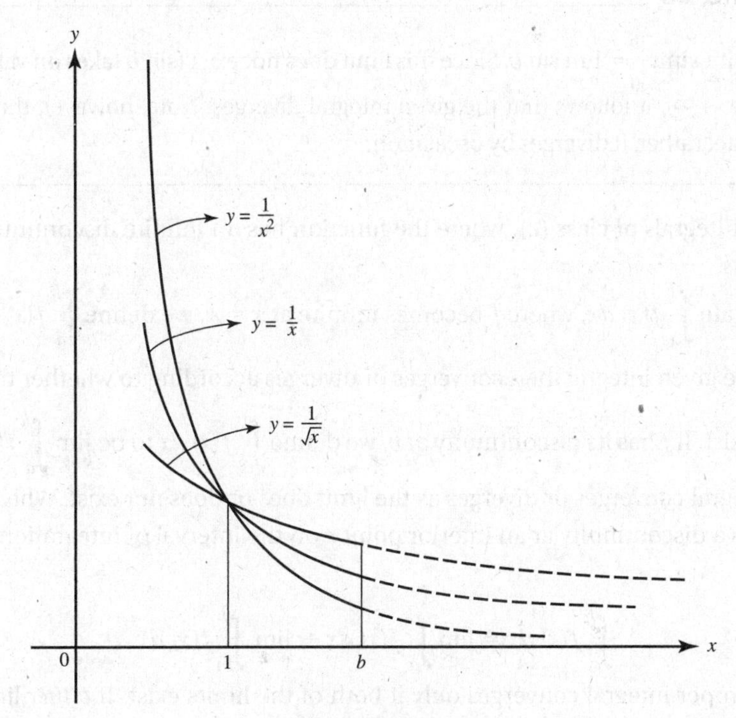

Figure N7–23

Example 19

$$\int_0^{\infty} \frac{dx}{x^2+9} = \lim_{b\to\infty} \int_0^b \frac{dx}{x^2+9} = \lim_{b\to\infty} \frac{1}{3}\tan^{-1}\frac{x}{3}\Big|_0^b = \lim_{b\to\infty} \frac{1}{3}\tan^{-1}\frac{b}{3} = \frac{1}{3}\cdot\frac{\pi}{2} = \frac{\pi}{6}.$$

Example 20

$$\int_0^{\infty} \frac{dy}{e^y} = \lim_{b\to\infty} \int_0^b e^{-y}\,dy = \lim_{b\to\infty} -(e^{-b}-1) = 1.$$

BC ONLY

➥ Example 21

$$\int_{-\infty}^{0} \frac{dz}{(z-1)^2} = \lim_{b\to-\infty} \int_{b}^{0} (z-1)^{-2}\,dz = \lim_{b\to-\infty} -\frac{1}{z-1}\Bigg|_{b}^{0}$$
$$= \lim_{b\to-\infty} -\left(-1 - \frac{1}{b-1}\right) = 1.$$

➥ Example 22

$$\int_{-\infty}^{0} e^{-x}\,dx = \lim_{b\to-\infty} -e^{-x}\Big|_{b}^{0} = \lim_{b\to-\infty} -(1 - e^{-b}) = +\infty.$$

Thus, this improper integral diverges.

➥ Example 23

$\int_{0}^{\infty} \cos x\,dx = \lim_{b\to\infty} \sin x\Big|_{0}^{b} = \lim_{b\to\infty} \sin b$. Since this limit does not exist ($\sin b$ takes on values between -1 and 1 as $b \to \infty$), it follows that the given integral diverges. Note, however, that it does not become infinite; rather, it diverges by oscillation.

Improper integrals of class (2), where the function has an infinite discontinuity, are handled as follows.

To investigate $\int_{a}^{b} f(x)\,dx$, where f becomes infinite at $x = a$, we define $\int_{a}^{b} f(x)\,dx$ to be $\lim_{k\to a^+} \int_{k}^{b} f(x)\,dx$. The given integral then converges or diverges according to whether the limit does or does not exist. If f has its discontinuity at b, we define $\int_{a}^{b} f(x)\,dx$ to be $\lim_{k\to b^-} \int_{a}^{k} f(x)\,dx$; again, the given integral converges or diverges as the limit does or does not exist. When, finally, the integrand has a discontinuity at an interior point c on the interval of integration ($a < c < b$), we let

$$\int_{a}^{b} f(x)\,dx = \lim_{k\to c^-} \int_{a}^{k} f(x)\,dx + \lim_{m\to c^+} \int_{m}^{b} f(x)\,dx.$$

Now the improper integral converges only if both of the limits exist. If *either* limit does not exist, the improper integral diverges.

The evaluation of improper integrals of class (2) is illustrated in Examples 24–31.

BC ONLY

➡ Example 24

Find $\int_0^1 \frac{dx}{\sqrt[3]{x}}$.

SOLUTION: $\int_0^1 \frac{dx}{\sqrt[3]{x}} = \lim_{k\to 0^+} \int_k^1 x^{-1/3}\,dx = \lim_{k\to 0^+} \frac{3}{2} x^{2/3} \Big|_k^1 = \lim_{k\to 0^+} \frac{3}{2}(1 - k^{2/3}) = \frac{3}{2}$.

In Figure N7–24 we interpret this integral as the first-quadrant area under $y = \frac{1}{\sqrt[3]{x}}$ and to the left of $x = 1$.

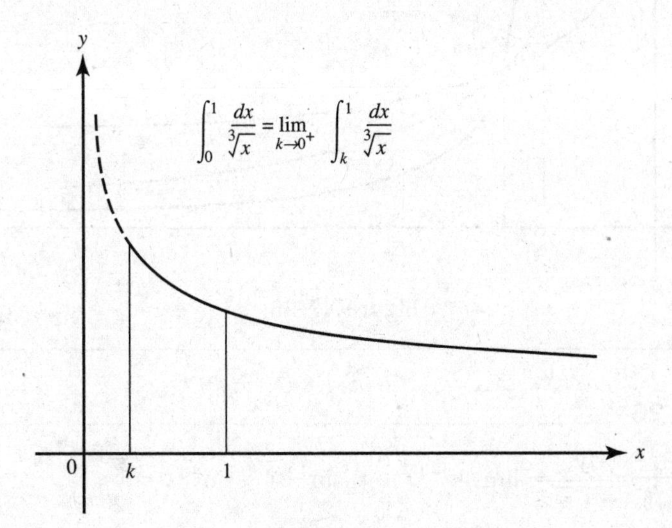

Figure N7–24

➡ Example 25

Does $\int_0^1 \frac{dx}{x^3}$ converge or diverge?

SOLUTION: $\int_0^1 \frac{dx}{x^3} = \lim_{k\to 0^+} \int_k^1 x^{-3}\,dx = \lim_{k\to 0^+} -\frac{1}{2x^2}\Big|_k^1 = \lim_{k\to 0^+} -\frac{1}{2}\left(1 - \frac{1}{k^2}\right) = \infty$.

Therefore, this integral diverges.

It can be shown that $\int_0^a \frac{dx}{x^p}$ $(a > 0)$ converges if $p < 1$ but diverges if $p \geqq 1$. Figure N7–25 shows an interpretation of $\int_0^1 \frac{dx}{x^p}$ in terms of areas where $p = \frac{1}{3}$, 1, and 3. Only the first-quadrant area under $y = \frac{1}{\sqrt[3]{x}}$ to the left of $x = 1$ exists. Note that

$$\int_0^1 \frac{dx}{x} = \lim_{k\to 0^+} \ln x \Big|_k^1 = \lim_{k\to 0^+} (\ln 1 - \ln k) = +\infty.$$

BC ONLY

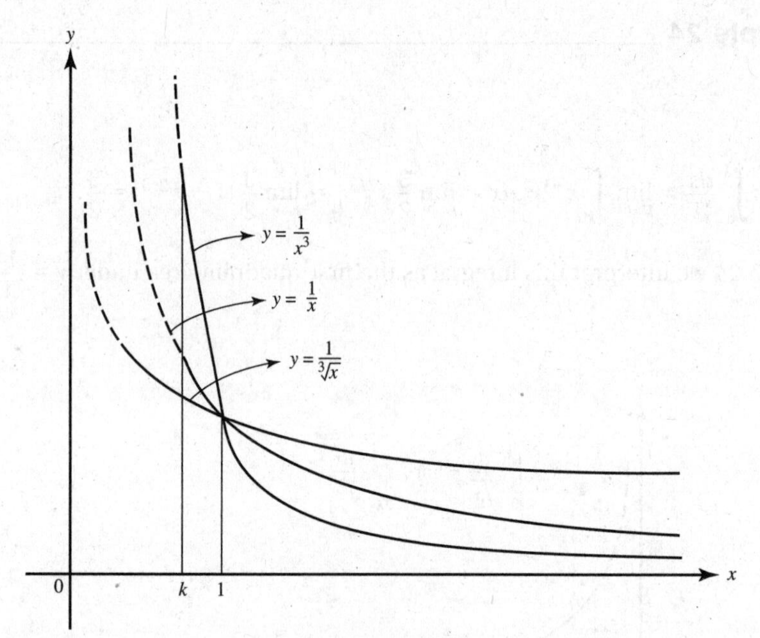

Figure N7–25

➡ Example 26

$$\int_0^2 \frac{dy}{\sqrt{4-y^2}} = \lim_{k\to 2^-} \int_0^k \frac{dy}{\sqrt{4-y^2}} = \lim_{k\to 2^-} \sin^{-1}\frac{y}{2}\Big|_0^k = \sin^{-1}1 - \sin^{-1}0 = \frac{\pi}{2}.$$

➡ Example 27

$$\int_2^3 \frac{dt}{(3-t)^2} = \lim_{k\to 3^-} -\int_2^k (3-t)^{-2}(-dt) = \lim_{k\to 3^-} \frac{1}{3-t}\Big|_2^k = +\infty.$$

This integral diverges.

➡ Example 28

$$\int_0^2 \frac{dx}{(x-1)^{2/3}} = \lim_{k\to 1^-} \int_0^k (x-1)^{-2/3}\,dx + \lim_{m\to 1^+} \int_m^2 (x-1)^{-2/3}\,dx$$

$$= \lim_{k\to 1^-} 3(x-1)^{1/3}\Big|_0^k + \lim_{m\to 1^+} 3(x-1)^{1/3}\Big|_m^2 = 3(0+1) + 3(1-0) = 6.$$

➡ Example 29

$$\int_{-2}^2 \frac{dx}{x^2} = \lim_{k\to 0^-} \int_{-2}^k x^{-2}\,dx + \lim_{m\to 0^+} \int_m^2 x^{-2}\,dx = \lim_{k\to 0^-} -\frac{1}{x}\Big|_{-2}^k + \lim_{m\to 0^+} -\frac{1}{x}\Big|_m^2.$$

Neither limit exists; the integral diverges.

BC ONLY

NOTE: This example demonstrates how careful one must be to notice a discontinuity at an interior point. If it were overlooked, one might proceed as follows:

$$\int_{-2}^{2} \frac{dx}{x^2} = -\frac{1}{x}\bigg|_{-2}^{2} = -\left(\frac{1}{2} + \frac{1}{2}\right) = -1.$$

Since this integrand is positive except at zero, the result obtained is clearly meaningless. Figure N7–26 shows the impossibility of this answer.

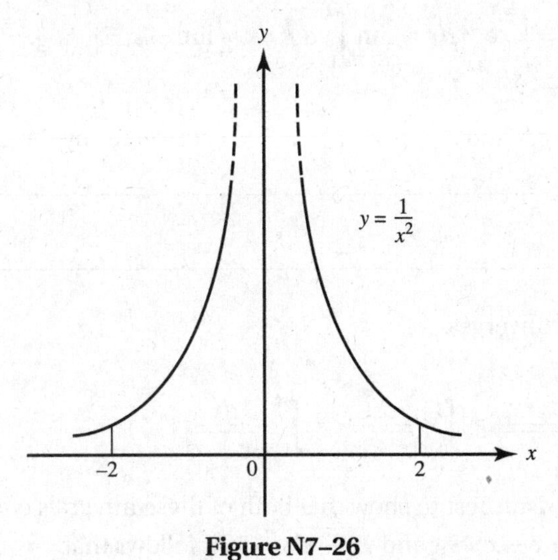

Figure N7–26

THE COMPARISON TEST

Comparison Test

We can often determine whether an improper integral converges or diverges by comparing it to a known integral on the same interval. This method is especially helpful when it is not easy to actually evaluate the appropriate limit by finding an antiderivative for the integrand. There are two cases.

(1) Convergence. If on the interval of integration $f(x) \le g(x)$ and $\int g(x)\,dx$ is known to converge, then $\int f(x)\,dx$ also converges. For example, consider $\int_1^{\infty} \frac{1}{x^3+1}\,dx$. We know that $\int_1^{\infty} \frac{1}{x^3}\,dx$ converges. Since $\frac{1}{x^3+1} < \frac{1}{x^3}$, the improper integral $\int_1^{\infty} \frac{1}{x^3+1}\,dx$ must also converge.

(2) Divergence. If on the interval of integration $f(x) \ge g(x)$ and $\int g(x)\,dx$ is known to diverge, then $\int f(x)\,dx$ also diverges. For example, consider $\int_0^1 \frac{\sec x}{x^3}\,dx$. We know that $\int_0^1 \frac{1}{x^3}\,dx$ diverges. Since $\sec x \ge 1$, it follows that $\frac{\sec x}{x^3} \ge \frac{1}{x^3}$; hence the improper integral $\int_0^1 \frac{\sec x}{x^3}\,dx$ must also diverge.

BC ONLY

Example 30

Determine whether or not $\int_1^{\infty} e^{-x^2}\,dx$ converges.

SOLUTION: Although there is no elementary function whose derivative is e^{-x^2}, we can still show that the given improper integral converges. Note, first, that if $x \geqq 1$ then $x^2 \geqq x$, so that $-x^2 \leqq -x$ and $e^{-x^2} \leqq e^{-x}$. Furthermore,

$$\int_1^{\infty} e^{-x}\,dx = \lim_{k\to\infty} \int_1^{k} e^{-x}\,dx = \lim_{k\to\infty} -e^{-x}\Big|_1^k = \frac{1}{e}.$$

Since $\int_1^{\infty} e^{-x}\,dx$ converges and $e^{-x^2} \leq e^{-x}$, $\int_1^{\infty} e^{-x^2}\,dx$ converges by the Comparison Test.

Example 31

Show that $\int_0^{\infty} \frac{dx}{\sqrt{x+x^4}}$ converges.

SOLUTION: $$\int_0^{\infty} \frac{dx}{\sqrt{x+x^4}} = \int_0^{1} \frac{dx}{\sqrt{x+x^4}} + \int_1^{\infty} \frac{dx}{\sqrt{x+x^4}};$$

we will use the Comparison Test to show that both of these integrals converge. Since if $0 < x \leqq 1$, then $x + x^4 > x$ and $\sqrt{x+x^4} > \sqrt{x}$, it follows that

$$\frac{1}{\sqrt{x+x^4}} < \frac{1}{\sqrt{x}} \quad (0 < x \leqq 1).$$

We know that $\int_0^{1} \frac{dx}{\sqrt{x}}$ converges; hence $\int_0^{1} \frac{dx}{\sqrt{x+x^4}}$ must converge. Further, if $x \geqq 1$ then $x + x^4 \geqq x^4$ and $\sqrt{x+x^4} \geqq \sqrt{x^4} = x^2$, so

$$\frac{1}{\sqrt{x+x^4}} \leqq \frac{1}{x^2} \quad (x \geqq 1).$$

We know that $\int_1^{\infty} \frac{1}{x^2}\,dx$ converges, hence $\int_1^{\infty} \frac{dx}{\sqrt{x+x^4}}$ also converges.

Thus the given integral, $\int_0^{\infty} \frac{dx}{\sqrt{x+x^4}}$, converges.

NOTE: Examples 32 and 33 involve finding the volumes of solids. Both lead to improper integrals.

BC ONLY

Example 32

Find the volume, if it exists, of the solid generated by rotating the region in the first quadrant bounded above by $y = \frac{1}{x}$, at the left by $x = 1$, and below by $y = 0$, about the x-axis.

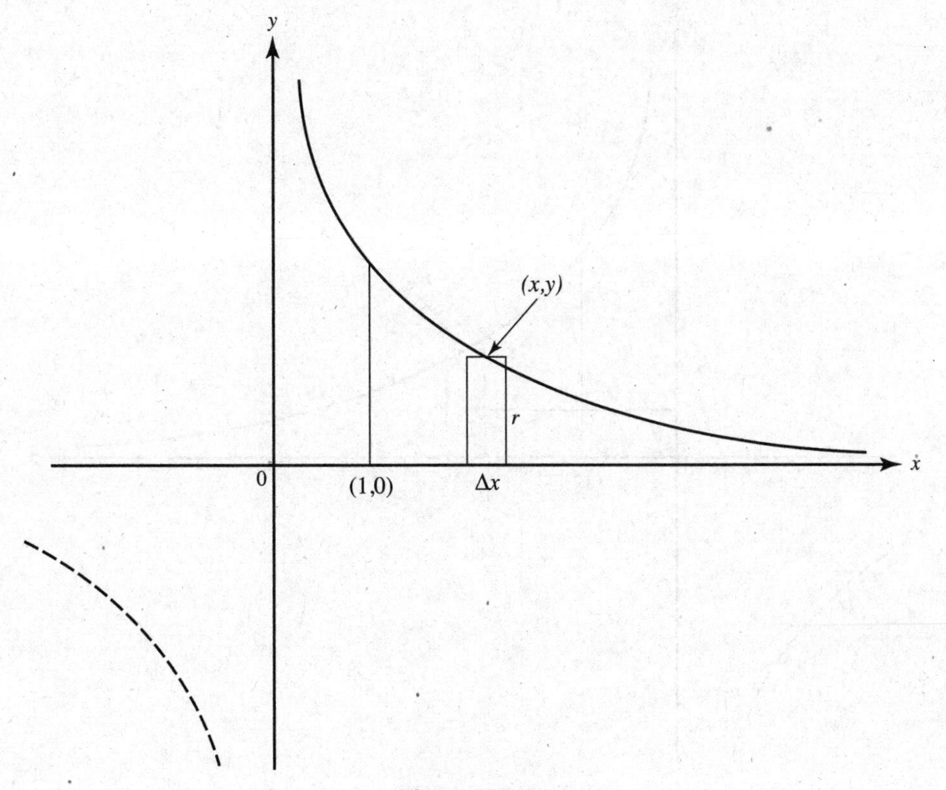

Figure N7–27

SOLUTION:

About the x-axis.

Disk.

$$\Delta V = \pi y^2 \Delta x.$$

$$V = \pi \int_1^{\infty} y^2\, dx = \pi \int_1^{\infty} \frac{1}{x^2}\, dx$$

$$= \pi \lim_{k\to\infty} \int_1^k \frac{1}{x^2}\, dx = \pi.$$

BC ONLY

➡ Example 33‡

Find the volume, if it exists, of the solid generated by rotating the region in the first quadrant bounded above by $y = \frac{1}{x}$, at the left by $x = 1$, and below by $y = 0$, about the y-axis.

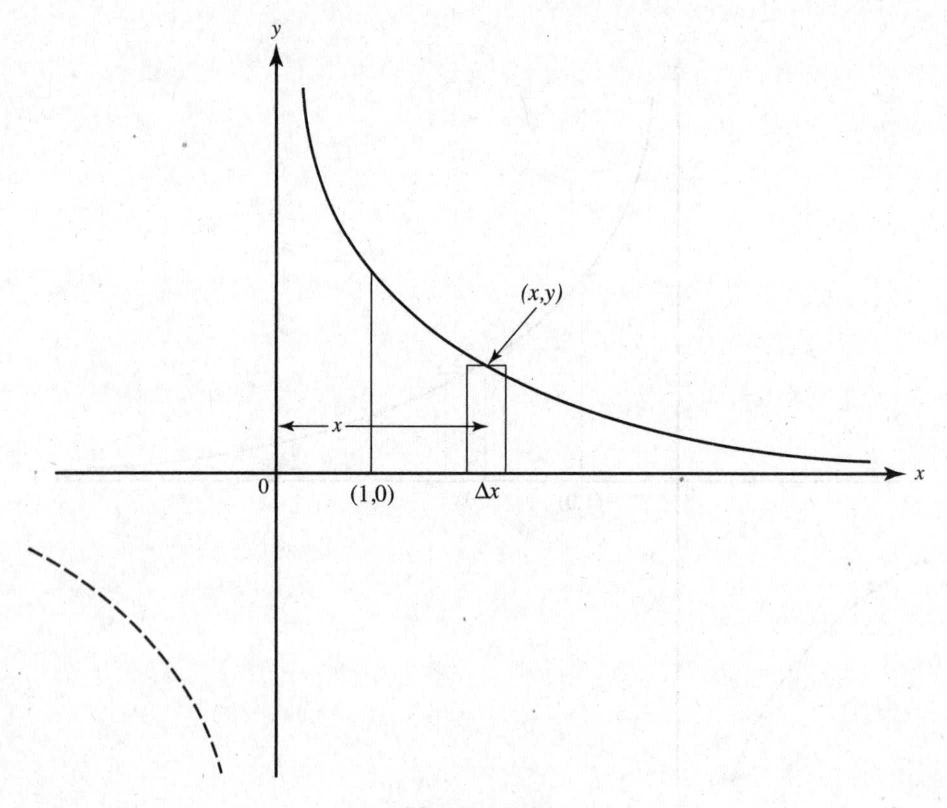

Figure N7–28

SOLUTION:

About the y-axis.

Shell.

$\Delta V = 2\pi xy\,\Delta x = 2\pi\,\Delta x.$

Note that $2\pi \int_{1}^{\infty} dx$ diverges to infinity.

CHAPTER SUMMARY

In this chapter, we have reviewed how to find areas and volumes using definite integrals. We've looked at area under a curve and between two curves. We've reviewed volumes of solids with known cross sections, and the methods of disks and washers for finding volumes of solids of revolution.

For BC Calculus students, we've applied these techniques to parametrically defined functions and polar curves and added methods for finding lengths of arc. We've also looked at improper integrals and tests for determining convergence and divergence.

‡No question requiring the use of shells will appear on the AP exam.

PRACTICE EXERCISES

Part A. Directions: Answer these questions *without* using your calculator.

AREA

In Questions 1–11, choose the alternative that gives the area of the region whose boundaries are given.

1. The curve of $y = x^2$, $y = 0$, $x = -1$, and $x = 2$.

   (A) $\frac{11}{3}$ (B) $\frac{7}{3}$ (C) 3 (D) 5 (E) 9

2. The parabola $y = x^2 - 3$ and the line $y = 1$.

   (A) $\frac{8}{3}$ (B) 32 (C) $\frac{32}{3}$ (D) $\frac{16}{3}$ (E) 12

3. The curve of $x = y^2 - 1$ and the y-axis.

   (A) $\frac{4}{3}$ (B) $\frac{2}{3}$ (C) $\frac{8}{3}$ (D) $\frac{1}{2}$ (E) 2

4. The parabola $y^2 = x$ and the line $x + y = 2$.

   (A) $\frac{5}{2}$ (B) $\frac{3}{2}$ (C) $\frac{11}{6}$ (D) $\frac{9}{2}$ (E) $\frac{29}{6}$

5. The curve of $y = \frac{4}{x^2 + 4}$, the x-axis, and the vertical lines $x = -2$ and $x = 2$.

   (A) $\frac{\pi}{4}$ (B) $\frac{\pi}{2}$ (C) 2π (D) π (E) 4π

6. The parabolas $x = y^2 - 5y$ and $x = 3y - y^2$.

   (A) $\frac{32}{3}$ (B) $\frac{139}{6}$ (C) $\frac{64}{3}$ (D) $\frac{128}{3}$ (E) 32

7. The curve of $y = \frac{2}{x}$ and $x + y = 3$.

   (A) $\frac{1}{2} - 2 \ln 2$ (B) $\frac{3}{2}$ (C) $\frac{1}{2} - \ln 4$

   (D) $\frac{5}{2}$ (E) $\frac{3}{2} - \ln 4$

8. In the first quadrant, bounded below by the x-axis and above by the curves of $y = \sin x$ and $y = \cos x$.

   (A) $2 - \sqrt{2}$ (B) $2 + \sqrt{2}$ (C) 2 (D) $\sqrt{2}$ (E) $2\sqrt{2}$

9. Bounded above by the curve $y = \sin x$ and below by $y = \cos x$ from $x = \frac{\pi}{4}$ to $x = \frac{5\pi}{4}$.

   (A) $2\sqrt{2}$ (B) $\frac{2}{\sqrt{2}}$ (C) $\frac{1}{2\sqrt{2}}$

   (D) $2(\sqrt{2} - 1)$ (E) $2(\sqrt{2} + 1)$

10. The curve $y = \cot x$, the line $x = \frac{\pi}{4}$, and the x-axis.

    (A) $\ln 2$ (B) $\frac{1}{2} \ln \frac{1}{2}$ (C) 1 (D) $\frac{1}{2} \ln 2$ (E) 2

11. The curve of $y = x^3 - 2x^2 - 3x$ and the x-axis.

    (A) $\frac{28}{3}$ (B) $\frac{79}{6}$ (C) $\frac{45}{4}$ (D) $\frac{71}{6}$ (E) $\frac{32}{3}$

12. The total area bounded by the cubic $x = y^3 - y$ and the line $x = 3y$ is equal to

(A) 4 (B) $\frac{16}{3}$ (C) 8 (D) $\frac{32}{3}$ (E) 16

13. The area bounded by $y = e^x$, $y = 2$, and the y-axis is equal to

(A) $3 - e$ (B) $e^2 - 1$ (C) $e^2 + 1$

(D) $2 \ln 2 - 1$ (E) $2 \ln 2 - 3$

BC ONLY

14. The area enclosed by the ellipse with parametric equations $x = 2 \cos \theta$ and $y = 3 \sin \theta$ equals

(A) 6π (B) $\frac{9}{2}\pi$ (C) 3π (D) $\frac{3}{2}\pi$ (E) $\frac{9}{4}\pi$

15. The area enclosed by one arch of the cycloid with parametric equations $x = \theta - \sin \theta$ and $y = 1 - \cos \theta$ equals

(A) $\frac{3\pi}{2}$ (B) 2π (C) 3π (D) 4π (E) 6π

16. The area enclosed by the curve $y^2 = x(1 - x)$ is given by

(A) $2\int_0^1 x\sqrt{1-x}\,dx$ (B) $2\int_0^1 \sqrt{x-x^2}\,dx$ (C) $4\int_0^1 \sqrt{x-x^2}\,dx$

(D) π (E) 2π

17. The figure below shows part of the curve of $y = x^3$ and a rectangle with two vertices at (0, 0) and (c, 0). What is the ratio of the area of the rectangle to the shaded part of it above the cubic?

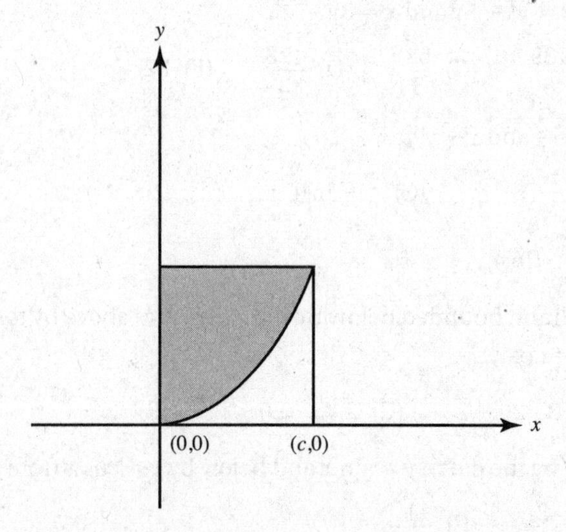

(A) 3:4 (B) 5:4 (C) 4:3 (D) 3:1 (E) 2:1

VOLUME

In Questions 18–24 the region whose boundaries are given is rotated about the line indicated. Choose the alternative that gives the volume of the solid generated.

18. $y = x^2$, $x = 2$, and $y = 0$; about the x-axis.

(A) $\frac{64\pi}{3}$ (B) 8π (C) $\frac{8\pi}{3}$ (D) $\frac{128\pi}{5}$ (E) $\frac{32\pi}{5}$

19. $y = x^2$, $x = 2$, and $y = 0$; about the y-axis.

(A) $\frac{16\pi}{3}$ (B) 4π (C) $\frac{32\pi}{5}$ (D) 8π (E) $\frac{8\pi}{3}$

20. The first quadrant region bounded by $y = x^2$, the y-axis, and $y = 4$; about the y-axis.

(A) 8π (B) 4π (C) $\frac{64\pi}{3}$ (D) $\frac{32\pi}{3}$ (E) $\frac{16\pi}{3}$

21. $y = x^2$ and $y = 4$; about the x-axis.

(A) $\frac{64\pi}{5}$ (B) $\frac{512\pi}{15}$ (C) $\frac{256\pi}{5}$

(D) $\frac{128\pi}{5}$ (E) $\frac{176\pi}{3}$

22. $y = x^2$ and $y = 4$; about the line $y = 4$.

(A) $\frac{256\pi}{15}$ (B) $\frac{256\pi}{5}$ (C) $\frac{512\pi}{5}$ (D) $\frac{512\pi}{15}$ (E) $\frac{64\pi}{3}$

23. An arch of $y = \sin x$ and the x-axis; about the x-axis.

(A) $\frac{\pi}{2}\left(\pi - \frac{1}{2}\right)$ (B) $\frac{\pi^2}{2}$ (C) $\frac{\pi^2}{4}$ (D) π^2 (E) $\pi(\pi - 1)$

24. A trapezoid with vertices at (2, 0), (2, 2), (4, 0), and (4, 4); about the x-axis.

(A) $\frac{56\pi}{3}$ (B) $\frac{128\pi}{3}$ (C) $\frac{92\pi}{3}$ (D) $\frac{112\pi}{3}$ (E) 20π

25. The base of a solid is a circle of radius a, and every plane section perpendicular to a diameter is a square. The solid has volume

(A) $\frac{8}{3}a^3$ (B) $2\pi a^3$ (C) $4\pi a^3$ (D) $\frac{16}{3}a^3$ (E) $\frac{8\pi}{3}a^3$

26. The base of a solid is the region bounded by the parabola $x^2 = 8y$ and the line $y = 4$, and each plane section perpendicular to the y-axis is an equilateral triangle. The volume of the solid is

(A) $\frac{64\sqrt{3}}{3}$ (B) $64\sqrt{3}$ (C) $32\sqrt{3}$ (D) 32 (E) $\frac{32\sqrt{3}}{3}$

27. The base of a solid is the region bounded by $y = e^{-x}$, the x-axis, the y-axis, and the line $x = 1$. Each cross section perpendicular to the x-axis is a square. The volume of the solid is

(A) $\frac{e^2}{2}$ (B) $e^2 - 1$ (C) $1 - \frac{1}{e^2}$ (D) $\frac{e^2 - 1}{2}$ (E) $\frac{1}{2}\left(1 - \frac{1}{e^2}\right)$

BC ONLY

ARC LENGTH

28. The length of the arc of the curve $y^2 = x^3$ cut off by the line $x = 4$ is

(A) $\frac{4}{3}(10\sqrt{10} - 1)$ (B) $\frac{8}{27}(10^{3/2} - 1)$ (C) $\frac{16}{27}(10^{3/2} - 1)$

(D) $\frac{16}{27}10\sqrt{10}$ (E) $8\sqrt{5}$

29. The length of the arc of $y = \ln \cos x$ from $x = \frac{\pi}{4}$ to $x = \frac{\pi}{3}$ equals

(A) $\ln \frac{\sqrt{3} + 2}{\sqrt{2} + 1}$ (B) 2 (C) $\ln(1 + \sqrt{3} - \sqrt{2})$

(D) $\sqrt{3} - 2$ (E) $\frac{\ln(\sqrt{3} + 2)}{\ln(\sqrt{2} + 1)}$

IMPROPER INTEGRALS

30. $\int_0^{\infty} e^{-x}\, dx =$

(A) 1 (B) $\frac{1}{e}$ (C) -1 (D) $-\frac{1}{e}$ (E) none of these

31. $\int_0^{e} \frac{du}{u} =$

(A) 1 (B) $\frac{1}{e}$ (C) $-\frac{1}{e^2}$ (D) -1 (E) none of these

32. $\int_1^{2} \frac{dt}{\sqrt[3]{t - 1}} =$

(A) $\frac{2}{3}$ (B) $\frac{3}{2}$ (C) 3 (D) 1 (E) none of these

33. $\int_2^{4} \frac{dx}{(x - 3)^{2/3}} =$

(A) 6 (B) $\frac{6}{5}$ (C) $\frac{2}{3}$ (D) 0 (E) none of these

34. $\int_2^{4} \frac{dx}{(x - 3)^2} =$

(A) 2 (B) -2 (C) 0 (D) $\frac{2}{3}$ (E) none of these

35. $\int_0^{\pi/2} \frac{\sin x}{\sqrt{1 - \cos x}}\, dx$

(A) -2 (B) $\frac{2}{3}$ (C) 2 (D) $\frac{1}{2}$ (E) none of these

In Questions 36–40, choose the alternative that gives the area, if it exists, of the region described.

36. In the first quadrant under the curve of $y = e^{-x}$.

(A) 1 (B) e (C) $\frac{1}{e}$ (D) 2 (E) none of these

37. In the first quadrant under the curve of $y = xe^{-x^2}$.

(A) 2 (B) $\frac{2}{e}$ (C) $\frac{1}{2}$ (D) $\frac{1}{2e}$ (E) none of these

38. In the first quadrant above $y = 1$, between the y-axis and the curve $xy = 1$.

(A) 1 (B) 2 (C) $\frac{1}{2}$ (D) 4 (E) none of these

39. Between the curve $y = \dfrac{4}{1 + x^2}$ and the x-axis.

BC ONLY

(A) 2π (B) 4π (C) 8π (D) π (E) none of these

40. Above the x-axis, between the curve $y = \dfrac{4}{\sqrt{1 - x^2}}$ and its asymptotes.

(A) $\dfrac{\pi}{2}$ (B) π (C) 2π (D) 4π (E) none of these

In Questions 41 and 42, choose the alternative that gives the volume, if it exists, of the solid generated.

41. $y = \dfrac{1}{x}$, at the left by $x = 1$, and below by $y = 0$; about the x-axis.

(A) $\dfrac{\pi}{2}$ (B) π (C) 2π (D) 4π (E) none of these

42. The first-quadrant region under $y = e^{-x}$; about the x-axis.

(A) $\dfrac{\pi}{2}$ (B) π (C) 2π (D) 4π (E) none of these

Part B. Directions: Some of the following questions require the use of a graphing calculator.

AREA

In Questions 43–47, choose the alternative that gives the area of the region whose boundaries are given.

43. The area bounded by the parabola $y = 2 - x^2$ and the line $y = x - 4$ is given by

(A) $\int_{-2}^{3} (6 - x - x^2)\, dx$ (B) $\int_{-2}^{1} (2 + x + x^2)\, dx$ (C) $\int_{-3}^{2} (6 - x - x^2)\, dx$

(D) $2\int_{0}^{\sqrt{2}} (2 - x^2)\, dx + \int_{-3}^{2} (4 - x)\, dx$ (E) $\int_{-7}^{2} \left[(y + 4) - \sqrt{2 - y}\right] dy$

BC ONLY

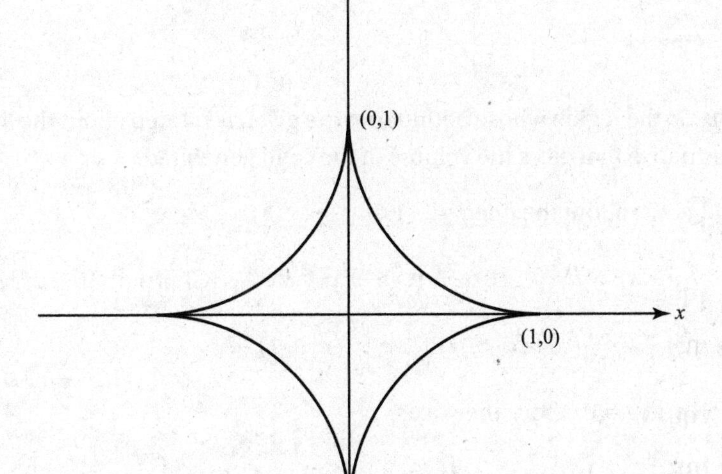

BC ONLY

44. The area enclosed by the hypocycloid (shown above) with parametric equations $x = \cos^3 t$ and $y = \sin^3 t$ as shown in the above diagram is

(A) $3\int_{\pi/2}^{0} \sin^4 t \cos^2 t\, dt$ (B) $4\int_{0}^{1} \sin^3 t\, dt$ (C) $-4\int_{\pi/2}^{0} \sin^6 t\, dt$

(D) $12\int_{0}^{\pi/2} \sin^4 t \cos^2 t\, dt$ (E) $6\int_{0}^{1} \sin^4 t \cos^2 t\, dt$

45. Suppose the following is a table of coordinates for $y = f(x)$, given that f is continuous on [1, 5]:

x	1	2	3	4	5
y	1.62	4.15	7.5	9.0	12.13

If a trapezoid sum in used, with $n = 4$, then the area under the curve, from $x = 1$ to $x = 5$, is equal, to two decimal places, to

(A) 6.88 (B) 13.76 (C) 20.30 (D) 25.73 (E) 27.53

BC ONLY

46. The area A enclosed by the four-leaved rose $r = \cos 2\theta$ equals, to three decimal places,

(A) 0.785 (B) 1.571 (C) 2.071 (D) 3.142 (E) 6.283

47. The area bounded by the small loop of the limaçon $r = 1 - 2\sin\theta$ is given by the definite integral

(A) $\int_{\pi/3}^{5\pi/3} \left[\frac{1}{2}(1 - 2\sin\theta)\right]^2 d\theta$

(B) $\int_{7\pi/6}^{3\pi/2} (1 - 2\sin\theta)^2\, d\theta$

(C) $\int_{\pi/6}^{\pi/2} (1 - 2\sin\theta)^2\, d\theta$

(D) $\int_{0}^{\pi/6} \left[\frac{1}{2}(1 - 2\sin\theta)\right]^2 d\theta + \int_{5\pi/6}^{\pi} \left[\frac{1}{2}(1 - 2\sin\theta)\right]^2 d\theta$

(E) $\int_{0}^{\pi/3} (1 - 2\sin\theta)^2\, d\theta$

VOLUME

In Questions 48–54 the region whose boundaries are given is rotated about the line indicated. Choose the alternative that gives the volume of the solid generated.

48. $y = x^2$ and $y = 4$; about the line $y = -1$.

(A) $4\pi\int_{-1}^{4} (y + 1)\sqrt{y}\, dy$ (B) $2\pi\int_{0}^{2} (4 - x^2)^2\, dx$ (C) $\pi\int_{-2}^{2} (16 - x^4)\, dx$

(D) $2\pi\int_{0}^{2} (24 - 2x^2 - x^4)\, dx$ (E) $2\pi\int_{0}^{2} (x^2 + 1)^2\, dx$

49. $y = 3x - x^2$ and $y = 0$; about the x-axis.

(A) $\pi\int_{0}^{3} (9x^2 + x^4)\, dx$ (B) $\pi\int_{0}^{3} (3x - x^2)^2\, dx$ (C) $\pi\int_{0}^{\sqrt{3}} (3x - x^2)\, dx$

(D) $2\pi\int_{0}^{3} y\sqrt{9 - 4y}\, dy$ (E) $\pi\int_{0}^{9/4} y^2\, dy$

50. $y = 3x - x^2$ and $y = x$; about the x-axis.

(A) $\pi \int_0^{3/2} \left[(3x - x^2)^2 - x^2\right] dx$ (B) $\pi \int_0^2 (9x^2 - 6x^3)\, dx$

(C) $\pi \int_0^2 \left[(3x - x^2)^2 - x^2\right] dx$ (D) $\pi \int_0^3 \left[(3x - x^2)^2 - x^4\right] dx$

(E) $\pi \int_0^3 (2x - x^2)^2\, dx$

51. $y = \ln x$, $y = 0$, $x = e$; about the line $x = e$.

(A) $\pi \int_1^e (e - x) \ln x\, dx$ (B) $\pi \int_0^1 (e - e^y)^2\, dy$ (C) $2\pi \int_1^e (e - \ln x)\, dx$

(D) $\pi \int_0^e \left(e^2 - 2e^{y+1} + e^{2y}\right) dy$ (E) $\pi \int_0^1 \left(e^2 - e^{2y}\right) dy$

BC ONLY

52. The curve with parametric equations $x = \tan\theta$, $y = \cos^2\theta$, and the lines $x = 0$, $x = 1$, and $y = 0$; about the x-axis.

(A) $\pi \int_0^{\pi/4} \cos^4\theta\, d\theta$ (B) $\pi \int_0^{\pi/4} \cos^2\theta \sin\theta\, d\theta$ (C) $\pi \int_0^{\pi/4} \cos^2\theta\, d\theta$

(D) $\pi \int_0^1 \cos^2\theta\, d\theta$ (E) $\pi \int_0^1 \cos^4\theta\, d\theta$

CHALLENGE

53. A sphere of radius r is divided into two parts by a plane at distance h $(0 < h < r)$ from the center. The volume of the smaller part equals

(A) $\frac{\pi}{3}\left(2r^3 + h^3 - 3r^2h\right)$ (B) $\frac{\pi h}{3}\left(3r^2 - h^2\right)$ (C) $\frac{4}{3}\pi r^3 + \frac{h^3}{3} - r^2h$

(D) $\frac{\pi}{3}\left(2r^3 + 3r^2h - h^3\right)$ (E) $\frac{\pi}{3}\left(r^3 - h^3\right)$

54. If the curves of $f(x)$ and $g(x)$ intersect for $x = a$ and $x = b$ and if $f(x) > g(x) > 0$ for all x on (a, b), then the volume obtained when the region bounded by the curves is rotated about the x-axis is equal to

(A) $\pi \int_a^b f^2(x)\, dx - \int_a^b g^2(x)\, dx$

(B) $\pi \int_a^b \left[f(x) - g(x)\right]^2 dx$

(C) $2\pi \int_a^b x\left[f(x) - g(x)\right] dx$

(D) $\pi \int_a^b \left[f^2(x) - g^2(x)\right] dx$

(E) $2\pi \int_a^b \left[f^2(x) - g^2(x)\right] dx$

ARC LENGTH

BC ONLY

55. The length of one arch of the cycloid $\begin{matrix} x = t - \sin t \\ y = 1 - \cos t \end{matrix}$ equals

(A) $\int_0^{\pi} \sqrt{1 - \cos t}\, dt$ (B) $\int_0^{2\pi} \sqrt{\frac{1 - \cos t}{2}}\, dt$ (C) $\int_0^{\pi} \sqrt{2 - 2\cos t}\, dt$

(D) $\int_0^{2\pi} \sqrt{2 - 2\cos t}\, dt$ (E) $2\int_0^{\pi} \sqrt{\frac{1 - \cos t}{2}}\, dt$

BC ONLY

56. The length of the arc of the parabola $4x = y^2$ cut off by the line $x = 2$ is given by the integral

(A) $\int_{-1}^{1} \sqrt{x^2+1}\,dx$ (B) $\frac{1}{2}\int_{0}^{2} \sqrt{4+y^2}\,dy$ (C) $\int_{-1}^{1} \sqrt{1+x}\,dx$

(D) $\int_{-1}^{1} \sqrt{4+y^2}\,dy$ (E) $\int_{-1}^{1} \sqrt{4+y^2}\,dy$

CHALLENGE

57. The length of $x = e^t \cos t$, $y = e^t \sin t$ from $t = 2$ to $t = 3$ is equal to

(A) $\sqrt{2}e^2\sqrt{e^2-1}$ (B) $\sqrt{2}\left(e^3-e^2\right)$ (C) $2\left(e^3-e^2\right)$

(D) $e^3(\cos 3+\sin 3)-e^2(\cos 2+\sin 2)$ (E) $2\sqrt{e^6-e^4}$

IMPROPER INTEGRALS

58. Which one of the following is an improper integral?

(A) $\int_{0}^{2} \frac{dx}{\sqrt{x+1}}$ (B) $\int_{-1}^{1} \frac{dx}{1+x^2}$ (C) $\int_{0}^{2} \frac{x\,dx}{1-x^2}$

(D) $\int_{0}^{\pi/3} \frac{\sin x\,dx}{\cos^2 x}$ (E) none of these

59. Which one of the following improper integrals diverges?

(A) $\int_{1}^{\infty} \frac{dx}{x^2}$ (B) $\int_{0}^{\infty} \frac{dx}{e^x}$ (C) $\int_{-1}^{1} \frac{dx}{\sqrt[3]{x}}$

(D) $\int_{-1}^{1} \frac{dx}{x^2}$ (E) none of these

60. Which one of the following improper integrals diverges?

(A) $\int_{0}^{\infty} \frac{dx}{1+x^2}$ (B) $\int_{0}^{1} \frac{dx}{x^{1/3}}$ (C) $\int_{0}^{\infty} \frac{dx}{x^3+1}$

(D) $\int_{0}^{\infty} \frac{dx}{e^x+2}$ (E) $\int_{0}^{\infty} \frac{dx}{x^{1/3}}$

Answer Key

1. **C**	13. **D**	25. **D**	37. **C**	49. **B**
2. **C**	14. **A**	26. **B**	38. **E**	50. **C**
3. **A**	15. **C**	27. **E**	39. **B**	51. **B**
4. **D**	16. **B**	28. **C**	40. **D**	52. **C**
5. **D**	17. **C**	29. **A**	41. **B**	53. **A**
6. **C**	18. **E**	30. **A**	42. **A**	54. **D**
7. **E**	19. **D**	31. **E**	43. **C**	55. **D**
8. **A**	20. **A**	32. **B**	44. **D**	56. **D**
9. **A**	21. **C**	33. **A**	45. **E**	57. **B**
10. **D**	22. **D**	34. **E**	46. **B**	58. **C**
11. **D**	23. **B**	35. **C**	47. **C**	59. **D**
12. **C**	24. **A**	36. **A**	48. **D**	60. **E**

Answers Explained

AREA

We give below, for each of Questions 1–17, a sketch of the region, and indicate a typical element of area. The area of the region is given by the definite integral. We exploit symmetry wherever possible.

1. **(C)**

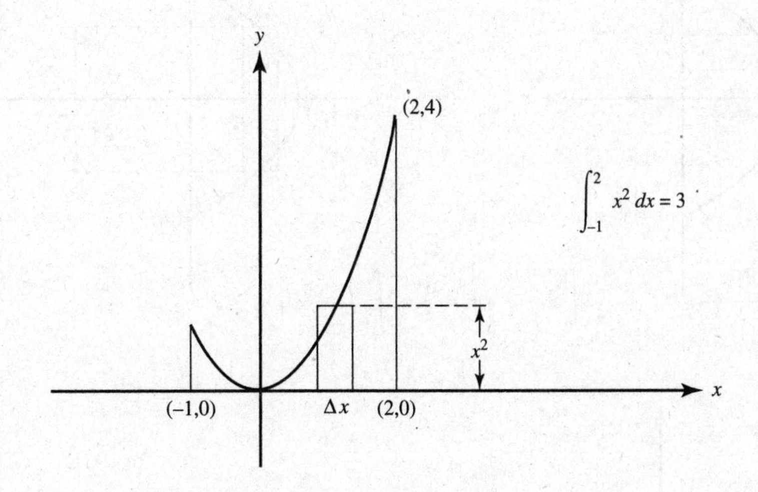

$$\int_{-1}^{2} x^2\,dx = 3$$

2. **(C)**

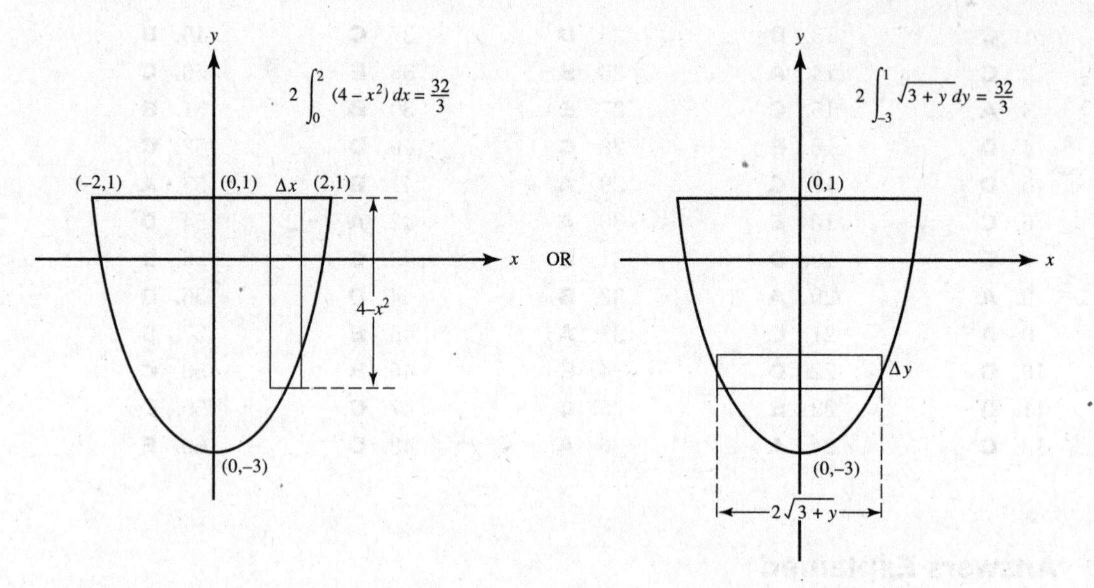

3. **(A)**

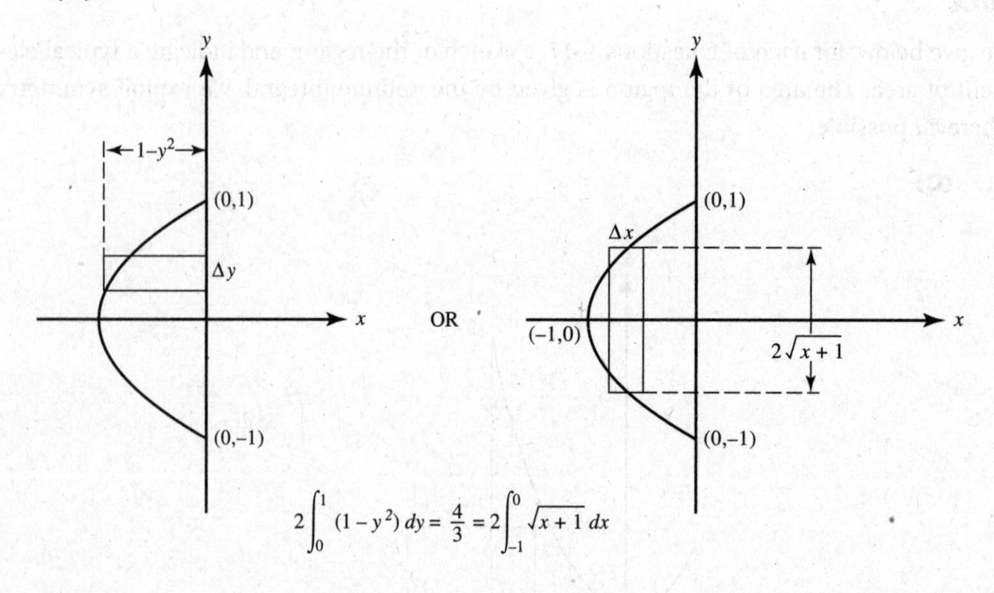

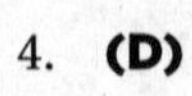

$$2\int_0^1 (1-y^2)\,dy = \frac{4}{3} = 2\int_{-1}^0 \sqrt{x+1}\,dx$$

4. **(D)**

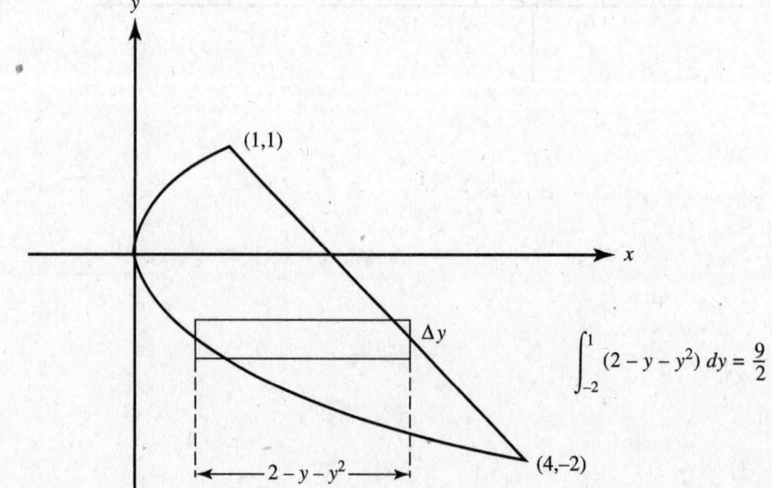

$$\int_{-2}^1 (2-y-y^2)\,dy = \frac{9}{2}$$

5. **(D)**

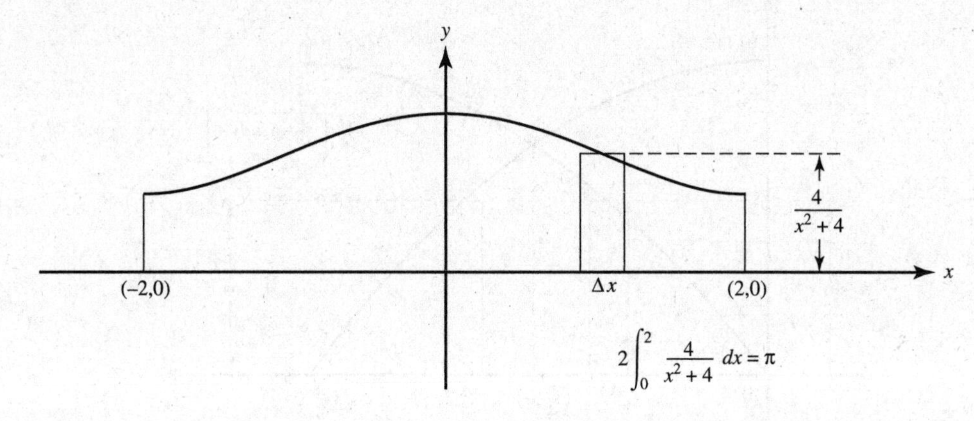

y
x
(–2,0)
Δx
(2,0)
$\frac{4}{x^2+4}$
$2\int_0^2 \frac{4}{x^2+4}\,dx = \pi$

6. **(C)**

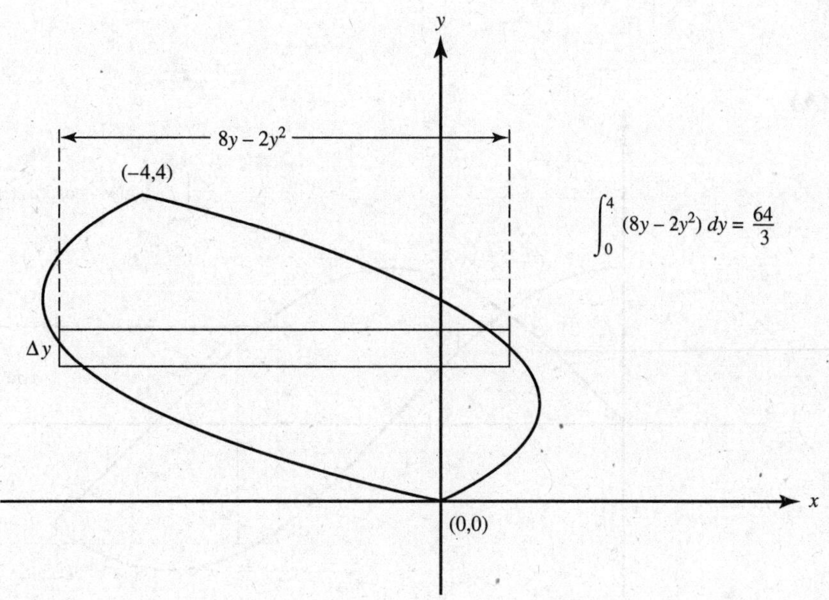

y
x
$8y - 2y^2$
(–4,4)
Δy
(0,0)
$\int_0^4 (8y - 2y^2)\,dy = \frac{64}{3}$

7. **(E)**

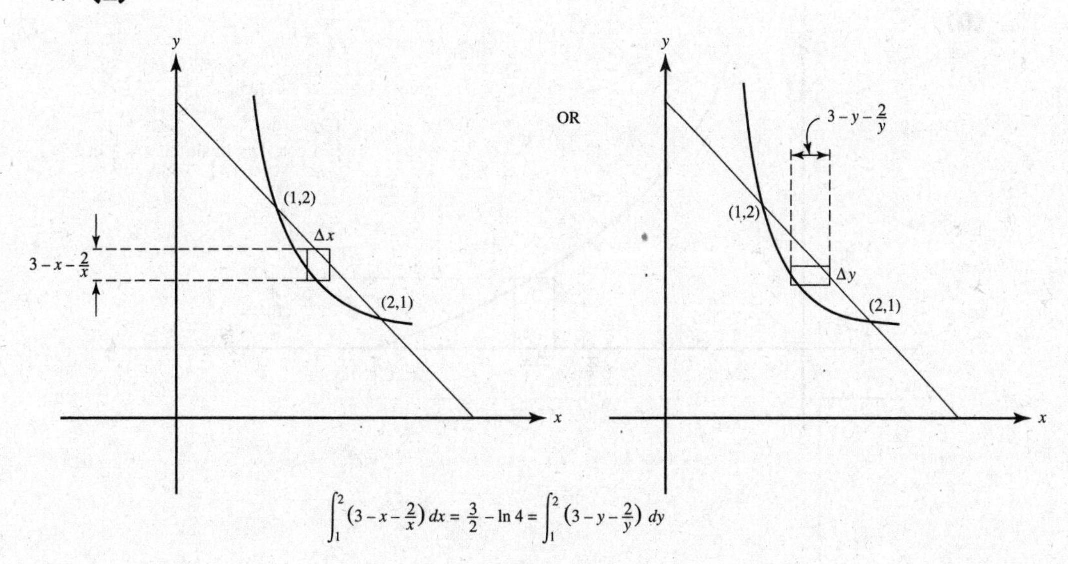

y
x
OR
$3 - x - \frac{2}{x}$
Δx
(1,2)
(2,1)
$3 - y - \frac{2}{y}$
Δy
$\int_1^2 \left(3 - x - \frac{2}{x}\right) dx = \frac{3}{2} - \ln 4 = \int_1^2 \left(3 - y - \frac{2}{y}\right) dy$

8. **(A)**

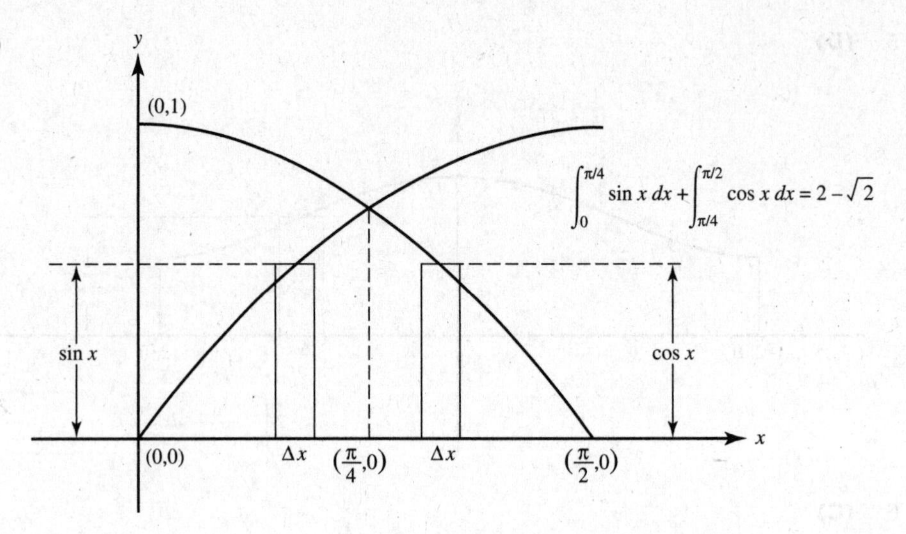

y
(0,1)
$\int_0^{\pi/4} \sin x\, dx + \int_{\pi/4}^{\pi/2} \cos x\, dx = 2 - \sqrt{2}$
sin x
cos x
(0,0)
Δx
$(\frac{\pi}{4},0)$
Δx
$(\frac{\pi}{2},0)$
x

9. **(A)**

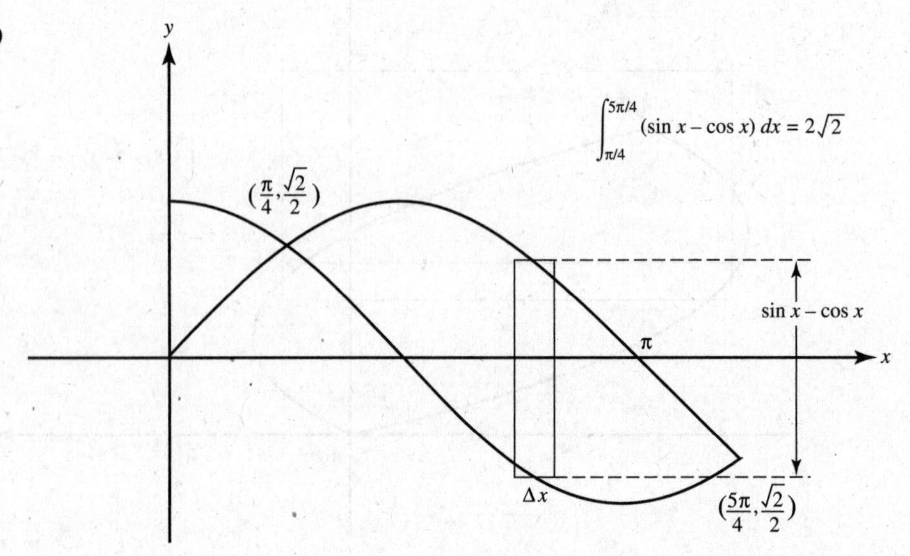

y
$\int_{\pi/4}^{5\pi/4} (\sin x - \cos x)\, dx = 2\sqrt{2}$
$(\frac{\pi}{4},\frac{\sqrt{2}}{2})$
sin x – cos x
π
x
Δx
$(\frac{5\pi}{4},\frac{\sqrt{2}}{2})$

10. **(D)**

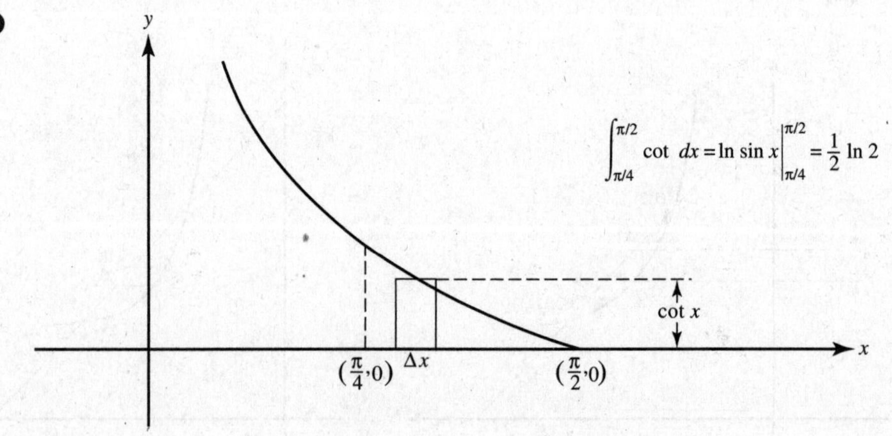

y
$\int_{\pi/4}^{\pi/2} \cot\, dx = \ln \sin x \Big|_{\pi/4}^{\pi/2} = \frac{1}{2} \ln 2$
cot x
$(\frac{\pi}{4},0)$
Δx
$(\frac{\pi}{2},0)$
x

11. **(D)**

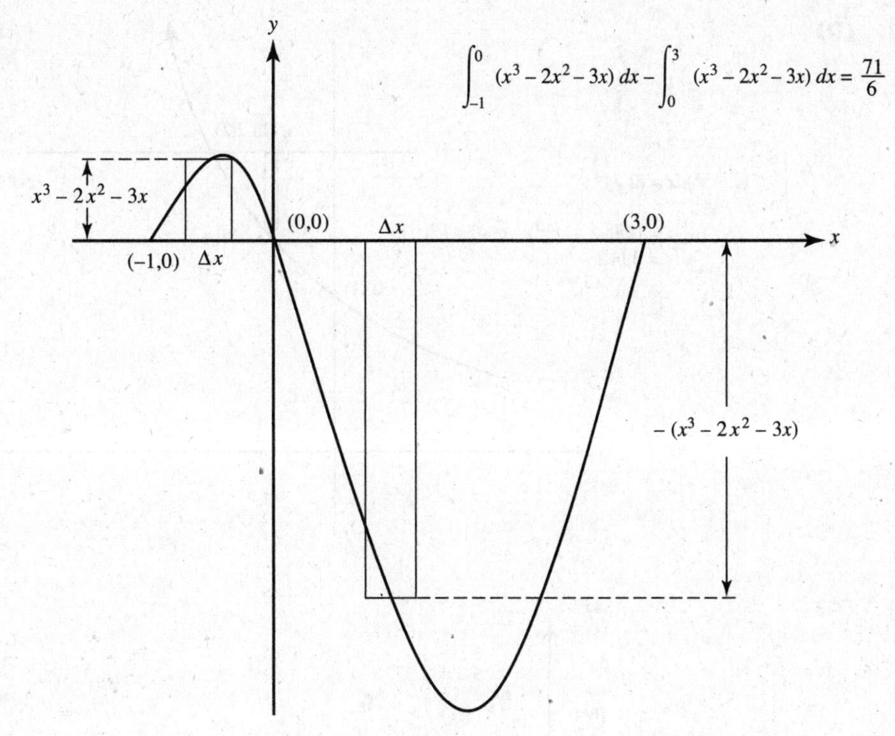

$\int_{-1}^{0} (x^3 - 2x^2 - 3x)\,dx - \int_{0}^{3} (x^3 - 2x^2 - 3x)\,dx = \frac{71}{6}$
y
x
$x^3 - 2x^2 - 3x$
(0,0)
Δx
(3,0)
(−1,0)
Δx
$-(x^3 - 2x^2 - 3x)$

12. **(C)**

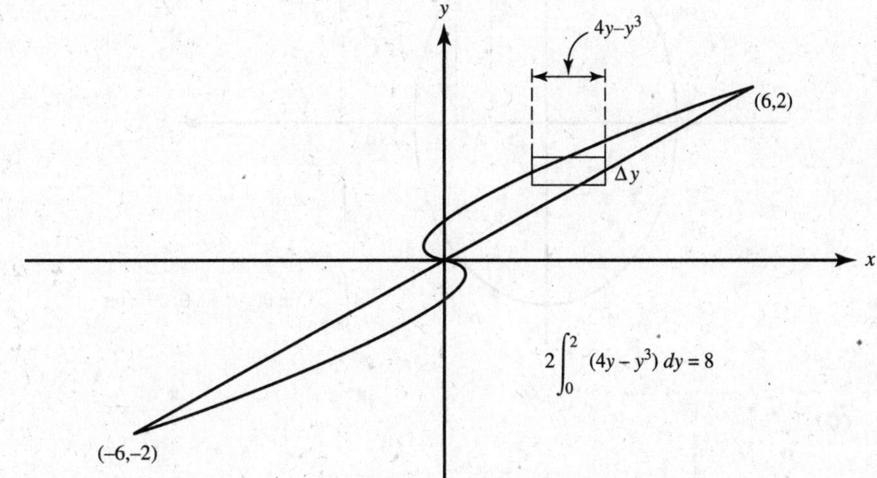

y
$4y-y^3$
(6,2)
Δy
x
$2\int_{0}^{2} (4y - y^3)\,dy = 8$
(−6,−2)

13. **(D)**

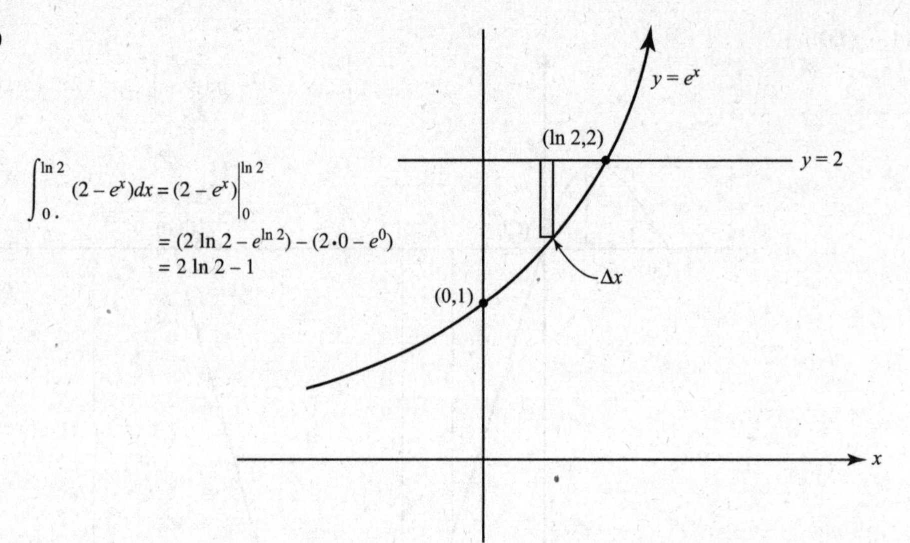

$$\int_{0}^{\ln 2} (2 - e^x)dx = (2x - e^x)\Big|_0^{\ln 2}$$
$$= (2 \ln 2 - e^{\ln 2}) - (2 \cdot 0 - e^0)$$
$$= 2 \ln 2 - 1$$

14. **(A)**

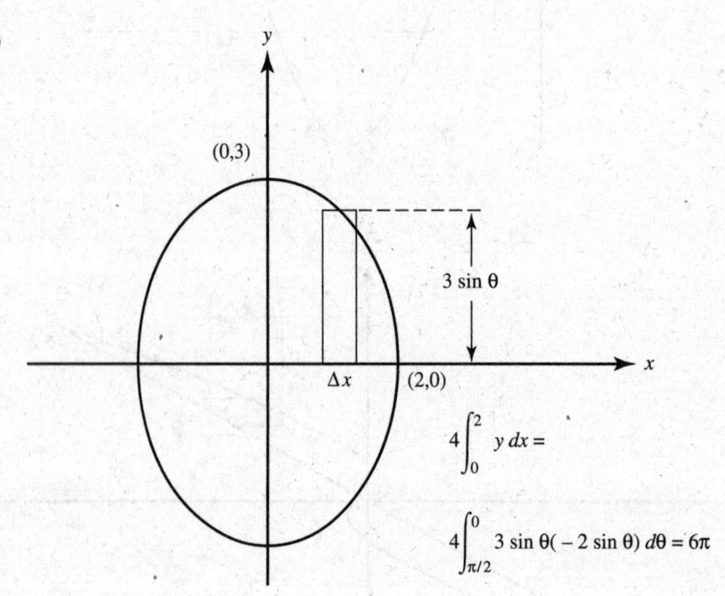

$$4\int_0^2 y\,dx =$$

$$4\int_{\pi/2}^{0} 3 \sin\theta(-2\sin\theta)\,d\theta = 6\pi$$

15. **(C)**

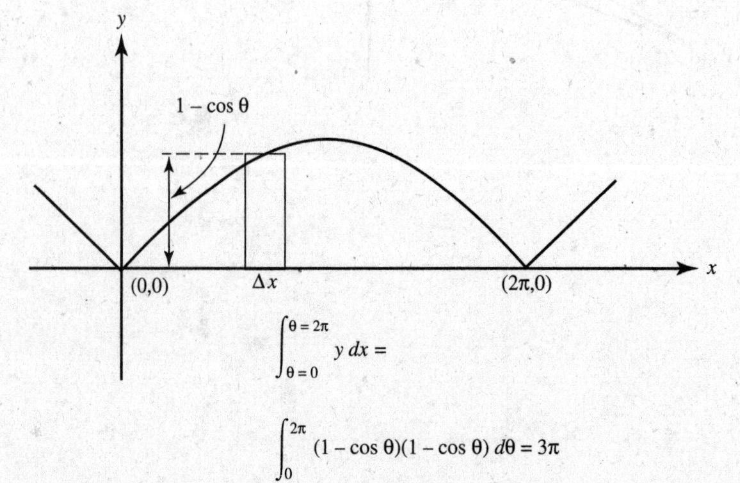

$$\int_{\theta=0}^{\theta=2\pi} y\,dx =$$

$$\int_0^{2\pi} (1 - \cos\theta)(1 - \cos\theta)\,d\theta = 3\pi$$

16. **(B)**

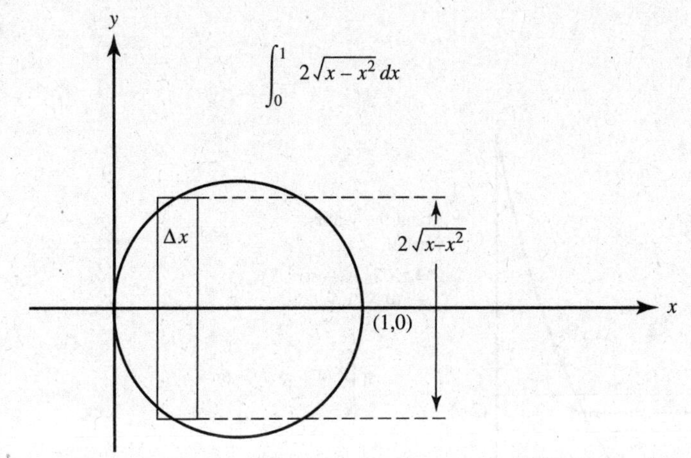

17. **(C)**

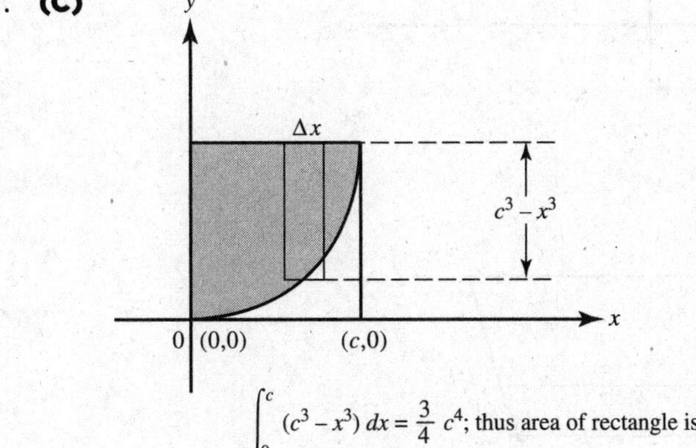

$\int_0^c (c^3 - x^3)\, dx = \frac{3}{4} c^4$; thus area of rectangle is to area of shaded region as 4 is to 3.

VOLUME

A sketch is given below, for each of Questions 18–27, in addition to the definite integral for each volume.

18. **(E)**

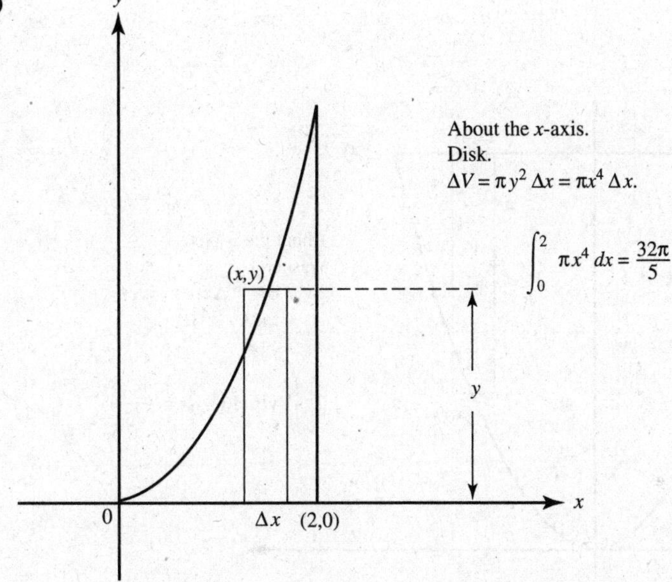

19. **(D)**

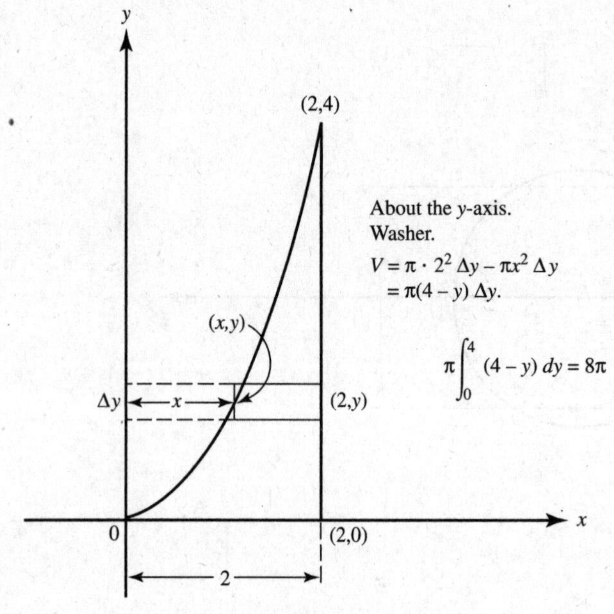

About the y-axis.
Washer.
$V = \pi \cdot 2^2 \, \Delta y - \pi x^2 \, \Delta y$
$= \pi(4 - y) \, \Delta y.$

$$\pi \int_0^4 (4 - y) \, dy = 8\pi$$

20. **(A)**

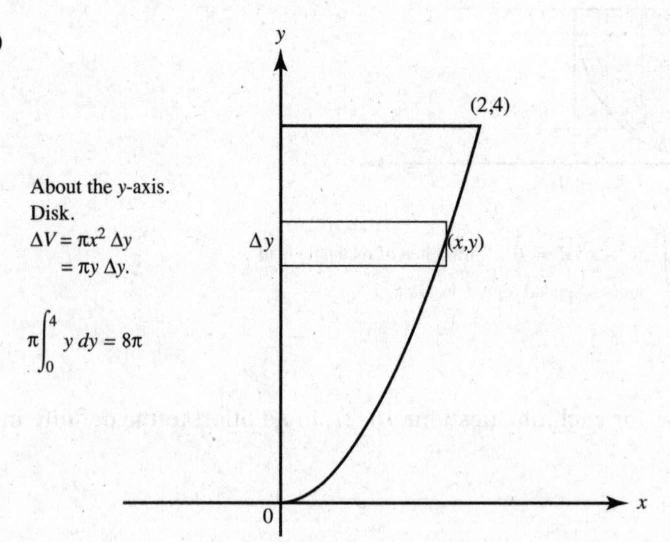

About the y-axis.
Disk.
$\Delta V = \pi x^2 \, \Delta y$
$= \pi y \, \Delta y.$

$$\pi \int_0^4 y \, dy = 8\pi$$

21. **(C)**

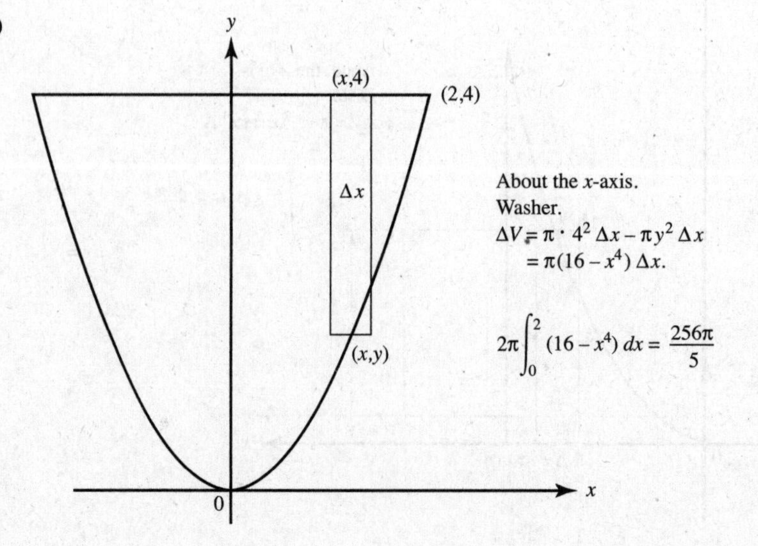

About the x-axis.
Washer.
$\Delta V = \pi \cdot 4^2 \, \Delta x - \pi y^2 \, \Delta x$
$= \pi(16 - x^4) \, \Delta x.$

$$2\pi \int_0^2 (16 - x^4) \, dx = \frac{256\pi}{5}$$

22. **(D)**

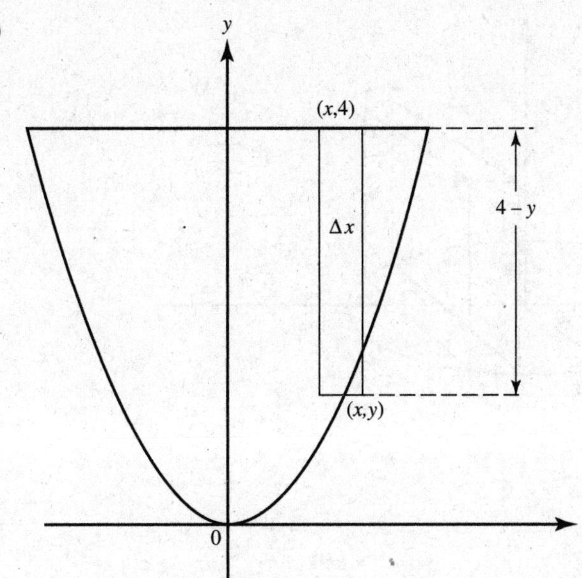

About $y = 4$.
Disk.
$\Delta V = \pi(4 - y)^2 \, \Delta x$
$= \pi(4 - x^2)^2 \, \Delta x.$

$$2\pi \int_0^2 (4 - x^2)^2 \, dx = \frac{512\pi}{15}$$

23. **(B)**

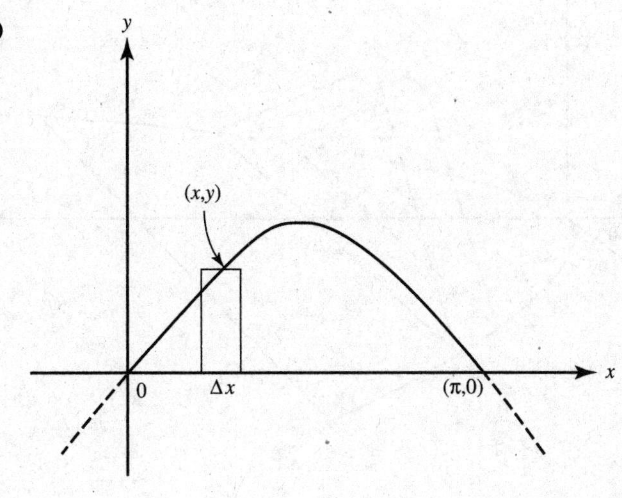

About the x-axis.
Disk.
$\Delta V = \pi y^2 \, \Delta x$
$= \pi \sin^2 x \, \Delta x.$

$$\pi \int_0^{\pi} \sin^2 x \; dx = \frac{\pi^2}{2}$$

24. **(A)**

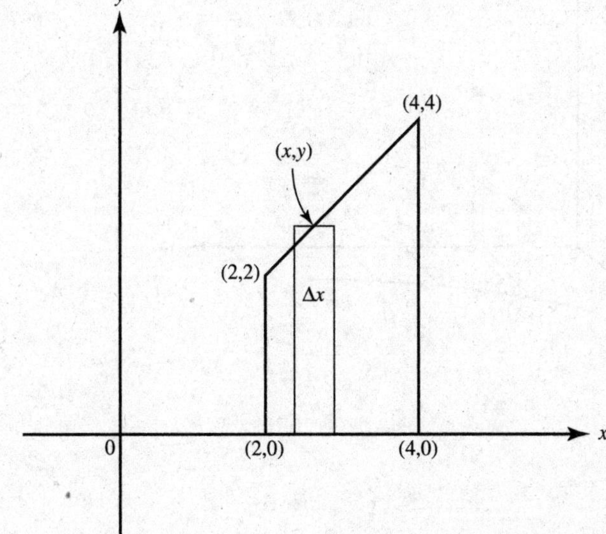

About the x-axis.
Disk.
$\Delta V = \pi y^2 \, \Delta x$

$$\pi \int_2^4 x^2 \, dx = \frac{56\pi}{3}$$

25. **(D)**

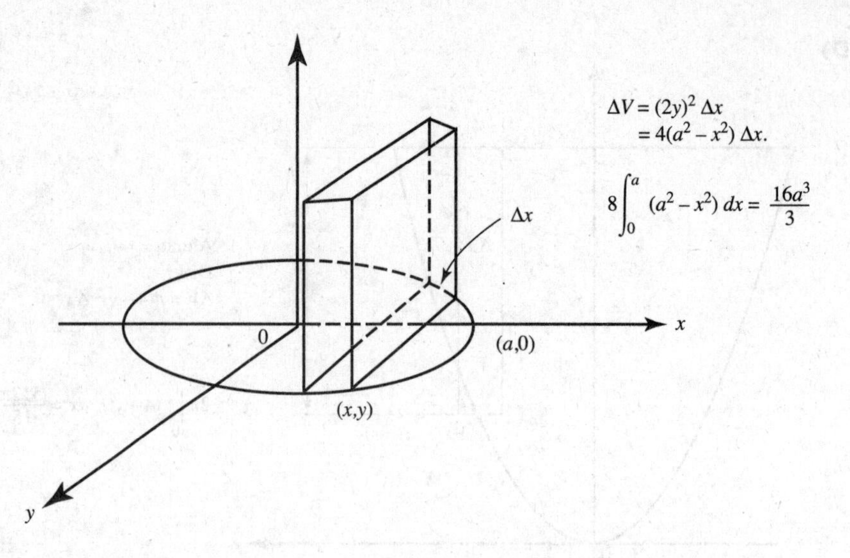

$$\Delta V = (2y)^2\,\Delta x = 4(a^2 - x^2)\,\Delta x.$$

$$8\int_0^a (a^2 - x^2)\,dx = \frac{16a^3}{3}$$

26. **(B)**

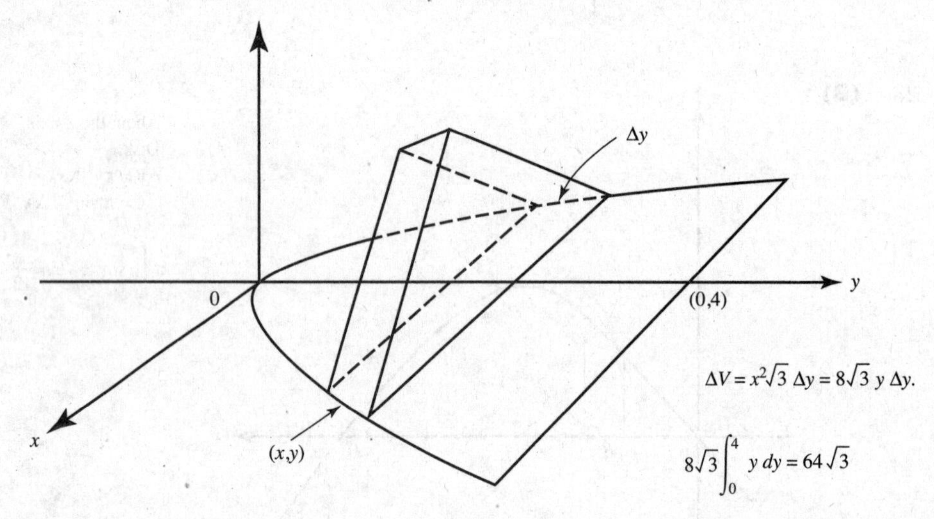

$$\Delta V = x^2\sqrt{3}\,\Delta y = 8\sqrt{3}\,y\,\Delta y.$$

$$8\sqrt{3}\int_0^4 y\,dy = 64\sqrt{3}$$

27. **(E)**

$$\int_0^1 (e^{-x})^2\,dx = \frac{1}{2}\left(1 - \frac{1}{e^2}\right)$$

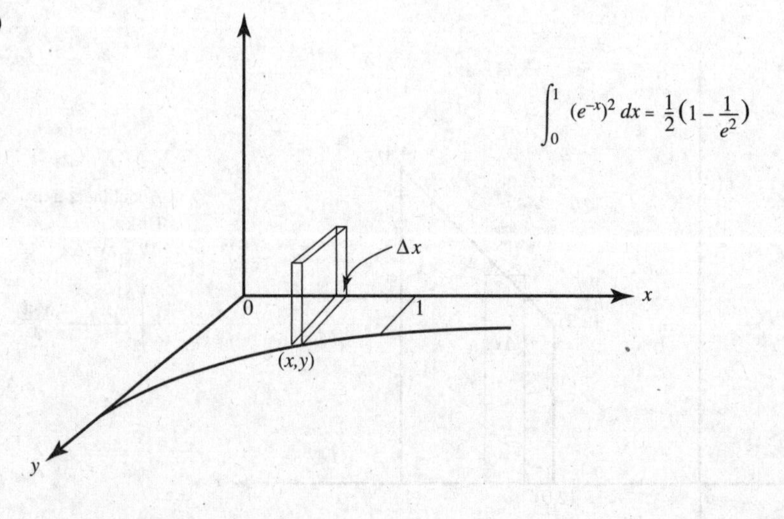

ARC LENGTH

28. **(C)** Note that the curve is symmetric to the x-axis. The arc length equals

$$2\int_0^4 \sqrt{1+\frac{9}{4}x}\,dx.$$

29. **(A)** Integrate $\int_{\pi/4}^{\pi/3} \sqrt{1+\tan^2 x}\,dx$. Replace the integrand by $\sec x$, and use formula (13) on page 212 to get $\ln|\sec x + \tan x|\Big|_{\pi/4}^{\pi/3}$.

IMPROPER INTEGRALS

30. **(A)** The integral equals $\lim\limits_{b\to\infty} -\frac{1}{e^x}\Big|_0^b = -(0-1)$.

31. **(E)** $\int_0^e \frac{du}{u} = \lim\limits_{h\to 0^+}\int_h^e \frac{du}{u} = \lim\limits_{h\to 0^+} \ln|u|\Big|_h^e = \lim\limits_{h\to 0^+} (\ln e - \ln h)$. So the integral diverges to infinity.

32. **(B)** Redefine as $\lim\limits_{k\to 1^+}\int_k^2 (t-1)^{-1/3}\,dt$.

33. **(A)** Rewrite as $\lim\limits_{k\to 3^-}\int_2^k (x-3)^{-2/3}\,dx + \lim\limits_{m\to 3^+}\int_m^4 (x-3)^{-2/3}\,dx$. Each integral converges to 3.

34. **(E)** $\int_2^4 \frac{dx}{(x-3)^2} = \int_2^3 \frac{dx}{(x-3)^2} + \int_3^4 \frac{dx}{(x-3)^2}$. Neither of the latter integrals converges; therefore the original integral diverges.

35. **(C)** Evaluate $\lim\limits_{k\to 0^+} 2\sqrt{1-\cos x}\Big|_k^{\pi/2}$.

36. **(A)**

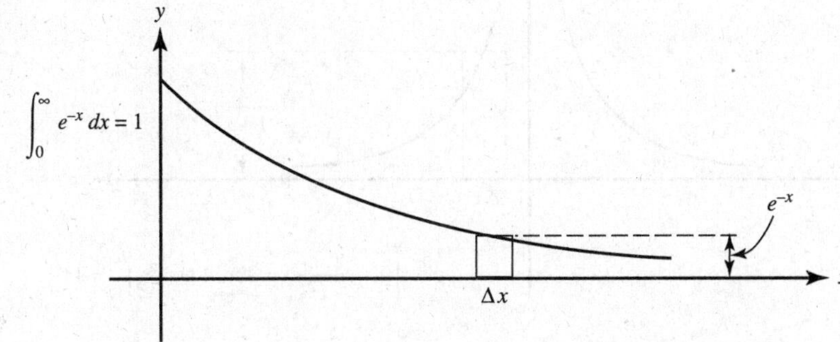

37. **(C)**

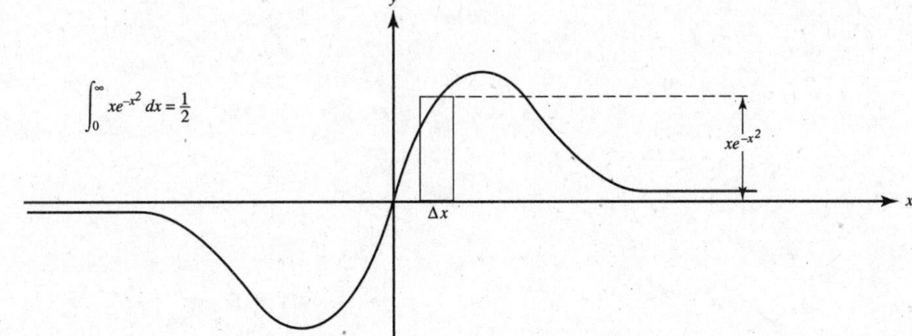

38. **(E)**

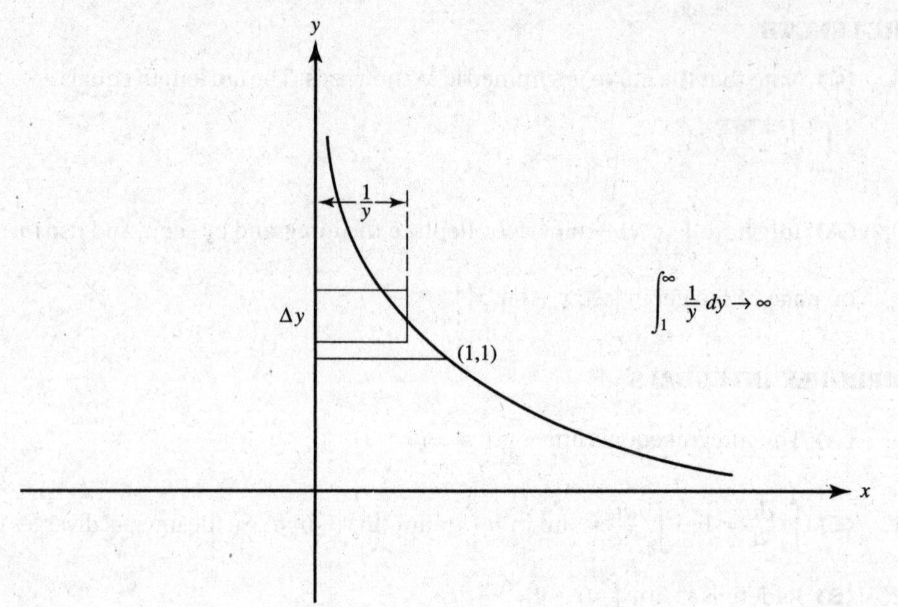

39. **(B)**

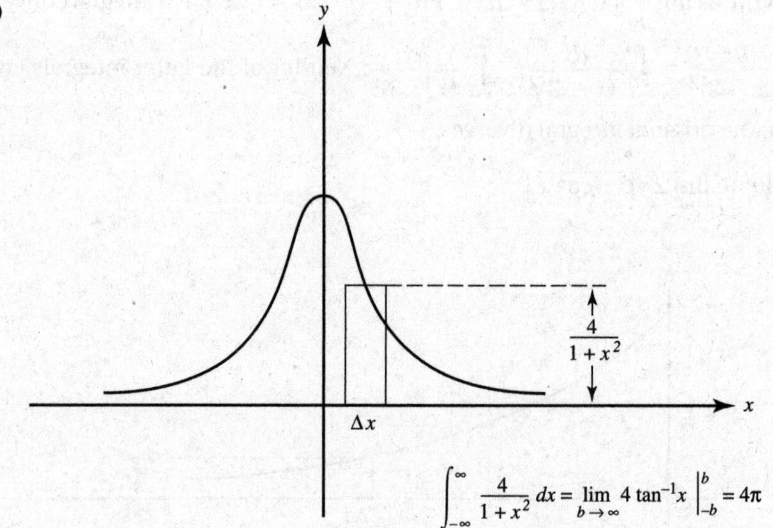

$$\int_{-\infty}^{\infty} \frac{4}{1+x^2}\,dx = \lim_{b\to\infty} 4\tan^{-1}x \Big|_{-b}^{b} = 4\pi$$

40. **(D)**

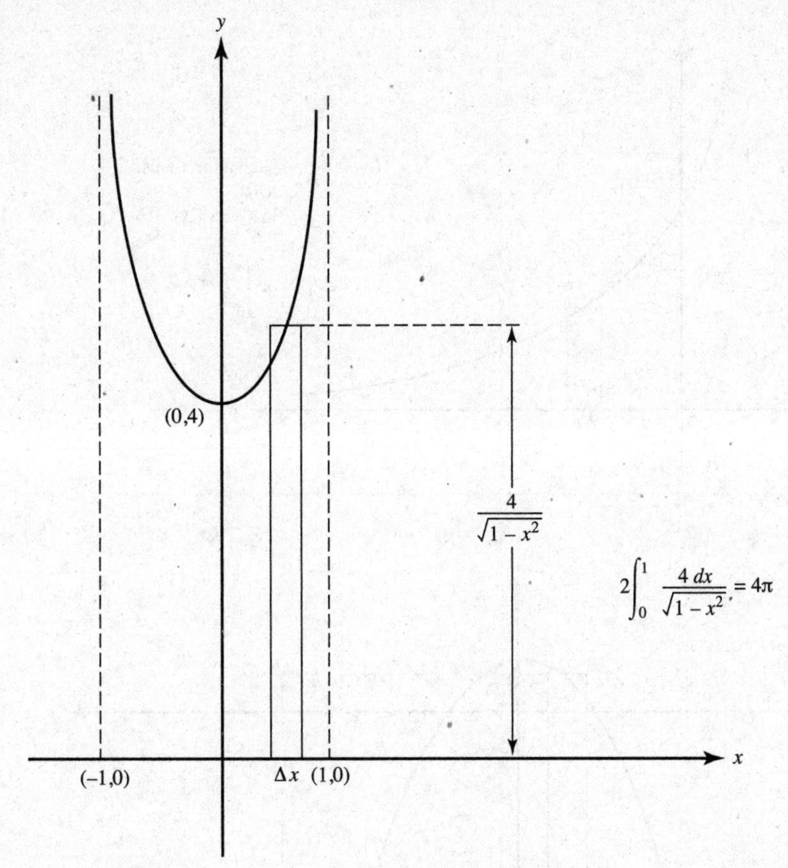

y
(0,4)
$\frac{4}{\sqrt{1-x^2}}$
$2\int_0^1 \frac{4\,dx}{\sqrt{1-x^2}} = 4\pi$
(−1,0)
Δx (1,0)
x

41. **(B)**

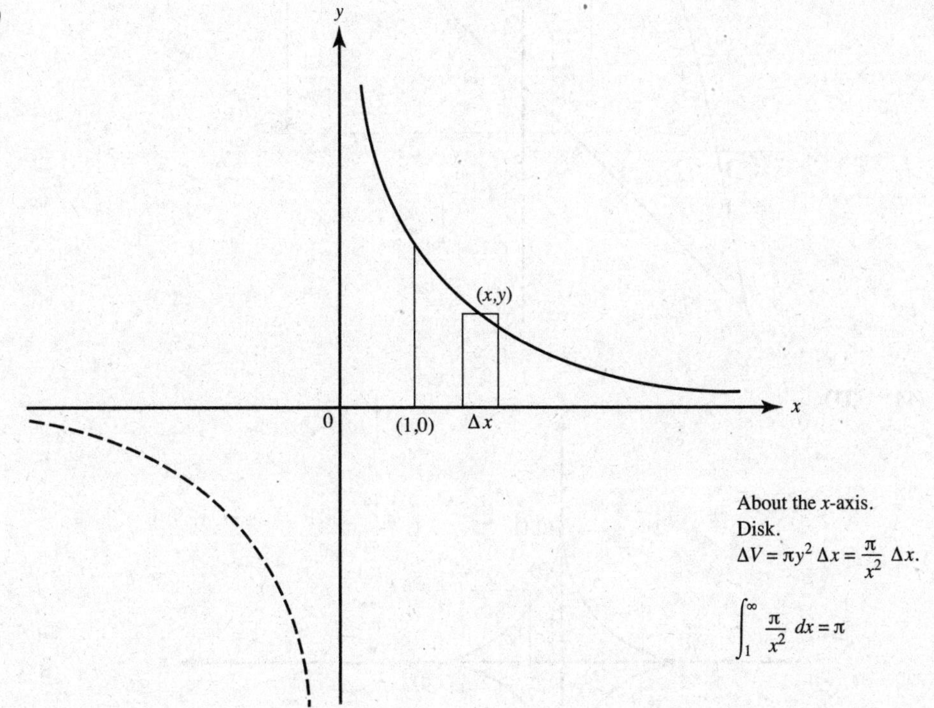

y
(x,y)
0
(1,0)
Δx
x
About the x-axis.
Disk.
$\Delta V = \pi y^2\,\Delta x = \frac{\pi}{x^2}\,\Delta x.$
$\int_1^\infty \frac{\pi}{x^2}\,dx = \pi$

42. **(A)**

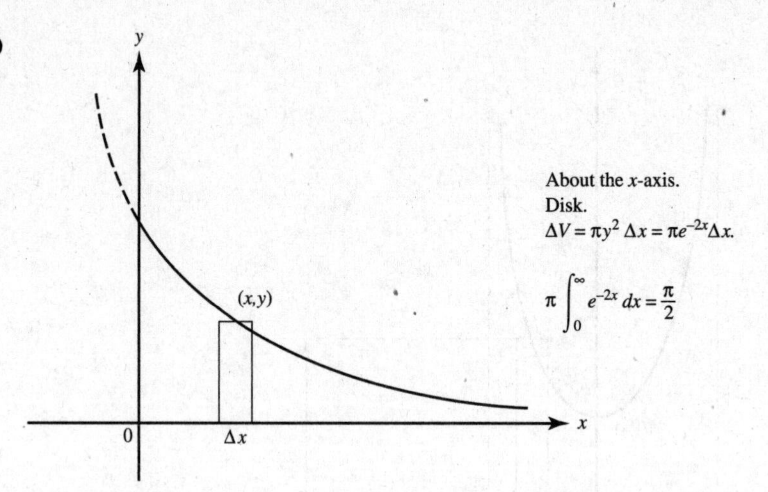

AREA

43. **(C)**

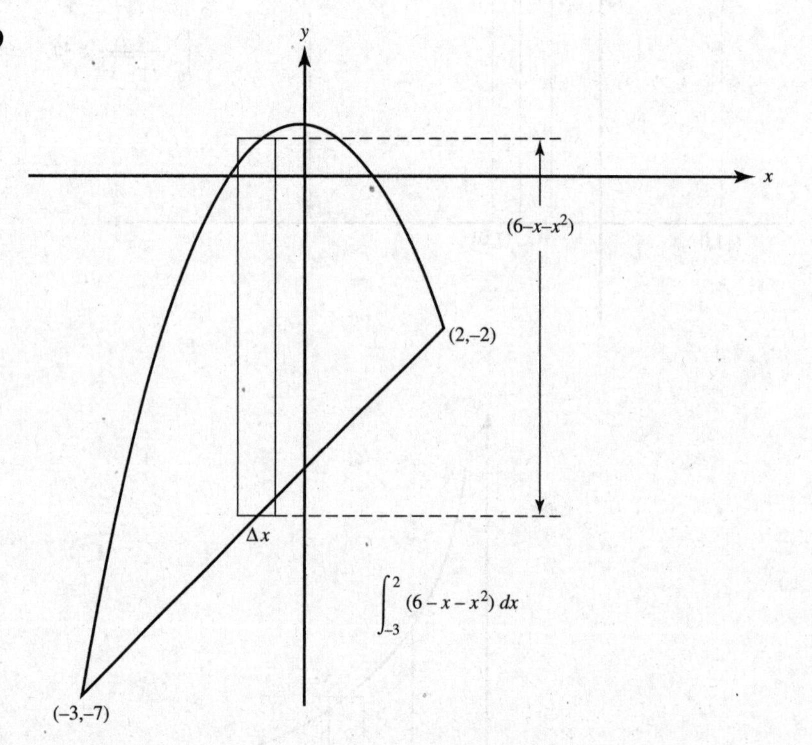

44. **(D)**

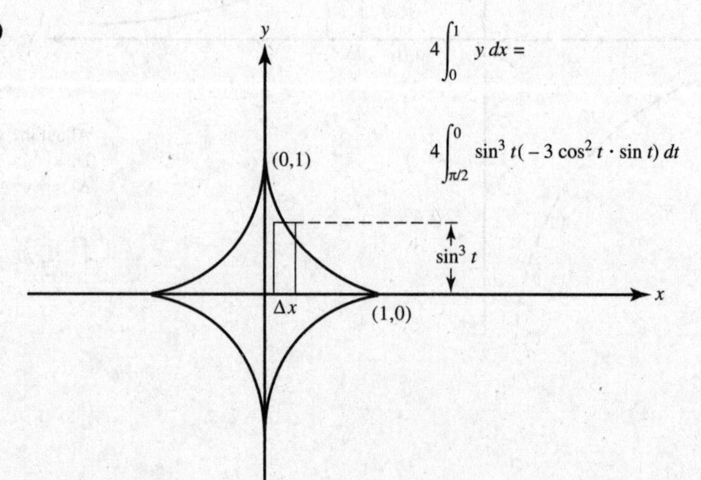

45. **(E)** $T(4) = \frac{1}{2}\left(1.62 + 2(4.15) + 2(7.5) + 2(9.0) + 12.13\right)$.

46. **(B)** $A = 8\int_0^{\pi/4} \frac{1}{2}\cos^2 2\theta\, d\theta = 1.571$, using a graphing calculator.

47. **(C)** The small loop is generated as θ varies from $\frac{\pi}{6}$ to $\frac{5\pi}{6}$. (C) uses the loop's symmetry.

VOLUME

48. **(D)**

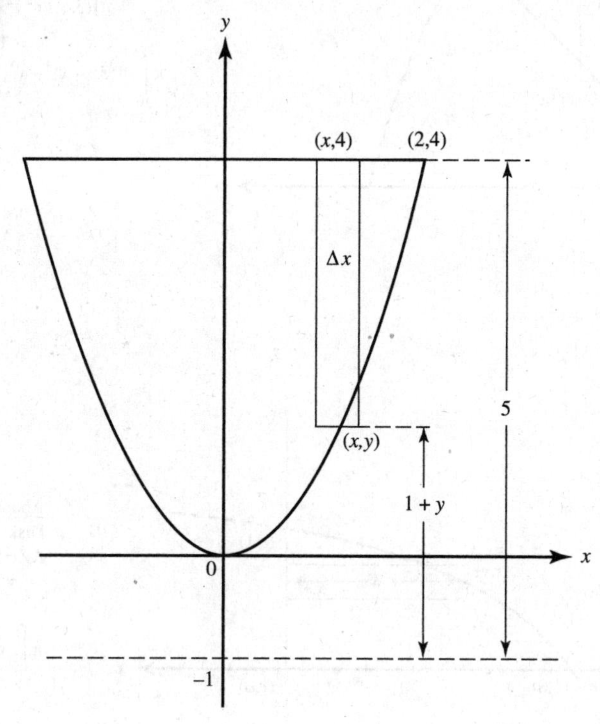

About $y = -1$.
Washer.
$\Delta V = \pi \cdot 5^2\,\Delta x - \pi(1+y)^2\,\Delta x$
$= \pi[25 - (1+y)^2]\,\Delta x.$

$$2\pi\int_0^2 (24 - 2x^2 - x^4)\,dx$$

49. **(B)**

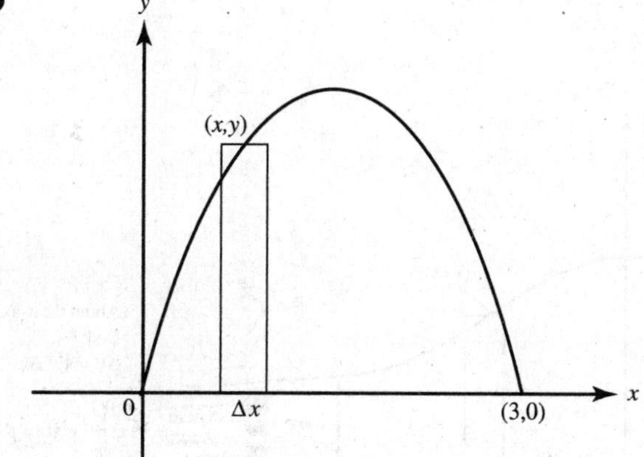

About the x-axis.
Disk.
$\Delta V = \pi y^2\,\Delta x$
$= \pi(3x - x^2)^2\,\Delta x.$

$$\pi\int_0^3 (3x - x^2)^2\,dx$$

50. **(C)**

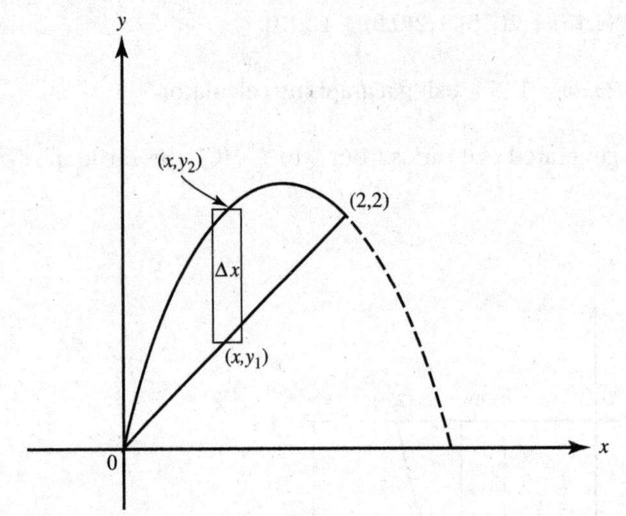

About the x-axis.
Washer.

$$\Delta V = \pi y_2^2\,\Delta x - \pi y_1^2\,\Delta x$$
$$= \pi[(3x - x^2)^2 - x^2]\,\Delta x.$$

$$\pi\int_0^2 [(3x - x)^2 - x^2]\,dx$$

51. **(B)**

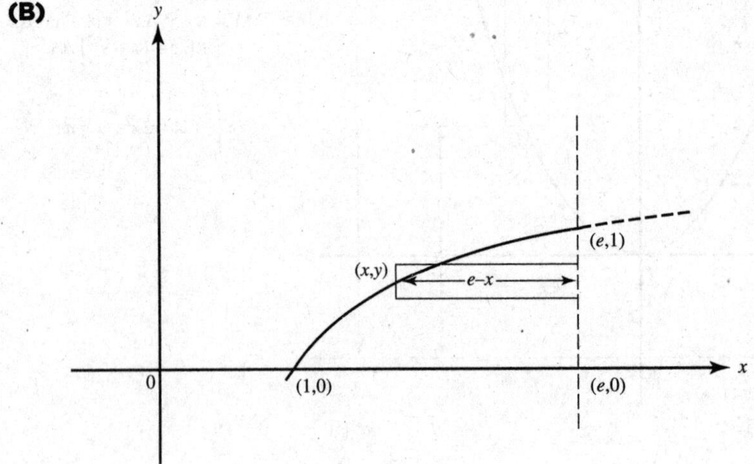

About $x = e$.
Disk.

$$\Delta V = \pi(e - x)^2\,\Delta y$$
$$= \pi(e - e^y)^2\,\Delta y.$$

$$\pi\int_0^1 (e - e^y)^2\,dy$$

52. **(C)**

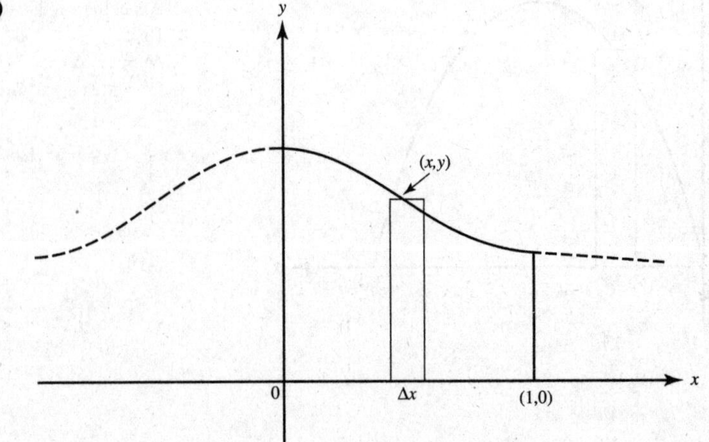

About the x-axis.
Disk.

$$\Delta V = \pi y^2\,\Delta x.$$

$$\pi\int_0^1 y^2\,dx = \pi\int_0^{\pi/4} \cos^2\theta\,d\theta$$

53. **(A)**

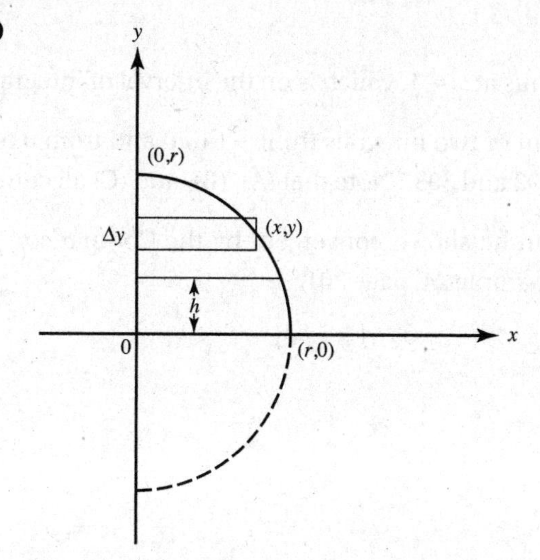

About the y-axis.
Disk.
$\Delta V = \pi x^2 \Delta y$
$= \pi(r^2 - y^2)\Delta y.$

$$\pi\int_h^r (r^2 - y^2)\,dy = \frac{\pi}{3}(2r^3 + h^3 - 3r^2h)$$

54. **(D)**

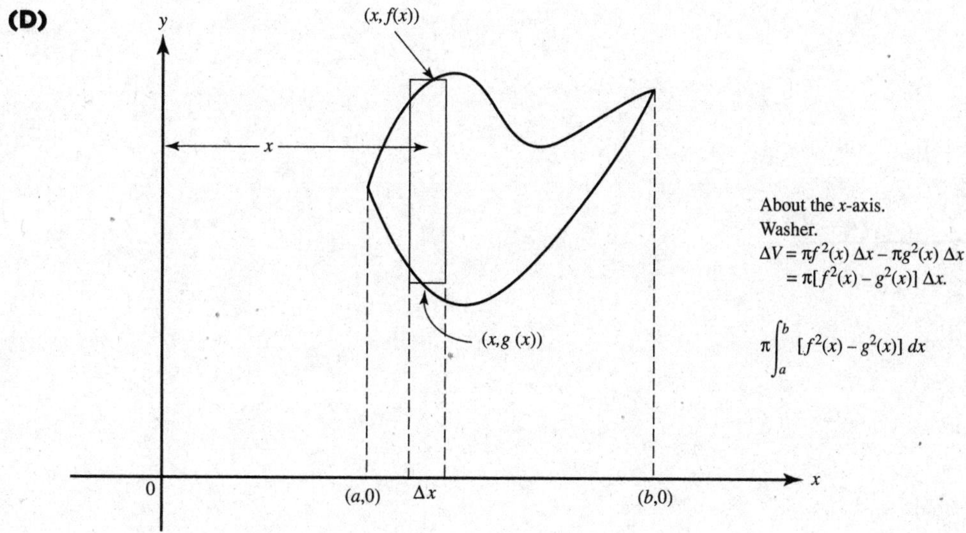

About the x-axis.
Washer.
$\Delta V = \pi f^2(x)\,\Delta x - \pi g^2(x)\,\Delta x$
$= \pi[f^2(x) - g^2(x)]\,\Delta x.$

$$\pi\int_a^b [f^2(x) - g^2(x)]\,dx$$

ARC LENGTH

55. **(D)** From (3) on page 295, we obtain the length:

$$\int_0^{2\pi}\sqrt{(1-\cos t)^2 + (\sin t)^2}\,dt = \int_0^{2\pi}\sqrt{2-\cos t}\,dt.$$

56. **(D)** Note that the curve is symmetric to the x-axis. Use (2) on page 295.

57. **(B)** Use (3) on page 295 to get the integral:

$$\int_2^3\sqrt{(-e^t\sin t + e^t\cos t)^2 + (e^t\cos t + e^t\sin t)^2}\,dt = \sqrt{2}e^t\Big|_2^3.$$

IMPROPER INTEGRALS

58. **(C)** The integrand is discontinuous at $x = 1$, which is on the interval of integration.

59. **(D)** The integral in (D) is the sum of two integrals from -1 to 0 and from 0 to 1. Both diverge (see Example 29, pages 302 and 303). Note that (A), (B), and (C) all converge.

60. **(E)** Choices (A), (C), and (D) can be shown convergent by the Comparison Test; the convergence of (B) is shown in Example 24, page 301.

Further Applications of Integration

8

CONCEPTS AND SKILLS

In this chapter, we will review many ways that definite integrals can be used to solve a variety of problems, notably distance traveled by an object in motion along a line. We'll see that in a variety of settings accumulated change can be expressed as a Riemann Sum whose limit becomes an integral of the rate of change.

For BC students, we'll expand our discussion of motion to include objects in motion in a plane along a parametrically defined curve.

A. MOTION ALONG A STRAIGHT LINE

If the motion of a particle P along a straight line is given by the equation $s = F(t)$, where s is the distance at time t of P from a fixed point on the line, then the velocity and acceleration of P at time t are given respectively by

$$v = \frac{ds}{dt} \quad \text{and} \quad a = \frac{dv}{dt} = \frac{d^2s}{dt^2}.$$

This topic was discussed as an application of differentiation on page 177. Here we will apply integration to find velocity from acceleration and distance from velocity.

Distance

If we know that particle P has velocity $v(t)$, where v is a continuous function, then the *distance* traveled by the particle during the time interval from $t = a$ to $t = b$ is the definite integral of its speed:

$$\int_a^b |v(t)|\, dt. \tag{1}$$

If $v(t) \geqq 0$ for all t on $[a, b]$ (i.e., P moves only in the positive direction), then (1) is equivalent to $\int_a^b v(t)\, dt$; similarly, if $v(t) \leqq 0$ on $[a, b]$ (P moves only in the negative direction), then (1) yields $-\int_a^b v(t)\, dt$. If $v(t)$ changes sign on $[a, b]$ (i.e., the direction of motion changes), then (1) gives the total distance traveled. Suppose, for example, that the situation is as follows:

$$\begin{array}{ll} a \leqq t \leqq c & v(t) \geqq 0; \\ c \leqq t \leqq d & v(t) \leqq 0; \\ d \leqq t \leqq b & v(t) \geqq 0. \end{array}$$

Then the total distance traveled during the time interval from $t = a$ to $t = b$ is exactly

$$\int_a^c v(t)\,dt - \int_c^d v(t)\,dt + \int_d^b v(t)\,dt.$$

Displacement

The *displacement* or *net change* in the particle's position from $t = a$ to $t = b$ is equal, by the Fundamental Theorem of Calculus (FTC), to

$$\int_a^b v(t)\,dt.$$

➡ Example 1

If a body moves along a straight line with velocity $v = t^3 + 3t^2$, find the distance traveled between $t = 1$ and $t = 4$.

SOLUTION: $\displaystyle\int_1^4 \left(t^3 + 3t^2\right) dt = \left(\frac{t^4}{4} + t^3\right)\Bigg|_1^4 = \frac{507}{4}.$

Note that $v > 0$ for all t on $[1, 4]$.

➡ Example 2

A particle moves along the x-axis so that its velocity at time t is given by $v(t) = 6t^2 - 18t + 12$.

(a) Find the total distance covered between $t = 0$ and $t = 4$.

(b) Find the displacement of the particle from $t = 0$ to $t = 4$.

SOLUTIONS:

(a) Since $v(t) = 6t^2 - 18t + 12 = 6(t - 1)(t - 2)$, we see that:

if $t < 1$, then $v > 0$;
if $1 < t < 2$, then $v < 0$;
if $2 < t$, then $v > 0$.

Thus, the total distance covered between $t = 0$ and $t = 4$ is

$$\int_0^1 v(t)\,dt - \int_1^2 v(t)\,dt + \int_2^4 v(t)\,dt. \qquad (2)$$

When we replace $v(t)$ by $6t^2 - 18t + 12$ in (2) and evaluate, we obtain 34 units for the total distance covered between $t = 0$ and $t = 4$. This can also be verified on your calculator by evaluating

$$\int_0^4 |v(t)|\,dt.$$

This example is the same as Example 26 on page 177, in which the required distance is computed by another method.

(b) To find the *displacement* of the particle from $t = 0$ to $t = 4$, we use the FTC, evaluating

$$\int_0^4 v(t)\,dt = \int_0^4 \left(6t^2 - 18t + 12\right) dt.$$

$$= \left(2t^3 - 9t^2 + 12t\right)\Big|_0^4 = 128 - 144 + 48 = 32.$$

This is the net change in position from $t = 0$ to $t = 4$, sometimes referred to as "position shift." Here it indicates the particle ended up 32 units to the right of its starting point.

Example 3

The acceleration of an object moving on a line is given at time t by $a = \sin t$; when $t = 0$ the object is at rest. Find the distance s it travels from $t = 0$ to $t = \frac{5\pi}{6}$.

SOLUTION: Since $a = \frac{d^2s}{dt^2} = \frac{dv}{dt} = \sin t$, it follows that

$$v(t) = \frac{ds}{dt} = \int \sin t\, dt; \qquad v(t) = -\cos t + C.$$

Also, $v(0) = 0$ yields $C = 1$. Thus $v(t) = 1 - \cos t$; and since $\cos t \leqq 1$ for all t we see that $v(t) \geqq 0$ for all t. Thus, the distance traveled is

$$\int_0^{5\pi/6} (1 - \cos t)\, dt = (t - \sin t)\Big|_0^{5\pi/6} = \frac{5\pi}{6} - \frac{1}{2}.$$

B. MOTION ALONG A PLANE CURVE

BC ONLY

In Chapter 4, §K, it was pointed out that, if the motion of a particle P along a curve is given parametrically by the equations $x = x(t)$ and $y = y(t)$, then at time t the position vector $\mathbf{R}$, the velocity vector $\mathbf{v}$, and the acceleration vector $\mathbf{a}$ are:

$$\mathbf{R} = \langle x, y \rangle;$$

$$\mathbf{v} = \frac{d\mathbf{R}}{dt} = \left\langle \frac{dx}{dt}, \frac{dy}{dt} \right\rangle = \langle v_x, v_y \rangle;$$

$$\mathbf{a} = \frac{d^2\mathbf{R}}{dt^2} = \frac{d\mathbf{v}}{dt} = \left\langle \frac{d^2x}{dt^2}, \frac{d^2y}{dt^2} \right\rangle = \langle a_x, a_y \rangle.$$

The components in the horizontal and vertical directions of $\mathbf{R}$, $\mathbf{v}$, and $\mathbf{a}$ are given, respectively, by the coordinates in the corresponding vector. The slope of $\mathbf{v}$ is $\frac{dy}{dx}$; its magnitude,

$$|\mathbf{v}| = \sqrt{\left(\frac{dx}{dt}\right)^2 + \left(\frac{dy}{dt}\right)^2},$$

is the *speed* of the particle, and the velocity vector is tangent to the path. The slope of $\mathbf{a}$ is $\frac{d^2y}{dt^2}\Big/\frac{d^2x}{dt^2}$. The distance the particle travels from time t_1 to t_2 is given by

$$\int_{t_1}^{t_2} |\mathbf{v}|\, dt = \int_{t_1}^{t_2} \sqrt{\left(\frac{dx}{dt}\right)^2 + \left(\frac{dy}{dt}\right)^2}\, dt$$

How integration may be used to solve problems of curvilinear motion is illustrated in the following examples.

BC ONLY

Example 4

Suppose a projectile is launched from the origin at an angle of elevation α and initial velocity v_0. Find the parametric equations for its flight path.

SOLUTION: We have the following initial conditions:

Position: $x(0) = 0; y(0) = 0.$

Velocity: $\frac{dx}{dt}(0) = v_0 \cos\alpha; \frac{dy}{dt}(0) = v_0 \sin\alpha$

We start with equations representing acceleration due to gravity and integrate each twice, determining the constants as shown:

Acceleration: $\frac{d^2x}{dt^2} = 0; \qquad \frac{d^2y}{dt^2} = -g;$

$$\frac{dx}{dt} = C_1 = v_0 \cos\alpha; \qquad \frac{dy}{dt} = -gt + C_2;$$

$$v_0 \sin\alpha = C_2;$$

$$x = (v_0 \cos\alpha)t + C_3; \qquad y = -\frac{1}{2}gt^2 + (v_0 \sin\alpha)t + C_4;$$

$$x(0) = 0 \text{ yields } C_3 = 0. \qquad y(0) = 0 \text{ yields } C_4 = 0.$$

Finally, then,

$$x = (v_0 \cos\alpha)t; \qquad y = -\frac{1}{2}gt^2 + (v_0 \sin\alpha)t.$$

If desired, t can be eliminated from this pair of equations to yield a parabola in rectangular coordinates.

Example 5

A particle $P(x, y)$ moves along a curve so that

$$\frac{dx}{dt} = 2\sqrt{x} \quad \text{and} \quad \frac{dy}{dt} = \frac{1}{x} \text{ at any time } t \geqq 0.$$

At $t = 0$, $x = 1$ and $y = 0$. Find the parametric equations of motion.

SOLUTION: Since $\frac{dx}{\sqrt{x}} = 2\,dt$, we integrate to get $2\sqrt{x} = 2t + C$, and use $x(0) = 1$ to find that $C = 2$. Therefore, $\sqrt{x} = t + 1$ and

$$x = (t+1)^2. \tag{1}$$

Then $\frac{dy}{dt} = \frac{1}{x} = \frac{1}{(t+1)^2}$ by (1), so $dy = \frac{dt}{(t+1)^2}$ and $y = -\frac{1}{t+1} + C'$. (2)

Since $y(0) = 0$, this yields $C' = 1$, and so (2) becomes

$$y = 1 - \frac{1}{t+1} = \frac{t}{t+1}.$$

Thus the parametric equations are

$$x = (t+1)^2 \quad \text{and} \quad y = \frac{t}{t+1}.$$

BC ONLY

➥ Example 6

The particle in Example 5 is in motion for 1 second, $0 \le t \le 1$. Find its position, velocity, speed, and acceleration at $t = 1$ and the distance it traveled.

SOLUTION: In Example 5 we derived the result $P(t) = \left\langle (t+1)^2, \frac{t}{t+1} \right\rangle$, the parametric representation of the particle's position. Hence its position at $t = 1$ is $P(1) = \left\langle 4, \frac{1}{2} \right\rangle$.

From $P(t)$ we write the velocity vector:

$$\mathbf{v} = \left\langle \frac{dx}{dt}, \frac{dy}{dt} \right\rangle = \left\langle 2(t+1), \frac{1}{(t+1)^2} \right\rangle$$

Hence, at $t = 1$ the particle's velocity is $\mathbf{v} = \left\langle 4, \frac{1}{4} \right\rangle$.

Speed is the magnitude of the velocity vector, so after 1 second the particle's speed is

$$|\mathbf{v}| = \sqrt{4^2 + \left(\frac{1}{4}\right)^2} \approx 4.008 \text{ units/sec.}$$

The particle's acceleration vector at $t = 1$ is

$$\mathbf{a} = \left\langle \frac{d^2x}{dt^2}, \frac{d^2y}{dt^2} \right\rangle = \left\langle 2, \frac{-2}{(t+1)^3} \right\rangle = \left\langle 2, -\frac{1}{4} \right\rangle.$$

On the interval $0 \le t \le 1$ the distance traveled by the particle is

$$\int_0^1 \sqrt{\left(\frac{dx}{dt}\right)^2 + \left(\frac{dy}{dt}\right)^2}\, dt = \int_0^1 \sqrt{[2(t+1)]^2 + \left[\frac{1}{(t+1)^2}\right]^2}\, dt \approx 3.057 \text{ units.}$$

➥ Example 7

A particle $P(x, y)$ moves along a curve so that its acceleration is given by

$$\mathbf{a} = \langle -4\cos 2t, -2\sin t \rangle \left(-\frac{\pi}{2} \le t \le \frac{\pi}{2} \right);$$

when $t = 0$, the particle is at (1, 0) with $\frac{dx}{dt} = 0$ and $\frac{dy}{dt} = 2$.

(a) Find the position vector $\mathbf{R}$ at any time t.

(b) Find a Cartesian equation for the path of the particle, and identify the conic on which P moves.

SOLUTIONS:

(a) $\mathbf{v} = \langle -2\sin 2t + c_1, 2\cos t + c_2 \rangle$, and since $\mathbf{v} = \langle 0,2 \rangle$ when $t = 0$, it follows that $c_1 = c_2 = 0$. So $\mathbf{v} = \langle -2\sin 2t, 2\cos t \rangle$. Also $\mathbf{R} = \langle \cos 2t + c_3, 2\sin t + c_4 \rangle$; and since $\mathbf{R} = \langle 1,0 \rangle$ when $t = 0$, we see that $c_3 = c_4 = 0$. Finally, then,

$$\mathbf{R} = \langle \cos 2t, 2\sin t \rangle.$$

BC ONLY

(b) From (a) the parametric equations of motion are

$$x = \cos 2t, \qquad y = 2 \sin t.$$

By a trigonometric identity,

$$x = 1 - 2\sin^2 t = 1 - \frac{y^2}{2}.$$

P travels in a counterclockwise direction along *part* of a parabola that has its vertex at (1, 0) and opens to the left. The path of the particle is sketched in Figure N8−1; note that $-1 \le x \le 1$, $-2 \le y \le 2$.

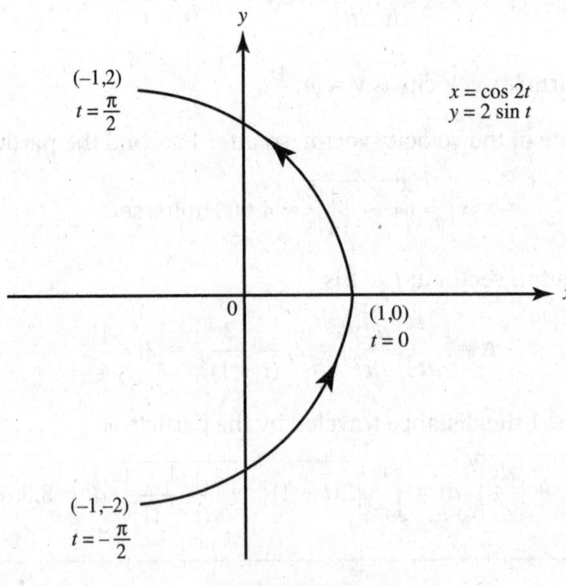

Figure N8−1

C. OTHER APPLICATIONS OF RIEMANN SUMS

We will continue to set up Riemann Sums to calculate a variety of quantities using definite integrals. In many of these examples, we will partition into *n* equal subintervals a given interval (or region or ring or solid or the like), approximate the quantity over each small subinterval (and assume it is constant there), then add up all these small quantities. Finally, as $n \to \infty$, we will replace the sum by its equivalent definite integral to calculate the desired quantity.

➡ Example 8

Amount of Leaking Water. Water is draining from a cylindrical pipe of radius 2 inches. At *t* seconds the water is flowing out with velocity $v(t)$ inches per second. Express the amount of water that has drained from the pipe in the first 3 minutes as a definite integral in terms of $v(t)$.

SOLUTION: We first express 3 min as 180 sec. We then partition [0,180] into *n* subintervals each of length Δt. In Δt sec, approximately $v(t)\ \Delta t$ in. of water have drained from the pipe.

Since a typical cross section has area 4π in.^{2} (Figure N8–2), in Δt sec the amount that has drained is

$$\left(4\pi \text{ in.}^2\right)(v(t) \text{ in./sec})(\Delta t \text{ sec}) = 4\pi v(t)\,\Delta t \text{ in.}^3.$$

The sum of the n amounts of water that drain from the pipe, as $n \to \infty$, is $\int_0^{180} 4\pi v(t)\,dt$; the units are cubic inches $\left(\text{in.}^3\right)$.

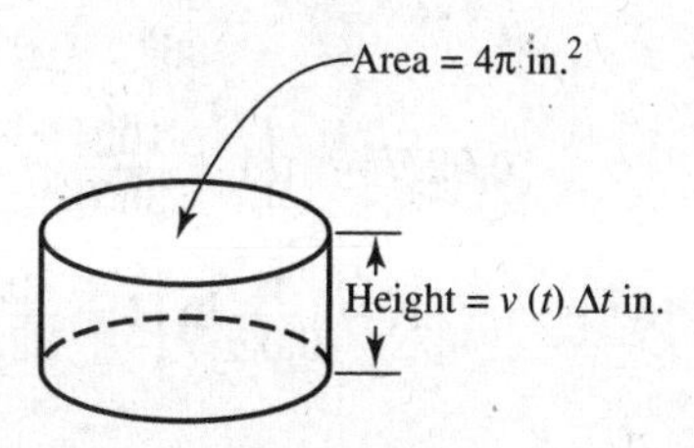

Figure N8–2

➡ Example 9

Traffic: Total Number of Cars. The density of cars (the number of cars per mile) on 10 miles of the highway approaching Disney World is equal approximately to $f(x) = 200[4 - \ln(2x + 3)]$, where x is the distance in miles from the Disney World entrance. Find the total number of cars on this 10-mile stretch.

SOLUTION: Partition the interval [0, 10] into n equal subintervals each of width Δx. In each subinterval the number of cars equals approximately the density of cars $f(x)$ times Δx, where $f(x) = 200[4 - \ln(2x + 3)]$. When we add n of these products we get $\sum f(x)\Delta x$, which is a Riemann Sum. As $n \to \infty$ (or as $\Delta x \to 0$), the Riemann Sum approaches the definite integral

$$\int_0^{10} 200[4 - \ln(2x + 3)]\,dx,$$

which, using our calculator, is approximately equal to 3118 cars.

➡ Example 10

Resource Depletion. In 2000 the yearly world petroleum consumption was about 77 billion barrels and the yearly exponential rate of increase in use was 2%. How many years after 2000 are the world's total estimated oil reserves of 1020 billion barrels likely to last?

SOLUTION: Given the yearly consumption in 2000 and the projected exponential rate of increase in consumption, the anticipated consumption during the Δtth part of a year (after 2000) is $77e^{0.02t}\,\Delta t$ billion barrels. The total to be used during the following N years is therefore $\int_0^N 77e^{0.02t}\,dt$. This integral must equal 1020 billion barrels.

We must now solve this equation for N. We get

$$3850e^{0.02t}\Big|_0^N = 1020,$$

$$3850\left(e^{0.02N} - 1\right) = 1020,$$

$$e^{0.02N} - 1 = \frac{1020}{3850},$$

$$e^{0.02N} = 1 + \frac{102}{385},$$

$$0.02N = \ln\left(1 + \frac{102}{385}\right),$$

$$N = \frac{1}{0.02}\ln\left(1 + \frac{102}{385}\right) \simeq 11.75 \text{ yr}.$$

Either more oil (or alternative sources of energy) must be found, or the world consumption must be sharply reduced.

D. FTC: DEFINITE INTEGRAL OF A RATE IS NET CHANGE

If f is continuous and $f(t) = \frac{dF}{dt}$, then we know from the FTC that

$$\int_a^b f(t)\,dt = F(b) - F(a).$$

The definite integral of the rate of change of a quantity over an interval is the *net change* or *net accumulation* of the quantity over that interval. Thus, $F(b) - F(a)$ is the net change in $F(t)$ as t varies from a to b.

We've already illustrated this principle many times. Here are more examples.

Example 11

Let $G(t)$ be the rate of growth of a population at time t. Then the increase in population between times $t = a$ and $t = b$ is given by $\int_a^b G(t)\,dt$. The population may consist of people, deer, fruit flies, bacteria, and so on.

Example 12

Suppose an epidemic is spreading through a city at the rate of $f(t)$ new people per week. Then

$$\int_0^4 f(t)\,dt$$

is the number of people who will become infected during the next 4 weeks (or the total change in the number of infected people).

Example 13

Suppose a rumor is spreading at the rate of $f(t) = 100e^{-0.2t}$ new people per day. Find the number of people who hear the rumor during the 5th and 6th days.

SOLUTION: $\int_4^6 100e^{-0.2t}\,dt = 74$ people.

If we let $F'(t) = f(t)$, then the integral above is the net change in $F(t)$ from $t = 4$ to $t = 6$, or the number of people who hear the rumor from the beginning of the 5th day to the end of the 6th.

Example 14

Economists define the *marginal cost of production* as the additional cost of producing one additional unit at a specified production level. It can be shown that if $C(x)$ is the cost at production level x then $C'(x)$ is the marginal cost at that production level.

If the marginal cost, in dollars, is $\frac{1}{x}$ per unit when x units are being produced, find the change in cost when production increases from 50 to 75 units.

SOLUTION: $\int_{50}^{75} \frac{1}{x}\,dx \simeq \$0.41$.

We replace "cost" above by "revenue" or "profit" to find total change in these quantities.

Example 15

After t minutes, a chemical is decomposing at the rate of $10e^{-t}$ grams per minute. Find the amount that has decomposed during the first 3 minutes.

SOLUTION: $\int_0^3 10e^{-t}\,dt \simeq 9.5$ g.

Example 16

An official of the Environmental Protection Agency estimates that t years from now the level of a particular pollutant in the air will be increasing at the rate of $(0.3 + 0.4t)$ parts per million per year (ppm/yr). Based on this estimate, find the change in the pollutant level during the second year.

SOLUTION: $\int_1^2 (0.3 + 0.4t)\,dt \simeq 0.9$ ppm.

CHAPTER SUMMARY

In this chapter we have reviewed how to find the distance traveled by an object in motion along a line and (for BC students) along a parametrically defined curve in a plane. We've also looked at a broad variety of applications of the definite integral to other situations where definite integrals of rates of change are used to determine accumulated change, using limits of Riemann Sums to create the integrals required.

PRACTICE EXERCISES

The aim of these questions is mainly to reinforce how to set up definite integrals, rather than how to integrate or evaluate them. Therefore we encourage using a graphing calculator wherever helpful.

1. A particle moves along a line in such a way that its position at time t is given by $s = t^3 - 6t^2 + 9t + 3$. Its direction of motion changes when

   (A) $t = 1$ only (B) $t = 2$ only (C) $t = 3$ only
   (D) $t = 1$ and $t = 3$ (E) $t = 1, 2$, and 3

2. A body moves along a straight line so that its velocity v at time t is given by $v = 4t^3 + 3t^2 + 5$. The distance the body covers from $t = 0$ to $t = 2$ equals

   (A) 34 (B) 55 (C) 24 (D) 44 (E) 49

3. A particle moves along a line with velocity $v = 3t^2 - 6t$. The total distance traveled from $t = 0$ to $t = 3$ equals

   (A) 2 (B) 4 (C) 8 (D) 9 (E) 16

4. The net change in the position of the particle in Question 3 is

   (A) 0 (B) 2 (C) 4 (D) 9 (E) 16

5. The acceleration of a particle moving on a straight line is given by $a = \cos t$, and when $t = 0$ the particle is at rest. The distance it covers from $t = 0$ to $t = 2$ is

   (A) $\sin 2$ (B) $1 - \cos 2$ (C) $\cos 2$ (D) $\sin 2 - 1$ (E) $-\cos 2$

6. During the worst 4-hr period of a hurricane the wind velocity, in miles per hour, is given by $v(t) = 5t - t^2 + 100$, $0 \leqslant t \leqslant 4$. The average wind velocity during this period (in mph) is

   (A) 10 (B) 100 (C) 102 (D) $104\frac{2}{3}$ (E) $108\frac{2}{3}$

7. A car accelerates from 0 to 60 mph in 10 sec, with constant acceleration. (Note that 60 mph = 88 ft/sec.) The acceleration (in ft/sec^2) is

   (A) 1.76 (B) 5.3 (C) 6 (D) 8 (E) 8.8

BC ONLY

For Questions 8–10 use the following information: The velocity **v** of a particle moving on a curve is given, at time t, by $\mathbf{v} = \langle t, -(1 - t)\rangle$. When $t = 0$, the particle is at point (0,1).

8. At time t the position vector **R** is

   (A) $\left\langle \frac{t^2}{2}, -\frac{(1 - t^2)}{2} \right\rangle$ (B) $\left\langle \frac{t^2}{2}, -\frac{(1 - t)^2}{2} \right\rangle$ (C) $\left\langle \frac{t^2}{2}, -\frac{t^2 - 2t}{2} \right\rangle$

   (D) $\left\langle \frac{t^2}{2}, -\frac{t^2 - 2t + 2}{2} \right\rangle$ (E) $\left\langle \frac{t^2}{2}, (1 - t)^2 \right\rangle$

9. The acceleration vector at time $t = 2$ is

   (A) $\langle 1, 1\rangle$ (B) $\langle 1, -1\rangle$ (C) $\langle 1, 2\rangle$ (D) $\langle 2, -1\rangle$ (E) $\left\langle 2, \frac{3}{2} \right\rangle$

BC ONLY

10. The speed of the particle is at a minimum when t equals

(A) 0 (B) $\frac{1}{2}$ (C) 1 (D) 1.5 (E) 2

11. A particle moves along a curve in such a way that its position vector and velocity vector are perpendicular at all times. If the particle passes through the point (4, 3), then the equation of the curve is

(A) $x^2 + y^2 = 5$ (B) $x^2 + y^2 = 25$ (C) $x^2 + 2y^2 = 34$
(D) $x^2 - y^2 = 7$ (E) $2x^2 - y^2 = 23$

12. The acceleration of an object in motion is given by the vector $\vec{a}(t) = \langle 2t, e^t \rangle$. If the object's initial velocity was $\vec{v}(0) = \langle 2,0 \rangle$, which is the velocity vector at any time t?

(A) $\vec{v}(t) = \langle t^2, e^t \rangle$ (B) $\vec{v}(t) = \langle t^2, e^t + 1 \rangle$ (C) $\vec{v}(t) = \langle t^2 + 2, e^t \rangle$
(D) $\vec{v}(t) = \langle t^2 + 2, e^t - 1 \rangle$ (E) $\vec{v}(t) = \langle 2, e^t - 1 \rangle$

13. The velocity of an object is given by $\vec{v}(t) = \langle 3\sqrt{t}, 4 \rangle$. If this object is at the origin when $t = 1$, where was it at $t = 0$?

(A) $(-3,-4)$ (B) $(-2,-4)$ (C) $(2,4)$ (D) $\left(\frac{3}{2}, 0\right)$ (E) $\left(-\frac{3}{2}, 0\right)$

14. Suppose the current world population is 6 billion and the population t years from now is estimated to be $P(t) = 6e^{0.024t}$ billion people. On the basis of this supposition, the average population of the world, in billions, over the next 25 years will be approximately

(A) 6.75 (B) 7.2 (C) 7.8 (D) 8.2 (E) 9.0

15. A beach opens at 8 A.M. and people arrive at a rate of $R(t) = 10 + 40t$ people per hour, where t represents the number of hours the beach has been open. Assuming no one leaves before noon, at what time will there be 100 people there?

(A) 9:45 (B) 10:00 (C) 10:15 (D) 10:30 (E) 10:45

16. A stone is thrown upward from the ground with an initial velocity of 96 ft/sec. Its average velocity (given that $a(t) = -32$ ft/sec^2) during the first 2 sec is

(A) 16 ft/sec (B) 32 ft/sec (C) 64 ft/sec (D) 80 ft/sec (E) 96 ft/sec

17. Suppose the amount of a drug in a patient's bloodstream t hr after intravenous administration is $30/(t + 1)^2$ mg. The average amount in the bloodstream during the first 4 hr is

(A) 6.0 mg (B) 11.0 mg (C) 16.6 mg (D) 24.0 mg (E) none of these

18. A rumor spreads through a town at the rate of $(t^2 + 10t)$ new people per day. Approximately how many people hear the rumor during the second week after it was first heard?

(A) 359 (B) 1535 (C) 1894 (D) 2000 (E) 2219

19. Oil is leaking from a tanker at the rate of $1000e^{-0.3t}$ gal/hr, where t is given in hours. The total number of gallons of oil that will leak out during the first 8 hr is approximately

(A) 1271 (B) 3031 (C) 3161 (D) 4323 (E) 11,023

20. Assume that the density of vehicles (number per mile) during morning rush hour, for the 20-mi stretch along the New York State Thruway southbound from the Tappan Zee Bridge, is given by $f(x)$, where x is the distance, in miles, south of the bridge. Which of the following gives the number of vehicles (on this 20-mi stretch) from the bridge to a point x mi south of the bridge?

(A) $\int_0^x f(t)\,dt$ (B) $\int_x^{20} f(t)\,dt$ (C) $\int_0^{20} f(x)\,dx$

(D) $\sum_{k=1}^{n} f(x_k)\,\Delta x$ (where the 20-mi stretch has been partitioned into n equal subintervals)

(E) none of these

21. The center of a city that we will assume is circular is on a straight highway. The radius of the city is 3 mi. The density of the population, in thousands of people per square mile, is given approximately by $f(r) = 12 - 2r$ at a distance r mi from the highway. The population of the city (in 1000s) is given by the integral

(A) $\int_0^3 (12 - 2r)\,dr$ (B) $2\int_0^3 (12 - 2r)\sqrt{9 - r^2}\,dr$

(C) $4\int_0^3 (12 - 2r)\sqrt{9 - r^2}\,dr$ (D) $\int_0^3 2\pi r(12 - 2r)\,dr$

(E) $2\int_0^3 2\pi r(12 - 2r)\,dr$

22. The population density of Winnipeg, which is located in the middle of the Canadian prairie, drops dramatically as distance from the center of town increases. This is shown in the following table:

x = distance (in mi) from the center	0	2	4	6	8	10
$f(x)$ = (density hundreds of people/mi^2)	50	45	40	30	15	5

Using a Riemann Sum, we can calculate the population living within a 10-mi radius of the center to be approximately

(A) 608,500 (B) 650,000 (C) 691,200 (D) 702,000 (E) 850,000

23. If a factory continuously dumps pollutants into a river at the rate of $\frac{\sqrt{t}}{180}$ tons per day, then the amount dumped after 7 weeks is approximately

(A) 0.07 ton (B) 0.90 ton (C) 1.55 tons (D) 1.9 tons (E) 1.27 tons

24. A roast at 160°F is put into a refrigerator whose temperature is 45°F. The temperature of the roast is cooling at time t at the rate of $(-9e^{-0.08t})$°F per minute. The temperature, to the nearest degree F, of the roast 20 min after it is put in the refrigerator is

(A) 45° (B) 70° (C) 81° (D) 90° (E) 115°

25. How long will it take to release 9 tons of pollutant if the rate at which pollutant is being released is $te^{-0.3t}$ tons per week?

(A) 10.2 weeks (B) 11.0 weeks (C) 12.1 weeks
(D) 12.9 weeks (E) none of these

26. What is the total area bounded by the curve $f(x) = x^3 - 4x^2 + 3x$ and the x-axis?

(A) -2.25 (B) 0.416 (C) 2.25 (D) 3 (E) 3.083

27. Water is leaking from a tank at the rate of $(-0.1t^2 - 0.3t + 2)$ gal/hr. The total amount, in gallons, that will leak out in the next 3 hr is approximately

(A) 1.00 (B) 2.08 (C) 3.13 (D) 3.48 (E) 3.75

28. A bacterial culture is growing at the rate of $1000e^{0.03t}$ bacteria in t hr. The total increase in bacterial population during the second hour is approximately

(A) 46 (B) 956 (C) 1046 (D) 1061 (E) 2046

29. A website went live at noon, and the rate of hits (visitors/hour) increased continuously for the first 8 hours, as shown in the graph below.

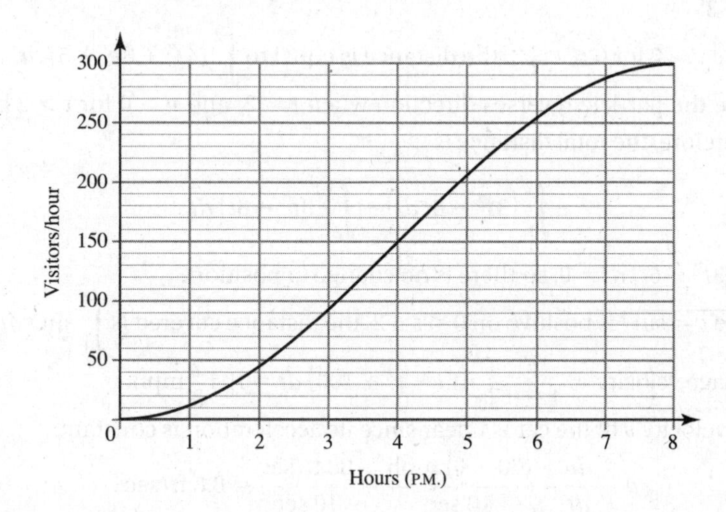

Approximately when did the 200th visitor go to this site?

(A) before 2 P.M. (B) between 2 and 3 P.M. (C) between 3 and 4 P.M.
(D) between 4 and 5 P.M. (E) after 5 P.M.

30. An observer recorded the velocity of an object in motion along the x-axis for 10 seconds. Based on the table below, use a trapezoidal approximation to estimate how far from its starting point the object came to rest at the end of this time.

t (sec)	0	2	3	5	7	10
$v(t)$ (units/sec)	2	3	1	-1	-2	0

(A) 0 units (B) 1 unit (C) 3 units (D) 4 units (E) 6 units

31. An 18-wheeler traveling at speed v mph gets about $(4 + 0.01v)$ mpg (miles per gallon) of diesel fuel. If its speed is $80\,\frac{t+1}{t+2}$ mph at time t, then the amount, in gallons, of diesel fuel used during the first 2 hr is approximately

(A) 20 (B) 21.5 (C) 23.1 (D) 24 (E) 25

Answer Key

1. **D**	9. **A**	17. **A**	25. **A**
2. **A**	10. **B**	18. **B**	26. **D**
3. **C**	11. **B**	19. **B**	27. **E**
4. **A**	12. **D**	20. **A**	28. **C**
5. **B**	13. **B**	21. **C**	29. **C**
6. **D**	14. **D**	22. **C**	30. **B**
7. **E**	15. **B**	23. **E**	31. **C**
8. **D**	16. **C**	24. **B**	

Answers Explained

1. **(D)** Velocity $v(t) = \frac{ds}{dt} = 3(t-1)(t-3)$, and changes sign both when $t = 1$ and when $t = 3$.
2. **(A)** Since $v > 0$ for $0 \leqq t \leqq 2$, the distance is equal to $\int_0^2 (4t^3 + 3t^2 + 5)\, dt$.
3. **(C)** Since the particle reverses direction when $t = 2$, and $v > 0$ for $t > 2$ but $v < 0$ for $t < 2$, therefore, the total distance is

$$-\int_0^2 (3t^2 - 6t\, dt) + \int_2^3 (3t^2 - 6t)\, dt.$$

4. **(A)** $\int_0^3 (3t^2 - 6t)\, dt = 0$, so there is no change in position.
5. **(B)** Since $v = \sin t$ is positive on $0 < t \leqq 2$, the distance covered is $\int_0^2 \sin t\, dt = 1 - \cos 2$.
6. **(D)** Average velocity $= \frac{1}{4-0}\int_0^4 (5t - t^2 + 100)\, dt = 104\frac{2}{3}$ mph.
7. **(E)** The velocity v of the car is linear since its acceleration is constant:

$$a = \frac{dv}{dt} = \frac{(60-0)\text{ mph}}{10\text{ sec}} = \frac{88\text{ ft/sec}}{10\text{ sec}} = 8.8\text{ ft/sec}^2$$

8. **(D)** $\mathbf{v} = \frac{d\mathbf{R}}{dt}$, so $\mathbf{R}(t) = \left\langle \frac{t^2}{2} + c_1, \frac{t^2}{2} - t + c_2 \right\rangle$. Since $\mathbf{R}(0) = \langle 0,1 \rangle$, $c_1 = 0$ and $c_2 = 1$.
9. **(A)** $\mathbf{a} = \mathbf{v}'(t) = \langle 1,1 \rangle$ for all t.
10. **(B)** $\mathbf{v} = \langle t, t-1 \rangle$. $|\mathbf{v}| = \sqrt{t^2 + (t-1)^2}$; $\frac{d|\mathbf{v}|}{dt} = \frac{2t-1}{|\mathbf{v}|}$; $\frac{d|\mathbf{v}|}{dt} = 0$ at $t = \frac{1}{2}$.
11. **(B)** Since $\mathbf{R} = \langle x, y \rangle$, its slope is $\frac{y}{x}$; since $\mathbf{v} = \left\langle \frac{dx}{dt}, \frac{dy}{dt} \right\rangle$, its slope is $\frac{dy}{dx}$.

    If $\mathbf{R}$ is perpendicular to $\mathbf{v}$, then $\frac{y}{x} \cdot \frac{dy}{dx} = -1$, so

$$\frac{y^2}{2} = -\frac{x^2}{2} + C \quad \text{and} \quad x^2 + y^2 = k\ (k > 0).$$

    Since (4, 3) is on the curve, the equation must be

$$x^2 + y^2 = 25.$$

12. **(D)** $\vec{v}(t) = \langle t^2 + c_1, e^t + c_2 \rangle$; since $\vec{v}(0) = \langle 2,0 \rangle$, $0^2 + c_1 = 2$ and $e^0 + c_2 = 0$; hence $c_1 = 2$ and $c_2 = -1$.
13. **(B)** The object's position is given by $x(t) = 2t^{\frac{3}{2}} + c_1$, $y(t) = 4t + c_2$. Since the object was at the origin at $t = 1$, $2 \cdot 1^{\frac{3}{2}} + c_1 = 0$ and $4 \cdot 1 + c_2 = 0$, making the position $x(t) = 2t^{\frac{3}{2}} - 2$, $y(t) = 4t - 4$. When $t = 0$, $x(0) = -2$, $y(0) = -4$.

14. **(D)** $\frac{1}{25-0}\int_0^{25} P(t)\,dt = \frac{1}{25}\int_0^{25} 6e^{0.024t} \simeq 8.2$ (billion people).

15. **(B)** We want the accumulated number of people to be 100:

$$\int_0^h (10 + 40t)dt = 100$$

$$\left(10t + 20t^2\right)\Big|_0^h = 100$$

$$20h^2 + 10h - 100 = 0$$

$$10(2h + 5)(h - 2) = 0$$

This occurs at $h = 2$ hours after 8 A.M.

16. **(C)** Average velocity $= \frac{1}{2-0}\int_0^2 (-32t + 96)\,dt = 64$ ft/sec.

17. **(A)** Average volume $= \frac{1}{4-0}\int_0^4 \frac{30}{(t+1)^2}dt \simeq 6$ mg.

18. **(B)** The number of new people who hear the rumor during the second week is

$$\int_7^{14} \left(t^2 + 10t\right) dt \simeq 1535.$$

Be careful with the units! The answer is the total change, of course, in $F(t)$ from $t = 7$ to $t = 14$ days, where $F'(t) = t^2 + 10t$.

19. **(B)** Total gallons $= \lim_{n\to\infty} \sum_{k=1}^{n} 1000e^{-0.3t_k}\,\Delta t = \int_0^8 1000e^{-0.3t}\,dt \approx 3031.$

20. **(A)** Be careful! The number of cars is to be measured over a distance of x (not 20) mi. The answer to the question is a function, not a number. Note that choice (C) gives the total number of cars on the entire 20-mi stretch.

21. **(C)** Since the strip of the city shown in the figure is at a distance r mi from the highway, it is $2\sqrt{9-r^2}$ mi long and its area is $2\sqrt{9-r^2}\,\Delta r$. The strip's population is approximately $2(12 - 2r)\sqrt{9-r^2}\,\Delta r$. The total population of the entire city is *twice* the integral $2\int_0^3 (12 - 2r)\sqrt{9-r^2}\,dr$ as it includes both halves of the city.

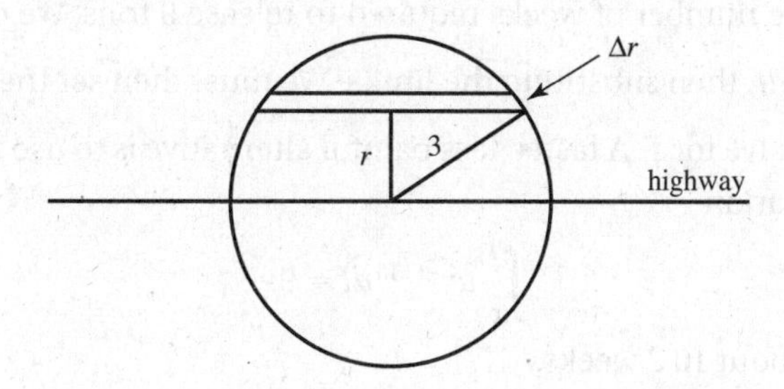

22. **(C)**

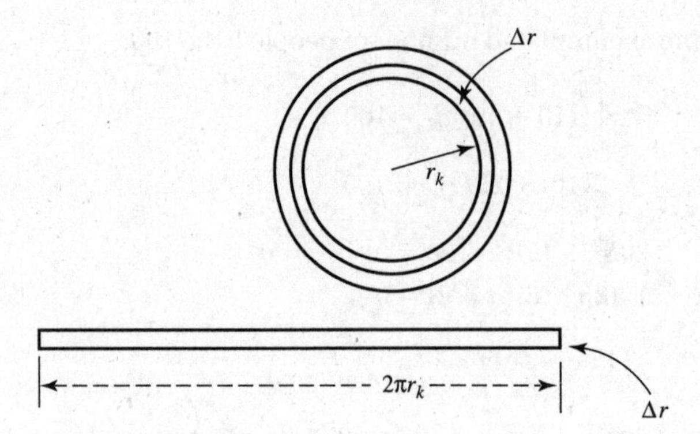

The population equals $\sum$ (area · density). We partition the interval [0,10] along a radius from the center of town into 5 equal subintervals each of width $\Delta r = 2$ mi. We will divide Winnipeg into 5 rings. Each has area equal to (circumference × width), so the area is $2\pi r_k \,\Delta r$ or $4\pi r_k$. The population in the ring is

$$\left(4\pi r_k\right)\cdot\left(\text{density at } r_k\right) = 4\pi r_k \cdot f\left(r_k\right).$$

A Riemann Sum, using left-hand values, is $4\pi\cdot 0\cdot 50 + 4\pi\cdot 2\cdot 45 + 4\pi\cdot 4\cdot 40 + 4\pi\cdot 6\cdot 30 + 4\pi\cdot 8\cdot 15 = 4\pi(90 + 160 + 180 + 120) \simeq 6912$ hundred people—or about 691,200 people.

23. **(E)** The total amount dumped after 7 weeks is

$$\int_0^{49} \frac{\sqrt{t}}{180}\,dt \simeq 1.27 \text{ tons.}$$

24. **(B)** The total change in temperature of the roast 20 min after it is put in the refrigerator is

$$\int_0^{20} -9e^{-0.08t}\,dt \qquad \text{or} \qquad -89.7°\text{F}.$$

Since its temperature was 160°F when placed in the refrigerator, then 20 min later it is (160 − 89.7)°F or about 70°F. Note that the temperature of the refrigerator (45°F) is not used in answering the question because it is already "built into" the cooling rate.

25. **(A)** Let T be the number of weeks required to release 9 tons. We can use parts to integrate $\int_0^T te^{-0.3t}\,dt$, then substitute the limits. We must then set the resulting expression equal to 9 and solve for T. A faster, less painful alternative is to use a graphing calculator to solve the equation

$$\int_0^T te^{-0.3t}\,dt = 9.$$

The answer is about 10.2 weeks.

26. **(D)** Note that the curve is above the x-axis on [0,1], but below on [1,3], and that the areas for $x < 0$ and $x > 3$ are unbounded.

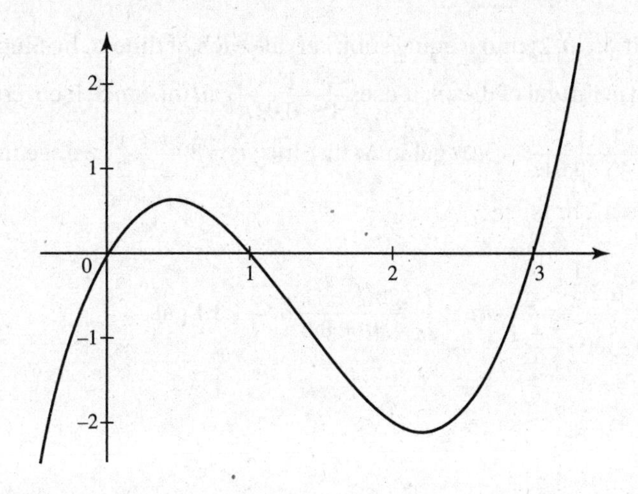

Using the calculator, we get

$$\int_0^3 |x^3 - 4x^2 + 3x|\, dx \approx 3.083.$$

27. **(E)** The FTC yields total change:

$$\int_0^3 \left(-0.1t^2 - 0.3t + 2\right) dt \approx 3.75 \text{ gal.}$$

28. **(C)** The total change (increase) in population during the second hour is given by $\int_1^2 1000e^{0.03t}\, dt$. The answer is 1046.

29. **(C)** Call the time in hours t and the function for visitors/hour $R(t)$. Then the area under the curve represents the number of visitors V. We will estimate the time k when $\int_0^k R(t)dt =$ 200 using a Riemann Sum.

The table shows one approach, based on the Midpoint Rule.

Hour	Visitors/hour (midpoint est.)	Total visitors since noon
Noon–1 P.M.	5	5
1–2 P.M.	25	30
2–3 P.M.	70	100
3–4 P.M.	120	220

We estimate that there had been about 100 visitors by 3 P.M. and 220 by 4 P.M., so the 200th visitor arrived between 3 and 4 P.M.

30. **(B)** $T = \left(\frac{2+3}{2}\right)2 + \left(\frac{3+1}{2}\right)1 + \left(\frac{1+(-1)}{2}\right)2 + \left(\frac{(-1)+(-2)}{2}\right)2 + \left(\frac{(-2)+0}{2}\right)3 = 1$ unit (to the right).

31. **(C)** We partition [0, 2] into n equal subintervals each of time Δt hr. Since the 18-wheeler gets $(4 + 0.01v)$ mi/gal of diesel, it uses $\frac{1}{4 + 0.01v}$ gal/mi. Since it covers $v \cdot \Delta t$ mi during Δt hr, it uses $\frac{1}{4 + 0.01v} \cdot v \cdot \Delta t$ gal in Δt hr. Since $v = 80\,\frac{t+1}{t+2}$, we see that the diesel fuel used in the first 2 hr is

$$\int_0^2 \frac{80 \cdot \frac{t+1}{t+2}}{4 + (0.01) \cdot \left(80 \cdot \frac{t+1}{t+2}\right)}\,dt = \int_0^2 \frac{80(t+1)}{4.8t + 8.8}\,dt \simeq 23.1 \text{ gal.}$$

Differential Equations 9

CONCEPTS AND SKILLS

In this chapter, we review how to write and solve differential equations, specifically,

- **writing differential equations to model dynamic situations;**
- **understanding a slope field as a graphical representation of a differential equation and its solutions;**
- **finding general and particular solutions of separable differential equations;**
- **and using differential equations to analyze growth and decay.**

We also review two additional BC Calculus topics:

- **Euler's method to estimate numerical solutions**
- **and using differential equations to analyze logistic growth and decay.**

A. BASIC DEFINITIONS

Differential equation

A *differential equation* (d.e.) is any equation involving a derivative. In §E of Chapter 5 we solved some simple differential equations. In Example 50, page 223, we were given the velocity at time t of a particle moving along the x-axis:

$$v(t) = \frac{dx}{dt} = 4t^3 - 3t^2. \qquad (1)$$

From this we found the antiderivative:

$$x(t) = t^4 - t^3 + C. \qquad (2)$$

If the initial position (at time $t = 0$) of the particle is $x = 3$, then

$$x(0) = 0 - 0 + C = 3,$$

and $C = 3$. So the solution to the initial-value problem is

$$x(t) = t^4 - t^3 + 3. \qquad (3)$$

Solution

General solution

Particular solution

A *solution* of a d.e. is any function that satisfies it. We see from (2) above that the d.e. (1) has an infinite number of solutions—one for each real value of C. We call the family of functions (2) the *general solution* of the d.e. (1). With the given initial condition $x(0) = 3$, we determined C, thus finding the unique solution (3). This is called the *particular solution*.

Domain

Note that the particular solution must not only satisfy the differential equation and the initial condition, but the function *must also be differentiable on an interval that contains the initial point.* Features such as vertical tangents or asymptotes restrict the domain of the solution. Therefore, even when they are defined by the same algebraic representation, particular solutions with different initial points may have different domains. Determining the proper domain is an important part of finding the particular solution.

In §A of Chapter 8 we solved more differential equations involving motion along a straight line. In §B we found parametric equations for the motion of a particle along a plane curve, given d.e.'s for the components of its acceleration and velocity.

Rate of Change

A differential equation contains derivatives. A derivative gives information about the rate of change of a function. For example:

(1) If P is the size of a population at time t, then we can interpret the d.e.

$$\frac{dP}{dt} = 0.0325P$$

as saying that at any time t the rate at which the population is growing is proportional (3.25%) to its size at that time.

(2) The d.e. $\frac{dQ}{dt} = -(0.000275)Q$ tells us that at any time t the rate at which the quantity Q is decreasing is proportional (0.0275%) to the quantity existing at that time.

(3) In psychology, one typical stimulus-response situation, known as *logarithmic response*, is that in which the response y changes at a rate inversely proportional to the strength of the stimulus x. This is expressed neatly by the differential equation

$$\frac{dy}{dx} = \frac{k}{x} \qquad (k \text{ a constant}).$$

If we suppose, further, that there is no response when $x = x_0$, then we have the condition $y = 0$ when $x = x_0$.

(4) We are familiar with the d.e.

$$a = \frac{d^2s}{dt^2} = -32 \text{ ft/sec}^2$$

for the acceleration due to gravity acting on an object at a height s above ground level at time t. The acceleration is the rate of change of the object's velocity.

B. SLOPE FIELDS

In this section we investigate differential equations by obtaining a *slope field* or calculator picture that approximates the general solution. We call the graph of a solution of a d.e. a *solution curve*.

The slope field of a d.e. is based on the fact that the d.e. can be interpreted as a statement about the slopes of its solution curves.

Example 1

The d.e. $\frac{dy}{dx} = y$ tells us that at any point (x, y) on a solution curve the slope of the curve is equal to its y-coordinate. Since the d.e. says that y is a function whose derivative is also y, we know that

$$y = e^x$$

is a solution. In fact, $y = Ce^x$ is a solution of the d.e. for every constant C, since $y' = Ce^x = y$. The d.e. $y' = y$ says that, at any point where $y = 1$, say $(0, 1)$ or $(1, 1)$ or $(5, 1)$, the slope of the solution curve is 1; at any point where $y = 3$, say $(0, 3)$, $(\ln 3, 3)$, or $(\pi, 3)$, the slope equals 3; and so on.

In Figure N9–1a we see some small line segments of slope 1 at several points where $y = 1$, and some segments of slope 3 at several points where $y = 3$. In Figure N9–1b we see the curve of $y = e^x$ with slope segments drawn in as follows:

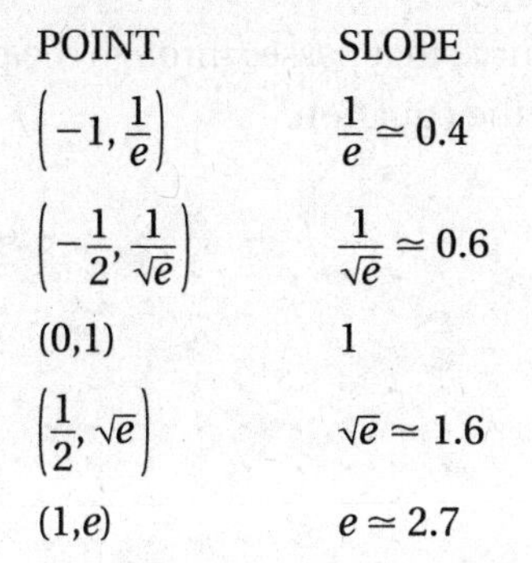

POINT	SLOPE
$\left(-1, \frac{1}{e}\right)$	$\frac{1}{e} \simeq 0.4$
$\left(-\frac{1}{2}, \frac{1}{\sqrt{e}}\right)$	$\frac{1}{\sqrt{e}} \simeq 0.6$
$(0,1)$	1
$\left(\frac{1}{2}, \sqrt{e}\right)$	$\sqrt{e} \simeq 1.6$
$(1,e)$	$e \simeq 2.7$

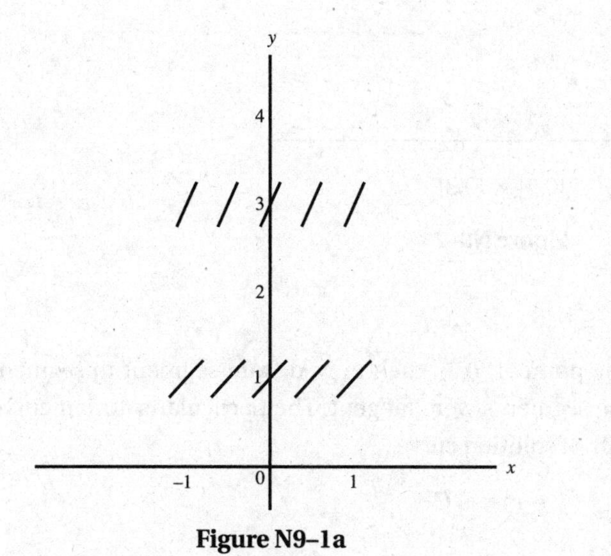

Figure N9–1a

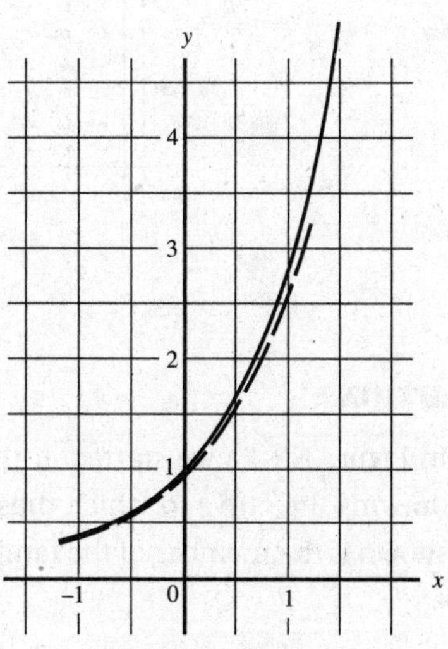

Figure N9–1b

Figure N9–1c is the slope field for the d.e. $\frac{dy}{dx} = y$. Slopes at many points are represented by small segments of the tangents at those points. The small segments approximate the solution curves. If we start at any point in the slope field and move so that the slope segments are always tangent to our motion, we will trace a solution curve. The slope field, as mentioned above, closely approximates the family of solutions.

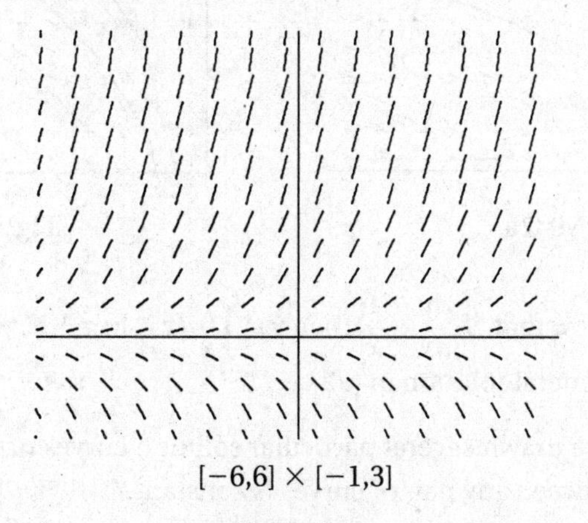

$[-6,6] \times [-1,3]$

Figure N9–1c

➡ Example 2

The slope field for the d.e. $\frac{dy}{dx} = \frac{1}{x}$ is shown in Figure N9–2.

(a) Carefully draw the solution curve that passes through the point (1, 0.5).

(b) Find the general solution for the equation.

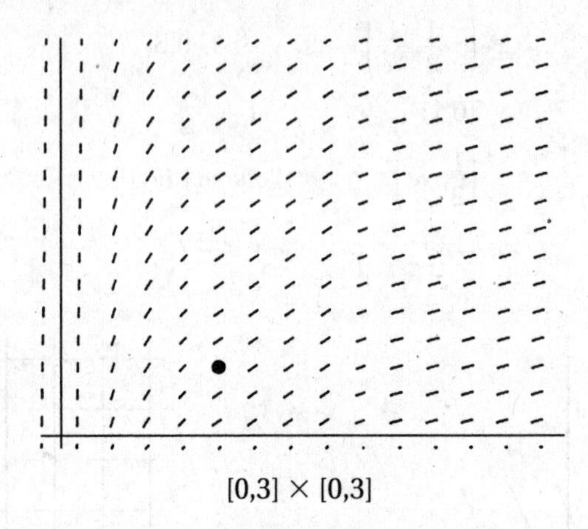

[0,3] × [0,3]

Figure N9–2

SOLUTIONS:

(a) In Figure N9–2a we started at the point (1, 0.5), then moved from segment to segment drawing the curve to which these segments were tangent. The particular solution curve shown is the member of the family of solution curves

$$y = \ln x + C$$

that goes through the point (1, 0.5).

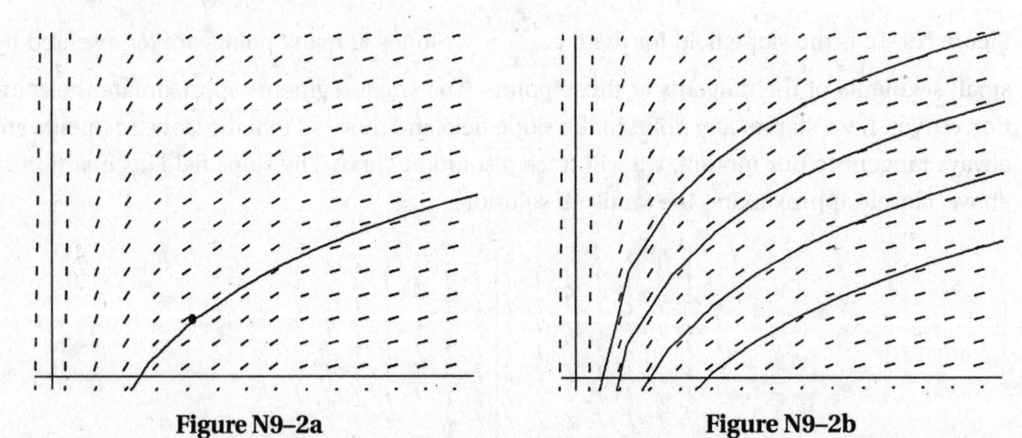

Figure N9–2a **Figure N9–2b**

(b) Since we already know that, if $\frac{dy}{dx} = \frac{1}{x}$, then $y = \int \frac{1}{x}dx = \ln x + C$, we are assured of having found the correct general solution in (a).

In Figure N9–2b we have drawn several particular solution curves of the given d.e. Note that the vertical distance between any pair of curves is constant.

➡ Example 3

Match each slope field in Figure N9–3 with the proper d.e. from the following set. Find the general solution for each d.e. The particular solution that goes through (0,0) has been sketched in.

(A) $y' = \cos x$ (B) $\frac{dy}{dx} = 2x$

(C) $\frac{dy}{dx} = 3x^2 - 3$ (D) $y' = -\frac{\pi}{2}$

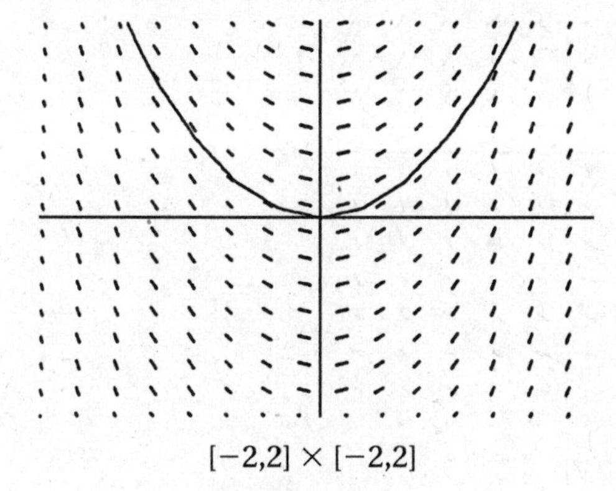

$[-2,2] \times [-2,2]$

Figure N9–3a

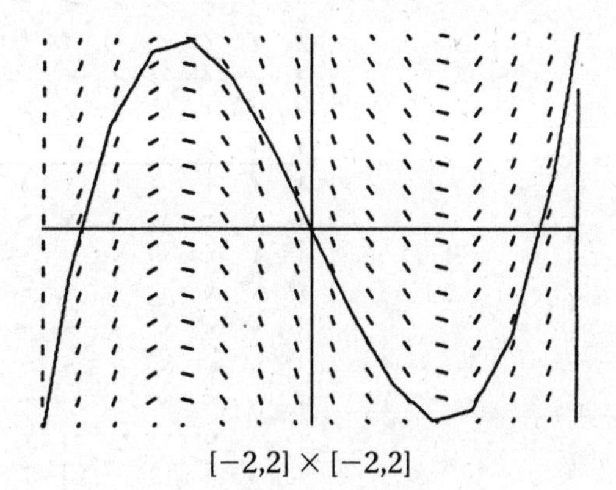

$[-2,2] \times [-2,2]$

Figure N9–3b

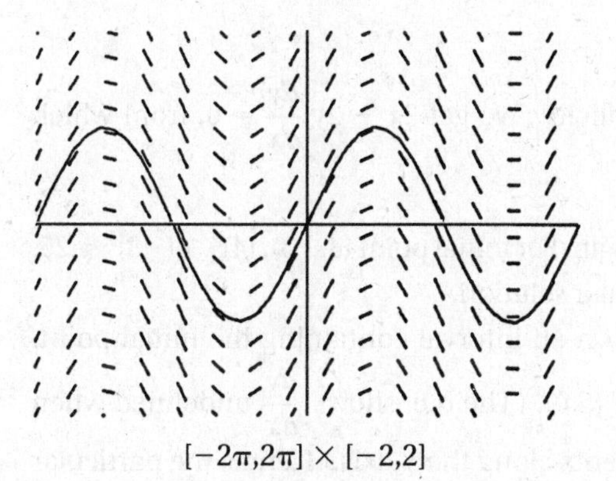

$[-2\pi,2\pi] \times [-2,2]$

Figure N9–3c

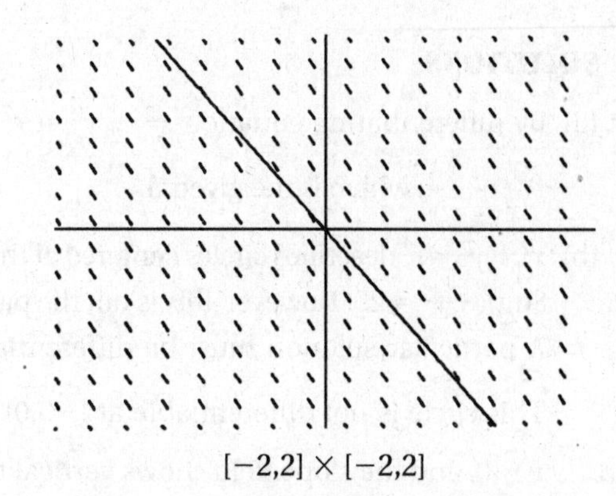

$[-2,2] \times [-2,2]$

Figure N9–3d

SOLUTIONS:

(A) goes with Figure N9–3c. The solution curves in the family $y = \sin x + C$ are quite obvious.
(B) goes with Figure N9–3a. The general solution is the family of parabolas $y = x^2 + C$.
For (C) the slope field is shown in Figure N9–3b. The general solution is the family of cubics $y = x^3 - 3x + C$.
(D) goes with Figure N9–3d; the general solution is the family of lines $y = -\frac{\pi}{2}x + C$.

Example 4

(a) Verify that relations of the form $x^2 + y^2 = r^2$ are solutions of the d.e. $\frac{dy}{dx} = -\frac{x}{y}$.

(b) Using the slope field in Figure N9–4 and your answer to (a), find the particular solution to the d.e. given in (a) that contains point $(4,-3)$.

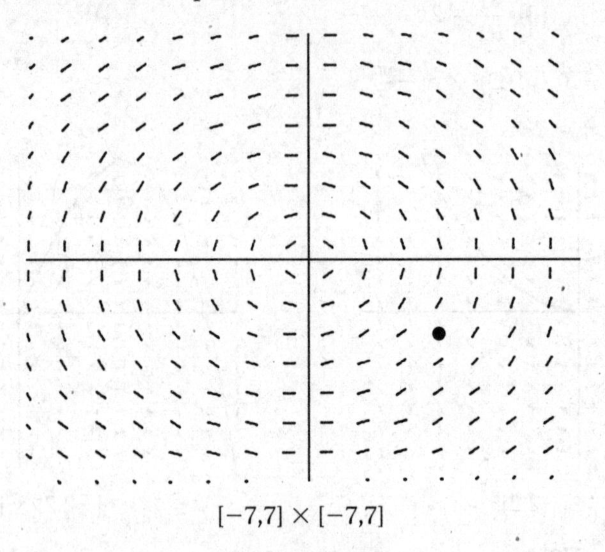

$[-7,7] \times [-7,7]$

Figure N9–4

SOLUTIONS:

(a) By differentiating equation $x^2 + y^2 = r^2$ implicitly, we get $2x + 2y\frac{dy}{dx} = 0$, from which $\frac{dy}{dx} = -\frac{x}{y}$, which is the given d.e.

(b) $x^2 + y^2 = r^2$ describes circles centered at the origin. For initial point $(4,-3)$, $(4)^2 + (-3)^2 = 25$. So $x^2 + y^2 = 25$. However, this is not the particular solution.

A particular solution must be differentiable on an interval containing the initial point. This circle is not differentiable at $(-5,0)$ and $(5,0)$. (The d.e. shows $\frac{dy}{dx}$ undefined when $y = 0$, and the slope field shows vertical tangents along the x-axis.) Hence, the particular solution includes only the semicircle in quadrants III and IV.

Solving $x^2 + y^2 = 25$ for y yields $y = \pm\sqrt{25 - x^2}$. The particular solution through point $(4,-3)$ is $y = -\sqrt{25 - x^2}$ with domain $-5 < x < 5$.

Derivatives of Implicitly Defined Functions

In Examples 2 and 3 above, each d.e. was of the form $\frac{dy}{dx} = f(x)$ or $y' = f(x)$. We were able to find the general solution in each case very easily by finding the antiderivative $y = \int f(x)\,dx$.

We now consider d.e.'s of the form $\frac{dy}{dx} = f(x,y)$, where $f(x,y)$ is an expression in x and y; that is, $\frac{dy}{dx}$ is an implicitly defined function. Example 4 illustrates such a differential equation. Here is another example.

Example 5

Figure N9–5 shows the slope field for

$$y' = x + y. \tag{1}$$

At each point (x,y) the slope is the sum of its coordinates. Three particular solutions have been added, through the points

(a) $(0,0)$ (b) $(0,-1)$ (c) $(0,-2)$

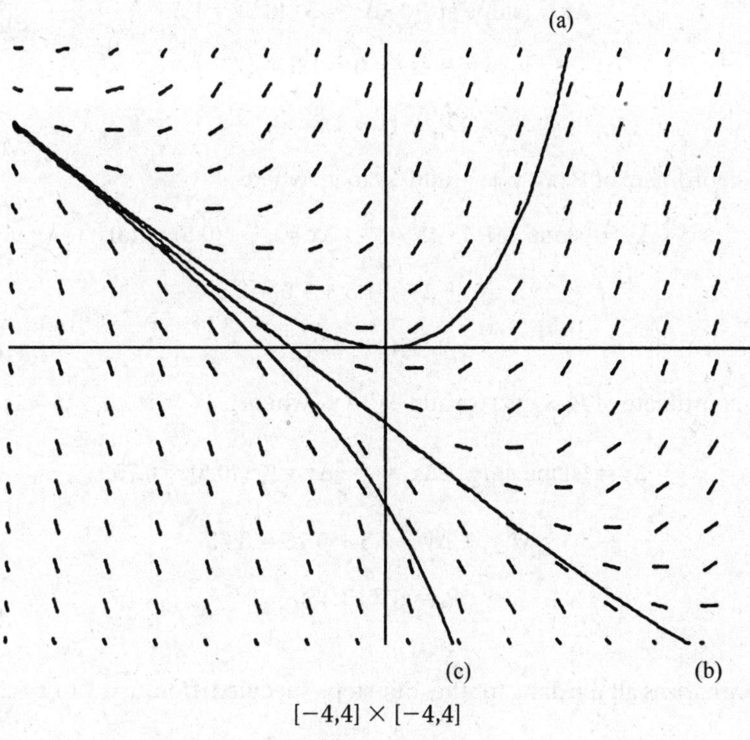

$[-4,4] \times [-4,4]$

Figure N9–5

C. EULER'S METHOD

BC ONLY

In §B we found solution curves to first-order differential equations *graphically,* using slope fields. Here we will find solutions *numerically,* using *Euler's method* to find points on solution curves.

When we use a slope field we start at an initial point, then move step by step so the slope segments are always tangent to the solution curve. With Euler's method we again select a starting point; but now we *calculate* the slope at that point (from the given d.e.), use the initial point and that slope to locate a new point, use the new point and calculate the slope at it (again from the d.e.) to locate still another point, and so on. The method is illustrated in Example 6.

BC ONLY

Example 6

Let $\frac{dy}{dx} = \frac{3}{x}$. Use Euler's method to approximate the y-values with four steps, starting at point $P_0(1, 0)$ and letting $\Delta x = 0.5$.

SOLUTION: The slope at $P_0 = (x_0, y_0) = (1, 0)$ is $\frac{dy}{dx} = \frac{3}{x_0} = \frac{3}{1} = 3$. To find the y-coordinate of $P_1(x_1, y_1)$, we add Δy to y_0. Since $\frac{dy}{dx} = \frac{\Delta y}{\Delta x}$, we estimate $\Delta y \approx \frac{dy}{dx} \cdot \Delta x$:

$$\Delta y = (\text{slope at } P_0) \cdot \Delta x = 3 \cdot (0.5) = 1.5.$$

Then $$y_1 = y_0 + \Delta y = 0 + 1.5 = 1.5$$

and $$P_1 = (1.5, 1.5).$$

To find the y-coordinate of $P_2(x_2, y_2)$ we add Δy to y_1, where

$$\Delta y = (\text{slope at } P_1) \cdot \Delta x = \frac{3}{x_1} \cdot \Delta x = \frac{3}{1.5} \cdot (0.5) = 1.0.$$

Then $$y_2 = y_1 + \Delta y = 1.5 + 1.0 = 2.5$$

and $$P_2 = (2.0, 2.5).$$

To find the y-coordinate of $P_3(x_3, y_3)$ we add Δy to y_2, where

$$\Delta y = (\text{slope at } P_2) \cdot \Delta x = \frac{3}{x_2} \cdot \Delta x = \frac{3}{2} \cdot (0.5) = 0.75.$$

Then $$y_3 = y_2 + \Delta y = 2.5 + 0.75 = 3.25,$$

$$P_3 = (2.5, 3.25),$$

and so on.

The table summarizes all the data, for the four steps specified, from $x = 1$ to $x = 3$:

TABLE FOR $\frac{dy}{dx} = \frac{3}{x}$

	x	y	(SLOPE) · (0.5)	=	Δy	TRUE y*
P_0	1	0	(3/1) · (0.5)	=	1.5	0
P_1	1.5	1.5	(3/1.5) · (0.5)	=	1.0	1.216
P_2	2.0	2.5	(3/2) · (0.5)	=	0.75	2.079
P_3	2.5	3.25	(3/2.5) · (0.5)	=	0.60	2.749
P_4	3.0	3.85	(3/3.0) · (0.5)	=	0.50	3.296

*To three decimal places.

The table gives us the numerical solution of $\frac{dy}{dx} = \frac{3}{x}$ using Euler's method. Figure N9–6a shows the graphical solution, which agrees with the data from the table, for x increasing from 1 to 3 by four steps with Δx equal to 0.5 Figure N9–6b shows this Euler graph and the particular solution of $\frac{dy}{dx} = \frac{3}{x}$ passing through the point (1, 0), which is $y = 3 \ln x$.

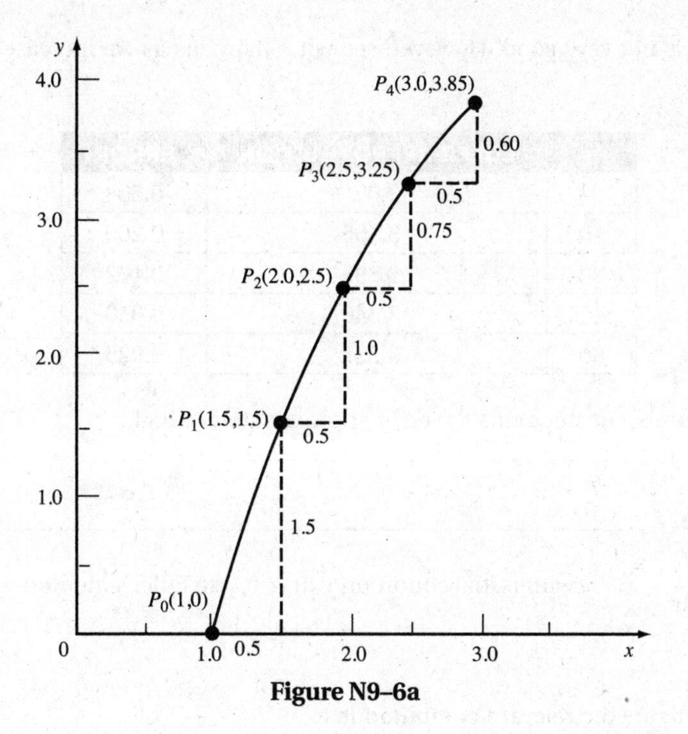

Figure N9–6a

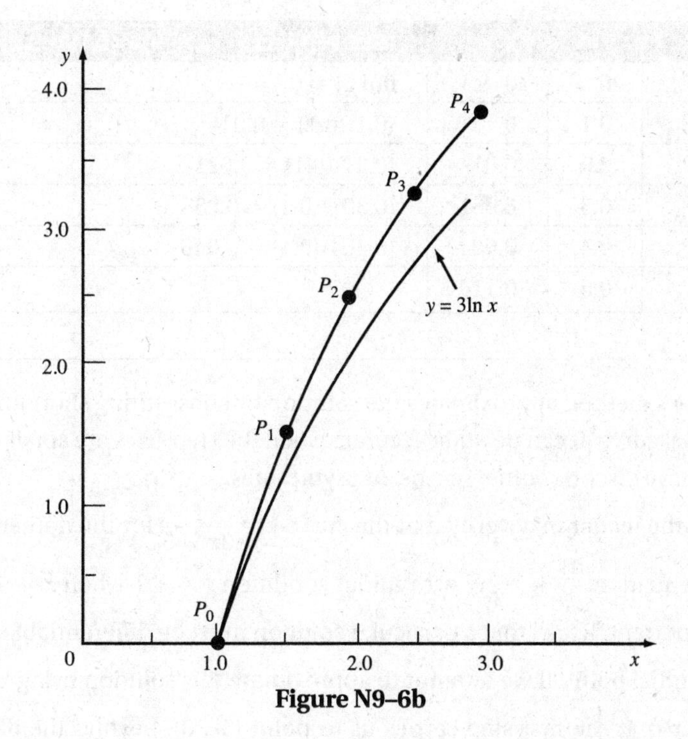

Figure N9–6b

We observe that, since y'' for $3 \ln x$ equals $-\frac{3}{x^2}$, the true curve is concave down and below the Euler graph.

The last column in the table on page 358 shows the true values (to three decimal places) of y. The Euler approximation for $3 \ln 3$ is 3.85; the true value is 3.296. The Euler approximation

BC ONLY

with four steps is not very good! However, see what happens as we increase the number n of steps:

n	EULER APPROXIMATION	ERROR
4	3.85	0.554
10	3.505	0.209
20	3.398	0.102
40	3.346	0.050
80	3.321	0.025

Doubling the number of steps cuts the error approximately in half.

Example 7

Given the d.e. $\frac{dy}{dx} = x + y$ with initial condition $y(0) = 0$, use Euler's method with $\Delta x = 0.1$ to estimate y when $x = 0.5$.

SOLUTION: Here are the relevant computations:

	x	y	(SLOPE) $\cdot \Delta x = (x + y) \cdot (0.1) = \Delta y$
P_0	0	0	$0(0.1) = 0$
P_1	0.1	0	$(0.1)(0.1) = 0.01$
P_2	0.2	0.01	$(0.21)(0.1) = 0.021$
P_3	0.3	0.031	$(0.331)(0.1) = 0.033$
P_4	0.4	0.064	$(0.464)(0.1) = 0.046$
P_5	0.5	0.110	

A Caution: Euler's method approximates the solution by substituting short line segments in place of the actual curve. It can be quite accurate when the step sizes are small, but only if the curve does not have discontinuities, cusps, or asymptotes.

For example, the reader may verify that the curve $y = \frac{1}{2x - 5}$ for the domain $x < \frac{5}{2}$ solves the differential equation $\frac{dy}{dx} = -2y^2$ with initial condition $y = -1$ when $x = 2$. The domain restriction is important. Recall that a particular solution must be differentiable on an interval containing the initial point. If we attempt to approximate this solution using Euler's method with step size $\Delta x = 1$, the first step carries us to point $(3, -3)$, beyond the discontinuity at $x = \frac{5}{2}$ and thus outside the domain of the solution. The accompanying graph (Figure N9–7, page 361) shows that this is nowhere near the solution curve with initial point $y = 1$ when $x = 3$ (and whose domain is $x > \frac{5}{2}$). Here, Euler's method fails because it leaps blindly across the vertical asymptote at $x = \frac{5}{2}$.

Always pay attention to the domain of any particular solution.

BC ONLY

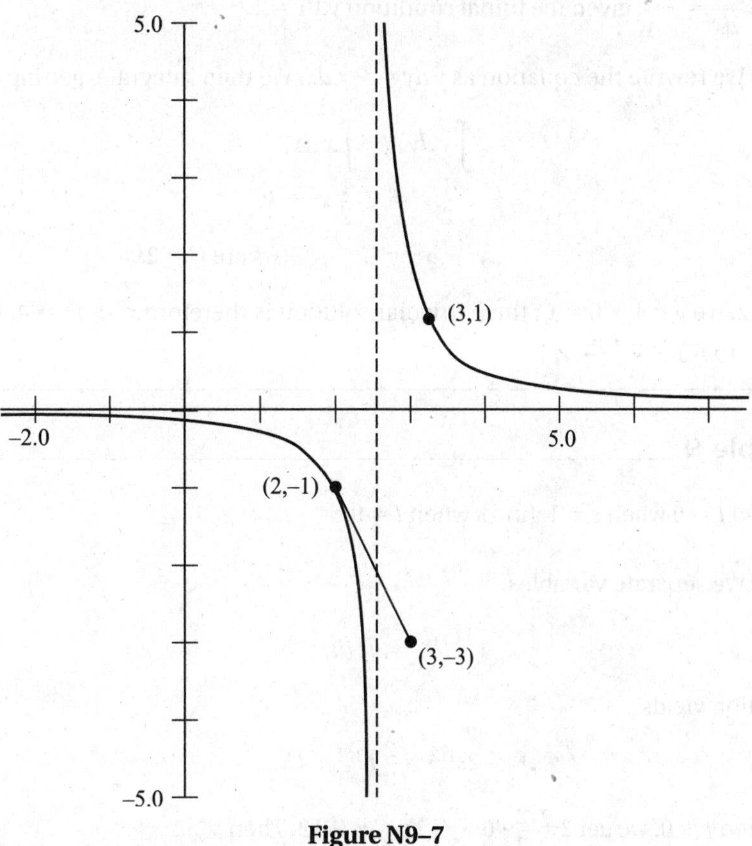

Figure N9–7

D. SOLVING FIRST-ORDER DIFFERENTIAL EQUATIONS ANALYTICALLY

In the preceding sections we solved differential equations graphically, using slope fields, and numerically, using Euler's method. Both methods yield approximations. In this section we review how to solve some differential equations *exactly.*

Separating Variables

Separable

A first-order d.e. in x and y is *separable* if it can be written so that all the terms involving y are on one side and all the terms involving x are on the other.

A differential equation has variables separable if it is of the form

$$\frac{dy}{dx} = \frac{f(x)}{g(y)} \quad \text{or} \quad g(y)\,dy - f(x)\,dx = 0.$$

The general solution is

$$\int g(y)\,dy - \int f(x)\,dx = C \quad (C \text{ arbitrary}).$$

➡ Example 8

Solve the d.e. $\frac{dy}{dx} = -\frac{x}{y}$, given the initial condition $y(0) = 2$.

SOLUTION: We rewrite the equation as $y\,dy = -x\,dx$. We then integrate, getting

$$\int y\,dy = -\int x\,dx,$$

$$\frac{1}{2}y^2 = -\frac{1}{2}x^2 + k,$$

$$y^2 + x^2 = C \qquad \text{(where } C = 2k\text{)}.$$

Since $y(0) = 2$, we get $4 + 0 = C$; the particular solution is therefore $x^2 + y^2 = 4$. (We need to specify above that $y > 0$. Why?)

➡ Example 9

If $\frac{ds}{dt} = \sqrt{st}$ and $t = 0$ when $s = 1$, find s when $t = 9$.

SOLUTION: We separate variables:

$$\frac{ds}{\sqrt{s}} = \sqrt{t}\,dt;$$

then integration yields

$$2s^{1/2} = \frac{2}{3}t^{3/2} + C.$$

Using $s = 1$ and $t = 0$, we get $2 = \frac{2}{3} \cdot 0 + C$, so $C = +12$. Then

$$2s^{1/2} = \frac{2}{3}t^{3/2} + 2 \qquad \text{or} \qquad s = \left(\frac{1}{3}t^{3/2} + 1\right)^2.$$

When $t = 9$, $s = \left(\frac{1}{3}(9^{3/2}) + 1\right)^2 = 100$.

➡ Example 10

If $(\ln y)\frac{dy}{dx} = \frac{y}{x}$, and $y = e$ when $x = 1$, find the value of y greater than 1 that corresponds to $x = e^4$.

SOLUTION: Separating, we get $\frac{\ln y}{y}\,dy = \frac{dx}{x}$. We integrate:

$$\frac{1}{2}\ln^2 y = \ln|x| + C$$

$$\ln y = \pm\sqrt{2\ln|x| + C}$$

$$y = e^{\pm\sqrt{2\ln|x| + C}}$$

Using $y = e$ when $x = 1$, we have $e = e^{\pm\sqrt{2\ln|1| + C}}$, or $e = e^{\pm\sqrt{C}}$. This is true only if $C = 1$, and we choose the positive square root. Hence, $y = e^{\sqrt{2\ln|x| + 1}}$. When $x = e^4$, $y = e^{\sqrt{2\ln|e^4| + 1}} = e^{\sqrt{2 \cdot 4 + 1}} = e^3$.

Example 11

Find the general solution of the differential equation $\frac{du}{dv} = e^{v-u}$.

SOLUTION: We rewrite $\frac{du}{dv} = \frac{e^v}{e^u}$ as $e^u\,du = e^v\,dv$.
Taking antiderivatives yields $e^u = e^v + C$, or $u = \ln(e^v + c)$.

E. EXPONENTIAL GROWTH AND DECAY

We now apply the method of separation of variables to three classes of functions associated with different rates of change. In each of the three cases, we describe the rate of change of a quantity, write the differential equation that follows from the description, then solve—or, in some cases, just give the solution of—the d.e. We list several applications of each case, and present relevant problems involving some of the applications.

Case I: Exponential Growth

An interesting special differential equation with wide applications is defined by the following statement: "A positive quantity y increases (or decreases) at a rate that at any time t is proportional to the amount present." It follows that the quantity y satisfies the d.e.

$$\frac{dy}{dt} = ky, \qquad (1)$$

Exponential growth

where $k > 0$ if y is increasing and $k < 0$ if y is decreasing.

From (1) it follows that

$$\frac{dy}{y} = k\,dt,$$

$$\int \frac{1}{y}\,dy = \int k\,dt,$$

$$\ln y = kt + C \qquad (C \text{ a constant}).$$

Then

$$y = e^{kt+C} = e^{kt} \cdot e^C$$
$$= ce^{kt} \qquad (\text{where } c = e^C).$$

If we are given an initial amount y, say y_0 at time $t = 0$, then

$$y_0 = c \cdot e^{k \cdot 0} = c \cdot 1 = c,$$

and our law of exponential change

$$y = ce^{kt} \qquad (2)$$

tells us that c is the initial amount of y (at time $t = 0$). If the quantity grows with time, then $k > 0$; if it decays (or diminishes, or decomposes), then $k < 0$. Equation (2) is often referred to as the *law of exponential growth or decay.*

The length of time required for a quantity that is decaying exponentially to be reduced by half is called its *half-life.*

Half-life

➡ Example 12

The population of a country is growing at a rate proportional to its population. If the growth rate per year is 4% of the current population, how long will it take for the population to double?

SOLUTION: If the population at time t is P, then we are given that $\frac{dP}{dt} = 0.04P$. Substituting in equation (2), we see that the solution is

$$P = P_0 e^{0.04t},$$

where P_0 is the initial population. We seek t when $P = 2P_0$:

$$2P_0 = P_0 e^{0.04t},$$
$$2 = e^{0.04t},$$
$$\ln 2 = 0.04t,$$
$$t = \frac{\ln 2}{0.04} \simeq 17.33 \text{ yr.}$$

➡ Example 13

The bacteria in a certain culture increase continuously at a rate proportional to the number present.

(a) If the number triples in 6 hours, how many will there be in 12 hours?
(b) In how many hours will the original number quadruple?

SOLUTIONS: We let N be the number at time t and N_0 the number initially. Then

$$\frac{dN}{dt} = kN, \qquad \frac{dN}{N} = k\,dt, \qquad \ln N = kt + C, \qquad \text{and} \qquad \ln N_0 = 0 + C,$$

hence, $C = \ln N_0$. The general solution is then $N = N_0 e^{kt}$, with k still to be determined.

Since $N = 3N_0$ when $t = 6$, we see that $3N_0 = N_0 e^{6k}$ and that $k = \frac{1}{6} \ln 3$. Thus

$$N = N_0 e^{(t \ln 3)/6}.$$

(a) When $t = 12$, $N = N_0 e^{2 \ln 3} = N_0 e^{\ln 3^2} = N_0 e^{\ln 9} = 9N_0$.
(b) We let $N = 4N_0$ in the centered equation above, and get

$$4 = e^{(t \ln 3)/6}, \qquad \ln 4 = \frac{t}{6} \ln 3, \qquad \text{and} \qquad t = \frac{6 \ln 4}{\ln 3} \simeq 7.6 \text{ hr.}$$

➡ Example 14

Radium-226 decays at a rate proportional to the quantity present. Its half-life is 1612 years. How long will it take for one quarter of a given quantity of radium-226 to decay?

SOLUTION: If $Q(t)$ is the amount present at time t, then it satisfies the equation

$$Q(t) = Q_0 e^{kt}, \tag{1}$$

where Q_0 is the initial amount and k is the (negative) factor of proportionality. Since it is given that $Q = \frac{1}{2}Q_0$ when $t = 1612$, equation (1) yields

$$\frac{1}{2}Q_0 = Q_0e^{k(1612)},$$
$$\frac{1}{2} = e^{1612k},$$
$$k = \frac{\ln \frac{1}{2}}{1612} = -0.00043$$

We now have

$$Q = Q_0e^{-0.00043t}. \tag{2}$$

When one quarter of Q_0 has decayed, three quarters of the initial amount remains. We use this fact in equation (2) to find t:

$$\frac{3}{4}Q_0 = Q_0e^{-0.00043t},$$
$$\frac{3}{4} = e^{-0.00043t},$$
$$t = \frac{\ln \frac{3}{4}}{-0.00043} \approx 669 \text{ yr.}$$

Applications of Exponential Growth

(1) A colony of bacteria may grow at a rate proportional to its size.

(2) Other populations, such as those of humans, rodents, or fruit flies, whose supply of food is unlimited may also grow at a rate proportional to the size of the population.

(3) Money invested at interest that is compounded continuously accumulates at a rate proportional to the amount present. The constant of proportionality is the interest rate.

(4) The demand for certain precious commodities (gas, oil, electricity, valuable metals) has been growing in recent decades at a rate proportional to the existing demand.

Each of the above quantities (population, amount, demand) is a function of the form ce^{kt} (with $k > 0$). (See Figure N9–7a.)

(5) Radioactive isotopes, such as uranium-235, strontium-90, iodine-131, and carbon-14, decay at a rate proportional to the amount still present.

(6) If P is the *present value* of a fixed sum of money A due t years from now, where the interest is compounded continuously, then P decreases at a rate proportional to the value of the investment.

(7) It is common for the concentration of a drug in the bloodstream to drop at a rate proportional to the existing concentration.

(8) As a beam of light passes through murky water or air, its intensity at any depth (or distance) decreases at a rate proportional to the intensity at that depth.

Each of the above four quantities (5 through 8) is a function of the form ce^{-kt} $(k > 0)$. (See Figure N9–7b.)

This is *exponential growth*. This is *exponential decay*.

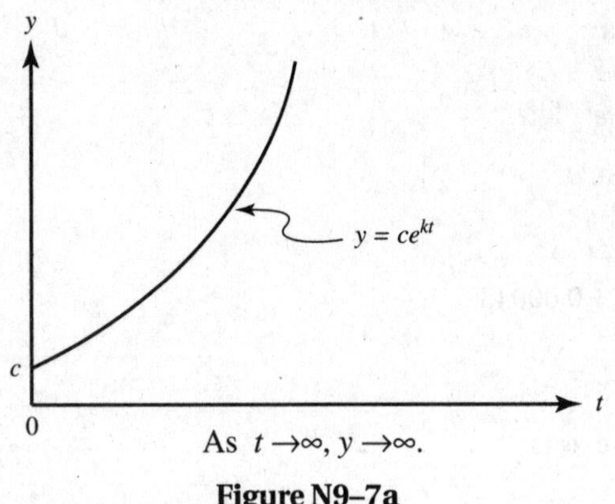

As $t \to \infty, y \to \infty$.

Figure N9–7a

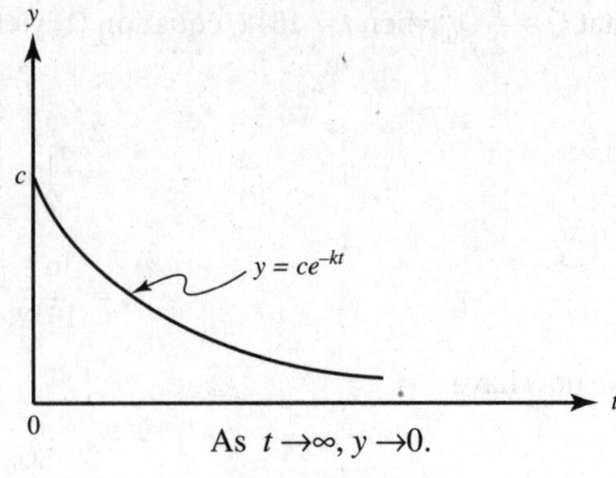

As $t \to \infty, y \to 0$.

Figure N9–7b

Example 15

At a yearly rate of 5% compounded continuously, how long does it take (to the nearest year) for an investment to triple?

SOLUTION: If P dollars are invested for t yr at 5%, the amount will grow to $A = Pe^{0.05t}$ in t yr. We seek t when $A = 3P$:

$$3 = e^{0.05t},$$

$$\frac{\ln 3}{0.05} = t \simeq 22 \text{ yr.}$$

Example 16

One important method of dating fossil remains is to determine what portion of the carbon content of a fossil is the radioactive isotope carbon-14. During life, any organism exchanges carbon with its environment. Upon death this circulation ceases, and the ^{14}C in the organism then decays at a rate proportional to the amount present. The proportionality factor is 0.012% per year.

When did an animal die, if an archaeologist determines that only 25% of the original amount of ^{14}C is still present in its fossil remains?

SOLUTION: The quantity Q of ^{14}C present at time t satisfies the equation

$$\frac{dQ}{dt} = -0.00012Q$$

with solution

$$Q(t) = Q_0 e^{-0.00012t}$$

(where Q_0 is the original amount). We are asked to find t when $Q(t) = 0.25Q_0$.

$$\begin{aligned} 0.25Q_0 &= Q_0 e^{-0.00012t}, \\ 0.25 &= e^{-0.00012t}, \\ \ln 0.25 &= -0.00012t, \\ -1.386 &= -0.00012t, \\ t &\simeq 11{,}550. \end{aligned}$$

Rounding to the nearest 500 yr, we see that the animal died approximately 11,500 yr ago.

➡ Example 17

In 1970 the world population was approximately 3.5 billion. Since then it has been growing at a rate proportional to the population, and the factor of proportionality has been 1.9% per year. At that rate, in how many years would there be one person per square foot of land? (The land area of Earth is approximately 200,000,000 mi^2, or about 5.5×10^{15} ft^2.)

SOLUTION: If $P(t)$ is the population at time t, the problem tells us that P satisfies the equation $\frac{dP}{dt} = 0.019P$. Its solution is the exponential growth equation

$$P(t) = P_0 e^{0.019t},$$

where P_0 is the initial population. Letting $t = 0$ for 1970, we have

$$3.5 \times 10^9 = P(0) = P_0 e^0 = P_0.$$

Then

$$P(t) = (3.5 \times 10^9)e^{0.019t}.$$

The question is: for what t does $P(t)$ equal 5.5×10^{15}? We solve

$$(3.5)(10^9)e^{0.019t} = (5.5)10^{15},$$
$$e^{0.019t} \simeq (1.6)10^6.$$

Taking the logarithm of each side yields

$$0.019t \simeq \ln 1.6 + 6 \ln 10 \simeq 14.3,$$
$$t \simeq 750 \text{ yr},$$

where it seems reasonable to round off as we have. Thus, if the human population continued to grow at the present rate, there would be one person for every square foot of land in the year 2720.

Case II: Restricted Growth

Restricted growth

The rate of change of a quantity $y = f(t)$ may be proportional, not to the amount present, but to a difference between that amount and a fixed constant. Two situations are to be distinguished: The rate of change is proportional to

(a) a fixed constant A minus the amount of the quantity present:

$$f'(t) = k\,[A - f(t)]$$

(b) the amount of the quantity present minus a fixed constant A:

$$f'(t) = -k[f(t) - A]$$

where (in both) $f(t)$ is the amount at time t and k and A are both positive. We may conclude that

(a) $f(t)$ is increasing (Fig. N9–8a):

$$f(t) = A - ce^{-kt}$$

(b) $f(t)$ is decreasing (Fig. N9–8b):

$$f(t) = A + ce^{-kt}$$

for some positive constant c.

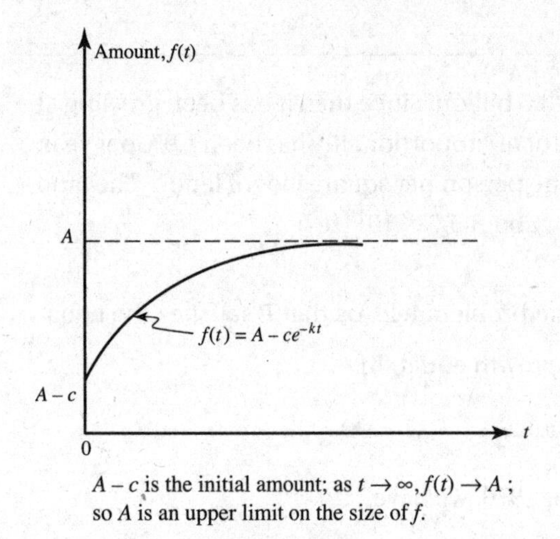

$A-c$ is the initial amount; as $t \to \infty, f(t) \to A$; so A is an upper limit on the size of f.

Figure N9–8a

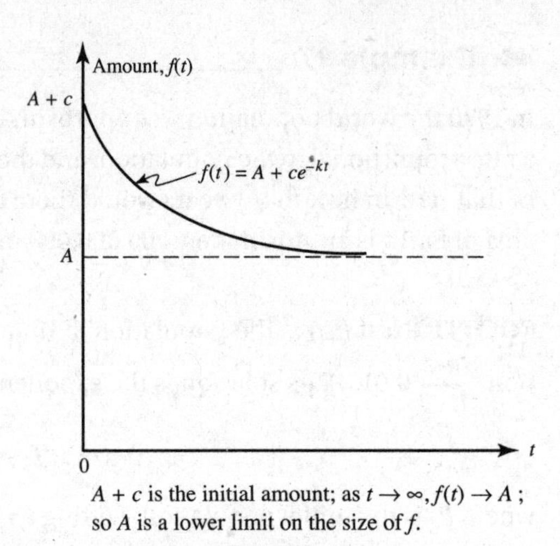

$A+c$ is the initial amount; as $t \to \infty, f(t) \to A$; so A is a lower limit on the size of f.

Figure N9–8b

Here is how we solve the d.e. for Case II(a), where $A - y > 0$. If the quantity at time t is denoted by y and k is the positive constant of proportionality, then

$$\begin{aligned} y' = \frac{dy}{dt} &= k(A-y), \\ \frac{dy}{A-y} &= k\,dt, \\ -\ln(A-y) &= kt + C, \\ \ln(A-y) &= -kt - C, \\ A - y &= e^{-kt} \cdot e^{-C} \\ &= ce^{-kt}, \text{ where } c = e^{-C}, \end{aligned}$$

and

$$y = A - ce^{-kt}.$$

Case II(b) can be solved similarly.

➡ Example 18

According to Newton's law of cooling, a hot object cools at a rate proportional to the difference between its own temperature and that of its environment. If a roast at room temperature 68°F is put into a 20°F freezer, and if, after 2 hours, the temperature of the roast is 40°F:
(a) What is its temperature after 5 hours?
(b) How long will it take for the temperature of the roast to fall to 21°F?

SOLUTIONS: This is an example of Case II(b) (the temperature is decreasing toward the limiting temperature 20°F).

(a) If $R(t)$ is the temperature of the roast at time t, then

$$\frac{dR(t)}{dt} = -k[R(t) - 20] \qquad \text{and} \qquad R(t) = 20 + ce^{-kt}.$$

Since $R(0) = 68$°F, we have

$$\begin{aligned} 68 &= 20 + c, \\ c &= 48, \\ R(t) &= 20 + 48e^{-kt}. \end{aligned}$$

Also, $R(2) = 40°F$, so	$40 = 20 + 48e^{-k \cdot 2}$
and	$e^{-k} \simeq 0.65,$
yielding	$R(t) = 20 + 48(0.65)^t \quad (*)$
and, finally,	$R(5) = 20 + 48(0.65)^5 \simeq 26°F.$

(b) Equation (*) in part (a) gives the roast's temperature at time t. We must find t when $R = 21$:

$$\begin{aligned} 21 &= 20 + 48\,(0.65)^t, \\ \frac{1}{48} &= (0.65)^t, \\ -\ln 48 &= t\ln(0.65), \\ t &\simeq 9 \text{ hr.} \end{aligned}$$

Example 19

Advertisers generally assume that the rate at which people hear about a product is proportional to the number of people who have not yet heard about it. Suppose that the size of a community is 15,000, that to begin with no one has heard about a product, but that after 6 days 1500 people know about it. How long will it take for 2700 people to have heard of it?

SOLUTION: Let $N(t)$ be the number of people aware of the product at time t. Then we are given that

$$N'(t) = k[15{,}000 - N(t)],$$

which is Case IIa. The solution of this d.e. is

$$N(t) = 15{,}000 - ce^{-kt}.$$

Since $N(0) = 0$, $c = 15{,}000$ and

$$N(t) = 15{,}000(1 - e^{-kt}).$$

Since 1500 people know of the product after 6 days, we have

$$\begin{aligned} 1500 &= 15{,}000(1 - e^{-6k}), \\ e^{-6k} &= 0.9, \\ k &= \frac{\ln 0.9}{-6} = 0.018. \end{aligned}$$

We now seek t when $N = 2700$:

$$\begin{aligned} 2700 &= 15{,}000(1 - e^{-0.018t}), \\ 0.18 &= 1 - e^{-0.018t}, \\ e^{-0.018t} &= 0.82, \\ t &\simeq 11 \text{ days.} \end{aligned}$$

Applications of Restricted Growth

(1) Newton's law of heating says that a cold object warms up at a rate proportional to the difference between its temperature and that of its environment. If you put a roast at 68°F into an oven of 400°F, then the temperature at time t is $R(t) = 400 - 332e^{-kt}$.

(2) Because of air friction, the velocity of a falling object approaches a limiting value L (rather than increasing without bound). The acceleration (rate of change of velocity) is proportional to the difference between the limiting velocity and the object's velocity. If initial velocity is zero, then at time t the object's velocity $V(t) = L(1 - e^{-kt})$.

(3) If a tire has a small leak, then the air pressure inside drops at a rate proportional to the difference between the inside pressure and the fixed outside pressure O. At time t the inside pressure $P(t) = O + ce^{-kt}$.

BC ONLY

Case III: Logistic Growth

The rate of change of a quantity (for example, a population) may be proportional both to the amount (size) of the quantity and to the difference between a fixed constant A and its amount (size). If $y = f(t)$ is the amount, then

$$y' = ky(A - y), \tag{1}$$

where k and A are both positive. Equation (1) is called the *logistic differential equation*; it is used to model logistic growth.

The solution of the d.e. (1) is

$$y = \frac{A}{1 + ce^{-Akt}} \tag{2}$$

for some positive constant c.

In most applications, $c > 1$. In these cases, the initial amount $A/(1 + c)$ is *less* than $A/2$. In all applications, since the exponent of e in the expression for $f(t)$ is negative for all positive t, therefore, as $t \to \infty$,

(1) $ce^{-Akt} \to 0$;

(2) the denominator of $f(t) \to 1$;

(3) $f(t) \to A$.

Thus, A is an upper limit of f in this growth model. When applied to populations, A is called the *carrying capacity* or the *maximum sustainable population*.

Shortly we will solve specific examples of the logistic d.e. (1), instead of obtaining the general solution (2), since the latter is algebraically rather messy. (It is somewhat less complicated to verify that y' in (1) can be obtained by taking the derivative of (2).)

Unrestricted Versus Restricted Growth

In Figures N9–9a and N9–9b we see the graphs of the growth functions of Cases I and III. The growth function of Case I is known as the *unrestricted* (or *uninhibited* or *unchecked*) model. It is not a very realistic one for most populations. It is clear, for example, that human populations cannot continue endlessly to grow exponentially. Not only is Earth's land area fixed, but also there are limited supplies of food, energy, and other natural resources. The growth function in Case III allows for such factors, which serve to check growth. It is therefore referred to as the *restricted* (or *inhibited*) model.

The two graphs are quite similar close to 0. This similarity implies that logistic growth is exponential at the start—a reasonable conclusion, since populations are small at the outset.

The S-shaped curve in Case III is often called a *logistic curve*. It shows that the rate of growth y':

(1) increases slowly for a while; i.e., $y'' > 0$;

(2) attains a maximum when $y = A/2$, at half the upper limit to growth;

(3) then decreases ($y'' < 0$), approaching 0 as y approaches its upper limit.

It is not difficult to verify these statements.

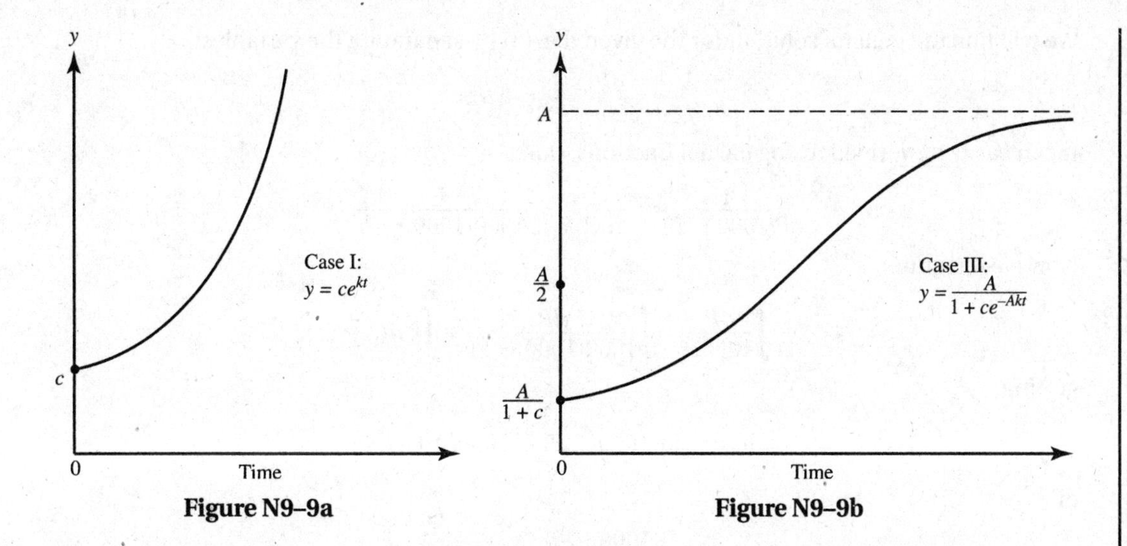

Figure N9–9a **Figure N9–9b**

BC ONLY

Applications of Logistic Growth

(1) Some diseases spread through a (finite) population P at a rate proportional to the number of people, $N(t)$, infected by time t and the number, $P - N(t)$, not yet infected. Thus $N'(t) = kN(P - N)$ and, for some positive c and k,

$$N(t) = \frac{P}{1 + ce^{-Pkt}}.$$

(2) A rumor (or fad or new religious cult) often spreads through a population P according to the formula in (1), where $N(t)$ is the number of people who have heard the rumor (acquired the fad, converted to the cult), and $P - N(t)$ is the number who have not.

(3) Bacteria in a culture on a Petri dish grow at a rate proportional to the product of the existing population and the difference between the maximum sustainable population and the existing population. (Replace bacteria on a Petri dish by fish in a small lake, ants confined to a small receptacle, fruit flies supplied with only a limited amount of food, yeast cells, and so on.)

(4) Advertisers sometimes assume that sales of a particular product depend on the number of TV commercials for the product and that the rate of increase in sales is proportional both to the existing sales and to the additional sales conjectured as possible.

(5) In an autocatalytic reaction a substance changes into a new one at a rate proportional to the product of the amount of the new substance present and the amount of the original substance still unchanged.

Example 20

Because of limited food and space, a squirrel population cannot exceed 1000. It grows at a rate proportional both to the existing population and to the attainable additional population. If there were 100 squirrels 2 years ago, and 1 year ago the population was 400, about how many squirrels are there now?

SOLUTION: Let P be the squirrel population at time t. It is given that

$$\frac{dP}{dt} = kP(1000 - P) \qquad (3)$$

with $P(0) = 100$ and $P(1) = 400$. We seek $P(2)$.

BC ONLY

We will find the general solution for the given d.e. (3) by separating the variables:

$$\frac{dP}{P(1000-P)} = k\,dt.$$

It can easily be verified, using partial fractions, that

$$\frac{1}{P(1000-P)} = \frac{1}{1000P} + \frac{1}{1000(1000-P)}.$$

Now we integrate:

$$\int\frac{dP}{1000P} + \int\frac{dP}{1000(1000-P)} = \int k\,dt,$$

getting

$$\ln P - \ln(1000-P) = 1000kt + C$$

or

$$\ln\frac{(1000-P)}{P} = -(1000kt + C),$$

$$\frac{1000-P}{P} = ce^{-1000kt} \quad \left(\text{where } c = e^{-C}\right),$$

$$\frac{1000}{P} - 1 = ce^{-1000kt},$$

$$\frac{1000}{P} = 1 + ce^{-1000kt},$$

$$\frac{P}{1000} = \frac{1}{1+ce^{-1000kt}},$$

and, finally (!),

$$P(t) = \frac{1000}{1+ce^{-1000kt}}. \tag{4}$$

Please note that this is precisely the solution "advertised" on page 370 in equation (2), with A equal to 1000.

Now, using our initial condition $P(0) = 100$ in (4), we get

$$\frac{100}{1000} = \frac{1}{1+c} \quad \text{and} \quad c = 9.$$

Using $P(1) = 400$, we get

$$400 = \frac{1000}{1+9e^{-1000k}},$$

$$1 + 9e^{-1000k} = 2.5,$$

$$e^{-1000k} = \frac{1.5}{9} = \frac{1}{6}. \tag{5}$$

Then the particular solution is

$$P(t) = \frac{1000}{1+9(1/6)^t} \tag{6}$$

and $P(2) \simeq 800$ squirrels.

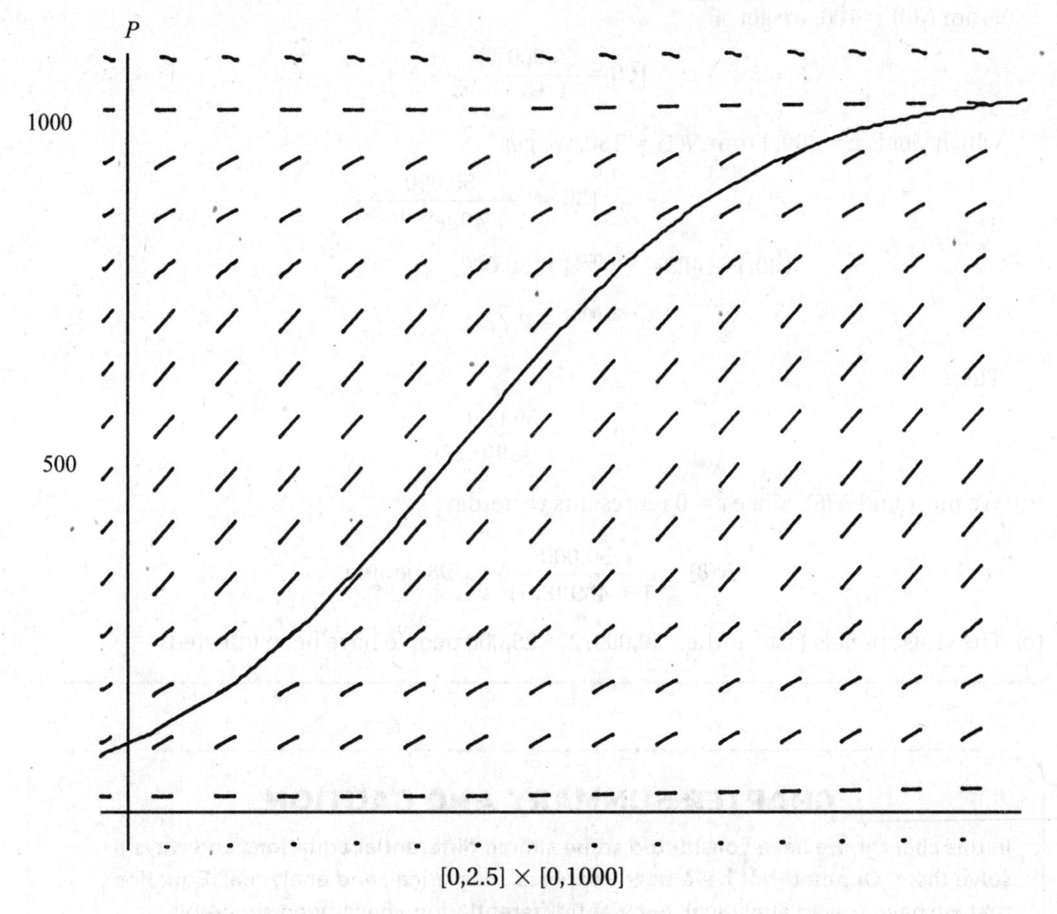

[0,2.5] × [0,1000]

Figure N9–10

Figure N9–10 shows the slope field for equation (3), with $k = 0.00179$, which was obtained by solving equation (5) above. Note that the slopes are the same along any horizontal line, and that they are close to zero initially, reach a maximum at $P = 500$, then diminish again as P approaches its limiting value, 1000. We have superimposed the solution curve for $P(t)$ that we obtained in (6) above.

➡ Example 21

Suppose a flu-like virus is spreading through a population of 50,000 at a rate proportional both to the number of people already infected and to the number still uninfected. If 100 people were infected yesterday and 130 are infected today:

(a) write an expression for the number of people $N(t)$ infected after t days;
(b) determine how many will be infected a week from today;
(c) indicate when the virus will be spreading the fastest.

SOLUTIONS:

(a) We are told that $N'(t) = k \cdot N \cdot (50{,}000 - N)$, that $N(0) = 100$, and that $N(1) = 130$. The d.e. describing logistic growth leads to

$$N(t) = \frac{50{,}000}{1 + ce^{-50{,}000kt}}.$$

From $N(0) = 100$, we get

$$100 = \frac{50{,}000}{1 + c},$$

which yields $c = 499$. From $N(1) = 130$, we get

$$130 = \frac{50{,}000}{1 + 499e^{-50{,}000k}},$$

$$130(1 + 499e^{-50{,}000k}) = 50{,}000,$$

$$e^{-50{,}000k} = 0.77.$$

Then

$$N(t) = \frac{50{,}000}{1 + 499(0.77)^t}.$$

(b) We must find $N(8)$. Since $t = 0$ represents yesterday:

$$N(8) = \frac{50{,}000}{1 + 499(0.77)^8} \simeq 798 \text{ people.}$$

(c) The virus spreads fastest when $50{,}000/2 = 25{,}000$ people have been infected.

CHAPTER SUMMARY AND CAUTION

In this chapter, we have considered some simple differential equations and ways to solve them. Our methods have been graphical, numerical, and analytical. Equations that we have solved analytically—by antidifferentiation—have been separable.

It is important to realize that, given a first-order differential equation of the type $\frac{dy}{dx} = F(x,y)$, it is the exception, rather than the rule, to be able to find the general solution by analytical methods. Indeed, a great many practical applications lead to d.e.'s for which no explicit algebraic solution exists.

PRACTICE EXERCISES

Part A. Directions: Answer these questions *without* using your calculator.

In Questions 1–10, $a(t)$ denotes the acceleration function, $v(t)$ the velocity function, and $s(t)$ the position or height function at time t. (The acceleration due to gravity is -32 ft/sec^2.)

1. If $a(t) = 4t - 1$ and $v(1) = 3$, then $v(t)$ equals

   (A) $2t^2 - t$ (B) $2t^2 - t + 1$ (C) $2t^2 - t + 2$ (D) $2t^2 + 1$ (E) $2t^2 + 2$

2. If $a(t) = 20t^3 - 6t$, $s(-1) = 2$, and $s(1) = 4$, then $v(t)$ equals

   (A) $t^5 - t^3$ (B) $5t^4 - 3t^2 + 1$ (C) $5t^4 - 3t^2 + 3$
   (D) $t^5 - t^3 + t + 3$ (E) $t^5 - t^3 + 1$

3. Given $a(t)$, $s(-1)$, and $s(1)$ as in Question 2, then $s(0)$ equals

   (A) 0 (B) 1 (C) 2 (D) 3 (E) 4

4. A stone is thrown straight up from the top of a building with initial velocity 40 ft/sec and hits the ground 4 sec later. The height of the building, in feet, is

   (A) 88 (B) 96 (C) 112 (D) 128 (E) 144

5. The maximum height is reached by the stone in Question 4 after

   (A) 4/5 sec (B) 4 sec (C) 5/4 sec (D) 5/2 sec (E) 2 sec

6. If a car accelerates from 0 to 60 mph in 10 sec, what distance does it travel in those 10 sec? (Assume the acceleration is constant and note that 60 mph $\doteq$ 88 ft/sec.)

   (A) 40 ft (B) 44 ft (C) 88 ft (D) 400 ft (E) 440 ft

7. A stone is thrown at a target so that its velocity after t sec is $(100 - 20t)$ ft/sec. If the stone hits the target in 1 sec, then the distance from the sling to the target is

   (A) 80 ft (B) 90 ft (C) 100 ft (D) 110 ft (E) 120 ft

8. What should the initial velocity be if you want a stone to reach a height of 100 ft when you throw it straight up?

   (A) 50 ft/sec (B) 80 ft/sec (C) 92 ft/sec
   (D) 96 ft/sec (E) 112 ft/sec

9. If the velocity of a car traveling in a straight line at time t is $v(t)$, then the difference in its odometer readings between times $t = a$ and $t = b$ is

   (A) $\int_a^b |v(t)|\,dt$

   (B) $\int_a^b v(t)\,dt$

   (C) the net displacement of the car's position from $t = a$ to $t = b$
   (D) the change in the car's position from $t = a$ to $t = b$
   (E) the car's average speed from $t = a$ to $t = b$.

10. If an object is moving up and down along the y-axis with velocity $v(t)$ and $s'(t) = v(t)$, then it is false that $\int_a^b v(t)\,dt$ gives

(A) $s(b) - s(a)$
(B) the net distance traveled by the object between $t = a$ and $t = b$
(C) the total change in $s(t)$ between $t = a$ and $t = b$
(D) the shift in the object's position from $t = a$ to $t = b$
(E) the total distance covered by the object from $t = a$ to $t = b$

11. Solutions of the differential equation $y\,dy = x\,dx$ are of the form

(A) $x^2 - y^2 = C$ (B) $x^2 + y^2 = C$ (C) $y^2 = Cx^2$
(D) $x^2 - Cy^2 = 0$ (E) $x^2 = C - y^2$

12. Find the domain of the particular solution to the differential equation in Question 11 that passes through point $(-2, 1)$.

(A) $x < 0$ (B) $-2 \le x < 0$ (C) $x < -\sqrt{3}$
(D) $|x| < \sqrt{3}$ (E) $|x| > \sqrt{3}$

13. If $\frac{dy}{dx} = \frac{y}{2\sqrt{x}}$ and $y = 1$ when $x = 4$, then

(A) $y^2 = 4\sqrt{x} - 7$ (B) $\ln y = 4\sqrt{x} - 8$ (C) $\ln y = \sqrt{x} - 2$
(D) $y = e^{\sqrt{x}}$ (E) $y = e^{\sqrt{x} - 2}$

14. If $\frac{dy}{dx} = e^y$ and $y = 0$ when $x = 1$, then

(A) $y = \ln|x|$ (B) $y = \ln|2 - x|$ (C) $e^{-y} = 2 - x$
(D) $y = -\ln|x|$ (E) $e^{-y} = x - 2$

15. If $\frac{dy}{dx} = \frac{x}{\sqrt{9 + x^2}}$ and $y = 5$ when $x = 4$, then y equals

(A) $\sqrt{9 + x^2} - 5$ (B) $\sqrt{9 + x^2}$ (C) $2\sqrt{9 + x^2} - 5$
(D) $\frac{\sqrt{9 + x^2} + 5}{2}$ (E) $\frac{\sqrt{9 + x^2}}{2}$

16. The general solution of the differential equation $x\,dy = y\,dx$ is a family of

(A) circles (B) hyperbolas (C) parallel lines
(D) parabolas (E) lines passing through the origin

17. The general solution of the differential equation $\frac{dy}{dx} = y$ is a family of

(A) parabolas (B) straight lines (C) hyperbolas
(D) ellipses (E) exponential curves

18. A function $f(x)$ that satisfies the equations $f(x)f'(x) = x$ and $f(0) = 1$ is

(A) $f(x) = \sqrt{x^2 + 1}$ (B) $f(x) = \sqrt{1 - x^2}$ (C) $f(x) = x$
(D) $f(x) = e^x$ (E) $f(x) = \frac{1}{2}x^2 + 1$

19. The curve that passes through the point (1, 1) and whose slope at any point (x, y) is equal to $\frac{3y}{x}$ has the equation

(A) $3x - 2 = y$ (B) $y^3 = x$ (C) $y = x^3$
(D) $3y^2 = x^2 + 2$ (E) $3y^2 - 2x = 1$

20. What is the domain of the particular solution in Question 19?

(A) all real numbers (B) $|x| \leq 1$ (C) $x \neq 0$ (D) $x < 0$ (E) $x > 0$

21. If $\frac{dy}{dx} = \frac{k}{x}$, k a constant, and if $y = 2$ when $x = 1$ and $y = 4$ when $x = e$, then, when $x = 2$, y equals

(A) 2 (B) 4 (C) ln 8 (D) ln 2 + 2 (E) ln 4 + 2

22. The slope field shown at the right is for the differential equation

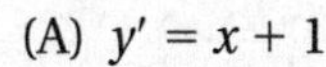

(A) $y' = x + 1$
(B) $y' = \sin x$
(C) $y' = -\sin x$
(D) $y' = \cos x$
(E) $y' = -\cos x$

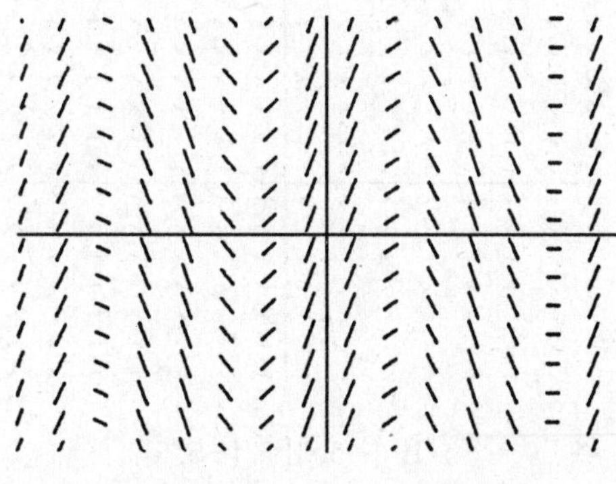

$[-2\pi, 2\pi] \times [-1.5, 1.5]$

23. The slope field at the right is for the differential equation

(A) $y' = 2x$
(B) $y' = 2x - 4$
(C) $y' = 4 - 2x$
(D) $y' = y$
(E) $y' = x + y$

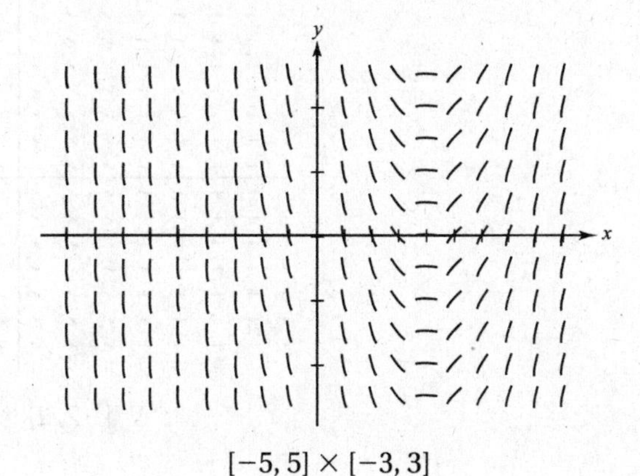

$[-5, 5] \times [-3, 3]$

24. A solution curve has been superimposed on the slope field shown at the right. The solution is for the differential equation and initial condition

(A) $y' = \tan x$; $y(0) = 0$
(B) $y' = \cot x$, $y(\pi/4) = 1$

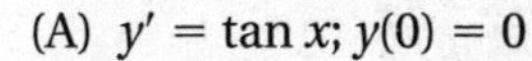

(C) $y' = 1 + x^2$; $y(0) = 0$
(D) $y' = \frac{1}{1 + x^2}$; $y\left(\frac{\pi}{4}\right) = 1$
(E) $y' = 1 + y^2$; $y(0) = 0$

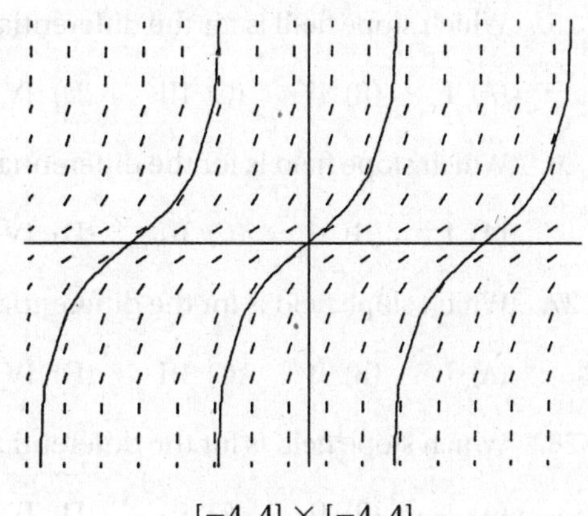

$[-4, 4] \times [-4, 4]$

The slope fields below are for Questions 25–30.

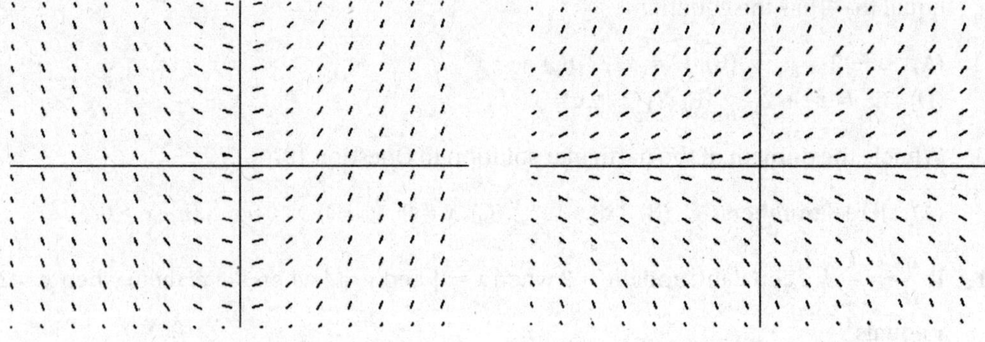

I $[-3, 3] \times [-3, 3]$

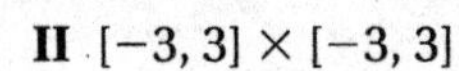

II $[-3, 3] \times [-3, 3]$

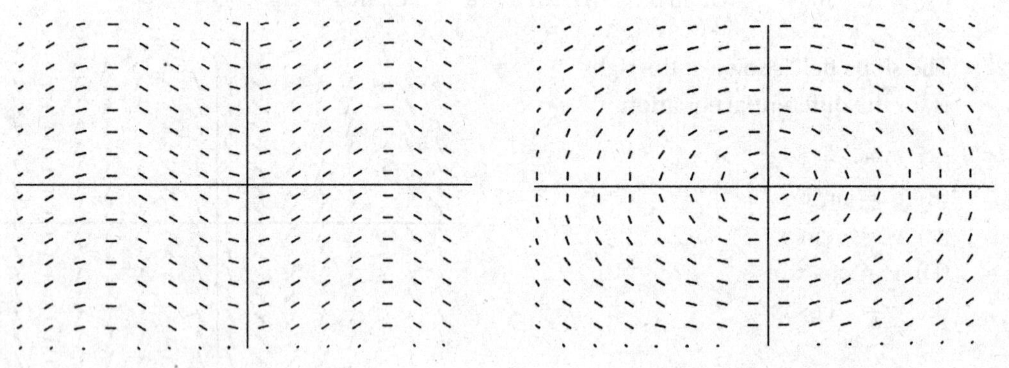

III $[-5, 5] \times [-5, 5]$

IV $[-3, 3] \times [-3, 3]$

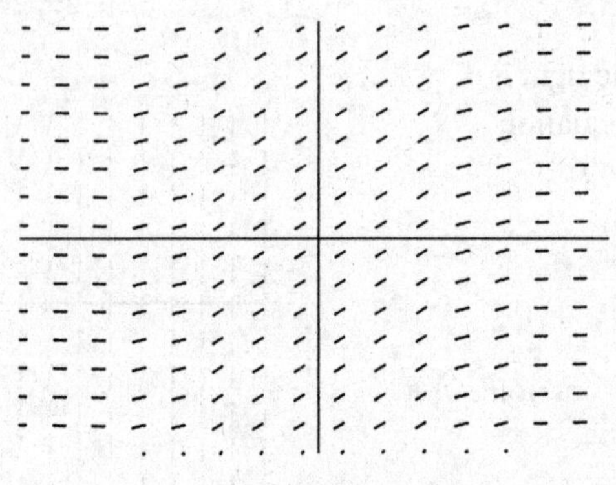

V $[-2, 2] \times [-2, 2]$

25. Which slope field is for the differential equation $y' = y$?

    (A) I (B) II (C) III (D) IV (E) V

26. Which slope field is for the differential equation $y' = -\frac{x}{y}$?

    (A) I (B) II (C) III (D) IV (E) V

27. Which slope field is for the differential equation $y' = \sin x$?

    (A) I (B) II (C) III (D) IV (E) V

28. Which slope field is for the differential equation $y' = 2x$?

    (A) I (B) II (C) III (D) IV (E) V

29. Which slope field is for the differential equation $y' = e^{-x^2}$?

(A) I (B) II (C) III (D) IV (E) V

30. A particular solution curve of a differential equation whose slope field is shown above in II passes through the point $(0, -1)$. The equation is

(A) $y = -e^{x}$ (B) $y = -e^{-x}$ (C) $y = x^2 - 1$

(D) $y = -\cos x$ (E) $y = -\sqrt{1 - x^2}$

31. If you use Euler's method with $\Delta x = 0.1$ for the d.e. $y' = x$, with initial value $y(1) = 5$, then, when $x = 1.2$, y is approximately

(A) 5.10 (B) 5.20 (C) 5.21 (D) 6.05 (E) 7.10

32. The error in using Euler's method in Question 31 is

(A) 0.005 (B) 0.010 (C) 0.050 (D) 0.500 (E) 0.720

33. Which differential equation has the slope field shown?

(A) $y' = y(y + 2)$
(B) $y' = x(y + 2)$
(C) $y' = xy + 2$
(D) $y' = \dfrac{x}{y + 2}$
(E) $y' = \dfrac{y}{y + 2}$

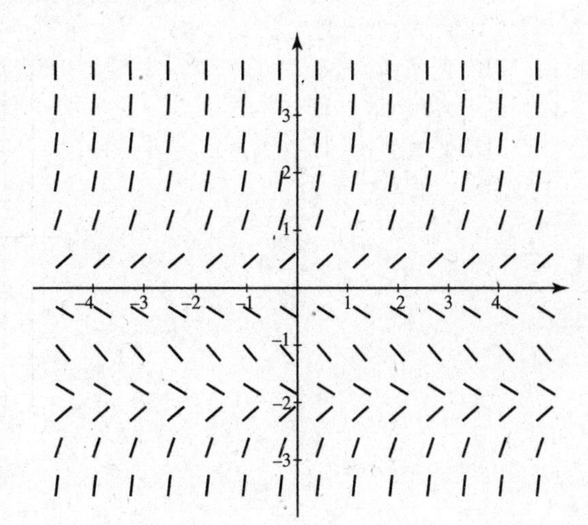

34. Which function is a possible solution of the slope field shown?

(A) $y = 1 - \frac{1}{x}$
(B) $y = 1 - \ln x$
(C) $y = 1 + \ln x$
(D) $y = 1 + e^{x}$
(E) $y = 1 + \tan x$

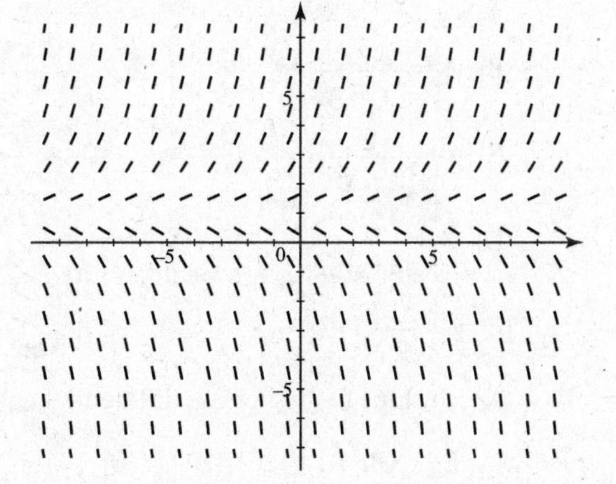

35. **Which differential equation has the slope field shown?**

(A) $y' = x^2 + 2y^2 + 1$
(B) $y' = 2y + 1$
(C) $y' = x - 2y + 1$
(D) $y' = x + 2y + 1$
(E) $y' = x + 1$

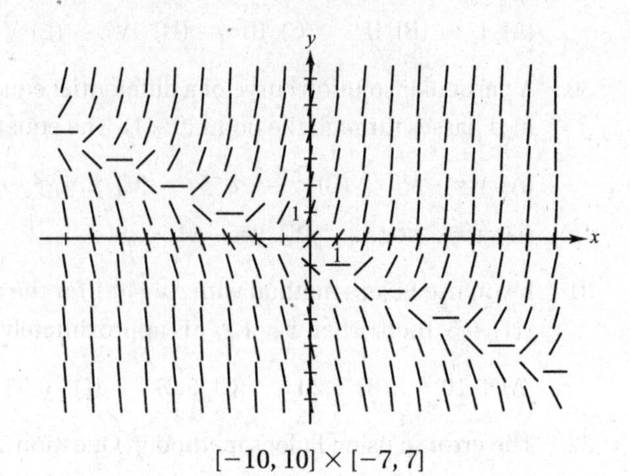

$[-10, 10] \times [-7, 7]$

36. **Which differential equation has the slope field shown?**

(A) $y' = x(x + y)$
(B) $y' = y(x + y)$
(C) $y' = x(x + 1)$
(D) $y' = y(y + 1)$
(E) $y' = x\left(x^2 + y\right)$

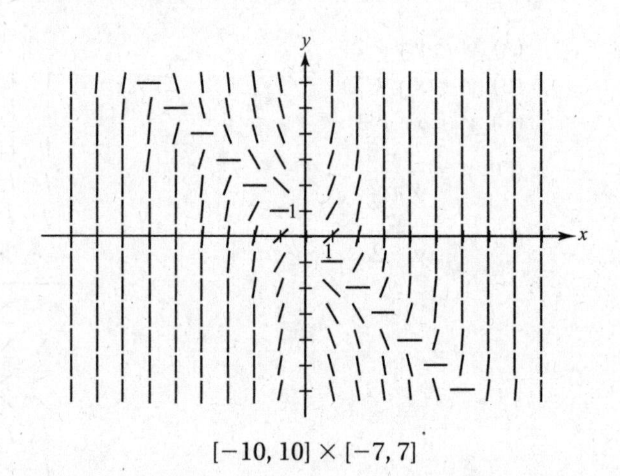

$[-10, 10] \times [-7, 7]$

37. **Which differential equation has the slope field shown?**

(A) $y' = \dfrac{2x}{2 + y^2}$

(B) $y' = \dfrac{2}{2x + y^2}$

(C) $y' = \dfrac{2}{2 + x^2}$

(D) $y' = \dfrac{2y}{2 + y^2}$

(E) $y' = \dfrac{2}{2 + y^2}$

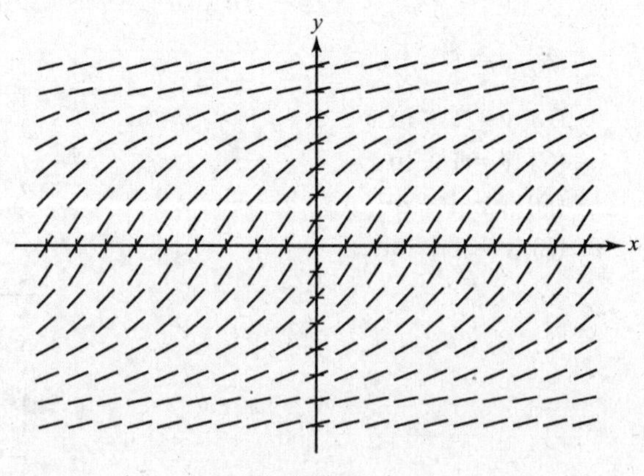

$[-10, 10] \times [-4, 4]$

Part B. Directions: Some of the following questions require the use of a graphing calculator.

38. If $\frac{ds}{dt} = \sin^2\left(\frac{\pi}{2}s\right)$ and if $s = 1$ when $t = 0$, then, when $s = \frac{3}{2}$, t is equal to

(A) $\frac{1}{2}$ (B) $\frac{\pi}{2}$ (C) 1 (D) $\frac{2}{\pi}$ (E) $-\frac{2}{\pi}$

39. If radium decomposes at a rate proportional to the amount present, then the amount R left after t yr, if R_0 is present initially and c is the negative constant of proportionality, is given by

(A) $R = R_0ct$ (B) $R = R_0e^{ct}$ (C) $R = R_0 + \frac{1}{2}ct^2$
(D) $R = e^{R_0ct}$ (E) $R = e^{R_0 + ct}$

40. The population of a city increases continuously at a rate proportional, at any time, to the population at that time. The population doubles in 50 yr. After 75 yr the ratio of the population P to the initial population P_0 is

(A) $\frac{9}{4}$ (B) $\frac{5}{2}$ (C) $\frac{4}{1}$ (D) $\frac{2\sqrt{2}}{1}$ (E) $\frac{3}{2}$

41. If a substance decomposes at a rate proportional to the amount of the substance present, and if the amount decreases from 40 g to 10 g in 2 hr, then the constant of proportionality is

(A) $-\ln 2$ (B) $-\frac{1}{2}$ (C) $-\frac{1}{4}$ (D) $\ln\frac{1}{4}$ (E) $\ln\frac{1}{8}$

42. If $\left(g'(x)\right)^2 = g(x)$ for all real x and $g(0) = 0$, $g(4) = 4$, then $g(1)$ equals

(A) $\frac{1}{4}$ (B) $\frac{1}{2}$ (C) 1 (D) 2 (E) 4

43. The solution curve of $y' = y$ that passes through point (2, 3) is

(A) $y = e^x + 3$ (B) $y = \sqrt{2x + 5}$ (C) $y = 0.406e^x$
(D) $y = e^x - \left(e^2 + 3\right)$ (E) $y = e^x/(0.406)$

44. At any point of intersection of a solution curve of the d.e. $y' = x + y$ and the line $x + y = 0$, the function y at that point

(A) is equal to 0 (B) is a local maximum (C) is a local minimum
(D) has a point of inflection (E) has a discontinuity

45. The slope field for $F'(x) = e^{-x^2}$ is shown at the right with the particular solution $F(0) = 0$ superimposed. With a graphing calculator, $\lim_{x\to\infty} F'(x)$ to three decimal places is

(A) 0.886 (B) 0.987
(C) 1.000 (D) 1.414
(E) ∞

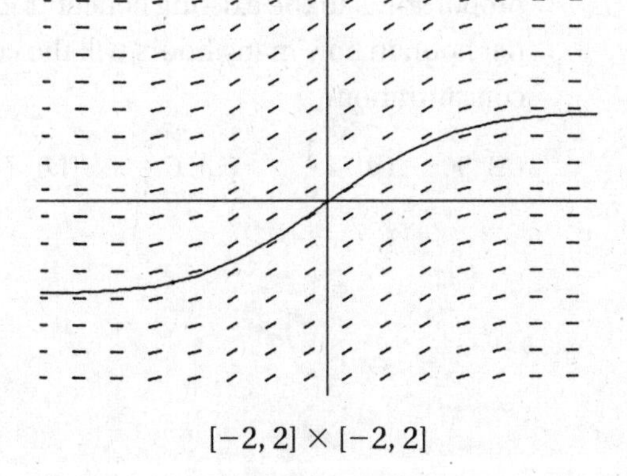

$[-2, 2] \times [-2, 2]$

BC ONLY

46. The graph displays logistic growth for a frog population F. Which differential equation could be the appropriate model?

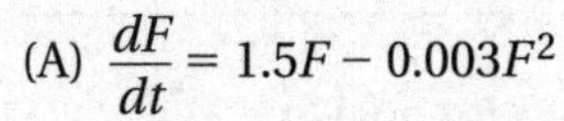

(A) $\frac{dF}{dt} = 1.5F - 0.003F^2$

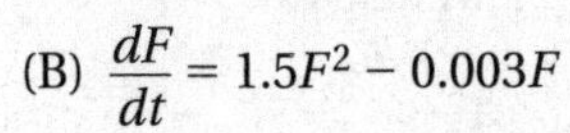

(B) $\frac{dF}{dt} = 1.5F^2 - 0.003F$

(C) $\frac{dF}{dt} = 3F - 0.003F^2$

(D) $\frac{dF}{dt} = 3F^2 - 0.003F$

(E) $\frac{dF}{dt} = 0.003F^2 - 3F$

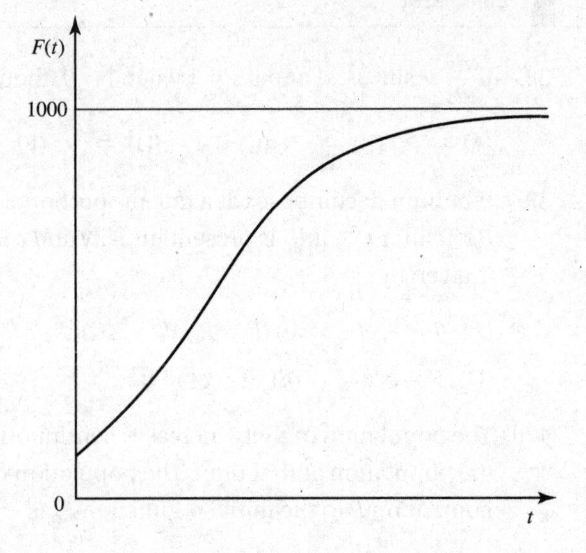

47. The table shows selected values of the derivative for a differentiable function f.

x	2	3	4	5	6	7
$f'(x)$	2.0	2.5	1.0	−0.5	−1.5	0.5

Given that $f(3) = 100$, use Euler's method with a step size of 2 to estimate $f(7)$.

(A) 101.5 (B) 102.5 (C) 103 (D) 104 (E) 104.5

48. A cup of coffee at temperature 180°F is placed on a table in a room at 68°F. The d.e. for its temperature at time t is $\frac{dy}{dt} = -0.11(y - 68)$; $y(0) = 180$. After 10 min the temperature (in °F) of the coffee is

(A) 96 (B) 100 (C) 105 (D) 110 (E) 115

49. Approximately how long does it take the temperature of the coffee in Question 48 to drop to 75°F?

(A) 10 min (B) 15 min (C) 18 min (D) 20 min (E) 25 min

50. The concentration of a medication injected into the bloodstream drops at a rate proportional to the existing concentration. If the factor of proportionality is 30% per hour, in how many hours will the concentration be one-tenth of the initial concentration?

(A) 3 (B) $4\frac{1}{3}$ (C) $6\frac{2}{3}$ (D) $7\frac{2}{3}$ (E) $13\frac{1}{2}$

BC ONLY

51. Which of the following statements characterize(s) the logistic growth of a population whose limiting value is L?

I. The rate of growth increases at first.

II. The growth rate attains a maximum when the population equals $\frac{L}{2}$.

III. The growth rate approaches 0 as the population approaches L.

(A) I only (B) II only (C) I and II only

(D) II and III only (E) I, II, and III

52. Which of the following d.e.'s is not logistic?

(A) $P' = P - P^2$ (B) $\frac{dy}{dt} = 0.01y(100 - y)$

(C) $\frac{dx}{dt} = 0.8x - 0.004x^2$ (D) $\frac{dR}{dt} = 0.16(350 - R)$

(E) $f'(t) = kf(t) \cdot [A - f(t)]$ (where k and A are constants)

53. Suppose $P(t)$ denotes the size of an animal population at time t and its growth is described by the d.e. $\frac{dP}{dt} = 0.002P(1000 - P)$. The population is growing fastest

(A) initially (B) when $P = 500$ (C) when $P = 1000$

(D) when $\frac{dP}{dt} = 0$ (E) when $\frac{d^2P}{dt^2} > 0$

54. According to Newton's law of cooling, the temperature of an object decreases at a rate proportional to the difference between its temperature and that of the surrounding air. Suppose a corpse at a temperature of 32°C arrives at a mortuary where the temperature is kept at 10°C. Then the differential equation satisfied by the temperature T of the corpse t hr later is

(A) $\frac{dT}{dt} = -k(T - 10)$ (B) $\frac{dT}{dt} = k(T - 32)$ (C) $\frac{dT}{dt} = 32e^{-kt}$

(D) $\frac{dT}{dt} = -kT(T - 10)$ (E) $\frac{dT}{dt} = kT(T - 32)$

55. If the corpse in Question 54 cools to 27°C in 1 hr, then its temperature (in °C) is given by the equation

(A) $T = 22e^{0.205t}$ (B) $T = 10e^{1.163t}$ (C) $T = 10 + 22e^{-0.258t}$

(D) $T = 32e^{-0.169t}$ (E) $T = 32 - 10e^{-0.093t}$

Answer Key

1. **C**	15. **B**	29. **E**	43. **C**
2. **B**	16. **E**	30. **A**	44. **C**
3. **D**	17. **E**	31. **C**	45. **A**
4. **B**	18. **A**	32. **B**	46. **C**
5. **C**	19. **C**	33. **A**	47. **D**
6. **E**	20. **E**	34. **D**	48. **C**
7. **B**	21. **E**	35. **D**	49. **E**
8. **B**	22. **D**	36. **A**	50. **D**
9. **A**	23. **B**	37. **E**	51. **E**
10. **E**	24. **E**	38. **D**	52. **D**
11. **A**	25. **B**	39. **B**	53. **B**
12. **C**	26. **D**	40. **D**	54. **A**
13. **E**	27. **C**	41. **A**	55. **C**
14. **C**	28. **A**	42. **A**	

Answers Explained

1. **(C)** $v(t) = 2t^2 - t + C$; $v(1) = 3$; so $C = 2$.

2. **(B)** If $a(t) = 20t^3 - 6t$, then

$$v(t) = 5t^4 - 3t^2 + C_1,$$

$$s(t) = t^5 - t^3 + C_1t + C_2,$$

Since

$$s(-1) = -1 + 1 - C_1 + C_2 = 2$$

and

$$s(1) = 1 - 1 + C_1 + C_2 = 4,$$

therefore

$$2C_2 = 6,\ C_2 = 3,$$
$$C_1 = 1.$$

So

$$v(t) = 5t^4 - 3t^2 + 1.$$

3. **(D)** From Answer 2, $s(t) = t^5 - t^3 + t + 3$, so $s(0) = C_2 = 3$.

4. **(B)** Since $a(t) = -32$, $v(t) = -32t + 40$, and the height of the stone $s(t) = -16t^2 + 40t + C$. When the stone hits the ground, 4 sec later, $s(t) = 0$, so

$$0 = -16(16) + 40(4) + C,$$
$$C = 96 \text{ ft}.$$

5. **(C)** From Answer 4

$$s(t) = -16t^2 + 40t + 96.$$

Then

$$s'(t) = -32t + 40,$$

which is zero if $t = 5/4$, and that yields maximum height, since $s''(t) = -32$.

6. **(E)** The velocity $v(t)$ of the car is linear, since its acceleration is constant and

$$a(t) = \frac{dv}{dt} = \frac{(60-0)\text{ mph}}{10\text{ sec}} = \frac{88\text{ ft/sec}}{10\text{ sec}} = 8.8\text{ ft/sec}^2$$

$$v(t) = 8.8t + C_1 \quad \text{and} \quad v(0) = 0, \quad \text{so } C_1 = 0;$$

$$s(t) = 4.4t^2 + C_2 \quad \text{and} \quad s(0) = 0, \quad \text{so } C_2 = 0;$$

$$s(10) = 4.4(10^2) = 440\text{ ft.}$$

7. **(B)** Since $v = 100 - 20t$, $s = 100t - 10t^2 + C$ with $s(0) = 0$. So $s(1) = 100 - 10 = 90$ ft.

8. **(B)** Since $v = -32t + v_0$ and $s = -16t^2 + v_0t$, we solve simultaneously:

$$0 = -32t + v_0,$$
$$100 = -16t^2 + v_0t.$$

These yield $t = 5/2$ and $v_0 = 80$ ft/sec.

9. **(A)** The odometer measures the total trip distance from time $t = a$ to $t = b$ (whether the car moves forward or backward or reverses its direction one or more times from $t = a$ to $t = b$). This total distance is given exactly by $\int_a^b |v(t)|\, dt$.

10. **(E)** (A), (B), (C), and (D) are all true. (E) is false: see Answer 9.

11. **(A)** Integrating yields $\frac{y^2}{2} = \frac{x^2}{2} + C$ or $y^2 = x^2 + 2C$ or $y = \pm\sqrt{x^2 + C'}$, where we have replaced the arbitrary constant $2C$ by C'.

12. **(C)** For initial point $(-2,1)$, $x^2 - y^2 = 3$. Rewriting the d.e. $y\,dy = x\,dx$ as $\frac{dy}{dx} = \frac{x}{y}$ reveals that the derivative does not exist when $y = 0$, which occurs at $x = \pm\sqrt{3}$. Since the particular solution must be differentiable in an interval containing $x = -2$, the domain is $x < -\sqrt{3}$.

13. **(E)** We separate variables. $\int \frac{dy}{y} = \frac{1}{2}\int x^{-\frac{1}{2}}\,dx$, so $\ln|y| = \sqrt{x} + c$. The initial point yields $\ln 1 = \sqrt{4} + c$; hence $c = -2$. With $y > 0$, the particular solution is $\ln y = \sqrt{x} - 2$, or $y = e^{\sqrt{x}-2}$.

14. **(C)** We separate variables. $\int e^{-y}dy = \int dx$, so $-e^{-y} = x + c$. The particular solution is $-e^{-y} = x - 2$.

15. **(B)** The general solution is $y = \frac{1}{2}\int (9 + x^2)^{-\frac{1}{2}}(2x\,dx) = \sqrt{9 + x^2} + C$; $y = 5$ when $x = 4$ yields $C = 0$.

16. **(E)** Since $\int \frac{dy}{y} = \int \frac{dx}{x}$, it follows that

$$\ln y = \ln x + C \quad \text{or} \quad \ln y = \ln x + \ln k;$$

so $y = kx$.

17. **(E)** $\int \frac{dy}{y} = \int dx$ yields $\ln|y| = x + c$; hence the general solution is $y = ke^x$, $k \neq 0$.

18. **(A)** We rewrite and separate variables, getting $y\frac{dy}{dx} = x$. The general solution is

$y^2 = x^2 + C$ or $f(x) = \pm\sqrt{x^2 + C}$.

19. **(C)** We are given that $\frac{dy}{dx} = \frac{3y}{x}$. The general solution is $\ln|y| = 3\ln|x| + C$.

Thus, $|y| = c\,|x^3|$; $y = \pm\, c\,x^3$. Since $y = 1$ when $x = 1$, we get $c = 1$.

20. **(E)** The d.e. $\frac{dy}{dx} = \frac{3y}{x}$ reveals that the derivative does not exist when $x = 0$.

Since the particular solution must be differentiable in an interval containing initial value $x = 1$, the domain is $x > 0$.

21. **(E)** The general solution is $y = k\ln|x| + C$, and the particular solution is $y = 2\ln|x| + 2$.

22. **(D)** We carefully(!) draw a curve for a solution to the d.e. represented by the slope field. It will be the graph of a member of the family $y = \sin x + C$. At the right we have superimposed the graph of the particular solution $y = \sin x - 0.5$.

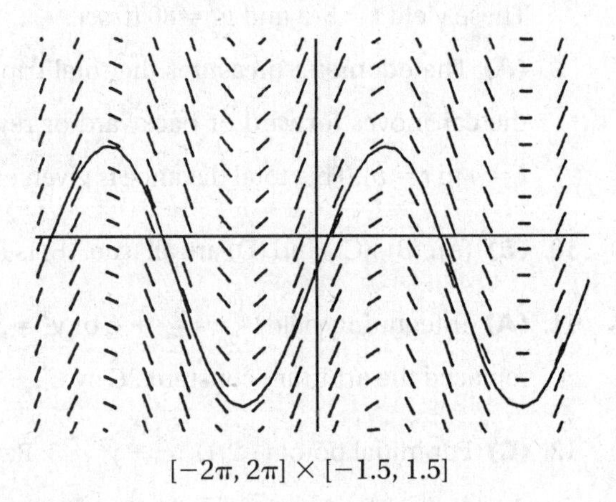

$[-2\pi, 2\pi] \times [-1.5, 1.5]$

23. **(B)**

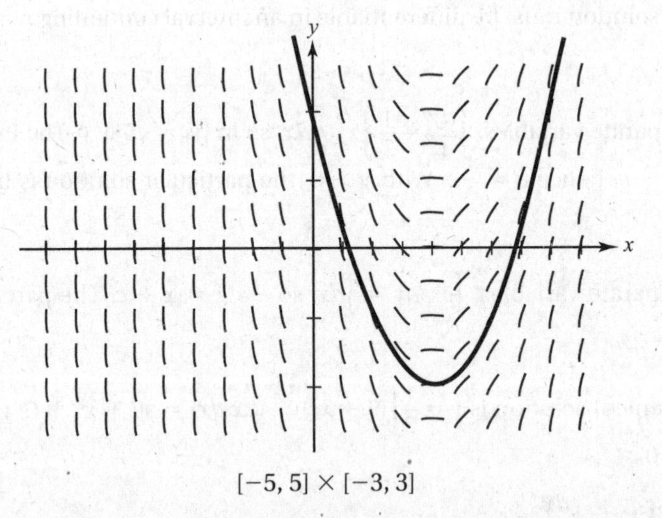

$[-5, 5] \times [-3, 3]$

It's easy to see that the answer must be choice (A), (B), or (C), because the slope field depends only on x: all the slope segments for a given x are parallel. Also, the solution curves in the slope field are all concave up, as they are only for choices (A) and (B). Finally, the solution curves all have a minimum at $x = 2$, which is true only for differential equation (B).

24. **(E)** The *solution curve* is $y = \tan x$, which we can obtain from the differential equation $y' = 1 + y^2$ with the condition $y(0) = 0$ as follows:

$$\frac{dy}{1+y^2} = dx, \qquad \tan^{-1} y = x, \qquad y = \tan x + C.$$

Since $y(0) = 0$, $C = 0$. Verify that (A) through (D) are incorrect.

NOTE: In matching slope fields and differential equations in Questions 25–29, keep in mind that if the slope segments along a vertical line are all parallel, signifying equal slopes for a fixed x, then the differential equation can be written as $y' = f(x)$. Replace "vertical" by "horizontal" and "x" by "y" in the preceding sentence to obtain a differential equation of the form $y' = g(y)$.

25. **(B)** The slope field for $y' = y$ must by II; it is the only one whose slopes are equal along a horizontal line.

26. **(D)** Of the four remaining slope fields, IV is the only one whose slopes are not equal along either a vertical or a horizontal line (the segments are *not* parallel). Its d.e. therefore cannot be either of type $y' = f(x)$ or $y' = g(y)$. The d.e. must be implicitly defined—that is, of the form $y' = F(x, y)$. So the answer here is IV.

27. **(C)** The remaining slope fields, I, III, and V, all have d.e.'s of the type $y' = f(x)$. The curves "lurking" in III are trigonometric curves—not so in I and V.

28. **(A)** Given $y' = 2x$, we immediately obtain the general solution, a family of parabolas, $y = x^2 + C$. (Trace the parabola in I through (0, 0), for example.)

29. **(E)** V is the only slope field still unassigned! Furthermore, the slopes "match" e^{-x^2}: the slopes are equal "about" the y-axis; slopes are very small when x is close to -2 and 2; and e^{-x^2} is a maximum at $x = 0$.

30. **(A)** From Answer 25, we know that the d.e. for slope field II is $y' = y$. The general solution is $y = ce^x$. For a solution curve to pass through point $(0, -1)$, we have $-1 = ce^0$ and $c = -1$.

31. **(C)** Euler's method for $y' = x$, starting at (1, 5), with $\Delta x = 0.1$, yields

x	y	(SLOPE)* $\cdot\ \Delta x = \Delta y$
1	5	$1 \cdot (0.1) = 0.1$
1.1	5.1	$(1.1) \cdot (0.1) = 0.11$
1.2	5.21	

*The slope is x.

32. **(B)** We want to compare the true value of $y(1.2)$ to the estimated value of 5.21 obtained using Euler's method in Solution 31. Solving the d.e. $\frac{dy}{dx} = x$ yields $y = \frac{x^2}{2} + C$, and initial condition $y(1) = 5$ means that $5 = \frac{1^2}{2} + C$, or $C = 4.5$. Hence $y(1.2) = \frac{1.2^2}{2} + 4.5 = 5.22$. The error is $5.22 - 5.21 = 0.01$.

33. **(A)** Slopes depend only on the value of y, and the slope field suggests that $y' = 0$ whenever $y = 0$ or $y = -2$.

34. **(D)** The slope field suggests that the solution function increases (or decreases) without bound as x increases, but approaches $y = 1$ as a horizontal asymptote as x decreases.

35. **(D)** The slope segments are not parallel in either the x or y direction, so the d.e. must include both x and y in the definition; this excludes (B) and (E). Next, (A) would result in all positive slopes. This is not the case, so (A) is eliminated. Finally, (C) will have a zero slope at the point (1, 1), and (D) will have a zero slope at the point (1, −1), thus, (D) will create this slope field.

36. **(A)** The slope segments are not parallel in either the x or y direction, so the d.e. must include both x and y in the definition; this excludes (C) and (D). (B) will have zero slopes along the x-axis, so we can eliminate (B). Finally, both (A) and (E) will have zero slopes along the y-axis, (E) will also have zero slopes at (1, −1) and (−1, −1), eliminating (E). (Note: (A) will have zero slopes at (1, −1) and (−1, 1).)

37. **(E)** The slope segments are parallel horizontally meaning that the slopes don't change as x varies; therefore, the d.e. is defined by the y-coordinate only. This excludes (A), (B), and (C). Choice (D) will have zero slopes along the x-axis, whereas (E) will never have zero slopes; thus (E) will create this slope field.

38. **(D)** We separate variables to get $\csc^2\left(\frac{\pi}{2}s\right)ds = dt$. We integrate:
$-\frac{2}{\pi}\cot\left(\frac{\pi}{2}s\right) = t + C$. With $t = 0$ and $s = 1$, $C = 0$. When $s = \frac{3}{2}$, we get $-\frac{2}{\pi}\cot\frac{3\pi}{4} = t$.

39. **(B)** Since $\frac{dR}{dt} = cR$, $\frac{dR}{R} = c\,dt$, and $\ln R = ct + C$. When $t = 0$, $R = R_0$; so $\ln R_0 = C$ or $\ln R = ct + \ln R_0$. Thus

$$\ln R - \ln R_0 = ct; \ln\frac{R}{R_0} = ct \quad \text{or} \quad \frac{R}{R_0} = e^{ct}.$$

40. **(D)** The question gives rise to the differential equation $\frac{dP}{dt} = kP$, where $P = 2P_0$ when $t = 50$. We seek $\frac{P}{P_0}$ for $t = 75$. We get $\ln\frac{P}{P_0} = kt$ with $\ln 2 = 50k$; then

$$\ln\frac{P}{P_0} = \frac{t}{50}\ln 2 \quad \text{or} \quad \frac{P}{P_0} = 2^{t/50}.$$

41. **(A)** We let S equal the amount present at time t; using $S = 40$ when $t = 0$ yields $\ln\frac{S}{40} = kt$. Since, when $t = 2$, $S = 10$, we get

$$k = \frac{1}{2}\ln\frac{1}{4} \quad \text{or} \quad \ln\frac{1}{2} \quad \text{or} \quad -\ln 2.$$

42. **(A)** We replace $g(x)$ by y and then solve the equation $\frac{dy}{dx} = \pm\sqrt{y}$. We use the constraints given to find the particular solution $2\sqrt{y} = x$ or $2\sqrt{g(x)} = x$.

43. **(C)** The general solution of $\frac{dy}{dx} = y$, or $\frac{dy}{y} = dx$ (with $y > 0$) is $\ln y = x + C$ or $y = ce^x$.

For a solution to pass through (2, 3), we have $3 = ce^2$ and $c = 3/e^2 \approx 0.406$.

44. **(C)** At a point of intersection, $y' = x + y$ and $x + y = 0$. So $y' = 0$, which implies that y has a critical point at the intersection. Since $y'' = 1 + y' = 1 + (x + y) = 1 + 0 = 1$, $y'' > 0$ and the function has a local minimum at the point of intersection. [See Figure N9–5, p. 357, showing the slope field for $y' = x + y$ and the curve $y = e^x - x - 1$ that has a local minimum at (0, 0).]

45. **(A)** Although there is no elementary function (one made up of polynomial, trigonometric, or exponential functions or their inverses) that is an antiderivative of $F'(x) = e^{-x^2}$, we know from the FTC, since $F(0) = 0$, that

$$F(x) = \int_0^x e^{-t^2}\, dt.$$

To approximate $\lim_{x\to\infty} F(x)$, use your graphing calculator.

For upper limits of integration $x = 50$ and $x = 60$, answers are identical to 10 decimal places. Rounding to three decimal places yields 0.886.

46. **(C)** Logistic growth is modeled by equations of the form $\frac{dP}{dt} = kP(L - P)$, where L is the upper limit. The graph shows $L = 1000$, so the differential equation must be $\frac{dF}{dt} = kF(1000 - F)\ 1000kF - kF^2$. Only equation C is of this form ($k = 0.003$).

47. **(D)** We start with $x = 3$ and $y = 100$. At $x = 3$, $\Delta y \approx \frac{dy}{dx} \cdot \Delta x = (2.5)(2) = 5$, moving us to $x = 3 + 2 = 5$ and $y = 100 + 5 = 105$. From there $\Delta y \approx \frac{dy}{dx} \cdot \Delta x = (-0.5)(2) = -1$, so when $x = 5 + 2 = 7$ we estimate $y = 105 + (-1) = 104$.

48. **(C)** We separate the variables in the given d.e., then solve:

$$\frac{dy}{y - 68} = -0.11dt,$$

$$\ln(y - 68) = -0.11t + c.$$

Since $y(0) = 180$, $\ln 112 = c$. Then

$$\ln\frac{y - 68}{112} = -0.11t,$$

$$y = 68 + 112e^{-0.11t}.$$

When $t = 10$, $y = 68 + 112e^{-1.1} \simeq 105$°F.

49. **(E)** The solution of the d.e. in Question 48, where y is the temperature of the coffee at time t, is

$$y = 68 + 112e^{-0.11t}.$$

We find t when $y = 75$°F:

$$75 = 68 + 112e^{-0.11t},$$

$$\frac{7}{112} = e^{-0.11t},$$

$$\frac{\ln 7 - \ln 112}{-0.11} = t \simeq 25 \text{ min}.$$

50. **(D)** If Q is the concentration at time t, then $\frac{dQ}{dt} = -0.30Q$. We separate variables and integrate:

$$\frac{dQ}{Q} = -0.30\, dt \quad\rightarrow\quad \ln Q = -0.30t + C.$$

We let $Q(0) = Q_0$. Then

$$\ln Q = -0.30t + \ln Q_0 \quad\rightarrow\quad \ln\frac{Q}{Q_0} = -0.30t \quad\rightarrow\quad \frac{Q}{Q_0}\ e^{-0.30t}.$$

We now find t when $Q = 0.1Q_0$:

$$0.1 = e^{-0.30t},$$

$$t = \frac{\ln 0.1}{-0.3} \simeq 7\frac{2}{3} \text{ hr.}$$

51. **(E)** See pages 370–371 for the characteristics of the logistic model.

52. **(D)** (A), (B), (C), and (E) are all of the form $y' = ky(a - y)$.

53. **(B)** The rate of growth, $\frac{dP}{dt}$, is greatest when its derivative is 0 and the curve of $y' = \frac{dP}{dt}$ is concave down. Since

$$\frac{dP}{dt} = 2P - 0.002P^2,$$

therefore

$$\frac{d^2P}{dt^2} = 2 - 0.004P,$$

which is equal to 0 if $y'' = \frac{2}{0.004}$, or 500, animals. The curve of y' is concave down for all P, since

$$\frac{d}{dt}\left(\frac{d^2P}{dt^2}\right) = -0.004,$$

so $P = 500$ is the maximum population.

54. **(A)** The description of temperature change here is an example of Case II (page 367): the rate of change is proportional to the amount or magnitude of the quantity present (i.e., the temperature of the corpse) minus a fixed constant (the temperature of the mortuary).

55. **(C)** Since (A) is the correct answer to Question 54, we solve the d.e. in (A) given the initial condition $T(0) = 32$:

$$\frac{dT}{T - 10} = -k\,dt,$$

$$\ln(T - 10) = -kt + C.$$

Using $T(0) = 32$, we get $\ln(22) = C$, so

$$\ln(T - 10) = -kt + \ln 22,$$

$$T - 10 = 22e^{-kt}.$$

To find k, we use the given information that $T(1) = 27$:

$$27 - 10 = 22e^{-k},$$

$$\frac{17}{22} = e^{-k},$$

$$k = 0.258.$$

Therefore $T = 10 + 22e^{-0.258t}$.

Sequences and Series

10

BC ONLY

CONCEPTS AND SKILLS

In this chapter, we review infinite series for BC Calculus students. Topics include

- **tests for determining convergence or divergence,**
- **functions defined as power series,**
- **Maclaurin and Taylor series,**
- **and estimates of errors.**

A. SEQUENCES OF REAL NUMBERS‡

An *infinite sequence* is a function whose domain is the set of positive integers, and is often denoted simply by a_n. The sequence defined, for example, by $a_n = \frac{1}{n}$ is the set of numbers $1, \frac{1}{2}, \frac{1}{3}, \ldots, \frac{1}{n}, \ldots$. The elements in this set are called the *terms* of the sequence, and the *nth* or *general* term of this sequence is $\frac{1}{n}$.

A sequence a_n *converges* to a finite number L if $\lim_{n\to\infty} a_n = L$. If a_n does not have a (finite) limit, we say the sequence is *divergent.*

Example 1

Does the sequence $a_n = \frac{1}{n}$ converge or diverge?

SOLUTION: $\lim_{n\to\infty} \frac{1}{n} = 0$; hence the sequence converges to 0.

Example 2

Does the sequence $a_n = \frac{3n^4 + 5}{4n^4 - 7n^2 + 9}$ converge or diverge?

SOLUTION: $\lim_{n\to\infty} \frac{3n^4 + 5}{4n^4 - 7n^2 + 9} = \frac{3}{4}$; hence the sequence converges to $\frac{3}{4}$.

‡Topic will not be tested on the AP examination, but some understanding of the notation and terminology is helpful.

BC ONLY

➡ Example 3

Does the sequence $a_n = 1 + \frac{(-1)^n}{n}$ converge or diverge?

SOLUTION: $\lim_{n\to\infty} 1 + \frac{(-1)^n}{n} = 1$; hence the sequence converges to 1. Note that the terms in the sequence $0, \frac{3}{2}, \frac{2}{3}, \frac{5}{4}, \frac{4}{5}, \frac{7}{6}, \ldots$ are alternately smaller and larger than 1. We say this sequence converges to 1 *by oscillation.*

➡ Example 4

Does the sequence $a_n = \frac{n^2 - 1}{n}$ converge or diverge?

SOLUTION: Since $\lim_{n\to\infty} \frac{n^2 - 1}{n} = \infty$, the sequence diverges (to infinity).

➡ Example 5

Does the sequence $a_n = \sin n$ converge or diverge?

SOLUTION: Because $\lim_{n\to\infty} \sin n$ does not exist, the sequence diverges. However, note that it does not diverge to infinity.

➡ Example 6

Does the sequence $a_n = (-1)^{n+1}$ converge or diverge?

SOLUTION: Because $\lim_{n\to\infty} (-1)^{n+1}$ does not exist, the sequence diverges. Note that the sequence $1, -1, 1, -1, \ldots$ diverges because it oscillates.

B. INFINITE SERIES

B1. Definitions

Infinite series

If a_n is a sequence of real numbers, then an *infinite series* is an expression of the form

$$\sum_{k=1}^{\infty} a_k = a_1 + a_2 + a_3 + \cdots + a_n + \cdots. \quad (1)$$

General term

The elements in the sum are called *terms*; a_n is the *n*th or *general term* of the series.

➡ Example 7

p-series

A series of the form $\sum_{k=1}^{\infty} \frac{1}{k^p}$ is called a *p-series.*

The *p*-series for $p = 2$ is $\sum_{k=1}^{\infty} \frac{1}{k^2} = \frac{1}{1^2} + \frac{1}{2^2} + \frac{1}{3^2} + \cdots + \frac{1}{n^2} + \cdots$.

BC ONLY

➡ Example 8

Harmonic series

The *p*-series with $p = 1$ is called the *harmonic series*:

$$\sum_{k=1}^{\infty} \frac{1}{k} = \frac{1}{1} + \frac{1}{2} + \frac{1}{3} + \cdots + \frac{1}{n} + \cdots.$$

➡ Example 9

Geometric series

A *geometric series* has a first term, *a*, and common ratio of terms, *r*:

$$\sum_{k=1}^{\infty} ar^{k-1} = a + ar + ar^2 + \cdots + ar^{n-1} + \cdots.$$

If there is a finite number *S* such that

$$\lim_{n\to\infty} \sum_{k=1}^{\infty} a_k = S,$$

Convergent

then we say that infinite series is *convergent*, or *converges to S*, or *has the sum S*, and we write, in this case,

$$\sum_{k=1}^{\infty} a_k = S.$$

When there is no source of confusion, the infinite series (1) may be indicated simply by

$$\sum a_k \quad \text{or} \quad \sum a_n.$$

➡ Example 10

Show that the geometric series $1 + \frac{1}{2} + \frac{1}{4} + \cdots + \frac{1}{2^n} + \cdots$ converges to 2.

SOLUTION: Let *S* represent the sum of the series; then:

$$S = \lim_{n\to\infty}\left(1 + \frac{1}{2} + \frac{1}{4} + \frac{1}{8} + \cdots + \frac{1}{2^n}\right);$$

$$\frac{1}{2}S = \lim_{n\to\infty}\left(\frac{1}{2} + \frac{1}{4} + \frac{1}{8} + \cdots + \frac{1}{2^{n+1}}\right).$$

Subtraction yields

$$\frac{1}{2}S = \lim_{n\to\infty}\left(1 - \frac{1}{2^{n+1}}\right).$$

Hence, $S = 2$.

BC ONLY

Example 11

Show that the *harmonic series* $1 + \frac{1}{2} + \frac{1}{3} + \frac{1}{4} + \cdots + \frac{1}{n} + \cdots$ diverges.

SOLUTION: The terms in the series can be grouped as follows:

$$1 + \frac{1}{2} + \left(\frac{1}{3} + \frac{1}{4}\right) + \left(\frac{1}{5} + \frac{1}{6} + \frac{1}{7} + \frac{1}{8}\right) + \left(\frac{1}{9} + \frac{1}{10} + \cdots + \frac{1}{16}\right) + \left(\frac{1}{17} + \cdots + \frac{1}{32}\right) + \cdots.$$

This sum clearly exceeds

$$1 + \frac{1}{2} + 2\left(\frac{1}{4}\right) + 4\left(\frac{1}{8}\right) + 8\left(\frac{1}{16}\right) + 16\left(\frac{1}{32}\right) + \cdots,$$

which equals

$$1 + \frac{1}{2} + \frac{1}{2} + \frac{1}{2} + \frac{1}{2} + \frac{1}{2} + \cdots.$$

Since that sum is not bounded, it follows that $\sum \frac{1}{n}$ diverges to ∞.

B2. Theorems About Convergence or Divergence of Infinite Series

The following theorems are important.

THEOREM 2a. If $\sum a_k$ converges, then $\lim_{n\to\infty} a_n = 0$.

This provides a convenient and useful test for divergence, since it is equivalent to the statement: If a_n does not approach zero, then the series $\sum a_k$ diverges. Note, however, particularly that the converse of Theorem 2a is *not* true. The condition that a_n approach zero is *necessary but not sufficient* for the convergence of the series. The harmonic series $\sum \frac{1}{n}$ is an excellent example of a series whose nth term goes to zero but that diverges (see Example 11 above). The series $\sum \frac{n}{n+1}$ diverges because $\lim_{n\to\infty} a_n = 1$, not zero; the series $\sum \frac{n}{n^2+1}$ does not converge (as will be shown shortly) even though $\lim_{n\to\infty} a_n = 0$.

THEOREM 2b. A finite number of terms may be added to or deleted from a series without affecting its convergence or divergence; thus

$$\sum_{k=1}^{\infty} a_k \quad \text{and} \quad \sum_{k=m}^{\infty} a_k$$

(where m is any positive integer) both converge or both diverge. (Note that the sums most likely will differ.)

THEOREM 2c. The terms of a series may be multiplied by a nonzero constant without affecting the convergence or divergence; thus

$$\sum_{k=1}^{\infty} a_k \quad \text{and} \quad \sum_{k=1}^{\infty} ca_k \ (c \neq 0)$$

both converge or both diverge. (Again, the sums will usually differ.)

THEOREM 2d. If $\sum a_n$ and $\sum b_n$ both converge, so does $\sum (a_n + b_n)$.

THEOREM 2e. If the terms of a convergent series are regrouped, the new series converges.

B3. Tests for Convergence of Infinite Series

BC ONLY

THE *n*th TERM TEST

*n*th Term Test

If $\lim_{n\to\infty} a_n \neq 0$, then $\sum a_n$ diverges.

NOTE: When working with series, it's a good idea to start by checking the *n*th Term Test. If the terms don't approach 0, the series cannot converge. This is often the quickest and easiest way to identify a divergent series.

(Because this is the contrapositive of Theorem 2a on page 394, it's always true. *But beware of the converse! Seeing that the terms do approach 0 does not guarantee that the series must converge.* It just means that you need to try other tests.)

Example 12

Does $\sum \frac{n}{2n+1}$ converge or diverge?

SOLUTION: Since $\lim_{n\to\infty} \frac{n}{2n+1} = \frac{1}{2} \neq 0$, the series $\sum \frac{n}{2n+1}$ diverges by the *n*th Term Test.

THE GEOMETRIC SERIES TEST

Geometric Series Test

A geometric series $\sum ar^n$ converges if and only if $|r| < 1$.

If $|r| < 1$, the sum of the series is $\frac{a}{1-r}$.

The series cannot converge unless it passes the *n*th Term Test; $\lim_{n\to\infty} ar^n = 0$ only if $|r| < 1$. As noted earlier, this is a necessary condition for convergence, but may not be sufficient. We now examine the sum using the same technique we employed in Example 10 on page 393:

$$
\begin{aligned}
S &= \lim_{n\to\infty}\left(a + ar + ar^2 + ar^3 + \cdots + ar^n\right);\\
rS &= \lim_{n\to\infty}\left(ar + ar^2 + ar^3 + \cdots + ar^n + ar^{n+1}\right);\\
(1-r)S &= \lim_{n\to\infty}\left(a - ar^{n+1}\right)\\
&= a - \lim_{n\to\infty} ar^{n+1} \left(\text{and remember: } |r| < 1\right)\\
&= a;\\
S &= \frac{a}{1-r}.
\end{aligned}
$$

Example 13

Does $0.3 + 0.03 + 0.003 + \cdots$ converge or diverge?

SOLUTION: The series $0.3 + 0.03 + 0.003 + \cdots$ is geometric with $a = 0.3$ and $r = 0.1$. Since $|r| < 1$, the series converges, and its sum is

$$S = \frac{a}{1-r} = \frac{0.3}{1-0.1} = \frac{0.3}{0.9} = \frac{1}{3}.$$

NOTE: $\frac{1}{3} = 0.333\ldots$, which is the given series.

BC ONLY

B4. Tests for Convergence of Nonnegative Series

The series $\sum a_n$ is called a *nonnegative series* if $a_n \geq 0$ for all n.

Integral Test

THE INTEGRAL TEST

Let $\sum a_n$ be a nonnegative series. If $f(x)$ is a continuous, positive, decreasing function and $f(n) = a_n$, then $\sum a_n$ converges if and only if the improper integral $\int_1^{\infty} f(x)\,dx$ converges.

Example 14

Does $\sum \frac{n}{n^2+1}$ converge?

SOLUTION: The associated improper integral is

$$\int_1^{\infty} \frac{x\,dx}{x^2+1},$$

which equals

$$\lim_{b\to\infty} \frac{1}{2}\ln\left(x^2+1\right)\Big|_1^b = \infty.$$

The improper integral and the infinite series both diverge.

Example 15

Test the series $\sum \frac{n}{e^n}$ for convergence.

SOLUTION:
$$\int_1^{\infty} \frac{x}{e^x}dx = \lim_{b\to\infty} \int_1^b xe^{-x}dx = \lim_{b\to\infty} - e^{-x}(1+x)\Big|_1^b$$
$$= -\lim_{b\to\infty}\left(\frac{1+b}{e^b} - \frac{2}{e}\right) = \frac{2}{e}$$

by an application of L'Hôpital's Rule. Thus $\sum \frac{n}{e^n}$ converges.

THE *p*-SERIES TEST

p-Series Test

A *p*-series $\sum_{n=1}^{\infty} \frac{1}{n^p}$ converges if $p > 1$, but diverges if $p \leq 1$.

This follows immediately from the Integral Test and the behavior of improper integrals of the form $\int_1^{\infty} \frac{1}{x^p}dx$.

Example 16

Does the series $1 + \frac{1}{2^3} + \frac{1}{3^3} + \cdots + \frac{1}{n^3} + \cdots$ converge or diverge?

SOLUTION: The series $1 + \frac{1}{2^3} + \frac{1}{3^3} + \cdots + \frac{1}{n^3} + \cdots$ is a *p*-series with $p = 3$; hence the series converges by the *p*-Series Test.

Example 17

Does the series $\sum \frac{1}{\sqrt{n}}$ converge or diverge?

SOLUTION: $\sum \frac{1}{\sqrt{n}}$ diverges, because it is a p-series with $p = \frac{1}{2}$.

BC ONLY

THE COMPARISON TEST

Comparison Test

We compare the general term of $\sum a_n$, the nonnegative series we are investigating, with the general term of a series, $\sum u_n$, known to converge or diverge.

(1) If $\sum u_n$ converges and $a_n \leqq u_n$, then $\sum a_n$ converges.

(2) If $\sum u_n$ diverges and $a_n \geqq u_n$, then $\sum a_n$ diverges.

Any known series can be used for comparison. Particularly useful are p-series, which converge if $p > 1$ but diverge if $p \leqq 1$, and geometric series, which converge if $|r| < 1$ but diverge if $|r| \geqq 1$.

Example 18

Does $\sum \frac{1}{1 + n^4}$ converge or diverge?

SOLUTION: Since $\frac{1}{1 + n^4} < \frac{1}{n^4}$ and the p-series $\sum \frac{1}{n^4}$ converges, $\sum \frac{1}{1 + n^4}$ converges by the Comparison Test.

Example 19

Does the series $\frac{1}{\sqrt{2}} + \frac{1}{\sqrt{5}} + \frac{1}{\sqrt{8}} + \cdots + \frac{1}{\sqrt{3n - 1}} + \cdots$ converge or diverge?

SOLUTION: $\frac{1}{\sqrt{2}} + \frac{1}{\sqrt{5}} + \frac{1}{\sqrt{8}} + \cdots + \frac{1}{\sqrt{3n - 1}} + \cdots$ diverges, since

$$\frac{1}{\sqrt{3n - 1}} > \frac{1}{\sqrt{3n}} = \frac{1}{\sqrt{3} \cdot n^{1/2}};$$

the latter is the general term of the divergent p-series $\sum \frac{c}{n^p}$, where $c = \frac{1}{\sqrt{3}}$ and $p = \frac{1}{2}$.

Remember in using the Comparison Test that you may either discard a finite number of terms or multiply each term by a nonzero constant without affecting the convergence of the series you are testing.

Example 20

Show that $\sum \frac{1}{n^n} = 1 + \frac{1}{2^2} + \frac{1}{3^3} + \cdots + \frac{1}{n^n} + \cdots$ converges.

SOLUTION: For $n > 2$, $\frac{1}{n^n} < \frac{1}{2^n}$ and $\sum \frac{1}{2^n}$ is a convergent geometric series with $r = \frac{1}{2}$.

BC ONLY

Limit Comparison Test

THE LIMIT COMPARISON TEST

Let $\sum a_n$ be a nonnegative series that we are investigating. Given $\sum b_n$, a nonnegative series known to be convergent or divergent:

(1) If $\lim_{n\to\infty} \frac{a_n}{b_n} = L$, where $0 < L < \infty$, then $\sum a_n$ and $\sum b_n$ both converge or diverge.

(2) If $\lim_{n\to\infty} \frac{a_n}{b_n} = 0$, and $\sum b_n$ converges, then $\sum a_n$ converges.

(3) If $\lim_{n\to\infty} \frac{a_n}{b_n} = \infty$, and $\sum b_n$ diverges, then $\sum a_n$ diverges.

Any known series can be used for comparison. Particularly useful are p-series, which converge if $p > 1$ but diverge if $p \le 1$, and geometric series, which converge if $|r| < 1$ but diverge if $|r| \ge 1$.

This test is useful when the direct comparisons required by the Comparison Test are difficult to establish or when the behavior of $\sum a_n$ is like that of $\sum b_n$, but the comparison of the individual terms is in the wrong direction necessary for the Comparison Test to be conclusive.

Example 21

Does $\sum \frac{1}{2n+1}$ converge or diverge?

SOLUTION: This series seems to be related to the divergent harmonic series, but $\frac{1}{2n+1} < \frac{1}{n}$, so the comparison fails. However, the Limit Comparison Test yields:

$$\lim_{n\to\infty} \frac{\frac{1}{2n+1}}{\frac{1}{n}} = \lim_{n\to\infty} \frac{n}{2n+1} = \frac{1}{2}.$$

Since $\sum \frac{1}{n}$ diverges, $\sum \frac{1}{2n+1}$ also diverges by the Limit Comparison Test.

Ratio Test

THE RATIO TEST

Let $\sum a_n$ be a nonnegative series, and let $\lim_{n\to\infty} \frac{a_{n+1}}{a_n} = L$, if it exists. Then $\sum a_n$ converges if $L < 1$ and diverges if $L > 1$.

If $L = 1$, this test is inconclusive; apply one of the other tests.

NOTE: It is good practice, when using the ratio test, to first write $\lim_{n\to\infty} \left|\frac{a_{n+1}}{a_n}\right|$; then, if it is known that the ratio is always nonnegative, you may rewrite the limit without the absolute value. However, when using the ratio test on a power series (see Examples 33–36), you must retain the absolute value throughout the limit process because it could be possible that $x < 0$.

Example 22

Does $\sum \frac{1}{n!}$, converge or diverge?

SOLUTION: $\lim_{n\to\infty} \frac{a_{n+1}}{a_n} = \lim_{n\to\infty} \frac{\frac{1}{(n+1)!}}{\frac{1}{n!}} = \lim_{n\to\infty} \frac{n!}{(n+1)!} = \lim_{n\to\infty} \frac{1}{n+1} = 0.$

Therefore this series converges by the Ratio Test.

BC ONLY

Example 23

Does $\sum \frac{n^n}{n!}$ converge or diverge?

SOLUTION: $$\frac{a_{n+1}}{a_n} = \frac{(n+1)^{n+1}}{(n+1)!} \cdot \frac{n!}{n^n} = \frac{(n+1)^n}{n^n}$$

and

$$\lim_{n\to\infty}\left(\frac{n+1}{n}\right)^n = \lim_{n\to\infty}\left(1+\frac{1}{n}\right)^n = e.$$

(See §E2, page 97.) Since $e > 1$, $\sum \frac{n^n}{n!}$ diverges by the Ratio Test.

Example 24

If the Ratio Test is applied to any p-series, $\sum \frac{1}{n^p}$, then

$$\frac{a_{n+1}}{a_n} = \frac{\frac{1}{(n+1)^p}}{\frac{1}{n^p}} = \left(\frac{n}{n+1}\right)^p \quad \text{and} \quad \lim_{n\to\infty}\left(\frac{n}{n+1}\right)^p = 1 \text{ for all } p.$$

But if $p > 1$ then $\sum \frac{1}{n^p}$ converges, while if $p \leqq 1$ then $\sum \frac{1}{n^p}$ diverges. This illustrates the failure of the Ratio Test to resolve the question of convergence when the limit of the ratio is 1.

Root Test

THE ROOT TEST

Let $\lim_{n\to\infty} \sqrt[n]{a_n} = L$, if it exists. Then $\sum a_n$ converges if $L < 1$ and diverges if $L > 1$.

If $L = 1$ this test is inconclusive; try one of the other tests.

The decision rule for this test is the same as that for the Ratio Test.

NOTE: The Root Test is not specifically tested on the AP Calculus Exam; however, we present it here because it may be helpful in determining convergence.

Example 25

The series $\sum \left(\frac{n}{2n+1}\right)^n$ converges by the Root Test, since

$$\lim_{n\to\infty} \sqrt[n]{\left(\frac{n}{2n+1}\right)^n} = \lim_{n\to\infty} \frac{n}{2n+1} = \frac{1}{2}.$$

B5. Alternating Series and Absolute Convergence

Alternating Series

Any test that can be applied to a nonnegative series can be used for a series all of whose terms are negative. We consider here only one type of series with mixed signs, the so-called *alternating series.* This has the form:

$$\sum_{k=1}^{\infty} (-1)^{k+1} a_k = a_1 - a_2 + a_3 - a_4 + \cdots + (-1)^{k+1} a_k + \cdots,$$

where $a_k > 0$. The series

$$1 - \frac{1}{2} + \frac{1}{3} - \frac{1}{4} + \cdots + (-1)^{n+1} \cdot \frac{1}{n} + \cdots$$

is the *alternating harmonic* series.

BC ONLY

Alternating Series Test

THE ALTERNATING SERIES TEST

An alternating series converges if:

(1) $a_{n+1} < a_n$ for all n, and
(2) $\lim_{n\to\infty} a_n = 0$.

➡ Example 26

Does the series $\sum \frac{(-1)^{n+1}}{n}$ converge or diverge?

SOLUTION: The alternating harmonic series $\sum \frac{(-1)^{n+1}}{n}$ converges, since

(1) $\frac{1}{n+1} < \frac{1}{n}$ for all n and
(2) $\lim_{n\to\infty} \frac{1}{n} = 0$.

➡ Example 27

Does the series $\frac{1}{2} - \frac{2}{3} + \frac{3}{4} - \cdots$ converge or diverge?

SOLUTION: The series $\frac{1}{2} - \frac{2}{3} + \frac{3}{4} - \cdots$ diverges, since we see that $\lim_{n\to\infty} a_n = \lim_{n\to\infty} \frac{n}{n+1}$ is 1, not 0. (By the nth Term Test, page 395, if a_n does not approach 0, then $\sum a_n$ does not converge.)

ABSOLUTE CONVERGENCE AND CONDITIONAL CONVERGENCE

Absolute convergence

A series with mixed signs is said to *converge absolutely* (or to be *absolutely convergent*) if the series obtained by taking the absolute values of its terms converges; that is, $\sum a_n$ converges absolutely if $\sum |a_n| = |a_1| + |a_2| + \cdots + |a_n| + \cdots$ converges.

Conditional convergence

A series that converges but not absolutely is said to *converge conditionally* (or to be *conditionally convergent*). The alternating harmonic series converges conditionally since it converges, but does not converge absolutely. (The harmonic series diverges.)

When asked to determine whether an alternating series is absolutely convergent, conditionally convergent, or divergent, it is often advisable to first consider the series of absolute values. Check first for divergence, using the nth Term Test. If that test shows that the series may converge, investigate further, using the tests for nonnegative series. If you find that the series of absolute values converges, then the alternating series is absolutely convergent. If, however, you find that the series of absolute values diverges, then you'll need to use the Alternating Series Test to see whether the series is conditionally convergent.

➡ Example 28

Determine whether $\sum \frac{(-1)^n n^2}{n^2+9}$ converges absolutely, converges conditionally, or diverges.

SOLUTION: We see that $\lim_{n\to\infty} \frac{n^2}{n^2+9} = 1$, not 0, so by the nth Term Test the series $\sum \frac{(-1)^n n^2}{n^2+9}$ is divergent.

➡ Example 29

BC ONLY

Determine whether $\sum \frac{\sin \frac{n\pi}{3}}{n^2}$ converges absolutely, converges conditionally, or diverges.

SOLUTION: Note that, since $\left|\sin \frac{n\pi}{3}\right| \leq 1$, $\lim_{n\to\infty} \frac{\sin \frac{n\pi}{3}}{n^2} = 0$; the series passes the nth Term Test.

Also, $\left|\frac{\sin \frac{n\pi}{3}}{n^2}\right| \leq \frac{1}{n^2}$ for all n.

But $\frac{1}{n^2}$ is the general term of a convergent p-series ($p = 2$), so by the Comparison Test the nonnegative series converges, and therefore the alternating series converges absolutely.

➡ Example 30

Determine whether $\sum \frac{(-1)^{n+1}}{\sqrt[3]{n+1}}$ converges absolutely, converges conditionally, or diverges.

SOLUTION: $\sum \frac{1}{\sqrt[3]{n+1}}$ is a p-series with $p = \frac{1}{3}$, so the nonnegative series diverges.

We see that $\frac{1}{\sqrt[3]{(n+1)+1}} < \frac{1}{\sqrt[3]{n+1}}$ and $\lim_{n\to\infty} \frac{1}{\sqrt[3]{n+1}} = 0$, so the alternating series converges; hence $\sum \frac{(-1)^{n+1}}{\sqrt[3]{n+1}}$ is conditionally convergent.

APPROXIMATING THE LIMIT OF AN ALTERNATING SERIES

Evaluating the sum of the first n terms of an alternating series, given by $\sum_{k=1}^{n} (-1)^{k+1} a_k$, yields an approximation of the limit, L. The error (the difference between the approximation and the true limit) is called the *remainder after n terms* and is denoted by R_n. When an alternating series is first shown to pass the Alternating Series Test, it's easy to place an upper bound on this remainder. Because the terms alternate in sign and become progressively smaller in magnitude, an alternating series converges on its limit by oscillation, as shown in Figure N10–1.

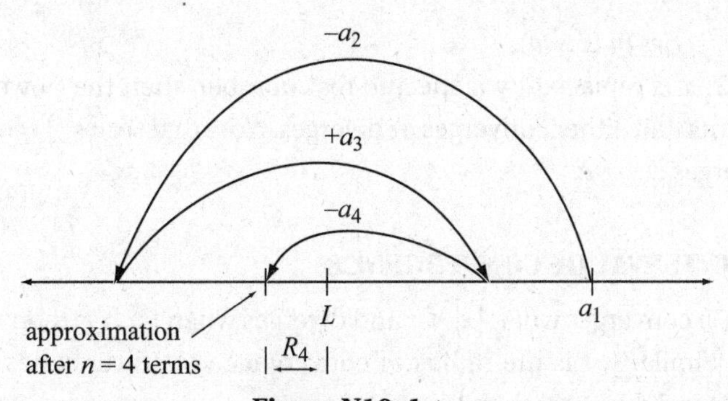

Figure N10–1

Because carrying out the approximation one more term would once more carry us beyond L, we see that the error is always less than that next term. Since $|R_n| < a_{n+1}$, the *alternating series error bound* for an alternating series is the first term omitted or dropped.

Alternating series error bound

BC ONLY

Example 31

The series $\sum_{k=1}^{\infty} \frac{(-1)^{k+1}}{k}$ passes the Alternating Series Test; hence its sum differs from the sum

$$\left(1 - \frac{1}{2} + \frac{1}{3} - \frac{1}{4} + \frac{1}{5} - \frac{1}{6}\right)$$

by less than $\frac{1}{7}$, which is the error bound.

Example 32

Use the alternating series error bound to determine how many terms must be summed to approximate to three decimal places the value of $1 - \frac{1}{4} + \frac{1}{9} - \frac{1}{16} + \cdots + \frac{(-1)^{n+1}}{n^2} + \cdots$?

SOLUTION: Since $\frac{1}{(n+1)^2} < \frac{1}{n^2}$ and $\lim_{n \to \infty} \frac{1}{n^2} = 0$, the series converges by the Alternating Series Test; therefore after summing a number of terms the remainder (alternating series error bound) will be less than the first omitted term.

We seek n such that $R_n = \frac{1}{(n+1)^2} < 0.001$. Thus n must satisfy $(n+1)^2 > 1000$, or $n > 30.623$. Therefore 31 terms are needed for the desired accuracy.

C. POWER SERIES

C1. Definitions; Convergence

An expression of the form

$$\sum_{k=0}^{\infty} a_k x^k = a_0 + a_1 x + a_2 x^2 + \cdots + a_n x^n + \cdots, \tag{1}$$

where the a's are constants, is called a *power series in x*; and

Power series

$$\sum_{k=0}^{\infty} a_k (x-a)^k = a_0 + a_1(x-a) + a_2(x-a)^2 + \cdots + a_n(x-a)^n + \cdots \tag{2}$$

is called a *power series in* $(x - a)$.

If in (1) or (2) x is replaced by a specific real number, then the power series becomes a series of constants that either converges or diverges. Note that series (1) converges if $x = 0$ and series (2) converges if $x = a$.

Radius of convergence

RADIUS AND INTERVAL OF CONVERGENCE

If power series (1) converges when $|x| < r$ and diverges when $|x| > r$, then r is called the *radius of convergence*. Similarly, r is the radius of convergence of power series (2) if (2) converges when $|x - a| < r$ and diverges when $|x - a| > r$.

Interval of convergence

The set of *all* values of x for which a power series converges is called its interval of convergence. To find the interval of convergence, first determine the radius of convergence by applying the Ratio Test to the series of absolute values. Then check each endpoint to determine whether the series converges or diverges there.

BC ONLY

➡ Example 33

Find all x for which the following series converges:

$$1 + x + x^2 + \cdots + x^n + \cdots. \qquad (3)$$

SOLUTION: By the Ratio Test, the series converges if

$$\lim_{n\to\infty}\left|\frac{u_{n+1}}{u_n}\right| = \lim_{n\to\infty}\left|\frac{x^{n+1}}{x^n}\right| = \lim_{n\to\infty}|x| = |x| < 1.$$

Thus, the radius of convergence is 1. The endpoints must be tested separately since the Ratio Test fails when the limit equals 1. When $x = 1$, (3) becomes $1 + 1 + 1 + \cdots$ and diverges; when $x = -1$, (3) becomes $1 - 1 + 1 - 1 + \cdots$ and diverges. Thus the interval of convergence is $-1 < x < 1$.

➡ Example 34

For what x does $\sum_{n=1}^{\infty} \frac{(-1)^{n-1}x^{n-1}}{n+1}$ converge?

SOLUTION: $$\lim_{n\to\infty}\left|\frac{u_{n+1}}{u_n}\right| = \lim_{n\to\infty}\left|\frac{x^n}{n+2}\cdot\frac{n+1}{x^{n-1}}\right| = \lim_{n\to\infty}|x| = |x| < 1.$$

The radius of convergence is 1. When $x = 1$, we have $\frac{1}{2} - \frac{1}{3} + \frac{1}{4} - \frac{1}{5} + \cdots$, an alternating convergent series; when $x = -1$, the series is $\frac{1}{2} + \frac{1}{3} + \frac{1}{4} + \cdots$, which diverges. Thus, the series converges if $-1 < x \leqq 1$.

➡ Example 35

For what values of x does $\sum_{n=1}^{\infty} \frac{x^n}{n!}$ converge?

SOLUTION: $$\lim_{n\to\infty}\left|\frac{u_{n+1}}{u_n}\right| = \lim_{n\to\infty}\left|\frac{x^{n+1}}{(n+1)!}\cdot\frac{n!}{x^n}\right| = \lim_{n\to\infty}\frac{|x|}{n+1} = 0,$$

which is always less than 1. Thus the series converges for all x.

➡ Example 36

Find all x for which the following series converges:

$$1 + \frac{x-2}{2^1} + \frac{(x-2)^2}{2^2} + \cdots + \frac{(x-2)^{n-1}}{2^{n-1}} + \cdots. \qquad (4)$$

SOLUTION: $$\lim_{n\to\infty}\left|\frac{u_{n+1}}{u_n}\right| = \lim_{n\to\infty}\left|\frac{(x-2)^n}{2^n}\cdot\frac{2^{n-1}}{(x-2)^{n-1}}\right| = \lim_{n\to\infty}\frac{|x-2|}{2} = \frac{|x-2|}{2},$$

which is less than 1 if $|x - 2| < 2$, that is, if $0 < x < 4$. Series (4) converges on this interval and diverges if $|x - 2| > 2$, that is, if $x < 0$ or $x > 4$. When $x = 0$, (4) is $1 - 1 + 1 - 1 + \cdots$ and diverges. When $x = 4$, (4) is $1 + 1 + 1 + \cdots$ and diverges. Thus, the interval of convergence is $0 < x < 4$.

BC ONLY

Example 37

Find all x for which the series $\sum_{n=1}^{\infty} n!\,x^n$ converges.

SOLUTION: $\sum_{n=1}^{\infty} n!\,x^n$ converges only at $x = 0$, since

$$\lim_{n\to\infty} \frac{u_{n+1}}{u_n} = \lim_{n\to\infty} (n + 1)\,x = \infty$$

unless $x = 0$.

C2. Functions Defined by Power Series

Let the function f be defined by

$$f(x) = \sum_{k=0}^{\infty} a_k\,(x - a)^k$$

$$= a_0 + a_1(x - a) + \cdots + a_n(x - a)^n + \cdots; \qquad (1)$$

its domain is the interval of convergence of the series.

Functions defined by power series behave very much like polynomials, as indicated by the following properties:

Property 2a. The function defined by (1) is continuous for each x in the interval of convergence of the series.

Property 2b. The series formed by differentiating the terms of series (1) converges to $f'(x)$ for each x within the radius of convergence of (1); that is,

$$f'(x) = \sum_{1}^{\infty} ka_k\,(x - a)^{k-1}$$

$$= a_1 + 2a_2(x - a) + \cdots + na_n\,(x - a)^{n-1} + \cdots. \qquad (2)$$

Note that power series (1) and its derived series (2) have the same radius of convergence but not necessarily the same interval of convergence.

Example 38

Let $f(x) = \sum_{k=1}^{\infty} \frac{x^k}{k(k+1)} = \frac{x}{1\cdot 2} + \frac{x^2}{2\cdot 3} + \cdots + \frac{x^n}{n(n+1)} + \cdots.$

Find the intervals of convergence of the power series for $f(x)$ and $f'(x)$.

SOLUTION:
$$\lim_{n\to\infty} \left| \frac{x^{n+1}}{(n+1)(n+2)} \cdot \frac{n(n+1)}{x^n} \right| = |x|;$$

also,

$$f(1) = \frac{1}{1\cdot 2} + \frac{1}{2\cdot 3} + \cdots + \frac{1}{n(n+1)} + \cdots;$$

and

$$f(-1) = -\frac{1}{1\cdot 2} + \frac{1}{2\cdot 3} - \cdots + \frac{(-1)^n}{n(n+1)} + \cdots.$$

Hence, the power series for f converges if $-1 \leqq x \leqq 1$.

For the derivative $f'(x) = \sum_{k=1}^{\infty} \frac{x^{k-1}}{k+1} = \frac{1}{2} + \frac{x}{3} + \frac{x^2}{4} + \cdots + \frac{x^{n-1}}{n+1} + \cdots$,

$$\lim_{n \to \infty} \left| \frac{x^n}{n+2} \cdot \frac{n+1}{x^{n-1}} \right| = |x|;$$

also,

$$f'(1) = \frac{1}{2} + \frac{1}{3} + \frac{1}{4} + \cdots;$$

and

$$f'(-1) = \frac{1}{2} - \frac{1}{3} + \frac{1}{4} - \cdots.$$

Hence, the power series for f' converges if $-1 \leqq x < 1$.

Thus, the series given for $f(x)$ and $f'(x)$ have the same radius of convergence, but their intervals of convergence differ.

Property 2c. The series obtained by integrating the terms of the given series (1) converges to $\int_a^x f(t)\, dt$ for each x within the interval of convergence of (1); that is,

$$\int_a^x f(t)dt = a_0 (x-a) + \frac{a_1 (x-a)^2}{2} + \frac{a_2 (x-a)^3}{3} + \cdots + \frac{a_n (x-a)^{n+1}}{n+1} + \cdots$$

$$= \sum_{k=0}^{\infty} \frac{a_k (x-a)^{k+1}}{k+1}.$$

Example 39

Let $f(x) = \frac{1}{(1-x)^2}$. Show that the power series for $\int f(x)dx$ converges for all values of x in the interval of convergence of the power series for $f(x)$.

SOLUTION: Obtain a series for $\frac{1}{(1-x)^2}$ by long division.

$$\begin{array}{r|l} & 1 + 2x + 3x^2 + 4x^3 + \cdots \\ \hline 1 - 2x + x^2 & 1 \\ & 1 - 2x + x^2 \\ \hline & 2x - x^2 \\ & 2x - 4x^2 + 2x^3 \\ \hline & +3x^2 - 2x^3 \\ & 3x^2 - 6x^3 + 3x^4 \\ \hline & 4x^3 - 3x^4 \end{array}$$

Then,

$$\frac{1}{(1-x)^2} = 1 + 2x + 3x^2 + \cdots + (n+1)x^n + \cdots.$$

It can be shown that the interval of convergence is $-1 < x < 1$.

Then by Property 2c

$$\int \frac{1}{(1-x)^2} dx = \int \left[1 + 2x + 3x^2 + \cdots + (n+1)x^n + \cdots\right] dx$$

$$\frac{1}{1-x} = c + x + x^2 + x^3 + \cdots + x^{n+1} + \cdots.$$

BC ONLY

Since when $x = 0$ we see that $c = 1$, we have

$$\frac{1}{1-x} = 1 + x + x^2 + x^3 + \cdots + x^n + \cdots.$$

Note that this is a geometric series with ratio $r = x$ and with $a = 1$; if $|x| < 1$, its sum is $\frac{a}{1-r} = \frac{1}{1-x}$.

C3. Finding a Power Series for a Function: Taylor and Maclaurin Series

If a function $f(x)$ is representable by a power series of the form

$$c_0 + c_1(x-a) + c_2(x-a)^2 + \cdots + c_n(x-a)^n + \cdots$$

on an interval $|x-a| < r$, then the coefficients are given by

$$c_n = \frac{f^{(n)}(a)}{n!}.$$

The series

Taylor Series

$$f(x) = f(a) + f'(a)(x-a) + \frac{f''(a)}{2!}(x-a)^2 + \cdots + \frac{f^{(n)}(a)}{n!}(x-a)^n + \cdots$$

is called the *Taylor series* of the function f about the number a. There is never more than one power series in $(x - a)$ for $f(x)$. It is required that the function and all its derivatives exist at $x = a$ if the function $f(x)$ is to generate a Taylor series expansion.

When $a = 0$ we have the special series

Maclaurin Series

$$f(x) = f(0) + f'(0)x + \frac{f''(0)}{2!}x^2 + \cdots + \frac{f^{(n)}(0)}{n!}x^n + \cdots,$$

called the *Maclaurin series* of the function f; this is the expansion of f about $x = 0$.

Example 40

Find the Maclaurin series for $f(x) = e^x$.

SOLUTION: Here $f'(x) = e^x, \ldots, f^{(n)}(x) = e^x, \ldots$, for all n. Then

$$f'(0) = 1, \ldots, f^{(n)}(0) = 1, \ldots,$$

for all n, making the coefficients $c_n = \frac{1}{n!}$:

$$e^x = 1 + x + \frac{x^2}{2!} + \frac{x^3}{3!} + \cdots + \frac{x^n}{n!} + \cdots.$$

Example 41

Find the Maclaurin expansion for $f(x) = \sin x$.

SOLUTION: $f(x) = \sin x$; $f(0) = 0$;

$f'(x) = \cos x$; $f'(0) = 1$;

$f''(x) = -\sin x$; $f''(0) = 0$;

$$f^{(3)}(x) = -\cos x; \qquad f^{(3)}(0) = -1;$$

$$f^{(4)}(x) = \sin x; \qquad f^{(4)}(0) = 0;$$

BC ONLY

Thus,

$$\sin x = x - \frac{x^3}{3!} + \frac{x^5}{5!} - \cdots + (-1)^{n-1}\frac{x^{2n-1}}{(2n-1)!} + \cdots.$$

➡ Example 42

Find the Maclaurin series for $f(x) = \frac{1}{1-x}$.

SOLUTION:

$$f(x) = (1-x)^{-1}; \qquad f(0) = 1;$$

$$f'(x) = (1-x)^{-2}; \qquad f'(0) = 1;$$

$$f''(x) = 2(1-x)^{-3}; \qquad f''(0) = 2;$$

$$f'''(x) = 3!(1-x)^{-4}; \qquad f'''(0) = 3!;$$

$$\vdots \qquad \vdots$$

$$f^{(n)}(x) = n!(1-x)^{-(n+1)}; \qquad f^{(n)}(0) = n!.$$

Then

$$\frac{1}{1-x} = 1 + x + x^2 + x^3 + \cdots + x^n + \cdots.$$

Note that this agrees exactly with the power series in x obtained by different methods in Example 39.

➡ Example 43

Find the Taylor series for the function $f(x) = \ln x$ about $x = 1$.

SOLUTION:

$$f(x) = \ln x; \qquad f(1) = \ln 1 = 0;$$

$$f'(x) = \frac{1}{x}; \qquad f'(1) = 1;$$

$$f''(x) = -\frac{1}{x^2}; \qquad f''(1) = -1;$$

$$f^{(3)}(x) = \frac{2}{x^3}; \qquad f^{(3)}(1) = 2;$$

$$f^{(4)}(x) = \frac{-3!}{x^4}; \qquad f^{(4)}(1) = -3!;$$

$$\vdots \qquad \vdots$$

$$f^{(n)}(x) = \frac{(-1)^{n-1}(n-1)!}{x^n}; \qquad f^{(n)}(1) = (-1)^{n-1}(n-1)!.$$

BC ONLY

Then

$$\ln x = (x-1) - \frac{(x-1)^2}{2} + \frac{(x-1)^3}{3} - \frac{(x-1)^4}{4} + \cdots + \frac{(-1)^{n-1}(x-1)^n}{n} + \cdots.$$

COMMON MACLAURIN SERIES

We list here for reference some frequently used Maclaurin series expansions together with their intervals of convergence:

FUNCTION	MACLAURIN SERIES	INTERVAL OF CONVERGENCE	
$\sin x$	$x - \frac{x^3}{3!} + \frac{x^5}{5!} - \cdots + \frac{(-1)^{n-1}x^{2n-1}}{(2n-1)!} + \cdots$	$-\infty < x < \infty$	(1)
$\cos x$	$1 - \frac{x^2}{2!} + \frac{x^4}{4!} - \cdots + \frac{(-1)^{n-1}x^{2n-2}}{(2n-2)!} + \cdots$	$-\infty < x < \infty$	(2)
e^x	$1 + x + \frac{x^2}{2!} + \frac{x^3}{3!} + \cdots + \frac{x^n}{n!} + \cdots$	$-\infty < x < \infty$	(3)
$\ln(1+x)$	$x - \frac{x^2}{2} + \frac{x^3}{3} - \frac{x^4}{4} + \cdots + \frac{(-1)^{n-1}x^n}{n} + \cdots$	$-1 < x \le 1$	(4)
$\tan^{-1} x$	$x - \frac{x^3}{3} + \frac{x^5}{5} - \frac{x^7}{7} + \cdots + \frac{(-1)^{n-1}x^{2n-1}}{2n-1} + \cdots$	$-1 \le x \le 1$	(5)

FUNCTIONS THAT GENERATE NO SERIES

Note that the following functions are among those that fail to generate a specific series in $(x - a)$ because the function and/or one or more derivatives do not exist at $x = a$:

FUNCTION	SERIES IT FAILS TO GENERATE
$\ln x$	about 0
$\ln(x-1)$	about 1
$\sqrt{x-2}$	about 2
$\sqrt{x-2}$	about 0
$\tan x$	about $\frac{\pi}{2}$
$\sqrt{1+x}$	about -1

BC ONLY

C4. Approximating Functions with Taylor and Maclaurin Polynomials

The function $f(x)$ at the point $x = a$ is approximated by a *Taylor polynomial* $P_n(x)$ of order n:

$$f(x) \simeq P_n(x) = f(a) + f'(a)(x - a) + \frac{f''(a)}{2!}(x - a)^2 + \cdots + \frac{f^{(n)}(a)}{n!}(x - a)^n.$$

The Taylor polynomial $P_n(x)$ and its first n derivatives all agree at a with f and its first n derivatives. The *order* of a Taylor polynomial is the order of the highest derivative, which is also the polynomial's last term.

In the special case where $a = 0$, the *Maclaurin polynomial* of order n that approximates $f(x)$ is

$$P_n(x) = f(0) + f'(0)x + \frac{f''(0)}{2!}x^2 + \cdots + \frac{f^{(n)}(0)}{n!}x^n.$$

The Taylor polynomial $P_1(x)$ at $x = 0$ is the tangent-line approximation to $f(x)$ near zero given by

$$f(x) \simeq P_1(x) = f(0) + f'(0)x.$$

It is the "best" linear approximation to f at 0, discussed at length in Chapter 4 §L.

A Note on Order and Degree

A Taylor polynomial has *degree n* if it has powers of $(x - a)$ up through the nth. If $f^{(n)}(a) = 0$, then the degree of $P_n(x)$ is less than n. Note, for instance, in Example 45, that the second-order polynomial $P_2(x)$ for the function $\sin x$ (which is identical with $P_1(x)$) is $x + 0 \cdot \frac{x^2}{2!}$, or just x, which has degree 1, not 2.

Example 44

Find the Taylor polynomial of order 4 at 0 for $f(x) = e^{-x}$. Use this to approximate $f(0.25)$.

SOLUTIONS: The first four derivatives are $-e^{-x}$, e^{-x}, $-e^{-x}$, and e^{-x}; at $a = 0$, these equal -1, 1, -1, and 1, respectively. The approximating Taylor polynomial of order 4 is therefore

$$e^{-x} \approx 1 - x + \frac{1}{2!}x^2 - \frac{1}{3!}x^3 + \frac{1}{4!}x^4.$$

With $x = 0.25$ we have

$$e^{-0.25} \approx 1 - 0.25 + \frac{1}{2!}(0.25)^2 - \frac{1}{3!}(0.25)^3 + \frac{1}{4!}(0.25)^4$$
$$\approx 0.7788.$$

This approximation of $e^{-0.25}$ is correct to four places.

BC ONLY

In Figure N10–2 we see the graphs of $f(x)$ and of the Taylor polynomials:

$$P_0(x) = 1;$$
$$P_1(x) = 1 - x;$$
$$P_2(x) = 1 - x + \frac{x^2}{2!};$$
$$P_3(x) = 1 - x + \frac{x^2}{2!} - \frac{x^3}{3!};$$
$$P_4(x) = 1 - x + \frac{x^2}{2!} - \frac{x^3}{3!} + \frac{x^4}{4!}.$$

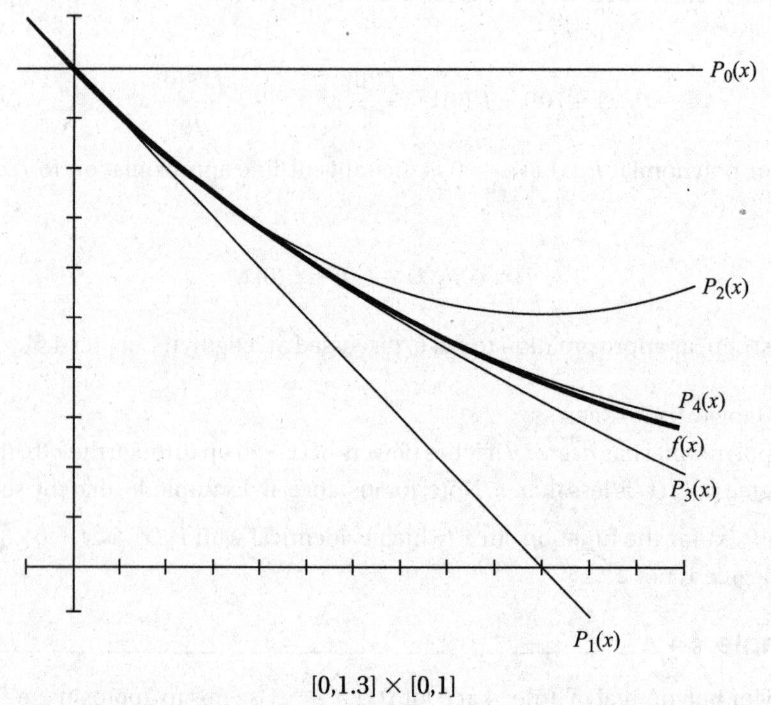

[0,1.3] × [0,1]

Figure N10–2

Notice how closely $P_4(x)$ hugs $f(x)$ even as x approaches 1. Since the series can be shown to converge for $x > 0$ by the Alternating Series Test, the error in $P_4(x)$ is less than the magnitude of the first omitted term, $\frac{x^5}{5!}$, or $\frac{1}{120}$ at $x = 1$. In fact, $P_4(1) = 0.375$ to three decimal places, close to $e^{-1} \approx 0.368$.

➡ Example 45

(a) Find the Taylor polynomials P_1, P_3, P_5, and P_7 at $x = 0$ for $f(x) = \sin x$.

(b) Graph f and all four polynomials in $[-2\pi, 2\pi] \times [-2,2]$.

(c) Approximate $\sin \frac{\pi}{3}$ using each of the four polynomials.

SOLUTIONS:

BC ONLY

(a) The derivatives of the sine function at 0 are given by the following table:

order of deriv	0	1	2	3	4	5	6	7
deriv of $\sin x$	$\sin x$	$\cos x$	$-\sin x$	$-\cos x$	$\sin x$	$\cos x$	$-\sin x$	$-\cos x$
deriv of $\sin x$ at 0	0	1	0	-1	0	1	0	-1

From the table we know that

$$P_1(x) = x;$$

$$P_3(x) = x - \frac{x^3}{3!};$$

$$P_5(x) = x - \frac{x^3}{3!} + \frac{x^5}{5!};$$

$$P_7(x) = x - \frac{x^3}{3!} + \frac{x^5}{5!} - \frac{x^7}{7!}.$$

(b) Figure N10–3a shows the graphs of $\sin x$ and the four polynomials. In Figure N10–3b we see graphs only of $\sin x$ and $P_7(x)$, to exhibit how closely P_7 "follows" the sine curve.

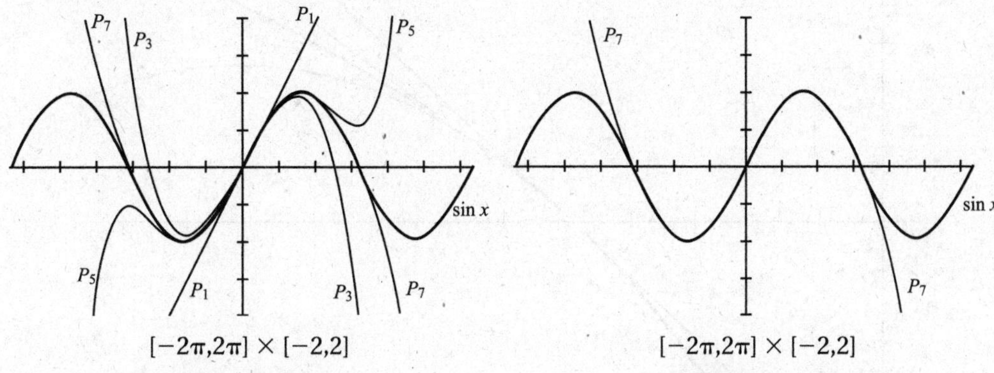

$[-2\pi, 2\pi] \times [-2, 2]$

Figure N10–3a

$[-2\pi, 2\pi] \times [-2, 2]$

Figure N10–3b

(c) To four decimal places, $\sin \frac{\pi}{3} = 0.8660$. Evaluating the polynomials at $\frac{\pi}{3}$, we get

$$P_1\left(\frac{\pi}{3}\right) = 1.0472, \quad P_3\left(\frac{\pi}{3}\right) = 0.8558, \quad P_5\left(\frac{\pi}{3}\right) = 0.8663, \quad P_7\left(\frac{\pi}{3}\right) = 0.8660.$$

We see that P_7 is correct to four decimal places.

➡ Example 46

(a) Find the Taylor polynomials of degrees 0, 1, 2, and 3 generated by $f(x) = \ln x$ at $x = 1$.

(b) Graph f and the four polynomials on the same set of axes.

(c) Using P_2, approximate $\ln 1.3$, and find a bound on the error.

BC ONLY

SOLUTIONS:

(a) The derivatives of $\ln x$ at $x = 1$ are given in the table:

order of deriv	0	1	2	3
deriv of $\ln x$	$\ln x$	$\frac{1}{x}$	$-\frac{1}{x^2}$	$\frac{2}{x^3}$
deriv at $x = 1$	0	1	-1	2

From the table we have

$$P_0(x) = 0;$$
$$P_1(x) = (x - 1);$$
$$P_2(x) = (x - 1) - \frac{(x - 1)^2}{2};$$
$$P_3(x) = (x - 1) - \frac{(x - 1)^2}{2} + \frac{(x - 1)^3}{3}.$$

(b) Figure N10–4 shows the graphs of $\ln x$ and the four Taylor polynomials above, in $[0,2.5] \times [-1,1]$.

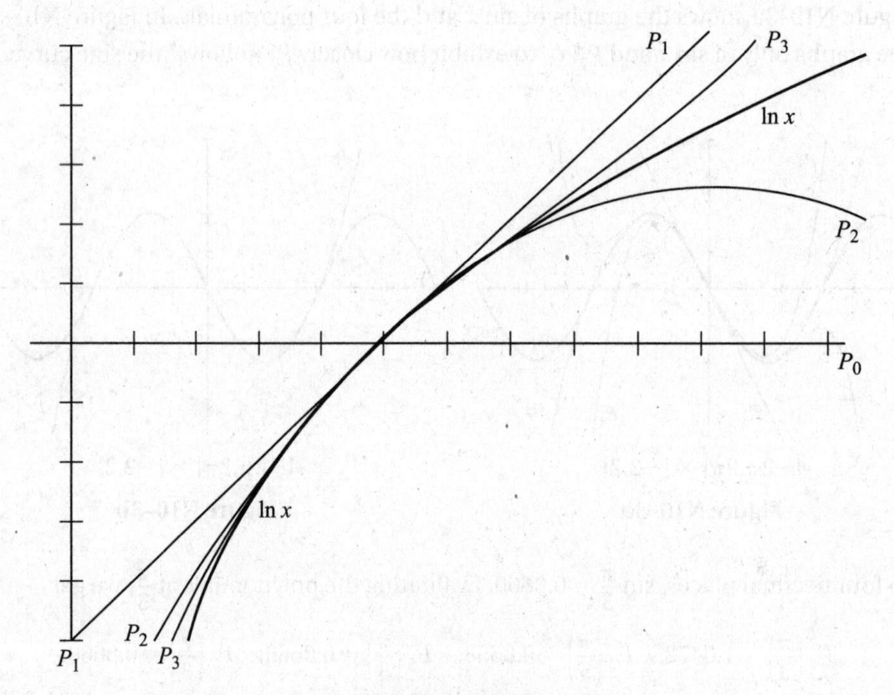

$[0,2.5] \times [-1,1]$

Figure N10–4

(c) $\ln 1.3 \approx P_2(1.3) = (1.3 - 1) - \frac{(1.3 - 1)^2}{2} = 0.3 - 0.045 = 0.255.$

For $x = 1.3$ the Taylor series converges by the Alternating Series Test, so the error is less than the magnitude of the first omitted term:

$$R_2(1.3) < \left|\frac{(1.3 - 1)^3}{3}\right| = 0.009.$$

BC ONLY

➡ Example 47

For what positive values of x is the approximate formula

$$\ln(1 + x) = x - \frac{x^2}{2}$$

correct to three decimal places?

SOLUTION: We can use series (4), page 408:

$$\ln(1 + x) = x - \frac{x^2}{2} + \frac{x^3}{3} - \cdots.$$

For $x > 0$, this is an alternating series with terms decreasing in magnitude and approaching 0, so the error committed by using the first two terms is less than $\frac{|x|^3}{3}$. If $\frac{|x|^3}{3} < 0.0005$, then the given approximation formula will yield accuracy to three decimal places. We therefore require that $|x|^3 < 0.0015$ or that $|x| < 0.114$.

C5. Taylor's Formula with Remainder; Lagrange Error Bound

When we approximate a function using a Taylor polynomial, it is important to know how large the remainder (error) may be. If at the desired value of x the Taylor series is alternating, this issue is easily resolved: the first omitted term serves as an upper bound on the error. However, when the approximation involves a nonnegative Taylor series, placing an upper bound on the error is more difficult. This issue is resolved by the Lagrange remainder.

TAYLOR'S THEOREM. If a function f and its first $(n + 1)$ derivatives are continuous on the interval $|x - a| < r$, then for each x in this interval

$$f(x) = f(a) + f'(a)(x - a) + \frac{f''(a)}{2!}(x - a)^2 + \cdots + \frac{f^{(n)}(a)}{n!}(x - a)^n + R_n(x),$$

where

$$R_n(x) = \frac{f^{(n+1)}(c)(x - a)^{n+1}}{(n + 1)!}$$

Lagrange remainder

and c is some number between a and x. $R_n(x)$ is called the *Lagrange remainder.*

Note that the equation above expresses $f(x)$ as the sum of the Taylor polynomial $P_n(x)$ and the error that results when that polynomial is used as an approximation for $f(x)$.

When we truncate a series after the $(n + 1)$st term, we can compute the error bound R_n, according to Lagrange, if we know what to substitute for c. In practice we find, not R_n exactly, but only an upper bound for it by assigning to c the value between a and x that determines the largest possible value of R_n. Hence:

$$R_n(x) < \max\left|\frac{f^{(n+1)}(c)(x - a)^{n+1}}{(n + 1)!}\right|,$$

Lagrange error bound

the *Lagrange error bound.*

BC ONLY

Example 48

Estimate the error in using the Maclaurin series generated by e^x to approximate the value of e.

SOLUTION: From Example 40 we know that $f(x) = e^x$ generates the Maclaurin series

$$e^x = 1 + x + \frac{x^2}{2!} + \cdots + \frac{x^n}{n!} + \cdots.$$

The Lagrange error bound is

$$R_n(x) < \max\left|\frac{e^c (x)^{n+1}}{(n+1)!}\right| \qquad (0 < c < x).$$

To estimate e, we use $x = 1$. For $0 < c < 1$, the maximum value of e^c is e. Thus:

$$e \approx 1 + 1 + \frac{1}{2!} + \frac{1}{3!} + \cdots + \frac{1}{n!} \quad \text{with error less than } \frac{e}{(n+1)!}.$$

Example 49

Find the Maclaurin series for ln (1 + x) and the associated Lagrange error bound.

SOLUTION:

$$f(x) = \ln(1 + x); \qquad f(0) = 0;$$

$$f'(x) = \frac{1}{1 + x}; \qquad f'(0) = 1;$$

$$f''(x) = -\frac{1}{(1 + x)^2}; \qquad f''(0) = -1;$$

$$f'''(x) = \frac{2}{(1 + x)^3}; \qquad f'''(0) = 2!;$$

$$\vdots \qquad\qquad \vdots$$

$$f^{(n)}(x) = \frac{(-1)^{n-1}(n-1)!}{(1 + x)^n}; \qquad f^{(n)}(0) = (-1)^{n-1}(n-1)!$$

$$f^{(n+1)}(x) = \frac{(-1)^n \cdot n!}{(1 + x)^{n+1}}.$$

Then

$$\ln(1 + x) = x - \frac{x^2}{2} + \frac{x^3}{3} - \frac{x^4}{4} + \cdots + (-1)^{n-1} \cdot \frac{x^n}{n} + R_n(x),$$

where the Lagrange error bound is

$$R_n(x) < \max\left|\frac{(-1)^n}{(1 + c)^{n+1}} \cdot \frac{x^{n+1}}{n+1}\right|.$$

NOTE: For $0 < x < 1$ the Maclaurin series is alternating, and the error bound simplifies to $R_n(x) < \frac{x^{n+1}}{n+1}$, the first omitted term. The more difficult Lagrange error bound applies for $-1 < x < 0$.

BC ONLY

➡ Example 50

Find the third-degree Maclaurin polynomial for $f(x) = \cos\left(x + \frac{\pi}{4}\right)$, and determine the upper bound on the error in estimating $f(0.1)$.

SOLUTION: We first make a table of the derivatives, evaluated at $x = 0$ and giving us the coefficients.

n	$f^{(n)}(x)$	$f^{(n)}(0)$	$C_n = \frac{f^{(n)}(0)}{n!}$
0	$\cos\left(x + \frac{\pi}{4}\right)$	$\frac{\sqrt{2}}{2}$	$\frac{\sqrt{2}}{2}$
1	$-\sin\left(x + \frac{\pi}{4}\right)$	$-\frac{\sqrt{2}}{2}$	$-\frac{\sqrt{2}}{2}$
2	$-\cos\left(x + \frac{\pi}{4}\right)$	$-\frac{\sqrt{2}}{2}$	$-\frac{\sqrt{2}}{2 \cdot 2!}$
3	$\sin\left(x + \frac{\pi}{4}\right)$	$\frac{\sqrt{2}}{2}$	$\frac{\sqrt{2}}{2 \cdot 3!}$

Thus $P_3(x) = \frac{\sqrt{2}}{2} - \frac{\sqrt{2}}{2}x - \frac{\sqrt{2}}{2 \cdot 2!}x^2 + \frac{\sqrt{2}}{2 \cdot 3!}x^3$.

Since this is not an alternating series for $x = 0.1$, we must use the Lagrange error bound:

$$R_3(x) < \max\left|\frac{f^{(4)}(c)}{4!}x^4\right|, \quad \text{where } x = 0.1 \text{ and } 0 < c < 0.1.$$

Note that $f^{(4)}(c) = \cos\left(c + \frac{\pi}{4}\right)$ is decreasing on the interval $0 < c < 0.1$, so its maximum value occurs at $c = 0$. Hence:

$$R_3(x) < \max\left|\frac{\cos\left(c + \frac{\pi}{4}\right)}{4!}x^4\right| = \frac{\cos\left(0 + \frac{\pi}{4}\right)}{4!}0.1^4 = \frac{\sqrt{2}}{2 \cdot 4! \cdot 10^4}.$$

C6. Computations with Power Series

The power series expansions of functions may be treated as any other functions for values of x that lie within their intervals of convergence. They may be added, subtracted, multiplied, divided (with division by zero to be avoided), differentiated, or integrated. These properties provide a valuable approach for many otherwise difficult computations. Indeed, power series are often very useful for approximating values of functions, evaluating indeterminate forms of limits, and estimating definite integrals.

➡ Example 51

Compute $\frac{1}{\sqrt{e}}$ to four decimal places.

SOLUTION: We can use the Maclaurin series,

$$e^x = 1 + x + \frac{x^2}{2!} + \frac{x^3}{3!} + \frac{x^4}{4!} + \cdots,$$

BC ONLY

and let $x = -\frac{1}{2}$ to get

$$\begin{aligned} e^{-1/2} &= 1 - \frac{1}{2} + \frac{1}{4 \cdot 2} - \frac{1}{8 \cdot 3!} + \frac{1}{16 \cdot 4} - \frac{1}{32 \cdot 5!} + R_5 \\ &= 1 - 0.50000 + 0.12500 - 0.02083 + 0.00260 - 0.00026 + R_5 \\ &= 0.60651 + R_5. \end{aligned}$$

Note that, since this series converges by the Alternating Series Test, R_5 is less than the first term dropped:

$$R_5 < \frac{1}{64 \cdot 6!} < 0.00003,$$

so $\frac{1}{\sqrt{e}} = 0.6065$, correct to four decimal places.

➡ Example 52

Estimate the error if the approximate formula

$$\sqrt{1+x} = 1 + \frac{x}{2}$$

is used and $|x| < 0.02$.

SOLUTION: We obtain the first few terms of the Maclaurin series generated by $f(x) = \sqrt{1+x}$:

$$\begin{aligned} f(x) &= \sqrt{1+x}; & f(0) &= 1; \\ f'(x) &= \frac{1}{2}(1+x)^{-1/2}; & f'(0) &= \frac{1}{2}; \\ f''(x) &= -\frac{1}{4}(1+x)^{-3/2}; & f''(0) &= -\frac{1}{4}; \\ f'''(x) &= \frac{3}{8}(1+x)^{-5/2}. & f'''(0) &= \frac{3}{8}. \end{aligned}$$

Then

$$\sqrt{1+x} = 1 + \frac{x}{2} - \frac{1}{4} \cdot \frac{x^2}{2} + \frac{3}{8} \cdot \frac{x^3}{6} - \cdots.$$

Note that for $x < 0$, the series is not alternating, so we must use the Lagrange error bound. Here R_1 is $\left|\frac{f''(c)x^2}{2!}\right|$, where $-0.02 < c < 0.02$. With $|x| < 0.02$, we see that the upper bound uses $c = -0.02$:

$$|R_1| < \frac{(0.02)^2}{8(1-0.02)^{3/2}} < 0.00005.$$

➡ Example 53

Use a series to evaluate $\lim_{x \to 0} \frac{\sin x}{x}$.

SOLUTION: From series (1), page 408,

$$\sin x = x - \frac{x^3}{3!} + \frac{x^5}{5!} - \cdots.$$

BC ONLY

Then

$$\lim_{x\to 0}\frac{\sin x}{x} = \lim_{x\to 0}\left(1 - \frac{x^2}{3!} + \frac{x^4}{5!} - \ldots\right) = 1,$$

a well-established result obtained previously.

Example 54

Use a series to evaluate $\lim_{x\to 0}\frac{\ln(x+1)}{3x}$.

SOLUTION: We can use series (4), page 408, and write

$$\lim_{x\to 0}\frac{x - \frac{x^2}{2} + \frac{x^3}{3} - \frac{x^4}{4} + \ldots}{3x} = \lim_{x\to 0}\frac{1}{3} - \frac{x}{6} + \frac{x^2}{9} - \ldots$$
$$= \frac{1}{3}.$$

Example 55

$$\lim_{x\to 0}\frac{e^{-x^2} - 1}{x^2} = \lim_{x\to 0}\frac{\left(1 - x^2 + \frac{x^4}{2!} - \frac{x^6}{3!} + \ldots\right) - 1}{x^2}$$

$$= \lim_{x\to 0}\frac{-x^2 + \frac{x^4}{2!} - \ldots}{x^2} = \lim_{x\to 0}\left(-1 + \frac{x^2}{2!} - \frac{x^4}{4!} + \ldots\right)$$

$$= -1.$$

Example 56

Show how a series may be used to evaluate π.

SOLUTION: Since $\frac{\pi}{4} = \tan^{-1} 1$, a series for $\tan^{-1} x$ may prove helpful. Note that

$$\tan^{-1} x = \int_0^x \frac{dt}{1+t^2}$$

and that a series for $\frac{1}{1+t^2}$ is obtainable easily by long division to yield

$$\frac{1}{1+t^2} = 1 - t^2 + t^4 - t^6 + \cdots.$$

If we integrate this series term by term and then evaluate the definite integral, we get

$$\tan^{-1} x = x - \frac{x^3}{3} + \frac{x^5}{5} - \frac{x^7}{7} + \cdots + \frac{(-1)^{n-1}x^{2n-1}}{2n-1} + \cdots.$$

BC ONLY

(Compare with series (5) on page 408 and note especially that this series converges on $-1 \leq x \leq 1$.)

For $x = 1$ we have:

$$\tan^{-1} 1 = 1 - \frac{1}{3} + \frac{1}{5} - \frac{1}{7} + \cdots.$$

Then

$$\frac{\pi}{4} = 1 - \frac{1}{3} + \frac{1}{5} - \frac{1}{7} + \cdots$$

and

$$\pi = 4\left(1 - \frac{1}{3} + \frac{1}{5} - \frac{1}{7} + \cdots\right).$$

Here are some approximations for π using this series:

number of terms	1	2	5	10	25	50	59	60
approximation	4	2.67	3.34	3.04	3.18	3.12	3.16	3.12

Since the series is alternating, the odd sums are greater, the even ones less, than the value of π. It is clear that several hundred terms may be required to get even two-place accuracy. There are series expressions for π that converge much more rapidly. (See Miscellaneous Free-Response Practice, Problem 12, page 468.)

Example 57

Use a series to evaluate $\int_0^{0.1} e^{-x^2}\,dx$ to four decimal places.

SOLUTION: Although $\int e^{-x^2}\,dx$ cannot be expressed in terms of elementary functions, we can write a series for e^u, replace u by $(-x^2)$, and integrate term by term. Thus,

$$e^{-x^2} = 1 - x^2 + \frac{x^4}{2!} - \frac{x^6}{3!} + \cdots,$$

so

$$\begin{aligned}\int_0^{0.1} e^{-x^2}\,dx &= x - \frac{x^3}{3} + \frac{x^5}{5\cdot 2!} - \frac{x^7}{7\cdot 3!} + \cdots \Big|_0^{0.1} \\ &= 0.1 - \frac{0.001}{3} + \frac{0.00001}{10} - \frac{0.0000001}{42} + \cdots \\ &= 0.1 - 0.00033 + 0.000001 + R_6 \\ &= 0.09967 + R_6.\end{aligned}$$

Since this is a convergent alternating series (with terms decreasing in magnitude and approaching 0), $|R_6| < \frac{10^{-7}}{42}$, which will not affect the fourth decimal place. Then, correct to four decimal places,

$$\int_0^{0.1} e^{-x^2}\,dx = 0.0997.$$

†C7. Power Series over Complex Numbers

BC ONLY

A *complex number* is one of the form $a + bi$, where a and b are real and $i^2 = -1$. If we allow complex numbers as replacements for x in power series, we obtain some interesting results.

Consider, for instance, the series

$$e^x = 1 + x + \frac{x^2}{2!} + \frac{x^3}{3!} + \frac{x^4}{4!} + \cdots + \frac{x^n}{n!} \cdots. \tag{1}$$

When $x = yi$, then (1) becomes

$$\begin{aligned} e^{yi} &= 1 + yi + \frac{(yi)^2}{2!} + \frac{(yi)^3}{3!} + \frac{(yi)^4}{4!} + \cdots \\ &= 1 + yi - \frac{y^2}{2!} - \frac{y^3 i}{3!} + \frac{y^4}{4!} + \cdots \\ &= \left(1 - \frac{y^2}{2!} + \frac{y^4}{4!} + \cdots\right) + i\left(y - \frac{y^3}{3!} + \frac{y^5}{5!} - \cdots\right). \end{aligned} \tag{2}$$

Then

$$e^{yi} = \cos y + i \sin y, \tag{3}$$

since the series within the parentheses of equation (2) converge respectively to $\cos y$ and $\sin y$. Equation (3) is called *Euler's formula.* It follows from (3) that

$$e^{i\pi} = -1,$$

and thus that

$$e^{i\pi} + 1 = 0,$$

sometimes referred to as *Euler's magic formula.*

CHAPTER SUMMARY

In this chapter, we have reviewed an important BC Calculus topic, infinite series. We have looked at a variety of tests to determine whether a series converges or diverges. We have worked with functions defined as power series, reviewed how to derive Taylor series, and looked at the Maclaurin series expansions for many commonly used functions. Finally, we have reviewed how to find bounds on the errors that arise when series are used for approximations.

† This is an optional topic not in the BC Course Description. We include it here because of the dramatic result.

BC ONLY

PRACTICE EXERCISES

Part A. Directions: Answer these questions *without* using your calculator.

NOTE: *No* questions on sequences will appear on the BC examination. We have nevertheless chosen to include the topic in Questions 1–5 because a series and its convergence are defined in terms of sequences. Review of sequences will enhance understanding of series.

1. Which sequence converges?

   (A) $a_n = n + \frac{3}{n}$ (B) $a_n = -1 + \frac{(-1)^n}{n}$ (C) $a_n = \sin\frac{n\pi}{2}$

   (D) $a_n = \frac{n!}{3^n}$ (E) $a_n = \frac{n}{\ln n}$

2. If $s_n = 1 + \frac{(-1)^n}{n}$, then

   (A) s_n diverges by oscillation (B) s_n converges to zero

   (C) $\lim_{n\to\infty} s_n = 1$ (D) s_n diverges to infinity

   (E) None of the above is true.

3. The sequence $a_n = \sin\frac{n\pi}{6}$

   (A) is unbounded (B) is monotonic

   (C) converges to a number less than 1 (D) is bounded

   (E) diverges to infinity

4. Which of the following sequences diverges?

   (A) $a_n = \frac{1}{n}$ (B) $a_n = \frac{(-1)^{n+1}}{n}$ (C) $a_n = \frac{2^n}{e^n}$

   (D) $a_n = \frac{n^2}{e^n}$ (E) $a_n = \frac{n}{\ln n}$

5. The sequence $\{r^n\}$ converges if and only if

   (A) $|r| < 1$ (B) $|r| \leqq 1$ (C) $-1 < r \leqq 1$

   (D) $0 < r < 1$ (E) $|r| > 1$

6. $\sum u_n$ is a series of constants for which $\lim_{n\to\infty} u_n = 0$. Which of the following statements is always true?

   (A) $\sum u_n$ converges to a finite sum. (B) $\sum u_n$ equals zero.

   (C) $\sum u_n$ does not diverge to infinity. (D) $\sum u_n$ is a positive series.

   (E) none of these

7. Note that $\frac{1}{n(n+1)} = \frac{1}{n} - \frac{1}{n+1}$ $(n \geqq 1)$. $\sum_{n=1}^{\infty} \frac{1}{n(n+1)}$ equals

   (A) 0 (B) 1 (C) $\frac{3}{2}$ (D) $\frac{3}{4}$ (E) ∞

BC ONLY

8. The sum of the geometric series $2 - 1 + \frac{1}{2} - \frac{1}{4} + \frac{1}{8} - \cdots$ is

(A) $\frac{4}{3}$ (B) $\frac{5}{4}$ (C) 1 (D) $\frac{3}{2}$ (E) $\frac{3}{4}$

9. Which of the following statements about series is true?

(A) If $\lim_{n\to\infty} u_n = 0$, then $\sum u_n$ converges.

(B) If $\lim_{n\to\infty} u_n \neq 0$, then $\sum u_n$ diverges.

(C) If $\sum u_n$ diverges, then $\lim_{n\to\infty} u_n \neq 0$.

(D) $\sum u_n$ converges if and only if $\lim_{n\to\infty} u_n = 0$.

(E) Statements A–D are all false.

10. Which of series A–D diverges?

(A) $\sum \frac{1}{n^2}$ (B) $\sum \frac{1}{n^2 + n}$ (C) $\sum \frac{n}{n^3 + 1}$

(D) $\sum \frac{n}{\sqrt{4n^2 - 1}}$ (E) Series A–D all converge.

11. Which of the following series diverges?

(A) $3 - 1 + \frac{1}{9} - \frac{1}{27} + \cdots$ (B) $\frac{1}{\sqrt{2}} - \frac{1}{\sqrt{3}} + \frac{1}{\sqrt{4}} - \cdots$

(C) $\frac{1}{2^2} - \frac{1}{3^2} + \frac{1}{4^2} - \cdots$ (D) $1 - 1.1 + 1.21 - 1.331 + \cdots$

(E) $\frac{1}{1 \cdot 2} - \frac{1}{2 \cdot 3} + \frac{1}{3 \cdot 4} - \frac{1}{4 \cdot 5} + \cdots$

12. Let $S = \sum_{n=1}^{\infty} \left(\frac{2}{3}\right)^n$; then S equals

(A) 1 (B) $\frac{3}{2}$ (C) $\frac{4}{3}$ (D) 2 (E) 3

13. Which of the following expansions is impossible?

(A) $\sqrt{x - 1}$ in powers of x (B) $\sqrt{x + 1}$ in powers of x

(C) $\ln x$ in powers of $(x - 1)$ (D) $\tan x$ in powers of $\left(x - \frac{\pi}{4}\right)$

(E) $\ln(1 - x)$ in powers of x

14. The series $\sum_{n=0}^{\infty} n!(x - 3)^n$ converges if and only if

(A) $x = 0$ (B) $2 < x < 4$ (C) $x = 3$
(D) $2 \leq x \leq 4$ (E) $x < 2$ or $x > 4$

15. Let $f(x) = \sum_{n=0}^{\infty} x^n$. The radius of convergence of $\int_0^x f(t)\, dt$ is

(A) 0 (B) 1 (C) 2 (D) e (E) ∞

16. The coefficient of x^4 in the Maclaurin series for $f(x) = e^{-x/2}$ is

(A) $-\frac{1}{24}$ (B) $\frac{1}{24}$ (C) $\frac{1}{96}$ (D) $-\frac{1}{384}$ (E) $\frac{1}{384}$

BC ONLY

17. If an appropriate series is used to evaluate $\int_0^{0.3} x^2e^{-x^2}\,dx$, then, correct to three decimal places, the definite integral equals

(A) 0.009 (B) 0.082 (C) 0.098 (D) 0.008 (E) 0.090

18. If the series $\tan^{-1} 1 = 1 - \frac{1}{3} + \frac{1}{5} - \frac{1}{7} + \cdots$ is used to approximate $\frac{\pi}{4}$ with an error less than 0.001, then the smallest number of terms needed is

(A) 100 (B) 200 (C) 300 (D) 400 (E) 500

19. Let f be the Taylor polynomial $P_7(x)$ of order 7 for $\tan^{-1} x$ about $x = 0$. Then it follows that, if $-0.5 < x < 0.5$,

(A) $f(x) = \tan^{-1} x$
(B) $f(x) \leqslant \tan^{-1} x$
(C) $f(x) \geqslant \tan^{-1} x$
(D) $f(x) > \tan^{-1} x$ if $x < 0$ but $< \tan^{-1} x$ if $x > 0$
(E) $f(x) < \tan^{-1} x$ if $x < 0$ but $> \tan^{-1} x$ if $x > 0$

20. Replace the first sentence in Question 19 by "Let f be the Taylor polynomial $P_9(x)$ of order 9 for $\tan^{-1} x$ about $x = 0$." Which choice given in Question 19 is now the correct one?

21. Which of the following series converges?

(A) $\sum \frac{1}{\sqrt[3]{n}}$ (B) $\sum \frac{1}{\sqrt{n}}$ (C) $\sum \frac{1}{n}$

(D) $\sum \frac{1}{10n-1}$ (E) $\sum \frac{2}{n^2-5}$

22. Which of the following series diverges?

(A) $\sum_{n=1}^{\infty} \frac{1}{n(n+1)}$ (B) $\sum_{n=1}^{\infty} \frac{n+1}{n!}$ (C) $\sum_{n=2}^{\infty} \frac{1}{n \ln n}$

(D) $\sum_{n=1}^{\infty} \frac{\ln n}{2^n}$ (E) $\sum_{n=1}^{\infty} \frac{n}{2^n}$

23. For which of the following series does the Ratio Test fail?

(A) $\sum \frac{1}{n!}$ (B) $\sum \frac{n}{2^n}$ (C) $1 + \frac{1}{2^{3/2}} + \frac{1}{3^{3/2}} + \frac{1}{4^{3/2}} + \cdots$

(D) $\frac{\ln 2}{2^2} + \frac{\ln 3}{2^3} + \frac{\ln 4}{2^4} + \cdots$ (E) $\sum \frac{n^n}{n!}$

24. Which of the following alternating series diverges?

(A) $\sum \frac{(-1)^{n-1}}{n}$ (B) $\sum \frac{(-1)^{n+1}(n-1)}{n+1}$ (C) $\sum \frac{(-1)^{n+1}}{\ln(n+1)}$

(D) $\sum \frac{(-1)^{n-1}}{\sqrt{n}}$ (E) $\sum \frac{(-1)^{n-1}(n)}{n^2+1}$

25. The power series $x + \frac{x^2}{2} + \frac{x^3}{3} + \cdots + \frac{x^n}{n} + \cdots$ converges if and only if

(A) $-1 < x < 1$ (B) $-1 \leqq x \leqq 1$ (C) $-1 \leqq x < 1$

(D) $-1 < x \leqq 1$ (E) $x = 0$

BC ONLY

26. The power series

$$(x+1)-\frac{(x+1)^2}{2!}+\frac{(x+1)^3}{3!}-\frac{(x+1)^4}{4!}+\cdots$$

diverges

(A) for no real x (B) if $-2 < x \leqq 0$ (C) if $x < -2$ or $x > 0$
(D) if $-2 \leqq x < 0$ (E) if $x \neq -1$

27. The series obtained by differentiating term by term the series

$$(x-2)+\frac{(x-2)^2}{4}+\frac{(x-2)^3}{9}+\frac{(x-2)^4}{16}+\cdots$$

converges for

(A) $1 \leqq x \leqq 3$ (B) $1 \leqq x < 3$ (C) $1 < x \leqq 3$

(D) $0 \leqq x \leqq 4$ (E) $0 < x < 4$

28. The Taylor polynomial of order 3 at $x = 0$ for $f(x) = \sqrt{1+x}$ is

(A) $1+\frac{x}{2}-\frac{x^2}{4}+\frac{3x^3}{8}$ (B) $1+\frac{x}{2}-\frac{x^2}{8}+\frac{x^3}{16}$ (C) $1-\frac{x}{2}+\frac{x^2}{8}-\frac{x^3}{16}$

(D) $1+\frac{x}{2}-\frac{x^2}{8}+\frac{x^3}{8}$ (E) $1-\frac{x}{2}+\frac{x^2}{4}-\frac{3x^3}{8}$

29. The Taylor polynomial of order 3 at $x = 1$ for e^x is

(A) $1+(x-1)+\frac{(x-1)^2}{2}+\frac{(x-1)^3}{3}$

(B) $e\left[1+(x-1)+\frac{(x-1)^2}{2}+\frac{(x-1)^3}{3}\right]$

(C) $e\left[1+(x+1)+\frac{(x+1)^2}{2!}+\frac{(x+1)^3}{3!}\right]$

(D) $e\left[1+(x-1)+\frac{(x-1)^2}{2!}+\frac{(x-1)^3}{3!}\right]$

(E) $e\left[1-(x-1)+\frac{(x-1)^2}{2!}+\frac{(x-1)^3}{3!}\right]$

30. The coefficient of $\left(x-\frac{\pi}{4}\right)^3$ in the Taylor series about $\frac{\pi}{4}$ of $f(x) = \cos x$ is

(A) $\frac{\sqrt{3}}{12}$ (B) $-\frac{1}{12}$ (C) $\frac{1}{12}$

(D) $\frac{1}{6\sqrt{2}}$ (E) $-\frac{1}{3\sqrt{2}}$

31. The coefficient of x^2 in the Maclaurin series for $e^{\sin x}$ is

(A) 0 (B) 1 (C) $\frac{1}{2!}$

(D) -1 (E) $\frac{1}{4}$

32. The coefficient of $(x-1)^5$ in the Taylor series for $x \ln x$ about $x = 1$ is

(A) $-\frac{1}{20}$ (B) $\frac{1}{5!}$ (C) $-\frac{1}{5!}$ (D) $\frac{1}{4!}$ (E) $-\frac{1}{4!}$

BC ONLY

33. The radius of convergence of the series $\sum_{n=1}^{\infty} \frac{x^n}{2^n} \cdot \frac{n^n}{n!}$ is

(A) 0 (B) 2 (C) $\frac{2}{e}$ (D) $\frac{e}{2}$ (E) ∞

34. The Taylor polynomial of order 3 at $x = 0$ for $(1 + x)^p$, where p is a constant, is

(A) $1 + px + p(p-1)x^2 + p(p-1)(p-2)x^3$

(B) $1 + px + \frac{p(p-1)}{2}x^2 + \frac{p(p-1)(p-2)}{3}x^3$

(C) $1 + px + \frac{p(p-1)}{2!}x^2 + \frac{p(p-1)(p-2)}{3!}x^3$

(D) $px + \frac{p(p-1)}{2!}x^2 + \frac{p(p-1)(p-2)}{3!}x^3$

(E) none of these

35. The Taylor series for $\ln(1 + 2x)$ about $x = 0$ is

(A) $2x - \frac{(2x)^2}{2} + \frac{(2x)^3}{3} - \frac{(2x)^4}{4} + \cdots$

(B) $2x - 2x^2 + 8x^3 - 16x^4 + \cdots$

(C) $2x - 4x^2 + 16x^3 + \cdots$

(D) $2x - x^2 + \frac{8}{3}x^3 - 4x^4 + \cdots$

(E) $2x - \frac{(2x)^2}{2!} + \frac{(2x)^3}{3!} - \frac{(2x)^4}{4!} + \cdots$

36. The set of all values of x for which $\sum_{n=1}^{\infty} \frac{n \cdot 2^n}{x^n}$ converges is

(A) only $x = 0$ (B) $|x| = 2$ (C) $-2 < x < 2$ (D) $|x| > 2$ (E) $|x| \geq 2$

37. The third-order Taylor polynomial $P_3(x)$ for $\sin x$ about $\frac{\pi}{4}$ is

(A) $\frac{1}{\sqrt{2}}\left(\left(x - \frac{\pi}{4}\right) - \frac{1}{3!}\left(x - \frac{\pi}{4}\right)^3\right)$

(B) $\frac{1}{\sqrt{2}}\left(1 + \left(x - \frac{\pi}{4}\right) - \frac{1}{2}\left(x - \frac{\pi}{4}\right)^2 + \frac{1}{3!}\left(x - \frac{\pi}{4}\right)^3\right)$

(C) $\frac{1}{\sqrt{2}}\left(1 + \left(x - \frac{\pi}{4}\right) - \frac{1}{2!}\left(x - \frac{\pi}{4}\right)^2 - \frac{1}{3!}\left(x - \frac{\pi}{4}\right)^3\right)$

(D) $1 + \left(x - \frac{\pi}{4}\right) - \frac{1}{2}\left(x - \frac{\pi}{4}\right)^2 - \frac{1}{6}\left(x - \frac{\pi}{4}\right)^3$

(E) $\frac{1}{\sqrt{2}}\left(1 + x - \frac{x^2}{2!} - \frac{x^3}{3!}\right)$

38. Let h be a function for which all derivatives exist at $x = 1$. If $h(1) = h'(1) = h''(1) = h'''(1) = 6$, which third-degree polynomial best approximates h there?

(A) $6 + 6x + 6x^2 + 6x^3$

(B) $6 + 6(x-1) + 6(x-1)^2 + 6(x-1)^3$

(C) $6 + 6x + 3x^2 + x^3$

(D) $6 + 6(x-1) + 3(x-1)^2 + (x-1)^3$

(E) $6 + 3(x-1) + 1(x-1)^2 + \frac{1}{4}(x-1)^3$

BC ONLY

Part B. Directions: Some of the following questions require the use of a graphing calculator.

NOTE: Because of the abilities of graphing calculators, Taylor Series and convergence are largely tested in a No Calculator environment; as such we offer only a few calculator-active multiple-choice questions here.

39. Which of the following statements about series is false?

(A) $\sum_{k=1}^{\infty} u_k = \sum_{k=m}^{\infty} u_k$, where m is any positive integer.

(B) If $\sum u_n$ converges, so does $\sum cu_n$ if $c \neq 0$.

(C) If $\sum a_n$ and $\sum b_n$ converge, so does $\sum (ca_n + b_n)$, where $c \neq 0$.

(D) If 1000 terms are added to a convergent series, the new series also converges.

(E) Rearranging the terms of a positive convergent series will not affect its convergence or its sum.

40. Which of the following statements is true?

(A) If $\sum u_n$ converges, then so does the series $\sum |u_n|$.

(B) If a series is truncated after the nth term, then the error is less than the first term omitted.

(C) If the terms of an alternating series decrease, then the series converges.

(D) If $r < 1$, then the series $\sum r^n$ converges.

(E) none of these

41. Which of the following series can be used to compute ln 0.8?

(A) $\ln (x - 1)$ expanded about $x = 0$

(B) $\ln x$ about $x = 0$

(C) $\ln x$ expanded about $x = 1$

(D) $\ln (x - 1)$ expanded about $x = 1$

(E) none of these

42. Let $f(x) = \sum_{n=0}^{\infty} a_n x^n$, $g(x) = \sum_{n=0}^{\infty} b_n x^n$. Suppose both series converge for $|x| < R$. Let x_0 be a number such that $|x_0| < R$. Which of statements A–D is false?

(A) $\sum_{n=0}^{\infty} (a_n + b_n)(x_0)^n$ converges to $f(x_0) + g(x_0)$.

(B) $\left[\sum_{n=0}^{\infty} a_n (x_0)^n\right]\left[\sum_{n=0}^{\infty} b_n (x_0)^n\right]$ converges to $f(x_0)g(x_0)$.

(C) $f(x) = \sum_{n=0}^{\infty} a_n x^n$ is continuous at $x = x_0$.

(D) $\sum_{n=1}^{\infty} n a_n x^{n-1}$ converges to $f'(x_0)$.

(E) Statements A–D are all true.

BC ONLY

43. If the approximate formula $\sin x = x - \frac{x^3}{3!}$ is used and $|x| < 1$ (radian), then the error is numerically less than

(A) 0.001 (B) 0.003 (C) 0.005 (D) 0.008 (E) 0.009

44. The function $f(x) = \sum_{n=0}^{\infty} a_n x^n$ and $f'(x) = -f(x)$ for all x. If $f(0) = 1$, then $f(0.2)$, correct to three decimal places, is

(A) 0.905 (B) 1.221 (C) 0.819 (D) 0.820 (E) 1.220

45. The sum of the series $\sum_{n=1}^{\infty} \left(\frac{\pi^3}{3^\pi}\right)^n$

(A) $= 0$ (B) $= 1$ (C) $= \frac{3^\pi}{\pi^3 - 3^\pi}$

(D) $= \frac{\pi^3}{3^\pi - \pi^3}$ (E) diverges

46. When $\sum_{1}^{\infty} \frac{(-1)^{n-1}}{3n-1}$ is approximated by the sum of its first 300 terms, the error is closest to

(A) 0.001 (B) 0.002 (C) 0.005 (D) 0.01 (E) 0.02

BC ONLY

Answer Key

1. **B**	13. **A**	25. **C**	37. **C**
2. **C**	14. **C**	26. **A**	38. **D**
3. **D**	15. **B**	27. **B**	39. **A**
4. **E**	16. **E**	28. **B**	40. **E**
5. **C**	17. **A**	29. **D**	41. **C**
6. **E**	18. **E**	30. **D**	42. **E**
7. **B**	19. **D**	31. **C**	43. **E**
8. **A**	20. **E**	32. **A**	44. **C**
9. **B**	21. **E**	33. **C**	45. **D**
10. **D**	22. **C**	34. **C**	46. **A**
11. **D**	23. **C**	35. **A**	
12. **D**	24. **B**	36. **D**	

Answers Explained

1. **(B)** $a_n = -1 + \frac{(-1)^n}{n}$ converges to -1.

2. **(C)** Note that $\lim_{n\to\infty} \frac{(-1)^n}{n} = 0$.

3. **(D)** The sine function varies continuously between -1 and 1 inclusive.

4. **(E)** Note that $a_n = \left(\frac{2}{e}\right)^n$ is a sequence of the type $s_n = r^n$ with $|r| < 1$; also that $\lim \frac{n^2}{e^n} = 0$ by repeated application of L'Hôpital's rule.

5. **(C)** $\lim_{n\to\infty} r_n = 0$ for $|r| < 1$; $\lim_{n\to\infty} r_n = 1$ for $r = 1$.

6. **(E)** The harmonic series $\sum_1^\infty \frac{1}{k}$ is a counterexample for (A), (B), and (C).

   $\sum_1^\infty \frac{(-1)^{k+1}}{k}$ shows that (D) does not follow.

7. **(B)** $\sum_1^\infty \frac{1}{n(n+1)} = \frac{1}{1\cdot 2} + \frac{1}{2\cdot 3} + \frac{1}{3\cdot 4} + \cdots + \frac{1}{n(n+1)} + \cdots$; so

   $s_n = 1 - \frac{1}{2} + \frac{1}{2} - \frac{1}{3} + \frac{1}{3} - \cdots + \frac{1}{n} - \frac{1}{n+1} = 1 - \frac{1}{n+1}$,

   and $\lim_{n\to\infty} s_n = 1$.

8. **(A)** $S = \frac{a}{1-r} = \frac{2}{1-\left(-\frac{1}{2}\right)} = \frac{4}{3}$.

9. **(B)** Find counterexamples for statements (A), (C), and (D).

10. **(D)** $\frac{n}{\sqrt{4n^2-1}} > \frac{n}{\sqrt{4n^2}} = \frac{1}{2}$, the general term of a divergent series.

11. **(D)** (A), (B), (C), and (E) all converge; (D) is the divergent geometric series with $r = -1.1$.

12. **(D)** $S = \frac{a}{1-r} = \frac{\frac{2}{3}}{1-\frac{2}{3}} = \frac{\frac{2}{3}}{\frac{1}{3}} = 2$.

13. **(A)** If $f(x) = \sqrt{x-1}$, then $f(0)$ is not defined.

14. **(C)** $\lim_{n\to\infty}(n+1)(x-3) = \infty$ unless $x = 3$.

BC ONLY

15. **(B)** The integrated series is $\sum_{n=0}^{\infty}\frac{x^{n+1}}{n+1}$ or $\sum_{n=1}^{\infty}\frac{x^n}{n}$. See Question 25.

16. **(E)** $e^{-x/2}=1+\left(-\frac{x}{2}\right)+\left(-\frac{x}{2}\right)^2\cdot\frac{1}{2!}+\left(-\frac{x}{2}\right)^3\cdot\frac{1}{3!}+\left(-\frac{x}{2}\right)^4\cdot\frac{1}{4!}+\cdots.$

17. **(A)** $\int_0^{0.3} x^2e^{-x^2}\,dx=\int_0^{0.3} x^2\left(1-x^2+\frac{x^4}{2!}-\ldots\right)dx$

$$=\int_0^{0.3}\left(x^2-x^4+\frac{x^6}{2!}-\ldots\right)dx$$

$$=\frac{x^3}{3}-\frac{x^5}{5}+\ldots\Big|_0^{0.3}$$

$$=0.009 \text{ to three decimal places.}$$

18. **(E)** The series satisfies the Alternating Series Test, so the error is less than the first term dropped, namely, $\frac{1}{2(n+1)-1}$, or $\frac{1}{2n+1}$ (see (5) on page 408), so $n\geqslant 500$.

19. **(D)** Note that the Taylor series for $\tan^{-1}x$ satisfies the Alternating Series Test and that $f(x)=x-\frac{x^3}{3}+\frac{x^5}{5}-\frac{x^7}{7}=P_7(x)$. If $x<0$, then the first omitted term, $\frac{x^9}{9}$, is negative. Hence $P_7(x)$ exceeds $\tan^{-1}x$.

20. **(E)** Now the first omitted term, $-\frac{x^{11}}{11}$, is positive for $x<0$. Hence $P_9(x)$ is less than $\tan^{-1}x$.

21. **(E)** Each series given is essentially a p-series. Only in (E) is $p>1$.

22. **(C)** Use the Integral Test on page 396.

23. **(C)** The limit of the ratio for the series $\sum\frac{1}{n^{3/2}}$ is 1, so this test fails; note for (E) that

$$\lim_{n\to\infty}\frac{u_{n+1}}{u_n}=\lim_{n\to\infty}\left(\frac{n+1}{n}\right)^n=\lim_{n\to\infty}\left(1+\frac{1}{n}\right)^n=e.$$

24. **(B)** $\lim_{n\to\infty}\frac{(-1)^{n+1}(n-1)}{n+1}$ does not equal 0.

25. **(C)** Since $\lim_{n\to\infty}\left|\frac{u_{n+1}}{u_n}\right|=|x|$, the series converges if $|x|<1$. We must test the endpoints: when $x=1$, we get the divergent harmonic series; $x=-1$ yields the convergent alternating harmonic series.

26. **(A)** $\lim_{n\to\infty}\left|\frac{x+1}{n+1}\right|=0$ for all $x\neq-1$; since the given series converges to 0 if $x=-1$, it therefore converges for *all* x.

27. **(B)** The differentiated series is $\sum_{n=1}^{\infty}\frac{(x-2)^{n-1}}{n}$; so

$$\lim_{n\to\infty}\left|\frac{u_{n+1}}{u_n}\right|=|x-2|.$$

28. **(B)** See Example 52 on page 416.

29. **(D)** Note that every derivative of e^x is e at $x=1$. The Taylor series is in powers of $(x-1)$ with coefficients of the form $c_n=\frac{e}{n!}$.

30. **(D)** For $f(x)=\cos x$ around $x=\frac{\pi}{4}$, $c_3=\frac{f^{(3)}\left(\frac{\pi}{4}\right)}{3!}=\frac{\sin\left(\frac{\pi}{4}\right)}{3!}=\frac{1/\sqrt{2}}{3!}$.

31. **(C)**

$$e^u = 1 + u + \frac{u^2}{2!} + \cdots; \text{ and } \sin x = x - \frac{x^3}{3!} + \cdots, \text{ so}$$

$$e^{\sin x} = 1 + \left(x - \frac{x^3}{3!} + \cdots\right) + \frac{1}{2!}\left(x - \frac{x^3}{3!} + \cdots\right)^2 + \cdots.$$

Or generate the Maclaurin series for $e^{\sin x}$.

BC ONLY

32. **(A)** $f(x) = x \ln x$,

$f'(x) = 1 + \ln x$,

$f''(x) = \frac{1}{x}$,

$f'''(x) = -\frac{1}{x^2}$,

$f^{(4)}(x) = \frac{2}{x^3}$,

$f^{(5)}(x) = -\frac{3 \cdot 2}{x^4}$; $\quad f^{(5)}(1) = -3 \cdot 2$.

So the coefficient of $(x - 1)^5$ is $-\frac{3 \cdot 2}{5!} = -\frac{1}{20}$.

33. **(C)** $\lim\limits_{n \to \infty} \left|\frac{u_{n+1}}{u_n}\right| = \lim\limits_{n \to \infty} \left|\frac{x^{n+1}(n+1)^{n+1}}{2^{n+1}(n+1)!} \cdot \frac{2^n \cdot n!}{x^n \cdot n^n}\right|$

$$= \lim_{n \to \infty} \left|\frac{x(n+1)^{n+1}}{2(n+1)n^n}\right| = \lim_{n \to \infty} \left|\frac{x}{2}\left(\frac{n+1}{n}\right)^n\right|$$

$$= \lim_{n \to \infty} \left|\frac{x}{2}\left(1 + \frac{1}{n}\right)^n\right| = \left|\frac{x}{2} \cdot e\right|.$$

Since the series converges when $\left|\frac{x}{2} \cdot e\right| < 1$, that is, when $|x| < \frac{2}{e}$, the radius of convergence is $\frac{x}{e}$.

34. **(C)** This polynomial is associated with the binomial series $(1 + x)^p$. Verify that $f(0) = 1$, $f'(0) = p$, $f''(0) = p(p - 1)$, $f'''(0) = p(p - 1)(p - 2)$.

35. **(A)** The fastest way to find the series for $\ln(1 + 2x)$ about $x = 0$ is to substitute $2x$ for x in the series

$$\ln(1 + x) = x - \frac{x^2}{2} + \frac{x^3}{3} - \frac{x^4}{4} + \cdots.$$

36. **(D)** $\lim\limits_{n \to \infty} \left|\frac{u_{n+1}}{u_n}\right| = \left|\frac{2}{x}\right|$. The series therefore converges if $\left|\frac{2}{x}\right| < 1$. If $x > 0$, $\left|\frac{2}{x}\right| = \frac{2}{x}$, which is less than 1 if $2 < x$. If $x < 0$, $\left|\frac{2}{x}\right| = -\frac{2}{x}$, which is less than 1 if $-2 > x$. Now for the endpoints:

$x = 2$ yields $1 + 1 + 1 + 1 + \ldots$, which diverges;
$x = -2$ yields $-1 + 1 - 1 + 1 - \ldots$, which diverges.

The answer is $|x| > 2$.

37. **(C)** The function and its first three derivatives at $\frac{\pi}{4}$ are $\sin \frac{\pi}{4} = \frac{1}{\sqrt{2}}$; $\cos \frac{\pi}{4} = \frac{1}{\sqrt{2}}$; $-\sin \frac{\pi}{4} = -\frac{1}{\sqrt{2}}$; and $-\cos \frac{\pi}{4} = -\frac{1}{\sqrt{2}}$. $P_3(x)$ is choice C.

BC ONLY

38. **(D)** $T_3(1) = \frac{h(1)}{0!} + \frac{h'(1)}{1!}(x-1) + \frac{h''(1)}{2!}(x-1)^2 + \frac{h'''(1)}{3!}(x-1)^3$

$$= \frac{6}{1} + \frac{6}{1}(x-1) + \frac{6}{2}(x-1)^2 + \frac{6}{6}(x-1)^3$$

39. **(A)** If $\sum_{k=1}^{\infty} u_k$ converges, so does $\sum_{k=m}^{\infty} u_k$, where m is any positive integer; but their *sums* are probably different.

40. **(E)** Note the following counterexamples:

(A) $\sum \frac{(-1)^{n-1}}{n}$ (B) $\sum \frac{1}{n}$ (C) $\sum \frac{(-1)^{n-1} \cdot n}{2n-1}$ (D) $\sum \left(-\frac{3}{2}\right)^{n-1}$

41. **(C)** Note that $\ln q$ is defined only if $q > 0$, and that the derivatives must exist at $x = a$ in the formula for the Taylor series on page 406.

42. **(E)** (A), (B), (C), and (D) are all true statements.

43. **(E)** The Maclaurin series $\sin x = x - \frac{x^3}{3!} + \frac{x^5}{5!} - \frac{x^7}{7} + \cdots$ converges by the Alternating Series Test, so the error $|R_4|$ is less than the first omitted term. For $x = 1$, we have $\frac{1}{5!} < 0.009$.

44. **(C)** $f(x) = a_0 + a_1x + a_2x^2 + a_3x^3 + \cdots$; if $f(0) = 1$, then $a_0 = 1$.

$f'(x) = a_1 + 2a_2x + 3a_3x^2 + 4a_4x^3 + \cdots$; $f'(0) = -f(0) = -1$,

so $a_1 = -1$. Since $f'(x) = -f(x)$, $f(x) = -f'(x)$:

$1 - x + a_2x^2 + a_3x^3 + \cdots = -\left(-1 + 2a_2x + 3a_3x^2 + 4a_4x^3 + \ldots\right)$

identically. Thus,

$$-2a_2 = -1, \qquad a_2 = \frac{1}{2},$$

$$-3a_3 = a_2, \qquad a_3 = -\frac{1}{3!},$$

$$-4a_4 = a_3, \qquad a_4 = \frac{1}{4!},$$

$$\vdots \qquad\qquad \vdots$$

It is clear, then, that

$$f(x) = 1 - x + \frac{x^2}{2!} - \frac{x^3}{3!} + \frac{x^4}{4!} - \cdots;$$

$$f(0.2) = 1 - 0.2 + \frac{(0.2)^2}{2!} - \frac{(0.2)^3}{3!} + R_3$$

$$= 0.819; |R_3| < 0.0005.$$

45. **(D)** Use a calculator to verify that the ratio $\frac{\pi^3}{3^\pi}$ (of the given geometric series) equals approximately 0.98. Since the ratio $r < 1$, the sum of the series equals

$$\frac{a}{1-r} \quad \text{or} \quad \frac{\pi^3}{3^\pi} \cdot \frac{1}{1-\frac{\pi^3}{3^\pi}}.$$

Simplify to get (D).

46. **(A)** Since the given series converges by the Alternating Series Test, the error is less in absolute value than the first term dropped, that is, less than $\left|\frac{(-1)^{300}}{903-1}\right| \simeq 0.0011.$

Choice (A) is closest to this approximation.

BC ONLY

Miscellaneous Multiple-Choice Practice Questions

11

These questions provide further practice for Parts A and B of Section I of the examination. Answers begin on page 448.

Part A. Directions: Answer these questions *without* using your calculator.

1. Which of the following functions is continuous at $x = 0$?

   (A) $f(x)\begin{cases} = \sin\frac{1}{x} & \text{for } x \neq 0 \\ = 0 & \text{for } x = 0 \end{cases}$

   (B) $f(x) = [x]$ (greatest-integer function)

   (C) $f(x)\begin{cases} = \frac{x}{x} & \text{for } x \neq 0 \\ = 0 & \text{for } x = 0 \end{cases}$

   (D) $f(x)\begin{cases} = x\sin\frac{1}{x} & \text{for } x \neq 0 \\ = 0 & \text{for } x = 0 \end{cases}$

   (E) $f(x) = \frac{x+1}{x}$

2. Which of the following statements about the graph of $y = \frac{x^2+1}{x^2-1}$ is *not* true?

   (A) The graph is symmetric to the y-axis.
   (B) The graph has two vertical asymptotes.
   (C) There is no y-intercept.
   (D) The graph has one horizontal asymptote.
   (E) There is no x-intercept.

3. $\lim\limits_{x\to 1^-}\left([x] - |x|\right)$

   (A) $= -1$ (B) $= 0$ (C) $= 1$ (D) $= 2$ (E) does not exist

4. The x-coordinate of the point on the curve $y = x^2 - 2x + 3$ at which the tangent is perpendicular to the line $x + 3y + 3 = 0$ is

   (A) $-\frac{5}{2}$ (B) $-\frac{1}{2}$ (C) $\frac{7}{6}$ (D) $\frac{5}{2}$ (E) $\frac{5}{6}$

5. $\lim\limits_{x\to 1}\frac{\frac{3}{x}-3}{x-1}$ is

   (A) -3 (B) -1 (C) 1 (D) 3 (E) nonexistent

6. For polynomial function p, $p''(2) = -6$, $p''(4) = 0$, and $p''(5) = 3$. Which is true?

   (A) p has an inflection point at $x = 4$.
   (B) p has a minimum at $x = 4$.
   (C) p has a root at $x = 4$.
   (D) p is increasing on [2,5].
   (E) A–D are all false.

7. $\int_0^6 |x-4|\, dx =$

(A) 6 (B) 8 (C) 10 (D) 11 (E) 12

8. $\lim_{x\to\infty} \frac{3+x-2x^2}{4x^2+9}$ is

(A) $-\frac{1}{2}$ (B) $\frac{1}{2}$ (C) 1 (D) 3 (E) nonexistent

9. The maximum value of the function $f(x) = x^4 - 4x^3 + 6$ on $[1, 4]$ is

(A) 1 (B) 0 (C) 3 (D) 6 (E) -27

10. Let $f(x) = \begin{cases} \frac{\sqrt{x+4}-3}{x-5} & \text{for } x \neq 5 \\ c & \text{for } x = 5 \end{cases}$, and let f be continuous at $x = 5$. Then $c =$

(A) $-\frac{1}{6}$ (B) 0 (C) $\frac{1}{6}$ (D) 1 (E) 6

11. $\int_0^{\pi/2} \cos^2 x \sin x\, dx =$

(A) -1 (B) $-\frac{1}{3}$ (C) 0 (D) $\frac{1}{3}$ (E) 1

12. If $\sin x = \ln y$ and $0 < x < \pi$, then, in terms of x, $\frac{dy}{dx}$ equals

(A) $e^{\sin x}\cos x$ (B) $e^{-\sin x}\cos x$ (C) $\frac{e^{\sin x}}{\cos x}$ (D) $e^{\cos x}$ (E) $e^{\sin x}$

13. If $f(x) = x\cos x$, then $f'\left(\frac{\pi}{2}\right)$ equals

(A) $\frac{\pi}{2}$ (B) 0 (C) -1 (D) $-\frac{\pi}{2}$ (E) 1

14. The equation of the tangent to the curve $y = e^x \ln x$, where $x = 1$, is

(A) $y = ex$ (B) $y = e^x + 1$ (C) $y = e(x-1)$ (D) $y = ex + 1$ (E) $y = x - 1$

15. If the displacement from the origin of a particle moving along the x-axis is given by $s = 3 + (t-2)^4$, then the number of times the particle reverses direction is

(A) 0 (B) 1 (C) 2 (D) 3 (E) 4

16. $\int_{-1}^{0} e^{-x}\, dx$ equals

(A) $1 - e$ (B) $\frac{1-e}{e}$ (C) $e - 1$ (D) $1 - \frac{1}{e}$ (E) $e + 1$

17. If $f(x) = \begin{cases} x^2 & \text{for } x \le 2 \\ 4x - x^2 & \text{for } x > 2 \end{cases}$, then $\int_{-1}^{4} f(x)\, dx$ equals

(A) 7 (B) $\frac{23}{3}$ (C) $\frac{25}{3}$ (D) 9 (E) $\frac{65}{3}$

18. If the position of a particle on a line at time t is given by $s = t^3 + 3t$, then the speed of the particle is decreasing when

(A) $-1 < t < 1$ (B) $-1 < t < 0$ (C) $t < 0$ (D) $t > 0$ (E) $|t| > 1$

19. A rectangle with one side on the x-axis is inscribed in the triangle formed by the lines $y = x$, $y = 0$, and $2x + y = 12$. The area of the largest such rectangle is

CHALLENGE

(A) 6 (B) 3 (C) $\frac{5}{2}$ (D) 5 (E) 7

20. The x-value of the first-quadrant point that is on the curve of $x^2 - y^2 = 1$ and closest to the point (3, 0) is

(A) 1 (B) $\frac{3}{2}$ (C) 2 (D) $\frac{5}{2}$ (E) 3

21. If $y = \ln(4x + 1)$, then $\frac{d^2y}{dx^2}$ is

(A) $\frac{1}{4}$ (B) $\frac{-1}{(4x+1)^2}$ (C) $\frac{-4}{(4x+1)^2}$ (D) $\frac{-16}{(4x+1)^2}$ (E) $\frac{-1}{16(4x+1)^2}$

22. The region bounded by the parabolas $y = x^2$ and $y = 6x - x^2$ is rotated about the x-axis so that a vertical line segment cut off by the curves generates a ring. The value of x for which the ring of largest area is obtained is

(A) 4 (B) 3 (C) $\frac{5}{2}$ (D) 2 (E) $\frac{3}{2}$

23. $\int \frac{dx}{x \ln x}$ equals

(A) $\ln |\ln x| + C$ (B) $-\frac{1}{\ln^2 x} + C$ (C) $\frac{(\ln x)^2}{2} + C$ (D) $\ln x + C$

24. The volume obtained by rotating the region bounded by $x = y^2$ and $x = 2 - y^2$ about the y-axis is equal to

(A) $\frac{16\pi}{3}$ (B) $\frac{32\pi}{3}$ (C) $\frac{32\pi}{15}$ (D) $\frac{64\pi}{15}$ (E) $\frac{8\pi}{3}$

25. The general solution of the differential equation $\frac{dy}{dx} = \frac{1 - 2x}{y}$ is a family of

(A) straight lines (B) circles (C) hyperbolas (D) parabolas (E) ellipses

26. Estimate $\int_0^4 \sqrt{25 - x^2}\, dx$ using the Left Rectangular Rule and two subintervals of equal width.

(A) $3 + \sqrt{21}$ (B) $5 + \sqrt{21}$ (C) $6 + 2\sqrt{21}$ (D) $8 + 2\sqrt{21}$ (E) $10 + 2\sqrt{21}$

27. $\int_0^7 \sin \pi x \, dx =$

(A) -2 (B) $-\frac{2}{\pi}$ (C) 0 (D) $\frac{1}{\pi}$ (E) $\frac{2}{\pi}$

28. $\lim_{x \to 0} \frac{\tan 3x}{2x} =$

(A) 0 (B) $\frac{1}{2}$ (C) $\frac{2}{3}$ (D) $\frac{3}{2}$ (E) ∞

29. $\lim_{h \to 0} \frac{\tan(\pi/4 + h) - 1}{h} =$

(A) 0 (B) $\frac{1}{2}$ (C) 1 (D) 2 (E) ∞

CHALLENGE

30. The number of values of k for which $f(x) = e^x$ and $g(x) = k \sin x$ have a common point of tangency is

(A) 0 (B) 1 (C) 2 (D) large but finite (E) infinite

31. The curve $2x^2 y + y^2 = 2x + 13$ passes through (3,1). Use the line tangent to the curve there to find the approximate value of y at $x = 2.8$.

(A) 0.5 (B) 0.9 (C) 0.95 (D) 1.1 (E) 1.4

32. $\int \cos^3 x \, dx =$

(A) $\frac{\cos^4 x}{4} + C$ (B) $\frac{\sin^4 x}{4} + C$ (C) $\sin x - \frac{\sin^3 x}{3} + C$

(D) $\sin x + \frac{\sin^3 x}{3} + C$ (E) $\cos x - \frac{\cos^3 x}{3} + C$

33. The region bounded by $y = \tan x$, $y = 0$, and $x = \frac{\pi}{4}$ is rotated about the x-axis. The volume generated equals

(A) $\pi - \frac{\pi^2}{4}$ (B) $\pi(\sqrt{2} - 1)$ (C) $\frac{3\pi}{4}$ (D) $\pi\left(1 + \frac{\pi}{4}\right)$ (E) $\frac{\pi}{3}$

34. $\lim_{h \to 0} \frac{a^h - 1}{h}$, for the constant $a > 0$, equals

(A) 1 (B) a (C) $\ln a$ (D) $\log_{10} a$ (E) $a \ln a$

35. Solutions of the differential equation whose slope field is shown here are most likely to be

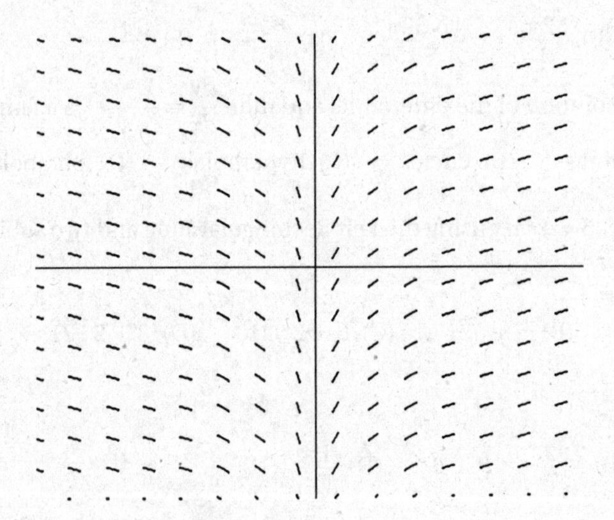

(A) quadratic (B) cubic (C) sinusoidal
(D) exponential (E) logarithmic

36. $\lim_{h \to 0} \frac{1}{h} \int_{\frac{\pi}{4}}^{\frac{\pi}{4} + h} \frac{\sin x}{x} \, dx =$

(A) 0 (B) 1 (C) $\frac{\sqrt{2}}{2}$ (D) $\frac{2\sqrt{2}}{\pi}$ (E) $\frac{2\sqrt{2}\,(\pi - 4)}{\pi^2}$

37. The graph of g, shown below, consists of the arcs of two quarter-circles and two straight-line segments. The value of $\int_0^{12} g(x)\,dx$ is

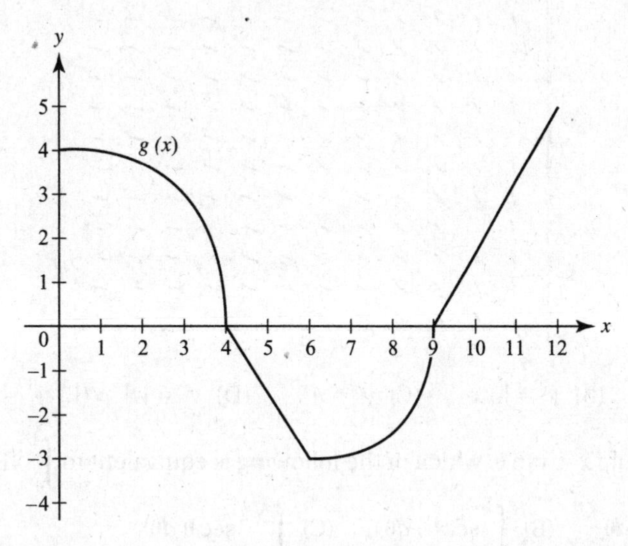

(A) $\pi + 2$ (B) $\frac{7\pi}{4} + \frac{9}{2}$ (C) $\frac{7\pi}{4} + 8$ (D) $7\pi + \frac{9}{2}$ (E) $\frac{25\pi}{4} + \frac{21}{2}$

38. Which of these could be a particular solution of the differential equation whose slope field is shown here?

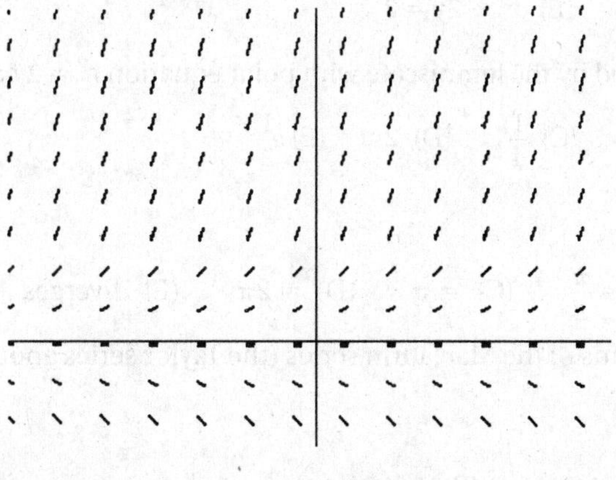

(A) $y = \frac{1}{x}$ (B) $y = \ln x$ (C) $y = e^x$ (D) $y = e^{-x}$ (E) $y = e^{x^2}$

39. What is the domain of the particular solution for $\frac{dy}{dx} = \frac{6x}{x^2 - 4}$ containing the point where $x = -1$?

(A) $x < 0$ (B) $x > -2$ (C) $-2 < x < 2$
(D) $x \neq \pm 2$ (E) none of these; no solution exists for $x = -1$

40. The slope field shown here is for the differential equation

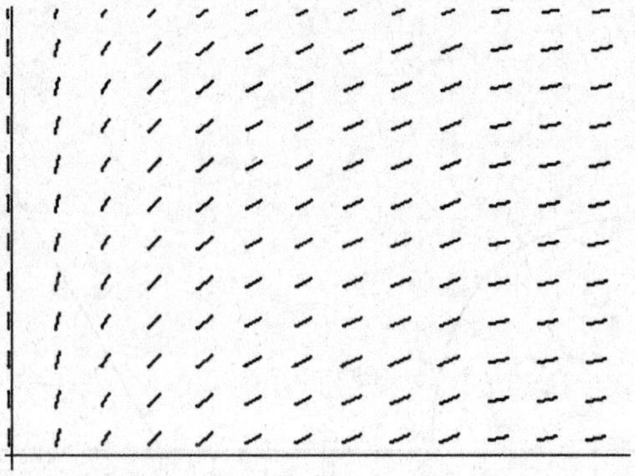

(A) $y' = \frac{1}{x}$ (B) $y' = \ln x$ (C) $y' = e^x$ (D) $y' = y$ (E) $y' = -y^2$

41. If we substitute $x = \tan\theta$, which of the following is equivalent to $\int_0^1 \sqrt{1+x^2}\,dx$?

(A) $\int_0^1 \sec\theta\,d\theta$ (B) $\int_0^1 \sec^3\theta\,d\theta$ (C) $\int_0^{\pi/4} \sec\theta\,d\theta$

(D) $\int_0^{\pi/4} \sec^3\theta\,d\theta$ (E) $\int_0^{\tan 1} \sec^3\theta\,d\theta$

BC ONLY

42. If $x = 2\sin u$ and $y = \cos 2u$, then a single equation in x and y is

(A) $x^2 + y^2 = 1$ (B) $x^2 + 4y^2 = 4$ (C) $x^2 + 2y = 2$
(D) $x^2 + y^2 = 4$ (E) $x^2 - 2y = 2$

43. The area bounded by the lemniscate with polar equation $r^2 = 2\cos 2\theta$ is equal to

(A) 4 (B) 1 (C) $\frac{1}{4}$ (D) 2 (E) $\frac{1}{2}$

44. $\int_{-\infty}^{\infty} \frac{dx}{x^2+1}$

(A) $= 0$ (B) $= \frac{\pi}{2}$ (C) $= \pi$ (D) $= 2\pi$ (E) diverges

45. The first four terms of the Maclaurin series (the Taylor series about $x = 0$) for $f(x) = \frac{1}{1-2x}$ are

(A) $1 + 2x + 4x^2 + 8x^3$ (B) $1 - 2x + 4x^2 - 8x^3$
(C) $-1 - 2x - 4x^2 - 8x^3$ (D) $1 - x + x^2 - x^3$
(E) $1 + x + x^2 + x^3$

46. $\int x^2 e^{-x}\,dx =$

(A) $\frac{1}{3}x^3e^{-x} + C$ (B) $-\frac{1}{3}x^3e^{-x} + C$ (C) $-x^2e^{-x} + 2xe^{-x} + C$

(D) $-x^2e^{-x} - 2xe^{-x} - 2e^{-x} + C$ (E) $-x^2e^{-x} + 2xe^{-x} - 2e^{-x} + C$

47. $\int_0^{\pi/2} \sin^2 x\,dx$ is equal to

(A) $\frac{1}{3}$ (B) $\frac{\pi}{4} - \frac{1}{4}$ (C) $\frac{\pi}{2}$ (D) $\frac{\pi}{2} - \frac{1}{3}$ (E) $\frac{\pi}{4}$

BC ONLY

48. A curve is given parametrically by the equations $x = t$, $y = 1 - \cos t$. The area bounded by the curve and the x-axis on the interval $0 \leqq t \leqq 2\pi$ is equal to

(A) $2(\pi + 1)$ (B) π (C) 4π (D) $\pi + 1$ (E) 2π

49. If $x = a \cot \theta$ and $y = a \sin^2 \theta$, then $\frac{dy}{dx}$, when $\theta = \frac{\pi}{4}$, is equal to

(A) $\frac{1}{2}$ (B) -1 (C) 2 (D) $-\frac{1}{2}$ (E) $-\frac{1}{4}$

50. Which of the following improper integrals diverges?

(A) $\int_0^{\infty} e^{-x^2}\,dx$ (B) $\int_{-\infty}^{0} e^x\,dx$ (C) $\int_0^1 \frac{dx}{x}$

(D) $\int_0^{\infty} e^{-x}\,dx$ (E) $\int_0^1 \frac{dx}{\sqrt{x}}$

51. $\int_2^4 \frac{du}{\sqrt{16 - u^2}}$ equals

(A) $\frac{\pi}{12}$ (B) $\frac{\pi}{6}$ (C) $\frac{\pi}{4}$ (D) $\frac{\pi}{3}$ (E) $\frac{2\pi}{3}$

52. $\lim\limits_{x \to 0^+} \left(\frac{1}{x}\right)^x$ is

(A) $-\infty$ (B) 0 (C) 1 (D) ∞ (E) nonexistent

53. A particle moves along the parabola $x = 3y - y^2$ so that $\frac{dy}{dt} = 3$ at all time t. The speed of the particle when it is at position (2,1) is equal to

(A) 0 (B) 3 (C) $3\sqrt{2}$ (D) $\sqrt{10}$ (E) $\sqrt{13}$

54. $\lim\limits_{x \to 0^+} \frac{\cot x}{\ln x} =$

(A) $-\infty$ (B) -1 (C) 0 (D) 1 (E) ∞

55. When rewritten as partial fractions, $\frac{3x + 2}{x^2 - x - 12}$ includes which of the following?

I. $\frac{1}{x + 3}$ II. $\frac{1}{x - 4}$ III. $\frac{2}{x - 4}$

(A) none (B) I only (C) II only (D) III only (E) I and III

56. Using two terms of an appropriate Maclaurin series, estimate $\int_0^1 \frac{1 - \cos x}{x}\,dx$.

(A) $\frac{1}{96}$ (B) $\frac{23}{96}$ (C) $\frac{1}{4}$ (D) $\frac{25}{96}$ (E) undefined; the integral is improper

57. The slope of the spiral $r = \theta$ at $\theta = \frac{\pi}{4}$ is

(A) $-\sqrt{2}$ (B) -1 (C) 1 (D) $\frac{4 + \pi}{4 - \pi}$ (E) undefined

Part B. Directions: Some of these questions require the use of a graphing calculator.

58. The graph of function h is shown here. Which of these statements is (are) true?

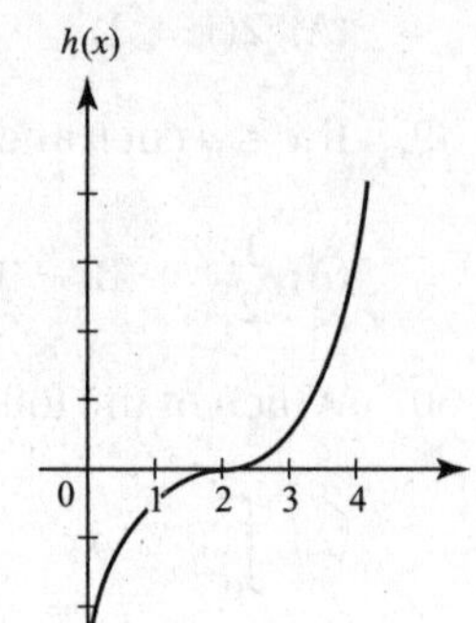

I. The first derivative is never negative.
II. The second derivative is constant.
III. The first and second derivatives equal 0 at the same point.

(A) I only (B) III only (C) I and II
(D) I and III (E) I, II, and III

59. Graphs of functions $f(x)$, $g(x)$, and $h(x)$ are shown below.

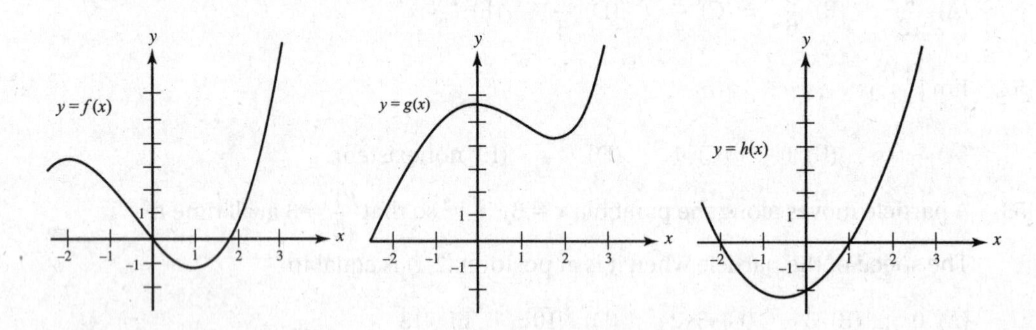

Consider the following statements:

I. $g(x) = f'(x)$
II. $f(x) = g'(x)$
III. $h(x) = g''(x)$

Which of these statements is (are) true?

(A) I only (B) II only (C) II and III only
(D) I, II, and III (E) none of I, II, or III

60. If $y = \int_3^x \frac{1}{\sqrt{3+2t}}\,dt$, then $\frac{d^2y}{dx^2} =$

(A) $-\frac{1}{(3+2x)^{\frac{3}{2}}}$ (B) $\frac{3}{(3+2x)^{\frac{5}{2}}}$ (C) $\frac{\sqrt{3+2x}}{2}$ (D) $\frac{(3+2x)^{\frac{3}{2}}}{3}$ (E) 0

61. If $\int_{-3}^{4} f(x)dx = 6$, then $\int_{-4}^{3} f(x+1)dx =$

(A) −6 (B) −5 (C) 5 (D) 6 (E) 7

62. At what point in the interval [1, 1.5] is the rate of change of $f(x) = \sin x$ equal to its average rate of change on the interval?

(A) 0.995 (B) 1.058 (C) 1.239 (D) 1.253 (E) 1.399

63. Suppose $f'(x) = x^2(x-1)$. Then $f''(x) = x(3x-2)$. Over which interval(s) is the graph of f both increasing and concave up?

I. $x < 0$ II. $0 < x < \frac{2}{3}$ III. $\frac{2}{3} < x < 1$ IV. $x > 1$

(A) I only (B) II only (C) II and IV (D) I and III (E) IV only

64. Which of the following statements is true about the graph of $f(x)$ in Question 62?

(A) The graph has no relative extrema.
(B) The graph has one relative extremum and one inflection point.
(C) The graph has two relative extrema and one inflection point.
(D) The graph has two relative extrema and two inflection points.
(E) None of the preceding statements is true.

65. The nth derivative of $\ln(x+1)$ at $x = 2$ equals

(A) $\frac{(-1)^{n-1}}{3^n}$ (B) $\frac{(-1)^n \cdot n!}{3^{n+1}}$ (C) $\frac{(-1)^{n-1}(n-1)!}{3^n}$
(D) $\frac{(-1)^{n+1} \cdot n!}{3^{n+1}}$ (E) $\frac{(-1)^{n+1}}{3^{n+1}}$

66. If $f(x)$ is continuous at the point where $x = a$, which of the following statements may be false?

(A) $\lim_{x \to a} f(x)$ exists. (B) $\lim_{x \to a} f(x) = f(a)$. (C) $f'(a)$ exists.

(D) $f(a)$ is defined. (E) $\lim_{x \to a^-} f(x) = \lim_{x \to a^+} f(x)$.

67. Suppose $\int_0^3 f(x+k)\,dx = 4$, where k is a constant. Then $\int_k^{3+k} f(x)\,dx$ equals

(A) 3 (B) $4-k$ (C) 4 (D) $4+k$ (E) $4k$

68. The volume, in cubic feet, of an "inner tube" with inner diameter 4 ft and outer diameter 8 ft is **CHALLENGE**

(A) $4\pi^2$ (B) $12\pi^2$ (C) $8\pi^2$ (D) $24\pi^2$ (E) $6\pi^2$

69. If $f(u) = \tan^{-1} u^2$ and $g(u) = e^u$, then the derivative of $f(g(u))$ is

(A) $\frac{2ue^u}{1+u^4}$ (B) $\frac{2ue^{u^2}}{1+u^4}$ (C) $\frac{2e^u}{1+4e^{2u}}$ (D) $\frac{2e^{2u}}{1+e^{4u}}$ (E) $\frac{2e^{2u}}{\sqrt{1-e^{4u}}}$

70. If $\sin(xy) = y$, then $\frac{dy}{dx}$ equals

(A) $\sec(xy)$ (B) $y\cos(xy) - 1$ (C) $\frac{1 - y\cos(xy)}{x\cos(xy)}$

(D) $\frac{y\cos(xy)}{1 - x\cos(xy)}$ (E) $\cos(xy)$

71. Let $x > 0$. Suppose $\frac{d}{dx}f(x) = g(x)$ and $\frac{d}{dx}g(x) = f(\sqrt{x})$; then $\frac{d^2}{dx^2}f(x^2) =$

(A) $f(x^4)$ (B) $f(x^2)$ (C) $2xg(x^2)$ (D) $\frac{1}{2x}f(x)$ (E) $2g(x^2) + 4x^2 f(x)$

72. The region bounded by $y = e^x$, $y = 1$, and $x = 2$ is rotated about the x-axis. The volume of the solid generated is given by the integral

(A) $\pi\int_0^2 e^{2x}dx$ (B) $2\pi\int_1^{e^2} (2 - \ln y)(y - 1)\, dy$ (C) $\pi\int_0^2 \left(e^{2x} - 1\right) dx$

(D) $2\pi\int_0^{e^2} y(2 - \ln y)\, dy$ (E) $\pi\int_0^2 \left(e^x - 1\right)^2 dx$

73. Suppose the function f is continuous on $1 \leqq x \leqq 2$, that $f'(x)$ exists on $1 < x < 2$, that $f(1) = 3$, and that $f(2) = 0$. Which of the following statements is *not* necessarily true?

(A) The Mean-Value Theorem applies to f on $1 \leqq x \leqq 2$.

(B) $\int_1^2 f(x)\, dx$ exists.

(C) There exists a number c in the closed interval [1, 2] such that $f'(c) = 0$.

(D) If k is any number between 0 and 3, there is a number c between 1 and 2 such that $f(c) = k$.

(E) If c is any number such that $1 < c < 2$, then $\lim_{x \to c} f(x)$ exists.

74. The region S in the figure is bounded by $y = \sec x$, the y-axis, and $y = 4$. What is the volume of the solid formed when S is rotated about the y-axis?

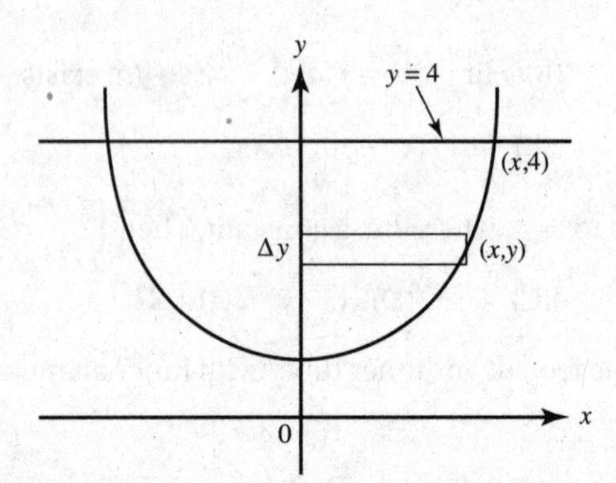

(A) 0.791 (B) 2.279 (C) 5.692 (D) 11.385 (E) 17.217

75. If 40 g of a radioactive substance decomposes to 20 g in 2 yr, then, to the nearest gram, the amount left after 3 yr is

(A) 10 (B) 12 (C) 14 (D) 16 (E) 17

76. An object in motion along a line has acceleration $a(t) = \pi t + \dfrac{2}{1 + t^2}$ and is at rest when $t = 1$. Its average velocity from $t = 0$ to $t = 2$ is

(A) 0.362 (B) 0.274 (C) 3.504 (D) 7.008 (E) 8.497

77. Find the area bounded by $y = \tan x$ and $x + y = 2$, and above the x-axis on the interval [0, 2].

(A) 0.919 (B) 0.923 (C) 1.013 (D) 1.077 (E) 1.494

78. An ellipse has major axis 20 and minor axis 10. Rounded off to the nearest integer, the maximum area of an inscribed rectangle is

(A) 50 (B) 79 (C) 80 (D) 82 (E) 100

79. The average value of $y = x \ln x$ on the interval $1 \leqq x \leqq e$ is

(A) 0.772 (B) 1.221 (C) 1.359 (D) 1.790 (E) 2.097

80. Let $f(x) = \int_0^x (1 - 2\cos^3 t)\,dt$ for $0 \leqq x \leqq 2\pi$. On which interval is f increasing?

(A) $0 < x < \pi$ (B) $0.654 < x < 5.629$ (C) $0.654 < x < 2\pi$
(D) $\pi < x < 2\pi$ (E) none of these

81. The table shows the speed of an object (in ft/sec) at certain times during a 6-sec period. Estimate its acceleration (in ft/sec^2) at $t = 2$ sec.

time, sec	0	1	3	6
speed, ft/sec	30	22	12	6

(A) -10 (B) -6 (C) -5 (D) -4 (E) $\frac{-10}{3}$

82. A maple-syrup storage tank 16 ft high hangs on a wall. The back is in the shape of the parabola $y = x^2$ and all cross sections parallel to the floor are squares. If syrup is pouring in at the rate of 12 ft^3/hr, how fast (in ft/hr) is the syrup level rising when it is 9 ft deep?

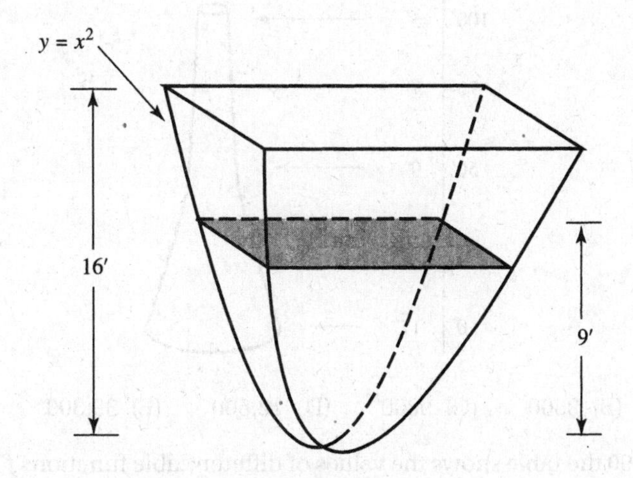

(A) $\frac{2}{27}$ (B) $\frac{1}{3}$ (C) $\frac{4}{3}$ (D) 36 (E) 162

83. In a protected area (no predators, no hunters), the deer population increases at a rate of $\frac{dP}{dt} = k(1000 - P)$, where $P(t)$ represents the population of deer at t yr. If 300 deer were originally placed in the area and a census showed the population had grown to 500 in 5 yr, how many deer will there be after 10 yr?

(A) 608 (B) 643 (C) 700 (D) 833 (E) 892

84. Shown is the graph of $f(x) = \dfrac{4}{x^2 + 1}$.

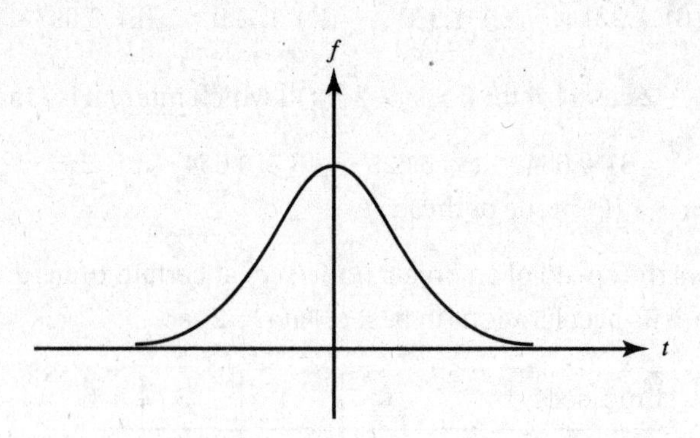

Let $H(x) = \int_0^x f(t)\,dt$. The local linearization of H at $x = 1$ is $H(x)$ equals

(A) $2x$ (B) $-2x - 4$ (C) $2x + \pi - 2$ (D) $-2x + \pi + 2$ (E) $2x + \ln 16 + 2$

85. A smokestack 100 ft tall is used to treat industrial emissions. The diameters, measured at 25-ft intervals, are shown in the table. Using the midpoint rule with 4 subintervals, estimate the volume of the smokestack to the nearest 100 ft^3.

ht	dia
100′	5′
75′	7′
50′	9′
25′	15′
0′	17′

(A) 8100 (B) 9500 (C) 9800 (D) 12,500 (E) 39,300

For Questions 86–90 the table shows the values of differentiable functions f and g.

x	f	f'	g	g'
1	2	$\frac{1}{2}$	−3	5
2	3	1	0	4
3	4	2	2	3
4	6	4	3	$\frac{1}{2}$

86. If $P(x) = \frac{f(x)}{g(x)}$, then $P'(3) =$

(A) -2 (B) $-\frac{8}{9}$ (C) $-\frac{1}{2}$ (D) $\frac{2}{3}$ (E) 2

87. If $H(x) = f(g(x))$, then $H'(3) =$

(A) 1 (B) 2 (C) 3 (D) 6 (E) 9

88. If $M(x) = f(x) \cdot g(x)$, then $M'(3) =$

(A) 2 (B) 6 (C) 8 (D) 14 (E) 16

89. If $K(x) = g^{-1}(x)$, then $K'(3) =$

(A) $-\frac{1}{2}$ (B) $-\frac{1}{3}$ (C) $\frac{1}{3}$ (D) $\frac{1}{2}$ (E) 2

90. If $R(x) = \sqrt{f(x)}$, then $R'(3) =$

(A) $\frac{1}{4}$ (B) $\frac{1}{2\sqrt{2}}$ (C) $\frac{1}{2}$ (D) $\sqrt{2}$ (E) 2

91. Water is poured into a spherical tank at a constant rate. If $W(t)$ is the rate of increase of the depth of the water, then W is

(A) constant (B) linear and increasing (C) linear and decreasing
(D) concave up (E) concave down

92. The graph of f' is shown below. If $f(7) = 3$ then $f(1) =$

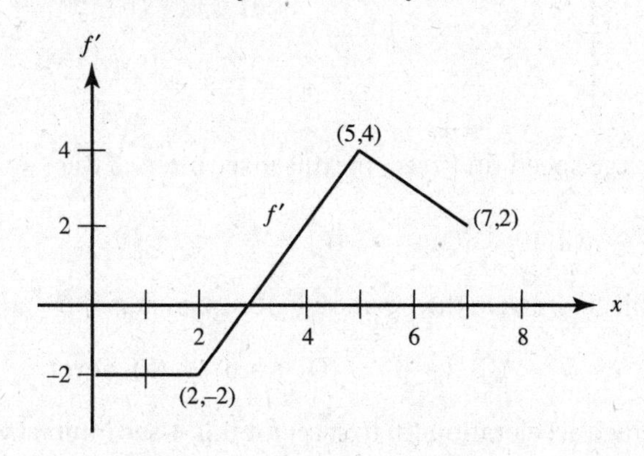

(A) -10 (B) -4 (C) -3 (D) 10 (E) 16

93. At an outdoor concert, the crowd stands in front of the stage filling a semicircular disk of radius 100 yd. The approximate density of the crowd x yd from the stage is given by

$$D(x) = \frac{20}{2\sqrt{x} + 1}$$

people per square yard. About how many people are at the concert?

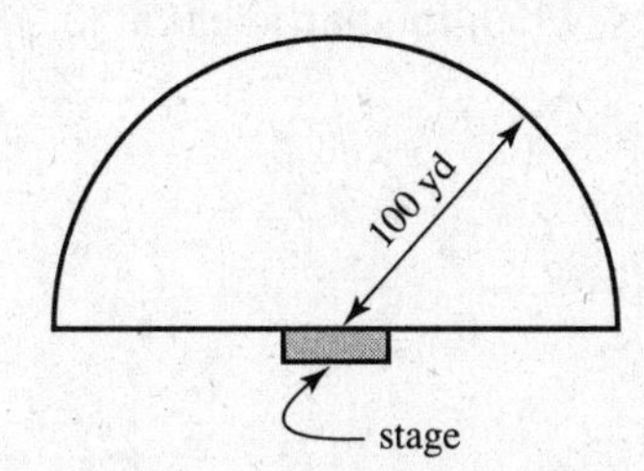

(A) 200 (B) 19,500 (C) 21,000 (D) 165,000 (E) 591,000

94. The Centers for Disease Control announced that, although more AIDS cases were reported this year, the rate of increase is slowing down. If we graph the number of AIDS cases as a function of time, the curve is currently

(A) increasing and linear
(B) increasing and concave down
(C) increasing and concave up
(D) decreasing and concave down
(E) decreasing and concave up

The graph below is for Questions 95–97. It shows the velocity, in feet per second, for $0 < t < 8$, of an object moving along a straight line.

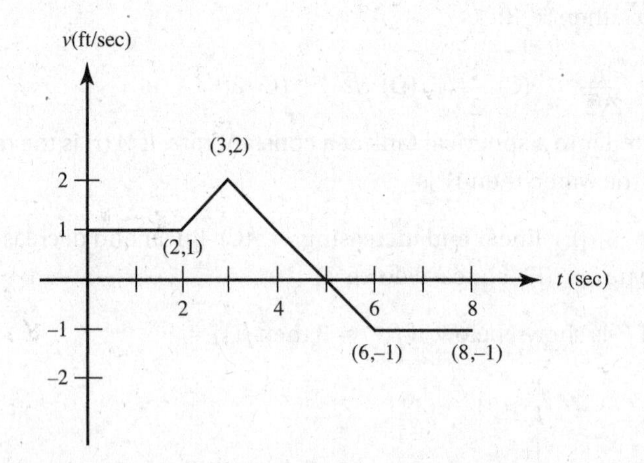

95. The object's average speed (in ft/sec) for this 8-sec interval was

(A) 0 (B) $\frac{3}{8}$ (C) 1 (D) $\frac{8}{3}$ (E) 8

96. When did the object return to the position it occupied at $t = 2$?

(A) $t = 4$ (B) $t = 5$ (C) $t = 6$ (D) $t = 8$ (E) never

97. The object's average acceleration (in ft/sec^2) for this 8-sec interval was

(A) -2 (B) $-\frac{1}{4}$ (C) 0 (D) $\frac{1}{4}$ (E) 1

98. If a block of ice melts at the rate of $\frac{72}{2t+3}$ cm^3/min, how much ice melts during the first 3 min?

(A) 8 cm^3 (B) 16 cm^3 (C) 21 cm^3 (D) 40 cm^3 (E) 79 cm^3

BC ONLY

99. A particle moves counterclockwise on the circle $x^2 + y^2 = 25$ with a constant speed of 2 ft/sec. Its velocity vector, **v**, when the particle is at (3, 4), equals

(A) $\left\langle -\frac{8}{5}, \frac{6}{5} \right\rangle$ (B) $\left\langle \frac{8}{5}, -\frac{6}{5} \right\rangle$ (C) $\left\langle -2\sqrt{3}, 2 \right\rangle$

(D) $\left\langle 2, -2\sqrt{3} \right\rangle$ (E) $\left\langle -2\sqrt{2}, 2\sqrt{2} \right\rangle$

BC ONLY

100. Let $\mathbf{R} = \langle a \cos kt, a \sin kt \rangle$ be the (position) vector from the origin to a moving point $P(x, y)$ at time t, where a and k are positive constants. The acceleration vector, **a**, equals

(A) $-k^2 \mathbf{R}$ (B) $a^2k^2\mathbf{R}$ (C) $-a\mathbf{R}$ (D) $-ak^2 \langle a \cos kt, a \sin kt \rangle$ (E) $-\mathbf{R}$

101. The length of the curve $y = 2^x$ between (0, 1) and (2, 4) is

(A) 3.141 (B) 3.664 (C) 4.823 (D) 5.000 (E) 7.199

102. The position of a moving object is given by $P(t) = \left(3t, e^t\right)$. Its acceleration is

(A) undefined
(B) constant in both magnitude and direction
(C) constant in magnitude only
(D) constant in direction only
(E) constant in neither magnitude nor direction

103. Suppose we plot a particular solution of $\frac{dy}{dx} = 4y$ from initial point (0, 1) using Euler's method. After one step of size $\Delta x = 0.1$, how big is the error?

(A) 0.09 (B) 1.09 (C) 1.49 (D) 1.90 (E) 2.65

104. We use the first three terms to estimate $\sum_{n=0}^{\infty} \frac{(-1)^n}{n^2+1}$. Which of the following statements is (are) true?

I. The estimate is 0.7.
II. The estimate is too low.
III. The estimate is off by less than 0.1.

(A) I only (B) III only (C) I and II (D) I and III (E) I, II, and III

105. Which of these diverges?

(A) $\sum_{n=1}^{\infty} \frac{2}{3^n}$ (B) $\sum_{n=1}^{\infty} \left(\frac{2}{3}\right)^n$ (C) $\sum_{n=1}^{\infty} \frac{2}{3n}$ (D) $\sum_{n=1}^{\infty} \frac{2}{n^3}$ (E) $\sum_{n=1}^{\infty} \frac{2n}{3^n}$

106. Find the radius of convergence of $\sum_{n=1}^{\infty} \frac{n!}{n^n} x^n$.

(A) 0 (B) $\frac{1}{e}$ (C) 1 (D) e (E) ∞

107. When we use $e^x \simeq 1 + x + \frac{x^2}{2}$ to estimate $\sqrt{e}$, the Lagrange remainder is no greater than

(A) 0.021 (B) 0.034 (C) 0.042 (D) 0.067 (E) 0.742

108. An object in motion along a curve has position $P(t) = (\tan t, \cos 2t)$ for $0 \leqslant t \leqslant 1$. How far does it travel?

(A) 0.96 (B) 1.73 (C) 2.10 (D) 2.14 (E) 3.98

Answer Key

1. **D**
2. **C**
3. **A**
4. **D**
5. **A**
6. **E**
7. **C**
8. **A**
9. **D**
10. **C**
11. **D**
12. **A**
13. **D**
14. **C**
15. **B**
16. **C**
17. **C**
18. **C**
19. **A**
20. **B**
21. **D**
22. **D**
23. **A**
24. **A**
25. **E**
26. **E**
27. **E**
28. **D**
29. **D**
30. **E**
31. **D**
32. **C**
33. **A**
34. **C**
35. **E**
36. **D**
37. **B**
38. **C**
39. **C**
40. **A**
41. **D**
42. **C**
43. **D**
44. **C**
45. **A**
46. **D**
47. **E**
48. **E**
49. **D**
50. **C**
51. **D**
52. **C**
53. **C**
54. **A**
55. **E**
56. **B**
57. **D**
58. **D**
59. **C**
60. **A**
61. **D**
62. **D**
63. **E**
64. **E**
65. **C**
66. **C**
67. **C**
68. **E**
69. **D**
70. **D**
71. **E**
72. **C**
73. **C**
74. **D**
75. **C**
76. **A**
77. **D**
78. **E**
79. **B**
80. **B**
81. **C**
82. **B**
83. **B**
84. **C**
85. **C**
86. **A**
87. **C**
88. **E**
89. **E**
90. **C**

9 . **D**

92. **B**
93. **B**
94. **B**
95. **C**
96. **E**
97. **B**
98. **D**
99. **A**
100. **A**
101. **B**
102. **D**
103. **A**
104. **D**
105. **C**
106. **D**
107. **B**
108. **D**

ANSWERS EXPLAINED

Part A

1. **(D)** If $f(x) = x \sin\frac{1}{x}$ for $x \neq 0$ and $f(0) = 0$ then,

$$\lim_{x\to 0} f(x) = 0 = f(0);$$

thus this function is continuous at 0. In (A), $\lim_{x\to 0} \sin\frac{1}{x}$ does not exist; in (B), f has a jump discontinuity; in (C), f has a removable discontinuity; and in (E), f has an infinite discontinuity.

2. **(C)** To find the y-intercept, let $x = 0$; $y = -1$.

3. **(A)** $\lim_{x\to 1^-} [x] - \lim_{x\to 1^-} |x| = 0 - 1 = -1$.

4. **(D)** The line $x + 3y + 3 = 0$ has slope $-\frac{1}{3}$; a line perpendicular to it has slope 3. The slope of the tangent to $y = x^2 - 2x + 3$ at any point is the derivative $2x - 2$. Set $2x - 2$ equal to 3.

5. **(A)** $\lim_{x\to 1} \frac{\frac{3}{x}-3}{x-1}$ is $f'(1)$, where $f(x) = \frac{3}{x}$, $f'(x) = -\frac{3}{x^2}$. Or simplify the given fraction to $\frac{3-3x}{x(x-1)} = \frac{3(1-x)}{x(x-1)} = \frac{-3}{x}$ $(x \neq 1)$.

6. **(E)** Because $p''(2) < 0$ and $p''(5) > 0$, p changes concavity somewhere in the interval $[2,5]$, but we cannot be sure p'' changes sign at $x = 4$.

7. **(C)** $\int_0^6 |x-4|\,dx = \int_0^4 (4-x)\,dx + \int_4^6 (x-4)\,dx = \left(4x - \frac{x^2}{2}\right)\Big|_0^4 + \left(\frac{x^2}{2} - 4x\right)\Big|_4^6$
$= 8 + [(18-24) - (8-16)] = 8 + (-6+8) = 10.$

Save time by finding the area under $y = |x-4|$ from a sketch!

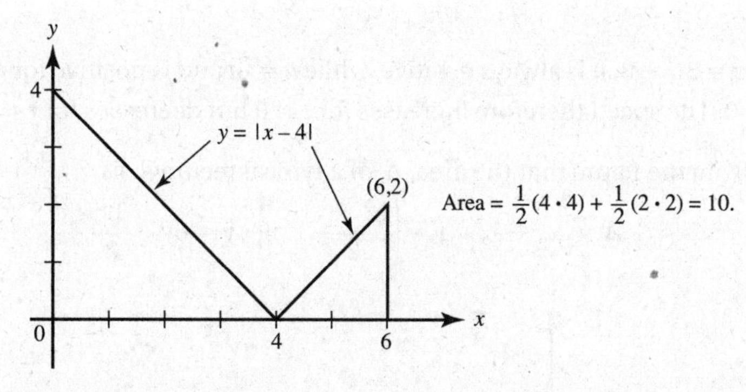

8. **(A)** Since the degrees of numerator and denominator are the same, the limit as $x \to \infty$ is the ratio of the coefficients of the terms of highest degree: $\frac{-2}{4}$.

9. **(D)** On the interval $[1, 4]$, $f'(x) = 0$ only for $x = 3$. Since $f(3)$ is a relative minimum, check the endpoints to find that $f(4) = 6$ is the absolute maximum of the function.

10. **(C)** To find $\lim f$ as $x \to 5$ (if it exists), multiply f by $\frac{\sqrt{x+4}+3}{\sqrt{x+4}+3}$.

$$f(x) = \frac{x-5}{(x-5)(\sqrt{x+4}+3)}$$

and if $x \neq 5$ this equals $\frac{1}{\sqrt{x+4}+3}$. So $\lim f(x)$ as $x \to 5$ is $\frac{1}{6}$. For f to be continuous at $x = 5$, $f(5)$ or c must also equal $\frac{1}{6}$.

11. **(D)** Evaluate $-\frac{1}{3}\cos^3 x \Big|_0^{\pi/2}$.

12. **(A)** $\cos x = \frac{1}{y}\frac{dy}{dx}$ and thus $\frac{dy}{dx} = y\cos x$. From the equation given, $y = e^{\sin x}$.

13. **(D)** If $f(x) = x\cos x$, then $f'(x) = -x\sin x + \cos x$, and

$$f'\left(\frac{\pi}{2}\right) = -\frac{\pi}{2}\cdot 1 + 0.$$

14. **(C)** If $y = e^x \ln x$, then $\frac{dy}{dx} = \frac{e^x}{x} + e^x \ln x$, which equals e when $x = 1$. Since also $y = 0$ when $x = 1$, the equation of the tangent is $y = e(x - 1)$.

15. **(B)** $v = 4(t - 2)^3$ and changes sign exactly once, when $t = 2$.

16. **(C)** Evaluate $-e^{-x}\Big|_{-1}^{0}$.

17. **(C)** $\int_{-1}^{4} f(x)\,dx = \int_{-1}^{2} x^2\,dx + \int_{2}^{4}\left(4x - x^2\right)dx = \frac{x^3}{3}\Big|_{-1}^{2} + \left(2x^2 - \frac{x^3}{3}\right)\Big|_{2}^{4}$.

18. **(C)** Since $v = 3t^2 + 3$, it is always positive, while $a = 6t$ and is positive for $t > 0$ but negative for $t < 0$. The speed therefore increases for $t > 0$ but decreases for $t < 0$.

19. **(A)** Note from the figure that the area, A, of a typical rectangle is

$$A = (x_2 - x_1)\cdot y = \left(\frac{12 - y}{2} - y\right)\cdot y = 6y - \frac{3y^2}{2}.$$

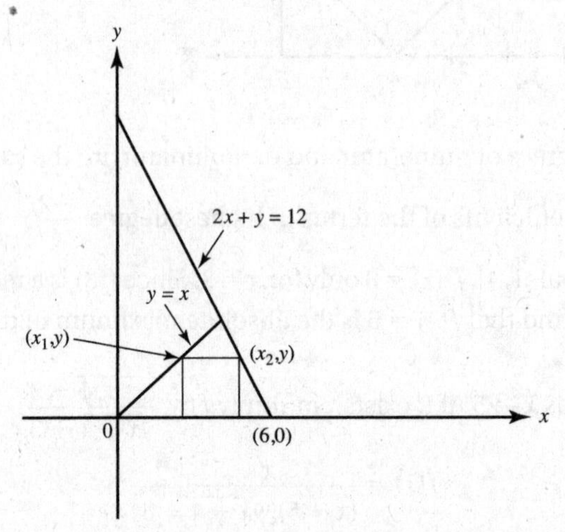

For $y = 2$, $\frac{dA}{dy} = 0$. Note that $\frac{d^2A}{dy^2}$ is always negative.

20. **(B)** If S represents the square of the distance from (3, 0) to a point (x, y) on the curve, then $S = (3 - x)^2 + y^2 = (3 - x)^2 + \left(x^2 - 1\right)$. Setting $\frac{ds}{dx} = 0$ yields the minimum distance at $x = \frac{3}{2}$.

21. **(D)** $\frac{dy}{dx} = \frac{4}{4x + 1} = 4(4x + 1)^{-1}$, so $\frac{d^2y}{dx^2} = 4\left(-1(4x + 1)^{-2}\cdot 4\right)$

22. **(D)** See the figure. Since the area, A, of the ring equals $\pi\left(y_2^2 - y_1^2\right)$,

$$A = \pi\left[(6x - x^2)^2 - x^4\right] = \pi\left[36x^2 - 12x^3 + x^4 - x^4\right]$$

and $\frac{dA}{dx} = \pi\left(72x - 36x^2\right) = 36\pi x\,(2 - x)$.

It can be verified that $x = 2$ produces the maximum area.

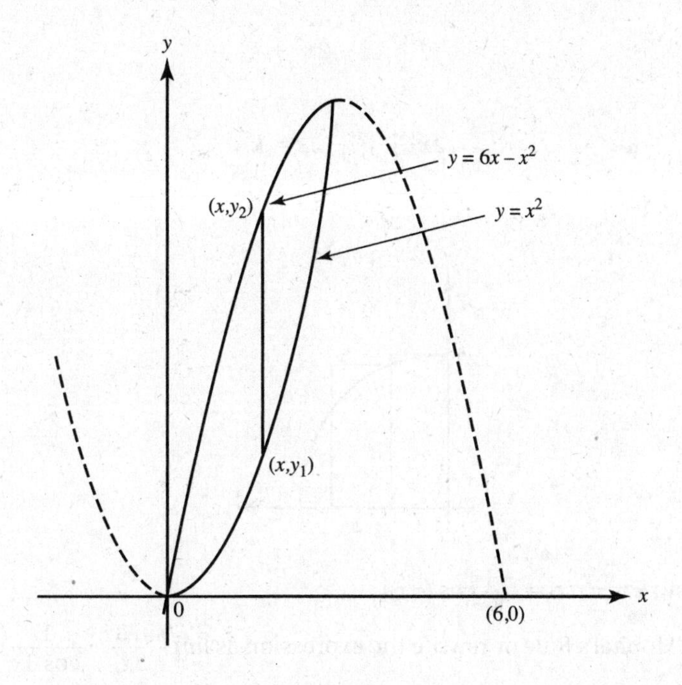

23. **(A)** This is of type $\int \frac{du}{u}$ with $u = \ln x$: $\int \frac{\frac{1}{x}dx}{\ln x}$.

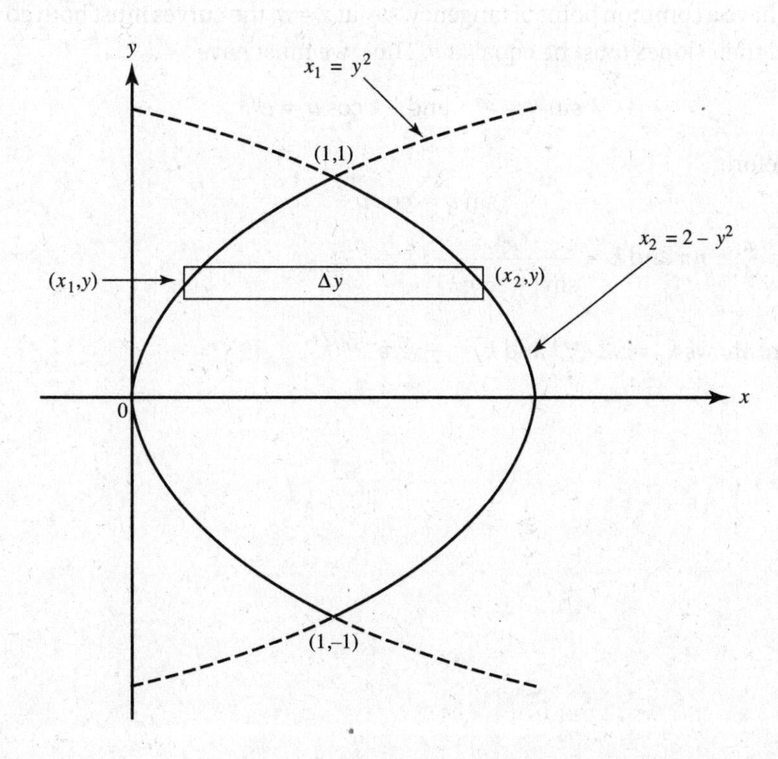

24. **(A)** About the y-axis; see the figure. Washer.

$$\Delta V = \pi\left(x_2^{\,2} - x_1^{\,2}\right)\Delta y, \text{ so } V = 2\pi \int_0^1 \left[(2 - y^2)^2 - y^4\right] dy = 2\pi \int_0^1 \left(4 - 4y^2\right) dy$$

25. **(E)** Separating variables, we get $y\,dy = (1 - 2x)\,dx$. Integrating gives

$$\frac{1}{2}y^2 = x - x^2 + C$$

or

$$y^2 = 2x - 2x^2 + k$$

or

$$2x^2 + y^2 - 2x = k.$$

26. **(E)** $2(5) + 2\sqrt{21}$.

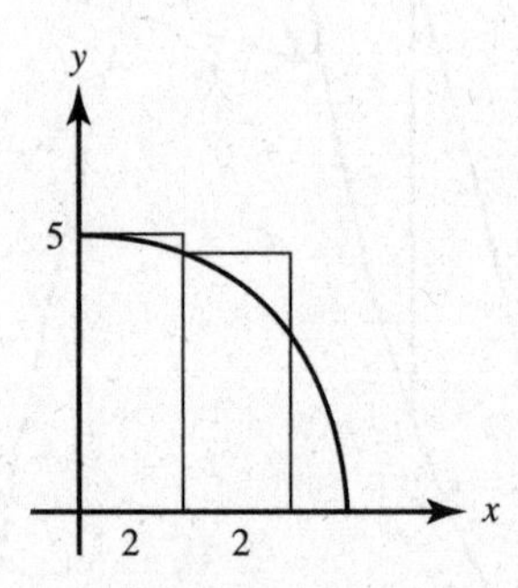

27. **(E)** $\frac{1}{\pi}\int_0^7 \sin \pi x\,(\pi\,dx) = \frac{-1}{\pi}\cos(\pi x)\Big|_0^7.$

28. **(D)** Use L'Hôpital's Rule or rewrite the expression as $\lim\limits_{x\to 0} \frac{\sin 3x}{3x} \cdot \frac{1}{\cos 3x} \cdot \frac{3}{2}$.

29. **(D)** For $f(x) = \tan x$, this is $f'\left(\frac{\pi}{4}\right) = \sec^2\left(\frac{\pi}{4}\right)$.

30. **(E)** The parameter k determines the amplitude of the sine curve. For $f = k \sin x$ and $g = e^x$ to have a common point of tangency, say at $x = q$, the curves must both go through (q, y) and their slopes must be equal at q. Thus, we must have

$$k \sin q = e^q \quad \text{and} \quad k \cos q = e^q,$$

and therefore

$$\sin q = \cos q.$$

Thus, $q = \frac{\pi}{4} \pm n\pi$ and $k = \dfrac{e^q}{\sin\left(\frac{\pi}{4} \pm n\pi\right)}$.

The figure shows $k_1 = \sqrt{2}\, e^{\pi/4}$ and $k_2 = -\sqrt{2}\, e^{-3\pi/4}$.

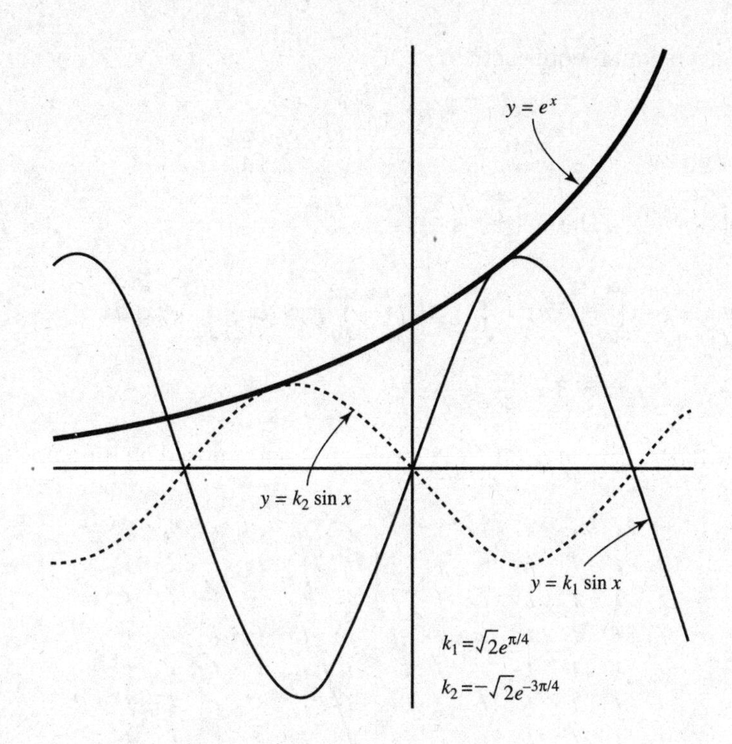

31. **(D)** We differentiate implicitly to find the slope $\frac{dy}{dx}$:

$$2\left(x^2\frac{dy}{dx} + 2xy\right) + 2y\frac{dy}{dx} = 2,$$

$$\frac{dy}{dx} = \frac{1 - 2xy}{x^2 + y}.$$

At (3, 1), $\frac{dy}{dx} = -\frac{1}{2}$. The linearization is $y \simeq -\frac{1}{2}(x - 3) + 1$.

32. **(C)** $\int \cos^3 x\, dx = \int \cos^2 x \cdot \cos x\, dx = \int \left(1 - \sin^2 x\right) \cos x\, dx$

$$= \int \cos x\, dx - \int \sin^2 x \cdot \cos x\, dx = \sin x - \frac{\sin^3 x}{3} + C.$$

33. **(A)** About the x-axis. Disk.

$$\Delta V = \pi y^2 \Delta x,$$

$$V = \pi \int_0^{\pi/4} \tan^2 x\, dx = \pi \int_0^{\pi/4} \left(\sec^2 x - 1\right) dx = \pi \left[\tan x - x\right]\Big|_0^{\pi/4}$$

$$= \pi\left(1 - \frac{\pi}{4}\right).$$

34. **(C)** Let $f(x) = a^x$; then $\lim_{h\to 0} \frac{a^h - 1}{h} = \lim_{h\to 0} \frac{a^{0+h} - a^0}{h} = f'(0) = a^0 \ln a = \ln a.$

35. **(E)** $\frac{dy}{dx}$ is a function of x alone; curves appear to be asymptotic to the y-axis and to increase more slowly as $|x|$ increases.

36. **(D)** The given limit is equivalent to

$$\lim_{h \to 0} \frac{F\left(\frac{\pi}{4} + h\right) - F\left(\frac{\pi}{4}\right)}{h} = F'\left(\frac{\pi}{4}\right),$$

where $F'(x) = \frac{\sin x}{x}$. The answer is $\frac{2\sqrt{2}}{\pi}$.

37. **(B)** $\int_0^{12} g(x)\,dx = \int_0^{4} g(x)\,dx + \int_4^{6} g(x)\,dx + \int_6^{9} g(x)\,dx + \int_9^{12} g(x)\,dx$

$$= 4\pi - 3 - \frac{9\pi}{4} + \frac{15}{2}.$$

38. **(C)** In the figure, the curve for $y = e^x$ has been superimposed on the slope field.

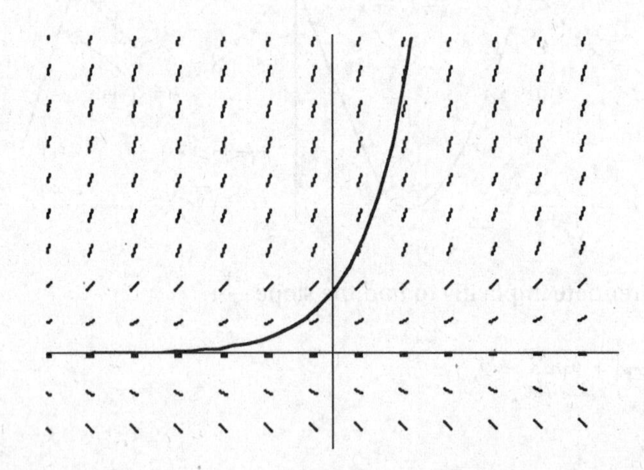

39. **(C)** The general solution is $y = 3 \ln|x^2 - 4| + C$. The differential equation $\frac{dy}{dx} = \frac{6x}{x^2 - 4}$ reveals that the derivative does not exist for $x = \pm 2$. The particular solution must be differentiable in an interval containing the initial value $x = -1$, so the domain is $-2 < x < 2$.

40. **(A)** The solution curve shown is $y = \ln x$, so the differential equation is $y' = \frac{1}{x}$.

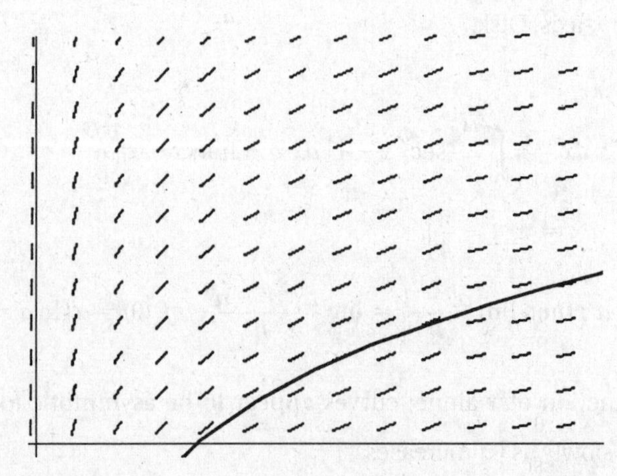

41. **(D)** $\sqrt{1 + \tan^2\theta} = \sec\theta$; $dx = \sec^2\theta$; $0 \le x \le 1$; so $0 \le \theta \le \frac{\pi}{4}$.

42. **(C)** The equations may be rewritten as $\frac{x}{2} = \sin u$ and $y = 1 - 2\sin^2 u$, giving $y = 1 - 2 \cdot \frac{x^2}{2}$.

43. **(D)** Use the formula for area in polar coordinates,

$$A = \frac{1}{2}\int_{\alpha}^{\beta} r^2 d\theta;$$

then the required area is given by

$$4 \cdot \frac{1}{2}\int_{0}^{\pi/4} 2\cos 2\theta.$$

(See polar graph 63 in the Appendix.)

44. **(C)** $\int_{-\infty}^{\infty} \frac{dx}{x^2+1} = \lim_{b\to\infty} \tan^{-1} x \Big|_{-b}^{b} = \frac{\pi}{2} - \left(-\frac{\pi}{2}\right) = \pi.$

45. **(A)** The first three derivatives of $\frac{1}{1-2x}$ are $\frac{2}{(1-2x)^2}$, $\frac{8}{(1-2x)^3}$, and $\frac{48}{(1-2x)^4}$.

The first four terms of the Maclaurin series (about $x = 0$) are 1, $+2x$, $+\frac{8x^2}{2!}$, and $+\frac{48x^3}{3!}$.

Note also that $\frac{1}{1-2x}$ represents the sum of an infinite geometric series with first term 1 and common ratio $2x$. Hence,

$$\frac{1}{1-2x} = 1 + 2x + (2x)^2 + (2x)^3 + \cdots$$

46. **(D)** We use parts, first letting $u = x^2$, $dv = e^{-x}dx$; then $du = 2x\,dx$, $v = -e^{-x}$ and

$$\int x^2e^{-x}dx = -x^2e^{-x} + 2\int xe^{-x}dx$$

Now we use parts again, letting $u = x$, $dv = e^{-x}dx$; then $du = dx$, $v = -e^{-x}$ and

$$-x^2e^{-x} + 2\int xe^{-x}dx = -x^2e^{-x} + 2\left(-xe^{-x} + \int e^{-x}dx\right)$$

Alternatively, we could use the Tic-Tac-Toe Method (see page 221):

u		dv
x^2	$+$	e^{-x}
$2x$	$-$	$-e^{-x}$
2	$+$	e^{-x}
0		$-e^{-x}$

Then $\int x^2e^{-x}dx = x^2(-e^{-x}) - (2x)e^{-x} + 2(-e^{-x}) + C$

47. **(E)** Use formula (20) in the Appendix to rewrite the integral as

$$\frac{1}{2}\int_0^{\pi/2} (1 - \cos 2x)\,dx = \frac{1}{2}\left(x - \frac{\sin 2x}{2}\right)\Bigg|_0^{\pi/2} = \frac{1}{2}\left(\frac{\pi}{2}\right).$$

48. **(E)** The area, A, is represented by $\int_0^{2\pi} (1 - \cos t) = 2\pi$.

49. **(D)** $\frac{dy}{dx} = \frac{\frac{dy}{d\theta}}{\frac{dx}{d\theta}} = \frac{2a\sin\theta\cos\theta}{-a\csc^2\theta} = -2\sin^3\theta\cos\theta.$

50. **(C)** Check to verify that each of the other improper integrals converges.

51. **(D)** Note that the integral is improper.

$$\lim_{k\to 4^-}\int_2^k \frac{du}{\sqrt{16-u^2}} = \lim_{k\to 4^-}\frac{1}{4}\int_2^k \frac{du}{\sqrt{1-\frac{u^2}{16}}} = \lim_{k\to 4^-}\frac{1}{4}\cdot 4\int_2^k \frac{\frac{1}{4}du}{\sqrt{1-\frac{u^2}{16}}} = \lim_{k\to 4^-}\sin^{-1}\frac{u}{4}\Bigg|_2^k = \frac{\pi}{3}$$

See Example 26, page 302.

52. **(C)** Let $y = \left(\frac{1}{x}\right)^x$. Then $\ln y = -x \ln x$ and

$$\lim_{x\to 0^+} \ln y = \lim_{x\to 0^+} \frac{-\ln x}{1/x}.$$

Now apply L'Hôpital's Rule:

$$\lim_{x\to 0^+} \ln y = \lim_{x\to 0^+} \frac{-1/x}{-1/x^2} = 0.$$

So, if $\lim_{x\to 0^+} \ln y = 0$, then $\lim_{x\to 0^+} y = 1$.

53. **(C)** The speed, $|\mathbf{v}|$, equals $\sqrt{\left(\frac{dx}{dt}\right)^2 + \left(\frac{dy}{dt}\right)^2}$, and since $x = 3y - y^2$,

$$\frac{dx}{dt} = (3-2y)\frac{dy}{dt} = (3-2y)\cdot 3.$$

Then $|\mathbf{v}|$ is evaluated, using $y = 1$, and equals $\sqrt{(3)^2 + (3)^2}$.

54. **(A)** This is an indeterminate form of type $\frac{\infty}{\infty}$; use L'Hôpital's Rule:

$$\lim_{x\to 0^+} \frac{\cot x}{\ln x} = \lim_{x\to 0^+} \frac{-\csc^2 x}{1/x} = \lim_{x\to 0^+} \frac{x}{\sin x}\cdot\frac{-1}{\sin x} = -\infty$$

55. **(E)** We find A and B such that $\frac{3x+2}{(x+3)(x-4)} = \frac{A}{x+3} + \frac{B}{x-4}$.

After multiplying by the common denominator, we have

$$3x + 2 = A(x-4) + B(x+3).$$

Substituting $x = -3$ yields $A = 1$, and $x = 4$ yields $B = 2$; hence,

$$\frac{3x+2}{x^2-x-12} = \frac{1}{x+3} + \frac{2}{x-4}.$$

56. **(B)** Since $\cos x \simeq 1 - \frac{x^2}{2!} + \frac{x^4}{4!}$, $\frac{1-\cos x}{x} \simeq \frac{x}{2} - \frac{x^3}{4!}$.

Then $\int_0^1 \frac{1-\cos x}{x}\,dx \simeq \left(\frac{x^2}{4} - \frac{x^4}{96}\right)\Bigg|_0^1 = \frac{1}{4} - \frac{1}{96}$.

Note that $\lim_{x\to 0^+} \frac{1-\cos x}{x} = 0$, so the integral is proper.

57. **(D)** We represent the spiral as $P(\theta) = (\theta\cos\theta, \theta\sin\theta)$. So

$$\frac{dy}{dx} = \frac{dy/d\theta}{dx/d\theta} = \frac{\theta\cos\theta + \sin\theta}{-\theta\sin\theta + \cos\theta} = \frac{\frac{\pi}{4}\cdot\frac{\sqrt{2}}{2} + \frac{\sqrt{2}}{2}}{-\frac{\pi}{4}\cdot\frac{\sqrt{2}}{2} + \frac{\sqrt{2}}{2}} = \frac{\pi/4 + 1}{-\pi/4 + 1}.$$

Part B

58. **(D)** Since h is increasing, $h' \geq 0$. The graph of h is concave downward for $x < 2$ and upward for $x > 2$, so h'' changes sign at $x = 2$, where it appears that $h' = 0$ also.

59. **(C)** I is false since, for example, $f'(-2) = f'(1) = 0$ but neither $g(-2)$ nor $g(1)$ equals zero.

    II is true. Note that $f = 0$ where g has relative extrema, and f is positive, negative, then positive on intervals where g increases, decreases, then increases.

    III is also true. Check the concavity of g: when the curve is concave down, $h < 0$; when up, $h > 0$.

60. **(A)** If $y = \int_3^x \frac{1}{\sqrt{3+2t}}\,dt$, then $\frac{dy}{dx} = \frac{1}{\sqrt{3+2x}}$, so $\frac{d^2y}{dx^2} = -\frac{1}{2}(3+2x)^{-3/2}(2)$.

61. **(D)** $\int_{-4}^{3} f(x+1)dx$ represents the area of the same region as $\int_{-3}^{4} f(x)dx$, translated one unit to the left.

62. **(D)** According to the Mean Value Theorem, there exists a number c in the interval $[1,1.5]$ such that $f'(c) = \frac{f(1.5) - f(1)}{1.5 - 1}$. Use your calculator to solve the equation $\cos c = \frac{\sin 1.5 - \sin 1}{0.5}$ for c (in radians).

63. **(E)** Here are the relevant sign lines:

    signs of $f'(x)$: $-$ (left of 0) | $-$ (between 0 and 1) | $+$ (right of 1)

    signs of $f''(x)$: $+$ (left of 0) | $-$ (between 0 and $\frac{2}{3}$) | $+$ (right of $\frac{2}{3}$)

    We see that f' and f'' are both positive only if $x > 1$.

64. **(E)** Note from the sign lines in Question 63 that f changes from decreasing to increasing at $x = 1$, so f has a local minimum there.
    Also, the graph of f changes from concave up to concave down at $x = 0$, then back to concave up at $x = \frac{2}{3}$; hence f has two points of inflection.

65. **(C)** The derivatives of $\ln(x+1)$ are $\frac{1}{x+1}, \frac{-1}{(x+1)^2}, \frac{+2!}{(x+1)^3}, \frac{-(3!)}{(x+1)^4}, \ldots$

    The nth derivative at $x = 2$ is $\frac{(-1)^{n-1}(n-1)!}{3^n}$.

66. **(C)** The absolute-value function $f(x) = |x|$ is continuous at $x = 0$, but $f'(0)$ does not exist.

67. **(C)** Let $F'(x) = f(x)$; then $F'(x+k) = f(x+k)$;

$$\int_0^3 f(x+k)\,dx = F(3+k) - F(k);$$
$$\int_k^{3+k} f(x)\,dx = F(3+k) - F(k).$$

    Or let $u = x + k$. Then $dx = du$; when $x = 0$, $u = k$; when $x = 3$, $u = 3 + k$.

68. **(E)** See the figure. The equation of the generating circle is $(x-3)^2+y^2=1$, which yields $x=3\pm\sqrt{1-y^2}$.

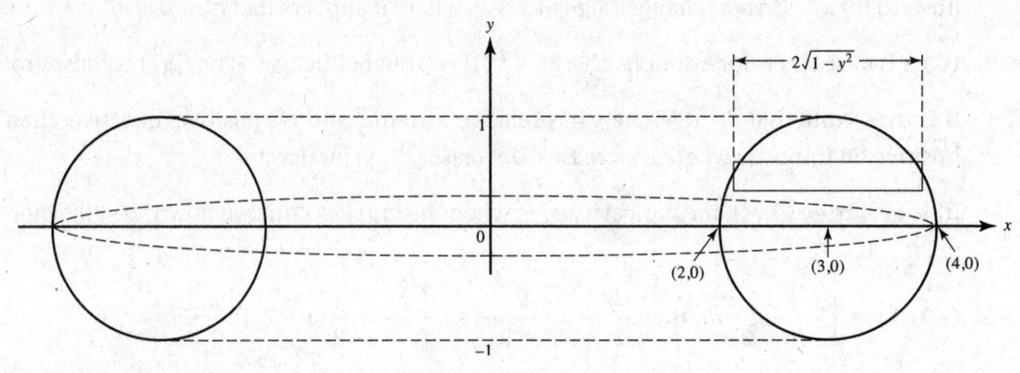

About the y-axis: $\Delta V = 2\pi\cdot 3\cdot 2\sqrt{1-y^2}\,\Delta y$.

$$\text{Thus, } V = 2\int_0^1 12\pi\sqrt{1-y^2}\,dy.$$

$$= 24\pi \text{ times the area of a quarter of a unit circle} = 6\pi^2.$$

69. **(D)** Note that $f(g(u)) = \tan^{-1}(e^{2u})$; then the derivative is $\frac{1}{1+(e^{2u})^2}(2e^{2u})$.

70. **(D)** Let $y' = \frac{dy}{dx}$. Then $\cos(xy)[xy'+y] = y'$. Solve for y'.

71. **(E)** $\frac{d^2}{dx^2}f(x^2) = \frac{d}{dx}\left[\frac{d}{dx}f(x^2)\right] = \frac{d}{dx}\left[\frac{d}{dx}f(x^2)\cdot\frac{dx^2}{dx}\right] = \frac{d}{dx}\left[g(x^2)\cdot 2x\right]$

$$= g(x^2)\frac{d}{dx}(2x) + 2x\frac{d}{dx}g(x^2) = g(x^2)\cdot 2 + 2x\frac{d}{dx^2}g(x^2)\frac{dx^2}{dx}$$

$$= 2g(x^2) + 2x\cdot f\left(\sqrt{x^2}\right)\cdot 2x = 2g(x^2) + 4x^2\cdot f(x).$$

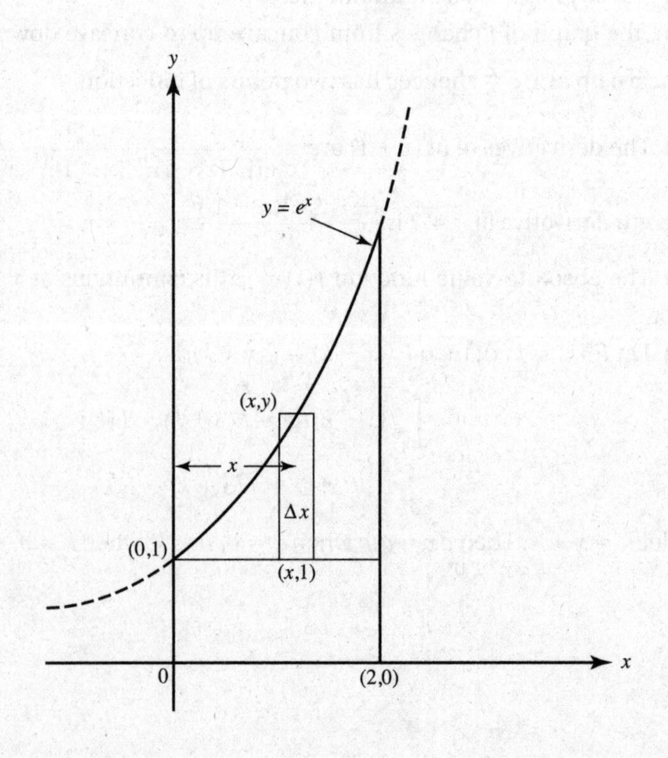

72. **(C)** About the x-axis; see the figure. Washer.

$$\Delta V = \pi\left(y^2 - 1^2\right)\Delta x,$$
$$V = \pi\int_0^2 (e^{2x} - 1)dx.$$

73. **(C)** By the Mean Value Theorem, there is a number c in [1, 2] such that

$$f'(c) = \frac{f(2) - f(1)}{2 - 1} = -3.$$

74. **(D)** The enclosed region, S, is bounded by $y = \sec x$, the y-axis, and $y = 4$. It is to be rotated about the y-axis.

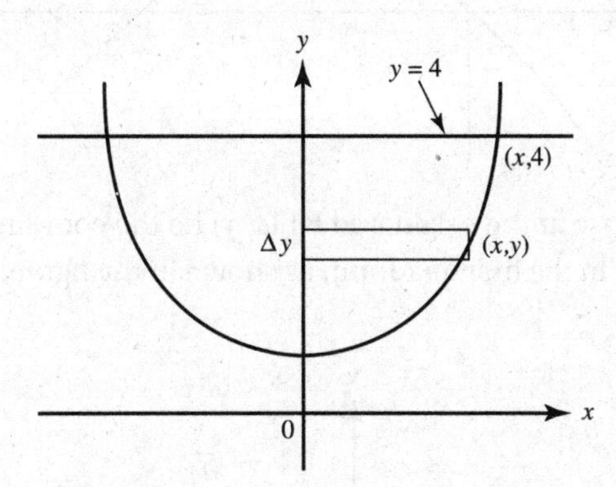

Use disks; then $\Delta V = \pi R^2 H = \pi(\text{arc sec } y)^2\,\Delta y$. Using the calculator, we find that

$$\pi\int_1^4 \left(\arccos\left(\frac{1}{y}\right)\right)^2 dx \simeq 11.385.$$

75. **(C)** If Q is the amount at time t, then $Q = 40e^{-kt}$. Since $Q = 20$ when $t = 2$, $k = -0.3466$. Now find Q when $t = 3$, from $Q = 40e^{-(0.3466)3}$, getting $Q = 14$ to the nearest gram.

76. **(A)** The velocity $v(t)$ is an antiderivative of $a(t)$, where $a(t) = \pi t + \dfrac{2}{1 + t^2}$. So $v(t) = \dfrac{\pi t^2}{2} + 2\arctan t + C$. Since $v(1) = 0$, $C = -\pi$.

$$\text{Required average velocity} = \frac{1}{2 - 0}\int_0^2 v(t)dt$$
$$= \frac{1}{2}\int_0^2 \left(\frac{\pi t^2}{2} + 2\arctan t - \pi\right) dt \simeq 0.362.$$

77. **(D)** Graph $y = \tan x$ and $y = 2 - x$ in $[-1, 3] \times [-1, 3]$ as shown on page 460. Note that

$$\Delta A = \left(x_{\text{line}} - x_{\text{curve}}\right)\Delta y$$
$$= (2 - y - \arctan y)\,\Delta y.$$

The limits are $y = 0$ and $y = b$, where b is the ordinate of the intersection of the curve and the line. Using the calculator, solve

$$\arctan y = 2 - y$$

and store the answer in memory as B. Evaluate the desired area:

$$\int_0^B (2 - y - \arctan y)\,dy \simeq 1.077.$$

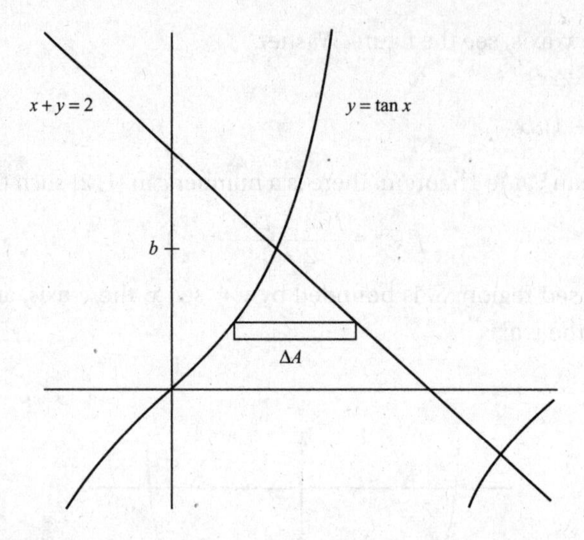

78. **(E)** Center the ellipse at the origin and let (x, y) be the coordinates of the vertex of the inscribed rectangle in the first quadrant, as shown in the figure.

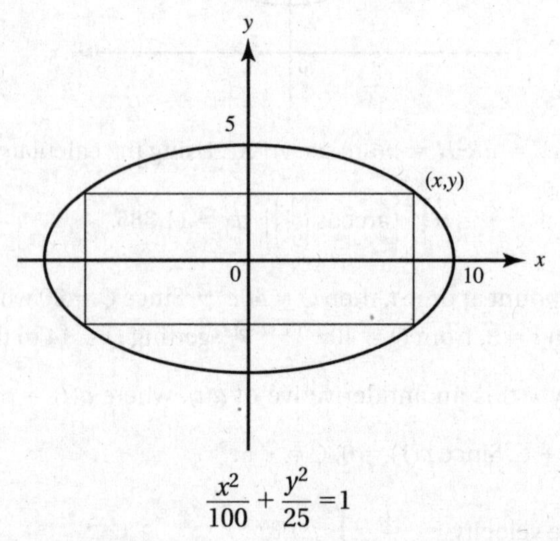

$$\frac{x^2}{100} + \frac{y^2}{25} = 1$$

To maximize the rectangle's area $A = 4xy$, solve the equation of the ellipse, getting

$$x = \sqrt{100 - 4y^2} = 2\sqrt{25 - y^2}.$$

So $A = 8y\sqrt{25 - y^2}$. Graph $y = 8x\sqrt{(25 - x^2)}$ in the window $[0,5] \times [0,150]$. The calculator shows that the maximum area (the y-coordinate) equals 100.

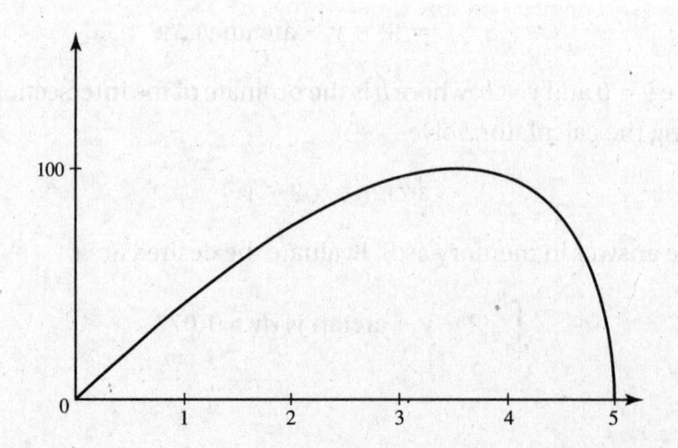

79. **(B)** $\dfrac{\int_1^e x \ln x \, dx}{e-1} \simeq 1.221.$

80. **(B)** When f' is positive, f increases. By the Fundamental Theorem of Calculus, $f'(x) = 1 - 2(\cos x)^3$. Graph f' in $[0, 2\pi] \times [-2, 4]$. It is clear that $f' > 0$ on the interval $a < x < b$. Using the calculator to solve $1 - 2(\cos x)^3 = 0$ yields $a = 0.654$ and $b = 5.629$.

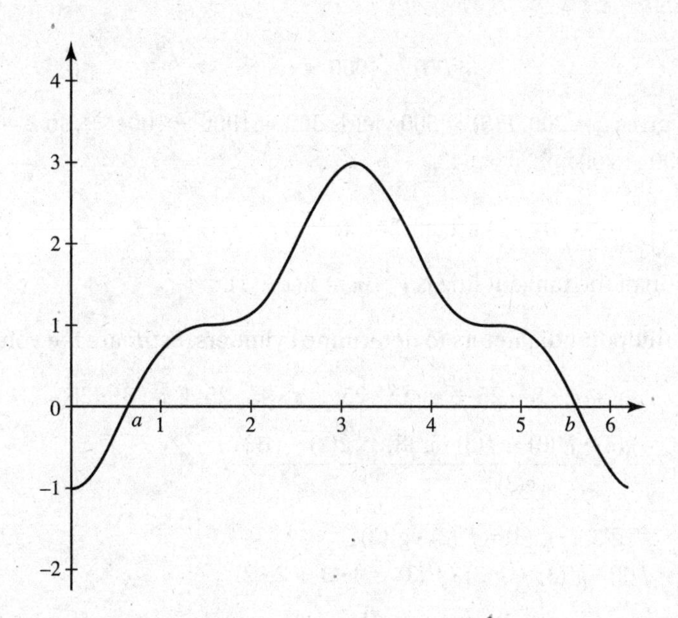

81. **(C)** $a(2) \simeq \dfrac{v(3) - v(1)}{3 - 1} = \dfrac{12 - 22}{2}.$

82. **(B)** The volume is composed of elements of the form $\Delta V = (2x)^2 \, \Delta y$. If h is the depth, in feet, then, after t hr,

$V(h) = 4\int_0^h y \, dy$ and $\dfrac{dV}{dt} = 4h\dfrac{dh}{dt}.$

Thus, $12 = 4\,(9)\,\dfrac{dh}{dt}$

and $\dfrac{dh}{dt} = \dfrac{1}{3}$ ft/hr.

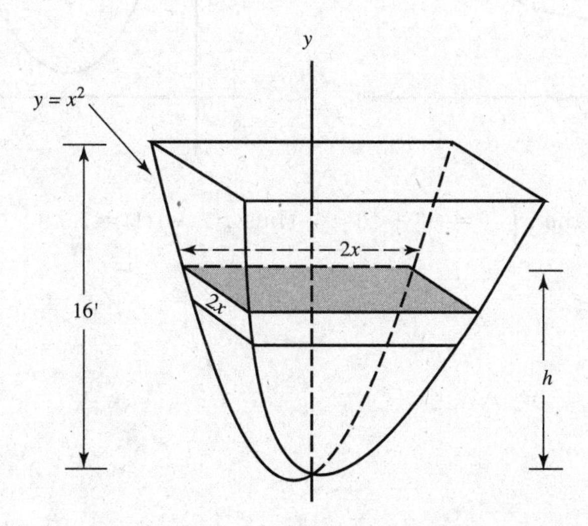

83. **(B)** Separating variables yields

$$\frac{dP}{1000 - P} = k\,dt,$$
$$-\ln(1000 - P) = kt + C,$$
$$1000 - P = ce^{-kt}.$$

Then

$$P(t) = 1000 - ce^{-kt}.$$

$P(0) = 300$ gives $c = 700$. $P(5) = 500$ yields $500 = 1000 - 700e^{-5k}$, so $k \simeq +0.0673$. Now $P(10) = 1000 - 700e^{-0.673} \simeq 643$.

84. **(C)** $H(1) = \int_0^1 \frac{4}{x^2+1}dx = 4\arctan 1 = \pi$. $H'(1) = f(1) = 2$.
The equation of the tangent line is $y - \pi = 2(x - 1)$.

85. **(C)** Using midpoint diameters to determine cylinders, estimate the volume to be

$$V \simeq \pi \cdot 8^2 \cdot 25 + \pi \cdot 6^2 \cdot 25 + \pi \cdot 4^2 \cdot 25 + \pi \cdot 3^2 \cdot 25.$$

86. **(A)** $\left(\frac{f}{g}\right)'(3) = \frac{g(3)\cdot f'(3) - f(3)\cdot g'(3)}{(g(3))^2} = \frac{2(2) - 4(3)}{2^2}$.

87. **(C)** $H'(3) = f'(g(3)) \cdot g'(3) = f'(2) \cdot g'(3)$.

88. **(E)** $M'(3) = f(3) \cdot g'(3) + g(3) \cdot f'(3) = 4 \cdot 3 + 2 \cdot 2$.

89. **(E)** $K'(3) = \frac{1}{g'(K(3))} = \frac{1}{g'(g^{-1}(3))} = \frac{1}{g'(4)} = \frac{1}{\frac{1}{2}}$.

90. **(C)** $R'(3) = \frac{1}{2}(f(3))^{-1/2} \cdot f'(3)$.

91. **(D)** Here are the pertinent curves, with d denoting the depth of the water:

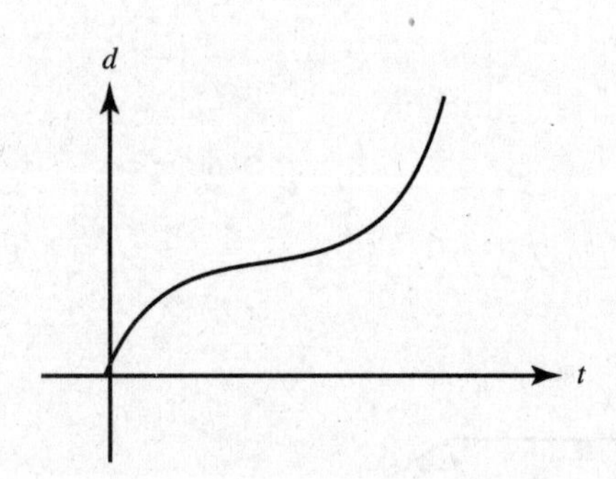

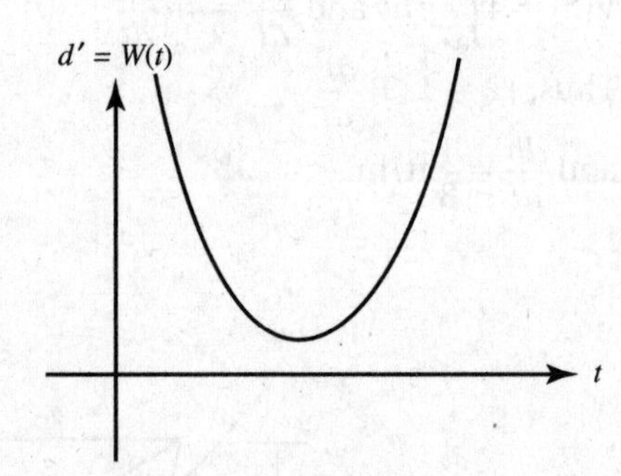

92. **(B)** Use areas; then $\int_1^7 f' = -3 + 10 = 7$. Thus, $f(7) - f(1) = 7$.

93. **(B)** The region x units from the stage can be approximated by the semicircular ring shown; its area is then the product of its circumference and its width.

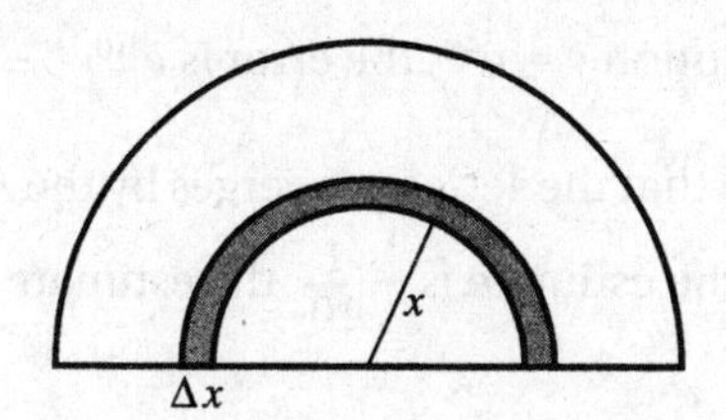

$\frac{1}{2}(2\pi x)$

Δx

The number of people standing in the region is the product of the area and the density:

$$\Delta P = (\pi x\,\Delta x)\left(\frac{20}{2\sqrt{x}+1}\right).$$

To find the total number of people, evaluate

$$20\pi \int_0^{100} \frac{x}{2\sqrt{x}+1}\,dx$$

94. **(B)** $\frac{dy}{dt}$ is positive, but decreasing; hence $\frac{dy^2}{dt^2} < 0$.

95. **(C)** Average speed $= \frac{\text{total distance}}{\text{elapsed time}} = \frac{\text{total area}}{8} = \frac{8}{8}$.

96. **(E)** On $2 \leqslant t \leqslant 5$, the object moved $3\frac{1}{2}$ ft to the right; then on $5 \leqslant t \leqslant 8$, it moved only $2\frac{1}{2}$ ft to the left.

97. **(B)** Average acceleration $= \frac{\Delta v}{\Delta t} = \frac{v(8) - v(0)}{8 - 0} = \frac{-1 - 1}{8}$.

98. **(D)** Evaluate $\int_0^3 \frac{72}{2t+3}\,dt = 36\ln(2t+3)\Big|_0^3 = 36\ln 3$.

99. **(A)** $2x\frac{dx}{dt} + 2y\frac{dy}{dt} = 0$ and $\frac{dy}{dt} = -\frac{3}{4}\frac{dx}{dt}$ at the point (3, 4).

Use, also, the facts that the speed is given by $|\mathbf{v}| = \sqrt{\left(\frac{dx}{dt}\right)^2 + \left(\frac{dy}{dt}\right)^2}$ and that the point moves counterclockwise; then $\left(\frac{dx}{dt}\right)^2 + \left(\frac{dy}{dt}\right)^2 = 4$, yielding $\frac{dx}{dt} = -\frac{8}{5}$ and $\frac{dy}{dt} = +\frac{6}{5}$ at the given point. The velocity vector, $\mathbf{v}$, at (3, 4) must therefore be $\left\langle -\frac{8}{5}, \frac{6}{5}\right\rangle$.

100. **(A)** $\mathbf{v} = \langle -ak\sin kt, ak\cos kt\rangle$, and

$$\mathbf{a} = \langle -ak^2\cos kt, ak^2\sin kt\rangle = -k^2\mathbf{R}.$$

101. **(B)** The formula for length of arc is

$$L = \int_a^b \sqrt{1 + \left(\frac{dy}{dx}\right)^2}\,dx.$$

Since $y = 2^x$, we find

$$L = \int_0^2 \sqrt{1 + (2^x \ln 2)^2}\,dx \simeq 3.664.$$

102. **(D)** $\mathbf{a}(t) = \langle 0, e^t \rangle$; the acceleration is always upward.

103. **(A)** At (0, 1), $\frac{dy}{dx} = 4$, so Euler's method yields (0.1, 1 + 0.1(4)) = (0.1, 1.4).

$\frac{dy}{dx} = 4y$ has particular solution $y = e^{4x}$; the error is $e^{4(0.1)} - 1.4$.

104. **(D)** $1 - \frac{1}{2} + \frac{1}{5} = 0.7$. Note that the series converges by the Alternating Series Test. Since the first term dropped in the estimate is $-\frac{1}{10}$, the estimate is too high, but within 0.1 of the true sum.

105. **(C)** $\sum_{n=1}^{\infty} \frac{2}{3n} = \frac{2}{3} \sum_{n=1}^{\infty} \frac{1}{n}$, which equals a constant times the harmonic series.

106. **(D)** We seek x such that

$$\lim_{n\to\infty} \left| \frac{(n+1)! \cdot x^{n+1}}{(n+1)^{n+1}} \cdot \frac{n^n}{n! \cdot x^n} \right| < 1$$

or such that $|x| \cdot \lim_{n\to\infty} \left(\frac{n}{n+1}\right)^n < 1$

or such that $|x| < \frac{1}{\lim_{n\to\infty} \left(\frac{n}{n+1}\right)^n}$.

The fraction equals $\lim_{n\to\infty} \left(\frac{n+1}{n}\right)^n \lim_{n\to\infty} \left(1 + \frac{1}{n}\right)^n = e$.

Then $|x| < e$ and the radius of convergence is e.

107. **(B)** The error is less than the maximum value of $\frac{e^c}{3!} x^3$ for $0 \le x \le \frac{1}{2}$.

This maximum occurs at $c = x = \frac{1}{2}$.

108. **(D)** Distance $= \int_0^1 \sqrt{\left(\frac{dx}{dt}\right)^2 + \left(\frac{dy}{dt}\right)^2}\, dt.$

$$= \int_0^1 \sqrt{(\sec^2 t)^2 + (-2\sin 2t)^2}\, dt.$$

Note that the curve is traced exactly once by the parametric equations from $t = 0$ to $t = 1$.

12 Miscellaneous Free-Response Practice Exercises

These problems provide further practice for both parts of Section II of the examination. Solutions begin on page 473.

Part A. Directions: A graphing calculator is required for some of these problems. See instructions on page 4.

x	2.5	3.2	3.5	4.0	4.6	5.0
$f(x)$	7.6	5.7	4.2	3.1	2.2	1.5

1. A function f is continuous, differentiable, and strictly decreasing on the interval [2.5,5]; some values of f are shown in the table above.

   (a) Estimate $f'(4.0)$ and $f'(4.8)$.
   (b) What does the table suggest may be true of the concavity of f? Explain.
   (c) Estimate $\int_{2.5}^{5} f(x)dx$ with a Riemann Sum using left endpoints.
   (d) Set up (but do not evaluate) a Riemann Sum that estimates the volume of the solid formed when f is rotated around the x-axis.

2. The equation of the tangent line to the curve $x^2y - x = y^3 - 8$ at the point (0,2) is $12y + x = 24$.

   (a) Given that the point $(0.3, y_0)$ is on the curve, find y_0 approximately, using the tangent line.
   (b) Find the true value of y_0.
   (c) What can you conclude about the curve near $x = 0$ from your answers to parts (a) and (b)?

3. A differentiable function f defined on $-7 < x < 7$ has $f(0) = 0$ and $f'(x) = 2x \sin x - e^{-x^2} + 1$. (Note: The following questions refer to f, not to f'.)

   (a) Describe the symmetry of f.
   (b) On what intervals is f decreasing?
   (c) For what values of x does f have a relative maximum? Justify your answer.
   (d) How many points of inflection does f have? Justify your answer.

4. Let C represent the piece of the curve $\sqrt[3]{64 - 16x^2}$ that lies in the first quadrant. Let S be the region bounded by C and the coordinate axes.

   (a) Find the slope of the line tangent to C at $y = 1$.
   (b) Find the area of S.
   (c) Find the volume generated when S is rotated about the x-axis.

5. Let R be the point on the curve of $y = x - x^2$ such that the line OR (where O is the origin) divides the area bounded by the curve and the x-axis into two regions of equal area. Set up (but do not solve) an integral to find the x-coordinate of R.

6. Suppose $f'' = \sin(2^x)$ for $-1 < x < 3.2$.

   (a) On what intervals is the graph of f concave downward? Justify your answer.
   (b) Find the x-coordinates of all relative minima of f'.
   (c) How many points of inflection does the graph of f' have? Justify your answer.

7. Let $f(x) = \cos x$ and $g(x) = x^2 - 1$.

   (a) Find the coordinates of any points of intersection of f and g.
   (b) Find the area bounded by f and g.

8. (a) In order to investigate mail-handling efficiency, at various times one morning a local post office checked the rate (letters/min) at which an employee was sorting mail. Use the results shown in the table and the trapezoid method to estimate the total number of letters he may have sorted that morning.

Time	8:00	8:30	10:00	11:20	12:00
Letters/min	10	12	8	9	11

   (b) Hoping to speed things up a bit, the post office tested a sorting machine that can process mail at the constant rate of 20 letters per minute. The graph shows the rate at which letters arrived at the post office and were dumped into this sorter.

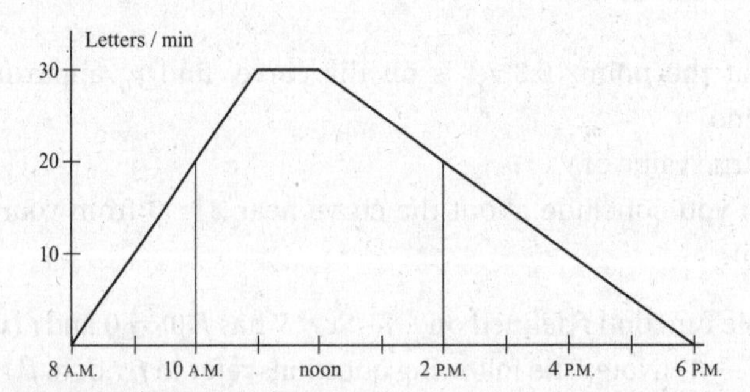

   (i) When did letters start to pile up?
   (ii) When was the pile the biggest?
   (iii) How big was it then?
   (iv) At about what time did the pile vanish?

9. Let R represent the region bounded by $y = \sin x$ and $y = x^4$. Find:

   (a) the area of R;
   (b) the volume of the solid whose base is R, if all cross sections perpendicular to the x-axis are isosceles triangles with height 3;
   (c) the volume of the solid formed when R is rotated around the x-axis.

10. The town of East Newton has a water tower whose tank is an ellipsoid, formed by rotating an ellipse about its minor axis. Since the tank is 20 feet tall and 50 feet wide, the equation of the ellipse is $\frac{x^2}{625} + \frac{y^2}{100} = 1$.

    (a) If there are 7.48 gallons of water per cubic foot, what is the capacity of this tank to the nearest thousand gallons?
    (b) East Newton imposes water rationing whenever the tank is only one-quarter full. Write an equation to find the depth of the water in the tank when rationing becomes necessary? (Do not solve.)

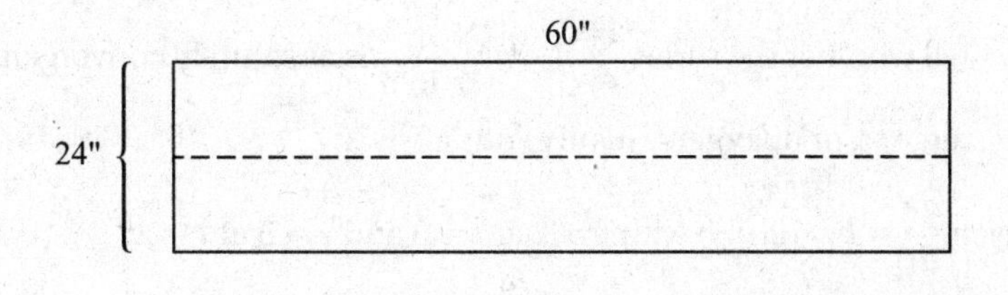

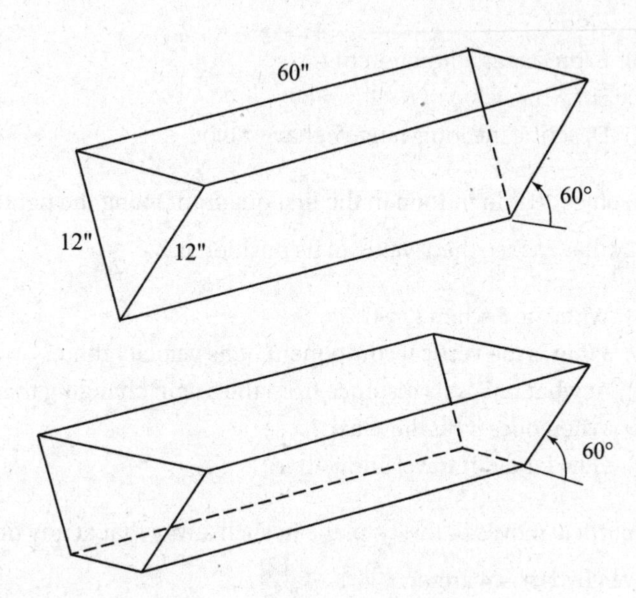

Note: Scales are different on the three figures.

11. The sides of a watering trough are made by folding a sheet of metal 24 inches wide and 5 feet (60 inches) long at an angle of 60°, as shown in the figure above. Ends are added, and then the trough is filled with water.

    (a) If water pours into the trough at the rate of 600 cubic inches per minute, how fast is the water level rising when the water is 4 inches deep?
    (b) Suppose, instead, the sheet of metal is folded twice, keeping the sides of equal height and inclined at an angle of 60°, as shown. Where should the folds be in order to maximize the volume of the trough? Justify your answer.

BC ONLY

12. (a) Using your calculator, verify that

$$\left(4\tan^{-1}(1/5)\right) - \left(\tan^{-1}(1/239)\right) \approx \frac{\pi}{4}.$$

(b) Use the Taylor polynomial of degree 7 about 0,

$$\tan^{-1} x \approx x - x^3/3 + x^5/5 - x^7/7,$$

to approximate $\tan^{-1} 1/5$, and the polynomial of degree 1 to approximate $\tan^{-1} 1/239$.

(c) Use part (b) to evaluate the expression in (a).

(d) Explain how the approximation for $\pi/4$ given here compares with that obtained using $\pi/4 = \tan^{-1} 1$.

13. (a) Show that the series $\sum_{n=1}^{\infty} (-1)^{n+1} \frac{1}{\ln(n+1)}$ converges.

(b) How many terms of the series are needed to get a partial sum within 0.1 of the sum of the whole series?

(c) Tell whether the series $\sum_{n=2}^{\infty} (-1)^n \frac{1}{n \ln n}$ is absolutely convergent, conditionally convergent, or divergent. Justify your answer.

14. Given $\frac{dy}{dt} = ky(10 - y)$ with $y = 2$ at $t = 0$ and $y = 5$ at $t = 2$:

(a) Find k.

(b) Express y as a function of t.

(c) For what value of t will $y = 8$?

(d) Describe the long-range behavior of y.

15. An object P is in motion in the first quadrant along the parabola $y = 18 - 2x^2$ in such a way that at t sec the x-value of its position is $x = \frac{1}{2}t$.

(a) Where is P when $t = 4$?

(b) What is the vertical component of its velocity there?

(c) At what rate is its distance from the origin changing then?

(d) When does it hit the x-axis?

(e) How far did it travel altogether?

16. A particle moves in the xy-plane in such a way that at any time $t \geq 0$ its position is given by $x(t) = 4 \arctan t$, $y(t) = \frac{12t}{t^2 + 1}$.

(a) Sketch the path of the particle, indicating the direction of motion.

(b) At what time t does the particle reach its highest point? Justify.

(c) Find the coordinates of that highest point, and sketch the velocity vector there.

(d) Describe the long-term behavior of the particle.

17. Let R be the region bounded by the curve $r = 2 + \cos 2\theta$, as shown.

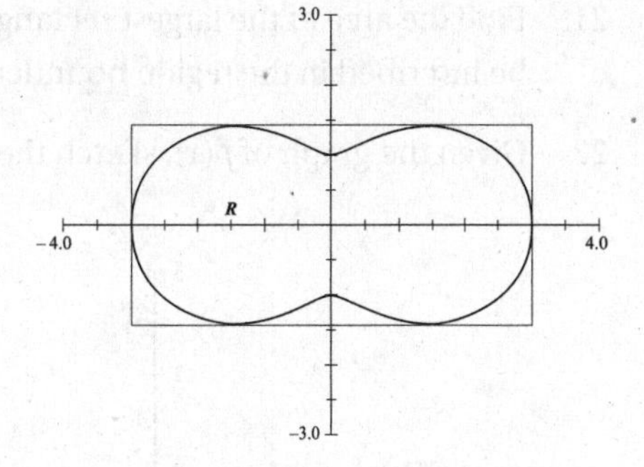

(a) Find the dimensions of the smallest rectangle that contains R and has sides parallel to the x- and y-axes.
(b) Find the area of R.

18. (a) For what *positive* values of x does $f(x) = \sum_{n=1}^{\infty} (-1)^{n+1} \frac{x^n}{\ln (n+1)}$ converge?
(b) How many terms are needed to estimate $f(0.5)$ to within 0.01?
(c) Would an estimate for $f(-0.5)$ using the same number of terms be more accurate, less accurate, or the same? Explain.

Part B. Directions: Answer these questions *without* using your calculator.

19. Draw a graph of $y = f(x)$, given that f satisfies all the following conditions:
(1) $f'(-1) = f'(1) = 0$.
(2) If $x < -1$, $f'(x) > 0$ but $f'' < 0$.
(3) If $-1 < x < 0$, $f'(x) > 0$ and $f'' > 0$.
(4) If $0 < x < 1$, $f'(x) > 0$ but $f'' < 0$.
(5) If $x > 1$, $f'(x) < 0$ and $f'' < 0$.

20. The figure below shows the graph of f', the derivative of f, with domain $-3 \le x \le 9$. The graph of f' has horizontal tangents at $x = 2$ and $x = 4$, and a corner at $x = 6$.

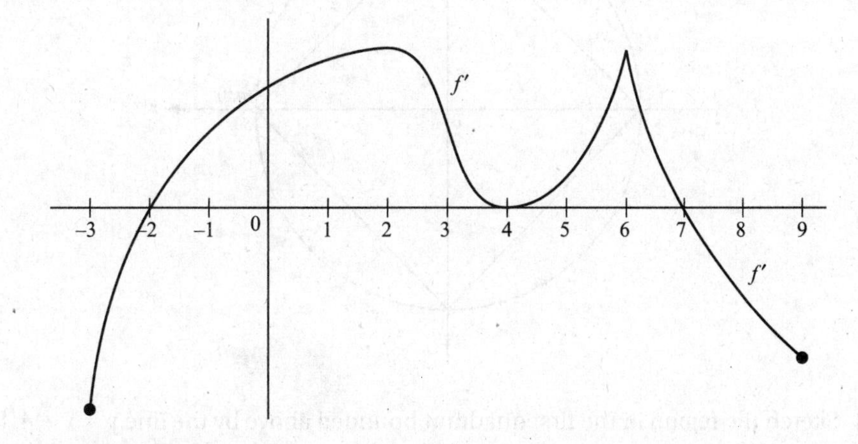

(a) Is f continuous? Explain.
(b) Find all values of x at which f attains a local minimum. Justify.
(c) Find all values of x at which f attains a local maximum. Justify.
(d) At what value of x does f attain its absolute maximum? Justify.
(e) Find all values of x at which the graph of f has a point of inflection. Justify.

21. Find the area of the largest rectangle (with sides parallel to the coordinate axes) that can be inscribed in the region bounded by the graphs of $f(x) = 8 - 2x^2$ and $g(x) = x^2 - 4$.

22. Given the graph of $f(x)$, sketch the graph of $f'(x)$.

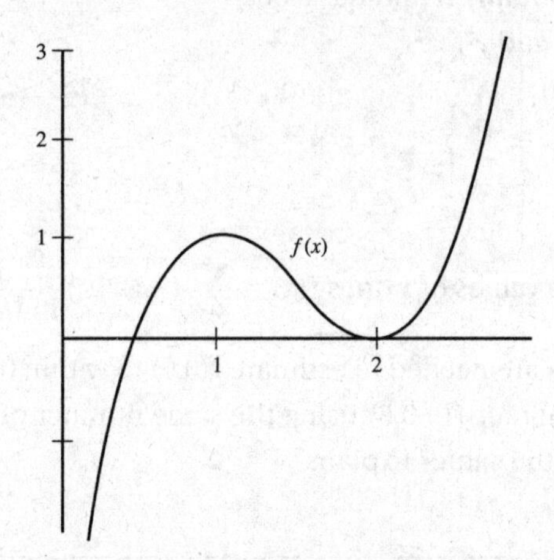

23. A cube is contracting so that its surface area decreases at the constant rate of 72 in.^{2}/sec. Determine how fast the volume is changing at the instant when the surface area is 54 ft^2.

24. A square is inscribed in a circle of radius a as shown in the diagram. Find the volume obtained if the region outside the square but inside the circle is rotated about a diagonal of the square.

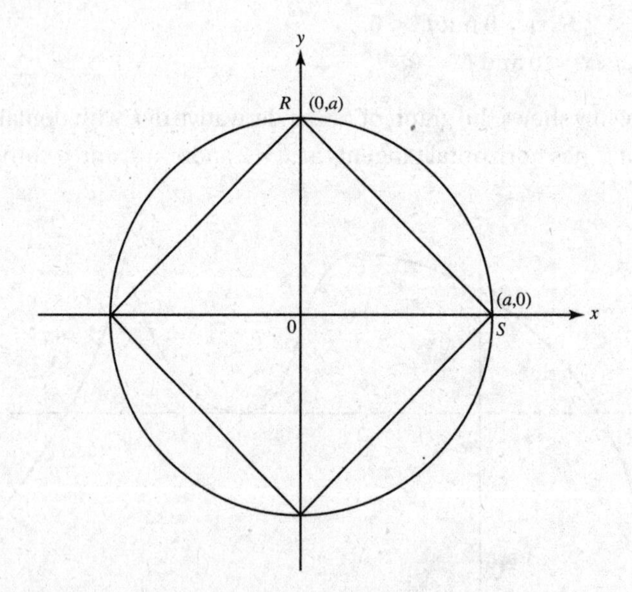

25. (a) Sketch the region in the first quadrant bounded above by the line $y = x + 4$, below by the line $y = 4 - x$, and to the right by the parabola $y = x^2 + 2$.
(b) Find the area of this region.

26. The graph shown below is based roughly on data from the U.S. Department of Agriculture.

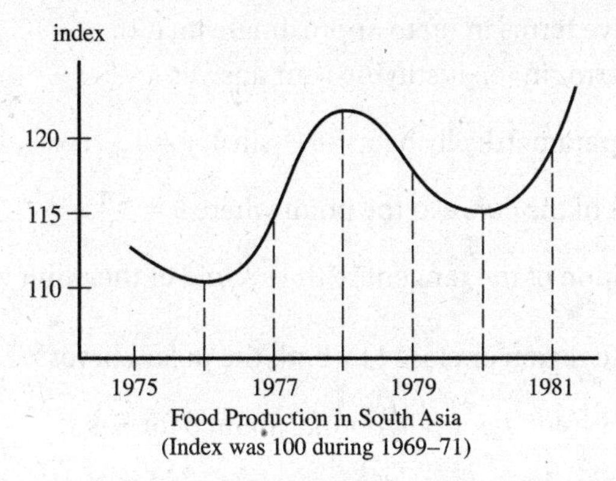

Food Production in South Asia
(Index was 100 during 1969–71)

(a) During which intervals did food production decrease in South Asia?
(b) During which intervals did the rate of change of food production increase?
(c) During which intervals did the increase in food production accelerate?

27. A particle moves along a straight line so that its acceleration at any time t is given in terms of its velocity v by $a = -2v$.

(a) Find v in terms of t if $v = 20$ when $t = 0$.
(b) Find the distance the particle travels while v changes from $v = 20$ to $v = 5$.

28. Let R represent the region bounded above by the parabola $y = 27 - x^2$ and below by the x-axis. Isosceles triangle AOB is inscribed in region R with its vertex at the origin O and its base $\overline{AB}$ parallel to the x-axis. Find the maximum possible area for such a triangle.

29. Newton's law of cooling states that the rate at which an object cools is proportional to the difference in temperature between the object and its surroundings.

It is 9:00 P.M., time for your milk and cookies. The room temperature is 68° when you pour yourself a glass of 40° milk and start looking for the cookie jar. By 9:03 the milk has warmed to 43°, and the phone rings. It's your friend, with a fascinating calculus problem. Distracted by the conversation, you forget about the glass of milk. If you dislike milk warmer than 60°, how long, to the nearest minute, do you have to solve the calculus problem and still enjoy acceptably cold milk with your cookies?

30. Let h be a function that is even and continuous on the closed interval $[-4,4]$. The function h and its derivatives have the properties indicated in the table below. Use this information to sketch a possible graph of h on $[-4,4]$.

x	$h(x)$	$h'(x)$	$h''(x)$
0	$-$	0	$+$
$0<x<1$	$-$	$+$	$+$
1	0	$+$	0
$1<x<2$	$+$	$+$	$-$
2	$+$	0	0
$2<x<3$	$+$	$+$	$+$
3	$+$	undefined	undefined
$3<x<4$	$+$	$-$	$-$

BC ONLY

31. (a) Find the Maclaurin series for $f(x) = \ln(1 + x)$.
(b) What is the radius of convergence of the series in (a)?
(c) Use the first five terms in (a) to approximate ln(1.2).
(d) Estimate the error in (c), justifying your answer.

32. A cycloid is given parametrically by $x = \theta - \sin\theta$, $y = 1 - \cos\theta$.
(a) Find the slope of the curve at the point where $\theta = \frac{2\pi}{3}$.
(b) Find the equation of the tangent to the cycloid at the point where $\theta = \frac{2\pi}{3}$.

33. Find the area of the region enclosed by both the polar curves $r = 4\sin\theta$ and $r = 4\cos\theta$.

34. (a) Find the 4th degree Taylor polynomial about 0 for cos x.
(b) Use part (a) to evaluate $\int_0^1 \cos x\,dx$.
(c) Estimate the error in (b), justifying your answer.

35. A particle moves on the curve of $y^3 = 2x + 1$ so that its distance from the x-axis is increasing at the constant rate of 2 units/sec. When $t = 0$, the particle is at (0,1).
(a) Find a pair of parametric equations $x = x(t)$ and $y = y(t)$ that describe the motion of the particle for nonnegative t.
(b) Find $|\mathbf{a}|$, the magnitude of the particle's acceleration, when $t = 1$.

36. Find the area of the region that the polar curves $r = 2 - \cos\theta$ and $r = 3\cos\theta$ enclose in common.

Answers Explained

Part A

1. **(a)** $f'(4.0) \simeq \dfrac{f(4.6) - f(4.0)}{4.6 - 4.0} = \dfrac{2.2 - 3.1}{0.6} = -1.5.$

$$f'(4.8) \simeq \frac{f(5) - f(4.6)}{5 - 4.6} = \frac{1.5 - 2.2}{0.4} = -1.75.$$

(b) It appears that the rate of change of f, while negative, is increasing. This implies that the graph of f is concave upward.

(c) $L = 7.6(0.7) + 5.7(0.3) + 4.2(0.5) + 3.1(0.6) + 2.2(0.4) = 11.87.$

(d) Using disks $\Delta V = \pi r^2 \Delta x$. One possible answer uses the left endpoints of the subintervals as values of r:

$$V \approx \pi(7.6)^2(0.7) + \pi(5.7)^2(0.3) + \pi(4.2)^2(0.5) + \pi(3.1)^2(0.6) + \pi(2.2)^2(0.4)$$

2. **(a)** $12y_0 + 0.3 = 24$ yields $y_0 \simeq 1.975.$

(b) Replace x by 0.3 in the equation of the curve:

$$(0.3)^2 y_0 - (0.3) = y_0^3 - 8 \text{ or}$$
$$y_0^3 - 0.09y_0 - 7.7 = 0.$$

The calculator's solution to three decimal places is $y_0 = 1.990$.

(c) Since the true value of y_0 at $x = 0.3$ exceeds the approximation, conclude that the given curve is concave up near $x = 0$. (Therefore, it is above the line tangent at $x = 0$.)

3. Graph $f'(x) = 2x \sin x - e^{(-x^2)} + 1$ in $[-7,7] \times [-10,10]$.

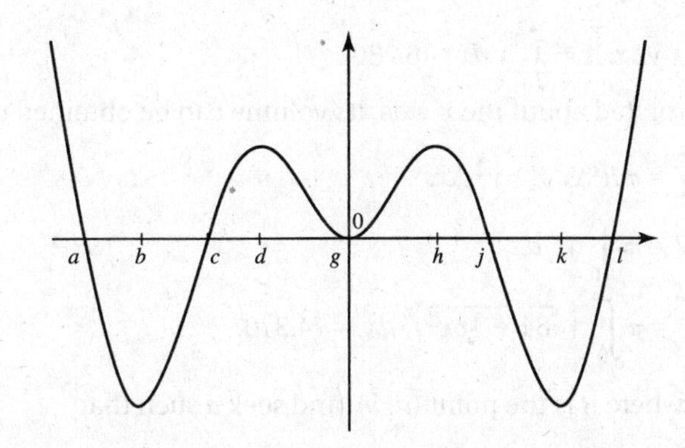

(a) Since f' is even and f contains (0, 0), f is odd and its graph is symmetric about the origin.

(b) Since f is decreasing when $f' < 0$, f decreases on the intervals (a, c) and (j, l). Use the calculator to solve $f'(x) = 0$. Conclude that f decreases on $-6.202 < x < -3.294$ and (symmetrically) on $3.294 < x < 6.202$.

(c) f has a relative maximum at $x = q$ if $f'(q) = 0$ and if f changes from increasing ($f' > 0$) to decreasing ($f' < 0$) at q. There are two relative maxima here: at $x = a = -6.202$ and at $x = j = 3.294$.

(d) f has a point of inflection when the graph of f changes its concavity; that is, when f' changes from increasing to decreasing, as it does at points d and h, or when f' changes from decreasing to increasing, as it does at points b, g, and k. So there are five points of inflection altogether.

4. In the graph below, C is the piece of the curve lying in the first quadrant. S is the region bounded by the curve C and the coordinate axes.

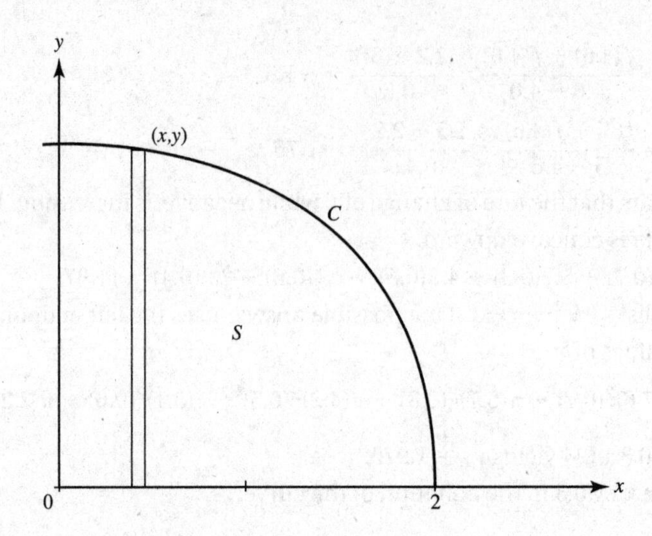

(a) Graph $y = \sqrt[3]{(64 - 16x^2)}$ in $[0,3] \times [0,5]$. Since you want dy/dx, the slope of the tangent, where $y = 1$, use the calculator to solve

$$\sqrt[3]{64 - 16x^2} = 1$$

(storing the answer at B). Then evaluate the slope of the tangent to C at $y = 1$:

$$f'(B) \approx -21.182.$$

(b) Since $\Delta A = y\Delta x$, $A = \int_0^2 y\,dx \approx 6.730.$

(c) When S is rotated about the x-axis, its volume can be obtained using disks:

$$\Delta V = \pi R^2 \Delta x = \pi y^2 \Delta x,$$

$$V = \pi \int_0^2 y^2\,dx$$

$$= \pi \int_0^2 \left(\sqrt[3]{64 - 16x^2}\right)^2 dx \approx 74.310.$$

5. See the figure, where R is the point (a, b), and seek a such that

$$\int_0^a \left(x - x^2 - \frac{b}{a} \cdot x\right) dx = \frac{1}{2}\int_0^1 (x - x^2)dx.$$

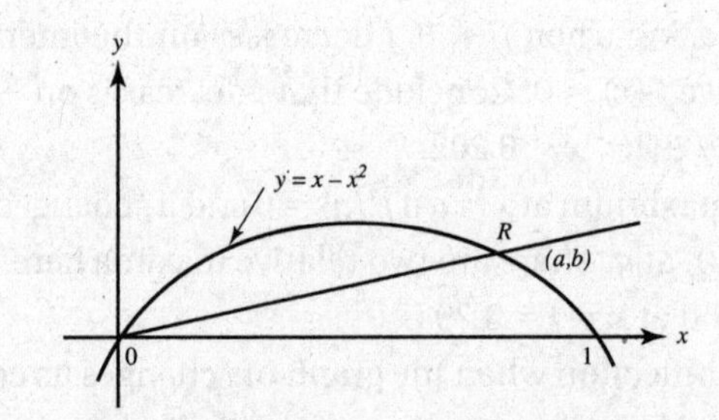

6. Graph $y = \sin 2^x$ in $[-1, 3.2] \times [-1, 1]$. Note that $y = f''$.

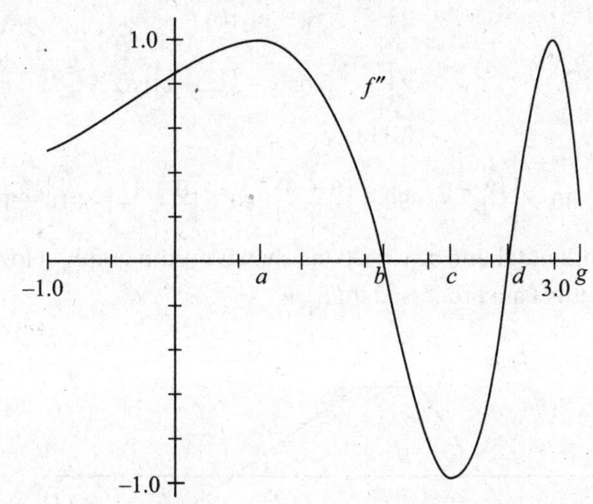

(a) The graph of f is concave downward where f'' is negative, namely, on (b, d). Use the calculator to solve $\sin 2^x = 0$, obtaining $b = 1.651$ and $d = 2.651$. The answer to (a) is therefore $1.651 < x < 2.651$.

(b) f' has a relative minimum at $x = d$, because f'' equals 0 at d, is less than 0 on (b, d), and is greater than 0 on (d, g). Thus f' has a relative minimum (from part a) at $x = 2.651$.

(c) The graph of f' has a point of inflection wherever its second derivative f''' changes from positive to negative or vice versa. This is equivalent to f'' changing from increasing to decreasing (as at a and g) or vice versa (as at c). Therefore, the graph of f' has three points of inflection on $[-1,3.2]$.

7. Graph $f(x) = \cos x$ and $g(x) = x^2 - 1$ in $[-2,2] \times [-2,2]$. Here, $y_1 = f$ and $y_2 = g$.

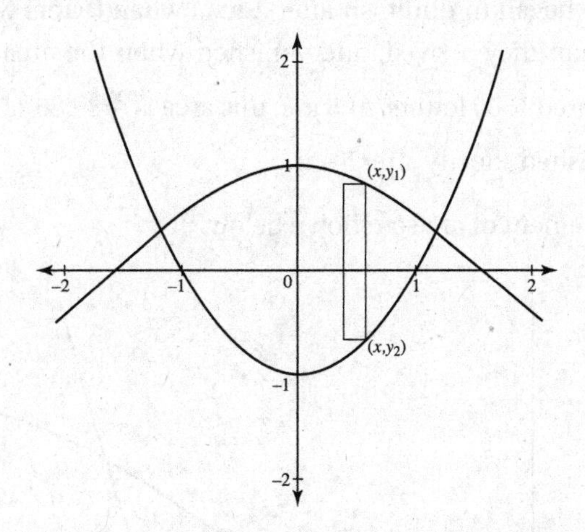

(a) Solve $\cos x = x^2 - 1$ to find the two points of intersection: $(1.177,0.384)$ and $(-1.177,0.384)$.

(b) Since $\Delta A = (y_1 - y_2)\,\Delta x = \left[f(x) - g(x)\right]\Delta x$, the area A bounded by the two curves is

$$\begin{aligned} A &= 2\int_0^{1.777} (y_1 - y_2)\,dx, \\ &= 2\int_0^{1.777} \left(\cos x - \left(x^2 - 1\right)\right) dx \\ &\approx 3.114. \end{aligned}$$

8. **(a)** $\left(\dfrac{10+12}{2}\right)\cdot 30 + \left(\dfrac{12+8}{2}\right)\cdot 90 + \left(\dfrac{8+9}{2}\right)\cdot 80 + \left(\dfrac{9+11}{2}\right)\cdot 40 = 2310$ letters

(b) Draw a horizontal line at $y = 20$ (as shown on the graph below), representing the rate at which letters are processed then.

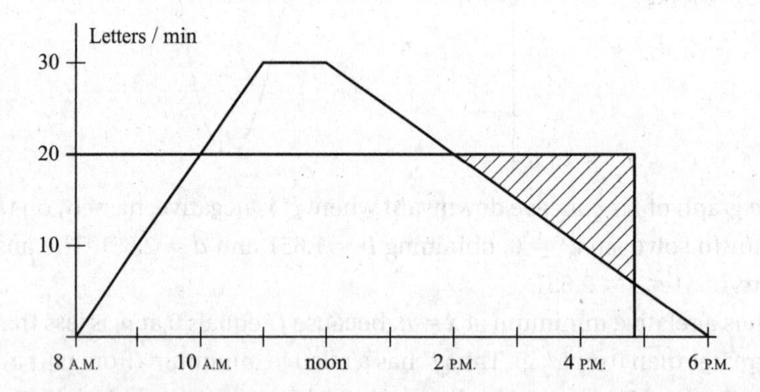

(i) Letters began to pile up when they arrived at a rate greater than that at which they were being processed, that is, at $t = 10$ A.M.

(ii) The pile was largest when the letters stopped piling up, at $t = 2$ P.M.

(iii) The number of letters in the pile is represented by the area of the small trapezoid above the horizontal line: $\frac{1}{2}(4\cdot 60 + 1\cdot 60)(10) = 1500$.

(iv) The pile began to diminish after 2 P.M., when letters were processed at a rate faster than they arrived, and vanished when the area of the shaded triangle represented 1500 letters. At 5 P.M. this area is $\frac{1}{2}(3\cdot 60)(15) = 1350$ letters, so the pile vanished shortly after 5 P.M.

9. Draw a vertical element of area as shown below.

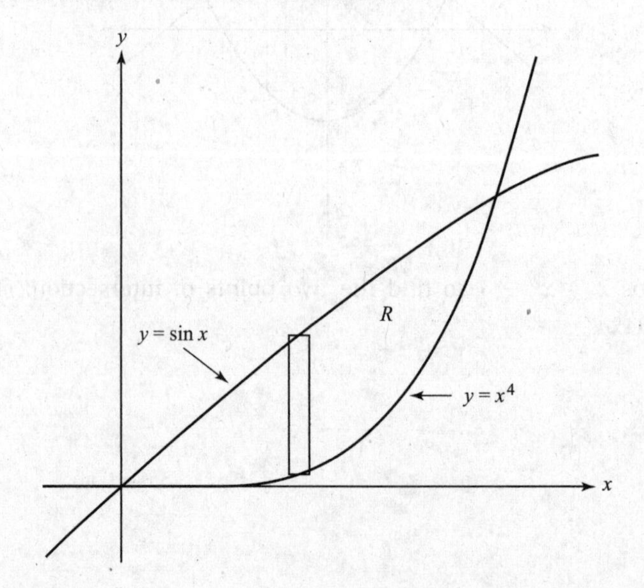

(a) Let a represent the x-value of the positive point of intersection of $y = x^4$ and $y = \sin x$. Solving $a^4 = \sin a$ with the calculator, we find $a = 0.9496$.

$$\Delta A = \left(y_{\text{top}} - y_{\text{bottom}}\right) \Delta x = \left(\sin x - x^4\right)\Delta x,$$
$$A = \int_0^a \left(\sin x - x^4\right) dx \simeq 0.264.$$

(b) Elements of volume are triangular prisms with height $h = 3$ and base $b = \left(\sin x - x^4\right)$, as shown below.

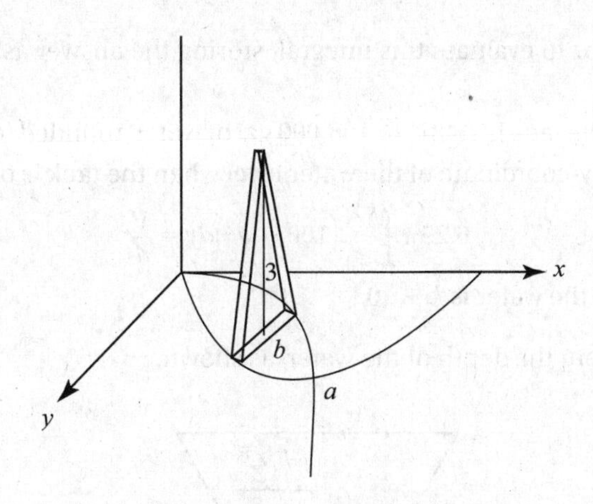

$$\Delta V = \frac{1}{2}\left(\sin x - x^4\right)(3)\Delta x,$$
$$V = \frac{3}{2}\int_0^a \left(\sin x - x^4\right) dx = 0.395.$$

(c) When R is rotated around the x-axis, the element generates washers. If r_1 and r_2 are the radii of the larger and smaller disks, respectively, then

$$\Delta V = \pi\left(r_1^2 - r_2^2\right)\Delta x = \pi\left(\left(\sin x\right)^2 - \left(x^4\right)^2\right)\Delta x,$$
$$V = \pi\int_0^a \left(\sin^2 x - x^8\right) dx = 0.529.$$

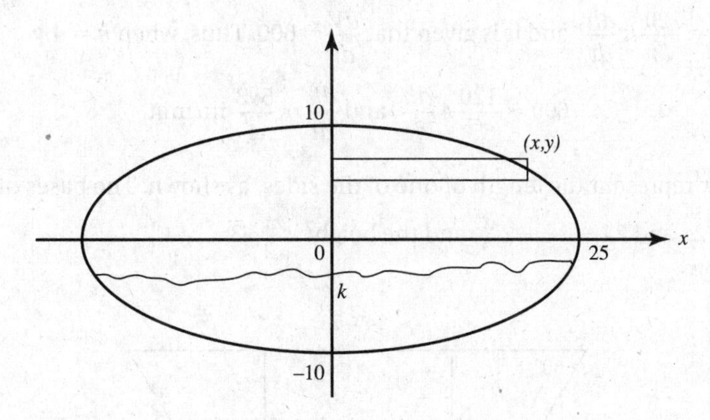

10. The figure above shows an elliptical cross section of the tank. Its equation is

$$\frac{x^2}{625} + \frac{y^2}{100} = 1.$$

(a) The volume of the tank, using disks, is $V = 2\pi \int_0^{10} x^2\, dy$, where the ellipse's symmetry about the x-axis has been exploited. The equation of the ellipse is equivalent to $x^2 = 6.25\left(100 - y^2\right)$, so

$$V = 12.5\pi \int_0^{10} \left(100 - y^2\right) dy.$$

Use the calculator to evaluate this integral, storing the answer as V to have it available for part (b).

The capacity of the tank is $7.48V$, or 196,000 gal of water, rounded to the nearest 1000 gal.

(b) Let k be the y-coordinate of the water level when the tank is one-fourth full. Then

$$6.25\pi \int_{-10}^{k} \left(100 - y^2\right) dy = \frac{V}{4}$$

and the depth of the water is $k + 10$.

11. **(a)** Let h represent the depth of the water, as shown.

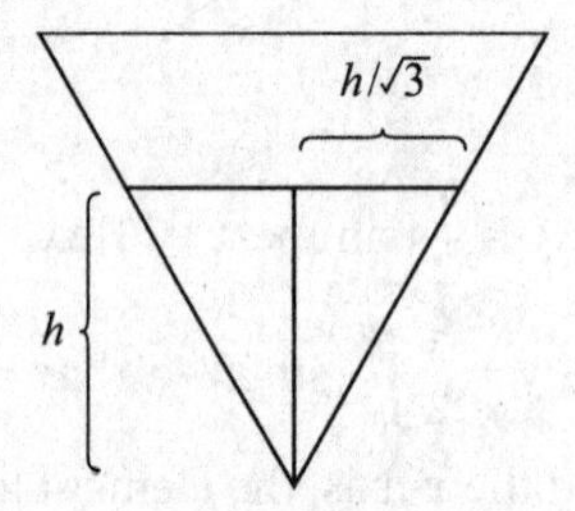

Then h is the altitude of an equilateral triangle, and the base $b = \frac{2h}{\sqrt{3}}$.

The volume of water is

$$V = \frac{1}{2}\left(\frac{2h}{\sqrt{3}}\right)h \cdot 60 = \frac{60h^2}{\sqrt{3}} \text{ in.}^3$$

Now $\frac{dV}{dt} = \frac{120}{\sqrt{3}} h \frac{dh}{dt}$, and it is given that $\frac{dV}{dt} = 600$. Thus, when $h = 4$,

$$600 = \frac{120}{\sqrt{3}} 4 \frac{dh}{dt}, \text{ and } \frac{dh}{dt} = \frac{5\sqrt{3}}{4} \text{ in/min.}$$

(b) Let x represent the length of one of the sides, as shown. The bases of the trapezoid are $24 - 2x$ and $24 - 2x + 2\frac{x}{2}$, and the height is $\frac{x}{2}\sqrt{3}$.

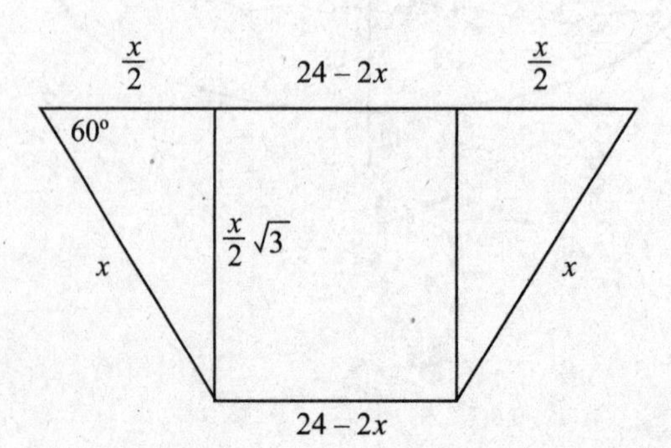

The volume of the trough (in in.^{3}) is given by

$$V = \frac{(24-2x)+(24-x)}{2}\cdot\frac{x}{2}\sqrt{3}\times 60 = 15\sqrt{3}\left(48x-3x^2\right) \qquad (0<x<12),$$

$$V' = 15\sqrt{3}(48-6x) = 0 \text{ when } x = 8.$$

Since $V'' = 15\sqrt{3}(-6) < 0$, the maximum volume is attained by folding the metal 8 inches from the edges.

12. **(a)** Both $\pi/4$ and the expression in brackets yield 0.7853981634, which is accurate to ten decimal places.

(b) $\tan^{-1}\frac{1}{5} = \frac{1}{5} - \frac{1}{3}\left(\frac{1}{5}\right)^3 + \frac{1}{5}\left(\frac{1}{5}\right)^5 - \frac{1}{7}\left(\frac{1}{5}\right)^7 = 0.197396.$

$$\tan^{-1}\frac{1}{239} = \frac{1}{239} = 0.004184.$$

(c) $4\tan^{-1}\frac{1}{5} - \tan^{-1}\frac{1}{239} = 0.7854$; this agrees with the value of $\frac{\pi}{4}$ to four decimal places.

(d) The series

$$\tan^{-1}1 = 1 - \frac{1}{3} + \frac{1}{5} - \frac{1}{7} + \cdots$$

converges *very* slowly. Example 56, page 417, evaluated the sum of 60 terms of the series for π (which equals $4\tan^{-1}1$). To four decimal places, we get $\pi = 3.1249$, which yields 0.7812 for $\pi/4$—not accurate even to two decimal places.

13. **(a)** The given series is alternating. Since $\lim_{n\to\infty}\ln(n+1) = \infty$, $\lim_{n\to\infty}\frac{1}{\ln(n+1)} = 0$. Since $\ln x$ is an increasing function,

$$\ln(n+1) > \ln n \quad \text{and} \quad \frac{1}{\ln(n+1)} < \frac{1}{\ln n}.$$

The series therefore converges.

(b) Since the series converges by the Alternating Series Test, the error in using the first n terms for the sum of the whole series is less than the absolute value of the $(n+1)$st term. Thus the error is less than $\frac{1}{\ln(n+1)}$. Solve for n using $\frac{1}{\ln(n+1)} < 0.1$:

$$\ln(n+1) > 10,$$
$$(n+1) > e^{10},$$
$$n > e^{10} - 1 > 22{,}025.$$

The given series converges very slowly!

(c) The series $\sum_{2}^{\infty}(-1)^n\frac{1}{n\ln n}$ is conditionally convergent. The given alternating series converges since the nth term approaches 0 and $\frac{1}{(n+1)\ln(n+1)} < \frac{1}{n\ln n}$. However, the *nonnegative* series diverges by the Integral Test, since

$$\int_2^{\infty}\frac{1}{x\ln x}\,dx = \lim_{b\to\infty}\ln(\ln x)\Big|_2^b = \infty.$$

14. **(a)** Solve by separation of variables:

$$\frac{dy}{y(10-y)} = k\,dt,$$

$$\frac{1}{10}\int\left(\frac{1}{y} + \frac{1}{10-y}\right)dy = \int k\,dt,$$

$$\frac{1}{10}\ln\left|\frac{y}{10-y}\right| = kt + C,$$

$$\ln\left|\frac{10-y}{y}\right| = -10\,(kt + C).$$

Let $c = e^{-10C}$; then

$$\frac{10-y}{y} = ce^{-10kt}.$$

Now use initial condition $y = 2$ at $t = 0$:

$$\frac{8}{2} = ce^{0} \text{ so } c = 4;$$

and the other condition, $y = 5$ at $t = 2$, gives

$$\frac{5}{5} = 4e^{-20k} \text{ or } k = \frac{1}{10}\ln 2.$$

(b) Since $c = 4$ and $k = \frac{1}{10}\ln 2$, then $\frac{10-y}{y} = 4e^{-10\left(\frac{1}{10}\ln 2\right)t}$.

Solving for y yields $y = \frac{10}{1 + 4\cdot 2^{-t}}$.

(c) $8 = \frac{10}{1 + 4\cdot 2^{-t}}$ means $1 + 4\cdot 2^{-t} = 1.25$, so $t = 4$.

(d) $\lim\limits_{t\to\infty}\frac{10}{1 + 4\cdot 2^{-t}} = 10$, so the value of y approaches 10.

15. **(a)** Since $x = \frac{1}{2}t$, $x(4) = \frac{1}{2}(4) = 2$. Since $y = 18 - 2\cdot 2^2 = 10$, P is at (2,10).

(b) Since $y = 18 - 2x^2$, $\frac{dy}{dt} = -4x\frac{dx}{dt}$. Since $x = \frac{1}{2}t$, $\frac{dx}{dt} = \frac{1}{2}$. Therefore

$\frac{dy}{dt} = -4x\frac{dx}{dt} = -4\cdot 2\cdot\frac{1}{2} = -4$ unit/sec.

(c) Let $D =$ the object's distance from the origin. Then

$D^2 = x^2 + y^2$, and at (2,10) $D = \sqrt{104}$.

$$2D\frac{dD}{dt} = 2x\frac{dx}{dt} + 2y\frac{dy}{dt},$$

$$2\sqrt{104}\,\frac{dD}{dt} = 2\cdot 2\cdot\frac{1}{2} + 2\cdot 10(-4),$$

$$\frac{dD}{dt} = \frac{-78}{2\sqrt{104}} = -3.824 \text{ unit/sec.}$$

(d) The object hits the x-axis when $y = 18 - 2x^2 = 0$, or $x = 3$. Since $x = \frac{1}{2}t = 3$, $t = 6$.

(e) The length of the arc of $y = 18 - 2x^2$ for $0 \le x \le 3$ is given by

$$L = \int\sqrt{1 + \left(\frac{dy}{dx}\right)^2}\,dx = \int_0^3 \sqrt{1 + (-4x)^2}\,dx = 18.460 \text{ units.}$$

16. **(a)** See graph.

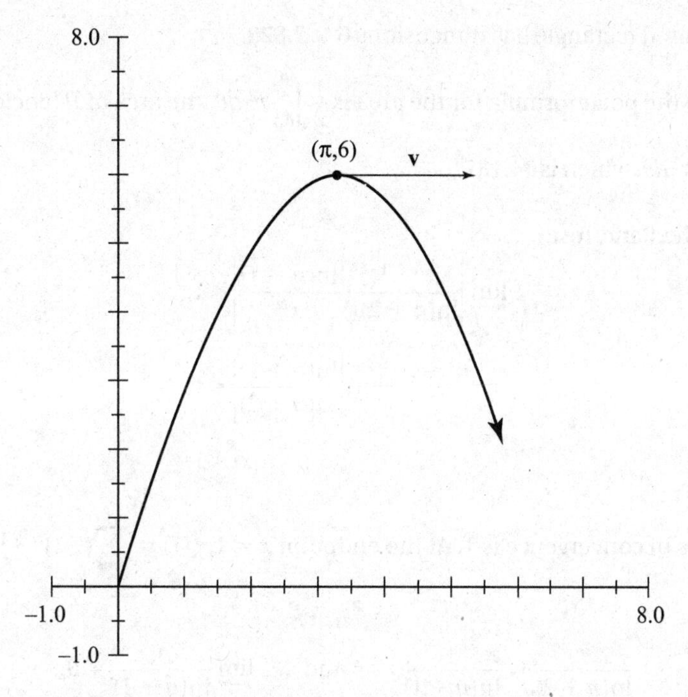

(b) You want to maximize $y(t) = \dfrac{12t}{t^2+1}$.

$$y'(t) = \frac{\left(t^2+1\right)(12) - 12t(2t)}{\left(t^2+1\right)^2} = \frac{12(1-t)(1+t)}{\left(t^2+1\right)^2}.$$

See signs analysis.

	0	Inc.	1	Dec.	t
y					
y'		+		−	

The maximum y occurs when $t = 1$, because y changes from increasing to decreasing there.

(c) Since $x(1) = 4\arctan 1 = \pi$ and $y(1) = \dfrac{12}{1+1} = 6$, the coordinates of the highest point are $(\pi,6)$.

Since $x'(t) = \dfrac{4}{1+t^2}$ and $y'(t) = \dfrac{12\left(1-t^2\right)}{\left(t^2+1\right)^2}$, so $\mathbf{v}(1) = \langle 2,0\rangle$. This vector is shown on the graph.

(d) $\lim\limits_{t\to\infty} x(t) = \lim\limits_{t\to\infty} 4\arctan t = 4\left(\dfrac{\pi}{2}\right) = 2\pi$, and $\lim\limits_{t\to\infty} y(t) = \lim\limits_{t\to\infty} \dfrac{12t}{t^2+1} = 0$. Thus the particle approaches the point $(2\pi,0)$.

17. **(a)** To find the smallest rectangle with sides parallel to the x- and y-axes, you need a rectangle formed by vertical and horizontal tangents as shown in the figure. The vertical tangents are at the x-intercepts, $x = \pm 3$. The horizontal tangents are at the points where y (not r) is a maximum. You need, therefore, to maximize

$$y = r\sin\theta = (2+\cos 2\theta)\sin\theta,$$

$$\frac{dy}{d\theta} = (2+\cos 2\theta)\cos\theta + \sin\theta(-2\sin 2\theta).$$

Use the calculator to find that $\frac{dy}{d\theta} = 0$ when $\theta = 0.7854$. Therefore, $y = 1.414$, so the desired rectangle has dimensions 6×2.828.

(b) Since the polar formula for the area is $\frac{1}{2}\int_{\theta_1}^{\theta_2} r^2\, d\theta$, the area of R (enclosed by r) is $4 \cdot \frac{1}{2}\int_0^{\pi/2} r^2 d\theta$, which is 14.137.

18. **(a)** Use the Ratio Test:

$$\lim_{n\to\infty}\left|\frac{x^{n+1}}{\ln(n+2)} \cdot \frac{\ln(n+1)}{x^n}\right| < 1,$$

$$|x| \lim_{n\to\infty}\left|\frac{\ln(n+1)}{\ln(n+2)}\right| < 1,$$

$$|x| < 1.$$

The radius of convergence is 1. At the endpoint $x = 1$, $f(1) = \sum_{n=1}^{\infty}(-1)^{n+1}\frac{1}{\ln(n+1)}$.

Since

$$\frac{1}{\ln(n+2)} < \frac{1}{\ln(n+1)} \qquad \text{and} \qquad \lim_{n\to\infty}\frac{1}{\ln(n+1)} = 0,$$

this series converges by the Alternating Series Test. Thus $f(x) = \sum_{n=1}^{\infty}(-1)^{n+1}\frac{x^n}{\ln(n+1)}$ converges for positive values $0 < x \le 1$.

(b) Because $f(0.5) = \sum_{n=1}^{\infty}(-1)^{n+1}\frac{(0.5)^n}{\ln(n+1)}$ satisfies the Alternating Series Test, the error in approximation after n terms is less than the magnitude of the next term.

The calculator shows that $\frac{(0.5)^{n+1}}{\ln(n+2)} < 0.01$ at $n = 5$ terms.

(c) $f(-0.5) = \sum_{n=1}^{\infty}(-1)^{n+1}\frac{(-0.5)^n}{\ln(n+1)} = \sum_{n=1}^{\infty}\frac{-(0.5)^n}{\ln(n+1)}$ is a negative series. Therefore the error will be larger than the magnitude of the first omitted term, and thus less accurate than the estimate for $f(0.5)$.

Part B

19. The graph shown below satisfies all five conditions. So do many others!

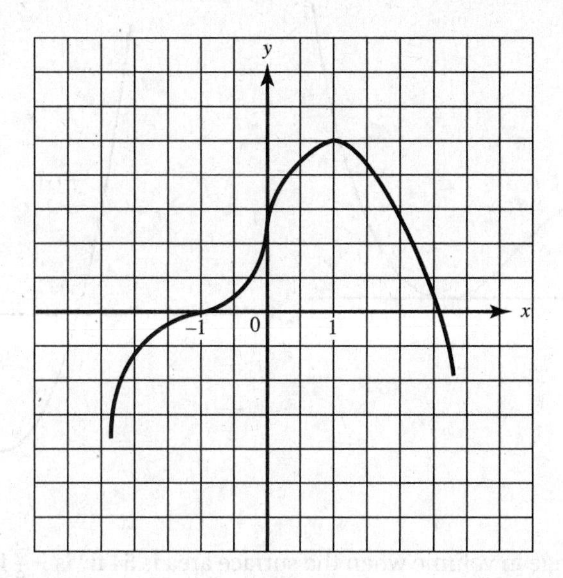

20. **(a)** f' is defined for all x in the interval. Since f is therefore differentiable, it must also be continuous.
(b) Because $f'(-2) = 0$ and f' changes from negative to positive there, f has a local minimum at $x = -2$. To the left of $x = 9$, f' is negative, so f is decreasing as it approaches that endpoint and reaches another local minimum there.
(c) Because f' is negative to the right of $x = -3$, f decreases from its left endpoint, indicating a local max there. Also, $f'(7) = 0$ and f' changes from positive to negative there, so f has a local relative maximum at $x = 7$.
(d) Note that $f(7) - f(-3) = \int_{-3}^{7} f'(x)\,dx$. Since there is more area above the x-axis than below the x-axis on $[-3,7]$, the integral is positive and $f(7) - f(-3) > 0$. This implies that $f(7) > f(-3)$, and that the absolute maximum occurs at $x = 7$.
(e) At $x = 2$ and also at $x = 6$, f' changes from increasing to decreasing, indicating that f changes from concave upward to concave downward at each. At $x = 4$, f' changes from decreasing to increasing, indicating that f changes from concave downward to concave upward there. Hence the graph of f has points of inflection at $x = 2$, 4, and 6.

21. Draw a sketch of the region bounded above by $y_1 = 8 - 2x^2$ and below by $y_2 = x^2 - 4$, and inscribe a rectangle in this region as described in the question. If (x, y_1) and (x, y_2) are the vertices of the rectangle in quadrants I and IV, respectively, then the area

$$A = 2x\left(y_1 - y_2\right) = 2x\left(12 - 3x^2\right), \quad \text{or} \quad A(x) = 24x - 6x^3.$$

Then $A'(x) = 24 - 18x^2 = 6\left(4 - 3x^2\right)$, which equals 0 when $x = \frac{2}{\sqrt{3}} = \frac{2\sqrt{3}}{3}$.

Check to verify that $A''(x) < 0$ at this point. This assures that this value of x yields maximum area, which is given by $\frac{4\sqrt{3}}{3} \times 8$.

22. The graph of $f'(x)$ is shown here.

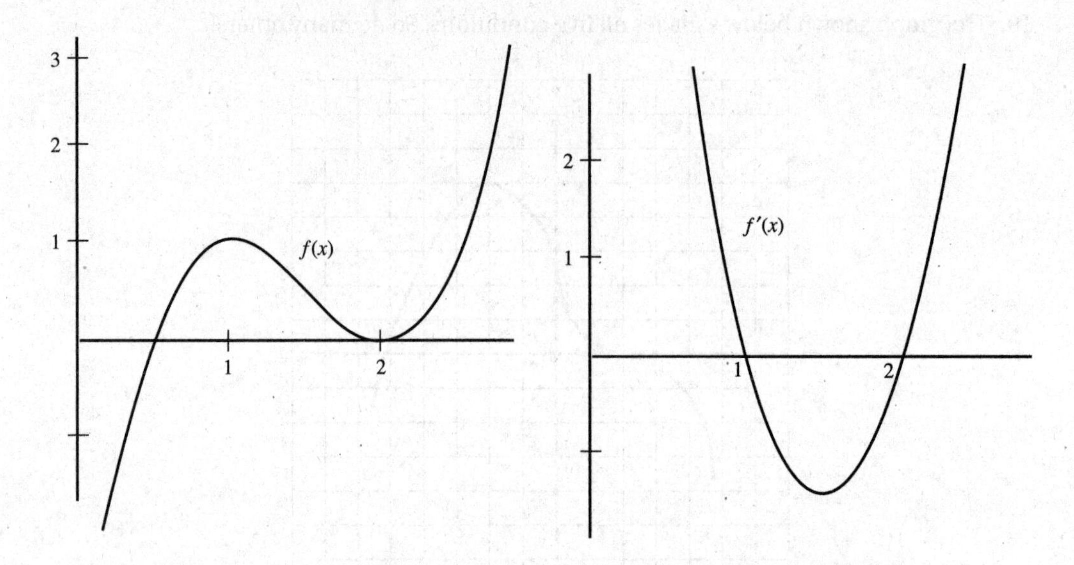

23. The rate of change in volume when the surface area is 54 ft³ is $-\frac{3}{8}$ ft³/sec.

24. See the figure. The equation of the circle is $x^2 + y^2 = a^2$; the equation of RS is $y = a - x$. If y_2 is an ordinate of the circle and y_1 of the line, then,

$$\Delta V = \pi {y_2}^2 \, \Delta x - \pi {y_1}^2 \, \Delta x,$$

$$V = 2\pi \int_0^a \left[(a^2 - x^2) - (a - x)^2 \right] dx = \frac{2}{3} \pi a^3.$$

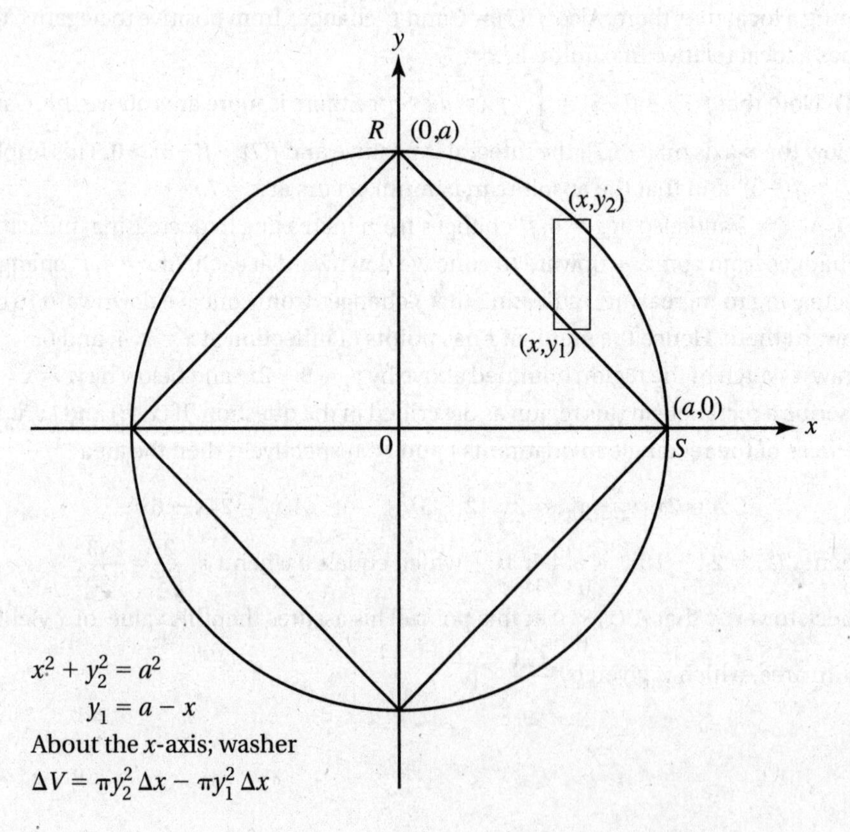

25. **(a)** The region is sketched in the figure. The pertinent points of intersection are labeled.

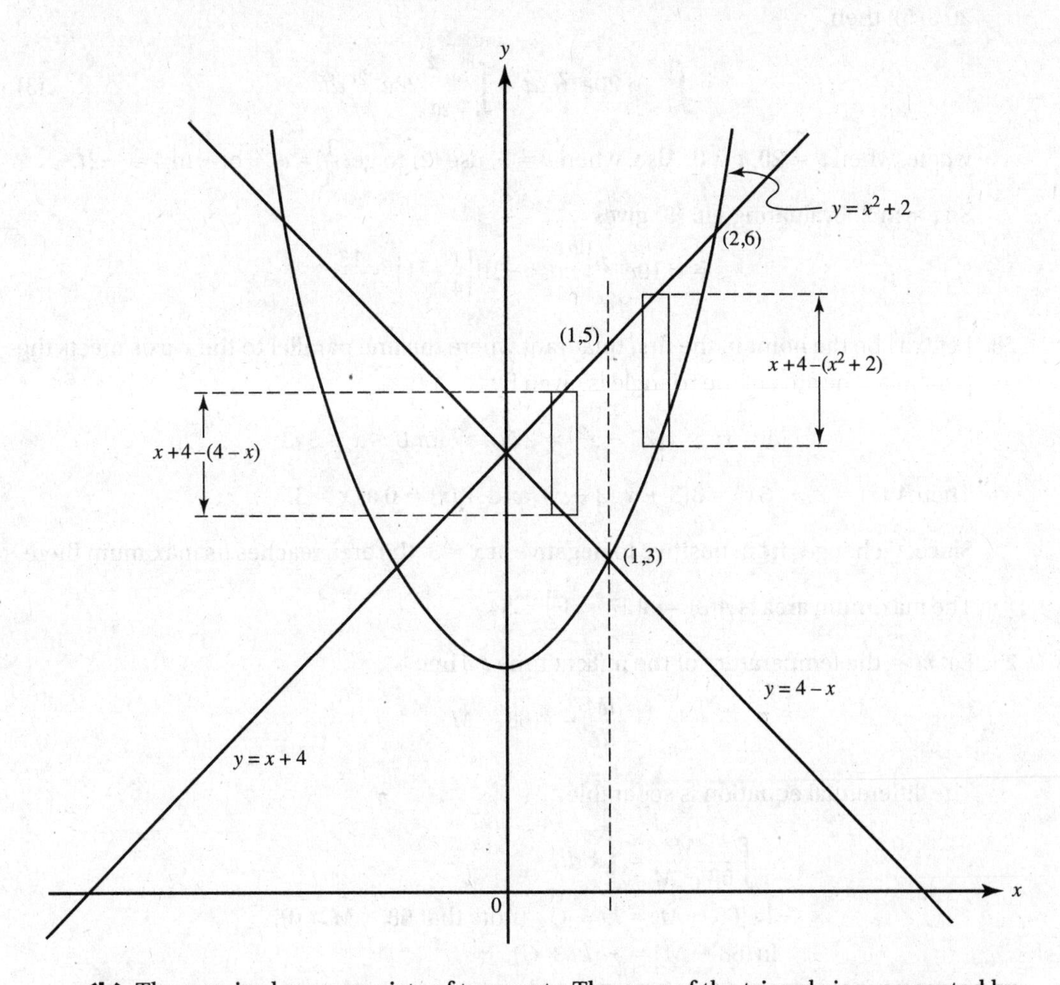

(b) The required area consists of two parts. The area of the triangle is represented by $\int_1^2 [(x+4)-(4-x)]\,dx$ and is equal to 1, while the area of the region bounded at the left by $x = 1$, above by $y = x + 4$, and at the right by the parabola is represented by $\int_1^2 \left[(x+4)-(x^2+2)\right] dx$. This equals

$$\int_1^2 (x + 2 - x^2)\,dx = \frac{x^2}{2} + 2x - \frac{x^3}{3}\bigg|_1^2 = \frac{7}{6}.$$

The required area, thus, equals $2\frac{1}{6}$ or $\frac{13}{6}$.

26. **(a)** 1975 to 1976 and 1978 to 1980.

(b) 1975 to 1977 and 1979 to 1981.

(c) 1976 to 1977 and 1980 to 1981.

27. **(a)** Since $a = \frac{dv}{dt} = -2v$, then, separating variables, $\frac{dv}{v} = -2dt$. Integrating gives

$$\ln v = -2t + C, \tag{1}$$

and, since $v = 20$ when $t = 0$, $C = \ln 20$. Then (1) becomes $\ln \frac{v}{20} = -2t$ or, solving for v,

$$v = 20e^{-2t}. \tag{2}$$

(b) Note that $v > 0$ for all t. Let s be the required distance traveled (as v decreases from 20 to 5); then

$$s = \int_{v=20}^{v=5} 20e^{-2t}\,dt = \int_{t=20}^{\ln=2} 20e^{-2t}\,dt, \tag{3}$$

where, when $v = 20$, $t = 0$. Also, when $v = 5$, use (2) to get $\frac{1}{4} = e^{-2t}$ or $-\ln 4 = -2t$.

So $t = \ln 2$. Evaluating s in (3) gives

$$s = -10e^{-2t}\Big|_0^{\ln 2} = -10\left(\frac{1}{4} - 1\right) = \frac{15}{2}.$$

28. Let (x,y) be the point in the first quadrant where the line parallel to the x-axis meets the parabola. The area of the triangle is given by

$$A = xy = x\left(27 - x^2\right) = 27x - x^3 \text{ for } 0 \le x \le 3\sqrt{3}.$$

Then $A'(x) = 27 - 3x^2 = 3(3 + x)(3 - x)$, and $A'(x) = 0$ at $x = 3$.

Since A' changes from positive to negative at $x = 3$, the area reaches its maximum there.

The maximum area is $A(3) = 3\left(27 - 3^2\right) = 54$.

29. Let $M =$ the temperature of the milk at time t. Then

$$\frac{dM}{dt} = k(68 - M).$$

The differential equation is separable:

$$\int \frac{dM}{68 - M} = \int k\,dt,$$

$$-\ln|68 - M| = kt + C \quad (\text{note that } 68 - M > 0),$$

$$\ln(68 - M) = -(kt + C),$$

$$68 - M = e^{-(kt + C)},$$

$$M = 68 - ce^{-kt},$$

where $c = e^{-C}$.

Find c, using the fact that $M = 40°$ when $t = 0$:

$$40 = 68 - ce^0 \qquad \text{means} \qquad c = 28.$$

Find k, using the fact that $M = 43°$ when $t = 3$:

$$43 = 68 - 28e^{-3k},$$

$$e^{-3k} = \frac{25}{28},$$

$$k = -\frac{1}{3}\ln\frac{25}{28}.$$

Hence $M = 68 - 28e^{\frac{1}{3}\ln\frac{25}{28}t}$.

Now find t when $M = 60°$:

$$60 = 68 - 28e^{\frac{1}{3}\ln\frac{25}{28}t},$$

$$e^{\frac{1}{3}\ln\frac{25}{28}t} = \frac{8}{28},$$

$$t = \frac{\ln \frac{8}{28}}{\frac{1}{3}\ln \frac{25}{28}} = 33.163.$$

Since the phone rang at $t = 3$, you have 30 min to solve the problem.

30. One possible graph of h is shown; it has the following properties:
    - continuity on $[-4,4]$,
    - symmetry about the y-axis,
    - roots at $x = -1, 1$,
    - horizontal tangents at $x = -2, 0, 2$,
    - points of inflection at $x = -3, -2, -1, 1, 2, 3$,
    - corners at $x = -3, 3$.

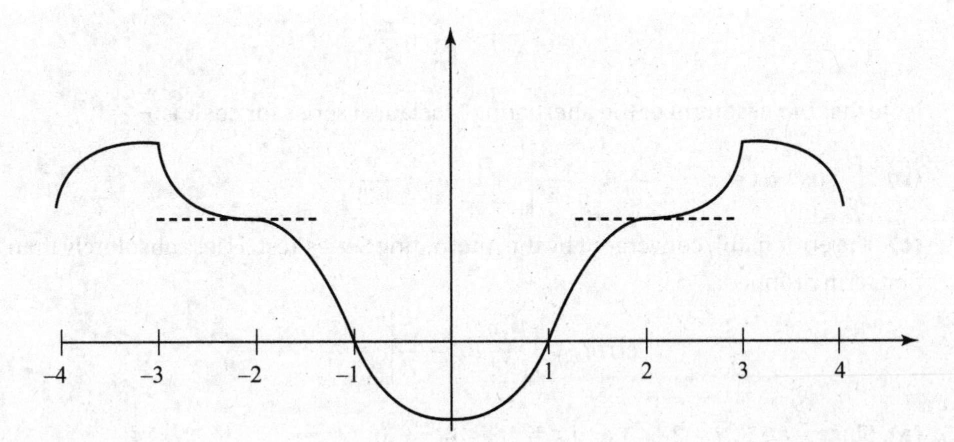

31. **(a)** Let $f(x) = \ln(1 + x)$. Then $f'(x) = \frac{1}{1+x}$, $f''(x) = -\frac{1}{(1+x)^2}$, $f'''(x) = \frac{2}{(1+x)^3}$, and $f^{(4)}(x) = -\frac{3!}{(1+x)^4}$, $f^{(5)}(x) = \frac{4!}{(1+x)^5}$. At $x = 0$, $f(0) = 0$, $f'(0) = 1$, $f''(0) = -1$, $f'''(0) = 2$, $f^{(4)}(0) = -(3!)$, and $f^{(5)}(0) = 4!$. So

$$\ln(1 + x) = x - \frac{x^2}{2} + \frac{x^3}{3} - \frac{x^4}{4} + \frac{x^5}{5} - \cdots.$$

(b) Using the Ratio Test, you know that the series converges when $\lim_{x\to\infty}\left|\frac{x^{n+1}}{n+1} \cdot \frac{n}{x^n}\right| < 1$, that is, when $|x| < 1$, or $-1 < x < 1$. Thus, the radius of convergence is 1.

(c) $\ln(1.2) = 0.2 - \frac{(0.2)^2}{2} + \frac{(0.2)^3}{3} + \frac{(0.2)^4}{4} + \frac{(0.2)^5}{5}$.

(d) Since the series converges by the Alternating Series Test, the error in the answer for (c) is less absolutely than $\frac{(0.2)^6}{6}$.

32. From the equations for x and y,

$$dx = (1 - \cos\theta)\,d\theta \qquad \text{and} \qquad dy = \sin\theta\,d\theta.$$

(a) The slope at any point is given by $\frac{dy}{dx}$, which here is $\frac{\sin\theta}{1 - \cos\theta}$. When $\theta = \frac{2\pi}{3}$, the slope is $\frac{\sqrt{3}}{3}$.

(b) When $\theta = \frac{2\pi}{3}$, $x = \frac{2\pi}{3} - \frac{\sqrt{3}}{2}$ and $y = 1 - \left(-\frac{1}{2}\right) = \frac{3}{2}$. The equation of the tangent is $9y - 3\sqrt{3} \cdot x = 18 - 2\pi\sqrt{3}$.

33. Both curves are circles with centers at, respectively, (2, 0) and $\left(2, \frac{\pi}{2}\right)$; the circles intersect at $\left(2\sqrt{2}, \frac{\pi}{4}\right)$. The common area is given by

$$\int_0^{\pi/4} (4 \sin \theta)^2 \, d\theta \quad \text{or} \quad \int_{\pi/4}^{\pi/2} (4 \cos \theta)^2 \, d\theta.$$

The answer is $2(\pi - 2)$.

34. **(a)** For $f(x) = \cos x$, $f'(x) = -\sin x$, $f''(x) = -\cos x$, $f'''(x) = \sin x$, $f^{(4)}(x) = \cos x$, $f^{(5)}(x) = -\sin x$, $f^{(6)}(x) = -\cos x$. The Taylor polynomial of order 4 about 0 is

$$\cos x = 1 - \frac{x^2}{2!} + \frac{x^4}{4!}.$$

Note that the next term of the alternating Maclaurin series for $\cos x$ is $-\frac{x^6}{6!}$.

(b) $\int_0^1 \cos x \, dx = x - \frac{x^3}{3 \cdot 2!} + \frac{x^5}{5 \cdot 4!}\Bigg|_0^1 = 1 - \frac{1}{6} + \frac{1}{120}$.

(c) The error in (b), convergent by the Alternating Series Test, is less absolutely than the first term dropped:

$$\text{error} < \int_0^1 \frac{x^6}{6!} \, dx = \frac{x^7}{7!}\Bigg|_0^1 = \frac{1}{7!}.$$

35. **(a)** Since $\frac{dy}{dx} = 2$, $y = 2t + 1$ and $x = 4t^3 + 6t^2 + 3t$.

(b) Since $\frac{d^2y}{dt^2} = 0$ and $\frac{d^2x}{dt^2} = 24t + 12$, then, when $t = 1$, $|\mathbf{a}| = 36$.

36. See the figure. The required area A is twice the sum of the following areas: that of the limaçon from 0 to $\frac{\pi}{3}$, and that of the circle from $\frac{\pi}{3}$ to $\frac{\pi}{2}$. Thus

$$A = 2\left[\frac{1}{2}\int_0^{\pi/3} (2 - \cos \theta)^2 \, d\theta + \frac{1}{2}\int_{\pi/3}^{\pi/2} (3\cos \theta)^2 \, d\theta\right]$$
$$= \frac{9\pi}{4} - 3\sqrt{3}$$

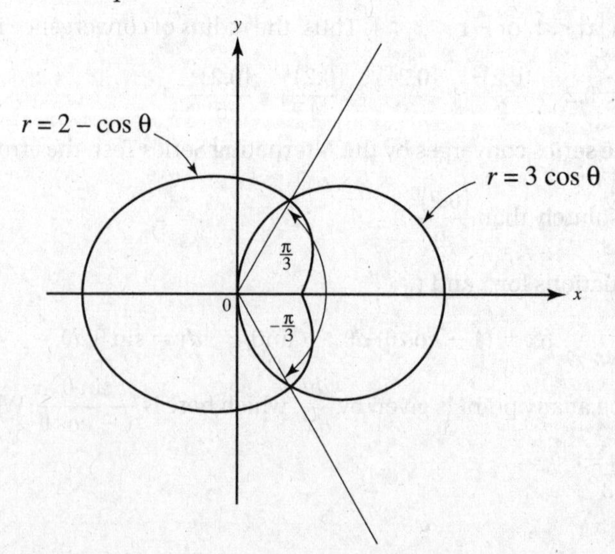

AB Practice Examinations

ANSWER SHEET
AB Practice Examination 1

Part A

1. Ⓐ Ⓑ Ⓒ Ⓓ
2. Ⓐ Ⓑ Ⓒ Ⓓ
3. Ⓐ Ⓑ Ⓒ Ⓓ
4. Ⓐ Ⓑ Ⓒ Ⓓ
5. Ⓐ Ⓑ Ⓒ Ⓓ
6. Ⓐ Ⓑ Ⓒ Ⓓ
7. Ⓐ Ⓑ Ⓒ Ⓓ
8. Ⓐ Ⓑ Ⓒ Ⓓ
9. Ⓐ Ⓑ Ⓒ Ⓓ
10. Ⓐ Ⓑ Ⓒ Ⓓ
11. Ⓐ Ⓑ Ⓒ Ⓓ
12. Ⓐ Ⓑ Ⓒ Ⓓ
13. Ⓐ Ⓑ Ⓒ Ⓓ
14. Ⓐ Ⓑ Ⓒ Ⓓ
15. Ⓐ Ⓑ Ⓒ Ⓓ
16. Ⓐ Ⓑ Ⓒ Ⓓ
17. Ⓐ Ⓑ Ⓒ Ⓓ
18. Ⓐ Ⓑ Ⓒ Ⓓ
19. Ⓐ Ⓑ Ⓒ Ⓓ
20. Ⓐ Ⓑ Ⓒ Ⓓ
21. Ⓐ Ⓑ Ⓒ Ⓓ
22. Ⓐ Ⓑ Ⓒ Ⓓ
23. Ⓐ Ⓑ Ⓒ Ⓓ
24. Ⓐ Ⓑ Ⓒ Ⓓ
25. Ⓐ Ⓑ Ⓒ Ⓓ
26. Ⓐ Ⓑ Ⓒ Ⓓ
27. Ⓐ Ⓑ Ⓒ Ⓓ
28. Ⓐ Ⓑ Ⓒ Ⓓ
29. Ⓐ Ⓑ Ⓒ Ⓓ
30. Ⓐ Ⓑ Ⓒ Ⓓ

Part B

31. Ⓐ Ⓑ Ⓒ Ⓓ
32. Ⓐ Ⓑ Ⓒ Ⓓ
33. Ⓐ Ⓑ Ⓒ Ⓓ
34. Ⓐ Ⓑ Ⓒ Ⓓ
35. Ⓐ Ⓑ Ⓒ Ⓓ
36. Ⓐ Ⓑ Ⓒ Ⓓ
37. Ⓐ Ⓑ Ⓒ Ⓓ
38. Ⓐ Ⓑ Ⓒ Ⓓ
39. Ⓐ Ⓑ Ⓒ Ⓓ
40. Ⓐ Ⓑ Ⓒ Ⓓ
41. Ⓐ Ⓑ Ⓒ Ⓓ
42. Ⓐ Ⓑ Ⓒ Ⓓ
43. Ⓐ Ⓑ Ⓒ Ⓓ
44. Ⓐ Ⓑ Ⓒ Ⓓ
45. Ⓐ Ⓑ Ⓒ Ⓓ

AB Practice Examination 1

SECTION I

Part A

TIME: 60 MINUTES

The use of calculators is not permitted for this part of the examination.
There are 30 questions in Part A, for which 60 minutes are allowed. Because there is no deduction for wrong answers, you should answer every question, even if you need to guess.

Directions: Choose the best answer for each question.

1. $\lim\limits_{x\to\infty} \dfrac{20x^2 - 13x + 5}{5 - 4x^3}$ is

   (A) -5 (B) 0 (C) 1 (D) ∞

2. $\lim\limits_{h\to 0} \dfrac{\ln(2+h) - \ln 2}{h}$ is

   (A) $\ln 2$ (B) $\frac{1}{2}$ (C) $\frac{1}{\ln 2}$ (D) ∞

3. If $y = e^{-x^2}$, then $y''(0)$ equals

   (A) 2 (B) 1 (C) 0 (D) -2

Questions 4 and 5. Use the following table, which shows the values of the differentiable functions f and g.

x	f	f'	g	g'
1	2	$\frac{1}{2}$	-3	5
2	3	1	0	4
3	4	2	2	3
4	6	4	3	$\frac{1}{2}$

4. The average rate of change of function f on [1,4] is

   (A) 7/6 (B) 4/3 (C) 15/8 (D) 15/4

5. If $h(x) = g(f(x))$ then $h'(3) =$

   (A) 1/2 (B) 1 (C) 4 (D) 6

GO ON TO THE NEXT PAGE

6. The derivative of a function f is given for all x by

$$f'(x) = x^2(x + 1)^3(x - 4)^2.$$

The set of x values for which f is a relative minimum is

(A) $\{0, -1, 4\}$ (B) $\{-1\}$ (C) $\{0, 4\}$ (D) $\{0, -1\}$

7. If $y = \dfrac{x - 3}{2 - 5x}$, then $\dfrac{dy}{dx}$ equals

(A) $\dfrac{17 - 10x}{(2 - 5x)^2}$ (B) $\dfrac{13}{(2 - 5x)^2}$ (C) $\dfrac{-13}{(2 - 5x)^2}$ (D) $\dfrac{17}{(2 - 5x)^2}$

8. The maximum value of the function $f(x) = xe^{-x}$ is

(A) $\dfrac{1}{e}$ (B) 1 (C) -1 (D) $-e$

9. Which equation has the slope field shown below?

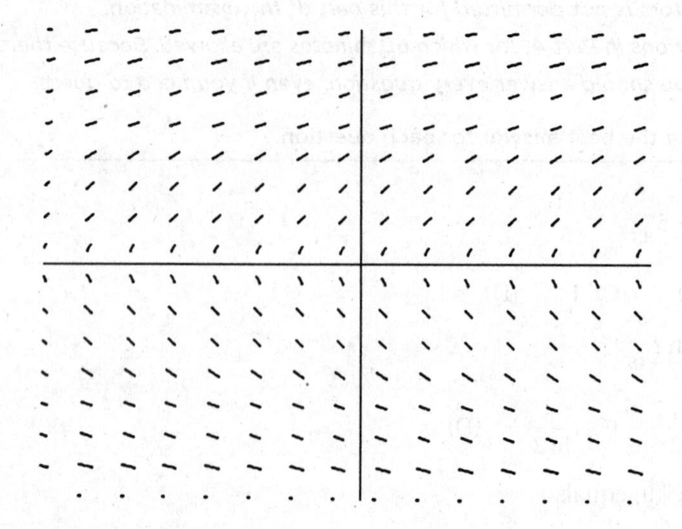

(A) $\dfrac{dy}{dx} = \dfrac{5}{y}$ (B) $\dfrac{dy}{dx} = \dfrac{5}{x}$ (C) $\dfrac{dy}{dx} = \dfrac{x}{y}$ (D) $\dfrac{dy}{dx} = 5y$

Questions 10–12. The graph below shows the velocity of an object moving along a line, for $0 \leqslant t \leqslant 9$.

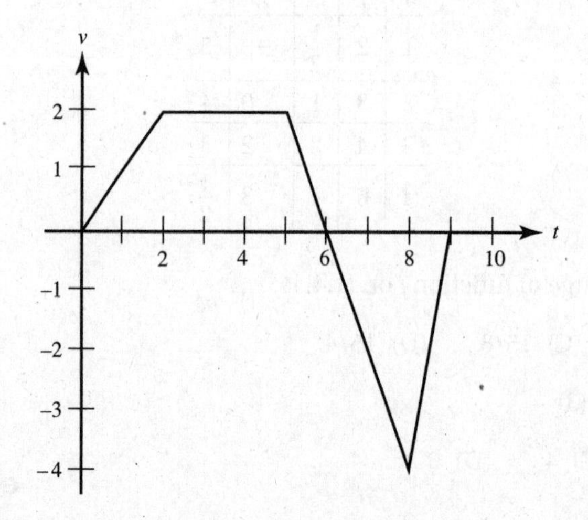

GO ON TO THE NEXT PAGE

10. At what time does the object attain its maximum acceleration?

(A) $2 < t < 5$ (B) $t = 6$ (C) $t = 8$ (D) $8 < t < 9$

11. The object is farthest from the starting point at $t =$

(A) 5 (B) 6 (C) 8 (D) 9

12. At $t = 8$, the object was at position $x = 10$. At $t = 5$, the object's position was $x =$

(A) 5 (B) 7 (C) 13 (D) 15

13. $\int_{\pi/4}^{\pi/2} \sin^3 \alpha \cos \alpha \, d\alpha$ is equal to

(A) $\frac{3}{16}$ (B) $\frac{1}{4}$ (C) $-\frac{1}{4}$ (D) $-\frac{3}{16}$

x	$f(x)$	$f'(x)$	$f''(x)$	$g(x)$	$g'(x)$	$g''(x)$
1	1	0	−7	1/3	−2	7

14. Given two twice-differentiable functions, $f(x)$ and $g(x)$. The table above gives values for $f(x)$ and $g(x)$ and their first and second derivatives at $x = 1$. Find $\lim\limits_{x \to 1} \frac{2f(x) - 6g(x)}{4x^2 - 4e^{3(x-1)}}$.

(A) −3 (B) 1 (C) 2 (D) nonexistent

15. A differentiable function has the values shown in this table:

x	2.0	2.2	2.3	2.7	2.8	3.0
$f(x)$	1.39	1.73	2.10	2.48	2.88	3.30

Estimate $f'(2.1)$.

(A) 0.34 (B) 1.56 (C) 1.70 (D) 1.91

16. If $A = \int_0^1 e^{-x} dx$ is approximated using various sums with the same number of subdivisions, and if L, R, and T denote, respectively, left Riemann Sum, right Riemann Sum, and trapezoid sum, then it follows that

(A) $R \leq A \leq T \leq L$ (B) $R \leq T \leq A \leq L$ (C) $L \leq T \leq A \leq R$ (D) $L \leq A \leq T \leq R$

17. The number of vertical tangents to the graph of $y^2 = x - x^3$ is

(A) 3 (B) 2 (C) 1 (D) 0

18. $\int_0^6 f(x - 1) dx =$

(A) $\int_1^7 f(x) dx$ (B) $\int_{-1}^5 f(x) dx$ (C) $\int_{-1}^5 f(x + 1) dx$ (D) $\int_1^7 f(x + 1) dx$

GO ON TO THE NEXT PAGE

19. The equation of the curve shown below is $y = \dfrac{4}{1 + x^2}$. What does the area of the shaded region equal?

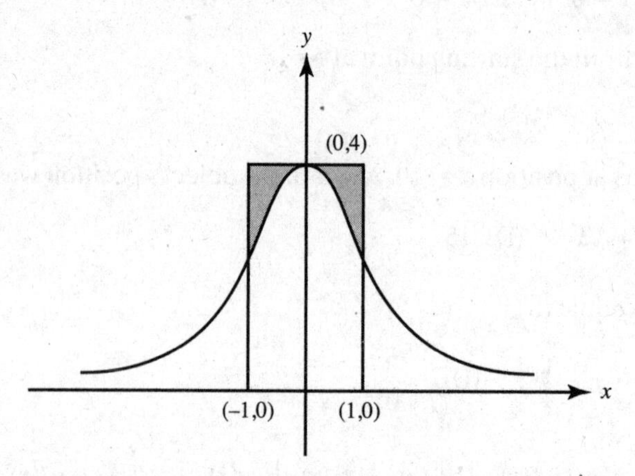

(A) $8 - \pi$ (B) $8 - 2\pi$ (C) $8 - 4\pi$ (D) $8 - 4 \ln 2$

20. Over the interval $0 \le x \le 10$, the average value of the function f shown below

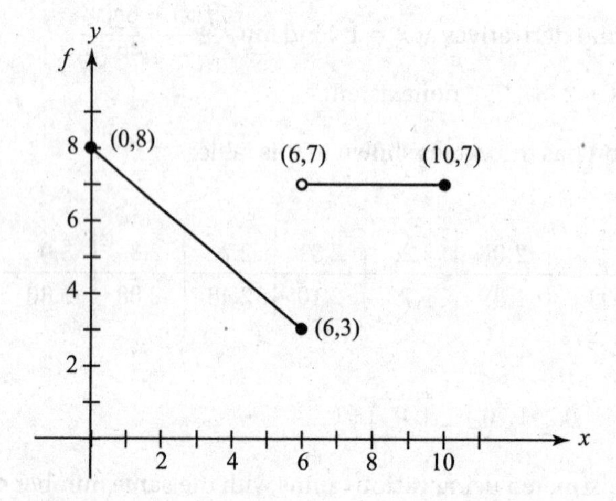

(A) is 6.10. (B) is 6.25.
(C) does not exist, because f is not continuous.
(D) does not exist, because f is not integrable.

21. If $f'(x) = 2f(x)$ and $f(2) = 1$, then $f(x) =$

(A) e^{2x-4} (B) $e^{2x} + 1 - e^4$ (C) e^{4-2x} (D) e^{x^2-4}

22. If $f(t) = \int_0^{t^2} \dfrac{1}{1 + x^2}\, dx$, then $f'(t)$ equals

(A) $\dfrac{1}{1 + t^2}$ (B) $\dfrac{2t}{1 + t^2}$ (C) $\dfrac{1}{1 + t^4}$ (D) $\dfrac{2t}{1 + t^4}$

GO ON TO THE NEXT PAGE

23. The curve $x^3+x\tan y = 27$ passes through (3,0). Use the tangent line there to estimate the value of y at $x = 3.1$. The value is

(A) -2.7 (B) -0.9 (C) 0 (D) 0.1

24. $\int (\sqrt{x} - 2)x^2 dx =$

(A) $\frac{2}{3}x^{3/2} - 2x + C$ (B) $\frac{5}{2}x^{3/2} - 4x + C$ (C) $\frac{2}{5}x^{5/2} - \frac{2}{3}x^3 + C$ (D) $\frac{2}{7}x^{7/2} - \frac{2}{3}x^3 + C$

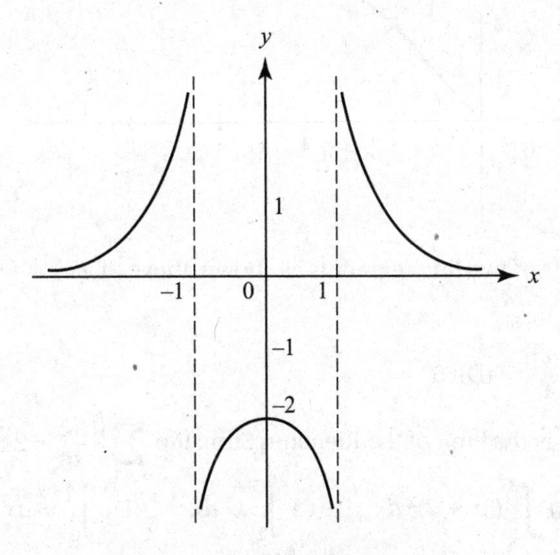

25. The graph of a function $y = f(x)$ is shown above. Which is true?

(A) $\lim_{x\to 1} f(x) = -\infty$ (B) $\lim_{x\to -\infty} f(x) = \pm 1$ (C) $\lim_{x\to -2} f(x) = 0$ (D) $\lim_{x\to \infty} f(x) = 0$

26. A function $f(x)$ equals $\frac{x^2 - x}{x - 1}$ for all x except $x = 1$. For the function to be continuous at $x = 1$, the value of $f(1)$ must be

(A) 0 (B) 1 (C) 2 (D) ∞

27. $\int_0^1 \frac{e^x}{(2 - e^x)^2} dx$ equals

(A) $-2\ln(2 - e)$ (B) $\frac{1 - e}{2 - e}$ (C) $\frac{1}{2 - e}$ (D) $\frac{e - 1}{2 - e}$

28. Suppose $f(x) = \int_0^x \frac{4 + t}{t^2 + 4} dt$. It follows that

(A) f increases for all x
(B) f has a critical point at $x = 0$
(C) f has a local min at $x = -4$
(D) f has a local max at $x = -4$

GO ON TO THE NEXT PAGE

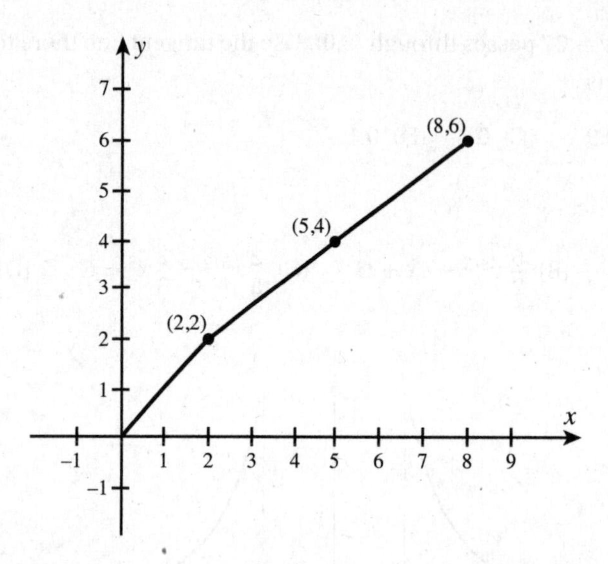

29. The graph of $f(x)$ consists of two line segments as shown above. If $g(x) = f^{-1}(x)$, the inverse function of $f(x)$, find $g'(4)$.

(A) $\frac{1}{5}$ (B) $\frac{2}{3}$ (C) $\frac{3}{2}$ (D) 5

30. Choose the integral that is the limit of the Riemann Sum: $\lim_{n\to\infty} \sum_{k=1}^{n} \left(\left(\frac{3k}{n} + 2\right)^2 \cdot \left(\frac{3}{n}\right)\right)$.

(A) $\int_2^5 (x+2)^2\,dx$ (B) $\int_0^3 (3x+2)^2\,dx$ (C) $\int_2^5 x^2\,dx$ (D) $\int_0^3 x^2\,dx$

STOP

END OF PART A, SECTION I

Part B

TIME: 45 MINUTES

> ***Some questions in this part of the examination require the use of a graphing calculator. There are 15 questions in Part B, for which 45 minutes are allowed. Because there is no deduction for wrong answers, you should answer every question, even if you need to guess.***
>
> **Directions: Choose the best answer for each question. If the exact numerical value of the correct answer is not listed as a choice, select the choice that is closest to the exact numerical answer.**

31. An object moving along a line has velocity $v(t) = t\cos t - \ln(t + 2)$, where $0 \leq t \leq 10$. The object achieves its maximum speed when t is approximately

(A) 5.107 (B) 6.419 (C) 7.550 (D) 9.538

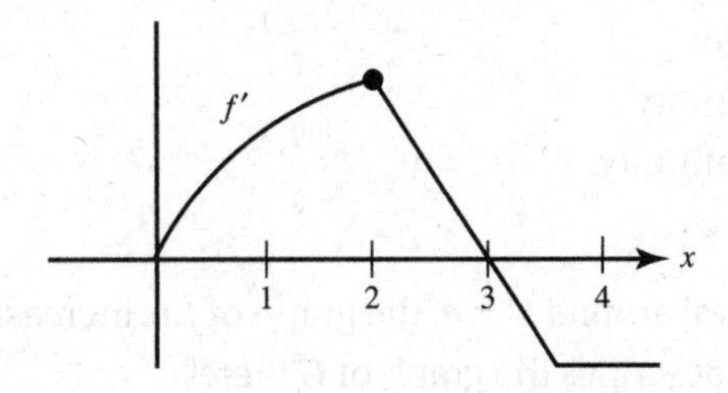

32. The graph of f', which consists of a quarter-circle and two line segments, is shown above. At $x = 2$ which of the following statements is true?

(A) f is not continuous.
(B) f is continuous but not differentiable.
(C) f has a local maximum.
(D) The graph of f has a point of inflection.

33. Let $H(x) = \int_0^x f(t)\,dt$, where f is the function whose graph appears below.

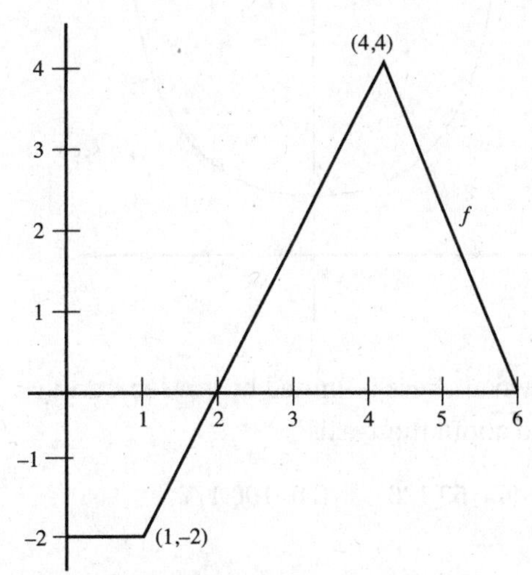

The local linearization of $H(x)$ near $x = 3$ is $H(x) \simeq$

(A) $-2x + 8$ (B) $2x - 4$ (C) $-2x + 4$ (D) $2x - 8$

GO ON TO THE NEXT PAGE

34. The table shows the speed of an object, in feet per second, at various times during a 12-second interval.

time (sec)	0	3	6	7	8	10	12
speed (ft/sec)	15	14	11	8	7	3	0

Estimate the distance the object travels, using the midpoint method with 3 subintervals.

(A) 100 ft (B) 110 ft (C) 112 ft (D) 114 ft

35. In a marathon, when the winner crosses the finish line many runners are still on the course, some quite far behind. If the density of runners x miles from the finish line is given by $R(x) = 20\left[1 - \cos\left(1 + 0.03x^2\right)\right]$ runners per mile, how many are within 8 miles of the finish line?

(A) 30 (B) 40 (C) 157 (D) 166

36. Which best describes the behavior of the function $y = \arctan\left(\frac{1}{\ln x}\right)$ at $x = 1$?

(A) It has a jump discontinuity.
(B) It has an infinite discontinuity.
(C) It has a removable discontinuity.
(D) It is continuous.

37. Let $G(x) = [f(x)]^2$. In an interval around $x = a$, the graph of f is increasing and concave downward, while G is decreasing. Which describes the graph of G there?

(A) concave downward (B) concave upward (C) point of inflection (D) quadratic

38. The value of c for which $f(x) = x + \frac{c}{x}$ has a local minimum at $x = 3$ is

(A) -9 (B) 0 (C) 6 (D) 9

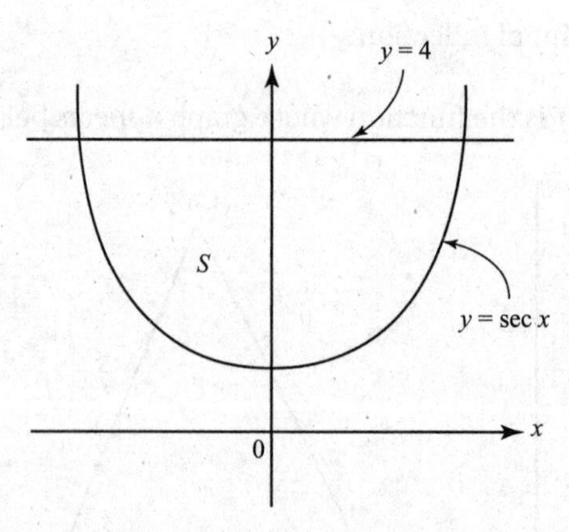

39. The region S in the figure shown above is bounded by $y = \sec x$ and $y = 4$. What is the volume of the solid formed when S is rotated about the x-axis?

(A) 11.385 (B) 23.781 (C) 53.126 (D) 108.177

GO ON TO THE NEXT PAGE

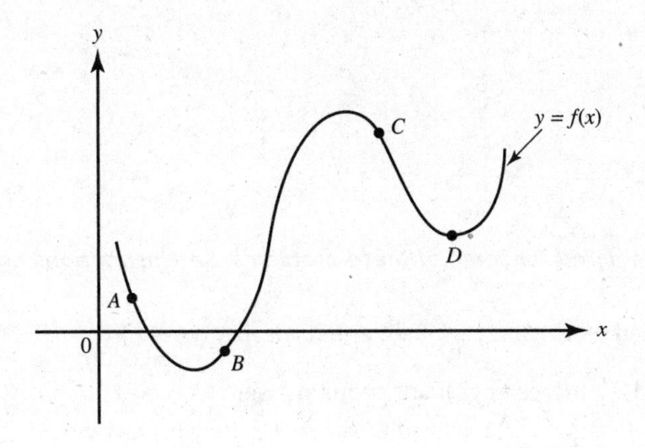

40. At which point on the graph of $y = f(x)$ shown above is $f'(x) < 0$ and $f''(x) > 0$?

(A) A (B) B (C) C (D) D

41. Let $f(x) = x^5 + 1$, and let g be the inverse function of f. What is the value of $g'(0)$?

(A) -1 (B) $\frac{1}{5}$ (C) $-\frac{1}{5}$ (D) $g'(0)$ does not exist.

42. The hypotenuse AB of a right triangle ABC is 5 feet, and one leg, AC, is decreasing at the rate of 2 feet per second. The rate, in square feet per second, at which the area is changing when $AC = 3$ is

(A) $\frac{7}{4}$ (B) $-\frac{3}{2}$ (C) $-\frac{7}{4}$ (D) $-\frac{7}{2}$

43. At how many points on the interval $[0,\pi]$ does $f(x) = 2\sin x + \sin 4x$ satisfy the Mean Value Theorem?

(A) 1 (B) 2 (C) 3 (D) 4

44. If the radius r of a sphere is increasing at a constant rate, then the rate of increase of the volume of the sphere is

(A) constant
(B) increasing
(C) decreasing
(D) decreasing for $r < 1$ and increasing for $r > 1$

45. The rate at which a purification process can remove contaminants from a tank of water is proportional to the amount of contaminant remaining. If 20% of the contaminant can be removed during the first minute of the process and 98% must be removed to make the water safe, approximately how long will the decontamination process take?

(A) 2 min (B) 5 min (C) 7 min (D) 18 min

STOP

END OF SECTION I

SECTION II

Part A

TIME: 30 MINUTES

2 PROBLEMS

A graphing calculator is required for some of these problems. See instructions on page 4.

1. A function f is defined on the interval [0,4], and its derivative is $f'(x) = e^{\sin x} - 2\cos 3x$.

   (a) On what interval is f increasing? Justify your answer.

   (b) At what value(s) of x does f have local maxima? Justify your answer.

   (c) How many points of inflection does the graph of f have? Justify your answer.

2. The rate of sales of a new software product is given by $S(t)$, where S is measured in hundreds of units per month and t is measured in months from the initial release date of January 1, 2012. The software company recorded these sales data:

t (months)	1	2	3	4	5	6	7
$S(t)$ (100s/mo)	1.54	1.88	2.32	3.12	3.78	4.90	6.12

   (a) Using a trapezoidal approximation, estimate the number of units the company sold during the second quarter (April 1, 2012, through June 30, 2012).

   (b) After looking at these sales figures, a manager suggests that the rate of sales can be modeled by assuming the rate to be initially 120 units/month and to double every 3 months. Write an equation for S based on this model.

   (c) Compare the model's prediction for total second quarter sales with your estimate from part a.

   (d) Use the model to predict the average value of $S(t)$ for the entire first year. Explain what your answer means.

STOP

END OF PART A, SECTION II

Part B

TIME: 60 MINUTES

4 PROBLEMS

No calculator is allowed for any of these problems.
If you finish Part B before time has expired, you may return to work on Part A, but you may not use a calculator.

3. The graph of function $y = f(x)$ passes through the point (1, 1) and satisfies the differential equation $\frac{dy}{dx} = \frac{6x^2 - 4}{y}$.

(a) Sketch the slope field for the differential equation at the 12 indicated points on the axes provided.

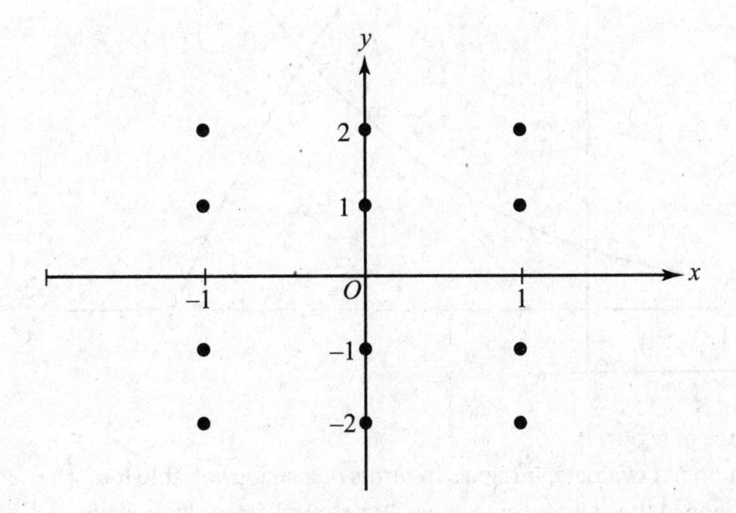

(b) Find the equation of the line tangent to $f(x)$ at the point (1, 1) and use the linear equation to estimate $f(1.2)$.

(c) Solve the differential equation, and find the particular solution for $y = f(x)$ that passes through the point (1, 1)

GO ON TO THE NEXT PAGE

4. Let R represent the first-quadrant region bounded by the y-axis and the curves $y = 2^x$ and $y = 8\cos\frac{\pi x}{6}$, as shown in the graph.

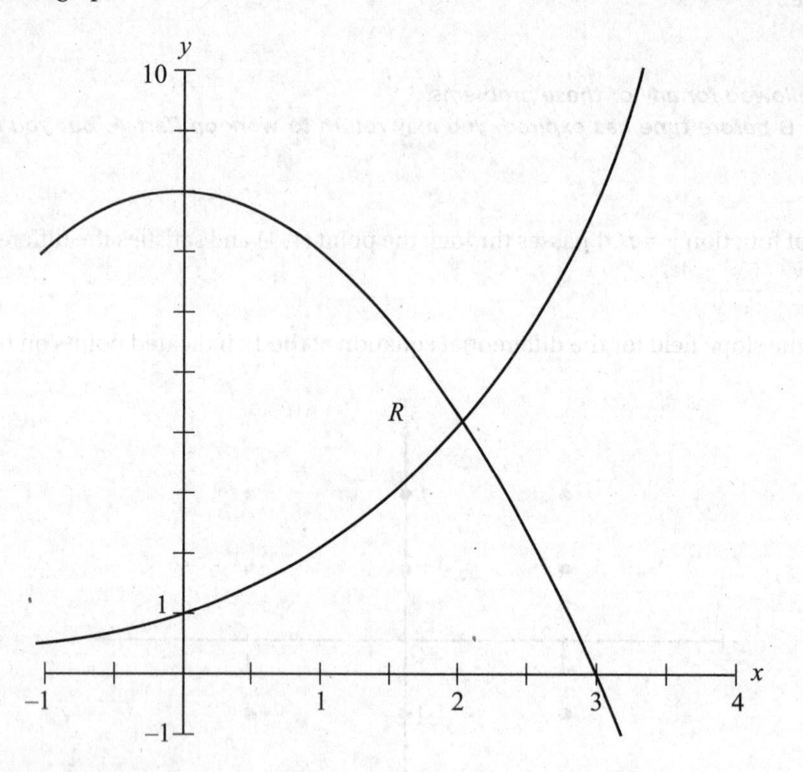

(a) Find the area of region R.

(b) Set up, but do not evaluate, integrals in terms of a single variable for:

(i) the volume of the solid formed when R is rotated around the x-axis,

(ii) the volume of the solid whose base is R, if all cross sections in planes perpendicular to the x-axis are squares.

5. Given the function $f(x) = e^{2x}(x^2 - 2)$:

(a) For what values of x is f decreasing?

(b) Does this decreasing arc reach a local or a global minimum? Justify your answer.

(c) Does f have a global maximum? Justify your answer.

6. (a) A spherical snowball melts so that its surface area shrinks at the constant rate of 10 square centimeters per minute. What is the rate of change of volume when the snowball is 12 centimeters in diameter?

(b) The snowball is packed most densely nearest the center. Suppose that, when it is 12 centimeters in diameter, its density x centimeters from the center is given by $d(x) = \frac{1}{1+\sqrt{x}}$ grams per cubic centimeter. Set up an integral for the total number of grams (mass) of the snowball then. Do not evaluate.

END OF TEST

ANSWER KEY
AB Practice Examination 1

Part A

1. **B**	9. **A**	17. **A**	25. **D**
2. **B**	10. **D**	18. **B**	26. **B**
3. **D**	11. **B**	19. **B**	27. **D**
4. **B**	12. **C**	20. **A**	28. **C**
5. **B**	13. **A**	21. **A**	29. **C**
6. **B**	14. **A**	22. **D**	30. **C**
7. **C**	15. **C**	23. **B**	
8. **A**	16. **A**	24. **D**	

Part B

31. **D**	35. **D**	39. **D**	43. **D**
32. **D**	36. **A**	40. **A**	44. **B**
33. **D**	37. **B**	41. **B**	45. **D**
34. **C**	38. **D**	42. **C**	

ANSWERS EXPLAINED

Section I Multiple-Choice

Part A

1. **(B)** Use the Rational Function Theorem on page 96.

2. **(B)** Note that $\lim_{h\to 0} \frac{\ln(2+h) - \ln 2}{h} = f'(2)$, where $f(x) = \ln x$.

3. **(D)** Since $y' = -2xe^{-x^2}$, therefore $y'' = -2(x \cdot e^{-x^2} \cdot (-2x) + e^{-x^2})$. Replace x by 0.

4. **(B)** $\frac{f(4) - f(1)}{4 - 1} = \frac{6 - 2}{4 - 1} = \frac{4}{3}$.

5. **(B)** $h'(3) = g'(f(3)) \cdot f'(3) = g'(4) \cdot f'(3) = \frac{1}{2} \cdot 2$.

6. **(B)** Since $f'(x)$ exists for all x, it must equal 0 for any x_0 for which f is a relative minimum, and it must also change sign from negative to positive as x increases through x_0. For the given derivative, this only occurs at $x = -1$.

7. **(C)** By the Quotient Rule (formula (6) on page 113),
$$\frac{dy}{dx} = \frac{(2-5x)(1) - (x-3)(-5)}{(2-5x)^2}.$$

8 **(A)** Here, $f'(x)$ is $e^{-x}(1 - x)$; f has maximum value when $x = 1$.

9. **(A)** Note that (1) on a horizontal line the slope segments are all parallel, so the slopes there are all the same and $\frac{dy}{dx}$ must depend only on y; (2) along the x-axis (where $y = 0$) the slopes are infinite; and (3) as y increases, the slope decreases.

10. **(D)** Acceleration is the derivative (the slope) of velocity v; v is largest on $8 < t < 9$.

11. **(B)** Velocity v is the derivative of position; because $v > 0$ until $t = 6$ and $v < 0$ thereafter, the position increases until $t = 6$ and then decreases; since the area bounded by the curve above the axis is larger than the area below the axis, the object is farthest from its starting point at $t = 6$.

12. **(C)** From $t = 5$ to $t = 8$, the displacement (the integral of velocity) can be found by evaluating definite integrals based on the areas of two triangles: $\frac{1}{2}(1)(2) - \frac{1}{2}(2)(4) = -3$. Thus, if K is the object's position at $t = 5$, then $K - 3 = 10$ at $t = 8$.

13. **(A)** The integral is of the form $\int u^3 du$; evaluate $\frac{1}{4}\sin^4 \alpha \Big|_{\pi/4}^{\pi/2}$.

14. **(A)** The limit $\lim_{x\to 1} \frac{2f(x) - 6g(x)}{4x^2 - 4e^{3(x-1)}}$ by substitution is of the form $\frac{2(1) - 6(1/3)}{4(1)^2 - 4e^{3(1-1)}} = \frac{0}{0}$, so apply L'Hôpital's Rule. You get $\lim_{x\to 2} \frac{2f'(x) - 6g'(x)}{8x - 12e^{3(x-1)}}$, which, by substitution, is $\frac{2(0) - 6(-2)}{8(1) - 12e^{3(1-1)}} = \frac{12}{-4} = -3$.

15. **(C)** $f'(2.1) \approx \frac{f(2.2) - f(2.0)}{2.2 - 2.0}$.

16. **(A)** $f(x) = e^{-x}$ is decreasing and concave upward.

17. **(A)** Implicit differentiation yields $2yy' = 1 - 3x^2$; so $\frac{dy}{dx} = \frac{1-3x^2}{2y}$. At a vertical tangent, $\frac{dy}{dx}$ is undefined; y must therefore equal 0 and the numerator be non-zero. The original equation with $y = 0$ is $0 = x - x^3$, which has three solutions.

18. **(B)** Let $t = x - 1$; then $t = -1$ when $x = 0$, $t = 5$ when $x = 6$, and $dt = dx$.

19. **(B)** The required area, A, is given by the integral

$$2\int_0^1 \left(4 - \frac{4}{1+x^2}\right) dx = 2\left(4x - 4\tan^{-1}x\right)\Big|_0^1 = 2\left(4 - 4\cdot\frac{\pi}{4}\right).$$

20. **(A)** The average value is $\frac{1}{10-0}\int_0^{10} f(x)dx$. The definite integral represents the sum of the areas of a trapezoid and a rectangle: $\frac{1}{2}(8+3)(6) = 4(7) = 61$.

21. **(A)** Solve the differential equation $\frac{dy}{dx} = 2y$ by separation of variables: $\frac{dy}{y} = 2dx$ yields $y = ce^{2x}$. The initial condition yields $1 = ce^{2\cdot 2}$; so $c = e^{-4}$ and $y = e^{2x-4}$.

22. **(D)** $\frac{d}{du}\int_0^u \frac{1}{1+x^2}dx = \frac{1}{1+u^2}$. When $u = t^2$,

$$\frac{d}{dt}\int_0^u \frac{1}{1+x^2}dx = \frac{1}{1+u^2}\frac{du}{dt} = \frac{1}{1+t^4}(2t).$$

23. **(B)** By implicit differentiation, $3x^2 + x\sec^2 y\frac{dy}{dx} + \tan y = 0$. At (3,0), $\frac{dy}{dx} = -9$; so the equation of the tangent line at (3,0) is $y = -9(x-3)$.

24. **(D)** $\int(\sqrt{x} - 2)x^2\,dx = \int(x^{5/2} - 2x^2)dx = \frac{2}{7}x^{7/2} - \frac{2}{3}x^3 + C.$

25. **(D)** The graph shown has the x-axis as a horizontal asymptote.

26. **(B)** Since $\lim_{x\to 1} f(x) = 1$, to render $f(x)$ continuous at $x = 1$ $f(1)$ must be defined to be 1.

27. **(D)** $-\int_0^1 (2 - e^x)^{-2}(-e^x dx) = \frac{1}{2-e^x}\Big|_0^1 = \frac{1}{2-e} - 1$

28. **(C)** Note that $f'(x) = \frac{4+x}{x^2+4}$, so f has a critical value at $x = -4$. As x passes through -4, the sign of f' changes from $-$ to $+$, so f has a local minimum at $x = -4$.

29. **(C)** Given the point (a,b) on function $f(x)$, $\left(f^{-1}\right)'(b) = \frac{1}{f'(a)}$. Note that the slope of the graph of $f(x)$ at $x = 5$ is $\frac{2}{3}$, so $g'(4) = \frac{1}{f'(5)} = \frac{1}{2/3} = \frac{3}{2}$.

30. **(C)** From the Riemann Sum, we see $\Delta x = \frac{3}{n}$, then $k\cdot\Delta x = \frac{3k}{n}$. Notice that the term involving k in the Riemann Sum is equal to $\frac{3k}{n}$. Thus, we try $x_k = \frac{3k}{n} + 2$, so, $a = 2$ and $\Delta x = \frac{b-2}{n} = \frac{3}{n}$, so $b = 5$. Since x_k replaces x, $f(x) = x^2$ giving the integral $\int_2^5 x^2\,dx$.

Part B

31. **(D)** Use your calculator to graph velocity against time. Speed is the absolute value of velocity. The greatest deviation from $v = 0$ is at $t = c$. With a calculator, $c = 9.538$.

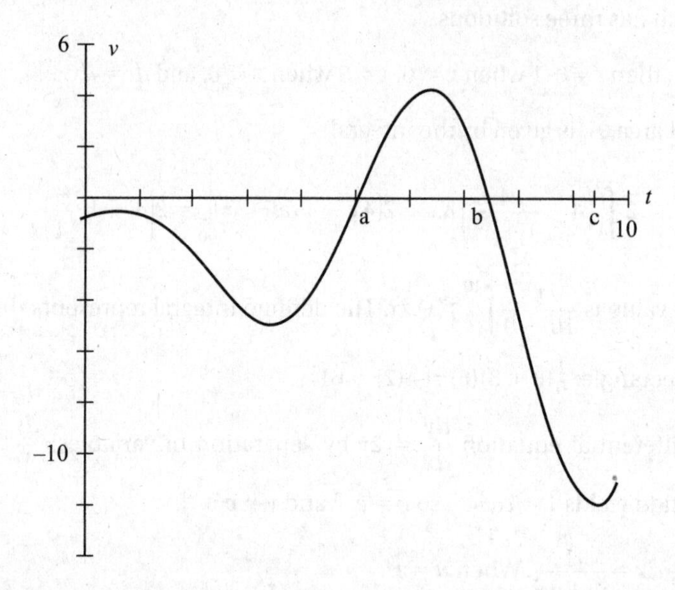

32. **(D)** Because f' changes from increasing to decreasing, f'' changes from positive to negative and thus the graph of f changes concavity.

33. **(D)** $H(3) = \int_0^3 f(t)dt$. We evaluate this definite integral by finding the area of a trapezoid (negative) and a triangle: $H(3) = -\frac{1}{2}(2 + 1)(2) + \frac{1}{2}(1)(2) = -2$, so the tangent line passes through the point $(3,-2)$. The slope of the line is $H'(3) = f(3) = 2$, so an equation of the line is $y - (-2) = 2(x - 3)$.

34. **(C)** The distance is approximately $14(6) + 8(2) + 3(4)$.

35. **(D)** $\int_0^8 R(x)\,dx = 166.396$

36. **(A)** Selecting an answer for this question from your calculator graph is unwise. In some windows the graph may appear continuous; in others there may seem to be cusps, or a vertical asymptote. Put the calculator aside. Find

$$\lim_{x\to 1^+}\left(\arctan\left(\frac{1}{\ln x}\right)\right) = \frac{\pi}{2} \text{ and } \lim_{x\to 1^-}\left(\arctan\left(\frac{1}{\ln x}\right)\right) = -\frac{\pi}{2}.$$

These limits indicate the presence of a jump discontinuity in the function at $x = 1$.

37. **(B)** We are given that (1) $f'(a) > 0$; (2) $f''(a) < 0$; and (3) $G'(a) < 0$. Since $G'(x) = 2f(x) \cdot f'(x)$, therefore $G'(a) = 2f(a) \cdot f'(a)$. Conditions (1) and (3) imply that (4) $f(a) < 0$. Since $G''(x) = 2[f(x) \cdot f''(x) + (f'(x))^2]$, therefore $G''(a) = 2[f(a)\,f''(a) + (f'(a))^2]$. Then the sign of $G''(a)$ is $2[(-) \cdot (-) + (+)]$ or positive, where the minus signs in the parentheses follow from conditions (4) and (2).

38. **(D)** Since $f'(x) = 1 - \frac{c}{x^2}$, it equals 0 for $x = \pm\sqrt{c}$. When $x = 3$, $c = 9$; this yields a minimum since $f''(3) > 0$.

39. **(D)** In the figure below, S is the region bounded by $y = \sec x$, the y axis, and $y = 4$.

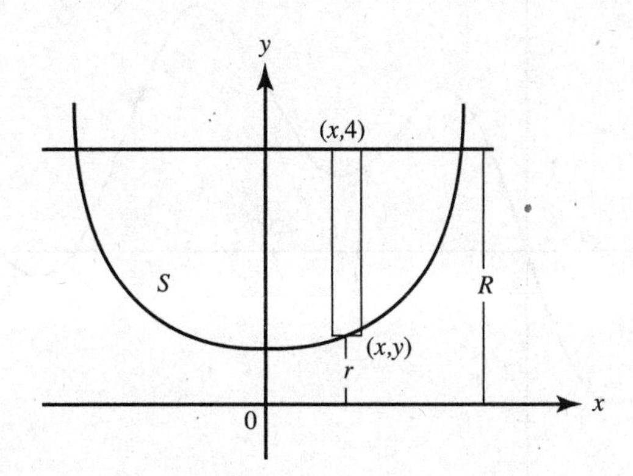

Send region S about the x-axis. Use washers; then $\Delta V = \pi(R^2 = r^2)\ \Delta x$. Symmetry allows you to double the volume generated by the first quadrant of S, so V is

$$2\pi \int_0^{\arccos\frac{1}{4}} \left(16 - \sec^2 x\right) dx.$$

A calculator yields 108.177.

40. **(A)** The curve falls when $f'(x) < 0$ and is concave up when $f''(x) > 0$.

41. **(B)** $g'(y) = \frac{1}{f'(x)} = \frac{1}{5x^4}$. To find $g'(0)$, find x such that $f(x) = 0$. By inspection, $x = -1$, so $g'(0) = \frac{1}{5(-1)^4} = \frac{1}{5}$.

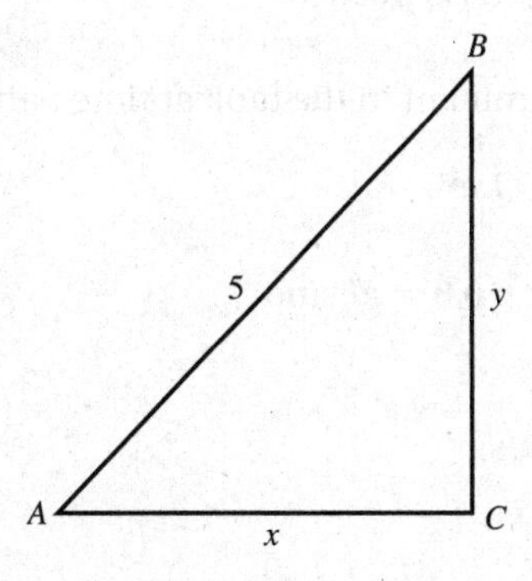

42. **(C)** It is given that $\frac{dx}{dt} = -2$; you want $\frac{dA}{dt}$, where $A = \frac{1}{2}xy$.

$$\frac{dA}{dt} = \frac{1}{2}\left(x\frac{dy}{dt} + y\frac{dx}{dy}\right) = \frac{1}{2}\left[3 \cdot \frac{dy}{dt} + y \cdot (-2)\right].$$

Since $y^2 = 25 - x^2$, it follows that $2y\frac{dy}{dt} = -2x\frac{dx}{dt}$ and, when $x = 3$, $y = 4$ and $\frac{dy}{dt} = \frac{3}{2}$.

Then $\frac{dA}{dt} = -\frac{7}{4}$.

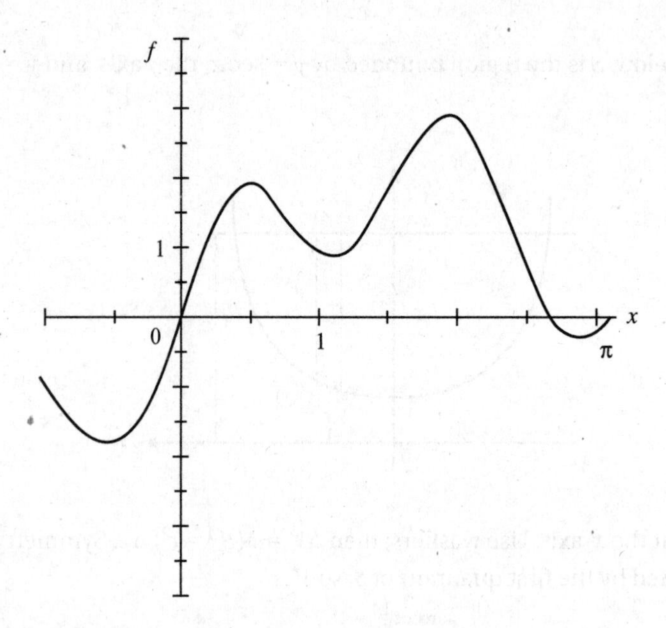

The function $f(x) = 2 \sin x + \sin 4x$ is graphed above.

43. (D) Since $f(0) = f(\pi)$ and f is both continuous and differentiable, Rolle's Theorem predicts at least one c in the interval such that $f'(c) = 0$.
There are four points in $[0,\pi]$ of the calculator graph above where the tangent is horizontal.

44. (B) Since $\frac{dr}{dt} = k$, a positive constant, $\frac{dV}{dt} = 4\pi r^2 \frac{dr}{dt} = 4\pi r^2\, k = cr^2$, where c is a positive constant.

Then $\frac{d^2V}{dt^2} = 2cr\frac{dr}{dt} = 2crk$, which is also positive.

45. (D) If $Q(t)$ is the amount of contaminant in the tank at time t and Q_0 is the initial amount, then

$$\frac{dQ}{dt} = kQ \text{ and } Q(t) = Q_0 e^{kt}.$$

Since $Q(1) = 0.8Q_0$, $0.8Q_0 = Q_0 e^{k \cdot 1}$, $0.8 = e^k$, and

$$Q(t) = Q_0(0.8)^t.$$

We seek t when $Q(t) = 0.02Q_0$. Thus,

$$0.02Q_0 = Q_0(0.8)^t$$

and

$$t \simeq 17.53 \text{ min.}$$

Section II Free-Response

Part A

AB/BC 1. This is the graph of $f'(x)$.

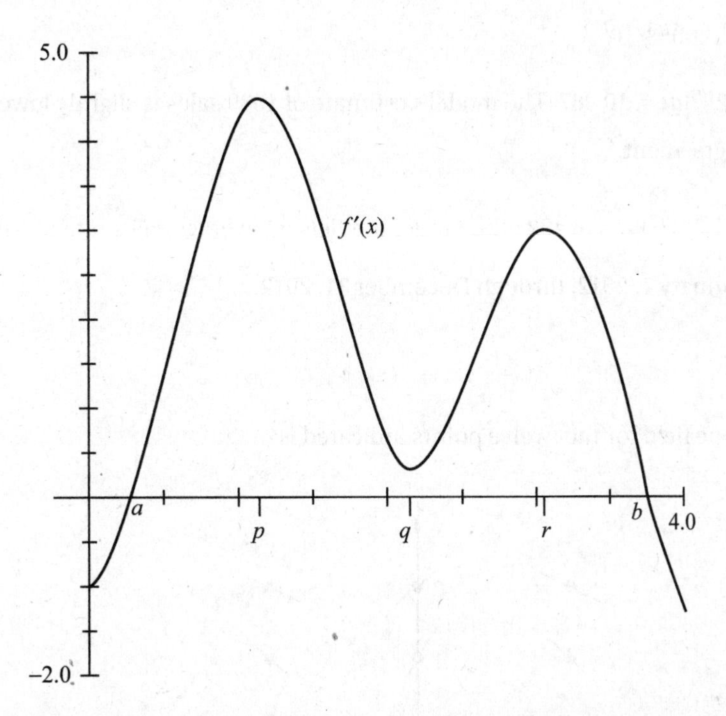

(a) f is increasing when $f'(x) > 0$. The graph shows this to be true in the interval $a < x < b$. Use the calculator to find a and b (where $e^x - 2\cos 3x = 0$); then $a = 0.283 < x < 3.760 = b$.

(b) See signs analysis.

	0		a		b		4
f		dec		inc		dec	
f'		–		+		–	

Since f decreases to the right of endpoint $x = 0$, f has a local maximum at $x = 0$. There is another local maximum at $x = 3.760$, because f changes from increasing to decreasing there.

(c) See signs analysis.

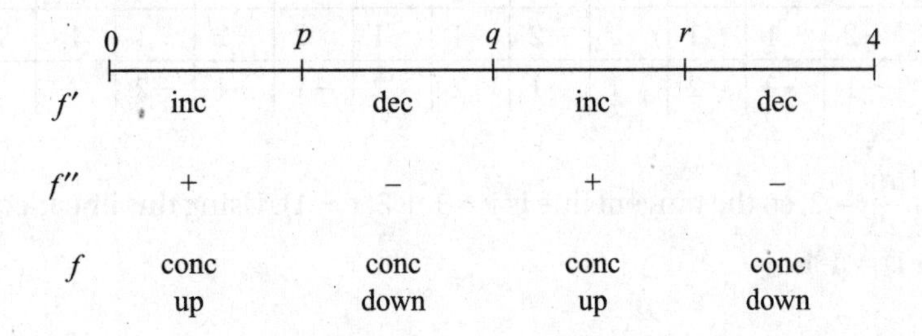

Since the graph of f changes concavity at p, q, and r, there are three points of inflection.

AB/BC 2. (a) Since April 1 is 3 months from January 1 and June 30 is 3 months later, we form the sum for the interval [3,6]:

$$\left(\frac{2.32+3.12}{2}\right)\cdot 1+\left(\frac{3.12+3.78}{2}\right)\cdot 1+\left(\frac{3.78+4.90}{2}\right)\cdot 1 = 10.51$$

We estimate the company sold 1051 software units during the second quarter.

(b) $S(t) = 1.2(2)^{t/3}$

(c) $\int_3^6 1.2(2^{t/3})dt = 10.387$. The model's estimate of 1039 sales is slightly lower, but the two are in close agreement.

(d) $\frac{1}{12}\int_0^{12} 1.2(2^{t/3})dt = 6.492$; the model predicts an average sales rate of 649.2 units per month from January 1, 2012, through December 31, 2012.

Part B

AB 3. (a) The slope field for the twelve points indicated is:

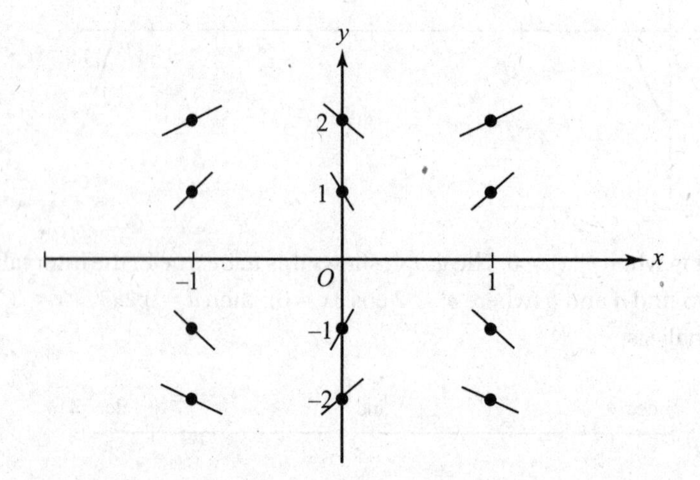

The slopes are given in the table below. Be sure that the segments you draw are correct relative to the other slopes in the slope field with respect to the steepness of the segments.

x	−1	−1	−1	−1	0	0	0	0	1	1	1	1
y	−2	−1	1	2	−2	−1	1	2	−2	−1	1	2
dy/dx	−1	−2	2	1	1	2	−2	−1	−1	−2	2	1

(b) At (1, 1), $\frac{dy}{dx} = 2$, so the tangent line is $y = 1 + 2(x - 1)$. Using this linear equation, $f(1.2) \approx 1 + 2(1.2 - 1) = 1.4$.

(c) The differential equation $\frac{dy}{dx} = \frac{6x^2 - 4}{y}$ is separable.

$$\int y\,dy = \int (6x^2 - 4)\,dx$$

$$\frac{y^2}{2} = 2x^3 - 4x + C_1$$

$$y = \pm\sqrt{4x^3 - 8x + C_2}\text{ ,where } C_2 = 2C_1$$

Since $f(x)$ passes through (1, 1), it must be true that $1 = \pm\sqrt{4(1)^3 - 8(1) + C_2}$.

Thus $C_2 = 5$, and the positive square root is used.

The solution is $f(x) = \sqrt{4x^3 - 8x + 5}$.

AB 4. (a) Draw a vertical element of area, as shown.

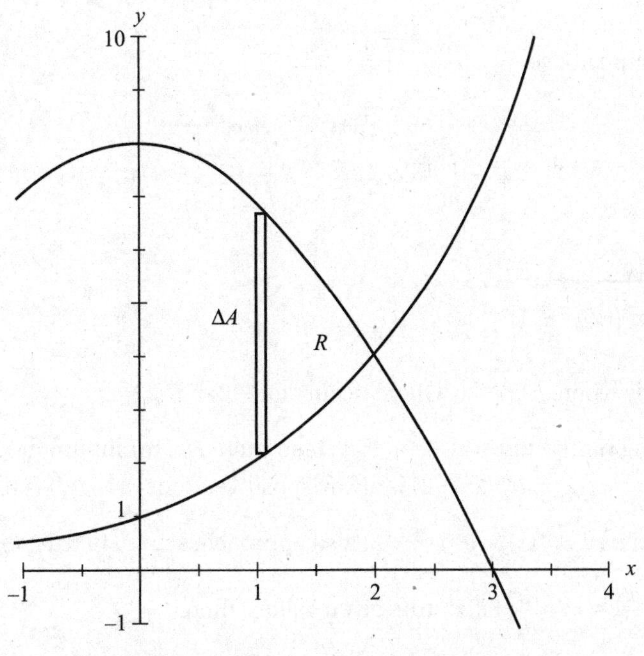

$$\Delta A = \left(y_{top} - y_{bottom}\right)\Delta x = \left(8\cos\frac{\pi x}{6} - 2^x\right)\Delta x,$$

$$A = \int_0^2 \left(8\cos\frac{\pi x}{6} - 2^x\right)dx$$

$$= \frac{6}{\pi}\cdot 8\int_0^2 \cos\frac{\pi x}{6}\,dx - \int_0^2 2^x\,dx$$

$$= \frac{48}{\pi}\cdot \sin\frac{\pi x}{6}\Big|_0^2 - \frac{2^x}{\ln 2}\Big|_0^2$$

$$= \frac{48}{\pi}\left(\sin\frac{\pi}{3} - \sin 0\right) - \left(\frac{2^2}{\ln 2} - \frac{2^0}{\ln 2}\right)$$

$$= \frac{24\sqrt{3}}{\pi} - \frac{3}{\ln 2}.$$

(b) (i) Use washers; then

$$\Delta V = (r_2^2 - r_1^2)\,\Delta x = \pi\,(y_{top}^2 - y_{bottom}^2)\,\Delta x,$$

$$V = \pi\int_0^2 \left[\left(8\cos\frac{\pi x}{6}\right)^2 - (2^x)^2\right]dx.$$

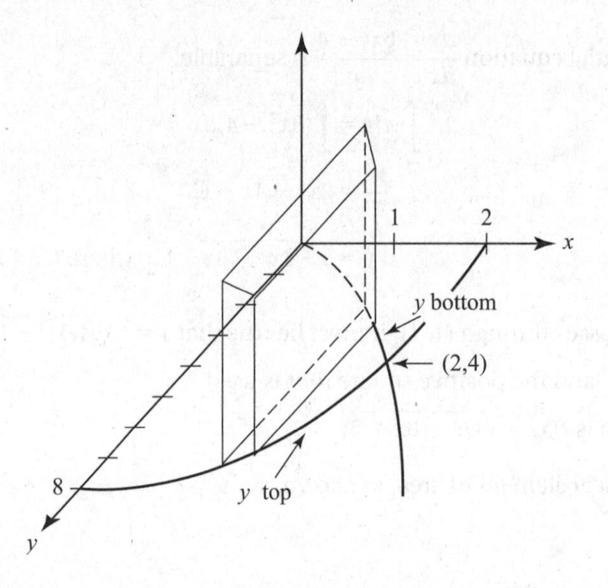

(ii) See the figure above.

$$\Delta V = s^2\,\Delta x = (y_{\text{top}} - y_{\text{bottom}})^2\,\Delta x,$$

$$V = \int_0^2 \left(8\cos\frac{\pi x}{6} - 2^x\right)^2 dx.$$

AB 5. (a)
$$\begin{aligned} f(x) &= e^{2x}(x^2-2),\\ f'(x) &= e^{2x}(2x) + 2e^{2x}(x^2-2)\\ &= 2e^{2x}(x+2)(x-1)\\ &= 0 \text{ at } x = -2, 1. \end{aligned}$$

f is decreasing where $f'(x) < 0$, which occurs for $-2 < x < 1$.

(b) f is decreasing on the interval $-2 < x < 1$, so there is a minimum at $(1, -e^2)$. Note that, as x approaches $\pm\infty$, $f(x) = e^{2x}(x^2-2)$ is always positive. Hence $(1, -e^2)$ is the global minimum.

(c) As x approaches $\pm\infty$, $f(x) = e^{2x}(x^2-2)$ also approaches $\pm\infty$. There is no global maximum.

AB/BC 6. (a) $S = 4\pi r^2$, so $\frac{dS}{dt} = 8\pi r\frac{dr}{dt}$. Substitute given values; then

$$-10 = 8\pi(6)\frac{dr}{dt}, \text{ so } \frac{dr}{dt} = -\frac{5}{24\pi} \text{ cm/min.}$$

Since $V = \frac{4}{3}\pi r^3$, therefore $\frac{dV}{dt} = 4\pi r^2\frac{dr}{dt}$. Substituting known values gives $\frac{dV}{dt} = 4\pi(6^2)\cdot\frac{-5}{24\pi} = -30\text{ cm}^3/\text{min}$.

(b) Regions of consistent density are concentric spherical shells. The volume of each shell is approximated by its surface area $(4\pi x^2)$ times its thickness (Δx). The weight of each shell is its density times its volume (g/cm^3·cm^3). If, when the snowball is 12 cm in diameter, ΔG is the weight of a spherical shell x cm from the center, then $\Delta G = \frac{1}{1+\sqrt{x}}\cdot 4\pi x^2\,\Delta x$, and the integral to find the weight of the snowball is

$$G = \int_0^6 \frac{1}{1+\sqrt{x}}\cdot 4\pi x^2 dx.$$

ANSWER SHEET
AB Practice Examination 2

Part A

1. Ⓐ Ⓑ Ⓒ Ⓓ
2. Ⓐ Ⓑ Ⓒ Ⓓ
3. Ⓐ Ⓑ Ⓒ Ⓓ
4. Ⓐ Ⓑ Ⓒ Ⓓ
5. Ⓐ Ⓑ Ⓒ Ⓓ
6. Ⓐ Ⓑ Ⓒ Ⓓ
7. Ⓐ Ⓑ Ⓒ Ⓓ
8. Ⓐ Ⓑ Ⓒ Ⓓ
9. Ⓐ Ⓑ Ⓒ Ⓓ
10. Ⓐ Ⓑ Ⓒ Ⓓ
11. Ⓐ Ⓑ Ⓒ Ⓓ
12. Ⓐ Ⓑ Ⓒ Ⓓ
13. Ⓐ Ⓑ Ⓒ Ⓓ
14. Ⓐ Ⓑ Ⓒ Ⓓ
15. Ⓐ Ⓑ Ⓒ Ⓓ
16. Ⓐ Ⓑ Ⓒ Ⓓ
17. Ⓐ Ⓑ Ⓒ Ⓓ
18. Ⓐ Ⓑ Ⓒ Ⓓ
19. Ⓐ Ⓑ Ⓒ Ⓓ
20. Ⓐ Ⓑ Ⓒ Ⓓ
21. Ⓐ Ⓑ Ⓒ Ⓓ
22. Ⓐ Ⓑ Ⓒ Ⓓ
23. Ⓐ Ⓑ Ⓒ Ⓓ
24. Ⓐ Ⓑ Ⓒ Ⓓ
25. Ⓐ Ⓑ Ⓒ Ⓓ
26. Ⓐ Ⓑ Ⓒ Ⓓ
27. Ⓐ Ⓑ Ⓒ Ⓓ
28. Ⓐ Ⓑ Ⓒ Ⓓ
29. Ⓐ Ⓑ Ⓒ Ⓓ
30. Ⓐ Ⓑ Ⓒ Ⓓ

Part B

31. Ⓐ Ⓑ Ⓒ Ⓓ
32. Ⓐ Ⓑ Ⓒ Ⓓ
33. Ⓐ Ⓑ Ⓒ Ⓓ
34. Ⓐ Ⓑ Ⓒ Ⓓ
35. Ⓐ Ⓑ Ⓒ Ⓓ
36. Ⓐ Ⓑ Ⓒ Ⓓ
37. Ⓐ Ⓑ Ⓒ Ⓓ
38. Ⓐ Ⓑ Ⓒ Ⓓ
39. Ⓐ Ⓑ Ⓒ Ⓓ
40. Ⓐ Ⓑ Ⓒ Ⓓ
41. Ⓐ Ⓑ Ⓒ Ⓓ
42. Ⓐ Ⓑ Ⓒ Ⓓ
43. Ⓐ Ⓑ Ⓒ Ⓓ
44. Ⓐ Ⓑ Ⓒ Ⓓ
45. Ⓐ Ⓑ Ⓒ Ⓓ

AB Practice Examination 2

SECTION I

Part A

TIME: 60 MINUTES

> ***The use of calculators is** not **permitted for this part of the examination.***
> ***There are 30 questions in Part A, for which 60 minutes are allowed. Because there is no deduction for wrong answers, you should answer every question, even if you need to guess.***
>
> **Directions:** Choose the best answer for each question.

1. $\lim\limits_{x \to 2} \dfrac{x^2 - 2}{4 - x^2}$ is

   (A) -1 (B) $-\dfrac{1}{2}$ (C) 0 (D) nonexistent

2. $\lim\limits_{x \to \infty} \dfrac{\sqrt{x} - 4}{4 - 3\sqrt{x}}$ is

   (A) $-\dfrac{1}{3}$ (B) -1 (C) 0 (D) ∞

3. If $y = \dfrac{e^{\ln u}}{u}$, then $\dfrac{dy}{du}$ equals

   (A) $e^{\ln u}$ (B) $\dfrac{e^{\ln u}}{u^2}$ (C) $\dfrac{e^{\ln u}(u-1)}{u^2}$ (D) 0

4. Using the line tangent to $f(x) = \sqrt{9 + \sin(2x)}$ at $x = 0$, an estimate of $f(0.06)$ is

   (A) 0.02 (B) 2.98 (C) 3.01 (D) 3.02

5. Air is escaping from a balloon at a rate of $R(t) = \dfrac{60}{1 + t^2}$ cubic feet per minute, where t is measured in minutes. How much air, in cubic feet, escapes during the first minute?

   (A) 15π (B) 30 (C) 45 (D) $30 \ln 2$

6. If $y = \sin^3(1 - 2x)$, then $\dfrac{dy}{dx}$ is

   (A) $3 \sin^2(1 - 2x)$ (B) $-2 \cos^3(1 - 2x)$ (C) $-6 \sin^2(1 - 2x)$
   (D) $-6 \sin^2(1 - 2x) \cos(1 - 2x)$

7. If $y = x^2 e^{1/x}$ $(x \neq 0)$, then $\dfrac{dy}{dx}$ is

   (A) $xe^{1/x}(x + 2)$ (B) $e^{1/x}(2x - 1)$ (C) $-\dfrac{2e^{1/x}}{x}$ (D) $e^{-x}(2x - x^2)$

8. A point moves along the curve $y = x^2 + 1$ so that the x-coordinate is increasing at the constant rate of $\frac{3}{2}$ units per second. The rate, in units per second, at which the distance from the origin is changing when the point has coordinates (1, 2) is equal to

(A) $\frac{7\sqrt{5}}{10}$ (B) $\frac{3\sqrt{5}}{2}$ (C) $\frac{3\sqrt{5}}{5}$ (D) $\frac{5}{2}$

9. $\lim_{h\to 0} \frac{\sqrt{25+h}-5}{h}$

(A) $= 0$ (B) $= \frac{1}{10}$ (C) $= 1$ (D) does not exist

10. The base of a solid is the first-quadrant region bounded by $y = \sqrt[4]{1-x^2}$. Each cross section perpendicular to the x-axis is a square with one edge in the xy-plane. The volume of the solid is

(A) $\frac{\pi}{4}$ (B) 1 (C) $\frac{\pi}{2}$ (D) π

11. $\int \frac{x}{\sqrt{9-x^2}}\,dx$ equals

(A) $-\frac{1}{2}\ln\sqrt{9-x^2}+C$ (B) $\sin^{-1}\frac{x}{3}+C$ (C) $-\sqrt{9-x^2}+C$ (D) $-\frac{1}{4}\sqrt{9-x^2}+C$

12. $\int \frac{(y-1)^2}{2y}\,dy$ equals

(A) $\frac{y^2}{4}-y+\frac{1}{2}\ln|y|+C$ (B) $y^2-y+\ln|2y|+C$ (C) $\frac{y^2}{4}-y+2\ln|2y|+C$ (D) $\frac{(y-1)^3}{3y^2}+C$

13. $\int_{\pi/6}^{\pi/2} \cot x\,dx$ equals

(A) 3 (B) $\ln\frac{1}{2}$ (C) $\ln 2$ (D) $\ln\frac{2}{\sqrt{3}}$

14. Given f' as graphed, which could be a graph of f?

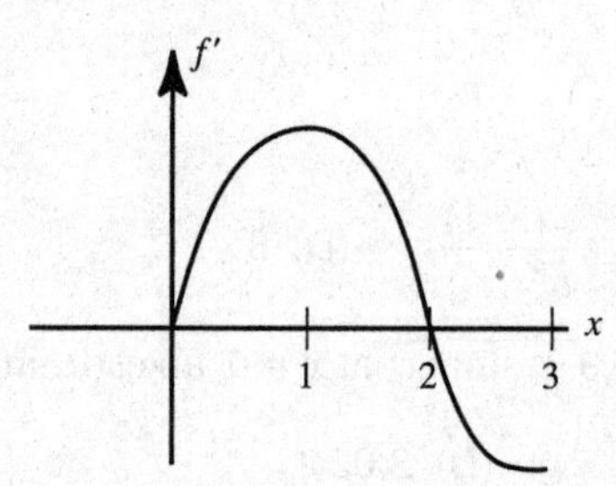

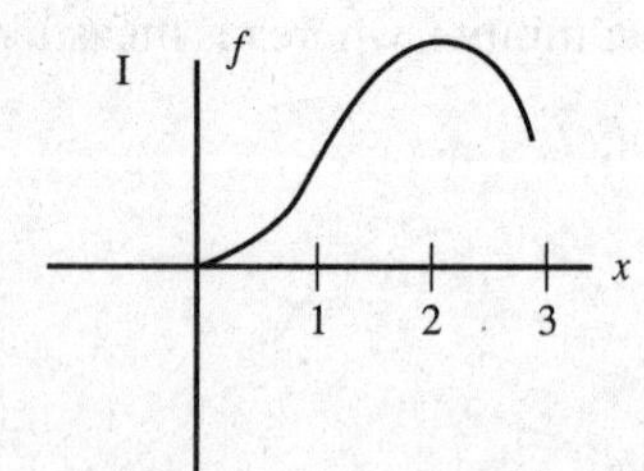

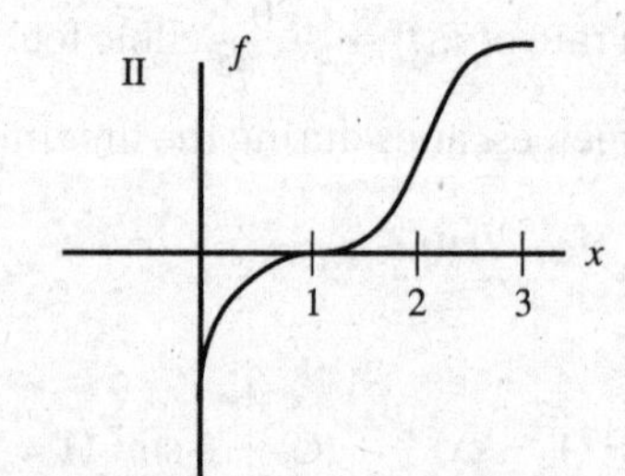

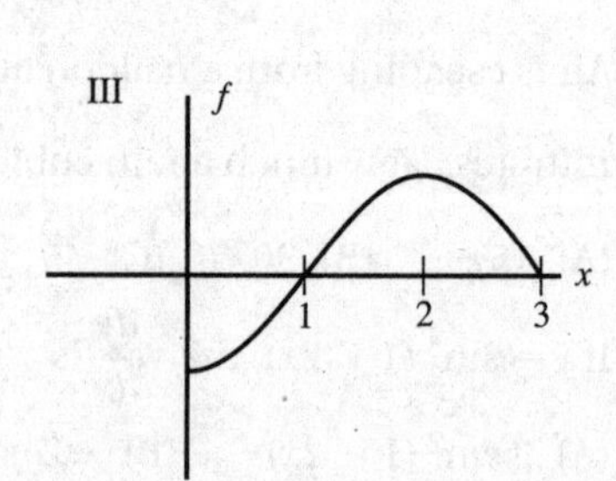

(A) I only (B) II only (C) III only (D) I and III

15. The first woman officially timed in a marathon was Violet Piercey of Great Britain in 1926. Her record of 3:40:22 stood until 1963, mostly because of a lack of women competitors. Soon after, times began dropping rapidly, but lately they have been declining at a much slower rate. Let $M(t)$ be the curve that best represents winning marathon times in year t. Which of the following is (are) positive for $t > 1963$?

I. $M(t)$
II. $M'(t)$
III. $M''(t)$

(A) I only (B) I and II only (C) I and III only (D) I, II, and III

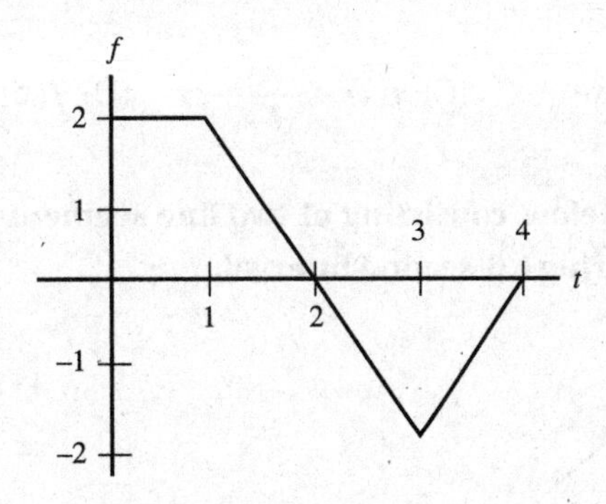

16. The graph of f is shown above. Let $G(x) = \int_0^x f(t)dt$ and $H(x) = \int_2^x f(t)\,dt$. Which of the following is true?

(A) $G(x) = H(x)$ (B) $G'(x) = H'(x + 2)$ (C) $G(x) = H(x + 2)$ (D) $G(x) = H(x) + 3$

17. The minimum value of $f(x) = x^2 + \frac{2}{x}$ on the interval $\frac{1}{2} \leq x \leq 2$ is

(A) 1 (B) 3 (C) $4\frac{1}{2}$ (D) 5

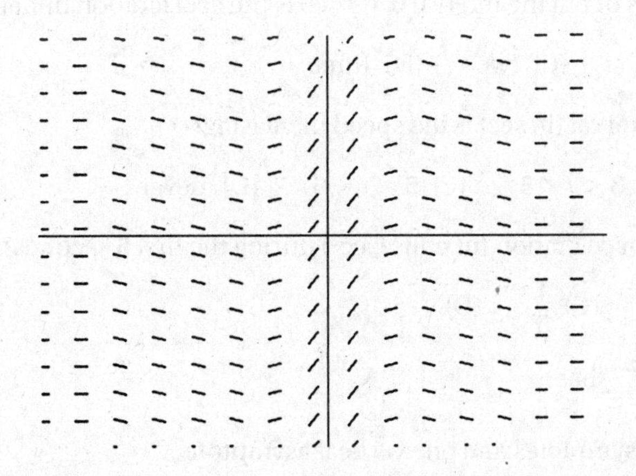

18. Which function could be a particular solution of the differential equation whose slope field is shown above?

(A) $y = \frac{2x}{x^2 + 1}$ (B) $y = \frac{x^2}{x^2 + 1}$ (C) $y = \sin x$ (D) $y = e^{-x^2}$

AB PRACTICE EXAMINATION 2

19. Which of the following functions could have the graph sketched below?

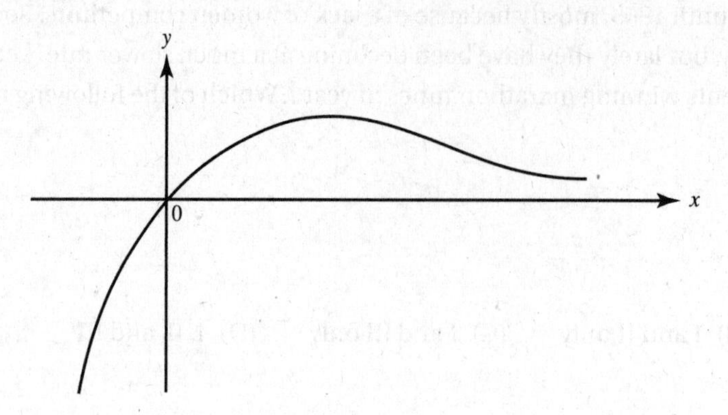

(A) $f(x) = xe^x$ (B) $f(x) = xe^{-x}$ (C) $f(x) = \frac{x}{x^2 + 1}$ (D) $f(x) = \frac{x^2}{x^3 + 1}$

Questions 20–22. Use the graph below, consisting of two line segments and a quarter-circle. The graph shows the velocity of an object during a 6-second interval.

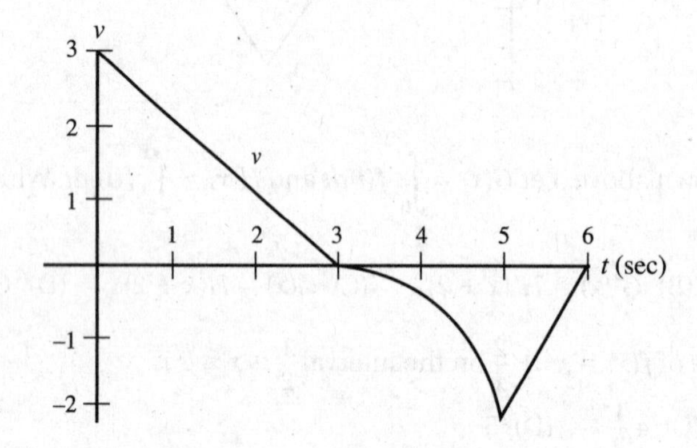

20. For how many values of t in the interval $0 < t < 6$ is the acceleration undefined?

(A) none (B) one (C) two (D) three

21. During what time interval (in sec) is the speed increasing?

(A) $0 < t < 3$ (B) $3 < t < 5$ (C) $5 < t < 6$ (D) never

22. What is the average acceleration (in units/sec^2) during the first 5 seconds?

(A) $-\frac{5}{2}$ (B) -1 (C) $\frac{1}{5}$ (D) $\frac{1}{2}$

23. The curve of $y = \frac{2x^2}{4 - x^2}$ has

(A) two horizontal asymptotes and one vertical asymptote
(B) two vertical asymptotes but no horizontal asymptote
(C) one horizontal and one vertical asymptote
(D) one horizontal and two vertical asymptotes

24. Suppose

$$f(x) = \begin{cases} x^2 & \text{if } x < -2, \\ 4 & \text{if } -2 < x \leq 1, \\ 6 - x & \text{if } x > 1. \end{cases}$$

AB PRACTICE EXAMINATION 2

Which statement is true?

(A) f is continuous everywhere.
(B) f is discontinuous only at $x = 1$.
(C) f is discontinuous at $x = -2$ and at $x = 1$.
(D) If $f(-2)$ is defined to be 4, then f will be continuous everywhere.

25. The function $f(x) = x^5 + 3x - 2$ passes through the point (1,2). Let f^{-1} denote the inverse of f. Then $(f^{-1})'(2)$ equals

(A) $\frac{1}{83}$ (B) $\frac{1}{8}$ (C) 8 (D) 83

26. $\int_1^e \frac{(\ln x)^3}{x}\,dx =$

(A) $\frac{1}{4}$ (B) 1 (C) $\frac{3}{2}$ (D) $\frac{e}{4}$

27. Which of the following statements is (are) true about the graph of $y = \ln(4 + x^2)$?

I. It is symmetric to the y-axis.
II. It has a local minimum at $x = 0$.
III. It has inflection points at $x = \pm 2$.

(A) I only (B) I and II only (C) II and III only (D) I, II, and III

28. Let $\int_0^x f(t)\,dt = x \sin \pi x$. Then $f(3) =$

(A) -3π (B) -1 (C) -3 (D) $-\pi$

29. Choose the integral that is the limit of the Riemann sum: $\lim_{n\to\infty} \sum_{k=1}^{n} \left(\left(\sqrt{\frac{2k}{n}} + 3 \right) \cdot \left(\frac{1}{n} \right) \right)$.

(A) $\int_3^4 \sqrt{2x}\,dx$ (B) $\int_0^1 \sqrt{2x+3}\,dx$ (C) $\int_0^1 \sqrt{2x}\,dx$ (D) $\int_3^4 \sqrt{2x+3}\,dx$

30. The region bounded by $y = e^x$, $y = 1$, and $x = 2$ is rotated about the x-axis. The volume of the solid generated is given by the integral:

(A) $\pi \int_0^2 e^{2x}\,dx$ (B) $\pi \int_0^2 (e^x - 1)^2\,dx$ (C) $\pi \int_0^2 (e^{2x} - 1)\,dx$ (D) $\pi \int_1^{e^2} \left(4 - \ln^2 y\right) dy$

STOP

END OF PART A, SECTION I

Part B

TIME: 45 MINUTES

Some questions in this part of the examination require the use of a graphing calculator. There are 15 questions in Part B, for which 45 minutes are allowed. Because there is no deduction for wrong answers, you should answer every question, even if you need to guess.

Directions: Choose the best answer for each question. If the exact numerical value of the correct answer is not listed as a choice, select the choice that is closest to the exact numerical answer.

31. A particle moves on a straight line so that its velocity at time t is given by $v = 12\sqrt{s}$, where s is its distance from the origin. If $s = 1$ when $t = 0$, then, when $t = 1$, s equals

(A) 6 (B) 7 (C) 36 (D) 49

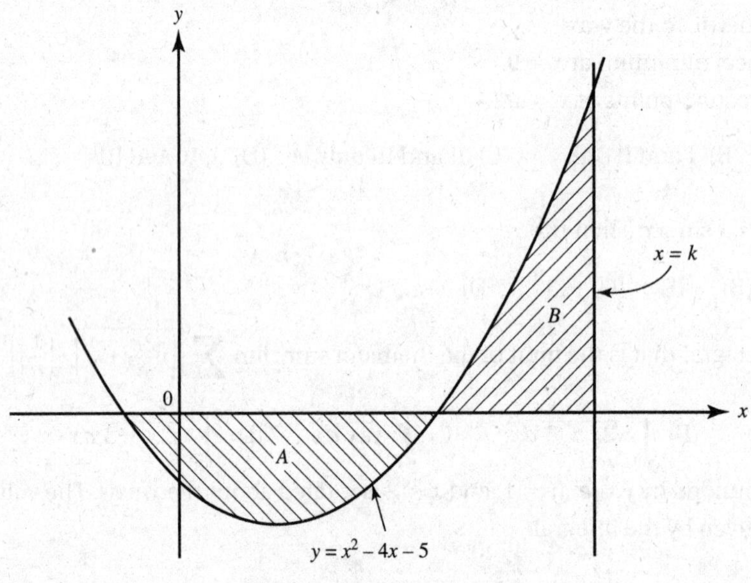

(This figure is not drawn to scale.)

32. The sketch shows the graphs of $f(x) = x^2 - 4x - 5$ and the line $x = k$. The regions labeled A and B have equal areas if $k =$

(A) 7.899 (B) 8 (C) 8.144 (D) 11

33. Bacteria in a culture increase at a rate proportional to the number present. An initial population of 200 triples in 10 hours. If this pattern of increase continues unabated, then the approximate number of bacteria after 1 full day is

(A) 1056 (B) 1440 (C) 2793 (D) 3240

34. When the substitution $x = 2t - 1$ is used, the definite integral $\int_3^5 t\sqrt{2t-1}\,dt$ may be expressed in the form $k\int_a^b (x+1)\sqrt{x}\,dx$, where $\{k, a, b\} =$

(A) $\left\{\frac{1}{4}, 2{,}3\right\}$ (B) $\left\{\frac{1}{4}, 5{,}9\right\}$ (C) $\left\{\frac{1}{2}, 2{,}3\right\}$ (D) $\left\{\frac{1}{2}, 5{,}9\right\}$

35. The curve defined by $x^3 + xy - y^2 = 10$ has a vertical tangent line when $x =$

(A) 1.037 (B) 1.087 (C) 2.074 (D) 2.096

Use the graph of f shown on [0,7] for Questions 36 and 37. Let $G(x) = \int_{2}^{3x-1} f(t)\,dt$.

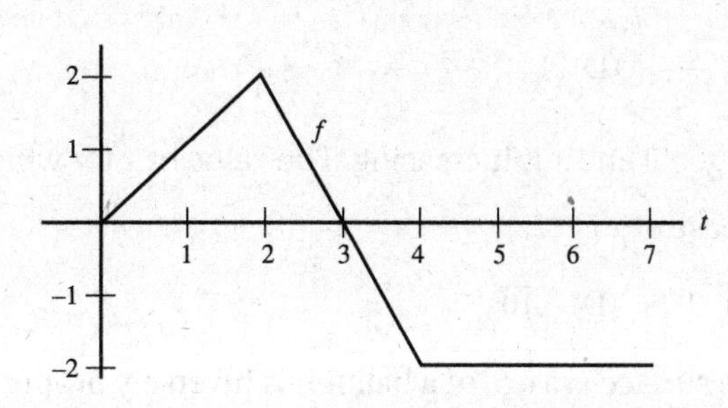

36. $G'(1)$ is

(A) 1 (B) 2 (C) 3 (D) 6

37. G has a local maximum at $x =$

(A) 1 (B) $\frac{4}{3}$ (C) 2 (D) 8

x	$f(x)$	$f'(x)$	$f''(x)$	$g(x)$	$g'(x)$	$g''(x)$
2	6	−1	−2	−2	1/3	−4/3

38. Given two twice-differentiable functions, $f(x)$ and $g(x)$. The table above gives values for $f(x)$ and $g(x)$ and their first and second derivatives at $x = 2$. Find $\lim_{x\to 2} \frac{f(x) + 3g(x)}{\frac{1}{2}x^2 - 2e^{x-2}}$.

(A) 0 (B) 1 (C) 6 (D) nonexistent

39. Using the left rectangular method and four subintervals of equal width, estimate $\int_0^8 |f(t)|\,dt$, where f is the function graphed below.

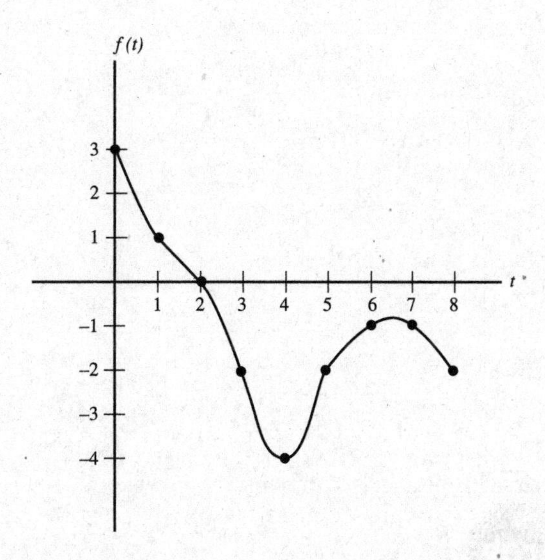

(A) 4 (B) 8 (C) 12 (D) 16

40. Suppose $f(3) = 2, f'(3) = 5$, and $f''(3) = -2$. Then $\frac{d^2}{dx^2}\left(f^2(x)\right)$ at $x = 3$ is equal to

(A) -20 (B) -4 (C) 10 (D) 42

41. The velocity of a particle in motion along a line (for $t \geq 0$) is $v(t) = \ln\left(2 - t^2\right)$. Find the acceleration when the object is at rest.

(A) -2 (B) 0 (C) $\frac{1}{2}$ (D) 1

42. Suppose $f(x) = \frac{1}{3}x^3 + x, x > 0$ and x is increasing. The value of x for which the rate of increase of f is 10 times the rate of increase of x is

(A) 1 (B) $\sqrt[3]{10}$ (C) 3 (D) $\sqrt{10}$

43. The rate of change of the surface area, S, of a balloon is inversely proportional to the square of the surface area. Which equation describes this relationship?

(A) $S(t) = \frac{k}{t^2}$ (B) $S(t) = \frac{k}{S^2}$ (C) $\frac{dS}{dt} = \frac{k}{S^2}$ (D) $\frac{dS}{dt} = \frac{k}{t^2}$

44. Two objects in motion from $t = 0$ to $t = 3$ seconds have positions $x_1(t) = \cos\left(t^2 + 1\right)$ and $x_2(t) = \frac{e^t}{2t}$, respectively. How many times during the 3 seconds do the objects have the same velocity?

(A) 0 (B) 1 (C) 3 (D) 4

45. After t years, $50e^{-0.015t}$ pounds of a deposit of a radioactive substance remain. The average amount per year *not* lost by radioactive decay during the second hundred years is

(A) 2.9 lb (B) 5.8 lb (C) 6.8 lb (D) 15.8 lb

STOP

END OF SECTION I

SECTION II

Part A

TIME: 30 MINUTES

2 PROBLEMS

A graphing calculator is required for some of these problems. See instructions on page 4.

1. Let function f be continuous and decreasing, with values as shown in the table:

x	2.5	3.2	3.5	4.0	4.6	5.0
$f(x)$	7.6	5.7	4.2	3.8	2.2	1.6

(a) Use the trapezoid method to estimate the area between f and the x-axis on the interval $2.5 \le x \le 5.0$.

(b) Find the average rate of change of f on the interval $2.5 \le x \le 5.0$.

(c) Estimate the instantaneous rate of change of f at $x = 2.5$.

(d) If $g(x) = f^{-1}(x)$, estimate the slope of g at $x = 4$.

2. The curve $y = \sqrt{8 \sin\left(\frac{\pi x}{6}\right)}$ divides a first quadrant rectangle into regions $\boldsymbol{A}$ and $\boldsymbol{B}$, as shown in the figure.

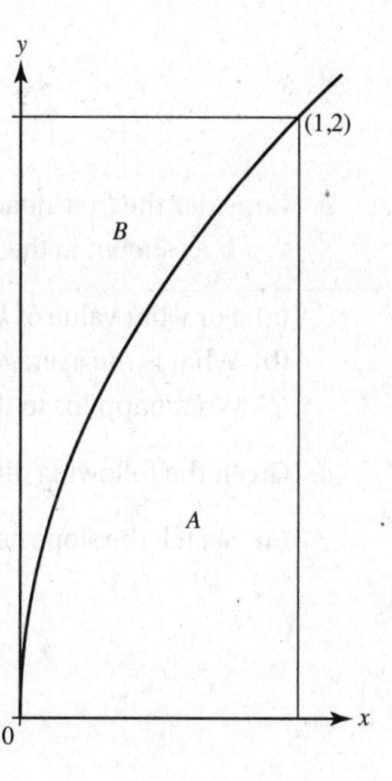

(a) Region $\boldsymbol{A}$ is the base of a solid. Cross sections of this solid perpendicular to the x-axis are rectangles. The height of each rectangle is 5 times the length of its base in region $\boldsymbol{A}$. Find the volume of this solid.

(b) The other region, $\boldsymbol{B}$, is rotated around the y-axis to form a different solid. Set up but do not evaluate an integral for the volume of this solid.

STOP

END OF PART A, SECTION II

AB PRACTICE EXAMINATION 2

Part B

TIME: 60 MINUTES
4 PROBLEMS

No calculator is allowed for any of these problems.
If you finish Part B before time has expired, you may return to work on Part A, but you may not use a calculator.

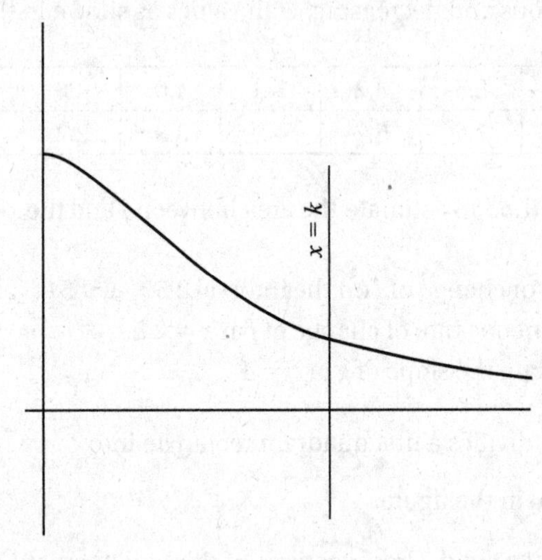

3. Consider the first-quadrant region bounded by the curve $y = \frac{18}{9 + x^2}$, the coordinate axes, and the line $x = k$, as shown in the figure above.

   (a) For what value of k will the area of this region equal π?
   (b) What is the average value of the function on the interval $0 \leq x \leq k$?
   (c) What happens to the area of the region as the value of k increases?

4. Given the following differential equation $\frac{dy}{dx} = x - 2y + 1$.

   (a) Sketch the slope field for the differential equation at the nine indicated points on the axes provided.

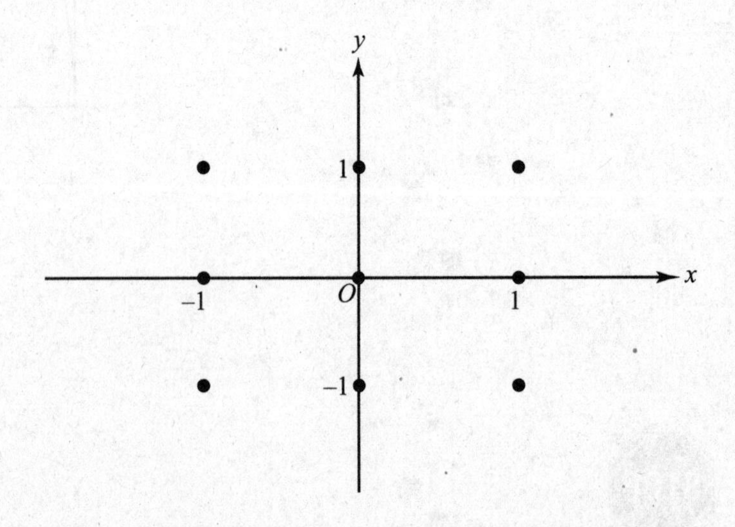

(b) Find the second derivative, $\frac{d^2y}{dx^2}$, in terms of x and y. The region in the xy-plane where all the solution curves to the differential equation are concave down can be expressed as a linear inequality. Find this region.

(c) The function $y = f(x)$ is the solution to the differential equation with initial condition $f(-1) = 0$. Determine whether f has a local maximum, local minimum, or neither at $x = -1$. Justify your answer.

(d) For which values of m and b is the line $y = mx + b$ a solution to the differential equation?

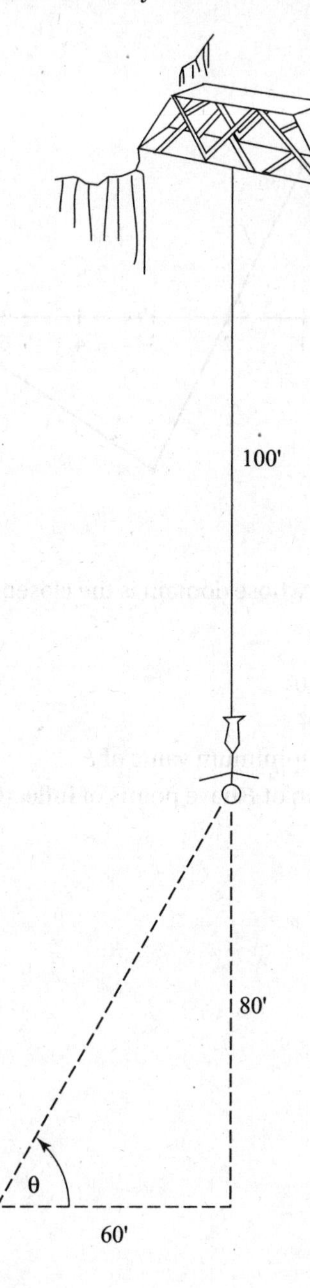

5. A bungee jumper has reached a point in her exciting plunge where the taut cord is 100 feet long with a 1/2-inch radius, and stretching. She is still 80 feet above the ground and is now falling at 40 feet per second. You are observing her jump from a spot on the ground 60 feet from the potential point of impact, as shown in the diagram above.

(a) Assuming the cord to be a cylinder with volume remaining constant as the cord stretches, at what rate is its radius changing when the radius is 1/2"?

(b) From your observation point, at what rate is the angle of elevation to the jumper changing when the radius is 1/2"?

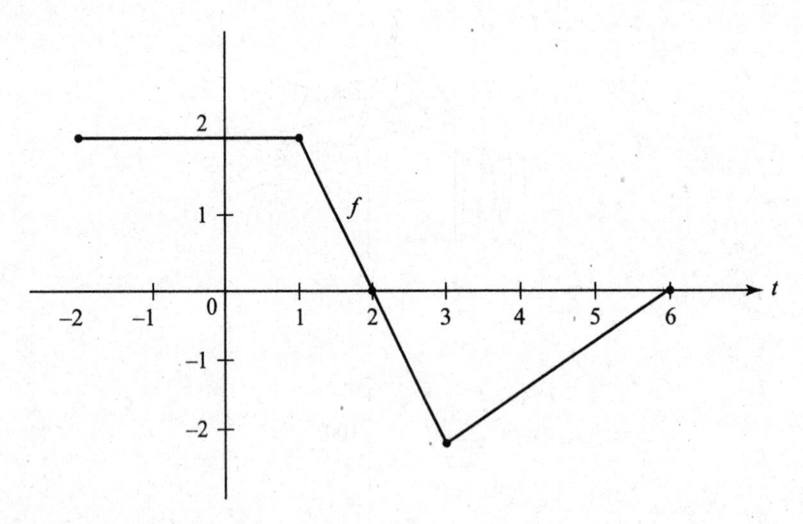

6. The figure above shows the graph of f, whose domain is the closed interval $[-2,6]$. Let $F(x) = \int_1^x f(t)\,dt$.

(a) Find $F(-2)$ and $F(6)$.

(b) For what value(s) of x does $F(x) = 0$?

(c) On what interval(s) is F increasing?

(d) Find the maximum value and the minimum value of F.

(e) At what value(s) of x does the graph of F have points of inflection? Justify your answer.

STOP

END OF TEST

ANSWER KEY
AB Practice Examination 2

Part A

1. **D**	9. **B**	17. **B**	25. **B**
2. **A**	10. **A**	18. **A**	26. **A**
3. **D**	11. **C**	19. **B**	27. **D**
4. **D**	12. **A**	20. **C**	28. **A**
5. **A**	13. **C**	21. **B**	29. **B**
6. **D**	14. **D**	22. **B**	30. **C**
7. **B**	15. **C**	23. **D**	
8. **B**	16. **D**	24. **C**	

Part B

31. **D**	35. **C**	39. **D**	43. **C**
32. **B**	36. **D**	40. **D**	44. **D**
33. **C**	37. **B**	41. **A**	45. **B**
34. **B**	38. **C**	42. **C**	

ANSWERS EXPLAINED

Section I Multiple-Choice

Part A

1. **(D)** $\frac{x^2-2}{4-x^2} \to +\infty$ as $x \to 2$.

2. **(A)** Divide both numerator and denominator by $\sqrt{x}$; $\lim_{x\to\infty} \frac{1-\frac{4}{\sqrt{x}}}{\frac{4}{\sqrt{x}}-3} = -\frac{1}{3}$.

3. **(D)** Since $e^{\ln u} = u$, $y = 1$.

4. **(D)** $f(0) = 3$, and $f'(x) = \frac{1}{2}(9 + \sin 2x)^{-1/2} \cdot (2\cos 2x)$, so $f'(0) = \frac{1}{3}$; $y \simeq \frac{1}{3}x + 3$.

5. **(A)** $\int_0^1 \frac{60}{1+t^2}\,dt = 60 \arctan t \Big|_0^1 = 60 \arctan 1 = 60 \cdot \frac{\pi}{4}$.

6. **(D)** Here $y' = 3\sin^2(1-2x)\cos(1-2x) \cdot (-2)$.

7. **(B)** $\frac{d}{dx}\left(x^2 e^{x^{-1}}\right) = x^2 e^{x^{-1}}\left(-\frac{1}{x^2}\right) + 2xe^{x^{-1}}$.

8. **(B)** Let s be the distance from the origin: then

$$s = \sqrt{x^2+y^2} \quad \text{and} \quad \frac{ds}{dt} = \frac{x\frac{dx}{dt} + y\frac{dy}{dt}}{\sqrt{x^2+y^2}}.$$

Since $\frac{dy}{dt} = 2x\frac{dx}{dt}$ and $\frac{dx}{dt} = \frac{3}{2}$, $\frac{dy}{dt} = 3x$. Substituting yields $\frac{ds}{dt} = \frac{3\sqrt{5}}{2}$.

9. **(B)** For $f(x) = \sqrt{x}$, this limit represents $f'(25)$.

10. **(A)** $V = \int_0^1 y^2\,dx = \int_0^1 \sqrt{1-x^2}\,dx$. This definite integral represents the area of a quadrant of the circle $x^2 + y^2 = 1$, hence $V = \frac{\pi}{4}$.

11. **(C)** $-\frac{1}{2}\int (9-x^2)^{-1/2}(-2x\,dx) = -\frac{1}{2}\frac{(9-x^2)^{\frac{1}{2}}}{1/2} + C$.

12. **(A)** The integral is rewritten as

$$\int \frac{(y-1)^2}{dy}\,dy = \frac{1}{2}\int \frac{y^2-2y+1}{y}\,dy,$$
$$= -\frac{1}{2}\int \left(y - 2 + \frac{1}{y}\right) dy,$$
$$= -\frac{1}{2}\left(\frac{y^2}{2} - 2y + \ln|y|\right) + C.$$

13. **(C)** $\int_{\pi/6}^{\pi/2} \cot x\,dx = \ln \sin x \Big|_{\pi/6}^{\pi/2} = 0 - \ln\frac{1}{2}$.

14. **(D)** Note:

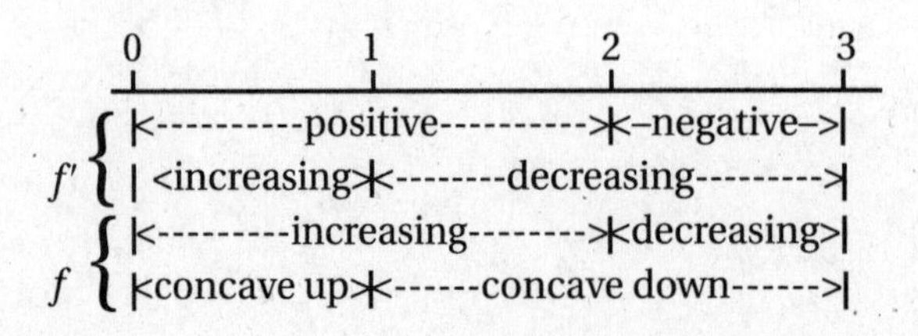

15. **(C)** The winning times are positive, decreasing, and concave upward.

16. **(D)** $G(x) = H(x) + \int_0^2 f(t)\,dt$, where $\int_0^2 f(t)dt$ represents the area of a trapezoid.

17. **(B)** $f'(x) = 0$ for $x = 1$ and $f''(1) > 0$.

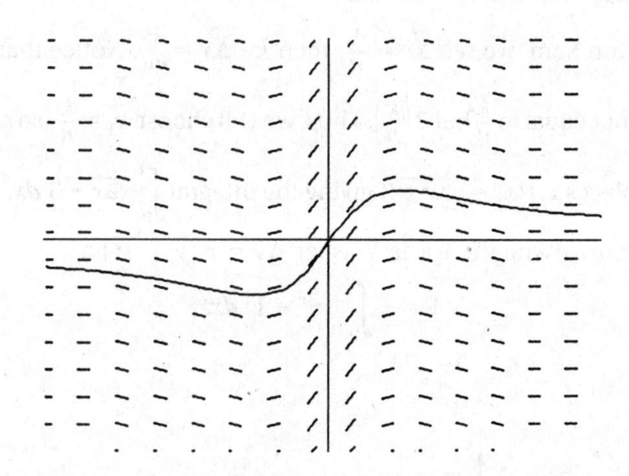

18. **(A)** Solution curves appear to represent odd functions with a horizontal asymptote. In the figure above, the curve in (B) of the question has been superimposed on the slope field.

19. **(B)** Note that

$$\lim_{x\to\infty} xe^x = \infty,\ \lim_{x\to\infty}\frac{e^x}{x} = \infty,\ \lim_{x\to-\infty}\frac{x}{x^2+1} = 0,\text{ and }\frac{x^2}{x^3+1} \geq 0 \text{ for } x > -1.$$

20. **(C)** v is not differentiable at $t = 3$ or $t = 5$.

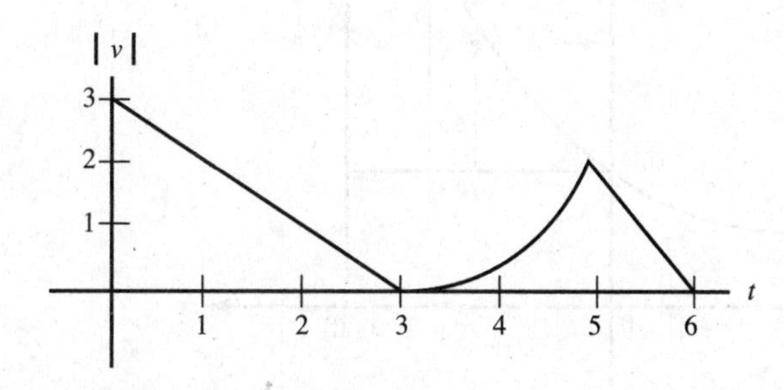

21. **(B)** Speed is the magnitude of velocity; its graph is shown above.

22. **(B)** The average rate of change of velocity is $\dfrac{v(5) - v(0)}{5 - 0} = \dfrac{-2 - 3}{5}$.

23. **(D)** The curve has vertical asymptotes at $x = 2$ and $x = -2$ and a horizontal asymptote at $y = -2$.

24. **(C)** The function is not defined at $x = -2$; $\lim_{x\to1^-} f(x) \neq \lim_{x\to1^+} f(x)$. Defining $f(-2) = 4$ will make f continuous at $x = -2$, but f will still be discontinuous at $x = 1$.

25. **(B)** Since $\left(f^{-1}\right)'(y) = \dfrac{1}{f'(x)}$,

$$\left(f^{-1}\right)'(y) = \frac{1}{5x^4+3} \text{ and } \left(f^{-1}\right)'(2) = \frac{1}{5\cdot1+3} = \frac{1}{8}.$$

AB PRACTICE EXAMINATION 2

26. **(A)** $\int_1^e \frac{\ln^3 x}{x}\,dx = \int_1^e (\ln x)^3\left(\frac{1}{x}\,dx\right) = \frac{1}{4}\ln^4 x\Big|_1^e = \frac{1}{4}\left(\ln^4 e - 0\right) = \frac{1}{4}.$

27. **(D)** $\ln\left(4 + x^2\right) = \ln\left(4 + (-x)^2\right); y' = \frac{2x}{4 + x^2}; y'' = \frac{-2\left(x^2 - 4\right)}{\left(4 + x^2\right)^2}.$

28. **(A)** $f(x) = \frac{d}{dx}(x \sin \pi x) = \pi x \cos \pi x + \sin \pi x.$

29. **(B)** From the Riemann Sum, we see $\Delta x = \frac{1}{n}$, then $k \cdot \Delta x = \frac{k}{n}$. Notice that the term involving k in the Riemann Sum is not equal to $\frac{k}{n}$ but $2\left(\frac{k}{n}\right)$. Thus, we will choose $x_k = \frac{k}{n}$, so $a = 0$ and $\Delta x = \frac{b-0}{n} = \frac{1}{n}$, so $b = 1$. Since x_k replaces x, $f(x) = \sqrt{2x+3}$ giving the integral $\int_0^1 \sqrt{2x+3}\,dx$.

30. **(C)** See the figure below. About the x-axis: Washer. $\Delta V = \pi\left(y^2 - 1^2\right)\Delta x$,

$$V = \pi \int_0^2 \left(e^{2x} - 1\right) dx.$$

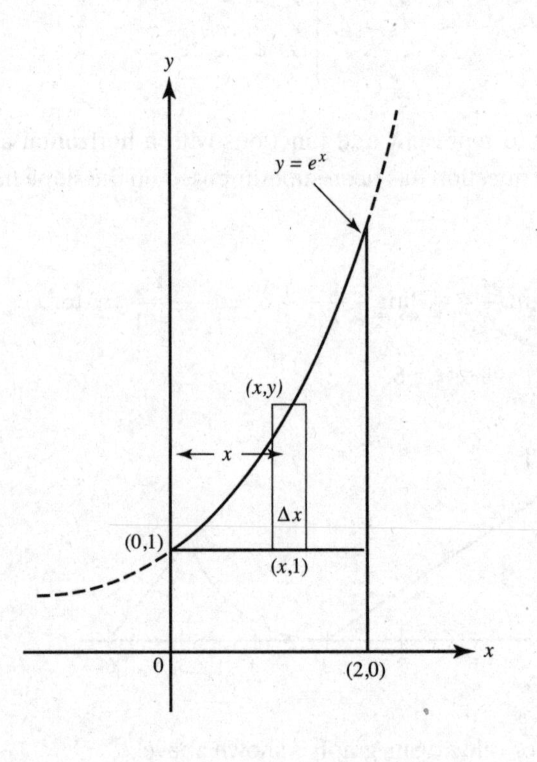

Part B

31. **(D)** We solve the differential equation $\frac{ds}{dt} = 12s^{\frac{1}{2}}$ by separation:

$$\int s^{-\frac{1}{2}}\, ds = 12 \int dt$$
$$2s^{\frac{1}{2}} = 12t + C$$
$$\sqrt{s} = 6t + C$$

If $s = 1$ when $t = 0$, we have $C = 1$; hence, $\sqrt{s} = 6t + 1$ so $\sqrt{s} = 7$ when $t = 1$.

32. **(B)**

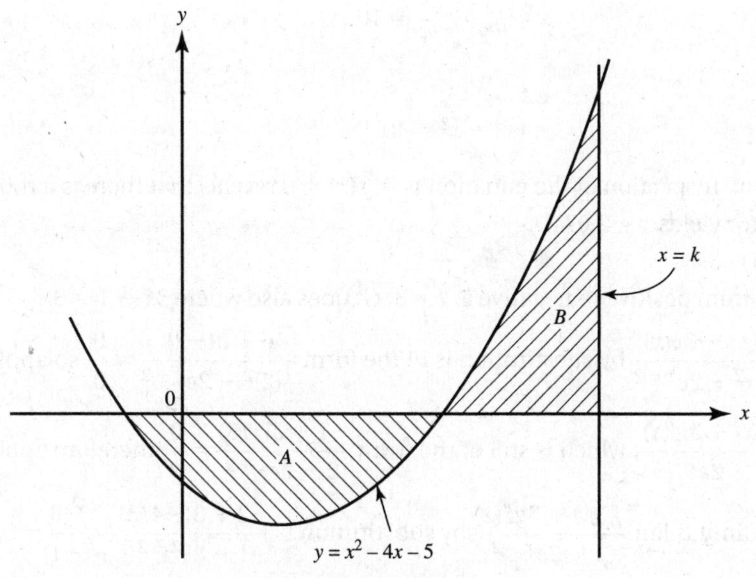

(This figure is not drawn to scale.)

The roots of $f(x) = x^2 - 4x - 5 = (x - 5)(x + 1)$ are $x = -1$ and 5. Since areas A and B are equal, therefore $\int_{-1}^{k} f(x)\, dx = 0$. Thus,

$$\left(\frac{x^3}{3} - 2x^2 - 5x\right)\Bigg|_{-1}^{k} = \left(\frac{k^3}{3} - 2k^2 - 5k\right) - \left(-\frac{1}{3} - 2 + 5\right)$$
$$= \frac{k^3}{3} - 2k^2 - 5k - \frac{8}{3} = 0.$$

Solving on a calculator gives k (or x) equal to 8.

33. **(C)** If N is the number of bacteria at time t, then $N = 200e^{kt}$. It is given that $3 = e^{10k}$. When $t = 24$, $N = 200e^{24k}$. Therefore $N = 200\left(e^{10k}\right)^{2.4} = 200(3)^{2.4} \approx 2793$ bacteria.

34. **(B)** Since $t = \frac{x+1}{2}$, $dt = \frac{1}{2}\, dx$. For $x = 2t - 1$, $t = 3$ yields $x = 5$ and $t = 5$ yields $x = 9$.

35. **(C)** Using implicit differentiation on the equation

$$x^3 + xy - y^2 = 10$$

AB PRACTICE EXAMINATION 2

yields

$$3x^2 + x\frac{dy}{dx} + y - 2y\frac{dy}{dx} = 0,$$

$$3x^2 + y = (2y - x)\frac{dy}{dx},$$

and

$$\frac{dy}{dx} = \frac{3x^2 + y}{2y - x}.$$

The tangent is vertical when $\frac{dy}{dx}$ is undefined; that is, when $2y - x = 0$.

Replacing y by $\frac{x}{2}$ in (1) gives

$$x^3 + \frac{x^2}{2} - \frac{x^2}{4} = 10$$

or

$$4x^3 + x^2 = 40.$$

Let $y_1 = 4x^3 + x^2 - 40$. Inspection of the equation $y_1 = f(x) = 0$ reveals that there is a root near $x = 2$. Solving on a calculator yields $x = 2.074$.

36. **(D)** $G'(x) = f(3x - 1) \cdot 3$.

37. **(B)** Since f changes from positive to negative at $t = 3$, G' does also where $3x - 1 = 3$.

38. **(C)** The limit $\lim_{x \to 2} \frac{f(x) + 3g(x)}{\frac{1}{2}x^2 - 2e^{x-2}}$ by substitution is of the form $\frac{6 + 3(-2)}{\frac{1}{2}(2)^2 - 2e^{2-2}} = \frac{0}{0}$, so apply L'Hopital's Rule. You get $\lim_{x \to 2} \frac{f'(x) + 3g'(x)}{x - 2e^{x-2}}$, which is still of the form $\frac{-1 + 3(1/3)}{2 - 2e^{2-2}} = \frac{0}{0}$, therefore apply L'Hopital's Rule again. The new limit is $\lim_{x \to 2} \frac{f''(x) + 3g''(x)}{1 - 2e^{x-2}}$ by substitution is $\frac{-2 + 3(-4/3)}{1 - 2e^{2-2}} = \frac{-6}{-1} = 6$.

39. **(D)** $2(3) + 2(0) + 2(4) + 2(1)$.

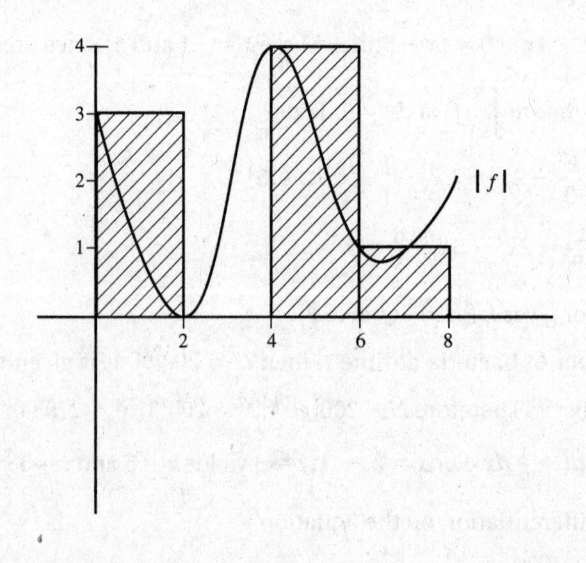

40. **(D)** $\frac{d}{dx}\left(f^2(x)\right) = 2f(x)f'(x),$

$$\frac{d^2}{dx^2}\left(f^2(x)\right) = 2[f(x)f''(x) + f'(x)f''(x)]$$

$$= 2[ff'' + (f')^2].$$

At $x = 3$, the answer is $2\left[2(-2) + 5^2\right] = 42.$

41. **(A)** The object is at rest when $v(t) = \ln\left(2 - t^2\right) = 0$; that occurs when $2 - t^2 = 1$, so $t = 1$.

The acceleration is $a(t) = v'(t) = \frac{-2t}{2 - t^2}$; $a(1) = \frac{-2(1)}{2 - 1^2}$.

42. **(C)** $\frac{df}{dt} = \left(x^2 + 1\right)\frac{dx}{dt}$. Find x when $\frac{df}{dt} = 10\frac{dx}{dt}$.

$$10\frac{dx}{dt} = \left(x^2 + 1\right)\frac{dx}{dt}$$

implies that $x = 3$.

43. **(C)** $\frac{dS}{dt}$ represents the rate of change of the surface area; if y is inversely proportional to x, then, $y = \frac{k}{x}$.

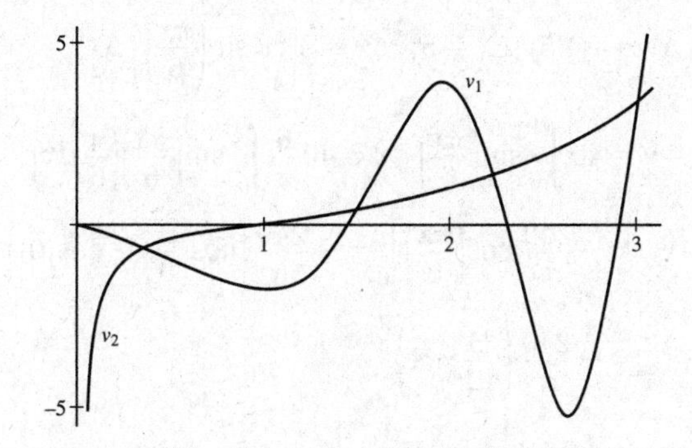

44. **(D)** The velocity functions are

$$v_1 = -2t\sin\left(t^2 + 1\right)$$

and

$$v_2 = \frac{2t\left(e^t\right) - 2e^t}{(2t)^2} = \frac{e^t(t - 1)}{2t^2}.$$

Graph both functions in $[0, 3] \times [-5, 5]$. The graphs intersect four times during the first 3 sec, as shown in the figure above.

45. **(B)** $\frac{\int_{100}^{200} 50e^{-0.015t}\,dt}{100} \simeq 5.778$ lb.

Section II Free-Response

Part A

AB/BC 1. (a) $T = \left(\frac{7.6+5.7}{2}\right)(0.7) + \left(\frac{5.7+4.2}{2}\right)(0.3) + \left(\frac{4.2+3.8}{2}\right)(0.5) +$

$\left(\frac{3.8+2.2}{2}\right)(0.6) + \left(\frac{2.2+1.6}{2}\right)(0.4) = 10.7.$

(b) $\frac{\Delta y}{\Delta x} = \frac{7.6-1.6}{2.5-5.0} = -2.4.$

(c) $f'(2.5) \approx \frac{5.7-7.6}{3.2-2.5} = -2.714.$

(d) To work with $g(x) = f^{-1}(x)$, interchange x and y:

x	7.6	5.7	4.2	3.8	2.2	1.6
$g(x)$	2.5	3.2	3.5	4.0	4.5	5.0

Now $g'(4) \approx \frac{4.0-3.5}{3.8-4.2} = -1.25$ OR $\frac{4.5-4.0}{2.2-3.8} = -0.313$ OR $\frac{4.5-3.5}{2.2-4.2} = -0.5.$

AB 2/BC 4. (a) The rectangular slices have base y, height $5y$, and thickness along the x-axis:

$$\Delta V = (y)(5y)\Delta x = 5y^2\Delta x = 5\left(\sqrt{8\sin\left(\frac{\pi x}{6}\right)}\right)\Delta x$$

$$V = 40\int_0^1 \sin\left(\frac{\pi x}{6}\right)dx = 40\,\frac{6}{\pi}\int_0^1 \sin\left(\frac{\pi x}{6}\right)\left(\frac{\pi}{6}\,dx\right)$$

$$= -\frac{240}{\pi}\cos\left(\frac{\pi x}{6}\right)\Big|_0^1 = -\frac{240}{\pi}\left(\cos\left(\frac{\pi}{6}\right) - \cos(0)\right)$$

$$= -\frac{240}{\pi}\left(\sqrt{\frac{3}{2}} - 1\right)$$

(b) The disks have radius x and thickness along the y-axis:

$\Delta V = \pi x^2 \Delta y$, so $V = \pi\int_0^2 x^2\,dy$

Now we solve for x in terms of y:

$y = \sqrt{8\sin\left(\frac{\pi x}{6}\right)}$, so $y^2 = 8\sin\left(\frac{\pi x}{6}\right)$ and $\frac{y^2}{8} = \sin\left(\frac{\pi x}{6}\right)$.

Then $\arcsin\frac{y^2}{8} = \left(\frac{\pi x}{6}\right)$, which gives us $x = \frac{6}{\pi}\arcsin\frac{y^2}{8}$.

Therefore $V = \pi\int_0^2 \left(\frac{6}{\pi}\arcsin\frac{y^2}{8}\right)^2 dy.$

(*NOTE*: Although the shells method is not a required AP topic, another correct integral for this volume is $V = 2\pi\int_0^1 x\left(2 - \sqrt{8\sin\frac{\pi x}{6}}\right)dx.$)

AB PRACTICE EXAMINATION 2

Part B

AB 3. (a)

$$\int_0^k \frac{18}{9+x^2}\,dx = \pi,$$

$$3 \cdot \frac{18}{9} \int_0^k \frac{\frac{1}{3}\,dx}{1+\left(\frac{x}{3}\right)^2} = \pi,$$

$$6 \arctan \frac{x}{3}\Big|_0^k = \pi,$$

$$6 \arctan \frac{k}{3} - 6 \arctan \frac{0}{3} = \pi,$$

$$\frac{k}{3} = \tan \frac{\pi}{6},$$

$$k = \sqrt{3},$$

See the figure on page 526.

(b) The average value of a function on an interval is the area under the graph of the function divided by the interval width, here $\frac{\pi}{\sqrt{3}}$.

(c) From part (a) you know that the area of the region is given by $\int_0^k \frac{18}{9+x^2}\,dx = 6 \arctan \frac{k}{3}$. Since $\lim_{k\to\infty} 6 \arctan \frac{k}{3} = 6\left(\frac{\pi}{2}\right) = 3\pi$, as k increases the area of the region approaches 3π.

AB 4. (a) The slope field for the nine points indicated is:

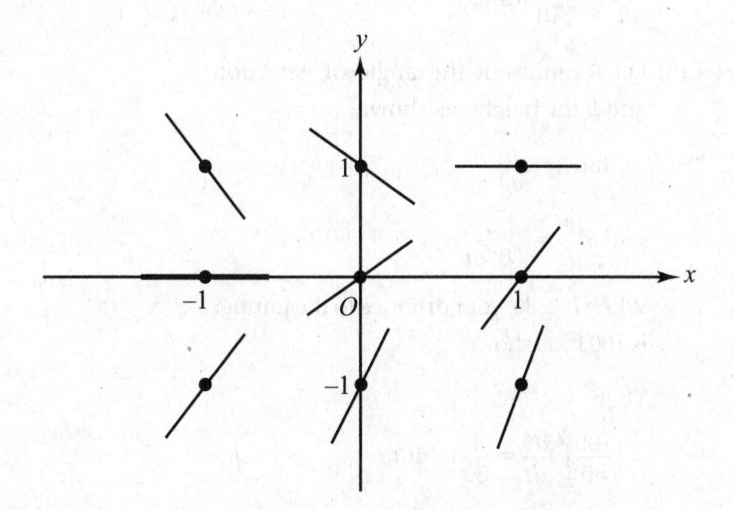

The slopes are given in the table below. Be sure that the segments you draw are correct relative to the other slopes in the slope field with respect to the steepness of the segments.

x	−1	−1	−1	0	0	0	1	1	1
y	−1	0	1	−1	0	1	−1	0	1
dy/dx	2	0	−2	3	1	−1	4	2	0

(b) $\frac{d^2y}{dx^2} = 1 - 2\frac{dy}{dx} = 1 - 2(x - 2y + 1) = 4y - 2x - 1$

The solution curves will be concave down when $\frac{d^2y}{dx^2} < 0.$

$4y - 2x - 1 < 0 \Rightarrow y < \frac{1}{2}x + \frac{1}{4}$

(c) $(-1,0)$ is a critical point because $\left.\frac{dy}{dx}\right|_{(-1,0)} = -1 - 2(0) + 1 = 0.$ And

$\left.\frac{d^2y}{dx^2}\right|_{(-1,0)} = 4(0) - 2(-1) - 1 = 1 > 0$, so by the second derivative test, the solution curve $f(x)$ has a local minimum.

(d) m is the slope $\left(\frac{dy}{dx}\right)$ so replace into the differential equation:

$m = x - 2(mx + b) + 1 = (1 - 2m)x + (1 - 2b)$

$1 - 2m = 0$ and $1 - 2b = m$, so $m = \frac{1}{2}$ and $b = \frac{1}{4}$.

AB/BC 5. (a) The volume of the cord is $V = \pi r^2 h$. Differentiate with respect to time, then substitute known values. (Be sure to use consistent units; here, all measurements have been converted to inches.)

$$\frac{dV}{dt} = \pi\left(r^2\frac{dh}{dt} + 2rh\frac{dr}{dt}\right),$$

$$0 = \pi\left(\left(\frac{1}{2}\right)^2 \cdot 480 + 2 \cdot \frac{1}{2} \cdot 1200\frac{dr}{dt}\right),$$

$$\frac{dr}{dt} = -\frac{1}{10} \text{ in/sec.}$$

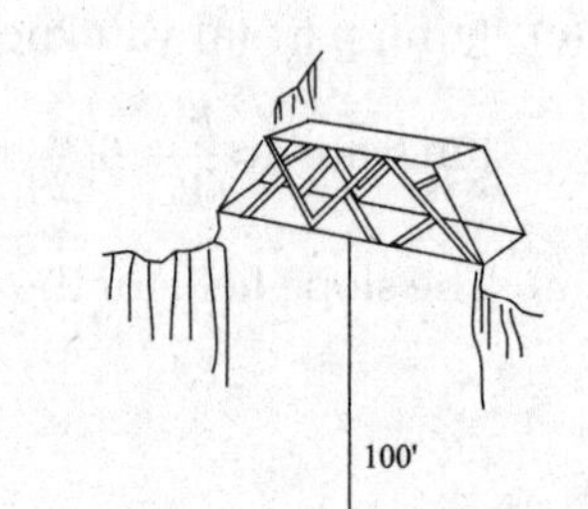

(b) Let θ represent the angle of elevation and h the height, as shown.

$$\tan\theta = \frac{h}{60}$$

$$\sec^2\theta\frac{d\theta}{dt} = \frac{1}{60}\frac{dh}{dt}$$

When $h = 80$, your distance to the jumper is 100 ft, as shown.

Then

$$\left(\frac{100}{60}\right)^2\frac{d\theta}{dt} = \frac{1}{60}(-40),$$

$$\frac{d\theta}{dt} = -\frac{6}{25} \text{ rad/sec.}$$

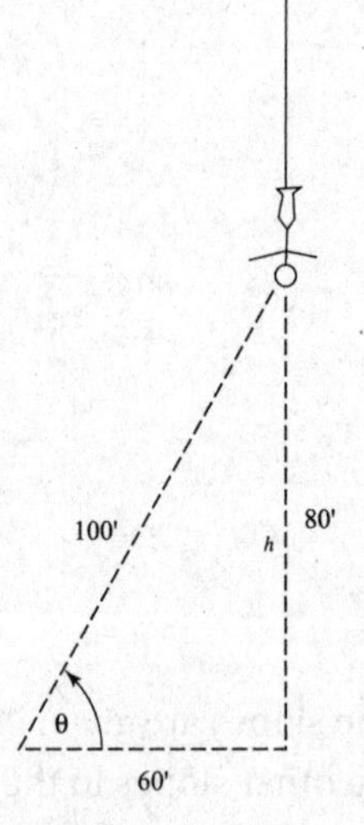

AB/BC 6. (a) $F(-2) = \int_1^{-2} f(t)\,dt = -\int_{-2}^{1} f(t)\,dt =$ the negative of the area of the shaded rectangle in the figure. Hence $F(-2) = -(3)(2) = -6$.

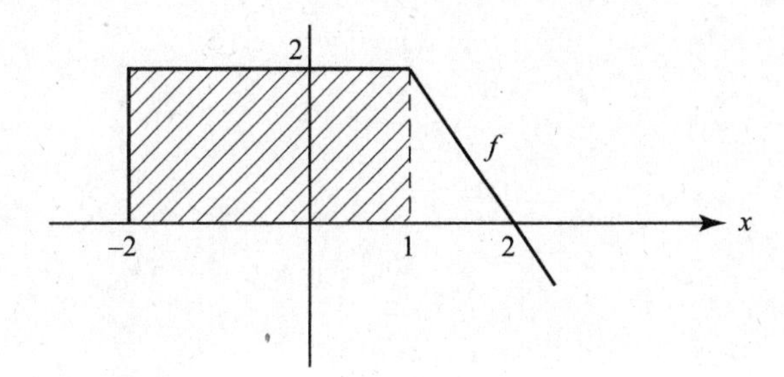

$F(6) = \int_1^6 f(t)\,dt$ is represented by the shaded triangles in the figure.

$$\int_1^6 f(t)\,dt = \int_1^2 f(t)\,dt + \int_2^6 f(t)\,dt$$
$$= \frac{1}{2}(1)(2) - \frac{1}{2}(4)(2) = -3.$$

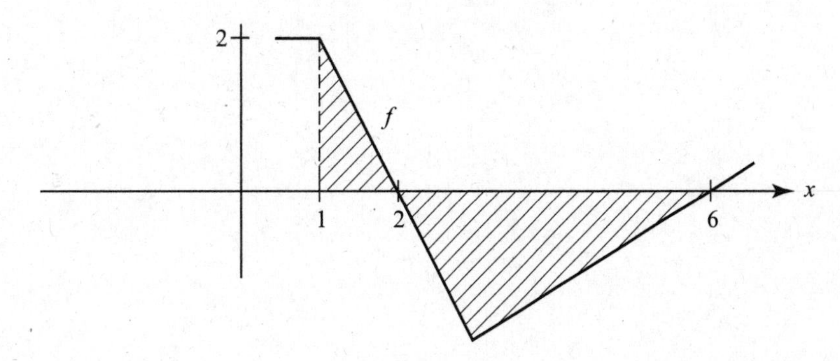

(b) $\int_1^1 f(t)\,dt = 0$, so $F(x) = 0$ at $x = 1$. $\int_1^3 f(t)\,dt = 0$ because the regions above and below the x-axis have the same area. Hence $F(x) = 0$ at $x = 3$.

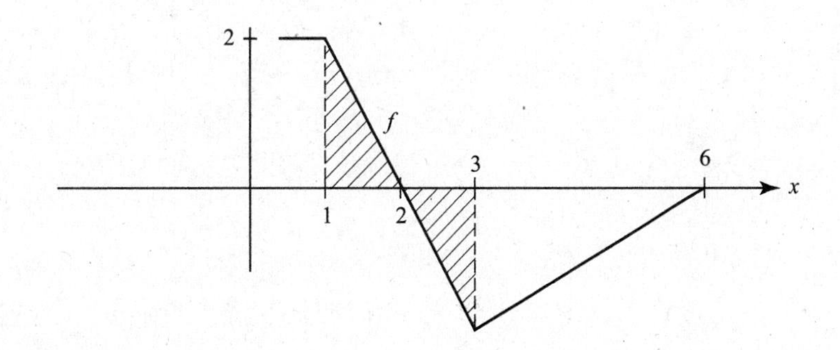

(c) F is increasing where $F' = f$ is positive: $-2 \le x < 2$.

(d) The maximum value of F occurs at $x = 2$, where $F' = f$ changes from positive to negative.
$F(2) = \int_1^2 f(t)\,dt = \frac{1}{2}(1)(2) = 1$.

The minimum value of F must occur at one of the endpoints. Since $F(-2) = -6$ and $F(6) = -3$, the minimum is at $x = -2$.

(e) F has points of inflection where F'' changes sign, as occurs where $F' = f$ goes from decreasing to increasing, at $x = 3$.

ANSWER SHEET
AB Practice Examination 3

Part A

1. Ⓐ Ⓑ Ⓒ Ⓓ	9. Ⓐ Ⓑ Ⓒ Ⓓ	17. Ⓐ Ⓑ Ⓒ Ⓓ	25. Ⓐ Ⓑ Ⓒ Ⓓ
2. Ⓐ Ⓑ Ⓒ Ⓓ	10. Ⓐ Ⓑ Ⓒ Ⓓ	18. Ⓐ Ⓑ Ⓒ Ⓓ	26. Ⓐ Ⓑ Ⓒ Ⓓ
3. Ⓐ Ⓑ Ⓒ Ⓓ	11. Ⓐ Ⓑ Ⓒ Ⓓ	19. Ⓐ Ⓑ Ⓒ Ⓓ	27. Ⓐ Ⓑ Ⓒ Ⓓ
4. Ⓐ Ⓑ Ⓒ Ⓓ	12. Ⓐ Ⓑ Ⓒ Ⓓ	20. Ⓐ Ⓑ Ⓒ Ⓓ	28. Ⓐ Ⓑ Ⓒ Ⓓ
5. Ⓐ Ⓑ Ⓒ Ⓓ	13. Ⓐ Ⓑ Ⓒ Ⓓ	21. Ⓐ Ⓑ Ⓒ Ⓓ	29. Ⓐ Ⓑ Ⓒ Ⓓ
6. Ⓐ Ⓑ Ⓒ Ⓓ	14. Ⓐ Ⓑ Ⓒ Ⓓ	22. Ⓐ Ⓑ Ⓒ Ⓓ	30. Ⓐ Ⓑ Ⓒ Ⓓ
7. Ⓐ Ⓑ Ⓒ Ⓓ	15. Ⓐ Ⓑ Ⓒ Ⓓ	23. Ⓐ Ⓑ Ⓒ Ⓓ	
8. Ⓐ Ⓑ Ⓒ Ⓓ	16. Ⓐ Ⓑ Ⓒ Ⓓ	24. Ⓐ Ⓑ Ⓒ Ⓓ	

Part B

31. Ⓐ Ⓑ Ⓒ Ⓓ	35. Ⓐ Ⓑ Ⓒ Ⓓ	39. Ⓐ Ⓑ Ⓒ Ⓓ	43. Ⓐ Ⓑ Ⓒ Ⓓ
32. Ⓐ Ⓑ Ⓒ Ⓓ	36. Ⓐ Ⓑ Ⓒ Ⓓ	40. Ⓐ Ⓑ Ⓒ Ⓓ	44. Ⓐ Ⓑ Ⓒ Ⓓ
33. Ⓐ Ⓑ Ⓒ Ⓓ	37. Ⓐ Ⓑ Ⓒ Ⓓ	41. Ⓐ Ⓑ Ⓒ Ⓓ	45. Ⓐ Ⓑ Ⓒ Ⓓ
34. Ⓐ Ⓑ Ⓒ Ⓓ	38. Ⓐ Ⓑ Ⓒ Ⓓ	42. Ⓐ Ⓑ Ⓒ Ⓓ	

AB Practice Examination 3

SECTION I

Part A

TIME: 60 MINUTES

The use of calculators is not permitted for this part of the examination.
There are 30 questions in Part A, for which 60 minutes are allowed. Because there is no deduction for wrong answers, you should answer every question, even if you need to guess.

Directions: Choose the best answer for each question.

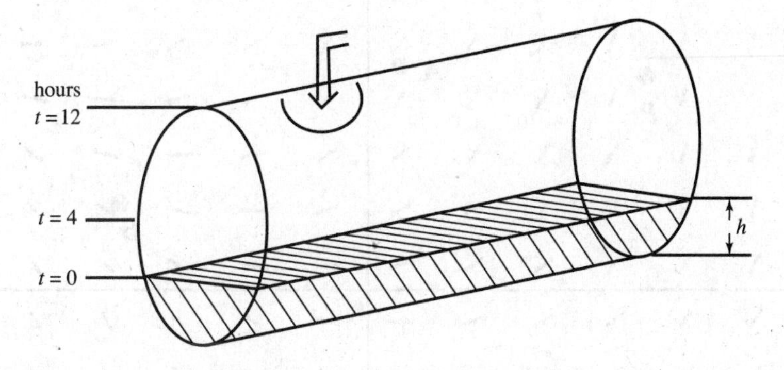

1. A cylindrical tank, shown in the figure above, is partially full of water at time $t = 0$, when more water begins flowing in at a constant rate. The tank becomes half full when $t = 4$, and is completely full when $t = 12$. Let h represent the height of the water at time t. During which interval is $\frac{dh}{dt}$ increasing?

   (A) $0 < t < 4$ (B) $0 < t < 8$ (C) $0 < t < 12$ (D) $4 < t < 12$

2. $\lim\limits_{h \to 0} \dfrac{\sin\left(\frac{\pi}{2} + h\right) - 1}{h}$ is

   (A) 1 (B) -1 (C) 0 (D) nonexistent

3. If $f(x) = x \ln x$, then $f'''(e)$ equals

   (A) $\frac{1}{e}$ (B) 2 (C) $-\frac{1}{e^2}$ (D) $\frac{2}{e^3}$

4. The equation of the tangent to the curve $2x^2 - y^4 = 1$ at the point $(-1, 1)$ is

   (A) $x + y = 0$ (B) $x + y = 2$ (C) $x - y = 0$ (D) $x - y = -2$

5. On which interval(s) does the function $f(x) = x^4 - 4x^3 + 4x^2 + 6$ increase?

(A) $x < 0$ and $1 < x < 2$ (B) $x > 2$ only (C) $0 < x < 1$ and $x > 2$ (D) $0 < x < 1$ only

6. $\int \frac{\cos x}{4 + 2 \sin x} dx$ equals

(A) $\ln |4 + 2 \sin x| + C$ (B) $-2 \ln |4 + 2 \sin x| + C$

(C) $\ln \sqrt{4 + 2 \sin x} + C$ (D) $2 \ln |4 + 2 \sin x| + C$

7. A local maximum value of the function $y = \frac{\ln x}{x}$ is

(A) 0 (B) $\frac{1}{e}$ (C) 1 (D) e

8. If a particle moves on a line according to the law $s = t^5 + 2t^3$, then the number of times it reverses direction is

(A) 3 (B) 2 (C) 1 (D) 0

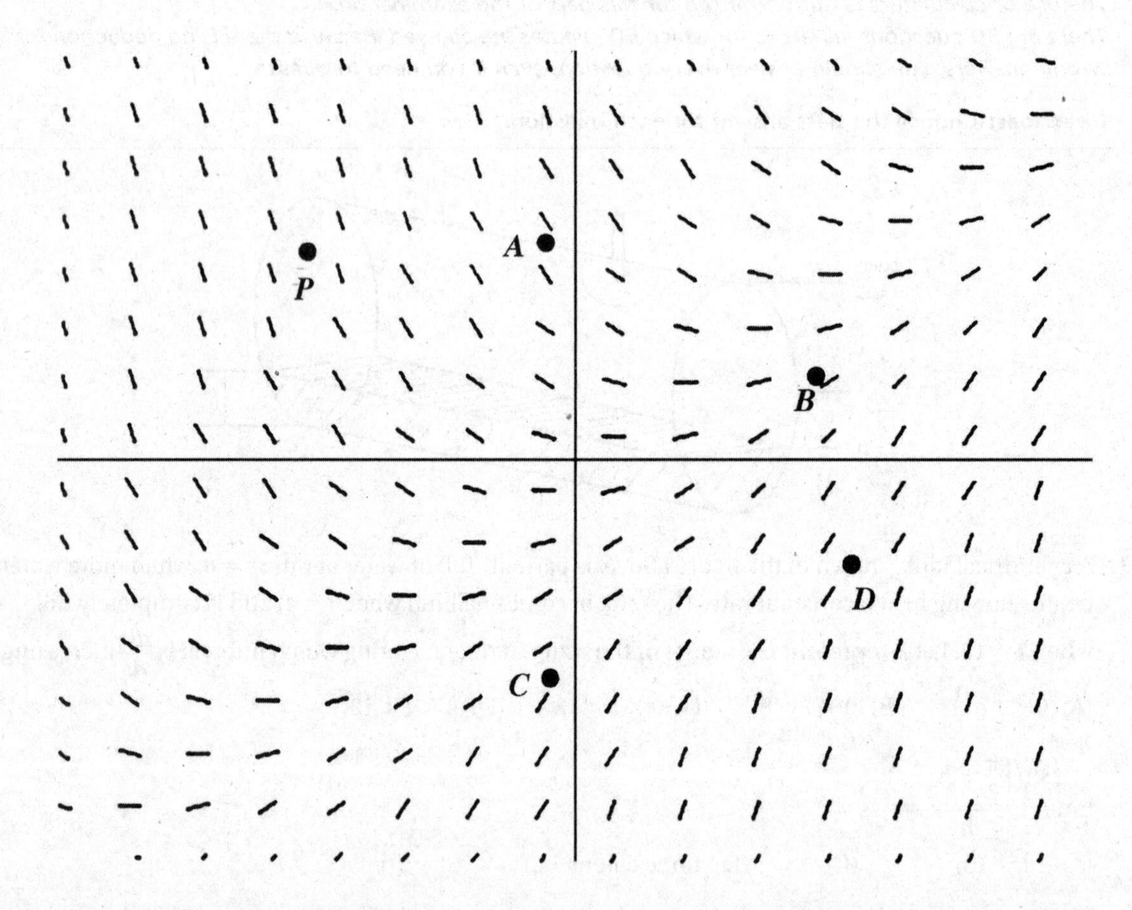

9. A particular solution of the differential equation whose slope field is shown above contains point P. This solution may also contain which other point?

(A) A (B) B (C) C (D) D

10. Let $F(x) = \int_5^x \frac{dt}{1-t^2}$. Which of the following statements is (are) true?

   I. The domain of F is $x \neq \pm 1$.
   II. $F(2) > 0$.
   III. The graph of F is concave upward.

   (A) I only (B) II only (C) III only (D) II and III only

11. As the tides change, the water level in a bay varies sinusoidally. At high tide today at 8 A.M., the water level was 15 feet; at low tide, 6 hours later at 2 P.M., it was 3 feet. How fast, in feet per hour, was the water level dropping at noon today?

   (A) $\frac{\pi}{2}$ (B) $\frac{\pi\sqrt{3}}{2}$ (C) $3\sqrt{3}$ (D) $6\sqrt{3}$

12. A smooth curve with equation $y = f(x)$ is such that its slope at each x equals x^2. If the curve goes through the point $(-1, 2)$, then its equation is

   (A) $y = x^3 + 3$ (B) $y = \frac{x^3}{3} + 7$ (C) $y = \frac{x^3+7}{3}$ (D) $y = 3x^3 + 5$

13. $\int \frac{e^u}{1+e^{2u}}\,du$ is equal to

   (A) $\ln(1 + e^{2u}) + C$ (B) $\frac{1}{2}\ln|1 + e^{2u}| + C$ (C) $\frac{1}{2}\tan^{-1}e^{2u} + C$ (D) $\tan^{-1}e^u + C$

14. Given $f(x) = \log_{10}x$ and $\log_{10}(102) \simeq 2.0086$, which is closest to $f'(100)$?

   (A) 0.0043 (B) 0.0086 (C) 0.01 (D) 1.0043

15. If $G(2) = 5$ and $G'(x) = \frac{10x}{9-x^2}$, then an estimate of $G(2.2)$ using a tangent-line approximation is

   (A) 4.4 (B) 5.8 (C) 6 (D) 11.6

16. The area bounded by the parabola $y = x^2$ and the lines $y = 1$ and $y = 9$ equals

   (A) $\frac{20}{3}$ (B) $\frac{52}{3}$ (C) 36 (D) $\frac{104}{3}$

17. Suppose $f(x) = \frac{x^2+x}{x}$ if $x \neq 0$ and $f(0) = 1$. Which of the following statements is (are) true of f?

   I. f is defined at $x = 0$.
   II. $\lim_{x\to 0} f(x)$ exists.
   III. f is continuous at $x = 0$.

   (A) I only (B) II only (C) I and II only (D) I, II, and III

18. Which function could have the graph shown below?

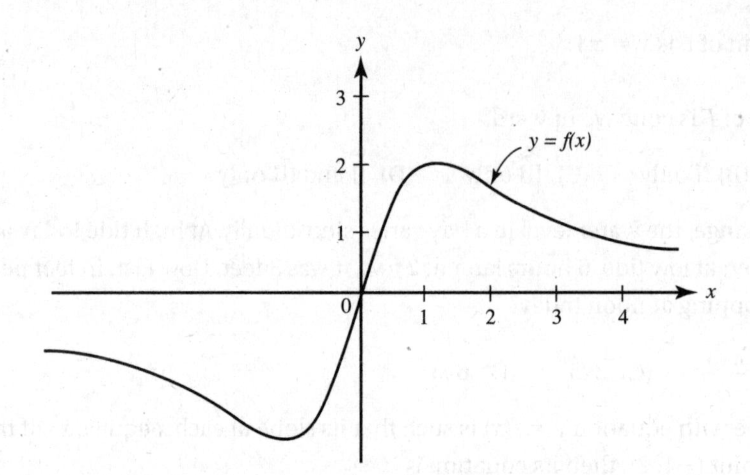

(A) $y = \dfrac{x}{x^2 + 1}$ (B) $y = \dfrac{4x}{x^2 + 1}$ (C) $y = \dfrac{x^2 + 3}{x^2 + 1}$ (D) $y = \dfrac{4x}{x + 1}$

19. Suppose the graph of f is both increasing and concave up on $a \leqslant x \leqslant b$. Then, using the same number of subdivisions, and with L, R, M, and T denoting, respectively, left, right, midpoint, and trapezoid sums, it follows that

(A) $R \leqslant T \leqslant M \leqslant L$ (B) $L \leqslant T \leqslant M \leqslant R$ (C) $R \leqslant M \leqslant T \leqslant L$ (D) $L \leqslant M \leqslant T \leqslant R$

20. $\lim\limits_{x \to 3} \dfrac{x + 3}{x^2 - 9}$ is

(A) $+\infty$ (B) 0 (C) $\dfrac{1}{6}$ (D) nonexistent

21. The only function that does not satisfy the Mean Value Theorem on the interval specified is

(A) $f(x) = x^2 - 2x$ on $[-3, 1]$

(B) $f(x) = \dfrac{1}{x}$ on $[1, 3]$

(C) $f(x) = \dfrac{x^3}{3} - \dfrac{x^2}{2} + x$ on $[-1, 2]$

(D) $f(x) = x + \dfrac{1}{x}$ on $[-1, 1]$

22. Suppose $f'(x) = x(x - 2)^2(x + 3)$. Which of the following is (are) true?

I. f has a local maximum at $x = -3$.
II. f has a local minimum at $x = 0$.
III. f has neither a local maximum nor a local minimum at $x = 2$.

(A) I only (B) II only (C) I and II only (D) I, II, and III

23. If $y = \ln \dfrac{x}{\sqrt{x^2 + 1}}$, then $\dfrac{dy}{dx}$ is

(A) $\dfrac{1}{x^2 + 1}$ (B) $\dfrac{1}{x(x^2 + 1)}$ (C) $\dfrac{2x^2 + 1}{x(x^2 + 1)}$ (D) $\dfrac{1}{x\sqrt{x^2 + 1}}$

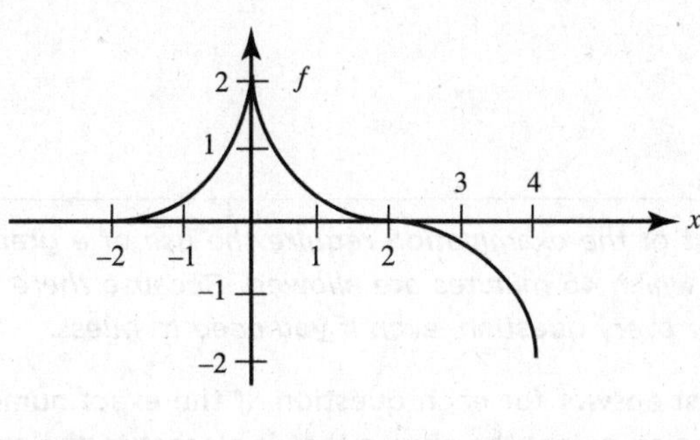

24. The graph of function f shown above consists of three quarter-circles. Which of the following is (are) equivalent to $\int_0^2 f(x)dx$?

I. $\frac{1}{2}\int_{-2}^{2} f(x)dx$

II. $\int_4^2 f(x)dx$

III. $\frac{1}{2}\int_0^4 f(x)dx$

(A) I only (B) II only (C) I and II only (D) I, II, and III

25. The base of a solid is the first-quadrant region bounded by $y = \sqrt[4]{4 - 2x}$, and each cross section perpendicular to the x-axis is a semicircle with a diameter in the xy-plane. The volume of the solid is

(A) $\frac{\pi}{2}\int_0^2 \sqrt{4 - 2x}\,dx$ (B) $\frac{\pi}{4}\int_0^2 \sqrt{4 - 2x}\,dx$ (C) $\frac{\pi}{8}\int_0^2 \sqrt{4 - 2x}\,dx$ (D) $\frac{\pi}{8}\int_{-2}^2 \sqrt{4 - 2x}\,dx$

26. The average value of $f(x) = 3 + |x|$ on the interval $[-2, 4]$ is

(A) $\frac{1}{3}$ (B) 2 (C) $\frac{14}{3}$ (D) 14

27. $\lim_{x\to\infty} \frac{3 + x - 2x^2}{4x^2 + 9}$ is

(A) $-\frac{1}{2}$ (B) $\frac{1}{2}$ (C) 1 (D) $\frac{3}{4}$

28. The area of the region in the xy-plane bounded by the curves $y = e^x$, $y = e^{-x}$, and $x = 1$ is equal to

(A) $e + \frac{1}{e} - 2$ (B) $e - \frac{1}{e}$ (C) $e + \frac{1}{e}$ (D) $2e - 2$

29. If $f(x) = \int_0^{x^2+2} \sqrt{1 + \cos t}\,dt$, then $f'(x) =$

(A) $2x\sqrt{1 + \cos(x^2 + 2)}$ (B) $2x\sqrt{1 - \sin x}$ (C) $\sqrt{1 + \cos(x^2 + 2)}$ (D) $\sqrt{[1 + \cos(x^2 + 2)]} \cdot 2x$

30. Choose the Riemann Sum whose limit is the integral: $\int_{-1}^{3} \ln(x^2 + 2)dx$.

(A) $\lim_{n\to\infty}\sum_{k=1}^{n}\left(\ln\left(\left(\frac{4k}{n} - 1\right)^2 + 2\right)\cdot\left(\frac{4}{n}\right)\right)$ (B) $\lim_{n\to\infty}\sum_{k=1}^{n}\left(\ln\left(\left(\frac{4k}{n} - 1\right)^2 + 2\right)\cdot\left(\frac{1}{n}\right)\right)$

(C) $\lim_{n\to\infty}\sum_{k=1}^{n}\left(\ln\left(\left(\frac{k}{n} - 1\right)^2 + 2\right)\cdot\left(\frac{1}{n}\right)\right)$ (D) $\lim_{n\to\infty}\sum_{k=1}^{n}\left(\ln\left(\left(\frac{k}{n} - 1\right)^2 + 2\right)\cdot\left(\frac{4}{n}\right)\right)$

STOP

END OF PART A, SECTION I

Part B

TIME: 45 MINUTES

Some questions in this part of the examination require the use of a graphing calculator. There are 15 questions in Part B, for which 45 minutes are allowed. Because there is no deduction for wrong answers, you should answer every question, even if you need to guess.

Directions: Choose the best answer for each question. If the exact numerical value of the correct answer is not listed as a choice, select the choice that is closest to the exact numerical answer.

31. A particle moves on a line according to the law $s = f(t)$ so that its velocity is $v = ks$, where k is a nonzero constant. The acceleration of the particle is

    (A) k^2v (B) k^2s (C) k (D) 0

32. A cup of coffee placed on a table cools at a rate of $\frac{dH}{dt} = -0.05(H - 70)$°F per minute, where H represents the temperature of the coffee and t is time in minutes. If the coffee was at 120°F initially, what will its temperature be to the nearest degree, 10 minutes later?

    (A) 73°F (B) 95°F (C) 100°F (D) 105°F

33. An investment of \$4000 grows at the rate of $320e^{0.08t}$ dollars per year after t years. Its value after 10 years is approximately

    (A) \$12,902 (B) \$8902 (C) \$7122 (D) \$4902

34. If $f(x) = (1 + e^x)$ then the domain of $f^{-1}(x)$ is

    (A) $(-\infty,\infty)$ (B) $(0,\infty)$ (C) $(1,\infty)$ (D) $(2,\infty)$

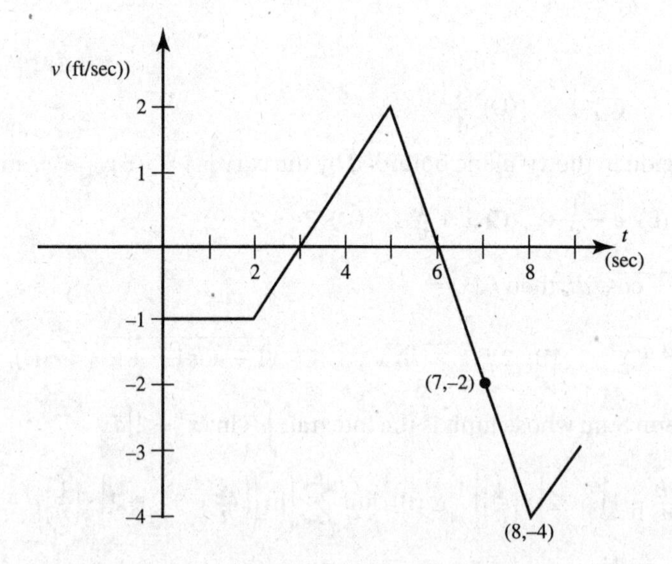

Questions 35 and 36. The graph shows the velocity of an object during the interval $0 \le t \le 9$.

35. The object attains its greatest speed at t =

    (A) 3 (B) 5 (C) 6 (D) 8

36. The object was at the origin at $t = 3$. It returned to the origin

(A) during $6 < t < 7$ (B) at $t = 7$ (C) during $7 < t < 8$ (D) at $t = 8$

37. When the region bounded by the y-axis, $y = e^x$, and $y = 2$ is rotated around the y-axis it forms a solid with volume

(A) 0.188 (B) 0.386 (C) 0.592 (D) 1.214

38. If $\sqrt{x-2}$ is replaced by u, then $\int_3^6 \frac{\sqrt{x-2}}{x}\,dx$ is equivalent to

(A) $\int_1^2 \frac{u\,du}{u^2+2}$ (B) $\int_1^2 \frac{2u^2}{u^2+2}\,du$ (C) $\int_3^6 \frac{2u^2}{u^2+2}\,du$ (D) $\int_3^6 \frac{u\,du}{u^2+2}$

39. The line tangent to the graph of function f at the point (8,1) intersects the y-axis at $y = 3$. Find $f'(8)$.

(A) $-\frac{1}{4}$ (B) 1 (C) $\frac{1}{4}$ (D) -4

40. How many points of inflection does the function f have on the interval $0 \leq x \leq 6$ if $f''(x) = 2 - 3\sqrt{x}\cos^3 x$?

(A) 1 (B) 2 (C) 3 (D) 4

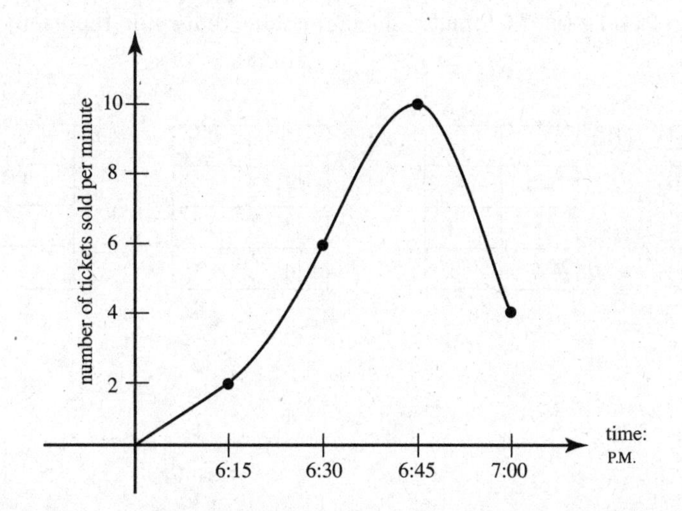

41. The graph shows the rate at which tickets were sold at a movie theater during the last hour before showtime. Using the right-rectangle method, estimate the size of the audience.

(A) 270 (B) 300 (C) 330 (D) 360

42. Find the x-coordinate where $f(x) = 4^{\sin x}$ and $g(x) = \ln\left(x^2\right)$ intersect and their derivatives have the same sign.

(A) -5.240 (B) -3.961 (C) -1.151 (D) 2.642

43. Which statement is true?

(A) If $f'(c) = 0$, then f has a local maximum or minimum at $\left(c, f(c)\right)$.
(B) If $f''(c) = 0$, then the graph of f has an inflection point at $\left(c, f(c)\right)$.
(C) If f is differentiable at $x = c$, then f is continuous at $x = c$.
(D) If f is continuous on (a,b), then f attains a maximum value on (a,b).

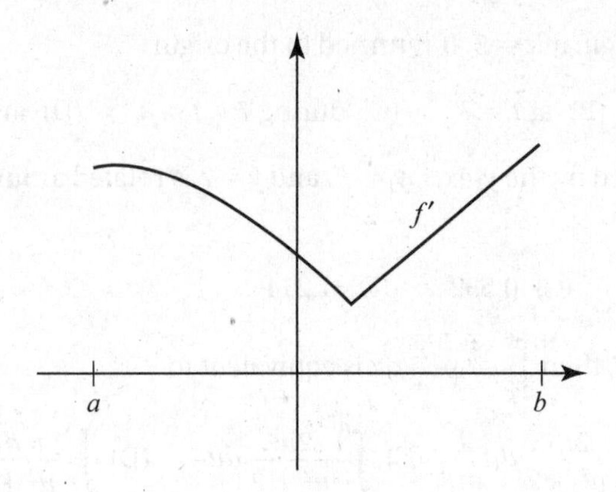

44. The graph of f' is shown above. Which statement(s) about f must be true for $a < x < b$?

I. f is increasing.
II. f is continuous.
III. f is differentiable.

(A) I only (B) II only (C) I and II only (D) I, II, and III

45. Given $\int_{-1}^{4} g(x)dx = 0$ and $g'(x) < 0$ for all x, select the table that could represent g on the interval $[-1,4]$.

(A)

x	y
−1	3
1	3
4	3

(B)

x	y
−1	7
1	−3
4	−1

(C)

x	y
−1	7
1	1
4	−3

(D)

x	y
−1	7
1	5
4	1

STOP

END OF SECTION I

SECTION II

Part A

TIME: 30 MINUTES

2 PROBLEMS

A graphing calculator is required for some of these problems. See instructions on page 4.

1. A curve is defined by $x^2y - 3y^2 = 48$.

   (a) Verify that $\frac{dy}{dx} = \frac{2xy}{6y - x^2}$.

   (b) Write an equation of the line tangent to this curve at (5,3).

   (c) Using your equation from part (a), estimate the y-coordinate of the point on the curve where $x = 4.93$.

   (d) Show that this curve has no horizontal tangent lines.

2. The table shows the depth of water, W, in a river, as measured at 4-hour intervals during a day-long flood. Assume that W is a differentiable function of time t.

t (hr)	0	4	8	12	16	20	24
$W(t)$ (ft)	32	36	38	37	35	33	32

   (a) Find the approximate value of $W'(16)$. Indicate units of measure.

   (b) Estimate the average depth of the water, in feet, over the time interval $0 \le t \le 24$ hours by using a trapezoidal approximation with subintervals of length $\Delta t = 4$ days.

   (c) Scientists studying the flooding believe they can model the depth of the water with the function $F(t) = 35 - 3\cos\left(\frac{t+3}{4}\right)$, where $F(t)$ represents the depth of the water, in feet, after t hours. Find $F'(16)$ and explain the meaning of your answer, with appropriate units, in terms of the river depth.

   (d) Use the function F to find the average depth of the water, in feet, over the time interval $0 \le t \le 24$ hours.

STOP

END OF PART A, SECTION II

AB PRACTICE EXAMINATION 3

Part B

TIME: 60 MINUTES

4 PROBLEMS

No calculator is allowed for any of these problems.

If you finish Part B before time has expired, you may return to work on Part A, but you may not use a calculator.

3. The region $\boldsymbol{R}$ is bounded by the curves $f(x) = \cos(\pi x) - 1$ and $g(x) = x(2 - x)$, as shown in the figure.

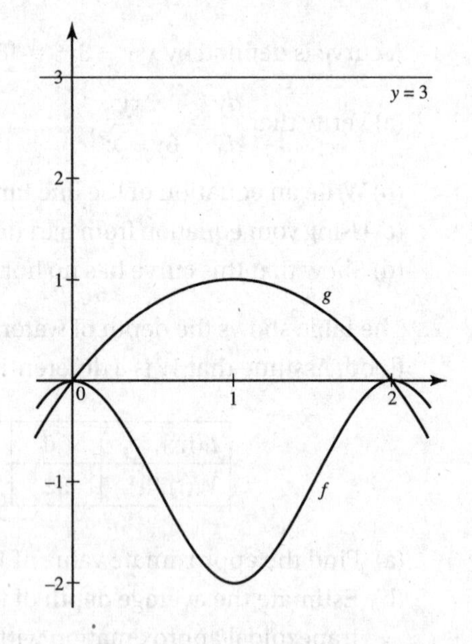

   (a) Find the area of $\boldsymbol{R}$.

   (b) A solid has base $\boldsymbol{R}$, and each cross section perpendicular to the x-axis is an isosceles right triangle whose hypotenuse lies in $\boldsymbol{R}$. Set up, but do not evaluate, an integral for the volume of this solid.

   (b) Set up, but do not evaluate, an integral for the volume of the solid formed when $\boldsymbol{R}$ is rotated around the line $y = 3$.

4. Two autos, P and Q, start from the same point and race along a straight road for 10 seconds. The velocity of P is given by $v_p(t) = 6\left(\sqrt{1 + 8t} - 1\right)$ feet per second. The velocity of Q is shown in the graph.

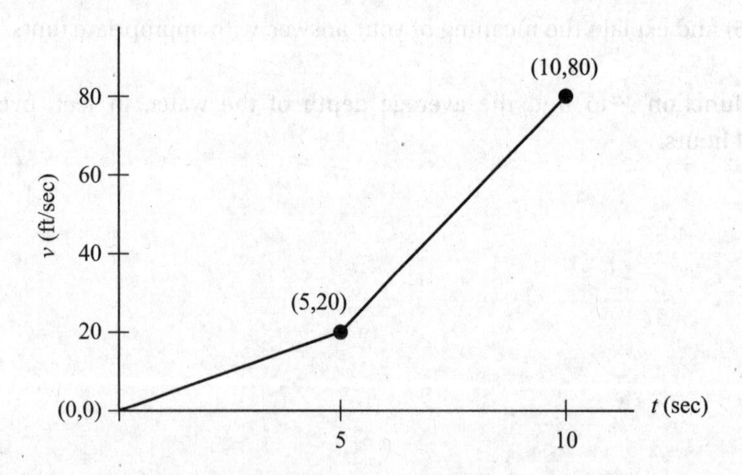

   (a) At what time is P's actual acceleration (in ft/sec^2) equal to its average acceleration for the entire race?

   (b) What is Q's acceleration (in ft/sec^2) then?

   (c) At the end of the race, which auto was ahead? Explain.

5. Given the differential equation $\frac{dy}{dx} = 2x\left(y^2 + 1\right)$

(a) Sketch the slope field for this differential equation at the points shown in the figure.

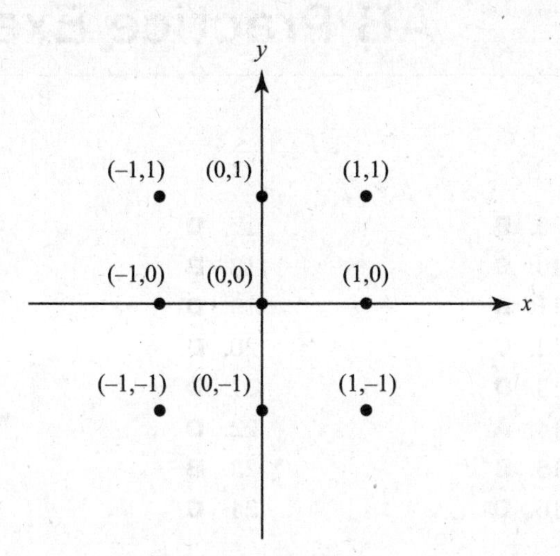

(b) Let f be the particular solution to the differential equation whose graph passes through (0,1). Express f as a function of x, and state its domain.

6. The graph shown is for $F(x) = \int_0^x f(t)dt$.

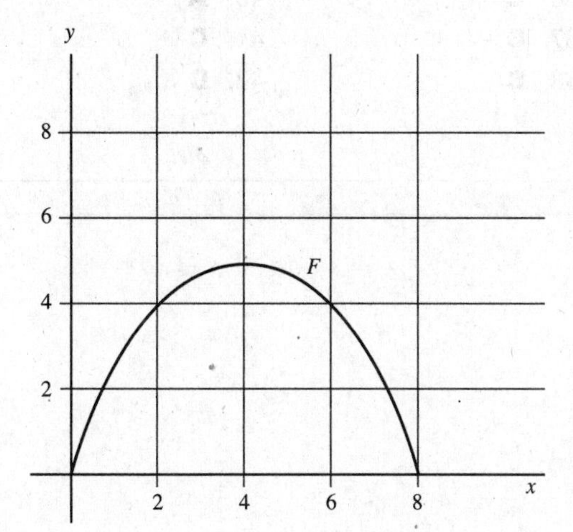

(a) What is $\int_0^2 f(t)dt$?

(b) What is $\int_2^7 f(t)dt$?

(c) At what value of x does $f(x) = 0$?

(d) Over what interval is $f'(x)$ negative?

(e) Let $G(x) = \int_2^x f(t)dt$. Sketch the graph of G on the same axes.

END OF TEST

ANSWER KEY
AB Practice Examination 3

Part A

1. **D**	9. **B**	17. **D**	25. **C**
2. **C**	10. **D**	18. **B**	26. **C**
3. **C**	11. **B**	19. **D**	27. **A**
4. **A**	12. **C**	20. **D**	28. **A**
5. **C**	13. **D**	21. **D**	29. **A**
6. **C**	14. **A**	22. **D**	30. **A**
7. **B**	15. **B**	23. **B**	
8. **D**	16. **D**	24. **C**	

Part B

31. **B**	35. **D**	39. **A**	43. **C**
32. **C**	36. **C**	40. **A**	44. **D**
33. **B**	37. **C**	41. **C**	45. **C**
34. **C**	38. **B**	42. **B**	

ANSWERS EXPLAINED

Section I Multiple-Choice

Part A

1. **(D)** $\frac{dh}{dt}$ will increase above the half-full level (that is, the height of the water will rise more rapidly) as the area of the surface of the water diminishes.

2. **(C)** The given limit equals $f'\left(\frac{\pi}{2}\right)$, where $f(x) = \sin x$.

3. **(C)** Since $f(x) = x \ln x$,

$$f'(x) = 1 + \ln x, \quad f''(x) = \frac{1}{x}, \quad \text{and} \quad f'''(x) = -\frac{1}{x^2}.$$

4. **(A)** Differentiate implicitly to get $4x - 4y^3 \frac{dy}{dx} = 0$. Substitute $(-1, 1)$ to find $\frac{dy}{dx} = -1$, the slope at this point, and write the equation of the tangent: $y - 1 = -1(x + 1)$.

5. **(C)** $f'(x) = 4x^3 - 12x^2 + 8x = 4x(x - 1)(x - 2)$. To determine the signs of $f'(x)$, inspect the sign at any point in each of the intervals $x < 0$, $0 < x < 1$, $1 < x < 2$, and $x > 2$. The function increases whenever $f'(x) > 0$.

6. **(C)** The integral is equivalent to $\frac{1}{2}\int \frac{2 \cos x\, dx}{4 + 2 \sin x} = \frac{1}{2}\int \frac{du}{u}$, where $u = 4 + 2\sin x$.

7. **(B)** Here $y' = \frac{1 - \ln x}{x^2}$, which is zero for $x = e$. Since the sign of y' changes from positive to negative as x increases through e, this critical value yields a local maximum. Note that $f(e) = \frac{1}{e}$.

8. **(D)** Since $v = \frac{ds}{dt} = 5t^4 + 6t^2 = t^2(5t^2 + 6)$ is always positive, there are no reversals in motion along the line.

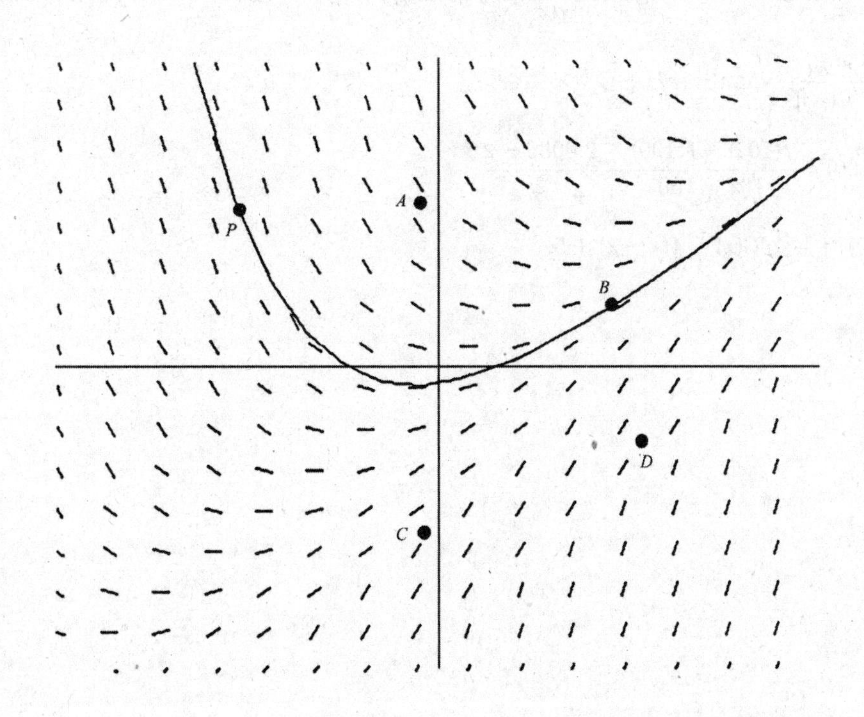

9. **(B)** The slope field suggests the curve shown above as a particular solution.

10. **(D)** Since $f(x) = F'(x) = \frac{1}{1-x^2}$, f is discontinuous at $x = 1$; the domain of F is therefore $x > 1$. On [2, 5] $f(x) < 0$, so $\int_5^2 f > 0$. $F''(x) = f'(x) = \frac{2x}{(1-x^2)^2}$, which is positive for $x > 1$.

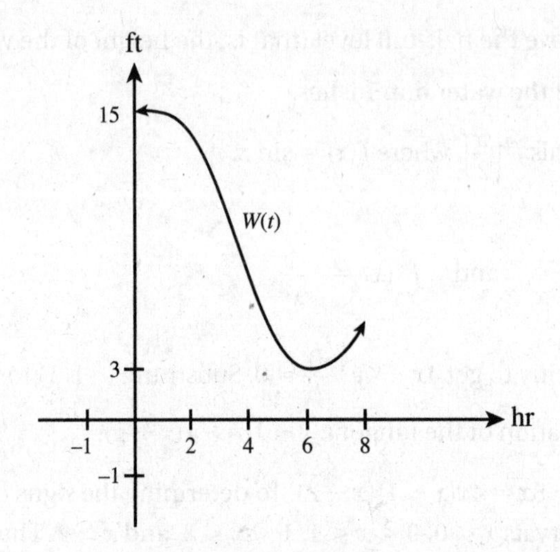

11. **(B)** In the graph above, $W(t)$, the water level at time t, is a cosine function with amplitude 6 ft and period 12 hr:

$$W(t) = 6\cos\left(\frac{\pi}{6}t\right) + 9 \text{ ft},$$

$$W'(t) = -\pi\sin\left(\frac{\pi}{6}t\right) \text{ ft/hr}.$$

Hence, $W'(4) = -\pi\sin\left(\frac{2\pi}{3}\right) = -\frac{\pi\sqrt{3}}{2}$ ft/hr.

12. **(C)** Solve the differential equation $\frac{dy}{dx} = x^2$, getting $y = \frac{x^3}{3} + C$. Use $x = -1$, $y = 2$ to determine $C = \frac{7}{3}$.

13. **(D)** $\int \frac{e^u du}{1 + (e^u)^2} = \tan^{-1}(e^u) + C$

14. **(A)** $f'(100) \simeq \frac{f(102) - f(100)}{102 - 100} = \frac{2.0086 - 2}{2}$.

15. **(B)** $G'(2) = 4$, so $G(x) \simeq 4(x - 2) + 5$.

16. **(D)** $A = 2\int_1^9 x\,dy = 2\int_1^9 \sqrt{y}\,dy = \frac{104}{3}$. See the figure below.

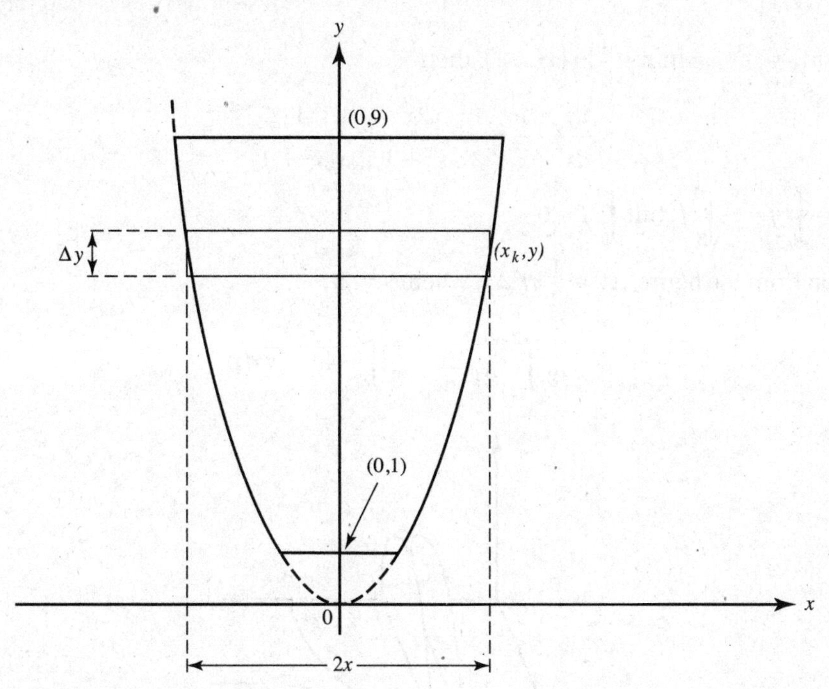

17. **(D)** Note that $\lim_{x\to 0} f(x) = f(0) = 1$.

18. **(B)** Note that (0, 0) is on the graph, as are (1, 2) and (−1, −2). So only (B) and (D) are possible. Since $\lim_{x\to\infty} y = \lim_{x\to -\infty} y = 0$, only (B) is correct.

19. **(D)** See the figure.

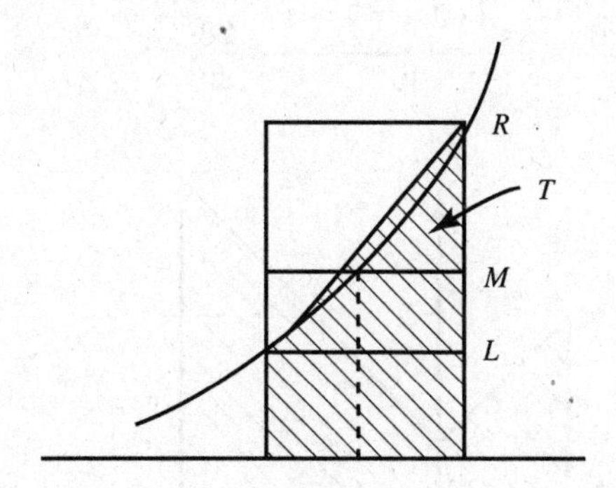

20. **(D)** $\frac{x+3}{x^2-9} = \frac{x+3}{(x+3)(x-3)} = \frac{1}{x-3}$; $\lim_{x\to 3^+} \frac{1}{x-3} = +\infty$; $\lim_{x\to 3^-} \frac{1}{x-3} = -\infty$.

21. **(D)** In (D), $f(x)$ is not defined at $x = 0$. Verify that each of the other functions satisfies both conditions of the Mean Value Theorem.

22. **(D)** The signs within the intervals bounded by the critical points are given below.

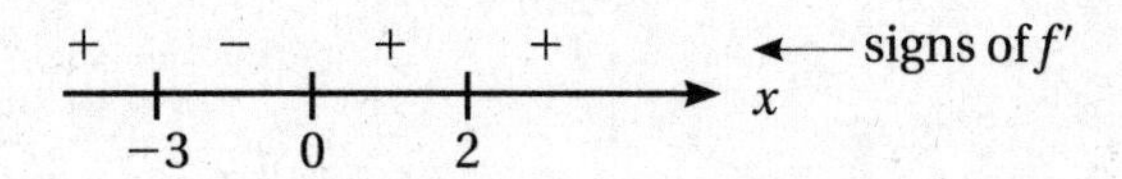

Since f changes from increasing to decreasing at $x = -3$, f has a local maximum at -3. Also, f has a local minimum at $x = 0$, because it is decreasing to the left of zero and increasing to the right.

23. **(B)** Since $\ln \dfrac{x}{\sqrt{x^2+1}} = \ln x - \dfrac{1}{2}\ln(x^2+1)$, then

$$\frac{dy}{dx} = \frac{1}{x} - \frac{1}{2} \cdot \frac{2x}{x^2+1} = \frac{1}{x(x^2+1)}.$$

24. **(C)** $\displaystyle\int_{-2}^{0} f = \int_{0}^{2} f = -\int_{2}^{4} f$, but $\displaystyle\int_{0}^{4} f = 0.$

25. **(C)** As seen from the figure, $\Delta V = \dfrac{1}{2}\pi r^2 \Delta x$, where $y = 2r$,

$$V = \frac{\pi}{2}\int_0^2 \left(\frac{y}{2}\right)^2 dx = \frac{\pi}{8}\int_0^2 \sqrt{4-2x}\, dx.$$

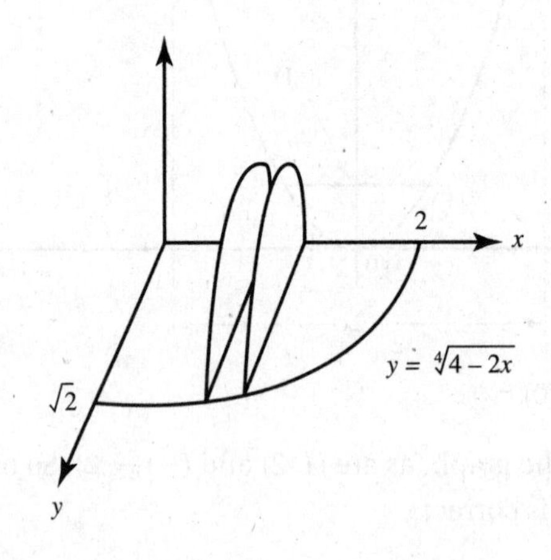

26. **(C)** From the figure below, $\dfrac{\int_{-2}^{4} f}{6} = \dfrac{\frac{5+3}{2}\cdot 2 + \frac{3+7}{2}\cdot 4}{6} = \dfrac{28}{6}.$

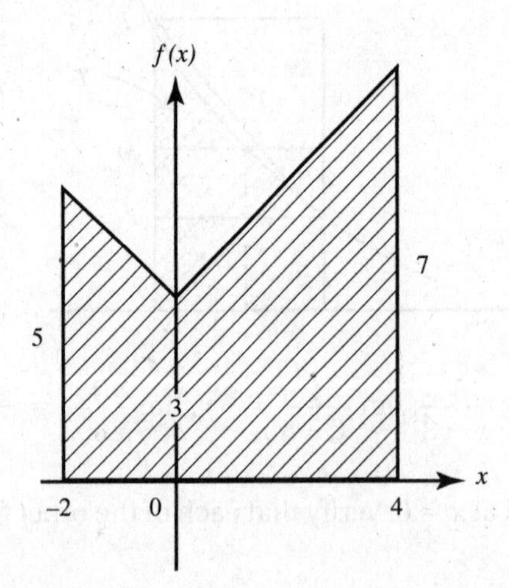

27. **(A)** Since the degrees of numerator and denominator are the same, the limit as $x \to \infty$ is the ratio of the coefficients of the terms of highest degree: $\frac{-2}{4}$.

28. **(A)** We see from the figure that $\Delta A = (y_2 - y_1)\Delta x$;

$$A = \int_0^1 (e^x - e^{-x})\,dx$$
$$= (e^x + e^{-x})\Big|_0^1 = e + \frac{1}{e} - 2.$$

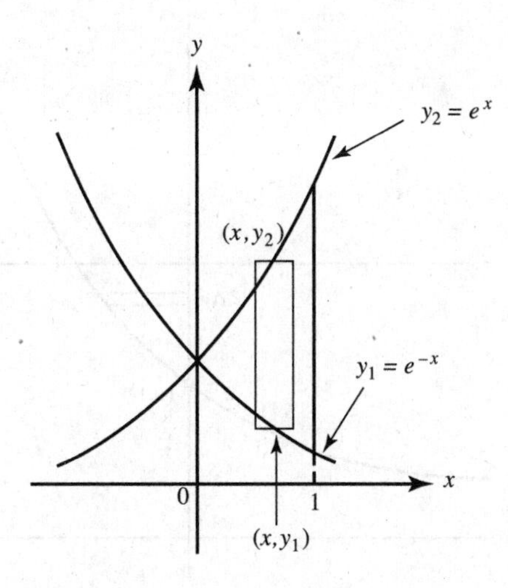

29. **(A)** Let $u = x^2 + 2$. Then

$$\frac{d}{du}\int_0^u \sqrt{1 + \cos t}\,dt = \sqrt{1 + \cos u}$$

and

$$\frac{d}{dx}\int_0^u \sqrt{1 + \cos t}\,dt = \sqrt{1 + \cos u}\,\frac{du}{dx} = \sqrt{1 + \cos\left(x^2 + 2\right)} \cdot (2x).$$

30. **(A)** From the integral we get $a = -1$, $b = 3$, so $\Delta x = \frac{3 - (-1)}{n} = \frac{4}{n}$ and $x_k = a + k \cdot \Delta x = -1 + \frac{4k}{n} = \frac{4k}{n} - 1$. Replace x with x_k and replace dx with Δx in the integrand to get the general term in the summation.

Part B

31. **(B)** Since $v = ks = \frac{ds}{dt}$, then $a = \frac{d^2s}{dt^2} = k\frac{ds}{dt} = kv = k^2s$.

32. **(C)** $\frac{dH}{H - 70} = -0.05\,dt$. $\quad \ln|H - 70| = -0.05t + C$

$$H - 70 = ce^{-0.05t}$$
$$H(x) = 70 + ce^{-0.05t}$$

The initial condition $H(0) = 120$ shows $c = 50$. Evaluate $H(10)$.

33. **(B)** Let P be the amount after t years. It is given that $P'(t) = 320\,e^{0.08t}$,

so $P(10) = P(0) + \int_0^{10} P'(t)\,dt = 8902.16$.

34. **(C)** The inverse of $y = 1 + e^x$ is $x = 1 + e^y$ or $y = \ln(x - 1)$; $(x - 1)$ must be positive.

35. **(D)** Speed is the magnitude of velocity: $|v(8)| = 4$.

36. **(C)** For $3 < t < 6$ the object travels to the right $\frac{1}{2}(3)(2) = 3$ units. At $t = 7$ it has returned 1 unit to the left; by $t = 8$, 4 units to the left.

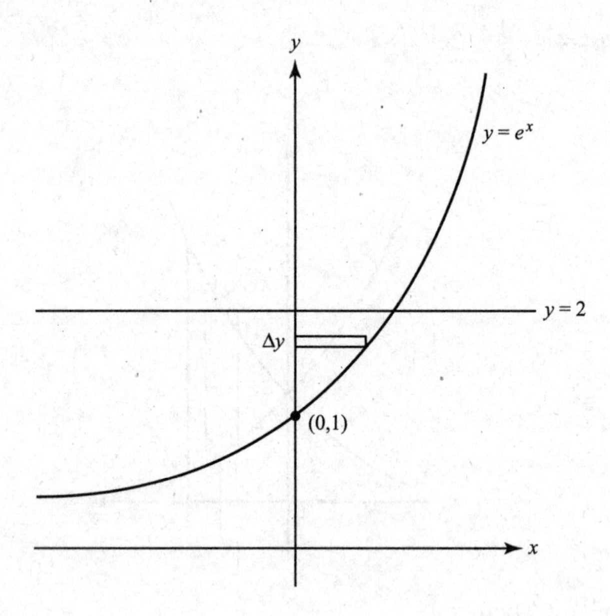

37. **(C)** Use disks: $\Delta V = \pi r^2 \Delta y = \pi x^2 \Delta y$, where $x = \ln y$. Use your calculator to evaluate $V = \pi \int_1^2 (\ln y)^2\, dy$.

38. **(B)** If $u = \sqrt{x - 2}$, then $u^2 = x - 2$, $x = u^2 + 2$, $dx = 2u\, du$. When $x = 3$, $u = 1$; when $x = 6$, $u = 2$.

39. **(A)** The tangent line passes through points (8,1) and (0,3). Its slope, $\frac{1 - 3}{8 - 0}$, is $f'(8)$.

40. **(A)** Graph f'' in $[0,6] \times [-5,10]$. The sign of f'' changes only at $x = a$, as seen in the figure.

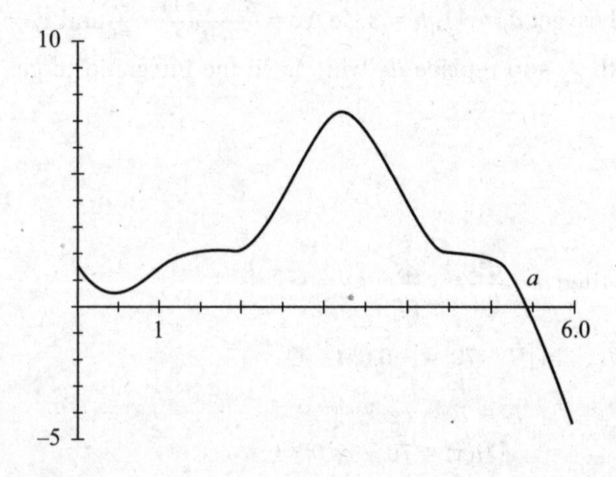

41. **(C)** In the graph below, the first rectangle shows 2 tickets sold per minute for 15 min, or 30 tickets. Similarly, the total is $2(15) + 6(15) + 10(15) + 4(15)$.

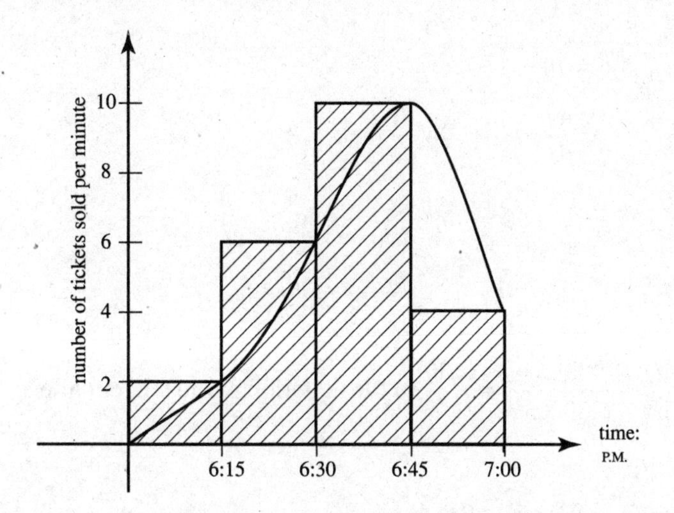

42. **(B)** Graph both functions in $[-8,8] \times [-5,5]$. The curves intersect at P, Q, R, and S (at T the curves are close but do not intersect). At point of intersection Q, both are decreasing. Use a graphing calculator to find that the intersection point at Q reveals an x-coordinate of -3.961.

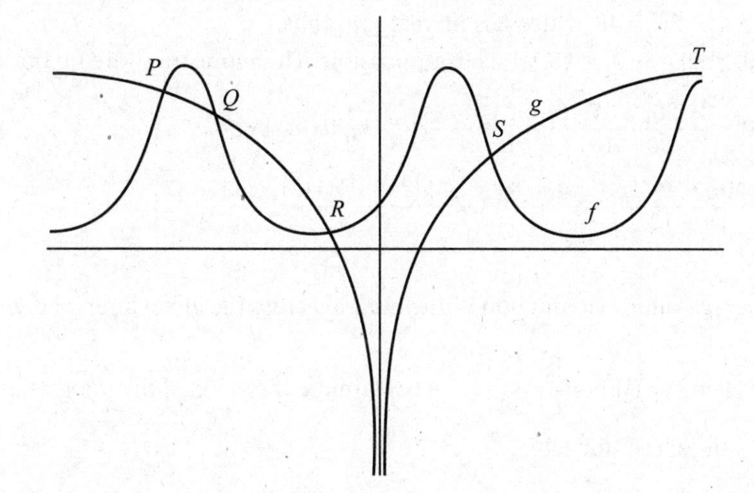

43. **(C)** Counterexamples are, respectively, for (A), $f(x) = x^3$, $c = 0$; for (B), $f(x) = x^4$, $c = 0$; for (D), $f(x) = x^2$ on $(-1, 1)$.

44. **(D)** $f'(x) > 0$; the curve shows that f' is defined for all $a < x < b$, so f is differentiable and therefore continuous.

45. **(C)** Since $g'(x) < 0$ for all x, the table must have decreasing values as x increases, and $\int_{-1}^{4} g(x)\,dx = 0$, so there must be at least one value above the x-axis and one value below the x-axis.

Section II Free-Response

Part A

AB 1. (a) Since $x^2y - 3y^2 = 48$,

$$x^2\frac{dy}{dx} + 2xy - 6y\frac{dy}{dx} = 0,$$

$$(x^2 - 6y)\frac{dy}{dx} = -2xy,$$

$$\frac{dy}{dx} = \frac{2xy}{6y - x^2}.$$

(b) At (5,3), $\frac{dy}{dx} = \frac{2(5)(3)}{6(3) - 5^2}, = -\frac{30}{7}$ so the equation of the tangent line is

$$y - 3 = -\frac{30}{7}(x - 5).$$

(c) $y - 3 = -\frac{30}{7}(4.93 - 5) = 0.3$, so $y = 3.3$.

(d) Horizontal tangent lines have $\frac{dy}{dx} = \frac{2xy}{6y - x^2} = 0$. This could happen only if

$2xy = 0$, which means that $x = 0$ or $y = 0$.
If $x = 0$, $0y - 3y^2 = 48$, which has no real solutions.
If $y = 0$, $x^2 \cdot 0 - 3 \cdot 0^2 = 48$, which is impossible. Therefore, there are no horizontal tangents.

AB/BC 2. (a) $W'(16) \approx \frac{W(20) - W(16)}{20 - 16} = \left(\frac{33 - 35}{4}\right) = -\frac{1}{2}$ ft/hr

$\left(\text{OR } \frac{W(16) - W(12)}{16 - 12} = \frac{35 - 37}{4} \text{ OR } \frac{W(20) - W(12)}{20 - 12} = \frac{33 - 37}{8}\right).$

(b) The average value of a function is the integral across the given interval divided by the interval width. Here Avg $(W) = \frac{\int_0^{24} W(t)dt}{24} - 0$. Estimate the value of the integral using trapezoid rule T with values from the table:

$$T = \left(\frac{32 + 36}{2}\right) \cdot 4 + \left(\frac{36 + 38}{2}\right) \cdot 4 + \left(\frac{38 + 37}{2}\right) \cdot 4 + \left(\frac{37 + 35}{2}\right) \cdot 4 + \left(\frac{35 + 33}{2}\right) \cdot 4 + \left(\frac{33 + 32}{2}\right) \cdot 4$$

$= 844$

Hence

$$\text{Avg } (W) \approx \frac{844}{24} = 35.167 \text{ ft.}$$

(c) For $F(t) = 35 - 3\cos\left(\frac{t + 3}{4}\right)$, use your calculator to evaluate

$F'(16) \approx -0.749$. After 16 hr, the river depth is dropping at the rate of 0.749 ft/hr.

(d) Avg $(F) = \frac{\int_0^{24} F(t)\,dt}{24 - 0} \approx 35.116$ ft.

Part B

AB/BC 3. (a) $\Delta A = \left(g(x) - f(x)\right)\Delta x$, so:

$$A = \int_0^2 \left(g(x) - f(x)\right)dx = \int_0^2 (x(2-x) - (\cos \pi x - 1)dx$$

$$= \int_0^2 \left(2x - x^2 - \cos \pi x + 1\right) dx = \left(x^2 - \frac{x^3}{3} - \frac{1}{\pi}\sin \pi x + x\right)\Bigg|_0^2$$

$$= \left(2^2 - \frac{2^3}{3} - \frac{1}{\pi}\sin 2\pi + 2\right) - \left(0^2 - \frac{0^3}{3} - \frac{1}{\pi}\sin 0 + 0\right)$$

$$= \frac{10}{3}.$$

(b) Let h = the hypotenuse of an isosceles right triangle, as shown in the figure. Then each leg of the triangle is $\frac{h}{\sqrt{2}}$ and

its area is $\frac{1}{2}\cdot\frac{h}{\sqrt{2}}\cdot\frac{h}{\sqrt{2}} = \frac{h^2}{4}$.

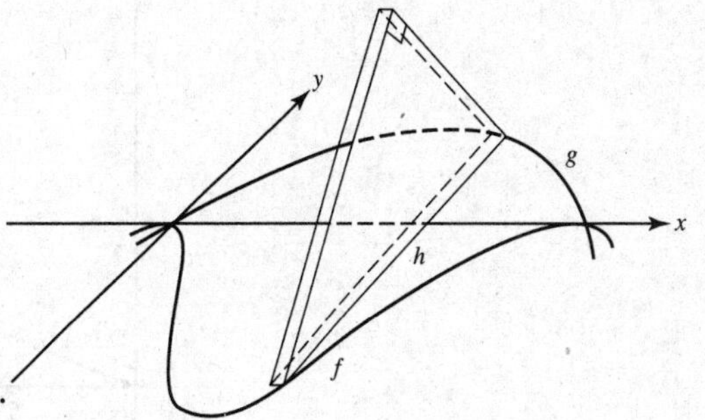

An element of volume is

$$\Delta V = \frac{h^2}{4}\Delta x = \frac{\left(g(x) - f(x)\right)^2}{4}\Delta x,$$

and thus $V = \frac{1}{4}\int_0^2 (x(2-x) - (\cos \pi x - 1))^2\, dx.$

(c) Washers; $\Delta V = \pi\left(r_1^2 - r_2^2\right)\Delta x$ where:

$$r_1 = 3 - f(x) = 3 - (\cos \pi x - 1),$$
$$r_2 = 3 - g(x) = 3 - x(2 - 1).$$

So $V = \pi\int_0^2 \left[(3 - (\cos \pi x - 1))^2 - (3 - x(2-x))^2\right] dx.$

AB/BC 4. (a) $v_p(t) = 6\left(\sqrt{1+8t} - 1\right)$, so $v(0) = 0$ and $v(10) = 48$.

The average acceleration is $\frac{\Delta v}{\Delta t} = \frac{48-0}{10-0} = \frac{24}{5}$ ft/sec^2.

Acceleration $a(t) = v'(t) = 6\cdot\frac{1}{2}(1+8t)^{-\frac{1}{2}}(8)$ft/sec^2.

$$\frac{24}{\sqrt{1+8t}} = \frac{24}{5} \text{ when } t = 3 \text{ sec.}$$

(b) Since Q's acceleration, for all t in $0 \le t \le 5$, is the slope of its velocity graph,

$a = \frac{20-0}{5-0} = 4$ft/sec^2.

(c) Find the distance each auto has traveled. For P, the distance is

$$\int_0^{10} 6\left(\sqrt{1+8t} - 1\right) =$$

$$6\left[\frac{1}{8}\int_0^{10}\sqrt{1+8t}\cdot 8\,dt - \int_0^{10} dt\right] =$$

$$6\left(\frac{1}{8}\cdot\frac{2}{3}(1+8t)^{\frac{3}{2}}-t\right)\Bigg|_0^{10}=$$

$$6\left(\frac{1}{12}\left(81^{\frac{3}{2}}-1^{\frac{3}{2}}\right)-10\right)=304 \text{ ft.}$$

For auto Q, the distance is the total area of the triangle and trapezoid under the velocity graph shown below, namely,

$$\frac{1}{2}(5)(20)+\frac{1}{2}(20+80)(5)=300 \text{ ft.}$$

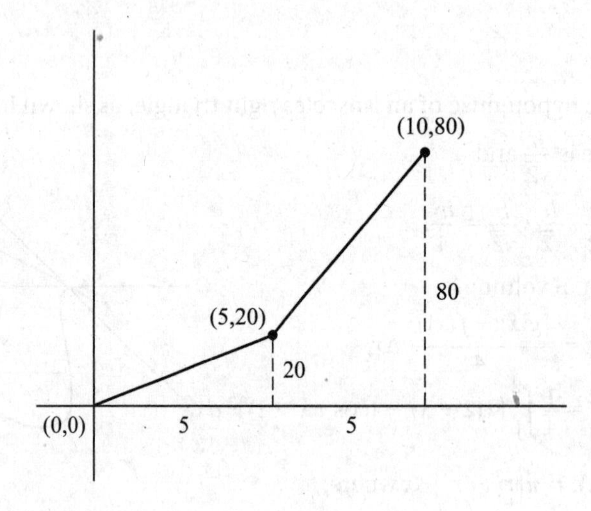

Auto P won the race.

AB 5. (a) Using the differential equation, evaluate the derivative at each point, then sketch a short segment having that slope. For example, at $(-1,-1)$, $\frac{dy}{dx}=2(-1)\left((-1)^2+1\right)-4$; draw a steeply decreasing segment at $(-1,-1)$.

Repeat this process at each of the other points. The result follows.

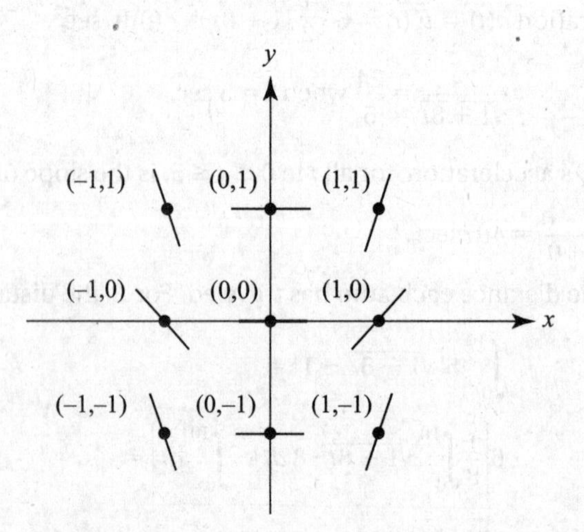

(b) The differential equation $\frac{dy}{dx} = 2x(y^2 + 1)$ is separable.

$$\int \frac{dy}{y^2+1} = \int 2x\,dx$$

$$\arctan(y) = x^2 + c$$

$$y = \tan(x^2 + c)$$

It is given that f passes through (0,1), so $1 = \tan(0^2 + c)$ and $c = \frac{\pi}{4}$.

The solution is $f(x) = \tan\left(x^2 + \frac{\pi}{4}\right)$.

The particular solution must be differentiable on an interval containing the initial point (0,1). The tangent function has vertical asymptotes at $x = \pm\frac{\pi}{2}$, hence:

$-\frac{\pi}{2} < x^2 + \frac{\pi}{4} < \frac{\pi}{2}$. (Since $x^2 \geq 0$, we ignore the left inequality.)

$$x^2 < \frac{\pi}{4}$$

$$|x| < \frac{\sqrt{\pi}}{2}$$

AB 6. (a) $\int_0^2 f(t)\,dt = F(2) = 4$.

(b) One estimate might be $\int_2^7 f(t)\,dt = F(7) - F(2) = 2 - 4 = -2$.

(c) $f(x) = F'(x)$; $F'(x) = 0$ at $x = 4$.

(d) $f'(x) = F''(x)$. F'' is negative when F is concave downward, which is true for the entire interval $0 < x < 8$.

(e) $$G(x) = \int_2^x f(t)dt$$

$$= \int_0^x f(t)dt - \int_0^2 f(t)dt$$

$$= F(x) - 4$$

Then the graph of G is the graph of F translated downward 4 units.

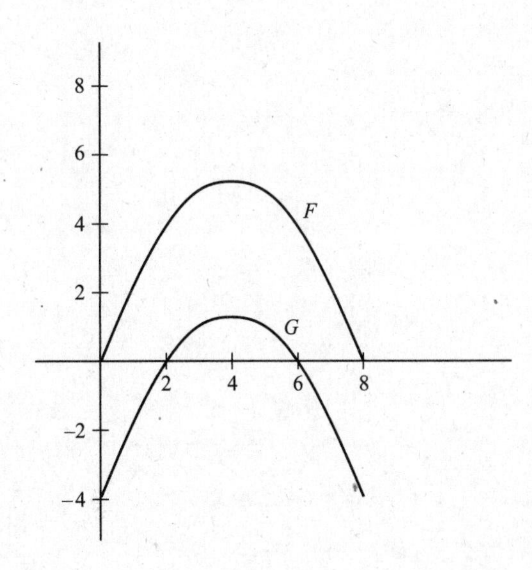

BC Practice Examinations

ANSWER SHEET
BC Practice Examination 1

Part A

1. Ⓐ Ⓑ Ⓒ Ⓓ
2. Ⓐ Ⓑ Ⓒ Ⓓ
3. Ⓐ Ⓑ Ⓒ Ⓓ
4. Ⓐ Ⓑ Ⓒ Ⓓ
5. Ⓐ Ⓑ Ⓒ Ⓓ
6. Ⓐ Ⓑ Ⓒ Ⓓ
7. Ⓐ Ⓑ Ⓒ Ⓓ
8. Ⓐ Ⓑ Ⓒ Ⓓ
9. Ⓐ Ⓑ Ⓒ Ⓓ
10. Ⓐ Ⓑ Ⓒ Ⓓ
11. Ⓐ Ⓑ Ⓒ Ⓓ
12. Ⓐ Ⓑ Ⓒ Ⓓ
13. Ⓐ Ⓑ Ⓒ Ⓓ
14. Ⓐ Ⓑ Ⓒ Ⓓ
15. Ⓐ Ⓑ Ⓒ Ⓓ
16. Ⓐ Ⓑ Ⓒ Ⓓ
17. Ⓐ Ⓑ Ⓒ Ⓓ
18. Ⓐ Ⓑ Ⓒ Ⓓ
19. Ⓐ Ⓑ Ⓒ Ⓓ
20. Ⓐ Ⓑ Ⓒ Ⓓ
21. Ⓐ Ⓑ Ⓒ Ⓓ
22. Ⓐ Ⓑ Ⓒ Ⓓ
23. Ⓐ Ⓑ Ⓒ Ⓓ
24. Ⓐ Ⓑ Ⓒ Ⓓ
25. Ⓐ Ⓑ Ⓒ Ⓓ
26. Ⓐ Ⓑ Ⓒ Ⓓ
27. Ⓐ Ⓑ Ⓒ Ⓓ
28. Ⓐ Ⓑ Ⓒ Ⓓ
29. Ⓐ Ⓑ Ⓒ Ⓓ
30. Ⓐ Ⓑ Ⓒ Ⓓ

Part B

31. Ⓐ Ⓑ Ⓒ Ⓓ
32. Ⓐ Ⓑ Ⓒ Ⓓ
33. Ⓐ Ⓑ Ⓒ Ⓓ
34. Ⓐ Ⓑ Ⓒ Ⓓ
35. Ⓐ Ⓑ Ⓒ Ⓓ
36. Ⓐ Ⓑ Ⓒ Ⓓ
37. Ⓐ Ⓑ Ⓒ Ⓓ
38. Ⓐ Ⓑ Ⓒ Ⓓ
39. Ⓐ Ⓑ Ⓒ Ⓓ
40. Ⓐ Ⓑ Ⓒ Ⓓ
41. Ⓐ Ⓑ Ⓒ Ⓓ
42. Ⓐ Ⓑ Ⓒ Ⓓ
43. Ⓐ Ⓑ Ⓒ Ⓓ
44. Ⓐ Ⓑ Ⓒ Ⓓ
45. Ⓐ Ⓑ Ⓒ Ⓓ

BC Practice Examination 1

SECTION I

Part A

TIME: 60 MINUTES

The use of calculators is not permitted for this part of the examination.

There are 30 questions in Part A, for which 60 minutes are allowed. Because there is no deduction for wrong answers, you should answer every question, even if you need to guess.

Directions: Choose the best answer for each question.

1. $\lim_{x\to\infty} \frac{20x^2 - 13x + 5}{5 - 4x^3}$ is

   (A) -5 (B) 0 (C) 1 (D) ∞

2. $\lim_{h\to 0} \frac{\ln(2+h) - \ln 2}{h}$ is

   (A) $\ln 2$ (B) $\frac{1}{2}$ (C) $\frac{1}{\ln 2}$ (D) ∞

3. If $x = \sqrt{1-t^2}$ and $y = \sin^{-1} t$, then $\frac{dy}{dx}$ equals

   (A) $-\frac{\sqrt{1-t^2}}{t}$ (B) $-t$ (C) 2 (D) $-\frac{1}{t}$

Questions 4 and 5. Use the following table, which shows the values of the differentiable functions *f* and *g*.

x	f	f'	g	g'
1	2	$\frac{1}{2}$	-3	5
2	3	1	0	4
3	4	2	2	3
4	6	4	3	$\frac{1}{2}$

4. The average rate of change of function f on [1,4] is

   (A) 7/6 (B) 4/3 (C) 15/8 (D) 15/4

5. If $h(x) = g(f(x))$ then $h'(3) =$

   (A) $\frac{1}{2}$ (B) 1 (C) 4 (D) 6

GO ON TO THE NEXT PAGE

6. $\int_1^2 (3x-2)^3\,dx$ is equal to

(A) $\frac{64}{3}$ (B) $\frac{63}{4}$ (C) $\frac{85}{4}$ (D) $\frac{255}{4}$

7. If $y = \frac{x-3}{2-5x}$, then $\frac{dy}{dx}$ equals

(A) $\frac{17-10x}{(2-5x)^2}$ (B) $\frac{13}{(2-5x)^2}$ (C) $\frac{-13}{(2-5x)^2}$ (D) $\frac{17}{(2-5x)^2}$

8. The maximum value of the function $f(x) = xe^{-x}$ is

(A) $\frac{1}{e}$ (B) 1 (C) -1 (D) $-e$

9. Which equation has the slope field shown below?

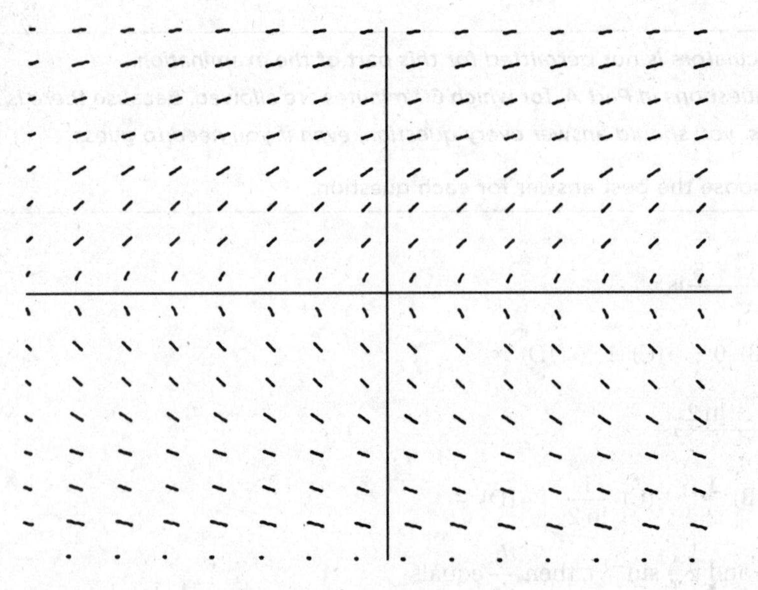

(A) $\frac{dy}{dx} = \frac{5}{y}$ (B) $\frac{dy}{dx} = \frac{5}{x}$ (C) $\frac{dy}{dx} = \frac{x}{y}$ (D) $\frac{dy}{dx} = 5y$

Questions 10–11. The graph below shows the velocity of an object moving along a line, for $0 \le t \le 9$.

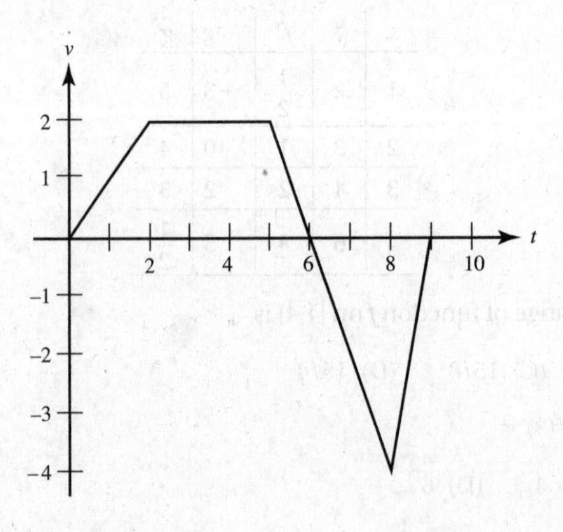

GO ON TO THE NEXT PAGE

10. At what time does the object attain its maximum acceleration?

(A) $2 < t < 5$ (B) $t = 6$ (C) $t = 8$ (D) $8 < t < 9$

11. The object is farthest from the starting point at $t =$

(A) 5 (B) 6 (C) 8 (D) 9

12. If $x = 2 \sin \theta$, then $\int_0^2 \frac{x^2 dx}{\sqrt{4 - x^2}}$ is equivalent to:

(A) $4 \int_0^2 \sin^2 \theta \, d\theta$ (B) $\int_0^{\pi/2} 4 \sin^2 \theta \, d\theta$

(C) $\int_0^{\pi/2} 2 \sin \theta \tan \theta \, d\theta$ (D) $\int_0^2 \frac{2 \sin^2 \theta}{\cos \theta} d\theta$

13. $\int_{-1}^{1} \left(1 - |x|\right) dx$

(A) $= 0$ (B) $= \frac{1}{2}$ (C) $= 1$ (D) does not exist

x	$f(x)$	$f'(x)$	$f''(x)$	$g(x)$	$g'(x)$	$g''(x)$
1	1	0	-7	1/3	-2	7

14. Given two twice-differentiable functions, $f(x)$ and $g(x)$. The table above gives values for $f(x)$ and $g(x)$ and their first and second derivatives at $x = 1$. Find $\lim_{x \to 1} \frac{2f(x) - 6g(x)}{4x^2 - 4e^{3(x-1)}}$.

(A) -3 (B) 1 (C) 2 (D) nonexistent

15. A differentiable function has the values shown in this table:

x	2.0	2.2	2.3	2.7	2.9	3.0
$f(x)$	1.39	1.73	2.10	2.48	2.88	3.30

Estimate $f'(2.1)$.

(A) 0.34 (B) 1.56 (C) 1.70 (D) 1.91

16. If $A = \int_0^1 e^{-x} dx$ is approximated using various sums with the same number of subdivisions, and if L, R, and T denote, respectively, left Riemann Sum, right Riemann Sum, and trapezoid sum, then it follows that

(A) $R \leq A \leq T \leq L$ (B) $R \leq T \leq A \leq L$ (C) $L \leq T \leq A \leq R$ (D) $L \leq A \leq T \leq R$

17. If $\frac{dy}{dx} = y \tan x$ and $y = 3$ when $x = 0$, then, when $x = \frac{\pi}{3}$, $y =$

(A) $2\sqrt{3}$ (B) $\frac{3}{2}$ (C) $\frac{3\sqrt{3}}{2}$ (D) 6

18. $\int_0^6 f(x - 1) dx =$

(A) $\int_1^7 f(x) dx$ (B) $\int_{-1}^{5} f(x) dx$ (C) $\int_{-1}^{5} f(x + 1) dx$ (D) $\int_1^7 f(x + 1) dx$

GO ON TO THE NEXT PAGE

BC PRACTICE EXAMINATION 1

19. The equation of the curve shown below is $y = \frac{4}{1+x^2}$. What does the area of the shaded region equal?

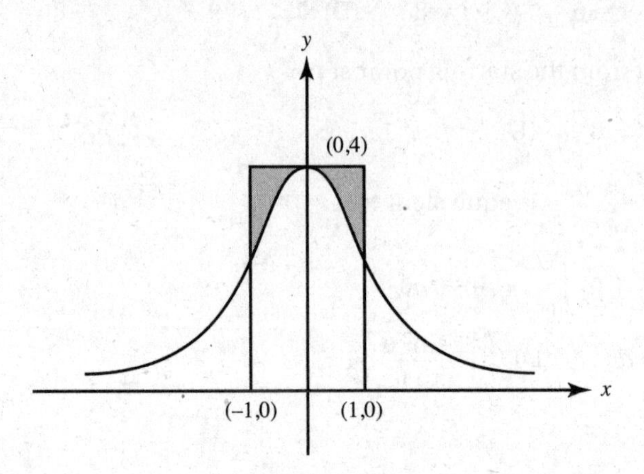

(A) $8 - \pi$ (B) $8 - 2\pi$ (C) $8 - 4\pi$ (D) $8 - 4\ln 2$

20. Find the slope of the curve $r = \cos 2\theta$ at $\theta = \frac{\pi}{6}$.

(A) $\frac{\sqrt{3}}{7}$ (B) $\frac{\sqrt{3}}{4}$ (C) $-\frac{\sqrt{3}}{4}$ (D) $-\sqrt{3}$

21. A particle moves along a line with velocity, in feet per second, $v = t^2 - t$. The total distance, in feet, traveled from $t = 0$ to $t = 2$ equals

(A) $\frac{1}{3}$ (B) $\frac{2}{3}$ (C) 2 (D) 1

22. If $f(t) = \int_0^{t^2} \frac{1}{1+x^2}\,dx$, then $f'(t)$ equals

(A) $\frac{1}{1+t^2}$ (B) $\frac{2t}{1+t^2}$ (C) $\frac{1}{1+t^4}$ (D) $\frac{2t}{1+t^4}$

23. The curve $x^3 + x \tan y = 27$ passes through (3,0). Use local linear approximation to estimate the value of y at $x = 3.1$. The value is

(A) −2.7 (B) −0.9 (C) 0 (D) 0.1

24. $\int x \cos x\, dx =$

(A) $x \sin x + \cos x + C$ (B) $x \sin x - \cos x + C$ (C) $\frac{x^2}{2} \sin x + C$ (D) $\frac{1}{2}\sin x^2 + C$

25. Which one of the following series converges?

(A) $\sum_{n=1}^{\infty} \frac{1}{\sqrt{n}}$ (B) $\sum_{n=1}^{\infty} \frac{1}{2n+1}$ (C) $\sum_{n=1}^{\infty} \frac{n}{n^2+1}$ (D) $\sum_{n=1}^{\infty} \frac{1}{n^2+1}$

26. The coefficient of the $(x-8)^2$ term in the Taylor polynomial for $y = x^{2/3}$ centered at $x = 8$ is

(A) $-\frac{1}{144}$ (B) $-\frac{1}{72}$ (C) $-\frac{1}{9}$ (D) $\frac{1}{144}$

27. If $f'(x) = h(x)$ and $g(x) = x^3$, then $\frac{d}{dx} f(g(x)) =$

(A) $h(x^3)$ (B) $3x^2 h(x)$ (C) $3x^2 h(x^3)$ (D) $h(3x^2)$

GO ON TO THE NEXT PAGE

28. $\int_0^{\infty} e^{-x/2}\,dx =$

(A) -2 (B) 1 (C) 2 (D) ∞

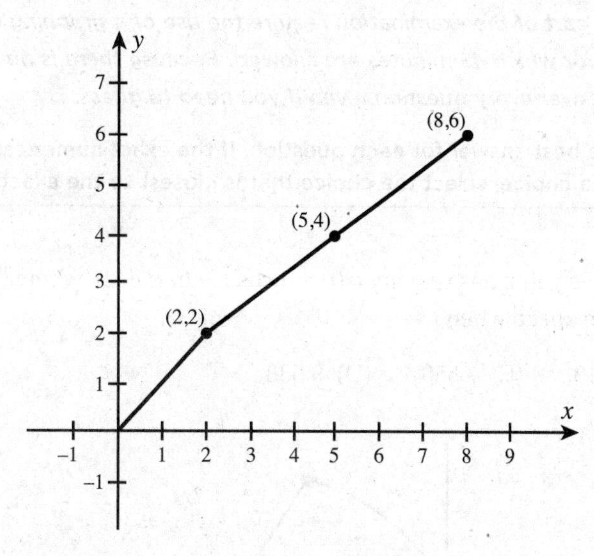

29. The graph of $f(x)$ consists of two line segments as shown above. If $g(x) = f^{-1}(x)$, the inverse function of $f(x)$, find $g'(4)$.

(A) $\frac{1}{5}$ (B) $\frac{2}{3}$ (C) $\frac{3}{2}$ (D) 5

30. Choose the integral that is the limit of the Riemann Sum: $\lim_{n\to\infty} \sum_{k=1}^{n} \left(\left(\frac{3k}{n} + 2 \right)^2 \cdot \left(\frac{3}{n} \right) \right)$.

(A) $\int_2^5 (x+2)^2\,dx$ (B) $\int_0^3 (3x+2)^2\,dx$ (C) $\int_2^5 x^2\,dx$ (D) $\int_0^3 x^2\,dx$

STOP

END OF PART A, SECTION I

BC PRACTICE EXAMINATION 1

Part B

TIME: 45 MINUTES

> *Some questions in this part of the examination require the use of a graphing calculator. There are 15 questions in Part B, for which 45 minutes are allowed. Because there is no deduction for wrong answers, you should answer every question, even if you need to guess.*
>
> **Directions:** Choose the best answer for each question. If the exact numerical value of the correct answer is not listed as a choice, select the choice that is closest to the exact numerical answer.

31. An object moving along a line has velocity $v(t) = t\cos t - \ln(t + 2)$, where $0 \leq t \leq 10$. The object achieves its maximum speed when $t =$

    (A) 5.107 (B) 6.419 (C) 7.550 (D) 9.538

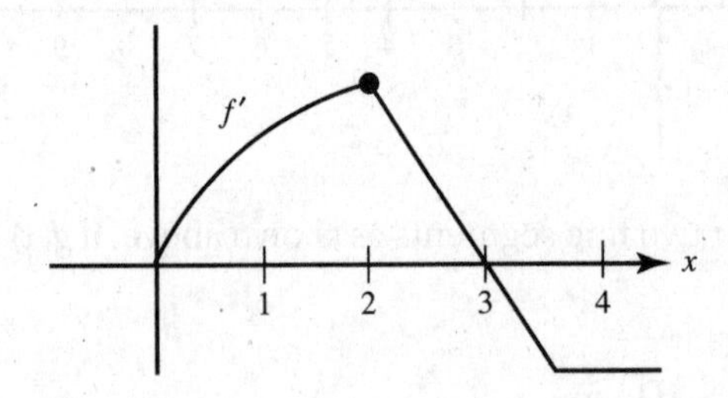

32. The graph of f', which consists of a quarter-circle and two line segments, is shown above. At $x = 2$ which of the following statements is true?

    (A) f is not continuous.
    (B) f is continuous but not differentiable.
    (C) f has a local maximum.
    (D) The graph of f has a point of inflection.

BC PRACTICE EXAMINATION 1

GO ON TO THE NEXT PAGE

33. Let $H(x) = \int_0^x f(t)dt$, where f is the function whose graph appears below.

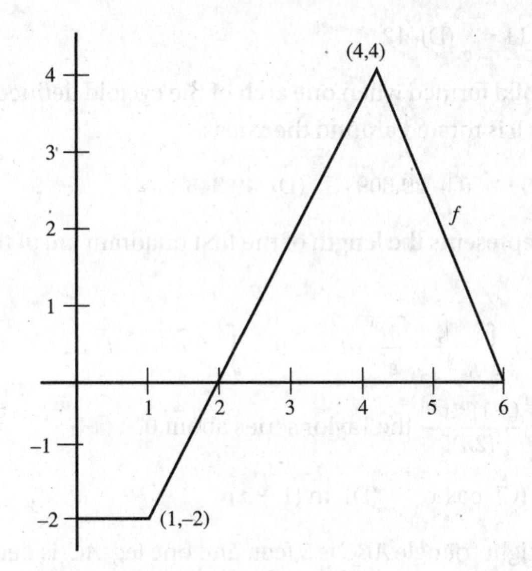

The tangent line approximating $H(x)$ near $x = 3$ is $H(x) \approx$.

(A) $-2x + 8$ (B) $2x - 4$ (C) $-2x + 4$ (D) $2x - 8$

34. The table shows the speed of an object, in feet per second, at various times during a 12-second interval.

time (sec)	0	3	6	7	8	10	12
speed (ft/sec)	15	14	11	8	7	3	0

Estimate the distance the object travels, using the midpoint method with 3 subintervals.

(A) 100 ft (B) 110 ft (C) 112 ft (D) 114 ft

35. In a marathon, when the winner crosses the finish line many runners are still on the course, some quite far behind. If the density of runners x miles from the finish line is given by $R(x) = 20\left[1 - \cos\left(1 + 0.03x^2\right)\right]$ runners per mile, how many are within 8 miles of the finish line?

(A) 30 (B) 40 (C) 157 (D) 166

36. Find the volume of the solid generated when the region bounded by the y-axis, $y = e^x$, and $y = 2$ is rotated around the y-axis.

(A) 0.592 (B) 1.214 (C) 2.427 (D) 3.998

37. The path of a satellite is given by the parametric equations

$$x = 4\cos t + \cos 12t,$$
$$y = 4\sin t + \sin 12t.$$

The upward velocity at $t = 1$ equals

(A) 3.073 (B) 3.999 (C) 12.287 (D) 12.666

GO ON TO THE NEXT PAGE

38. You wish to estimate e^x, over the interval $|x| < 2$, with an error less than 0.001. The Lagrange error term suggests that you use a Taylor polynomial at 0 with degree at least

(A) 9 (B) 10 (C) 11 (D) 12

39. Find the volume of the solid formed when one arch of the cycloid defined parametrically by $x = \theta - \sin\theta$, $y = 1 - \cos\theta$ is rotated around the x-axis.

(A) 15.708 (B) 16.755 (C) 29.609 (D) 49.348

40. Which definite integral represents the length of the first quadrant arc of the curve defined by $x(t) = e^t$, $y(t) = 1 - t^2$?

(A) $\int_{-1}^{1} \sqrt{1 + \frac{4t^2}{e^{2t}}}\, dt$ (B) $\int_{1/e}^{e} \sqrt{1 + \frac{4t^2}{e^{2t}}}\, dt$ (C) $\int_{-1}^{1} \sqrt{e^{2t} + 4t^2}\, dt$ (D) $\int_{1/e}^{e} \sqrt{e^{2t} + 4t^2}\, dt$

41. For which function is $\sum_{n=0}^{\infty} \frac{(-1)^n x^{2n}}{(2n)!}$ the Taylor series about 0?

(A) e^{-x} (B) $\sin x$ (C) $\cos x$ (D) $\ln(1 + x)$

42. The hypotenuse AB of a right triangle ABC is 5 feet, and one leg, AC, is decreasing at the rate of 2 feet per second. The rate, in square feet per second, at which the area is changing when $AC = 3$ is

(A) $\frac{7}{4}$ (B) $-\frac{3}{2}$ (C) $-\frac{7}{4}$ (D) $-\frac{7}{2}$

43. At how many points on the interval $[0,\pi]$ does $f(x) = 2\sin x + \sin 4x$ satisfy the Mean Value Theorem?

(A) 1 (B) 2 (C) 3 (D) 4

44. As a cup of hot chocolate cools, its temperature after t minutes is given by $H(t) = 70 + ke^{-0.4t}$. If its initial temperature was 120°F, what was its average temperature (in °F) during the first 10 minutes?

(A) 79.1 (B) 82.3 (C) 95.5 (D) 99.5

45. The rate at which a purification process can remove contaminants from a tank of water is proportional to the amount of contaminant remaining. If 20% of the contaminant can be removed during the first minute of the process and 98% must be removed to make the water safe, approximately how long will the decontamination process take?

(A) 2 min (B) 5 min (C) 7 min (D) 18 min

STOP

END OF SECTION I

SECTION II

Part A

TIME: 30 MINUTES

2 PROBLEMS

A graphing calculator is required for some of these problems. See instructions on page 4.

1. A function f is defined on the interval $[0,4]$, and its derivative is $f'(x) = e^{\sin x} - 2\cos 3x$.

   (a) On what interval is f increasing?

   (b) At what value(s) of x does f have local maxima? Justify your answer.

   (c) How many points of inflection does the graph of f have? Justify your answer.

2. The rate of sales of a new software product is given by $S(t)$, where S is measured in hundreds of units per month and t is measured in months from the initial release date of January 1, 2012. The software company recorded these sales data:

t (months)	1	2	3	4	5	6	7
$S(t)$ (100s/month)	1.54	1.88	2.32	3.12	3.78	4.90	6.12

   (a) Using a trapezoidal approximation, estimate the number of units the company sold during the second quarter (April 1, 2012, through June 30, 2012).

   (b) After looking at these sales figures, a manager suggests that the rate of sales can be modeled by assuming the rate to be initially 120 units/month and to double every 3 months. Write an equation for S based on this model.

   (c) Compare the model's prediction for total second quarter sales with your estimate from part (a).

   (d) Use the model to predict the average value of $S(t)$ for the entire first year. Explain what your answer means.

STOP

END OF PART A, SECTION II

BC PRACTICE EXAMINATION 1

Part B

TIME: 60 MINUTES

4 PROBLEMS

No calculator is allowed for any of these problems.

If you finish Part B before time has expired, you may return to work on Part A, but you may not use a calculator.

3. The velocity of an object in motion in the plane for $0 \le t \le 1$ is given by the vector

   $$\mathbf{v}(t) = \left\langle \frac{1}{\sqrt{4-t^2}}, \frac{t}{\sqrt{4-t^2}} \right\rangle.$$

   (a) When is this object at rest?

   (b) If this object was at the origin when $t = 0$, what are its speed and position when $t = 1$?

   (c) Find an equation of the curve the object follows, expressing y as a function of x.

4. (a) Write the first four terms and the general term of the Maclaurin series for $f(x) = \ln(e + x)$.

   (b) What is the radius of convergence?

   (c) Use the first three terms of that series to write an expression that estimates the value of $\int_0^1 \ln(e + x^2)dx$.

5. After pollution-abatement efforts, conservation researchers introduce 100 trout into a small lake. The researchers predict that after m months the rate of growth, F, of the trout population will be modeled by the differential equation $\frac{dF}{dm} = 0.0002F(600 - F)$.

   (a) How large is the trout population when it is growing the fastest?

   (b) Solve the differential equation, expressing F as a function of m.

   (c) How long after the lake was stocked will the population be growing the fastest?

6. (a) A spherical snowball melts so that its surface area shrinks at the constant rate of 10 square centimeters per minute. What is the rate of change of volume when the snowball is 12 centimeters in diameter?

   (b) The snowball is packed most densely nearest the center. Suppose that, when it is 12 centimeters in diameter, its density x centimeters from the center is given by $d(x) = \frac{1}{1 + \sqrt{x}}$ grams per cubic centimeter. Set up an integral for the total number of grams (mass) of the snowball then. Do not evaluate.

END OF TEST

ANSWER KEY
BC Practice Examination 1

Part A

1. **B**	9. **A**	17. **D**	25. **D**
2. **B**	10. **D**	18. **B**	26. **A**
3. **D**	11. **B**	19. **B**	27. **C**
4. **B**	12. **B**	20. **A**	28. **C**
5. **B**	13. **C**	21. **D**	29. **C**
6. **C**	14. **A**	22. **D**	30. **C**
7. **C**	15. **C**	23. **B**	
8. **A**	16. **A**	24. **A**	

Part B

31. **D**	35. **D**	39. **D**	43. **D**
32. **D**	36. **A**	40. **C**	44. **B**
33. **D**	37. **C**	41. **C**	45. **D**
34. **C**	38. **B**	42. **C**	

ANSWERS EXPLAINED

The explanations for questions not given below will be found in the answer section for AB Practice Examination 1 on pages 506–510. Identical questions in Section I of Practice Examinations AB 1 and BC 1 have the same number. For example, explanations of the answers for Question 1, not given below, will be found in Section I of Calculus AB Practice Examination 1, Answer 1, page 506.

Section I Multiple-Choice

Part A

3. **(D)** Here,

$$\frac{dy}{dx} = \frac{\frac{dy}{dt}}{\frac{dx}{dt}} = \frac{\frac{1}{\sqrt{1-t^2}}}{\frac{1}{2}\frac{(-2t)}{\sqrt{1-t^2}}} = -\frac{1}{t}.$$

6. **(C)** $\frac{1}{3}\int_1^2 (3x-2)^3(3dx) = \frac{1}{12}(3x-2)^4\Big|_1^2.$

12. **(B)** Note that, when $x = 2\sin\theta$, $x^2 = 4\sin^2\theta$, $dx = 2\cos\theta\, d\theta$, and $\sqrt{4-x^2} = 2\cos\theta$. Also,

$$\text{when } x = 0, \theta = 0;$$

$$\text{when } x = 2, \frac{\pi}{2}.$$

13. **(C)** The given integral is equivalent to $\int_{-1}^{0} (1+x)\,dx + \int_0^1 (1-x)\,dx.$

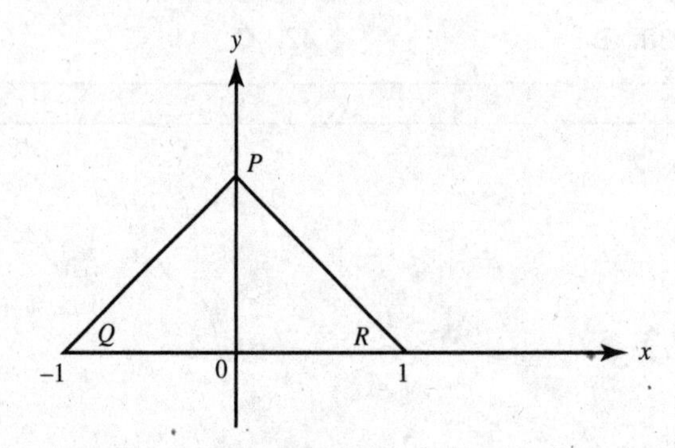

The figure shows the graph of $f(x) = 1 - |x|$ on $[-1,1]$.
The area of triangle PQR is equal to $\int_{-1}^{1} \left(1 - |x|\right) dx.$

17. **(D)** Separating variables yields $\frac{dy}{y} = \tan x$, so $\ln y = -\ln\cos x + C$. With $y = 3$ when $x = 0$, $C = \ln 3$. The general solution is therefore $(\cos x)\, y = 3$. When $x = \frac{\pi}{3}$,

$$\cos x = \frac{1}{2} \quad \text{and} \quad y = 6.$$

20. **(A)** Represent the coordinates parametrically as $(r\cos\theta, r\sin\theta)$. Then

$$\frac{dy}{dx} = \frac{\frac{dy}{d\theta}}{\frac{dx}{d\theta}} = \frac{r\cos\theta + \frac{dr}{d\theta}\cdot\sin\theta}{-r\sin\theta + \frac{dx}{d\theta}\cdot\cos\theta}.$$

Note that $\frac{dr}{d\theta} = -2\sin 2\theta$, and evaluate $\frac{dy}{dx}$ at $\theta = \frac{\pi}{6}$. (Alternatively, write $x = \cos 2\theta\cos\theta$ and $y = \cos 2\theta\sin\theta$ to find $\frac{dy}{dx}$ from $\frac{dy/d\theta}{dx/d\theta}$.)

21. **(D)** Note that v is negative from $t = 0$ to $t = 1$, but positive from $t = 1$ to $t = 2$. Thus the distance traveled is given by

$$-\int_0^1 (t^2 - t)\,dt + \int_1^2 (t^2 - t)\,dt.$$

24. **(A)** Use parts; then $u = x$, $dv = \cos x\,dx$; $du = dx$, $v = \sin x$. Thus,

$$\int x\cos x\,dx = x\sin x - \int \sin x\,dx.$$

25. **(D)** (A), a p-series with $p = 1/2$, diverges. We would like to compare (B) to $\sum_{n=1}^{\infty}\frac{1}{n}$, but $\frac{1}{2n+1} < \frac{1}{n}$, so we use the Limit Comparison Test $\lim_{n\to\infty}\frac{\frac{1}{2n+1}}{1/n} = \lim_{n\to\infty}\frac{n}{2n+1} = \frac{1}{2}$; since $\sum_{n=1}^{\infty}\frac{1}{n}$ diverges, (B) diverges. For (C), we need to use LCT for the same reason as (B), $\lim_{n\to\infty}\frac{\frac{n}{n^2+1}}{1/n} = \lim_{n\to\infty}\frac{n^2}{n^2+1} = 1$; again since $\sum_{n=1}^{\infty}\frac{1}{n}$ diverges, (C) diverges. For (D), we would like to compare to $\sum_{n=1}^{\infty}\frac{1}{n^2}$ and $\frac{1}{n^2+1} < \frac{1}{n^2}$, so by the Comparison Test (D) converges.

26. **(A)** By Taylor's Theorem, the coefficient is $\frac{f''(8)}{2!}$. For $f(x) = x^{\frac{2}{3}}$, $f'(x) = \frac{2}{3}x^{-\frac{1}{3}}$ and $f''(x) = -\frac{2}{9}x^{-\frac{4}{3}}$; hence $f''(8) = -\frac{2}{9}(8)^{-\frac{4}{3}} = -\frac{2}{9}\cdot\frac{1}{16} = -\frac{1}{72}$, making the coefficient $-\frac{1}{72}\cdot\frac{1}{2!}$.

27. **(C)** Here,

$$\frac{d}{dx}f\big(g(x)\big) = f'\big(g(x)\big)g'(x) = h\big(g(x)\big)g'(x) = h\big(x^3\big)\cdot 3x^2.$$

28. **(C)** Evaluate

$$\lim_{b\to\infty}\int_0^b e^{-x/2}\,dx = -\lim_{b\to\infty} 2e^{-x/2}\Big|_0^b = -2(0-1).$$

Part B

36. **(A)**

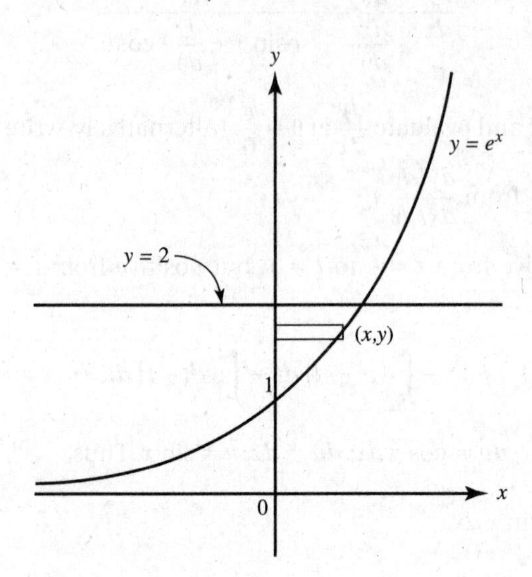

Using disks, $\Delta V = \pi R^2 \Delta y = \pi(\ln y)^2\, \Delta y$. Evaluate

$$V = \int_1^2 \pi(\ln y)^2\, dy.$$

The required volume is 0.592.

37. **(C)** The vertical component of velocity is

$$\frac{dy}{dt} = 4 \cos t + 12 \cos 12t.$$

Evaluate at $t = 1$.

38. **(B)** The Maclaurin expansion is

$$e^x = 1 + x + \frac{x^2}{2!} + \frac{x^3}{3!} + \cdots + \frac{x^n}{n!} + \cdots.$$

The Lagrange remainder R, after n terms, for some c in the interval $|x| \leqq 2$, is

$$R = \frac{f^{(n+1)}(c) \cdot c^{n+1}}{(n+1)!} = \frac{e^c\, c^{n+1}}{(n+1)!}.$$

Since R is greatest when $c = 2$, n needs to satisfy the inequality

$$\frac{e^2\, 2^{n+1}}{(n+1)!} < 0.001.$$

Using a calculator to evaluate $y = \dfrac{e^2\, 2^{x+1}}{(x+1)!}$ successively at various integral values of x gives $y(8) > 0.01$, $y(9) > 0.002$, $y(10) < 3.8 \times 10^{-4} < 0.0004$. Thus we achieve the desired accuracy with a Taylor polynomial at 0 of degree at least 10.

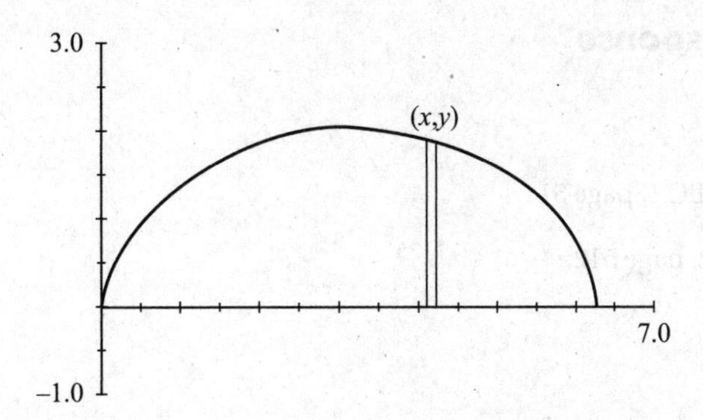

39. **(D)** On your calculator, graph one arch of the cycloid for t in $[0,2\pi]$ and (x,y) in $[0,7] \times [-1,3]$. Use disks; then the desired volume is

$$\begin{aligned} V &= \pi \int_{t=0}^{t=2\pi} y^2\, dx \\ &= \pi \int_0^{2\pi} (1 - \cos t)^2 (1 - \cos t)\, dt \\ &= \pi \int_0^{2\pi} (1 - \cos t)^3\, dt \\ &\simeq 49.348. \end{aligned}$$

40. **(C)** In the first quadrant, both x and y must be positive; $x(t) = e^t$ is positive for all t, but $y(t) = 1 - t^2$ is positive only for $-1 < t < 1$. The arc length is

$$\int_{t_1}^{t_2} \sqrt{\left(\frac{dx}{dt}\right)^2 + \left(\frac{dy}{dt}\right)^2}\, dt = \int_{-1}^{1} \sqrt{(e^t)^2 + (-2t)^2}\, dt.$$

41. **(C)** See series (2) on page 408.

44. **(B)** At $t = 0$, we know $H = 120$, so $120 = 70 + ke^{-0.4(0)}$, and thus $k = 50$. The average temperature for the first 10 minutes is $\frac{1}{10 - 0} \int_0^{10} \left(70 + 50e^{-0.4t}\right) dt$.

Section II Free-Response

Part A

1. See solution for AB/BC 1, page 511.
2. See solution for AB 2, page 512.

Part B

3. (a) Because $\frac{dx}{dt} = \frac{1}{\sqrt{4-t^2}}$, which never equals zero, the object is never at rest.

(b) $\mathbf{v}(1) = \left\langle \frac{1}{\sqrt{4-1^2}}, \frac{1}{\sqrt{4-1^2}} \right\rangle = \left\langle \frac{1}{\sqrt{3}}, \frac{1}{\sqrt{3}} \right\rangle$, so the object's speed is

$$|\mathbf{v}(1)| = \sqrt{\left(\frac{1}{\sqrt{3}}\right)^2 + \left(\frac{1}{\sqrt{3}}\right)^2} = \sqrt{\frac{2}{3}}.$$

Position is the antiderivative of velocity $\mathbf{v}(t) = \left\langle \frac{1}{\sqrt{4-t^2}}, \frac{t}{\sqrt{4-t^2}} \right\rangle$.

$$y = \int \frac{1}{\sqrt{4-t^2}}\,dt = 2 \cdot \frac{1}{2} \int \frac{\frac{1}{2}\,dt}{\sqrt{1-\left(\frac{t}{2}\right)^2}} = \arcsin\frac{t}{2} + c.$$

Since $P(0) = (0,0)$, $\arcsin\frac{0}{2} + c = 0$,and thus $c = 0$.

$$y = \int \frac{1}{\sqrt{4-t^2}}\,dt = -\frac{1}{2}\int (4-t^2)^{-1/2}(-2t\,dt) - \sqrt{4-t^2} + c.$$

Since $P(0) = (0,0)$, $-\sqrt{4-0^2} + c = 0$ and thus $c = 2$.

Then

$$P(t) = \left\langle \arcsin\frac{t}{2}, 2 - \sqrt{4-t^2} \right\rangle,$$

$$P(1) = \left\langle \arcsin\frac{1}{2}, 2 - \sqrt{4-1} \right\rangle = \left\langle \frac{\pi}{6}, 2 - \sqrt{3} \right\rangle.$$

(c) Solving $x = \arcsin\frac{t}{2}$ for t yields $t = 2\sin x$. Therefore

$$y = 2 - \sqrt{4 - (2\sin x)^2} = 2 - 2\sqrt{1 - \sin^2 x} = 2 - 2|\cos x|.$$

Since $0 \le t \le 1$ means $0 \le x \le \frac{\pi}{6}$, then $\cos x > 0$, so $y = 2 - 2\cos x$.

4. (a) To write the Maclaurin series for $f(x) = \ln(e + x)$, use Taylor's Theorem at $x = 0$.

n	$f^{(n)}(x)$	$f^{(n)}(0)$	$a_n = \frac{f^{(n)}(0)}{n!}$
0	$\ln(e + x)$	1	1
1	$\frac{1}{e+x}$	$\frac{1}{e}$	$\frac{1}{e}$
2	$-(e+x)^{-2}$	$-\frac{1}{e^2}$	$-\frac{1}{2e^2}$
3	$2(e+x)^{-3}$	$\frac{2}{e^3}$	$\frac{1}{3e^3}$

$$f(x) = 1 + \frac{x}{e} - \frac{x^2}{2e^2} + \frac{x^3}{3e^3} \cdots + \frac{(-1)^{n+1}x^n}{ne^n} + \cdots \text{ (for } n \geq 1\text{)}.$$

(b) By the Ratio Test, the series converges when

$$\lim_{n\to\infty}\left|\frac{x^{n+1}}{(n+1)e^{n+1}} \cdot \frac{ne^n}{x^n}\right| < 1,$$

$$|x| \lim_{n\to\infty} \frac{1}{e} < 1,$$

$$|x| < e.$$

Thus, the radius of convergence is e.

(c) $\int_0^1 \ln\left(e + x^2\right) dx \approx \int_0^1 1 + \frac{x^2}{e} - \frac{(x^2)^2}{2e^2} dx,$

$$\int_0^1 \ln\left(e + x^2\right) dx \approx \left(x + \frac{x^3}{3e} - \frac{x}{5 \cdot 2e^2}\right)\Bigg|_0^1 = 1 + \frac{1}{3e} - \frac{1}{10e^2}$$

5. (a) To find the maximum rate of growth, first find the derivative of $\frac{dF}{dm} = 0.0002F(600 - F) = 0.0002\left(600F - F^2\right)$. $\frac{d^2F}{dm^2} = 0.0002(600 - 2F)$, which equals 0 when $F = 300$. A signs analysis shows that $\frac{d^2F}{dm^2}$ changes from positive to negative there, confirming that $\frac{dF}{dm}$ is at its maximum when there are 300 trout.

BC PRACTICE EXAMINATION 1

(b) The differential equation $\frac{dF}{dm} = 0.0002F(600 - F)$ is separable.

$$\int \frac{dF}{F(600 - F)} = 0.0002 \int dm.$$

To integrate the left side of this equation, use the method of partial fractions.

$$\frac{1}{F(600 - F)} = \frac{A}{F} + \frac{B}{600 - F},$$

$$1 = A(600 - F) + B(F).$$

Let $F = 0$; then $A = \frac{1}{600}$.

Let $F = 600$; then $B = \frac{1}{600}$.

$$\frac{1}{600} \int \left(\frac{1}{F} + \frac{1}{(600 - F)} \right) dF = 0.0002 \int dm,$$

$$\int \frac{1}{F} dF + (-1) \int \frac{-1}{(600 - F)} dF = 0.12 \int dm,$$

$$\ln F - \ln(600 - F) = 0.12m + C,$$

$$\ln \left(\frac{F}{600 - F} \right) = 0.12m + C,$$

$$\frac{F}{600 - F} = e^{0.12m + C}$$

$$= ce^{0.12m}, \text{ where } c = e^C,$$

$$F = \frac{600}{1 + ce^{-0.12m}}.$$

Since $F = 100$ when $m = 0$, $100 = \frac{600}{1 + c}$, so $c = 5$.

The solution is $F = \frac{600}{1 + 5e^{-0.12m}}$.

(c) In (a) the population was found to be growing the fastest when $F = 300$. Then:

$$300 = \frac{600}{1 + 5e^{-0.12m}},$$

$$e^{-0.12m} = \frac{1}{5},$$

$$m = \frac{\ln \frac{1}{5}}{-0.12} \text{ months.}$$

6. See solution for AB/BC 6, page 514.

ANSWER SHEET
BC Practice Examination 2

Part A

1. Ⓐ Ⓑ Ⓒ Ⓓ
2. Ⓐ Ⓑ Ⓒ Ⓓ
3. Ⓐ Ⓑ Ⓒ Ⓓ
4. Ⓐ Ⓑ Ⓒ Ⓓ
5. Ⓐ Ⓑ Ⓒ Ⓓ
6. Ⓐ Ⓑ Ⓒ Ⓓ
7. Ⓐ Ⓑ Ⓒ Ⓓ
8. Ⓐ Ⓑ Ⓒ Ⓓ
9. Ⓐ Ⓑ Ⓒ Ⓓ
10. Ⓐ Ⓑ Ⓒ Ⓓ
11. Ⓐ Ⓑ Ⓒ Ⓓ
12. Ⓐ Ⓑ Ⓒ Ⓓ
13. Ⓐ Ⓑ Ⓒ Ⓓ
14. Ⓐ Ⓑ Ⓒ Ⓓ
15. Ⓐ Ⓑ Ⓒ Ⓓ
16. Ⓐ Ⓑ Ⓒ Ⓓ
17. Ⓐ Ⓑ Ⓒ Ⓓ
18. Ⓐ Ⓑ Ⓒ Ⓓ
19. Ⓐ Ⓑ Ⓒ Ⓓ
20. Ⓐ Ⓑ Ⓒ Ⓓ
21. Ⓐ Ⓑ Ⓒ Ⓓ
22. Ⓐ Ⓑ Ⓒ Ⓓ
23. Ⓐ Ⓑ Ⓒ Ⓓ
24. Ⓐ Ⓑ Ⓒ Ⓓ
25. Ⓐ Ⓑ Ⓒ Ⓓ
26. Ⓐ Ⓑ Ⓒ Ⓓ
27. Ⓐ Ⓑ Ⓒ Ⓓ
28. Ⓐ Ⓑ Ⓒ Ⓓ
29. Ⓐ Ⓑ Ⓒ Ⓓ
30. Ⓐ Ⓑ Ⓒ Ⓓ

Part B

31. Ⓐ Ⓑ Ⓒ Ⓓ
32. Ⓐ Ⓑ Ⓒ Ⓓ
33. Ⓐ Ⓑ Ⓒ Ⓓ
34. Ⓐ Ⓑ Ⓒ Ⓓ
35. Ⓐ Ⓑ Ⓒ Ⓓ
36. Ⓐ Ⓑ Ⓒ Ⓓ
37. Ⓐ Ⓑ Ⓒ Ⓓ
38. Ⓐ Ⓑ Ⓒ Ⓓ
39. Ⓐ Ⓑ Ⓒ Ⓓ
40. Ⓐ Ⓑ Ⓒ Ⓓ
41. Ⓐ Ⓑ Ⓒ Ⓓ
42. Ⓐ Ⓑ Ⓒ Ⓓ
43. Ⓐ Ⓑ Ⓒ Ⓓ
44. Ⓐ Ⓑ Ⓒ Ⓓ
45. Ⓐ Ⓑ Ⓒ Ⓓ

BC Practice Examination 2

SECTION I

Part A

TIME: 60 MINUTES

The use of calculators is not permitted for this part of the examination.
There are 30 questions in Part A, for which 60 minutes are allowed. Because there is no deduction for wrong answers, you should answer every question, even if you need to guess.

Directions: Choose the best answer for each question.

1. A function $f(x)$ equals $\frac{x^2 - x}{x - 1}$ for all x except $x = 1$. If $f(1) = k$, for what value of k would the function be continuous at $x = 1$?

   (A) 0 (B) 1 (C) 2 (D) No such k exists.

2. $\lim\limits_{x \to 0} \frac{\sin \frac{x}{2}}{x}$ is

   (A) 2 (B) 0 (C) $\frac{1}{2}$ (D) nonexistent

3. The first four terms of the Taylor series about $x = 0$ of $\sqrt{1 + x}$ are

   (A) $1 - \frac{x}{2} + \frac{x^2}{4 \cdot 2} - \frac{3x^3}{8 \cdot 6}$ (B) $x + \frac{x^2}{2} + \frac{x^3}{8} + \frac{x^4}{48}$ (C) $1 + \frac{x}{2} - \frac{x^2}{8} + \frac{x^3}{16}$

   (D) $-1 + \frac{x}{2} - \frac{x^2}{8} + \frac{x^3}{16}$

4. Using the line tangent to $f(x) = \sqrt{9 + \sin(2x)}$ at $x = 0$, an estimate of $f(0.06)$ is

   (A) 0.02 (B) 2.98 (C) 3.01 (D) 3.02

5. Air is escaping from a balloon at a rate of $R(t) = \frac{60}{1 + t^2}$ cubic feet per minute, where t is measured in minutes. How much air, in cubic feet, escapes during the first minute?

   (A) 15π (B) 30 (C) 45 (D) $30 \ln 2$

GO ON TO THE NEXT PAGE

6. The motion of a particle in a plane is given by the pair of equations $x = \cos 2t$, $y = \sin 2t$. The magnitude of its acceleration at any time t equals

(A) 1 (B) 2 (C) 4 (D) 16

7. Let $f(x) = (x-1) + \frac{(x-1)^2}{4} + \frac{(x-1)^3}{9} + \frac{(x-1)^4}{16} + \cdots$

The interval of convergence of $f'(x)$ is

(A) $0 \leqq x \leqq 2$ (B) $0 \leqq x < 2$ (C) $0 < x \leqq 2$ (D) $0 < x < 2$

8. A point moves along the curve $y = x^2 + 1$ so that the x-coordinate is increasing at the constant rate of $\frac{3}{2}$ units per second. The rate, in units per second, at which the distance from the origin is changing when the point has coordinates (1, 2) is equal to

(A) $\frac{7\sqrt{5}}{0}$ (B) $\frac{3\sqrt{5}}{2}$ (C) $\frac{3\sqrt{5}}{5}$ (D) $\frac{5}{2}$

9. $\lim\limits_{h \to 0} \frac{\sqrt{25+h} - 5}{h}$

(A) $= 0$ (B) $= \frac{1}{10}$ (C) $= 1$ (D) does not exist

10. $\int_0^{\pi/3} \sec^2 x \tan^2 x \, dx$ equals

(A) $\frac{\sqrt{3}}{27}$ (B) $\frac{\sqrt{3}}{3}$ (C) $\sqrt{3}$ (D) $3\sqrt{3}$

11. $\int_1^e \ln x \, dx$ equals

(A) $e - 1$ (B) $e + 1$ (C) 1 (D) -1

12. $\int \frac{(y-1)^2}{2y} \, dy$ equals

(A) $\frac{y^2}{4} - y + \frac{1}{2} \ln |y| + C$ (B) $y^2 - y + \ln|2y| + C$ (C) $\frac{y^2}{4} - y + 2 \ln|2y| + C$

(D) $\frac{(y-1)^3}{3y^2} + C$

13. $\int \frac{x-6}{x^2 - 3x} \, dx =$

(A) $\ln |x^2 (x-3)| + C$ (B) $-\ln |x^2 (x-3)| + C$ (C) $\ln \left| \frac{x^2}{(x-3)} \right| + C$

(D) $\ln \left| \frac{x-3}{x^2} \right| + C$

GO ON TO THE NEXT PAGE

14. Given f' as graphed, which could be a graph of f?

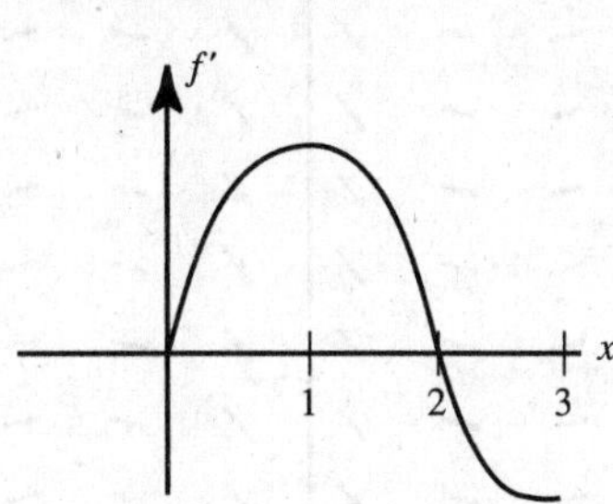

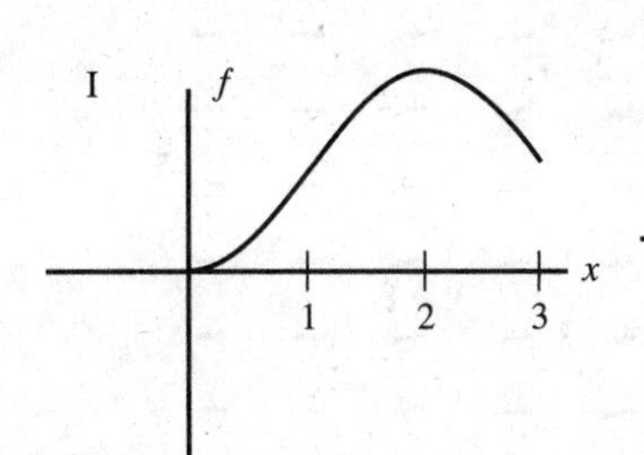

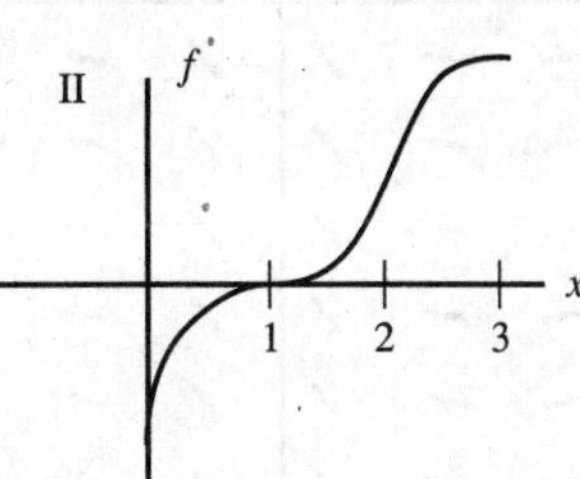

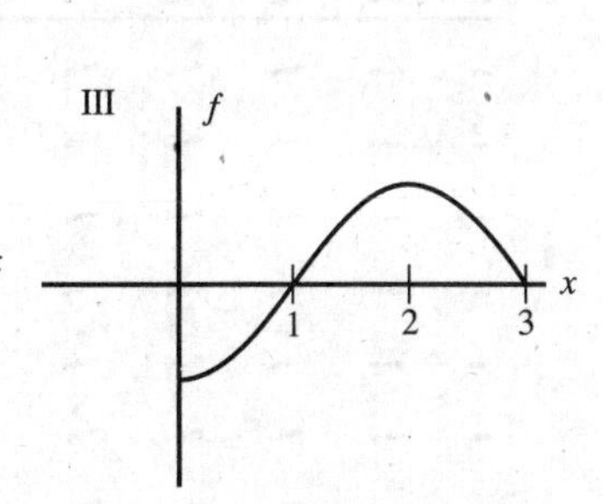

(A) I only (B) II only (C) III only (D) I and III only

15. The first woman officially timed in a marathon was Violet Piercey of Great Britain in 1926. Her record of 3:40:22 stood until 1963, mostly because of a lack of women competitors. Soon after, times began dropping rapidly, but lately they have been declining at a much slower rate. Let $M(t)$ be the curve that best represents winning marathon times in year t. Which of the following is positive for $t > 1963$?

I. $M(t)$
II. $M'(t)$
III. $M''(t)$

(A) I only (B) I and II only (C) I and III only (D) I, II, and III

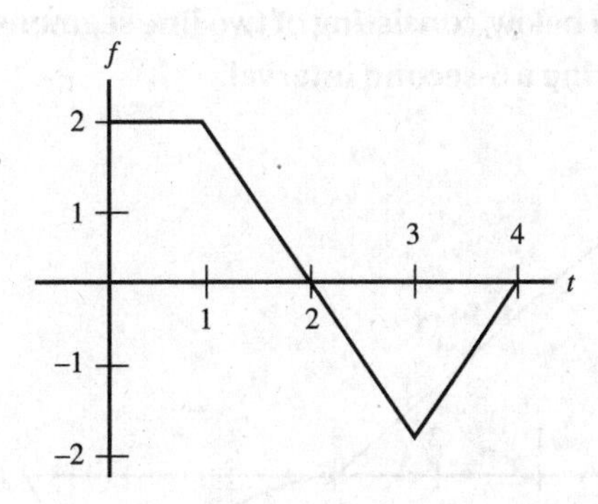

16. The graph of f is shown above. Let $G(x) = \int_0^x f(t)dt$ and $H(x) = \int_2^x f(t)dt$. Which of the following is true?

(A) $G(x) = H(x)$ (B) $G(x) = H'(x + 2)$ (C) $G(x) = H(x + 2)$ (D) $G(x) = H(x) + 3$

17. The minimum value of $f(x) = x^2 + \frac{2}{x}$ on the interval $\frac{1}{2} \leq x \leq 2$ is

(A) 1 (B) 3 (C) $4\frac{1}{2}$ (D) 5

GO ON TO THE NEXT PAGE

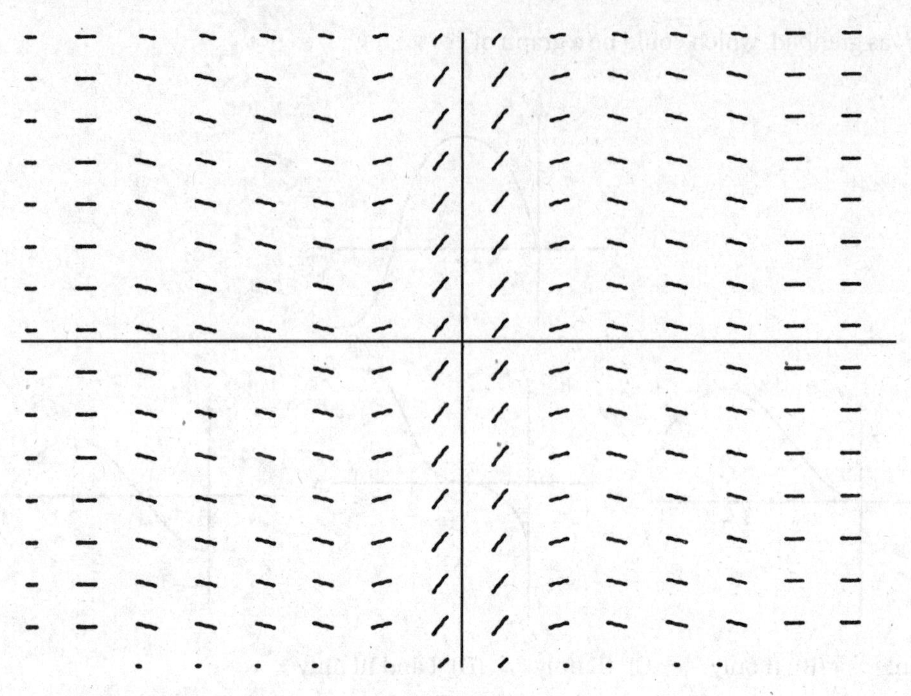

18. Which function could be a particular solution of the differential equation whose slope field is shown above?

(A) $y = \frac{2x}{x^2+1}$ (B) $y = \frac{x^2}{x^2+1}$ (C) $y = \sin x$ (D) $y = e^{-x^2}$

19. A particular solution of the differential equation $\frac{dy}{dx} = x + y$ passes through the point (2,1). Using Euler's method with $\Delta x = 0.1$, estimate its y-value at $x = 2.2$.

(A) 1.30 (B) 1.34 (C) 1.60 (D) 1.64

Questions 20 and 21. Use the graph below, consisting of two line segments and a quarter-circle. The graph shows the velocity of an object during a 6-second interval.

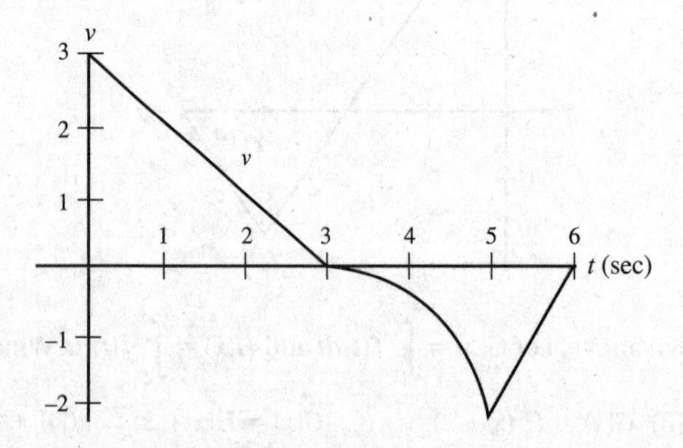

20. For how many values of t in the interval $0 < t < 6$ is the acceleration undefined?

(A) zero (B) one (C) two (D) three

21. During what time interval (in sec) is the speed increasing?

(A) $0 < t < 3$ (B) $3 < t < 5$ (C) $5 < t < 6$ (D) never

GO ON TO THE NEXT PAGE

22. If $\frac{dy}{dx} = \frac{y}{x}$ $(x > 0, y > 0)$ and $y = 3$ when $x = 1$, then

(A) $x^2 + y^2 = 10$ (B) $y = x + 2$ (C) $y^2 - x^2 = 8$ (D) $y = 3x$

23. A solid is cut out of a sphere of radius 2 by two parallel planes each 1 unit from the center. The volume of this solid is

(A) $\frac{10\pi}{3}$ (B) $\frac{32\pi}{3}$ (C) $\frac{25\pi}{3}$ (D) $\frac{22\pi}{3}$

24. Which one of the following improper integrals converges?

(A) $\int_{-1}^{1} \frac{dx}{(x+1)^2}$ (B) $\int_{1}^{\infty} \frac{dx}{\sqrt{x}}$ (C) $\int_{0}^{\infty} \frac{dx}{(x^2+1)}$ (D) $\int_{1}^{3} \frac{dx}{(2-x)^3}$

25. The function $f(x) = x^5 + 3x - 2$ passes through the point (1,2). Let f^{-1} denote the inverse of f. Then $(f^{-1})'(2)$ equals

(A) $\frac{1}{83}$ (B) $\frac{1}{8}$ (D) 8 (E) 83

26. Find the domain of the particular solution of $\frac{dy}{dx} = 1 + y^2$ that passes through the origin.

(A) all x (B) $x \geq 0$ (C) $|x| < \frac{\pi}{2}$ (D) $0 \leq x < \frac{\pi}{2}$

27. Which of the following statements is (are) true about the graph of $y = \ln(4 + x^2)$?

I. It is symmetric to the y-axis.
II. It has a local minimum at $x = 0$.
III. It has inflection points at $x = \pm 2$.

(A) I only (B) I and II only (C) II and III only (D) I, II, and III

28. $\int_{1}^{2} \frac{dx}{\sqrt{4-x^2}}$ is

(A) $-\frac{\pi}{3}$ (B) $\frac{\pi}{6}$ (C) $\frac{\pi}{3}$ (D) nonexistent

29. Choose the integral that is the limit of the Riemann Sum: $\lim_{n\to\infty} \sum_{k=1}^{n} \left(\left(\sqrt{\frac{2k}{n}} + 3 \right) \cdot \left(\frac{1}{n} \right) \right)$.

(A) $\int_{3}^{4} \sqrt{2x}\, dx$ (B) $\int_{0}^{1} \sqrt{2x+3}\, dx$ (C) $\int_{0}^{1} \sqrt{2x}\, dx$ (D) $\int_{3}^{4} \sqrt{2x+3}\, dx$

30. Which infinite series converge(s)?

I. $\sum_{n=1}^{\infty} \frac{3^n}{n!}$ II. $\sum_{n=1}^{\infty} \frac{3^n}{n^3}$ III. $\sum_{n=1}^{\infty} \frac{n^2}{n^3+1}$

(A) I only (B) II only (C) III only (D) I and III only

STOP

END OF PART A, SECTION I

Part B

TIME: 45 MINUTES

Some questions in this part of the examination require the use of a graphing calculator. There are 15 questions in Part B, for which 45 minutes are allowed. Because there is no deduction for wrong answers, you should answer every question, even if you need to guess.

Directions: Choose the best answer for each question. If the exact numerical value of the correct answer is not listed as a choice, select the choice that is closest to the exact numerical answer.

31. Find the area bounded by the spiral $r = \ln \theta$ on the interval $\pi \leqq \theta \leqq 2\pi$.

    (A) 2.405 (B) 3.743 (C) 4.810 (D) 7.487

32. Write an equation for the line tangent to the curve defined by $\mathbf{F}(t) = \langle t^2 + 1, 2^t \rangle$ at the point where $y = 4$.

    (A) $y - 4 = (\ln 2)(x - 2)$ (B) $y - 4 = (4 \ln 2)(x - 2)$

    (C) $y - 4 = (\ln 2)(x - 5)$ (D) $y - 4 = (4 \ln 2)(x - 5)$

33. Bacteria in a culture increase at a rate proportional to the number present. An initial population of 200 triples in 10 hours. If this pattern of increase continues unabated, then the approximate number of bacteria after 1 full day is

    (A) 1056 (B) 1440 (C) 2793 (D) 3240

34. When the substitution $x = 2t - 1$ is used, the definite integral $\int_3^5 t\sqrt{2t-1}\,dt$ may be expressed in the form $k\int_a^b (x+1)\sqrt{x}\,dx$, where $\{k, a, b\} =$

    (A) $\left\{\frac{1}{4}, 2, 3\right\}$ (B) $\left\{\frac{1}{4}, 5, 9\right\}$ (C) $\left\{\frac{1}{2}, 2, 3\right\}$ (D) $\left\{\frac{1}{2}, 5, 9\right\}$

35. The curve defined by $x^3 + xy - y^2 = 10$ has a vertical tangent line when $x =$

    (A) 1.037 (B) 1.087 (C) 2.074 (D) 2.096

Questions 36 and 37. Use the graph of *f* shown on [0,7]. Let $G(x) = \int_2^{3x-1} f(t)\,dt$.

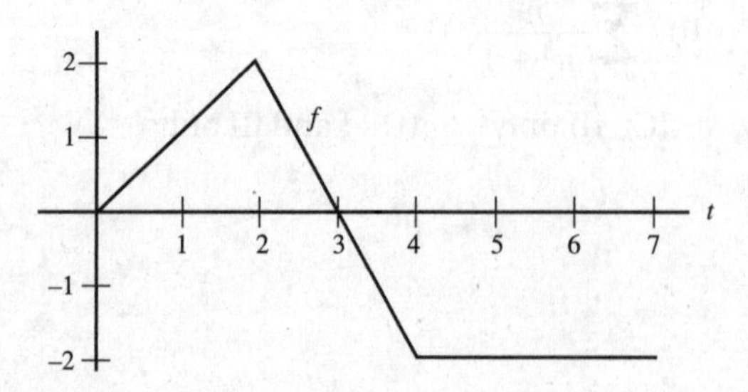

36. $G'(1)$ is

    (A) 1 (B) 2 (C) 3 (D) 6

GO ON TO THE NEXT PAGE

37. G has a local maximum at $x =$

(A) 1 (B) $\frac{4}{3}$ (C) 2 (D) 8

x	$f(x)$	$f'(x)$	$f''(x)$	$g(x)$	$g'(x)$	$g''(x)$
2	6	−1	−2	−2	1/3	−4/3

38. Given two twice-differentiable functions, $f(x)$ and $g(x)$. The table above gives values for $f(x)$ and $g(x)$ and their first and second derivatives at $x = 2$. Find $\lim_{x \to 2} \frac{f(x) + 3g(x)}{\frac{1}{2}x^2 - 2e^{x-2}}$.

(A) 0 (B) 1 (C) 6 (D) nonexistent

39. Using the left rectangular method and four subintervals of equal width, estimate $\int_0^8 |f(t)|\, dt$, where f is the function graphed below.

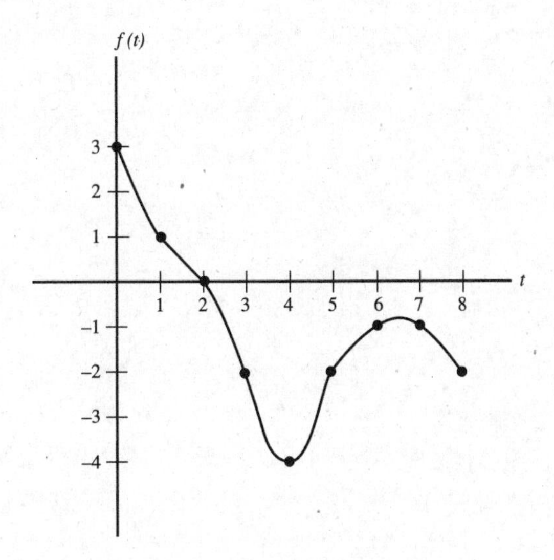

(A) 4 (B) 8 (C) 12 (D) 16

40. Given the function $f(x) = 2\cos(3x - 1) - 0.5e^{0.5x}$, find all values of x that satisfy the result of the Mean Value Theorem for the function f on the interval $[-3, -1]$. Note: The derivative of f is $f'(x) = -6\sin(3x - 1) - 0.25e^{0.5x}$.

(A) −2.800 and −1.772 (B) −2.242 and −1.296
(C) −2.812 and −1.760 (D) −2.843 and −1.729

41. The base of a solid is the region bounded by $x^2 = 4y$ and the line $y = 2$, and each plane section perpendicular to the y-axis is a square. The volume of the solid is

(A) 8 (B) 16 (C) 32 (D) 64

42. An object initially at rest at (3,3) moves with acceleration $a(t) = \langle 2, e^{-t} \rangle$. Where is the object at $t = 2$?

(A) (4, −0.865) (B) (4, 1.135) (C) (7, 2.135) (D) (7, 4.135)

GO ON TO THE NEXT PAGE

43. Find the length of the curve $y = \ln x$ between the points where $y = \frac{1}{2}$ and $y = 1$.

(A) 0.531 (B) 0.858 (C) 1.182 (D) 1.356

44. Using the first two terms in the Maclaurin series for $y = \cos x$ yields accuracy to within 0.001 over the interval $|x| < k$ when $k =$

(A) 0.394 (B) 0.707 (C) 0.786 (D) 0.788

45. After t years, $50e^{-0.015t}$ pounds of a deposit of a radioactive substance remain. The average amount per year *not* lost by radioactive decay during the second hundred years is

(A) 2.9 lb (B) 5.8 lb (C) 6.8 lb (D) 15.8 lb

STOP

END OF SECTION I

SECTION II

Part A

TIME: 30 MINUTES

2 PROBLEMS

A graphing calculator is required for some of these problems.
See instructions on page 4.

1. Let function f be continuous and decreasing, with values as shown in the table:

x	2.5	3.2	3.5	4.0	4.6	5.0
$f(x)$	7.6	5.7	4.2	3.8	2.2	1.6

(a) Use the trapezoid method to estimate the area between f and the x-axis on the interval $2.5 \le x \le 5.0$.
(b) Find the average rate of change of f on the interval $2.5 \le x \le 5.0$.
(c) Estimate the instantaneous rate of change of f at $x = 2.5$.
(d) If $g(x) = f^{-1}(x)$, estimate the slope of g at $x = 4$.

2. An object starts at point (1,3), and moves along the parabola $y = x^2 + 2$ for $0 \le t \le 2$, with the horizontal component of its velocity given by $\frac{dx}{dt} = \frac{4}{t^2 + 4}$.

(a) Find the object's position at $t = 2$.
(b) Find the object's speed at $t = 2$.
(c) Find the distance the object traveled during this interval.

STOP

END OF PART A, SECTION II

BC PRACTICE EXAMINATION 2

Part B

TIME: 60 MINUTES

4 PROBLEMS

No calculator is allowed for any of these problems.
If you finish Part B before time has expired, you may return to work on Part A, but you may not use a calculator.

3. Given a function f such that $f(3) = 1$ and $f^{(n)}(3) = \frac{(-1)^n n!}{(2n+1)2^n}$.

   (a) Write the first four nonzero terms and the general term of the Taylor series for f around $x = 3$.
   (b) Find the radius of convergence of the Taylor series.
   (c) Show that the third-degree Taylor polynomial approximates $f(4)$ to within 0.01.

4. The curve $y = \sqrt{8 \sin\left(\frac{\pi x}{6}\right)}$ divides a first quadrant rectangle into regions $\boldsymbol{A}$ and $\boldsymbol{B}$, as shown in the figure.

   (a) Region $\boldsymbol{A}$ is the base of a solid. Cross sections of this solid perpendicular to the x-axis are rectangles. The height of each rectangle is 5 times the length of its base in region $\boldsymbol{A}$. Find the volume of this solid.
   (b) The other region, $\boldsymbol{B}$, is rotated around the y-axis to form a different solid. Set up but do not evaluate an integral for the volume of this solid.

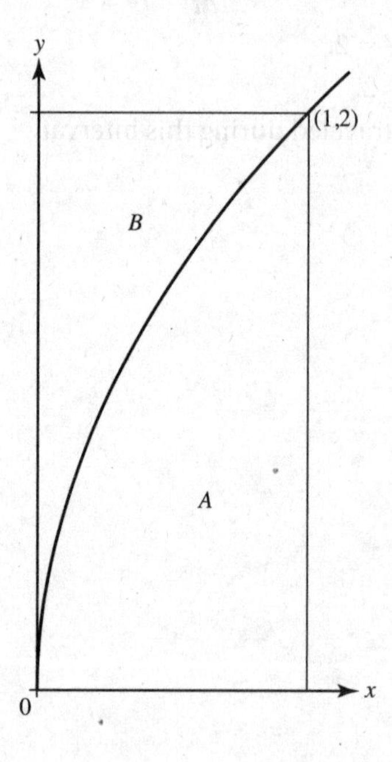

GO ON TO THE NEXT PAGE

5. A bungee jumper has reached a point in her exciting plunge where the taut cord is 100 feet long with a 1/2-inch radius, and stretching. She is still 80 feet above the ground and is now falling at 40 feet per second. You are observing her jump from a spot on the ground 60 feet from the potential point of impact, as shown in the diagram below.

   (a) Assuming the cord to be a cylinder with volume remaining constant as the cord stretches, at what rate is its radius changing when the radius is 1/2″?

   (b) From your observation point, at what rate is the angle of elevation to the jumper changing when the radius is 1/2″?

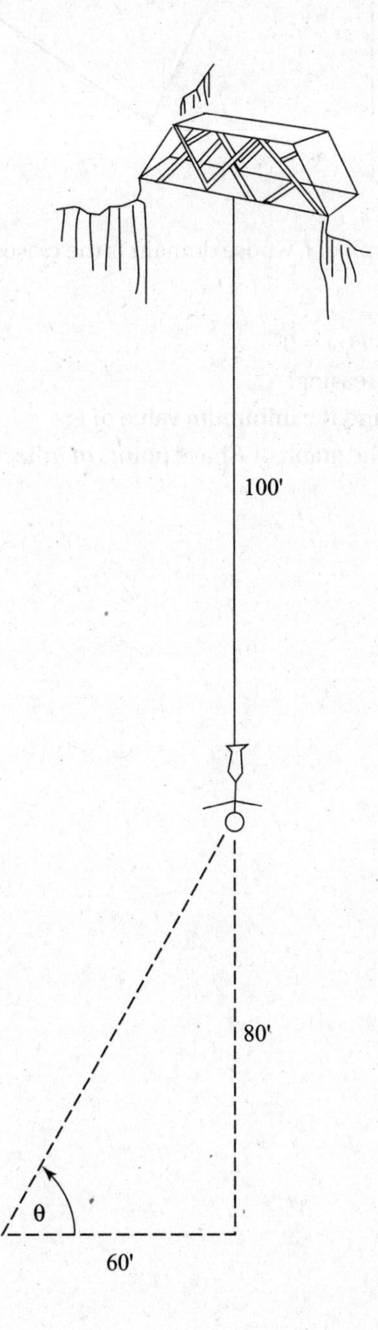

GO ON TO THE NEXT PAGE

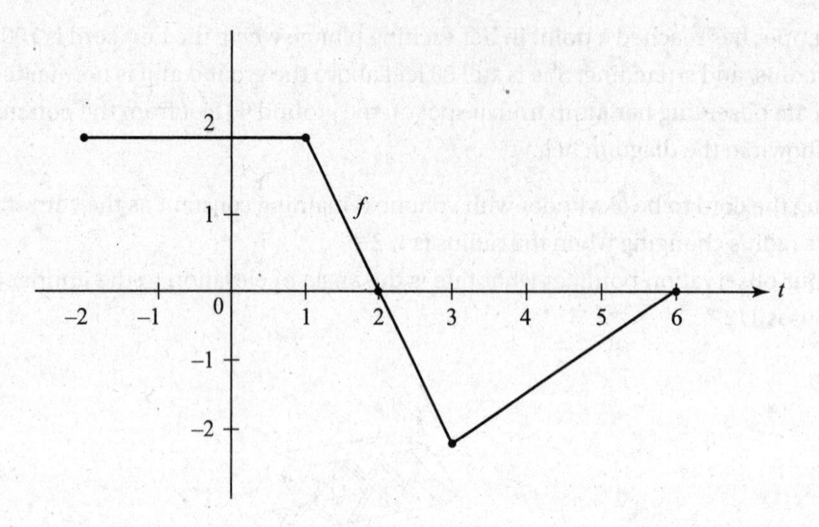

6. The figure above shows the graph of f, whose domain is the closed interval $[-2,6]$. Let $F(x) = \int_1^x f(t)dt$.

(a) Find $F(-2)$ and $F(6)$.

(b) For what value(s) of x does $F(x) = 0$?

(c) On what interval(s) is F increasing?

(d) Find the maximum value and the minimum value of F.

(e) At what value(s) of x does the graph of F have points of inflection? Justify your answer.

STOP

END OF TEST

ANSWER KEY
BC Practice Examination 2

Part A

1. **B**
2. **C**
3. **C**
4. **D**
5. **A**
6. **C**
7. **B**
8. **B**
9. **B**
10. **C**
11. **C**
12. **A**
13. **C**
14. **D**
15. **C**
16. **D**
17. **B**
18. **A**
19. **D**
20. **C**
21. **B**
22. **D**
23. **D**
24. **C**
25. **B**
26. **C**
27. **D**
28. **C**
29. **B**
30. **A**

Part B

31. **B**
32. **C**
33. **C**
34. **B**
35. **C**
36. **D**
37. **B**
38. **C**
39. **D**
40. **A**
41. **C**
42. **D**
43. **C**
44. **A**
45. **B**

ANSWERS EXPLAINED

The explanations for questions not given below will be found in the answer section for AB Practice Exam 2 on pages 530–535. Identical questions in Section I of Practice Examinations AB 2 and BC 2 have the same number. For example, explanations of the answers for Questions 4 and 5, not given below, will be found in Section I of Calculus AB Practice Exam 2, Answers 4 and 5, page 530.

Section I Multiple-Choice

Part A

1. **(B)** Since $\lim_{x\to 1} f(x) = 1$, to render $f(x)$ continuous at $x = 1$, define $f(1)$ to be 1.

2. **(C)** The limit $\lim_{x\to 0} \frac{\sin\left(\frac{x}{2}\right)}{x}$ is of the form $\frac{0}{0}$ so after applying L'Hôpital's Rule, we get $\lim_{x\to 0} \frac{\frac{1}{2}\cos\left(\frac{x}{2}\right)}{1} = \frac{1}{2}$. Or, using the substitution $\theta = \frac{x}{2}$, we can rewrite the limit as $\lim_{x\to 0} \frac{\sin\left(\frac{x}{2}\right)}{x} = \lim_{x\to 0} \frac{\sin\theta}{2\theta} = \frac{1}{2}\lim_{x\to 0} \frac{\sin\theta}{\theta} = \frac{1}{2}\cdot 1 = \frac{1}{2}$.

3. **(C)** Obtain the first few terms of the Maclaurin series generated by $f(x) = \sqrt{1+x}$:

$$f(x) = \sqrt{1+x}; \qquad f(0) = 1;$$

$$f'(x) = \frac{1}{2}(1+x)^{-1/2}; \qquad f'(0) = \frac{1}{2};$$

$$f''(x) = -\frac{1}{4}(1+x)^{-3/2}; \qquad f''(0) = -\frac{1}{4};$$

$$f'''(x) = \frac{3}{8}(1+x)^{-5/2}; \qquad f'''(0) = \frac{3}{8};$$

So $\sqrt{1+x} = 1 + \frac{x}{2} - \frac{1}{4}\cdot\frac{x^2}{2} + \frac{3}{8}\cdot\frac{x^3}{6} - \cdots$.

6. **(C)** Here,

$$\frac{dx}{dt} = -2\sin 2t, \quad \frac{dy}{dt} = 2\cos 2t,$$

and

$$\frac{d^2x}{dt^2} = -4\cos 2t, \quad \frac{d^2y}{dt^2} = -4\sin 2t;$$

and the magnitude of the acceleration, $|\mathbf{a}|$, is given by

$$|\mathbf{a}| = \sqrt{\left(\frac{d^2x}{dt^2}\right)^2 + \left(\frac{d^2y}{dt^2}\right)^2}$$

$$= \sqrt{(-4\cos 2t)^2 + (-4\sin 2t)^2}$$

$$= \sqrt{16\left(\cos^2 2t + \sin^2 2t\right)} = 4.$$

7. **(B)** $f'(x) = 1 + \frac{(x-1)}{2} + \frac{(x-1)^2}{3} + \frac{(x-1)^3}{4} + \cdots$

By the Ratio Test (page 398) the series converges when

$$\lim_{n\to\infty}\left|\frac{(x-1)^{n+1}}{n+2}\cdot\frac{n+1}{(x-1)^n}\right| < 1$$

$$|x-1| \lim_{n\to\infty}\frac{n+1}{n+2} < 1$$

$$|x-1| < 1.$$

Checking the endpoints, we find:

$f'(0) = 1 - \frac{1}{2} + \frac{1}{3} - \frac{1}{4} + \cdots$ is the alternating harmonic series, which converges.

$f'(2) = 1 + \frac{1}{2} + \frac{1}{3} + \frac{1}{4} + \cdots$ is the harmonic series, which diverges.

Hence the interval of convergence is $0 \le x < 2$.

10. **(C)** $\int_0^{\pi/3} (\tan x)^2 \left(\sec^2 x\, dx\right) = \left.\frac{\tan^3 x}{3}\right|_0^{\pi/3} = \frac{1}{3}\left(3\sqrt{3} - 0\right).$

11. **(C)** We integrate by parts using $u = \ln x$, $dv = dx$; then $du = \frac{1}{x}\,dx$, $v = x$, and

$$\int_1^e \ln x\, dx = \left.\left(x \ln x - \int x \frac{1}{x}\, dx\right)\right|_1^e = \left.(x \ln x - x)\right|_1^e$$

$$= e \ln e - e - (1 \ln 1 - 1) = e - e - (0 - 1) = 1.$$

13. **(C)** Use the method of partial fractions, letting

$$\frac{x-6}{x(x-3)} = \frac{A}{x} + \frac{B}{x-3}.$$

$$x - 6 = A(x-3) + Bx$$

Letting $x = 0$, we find $A = 2$, and letting $x = 3$ yields $B = -1$.

Now $\int \left(\frac{2}{x} - \frac{1}{x-3}\right) dx = 2 \ln|x| - \ln|x-3| + C.$

19. **(D)** At (2,1), $\frac{dy}{dx} = 3$. Use $\Delta x = 0.1$; then Euler's method moves to $(2.1, 1 + 3(0.1))$.

At (2.1, 1.3), $\frac{dy}{dx} = 3.4$, so the next point is $(2.2, 1.3 + 3.4(0.1))$.

22. **(D)** Separate variables to get $\frac{dy}{y} = \frac{dx}{x}$, and integrate to get $\ln y = \ln x + C$.

Since $y = 3$ when $x = 1$, $C = \ln 3$. Then $y = e^{(\ln x + \ln 3)} = e^{\ln x} \cdot e^{\ln 3} = 3x$.

23. **(D)** The generating circle has equation $x^2 + y^2 = 4$. Using disks, the volume, V, is given by

$$V = \pi \int_{-1}^{1} x^2\, dy = 2\pi \int_0^1 \left(4 - y^2\right) dy = 2\pi \left.\left(4y - \frac{y^3}{3}\right)\right|_0^1.$$

24. **(C)** $\int_0^\infty \frac{dx}{x^2+1} = \lim_{b\to\infty} \left.\tan^{-1} x\right|_0^b = \frac{\pi}{2}$. The integrals in (A), (B), and (D) all diverge to infinity.

26. **(C)** Using separation of variables:

$$\frac{dy}{dx} = 1 + y^2$$

$$\int \frac{dy}{1 + y^2} = \int dx$$

$$\arctan y = x + C$$

$$y = \tan(x + C)$$

Given initial point (0,0), we have $0 = \tan(0 + C)$; hence $C = 0$ and the particular solution is $y = \tan(x)$. Because this function has vertical asymptotes at $x = \pm\frac{\pi}{2}$ and the particular solution must be differentiable in an interval containing the initial point $x = 0$, the domain is $|x| < \frac{\pi}{2}$.

28. **(C)** $\displaystyle\int_1^2 \frac{1}{\sqrt{4 - x^2}}\,dx = \lim_{h \to 2^-} \int_1^h \frac{1}{\sqrt{4 - x^2}}\,dx = \lim_{h \to 2^-} \int_1^h \frac{\left(\frac{1}{2}dx\right)}{\sqrt{1 - \left(\frac{x}{2}\right)^2}} = \lim_{h \to 2^-} \sin^{-1}\frac{x}{2}\Big|_1^h$

$$= \lim_{h \to 2^-}\left(\sin^{-1}\frac{h}{2} - \sin^{-1}\frac{1}{2}\right) = \frac{\pi}{2} - \frac{\pi}{6} = \frac{\pi}{3}.$$

29. **(B)** From the Riemann Sum, we see $\Delta x = \frac{1}{n}$, then $k \cdot \Delta x = \frac{k}{n}$. Notice that the term involving k in the Riemann Sum is not equal to $\frac{k}{n}$ but $2\left(\frac{k}{n}\right)$. Thus, the only choice for x_k is $x_k = \frac{k}{n}$, so $a = 0$ and $\Delta x = \frac{b - 0}{n} = \frac{1}{n}$, so $b = 1$. Since x_k replaces x, $f(x) = \sqrt{2x + 3}$ giving the integral $\int_0^1 \sqrt{2x + 3}\,dx$.

30. **(A)** I. $\displaystyle\sum_{n=1}^{\infty} \frac{3^n}{n!}$ converges by the Ratio Test: $\displaystyle\lim_{n \to \infty} \left|\frac{3^{n+1}}{(n+1)!} \cdot \frac{n!}{3^n}\right| = \lim_{n \to \infty} \left|\frac{3}{n+1}\right| = 0 < 1.$

II. $\displaystyle\sum_{n=1}^{\infty} \frac{3^n}{n^3}$ diverges by the nth Term Test: $\displaystyle\lim_{n \to \infty} \frac{3^n}{n^3} = \infty.$

III. $\displaystyle\sum_{n=1}^{\infty} \frac{n^2}{n^3 + 1}$ diverges by the Limit Comparison Test: $\displaystyle\lim_{n \to \infty} \frac{\frac{n^2}{n^3+1}}{1/n} = \lim_{n \to \infty} \frac{n^3}{n^3 + 1} = 1$; therefore both series diverge since we know that the harmonic series, $\displaystyle\sum_{n=1}^{\infty} \frac{1}{n}$, diverges. Note that the Comparison Test is not appropriate for this series because $\frac{n^2}{n^3 + 1} < \frac{1}{n}$.

Part B

31. **(B)** Since the equation of the spiral is $r = \ln\theta$, use the polar mode. The formula for area in polar coordinates is

$$\frac{1}{2}\int_{\theta_1}^{\theta_2} r^2\,d\theta.$$

Therefore, calculate

$$0.5\int_{\pi}^{2\pi} \ln^2\theta\,d\theta.$$

The result is 3.743.

32. **(C)** When $y = 2^t = 4$, we have $t = 2$, so the line passes through point $F(2) = (5,4)$.

Also $\dfrac{dy}{dx} = \dfrac{\frac{dy}{dt}}{\frac{dx}{dt}} = \dfrac{2^t \ln 2}{2t}$, so at $t = 2$ the slope of the tangent line is

$\dfrac{dy}{dx} = \ln 2.$

An equation for the tangent line is $y - 4 = \ln 2(x - 5)$.

38. **(C)** The limit $\lim_{x\to 2} \dfrac{f(x) + 3g(x)}{\frac{1}{2}x^2 - 2e^{x-2}}$ by substitution is of the form $\dfrac{6 + 3(-2)}{\frac{1}{2}(2)^2 - 2e^{2-2}} = \dfrac{0}{0}$, so apply

L'Hôpital's Rule. You get $\lim_{x\to 2} \dfrac{f'(x) + 3g'(x)}{x - 2e^{x-2}}$, which is still of the form $\dfrac{-1 + 3(1/3)}{2 - 2e^{2-2}} = \dfrac{0}{0}$; therefore apply

L'Hôpital's Rule again. The new limit is $\lim_{x\to 2} \dfrac{f''(x) + 3g''(x)}{1 - 2e^{x-2}}$ by substitution is $\dfrac{-2 + 3(-4/3)}{1 - 2e^{2-2}} = \dfrac{-6}{-1} = 6.$

40. **(A)** $f(x)$ is differentiable on $[-3, -1]$ so MVT guarantees that there exists at least one x value such that

$f'(x) = \dfrac{f(-1) - f(-3)}{-1 - (-3)} = 0.08957778$; solving this equation on your graphing calculator yields

$x = -2.800, -1.772$ on $[-3, -1]$.

41. **(C)** See figure below.

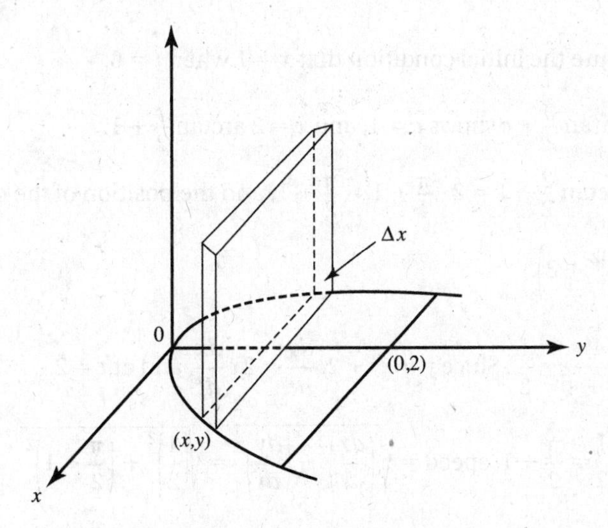

$$\Delta V = (2x)^2\,\Delta y = 4x^2\,\Delta y,$$

$$= 16y\,\Delta y$$

$$V = \int_0^2 16y\,dy.$$

42. **(D)** At $t = 0$, $\mathbf{R}(0) = \langle 3,3\rangle$ and $\mathbf{v}(0) = \langle 0,0\rangle$. Given $a(t) = \langle 2, e^{-t}\rangle$, using the FTC

$\mathbf{v}(t) = \left\langle 0 + \int_0^t 2du,\ 0 + \int_0^t e^{-u}\,du\right\rangle = \left\langle 2u\Big|_0^t,\ -e^{-u}\Big|_0^t\right\rangle = \langle 2t, -e^{-t} + 1\rangle$ and

$\mathbf{R}(2) = \left\langle 3 + \int_0^2 2tdt,\ 3 + \int_0^2 (-e^{-t} + 1)\,dt\right\rangle = \langle 7, 4.135\rangle$

43. **(C)** The endpoints of the arc are $\left(\sqrt{e}, \frac{1}{2}\right)$ and $(e,1)$. The arc length is given by

$$\int_{x_1}^{x_2} \sqrt{1+\left(\frac{dy}{dx}\right)^2}\,dx = \int_{\sqrt{e}}^{e} \sqrt{1+\left(\frac{1}{x}\right)^2}\,dx \text{ or } \int_{y_1}^{y_2} \sqrt{1+\left(\frac{dx}{dy}\right)^2}\,dy = \int_{1/2}^{1} \sqrt{1+(e^y)^2}\,dy.$$

44. **(A)** Find k such that $\cos x$ will differ from $\left(1-\frac{x^2}{2}\right)$ by less than 0.001 at $x = k$.

Solve

$$0 = \cos x - \left(1-\frac{x^2}{2}\right) - 0.001,$$

which yields x or $k = 0.394$.

Section II Free-Response

Part A

1. See solution for AB/BC 1, page 536.

2. (a) Position is the antiderivative of: $\frac{dx}{dt} = \frac{4}{t^2+4}$:

$$x = \int \frac{4}{t^2+4}\,dt = \frac{4}{4}\int \frac{1}{\left(\frac{t}{2}\right)^2+1}\,dt = 2\int \frac{\frac{1}{2}\,dt}{\left(\frac{t}{2}\right)^2+1}$$

$$= 2\arctan\frac{t}{2} + c.$$

To find c, substitute the initial condition that $x = 1$ when $t = 0$:

$$1 = 2\arctan\frac{0}{2} + c \text{ shows } c = 1, \text{ and } x = 2\arctan\frac{t}{2} + 1.$$

At $t = 2$, $x = 2\arctan\frac{2}{2} + 1 = 2\cdot\frac{\pi}{4} + 1 = \frac{\pi}{2} + 1$, and the position of the object is $\left(\frac{\pi}{2}+1, \left(\frac{\pi}{2}+1\right)^2 + 2\right)$.

(b) At $t = 2$, $\frac{dx}{dt} = \frac{4}{2^2+4} = \frac{1}{2}$. Since $y = x^2 + 2$, $\frac{dy}{dt} = 2x\frac{dx}{dt}$, and at $t = 2$,

$$\frac{dy}{dt} = 2\left(\frac{\pi}{2}+1\right)\cdot\frac{1}{2} = \frac{\pi}{2} + 1, \text{ speed} = \sqrt{\left(\frac{dx}{dt}\right)^2 + \left(\frac{dy}{dt}\right)^2} = \sqrt{\left(\frac{1}{2}\right)^2 + \left(\frac{\pi}{2}+1\right)^2}.$$

(c) The distance traveled is the length of the arc of $y = x^2 + 2$ in the interval $1 < x < \frac{\pi}{2} + 1$:

$$L = \int \sqrt{1+\left(\frac{dy}{dx}\right)^2}\,dx = \int_1^{\pi/2+1} \sqrt{1+(2x)^2}\,dx = 5.839.$$

Part B

3. (a)

n	$f^{(n)}(3) = \frac{(-1)^n n!}{(2n+1)2^n}$	$a_n = \frac{f^{(n)}(3)}{n!}$
0	1	1
1	$\frac{-1}{3 \cdot 2}$	$\frac{-1}{3 \cdot 2}$
2	$\frac{2!}{5 \cdot 2^2}$	$\frac{1}{5 \cdot 2^2}$
3	$\frac{-3!}{7 \cdot 2^3}$	$\frac{-1}{7 \cdot 2^3}$

$$f(x) = 1 - \frac{1}{6}(x-3) + \frac{1}{20}(x-3)^2 - \frac{1}{56}(x-3)^3 + \cdots + \frac{(-1)^n (x-3)^n}{(2n+1) \cdot 2^n} + \cdots.$$

(b) By the Ratio Test, the series converges when

$$\lim_{x \to \infty} \left| \frac{(x-3)^{n+1}}{(2n+3) \cdot 2^{n+1}} \cdot \frac{(2n+1) \cdot 2^n}{(x-3)^n} \right| < 1,$$

$$|x-3| \lim_{x \to \infty} \left| \frac{(2n+1)}{(2n+3) \cdot 2} \right| < 1,$$

$$|x-3| < 2.$$

Thus, the radius of convergence is 2.

(c) $f(4) = \sum_{n=0}^{\infty} \frac{(-1)^n}{(2n+1)2^n}$ is an alternating series. Since $\frac{1}{(2n+3)2^{n+1}} < \frac{1}{(2n+1)2^n}$ and $\lim_{n \to \infty} \frac{1}{(2n+1)2^n} = 0$ it converges by the Alternating Series Test. Therefore the error is less than the magnitude of the first omitted term:

$$\text{error} < \frac{(4-3)^4}{(2 \cdot 4 + 1) \cdot 2^4} = \frac{1}{144} < 0.01.$$

4. See solution for AB 2/BC 4, page 536.

5. See solution for AB/BC 5, page 538.

6. See solution for AB/BC 6, page 539.

ANSWER SHEET
BC Practice Examination 3

Part A

1. Ⓐ Ⓑ Ⓒ Ⓓ	9. Ⓐ Ⓑ Ⓒ Ⓓ	17. Ⓐ Ⓑ Ⓒ Ⓓ	25. Ⓐ Ⓑ Ⓒ Ⓓ
2. Ⓐ Ⓑ Ⓒ Ⓓ	10. Ⓐ Ⓑ Ⓒ Ⓓ	18. Ⓐ Ⓑ Ⓒ Ⓓ	26. Ⓐ Ⓑ Ⓒ Ⓓ
3. Ⓐ Ⓑ Ⓒ Ⓓ	11. Ⓐ Ⓑ Ⓒ Ⓓ	19. Ⓐ Ⓑ Ⓒ Ⓓ	27. Ⓐ Ⓑ Ⓒ Ⓓ
4. Ⓐ Ⓑ Ⓒ Ⓓ	12. Ⓐ Ⓑ Ⓒ Ⓓ	20. Ⓐ Ⓑ Ⓒ Ⓓ	28. Ⓐ Ⓑ Ⓒ Ⓓ
5. Ⓐ Ⓑ Ⓒ Ⓓ	13. Ⓐ Ⓑ Ⓒ Ⓓ	21. Ⓐ Ⓑ Ⓒ Ⓓ	29. Ⓐ Ⓑ Ⓒ Ⓓ
6. Ⓐ Ⓑ Ⓒ Ⓓ	14. Ⓐ Ⓑ Ⓒ Ⓓ	22. Ⓐ Ⓑ Ⓒ Ⓓ	30. Ⓐ Ⓑ Ⓒ Ⓓ
7. Ⓐ Ⓑ Ⓒ Ⓓ	15. Ⓐ Ⓑ Ⓒ Ⓓ	23. Ⓐ Ⓑ Ⓒ Ⓓ	
8. Ⓐ Ⓑ Ⓒ Ⓓ	16. Ⓐ Ⓑ Ⓒ Ⓓ	24. Ⓐ Ⓑ Ⓒ Ⓓ	

Part B

31. Ⓐ Ⓑ Ⓒ Ⓓ	35. Ⓐ Ⓑ Ⓒ Ⓓ	39. Ⓐ Ⓑ Ⓒ Ⓓ	43. Ⓐ Ⓑ Ⓒ Ⓓ
32. Ⓐ Ⓑ Ⓒ Ⓓ	36. Ⓐ Ⓑ Ⓒ Ⓓ	40. Ⓐ Ⓑ Ⓒ Ⓓ	44. Ⓐ Ⓑ Ⓒ Ⓓ
33. Ⓐ Ⓑ Ⓒ Ⓓ	37. Ⓐ Ⓑ Ⓒ Ⓓ	41. Ⓐ Ⓑ Ⓒ Ⓓ	45. Ⓐ Ⓑ Ⓒ Ⓓ
34. Ⓐ Ⓑ Ⓒ Ⓓ	38. Ⓐ Ⓑ Ⓒ Ⓓ	42. Ⓐ Ⓑ Ⓒ Ⓓ	

BC Practice Examination 3

SECTION I

Part A

TIME: 60 MINUTES

The use of calculators is not permitted for this part of the examination.
There are 30 questions in Part A, for which 60 minutes are allowed. Because there is no deduction for wrong answers, you should answer every question, even if you need to guess.

Directions: Choose the best answer for each question.

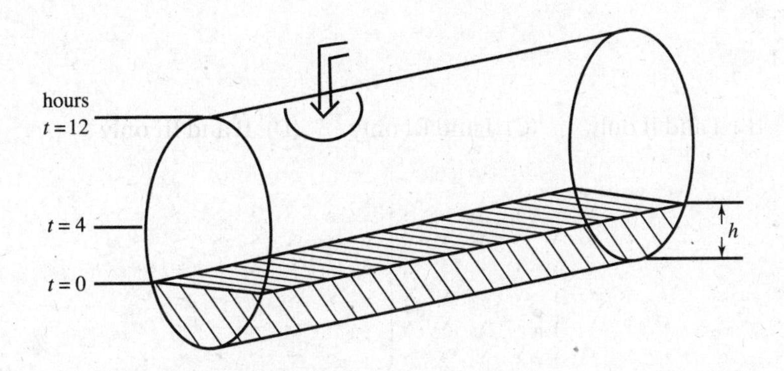

1. A cylindrical tank, shown in the figure above, is partially full of water at time $t = 0$, when more water begins flowing in at a constant rate. The tank becomes half full when $t = 4$, and is completely full when $t = 12$. Let h represent the height of the water at time t. During which interval is $\frac{dh}{dt}$ increasing?

   (A) $0 < t < 4$ (B) $0 < t < 8$ (C) $0 < t < 12$ (D) $4 < t < 12$

2. $\lim_{h \to 0} \frac{\sin\left(\frac{\pi}{2} + h\right) - 1}{h}$ is

   (A) 1 (B) -1 (C) 0 (D) nonexistent

3. $\sum_{n=0}^{\infty} \frac{3n}{4^{n+1}}$

   (A) $= 1.$ (B) $= 3.$ (C) $= 4.$ (D) diverges.

GO ON TO THE NEXT PAGE

BC PRACTICE EXAMINATION 3

4. The equation of the tangent to the curve $2x^2 - y^4 = 1$ at the point $(-1, 1)$ is

(A) $x + y = 0$ (B) $x + y = 2$ (C) $x - y = 0$ (D) $x - y = -2$

5. The nth term of the Taylor series expansion about $x = 0$ of the function $f(x) = \dfrac{1}{1 + 2x}$ is

(A) $(2x)^n$ (B) $2x^n$ (C) $(-1)^{n-1}(2x)^{n-1}$ (D) $(-1)^n (2x)^{n-1}$

6. When the method of partial fractions is used to decompose $\dfrac{2x+3}{x^2 - 3x + 2}$, one of the fractions obtained is

(A) $-\dfrac{7}{x-1}$ (B) $-\dfrac{5}{x-1}$ (C) $\dfrac{5}{x-1}$ (D) $\dfrac{7}{x-1}$

7. A local maximum value of the function $y = \dfrac{\ln x}{x}$ is

(A) 0 (B) $\dfrac{1}{e}$ (C) 1 (D) e

8. Which of the following series converge?

I. $\sum_{n=1}^{\infty} n^2 e^{-n}$

II. $\sum_{n=1}^{\infty} (-1)^n 3^{1/n}$

III. $\sum_{n=1}^{\infty} \dfrac{n^2 - 1}{n^3 + 1}$

(A) I only (B) I and II only (C) I and III only (D) II and III only

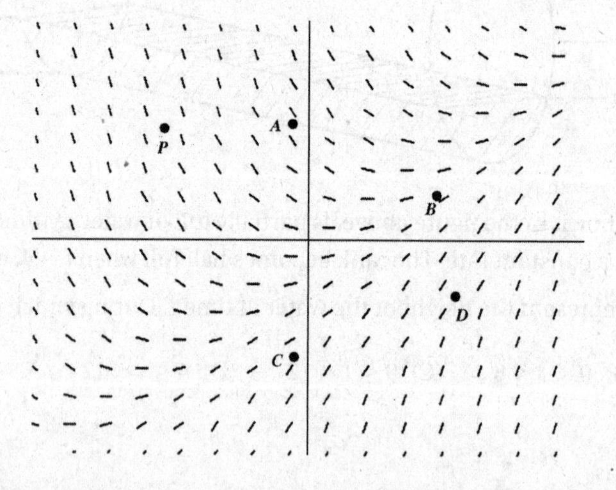

9. A particular solution of the differential equation whose slope field is shown above contains point P. This solution may also contain which other point?

(A) A (B) B (C) C (D) D

GO ON TO THE NEXT PAGE

BC PRACTICE EXAMINATION 3

10. Let $F(x) = \int_5^x \frac{dt}{1-t^2}$. Which of the following statements is (are) true?

I. The domain of F is $x \neq \pm 1$.

II. $F(2) > 0$.

III. The graph of F is concave upward.

(A) I only (B) II only (C) III only (D) II and III only

11. As the tides change, the water level in a bay varies sinusoidally. At high tide today at 8 A.M., the water level was 15 feet; at low tide, 6 hours later at 2 P.M., it was 3 feet. How fast, in feet per hour, was the water level dropping at noon today?

(A) $\frac{\pi}{2}$ (B) $\frac{\pi\sqrt{3}}{2}$ (C) $3\sqrt{3}$ (D) $6\sqrt{3}$

12. Let $\int_0^x f(t)\,dt = x \sin \pi x$. Then $f(3)=$

(A) -3π (B) -1 (C) 1 (D) 3π

13. $\int \frac{e^u}{1+e^{2u}}\,du$ is equal to

(A) $\ln(1+e^{2u})+C$ (B) $\frac{1}{2}\ln|1+e^{2u}|+C$ (C) $\frac{1}{2}\tan^{-1}e^{2u}+C$ (D) $\tan^{-1}e^u+C$

14. Given $f(x) = \log_{10}x$ and $\log_{10}(102) \simeq 2.0086$, which is closest to $f'(100)$?

(A) 0.0043 (B) 0.0086 (C) 0.01 (D) 1.0043

15. If $G(2) = 5$ and $G'(x) = \frac{10x}{9-x^2}$, then an estimate of $G(2.2)$ using a tangent-line approximation is

(A) 4.4 (B) 5.8 (C) 6 (D) 11.6

16. The area bounded by the parabola $y = x^2$ and the lines $y = 1$ and $y = 9$ equals

(A) $\frac{20}{3}$ (B) $\frac{52}{3}$ (C) 36 (D) $\frac{104}{3}$

17. The first-quadrant region bounded by $y = \frac{1}{\sqrt{x}}$, $y = 0$, $x = q$ $(0 < q < 1)$, and $x = 1$ is rotated about the x-axis. The volume obtained as $q \to 0^+$

(A) $=\frac{2\pi}{3}$ (B) $=\frac{4\pi}{3}$ (C) $=2\pi$ (D) diverges

18. A curve is given parametrically by the equations

$$x = 3 - 2\sin t \text{ and } y = 2\cos t - 1.$$

The length of the arc from $t = 0$ to $t = \pi$ is

(A) $\pi\sqrt{2}$ (B) π (D) 2π (D) 4π

19. Suppose the graph of f is both increasing and concave up on $a \leqslant x \leqslant b$. Then, using the same number of subdivisions, and with L, R, M, and T denoting, respectively, left, right, midpoint, and trapezoid sums, it follows that

(A) $R \leqslant T \leqslant M \leqslant L$ (B) $L \leqslant T \leqslant M \leqslant R$ (C) $R \leqslant M \leqslant T \leqslant L$ (D) $L \leqslant M \leqslant T \leqslant R$

GO ON TO THE NEXT PAGE

20. Which of the following statements about the graph of $y = \frac{x^2}{x-2}$ is (are) true?

I. The graph has no horizontal asymptote.
II. The line $x = 2$ is a vertical asymptote.
III. The line $y = x + 2$ is an oblique asymptote.

(A) I only (B) II only (C) I and III only (D) I, II, and III

21. The only function that does not satisfy the Mean Value Theorem on the interval specified is

(A) $f(x) = x^2 - 2x$ on $[-3, 1]$

(B) $f(x) = \frac{1}{x}$ on $[1, 3]$

(C) $f(x) = \frac{x^3}{3} - \frac{x^2}{2} + x$ on $[-1, 2]$

(D) $f(x) = x + \frac{1}{x}$ on $[-1, 1]$

22. $\int_0^1 x^2 e^x dx$

(A) $-3e - 2$ (B) $e - 2$ (C) $3e$ (D) $4e - 1$

23. Suppose $f(3) = 2$, $f'(3) = 5$, and $f''(3) = -2$. Then $\frac{d^2}{dx^2}\left(f^2(x)\right)$ at $x = 3$ is equal to

(A) -20 (B) 10 (C) 20 (D) 42

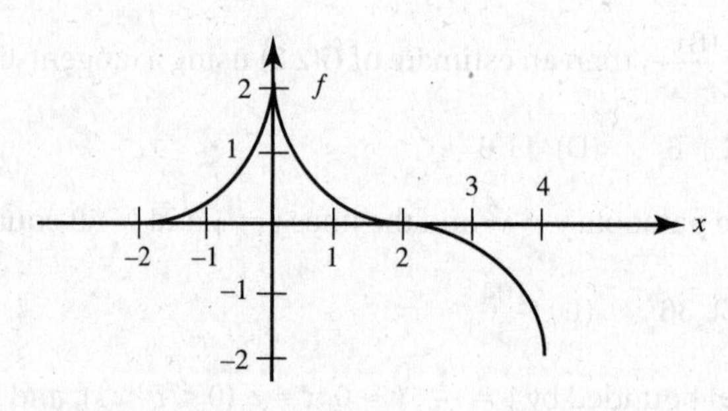

24. The graph of function f shown above consists of three quarter-circles.

Which of the following is (are) equivalent to $\int_0^2 f(x)dx$?

I. $\frac{1}{2}\int_{-2}^{2} f(x)dx$

II. $\int_4^2 f(x)dx$

III. $\frac{1}{2}\int_0^4 f(x)dx$

(A) I only (B) II only (C) I and II only (D) I, II, and III

25. The base of a solid is the first-quadrant region bounded by $y = \sqrt[4]{4 - 2x}$, and each cross section perpendicular to the x-axis is a semicircle with a diameter in the xy-plane. The volume of the solid is

(A) $\frac{\pi}{2}\int_0^2 \sqrt{4 - 2x}\,dx$ (B) $\frac{\pi}{4}\int_0^2 \sqrt{4 - 2x}\,dx$ (C) $\frac{\pi}{8}\int_0^2 \sqrt{4 - 2x}\,dx$ (D) $\frac{\pi}{8}\int_{-2}^2 \sqrt{4 - 2x}\,dx$

GO ON TO THE NEXT PAGE

26. The average value of $f(x) = 3 + |x|$ on the interval $[-2, 4]$ is

(A) $\frac{1}{3}$ (B) 2 (C) $\frac{14}{3}$ (D) 14

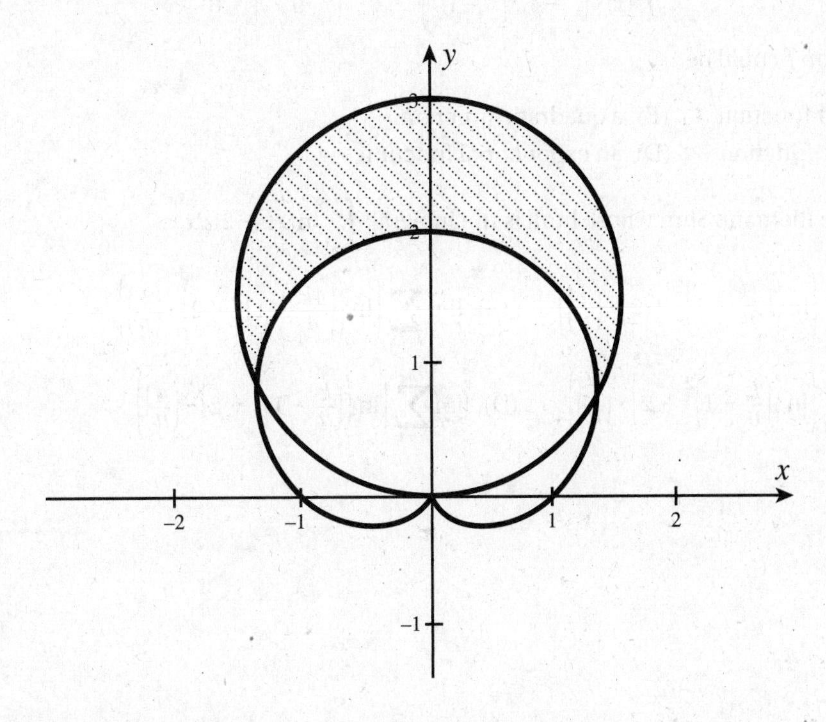

27. The area inside the circle $r = 3\sin\theta$ and outside the cardioid $r = 1 + \sin\theta$, shaded in the figure above, is given by

(A) $\int_{\pi/6}^{\pi/2} \left[9\sin^2\theta - (1 + \sin\theta)^2\right] d\theta$ (B) $\int_{\pi/6}^{\pi/2} (2\sin\theta - 1)^2\, d\theta$

(C) $\frac{1}{2}\int_{\pi/6}^{5\pi/6} (8\sin^2\theta - 1)\, d\theta$ (D) $\frac{9\pi}{4} - \frac{1}{2}\int_{\pi/6}^{5\pi/6} (1 + \sin\theta)^2\, d\theta$

28. Let $f(x) = \begin{cases} \dfrac{x^2 - 36}{x - 6} & \text{if } x \neq 6, \\ 12 & \text{if } x = 6. \end{cases}$

Which of the following statements is (are) true?

I. f is defined at $x = 6$.
II. $\lim_{x \to 6} f(x)$ exists.
III. f is continuous at $x = 6$.

(A) I only (B) II only (C) I and II only (D) I, II, and III

GO ON TO THE NEXT PAGE

29. The table below shows values of $f''(x)$ for various values of x:

x	-1	0	1	2	3
$f''(x)$	-4	-1	2	5	8

The function f could be

(A) a linear function (B) a quadratic function
(C) a cubic function (D) an exponential function

30. Choose the Riemann Sum whose limit is the integral: $\int_{-1}^{3} \ln(x^2 + 2)dx$.

(A) $\lim_{n\to\infty} \sum_{k=1}^{n} \left(\ln\left(\left(\frac{4k}{n} - 1 \right)^2 + 2 \right) \cdot \left(\frac{4}{n} \right) \right)$ (B) $\lim_{n\to\infty} \sum_{k=1}^{n} \left(\ln\left(\left(\frac{4k}{n} - 1 \right)^2 + 2 \right) \cdot \left(\frac{1}{n} \right) \right)$

(C) $\lim_{n\to\infty} \sum_{k=1}^{n} \left(\ln\left(\left(\frac{k}{n} - 1 \right)^2 + 2 \right) \cdot \left(\frac{1}{n} \right) \right)$ (D) $\lim_{n\to\infty} \sum_{k=1}^{n} \left(\ln\left(\left(\frac{k}{n} - 1 \right)^2 + 2 \right) \cdot \left(\frac{4}{n} \right) \right)$

STOP

END OF PART A, SECTION I

Part B

TIME: 45 MINUTES

Some questions in this part of the examination require the use of a graphing calculator. There are 15 questions in Part B, for which 45 minutes are allowed. Because there is no deduction for wrong answers, you should answer every question, even if you need to guess.

Directions: Choose the best answer for each question. If the exact numerical value of the correct answer is not listed as a choice, select the choice that is closest to the exact numerical answer.

31. Where, in the first quadrant, does the rose $r = \sin 3\theta$ have a vertical tangent?

(A) $\theta = 0.393$ (B) $\theta = 0.468$ (C) $\theta = 0.598$ (D) $\theta = 0.659$

32. A cup of coffee placed on a table cools at a rate of $\frac{dH}{dt} = -0.05(H - 70)$°F per minute, where H represents the temperature of the coffee and t is time in minutes. If the coffee was at 120°F initially, what will its temperature be 10 minutes later?

(A) 73°F (B) 95°F (C) 100°F (D) 105°F

33. An investment of \$4000 grows at the rate of $320e^{0.08t}$ dollars per year after t years. Its value after 10 years is approximately

(A) \$12,902 (B) \$8902 (C) \$7122 (D) \$4902

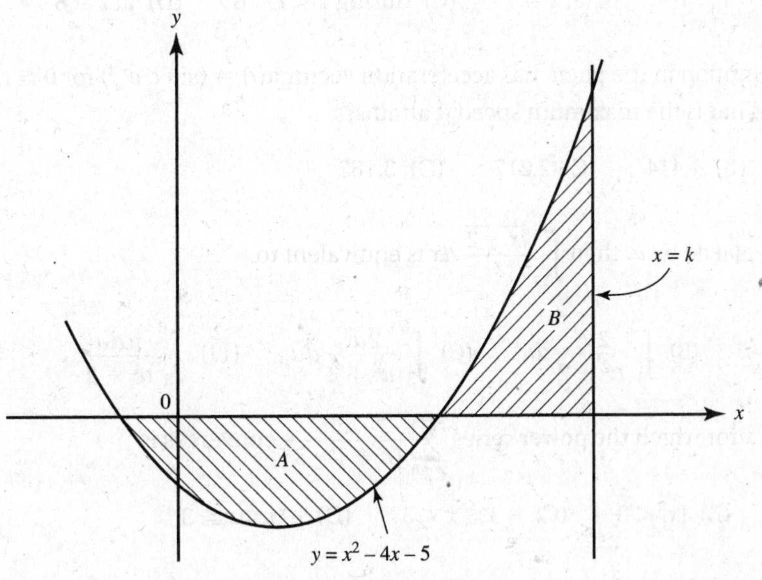

(This figure is not drawn to scale.)

34. The sketch shows the graphs of $f(x) = x^2 - 4x - 5$ and the line $x = k$. The regions labeled A and B have equal areas if $k =$

(A) 7.766 (B) 7.899 (C) 8 (D) 11

GO ON TO THE NEXT PAGE

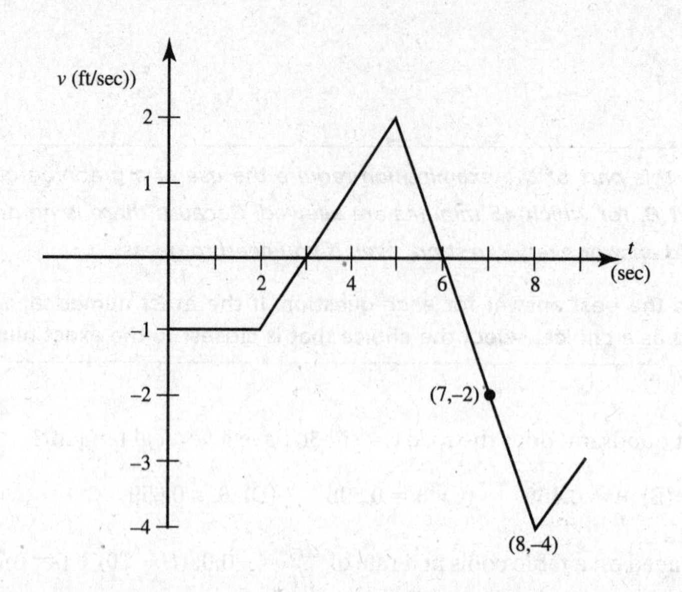

Questions 35 and 36. The graph shows the velocity of an object during the interval $0 \leq t \leq 9$.

35. The object attains its greatest speed at $t =$

(A) 3 sec (B) 5 sec (C) 6 sec (D) 8 sec

36. The object was at the origin at $t = 3$. It returned to the origin

(A) during $6 < t < 7$ (B) at $t = 7$ (C) during $7 < t < 8$ (D) at $t = 8$

37. An object in motion in the plane has acceleration vector $\mathbf{a}(t) = \langle \sin t, e^{-t} \rangle$ for $0 \leqq t \leqq 5$. It is at rest when $t = 0$. What is the maximum speed it attains?

(A) 1.022 (B) 1.414 (C) 2.217 (D) 3.162

38. If $\sqrt{x-2}$ is replaced by u, then $\int_3^6 \frac{\sqrt{x-2}}{x}\, dx$ is equivalent to

(A) $\int_1^2 \frac{u\,du}{u^2+2}$ (B) $\int_1^2 \frac{2u^2}{u^2+2}\,du$ (C) $\int_3^6 \frac{2u^2}{u^2+2}\,du$ (D) $\int_3^6 \frac{u\,du}{u^2+2}$

39. The set of all x for which the power series $\sum_{n=0}^{\infty} \frac{x^n}{(n+1)\cdot 3^n}$ converges is

(A) $\{-3, 3\}$ (B) $|x| < 3$ (C) $-3 \leqq x < 3$ (D) $-3 < x \leqq 3$

40. A particle moves along a line with acceleration $a = 6t$. If, when $t = 0$, $v = 1$, then the total distance traveled between $t = 0$ and $t = 3$ equals

(A) 30 (B) 28 (C) 27 (D) 26

GO ON TO THE NEXT PAGE

41. The definite integral $\int_1^{10} \sqrt{1 + \frac{9}{x^2}}\, dx$ represents the length of an arc of an increasing function $f(x)$. If one end of the arc is at the point (1,2), then an equation describing the curve is

(A) $y = 3 \ln x + 2$ (B) $y = 3 \ln x$ (C) $y = 11 - \frac{9}{x}$ (D) $y = 9 - \frac{9}{x}$

42. Two objects in motion from $t = 0$ to $t = 3$ seconds have positions $x_1(t) = \cos\left(t^2 + 1\right)$ and $x_2(t) = \frac{e^t}{2t}$, respectively. How many times during the 3 seconds do the objects have the same velocity?

(A) 0 (B) 3 (C) 4 (D) 6

43. Which statement is true?

(A) If $f'(c) = 0$, then f has a local maximum or minimum at $(c, f(c))$.
(B) If $f''(c) = 0$, then the graph of f has an inflection point at $(c, f(c))$.
(C) If f is differentiable at $x = c$, then f is continuous at $x = c$.
(D) If f is continuous on (a, b), then f maintains a maximum value on (a, b).

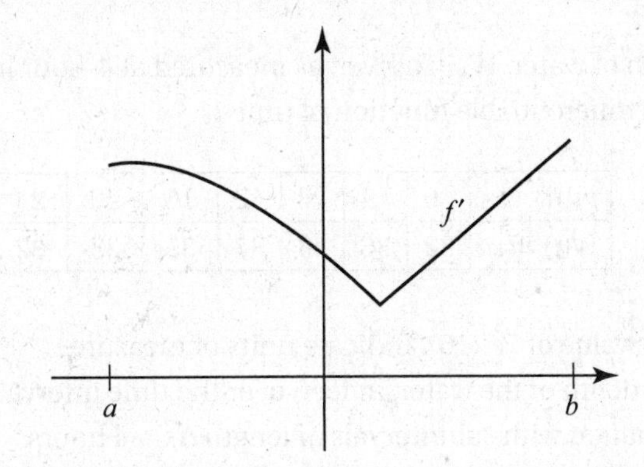

44. The graph of f' is shown above. Which statements about f must be true for $a < x < b$?

I. f is increasing.
II. f is continuous.
III. f is differentiable.

(A) I only (B) II only (C) I and II only (D) I, II, and III

45. Given $\int_{-1}^{4} g(x)dx = 0$ and $g'(x) < 0$ for all x, select the table that could represent g on the interval $[-1,4]$.

(A)

x	y
−1	3
1	3
4	3

(B)

x	y
−1	7
1	−3
4	−1

(C)

x	y
−1	7
1	1
4	−3

(D)

x	y
−1	7
1	5
4	1

STOP

END OF SECTION I

SECTION II

Part A

TIME: 30 MINUTES

2 PROBLEMS

A graphing calculator is required for some of these problems. See instructions on page 4.

1. The Boston Red Sox play in Fenway Park, notorious for its Green Monster, a wall 37 feet tall and 315 feet from home plate at the left-field foul line. Suppose a batter hits a ball 2 feet above home plate, driving the ball down the left-field line at an initial angle of 30° above the horizontal, with initial velocity of 120 feet per second. (Since Fenway is near sea level, assume that the acceleration due to gravity is -32.172 ft/sec^2.)

   (a) Write the parametric equations for the location of the ball t seconds after it has been hit.
   (b) At what elevation does the ball hit the wall?
   (c) How fast is the ball traveling when it hits the wall?

2. The table shows the depth of water, W, in a river, as measured at 4-hour intervals during a day-long flood. Assume that W is a differentiable function of time t.

t (hr)	0	4	8	12	16	20	24
$W(t)$ (ft)	32	36	38	37	35	33	32

   (a) Find the approximate value of $W'(16)$. Indicate units of measure.
   (b) Estimate the average depth of the water, in feet, over the time interval $0 \le t \le 24$ hours by using a trapezoidal approximation with subintervals of length $\Delta t = 4$ hours.
   (c) Scientists studying the flooding believe they can model the depth of the water with the function $F(t) = 35 - 3\cos\left(\frac{t+3}{4}\right)$, where $F(t)$ represents the depth of the water, in feet, after t hours. Find $F'(16)$ and explain the meaning of your answer, with appropriate units, in terms of the river depth.
   (d) Use the function F to find the average depth of the water, in feet, over the time interval $0 \le t \le 24$ hours.

STOP

END OF PART A, SECTION II

Part B

TIME: 60 MINUTES

4 PROBLEMS

No calculator is allowed for any of these problems.
If you finish Part B before time has expired, you may return to work on Part A, but you may not use a calculator.

3. The region **R** is bounded by the curves $f(x) = \cos(\pi x) - 1$ and $g(x) = x(2 - x)$, as shown in the figure.
   (a) Find the area of **R**.
   (b) A solid has base **R**, and each cross section perpendicular to the x-axis is an isosceles right triangle whose hypotenuse lies in **R**. Set up, but do not evaluate, an integral for the volume of this solid.
   (c) Set up, but do not evaluate, an integral for the volume of the solid formed when **R** is rotated around the line $y = 3$.

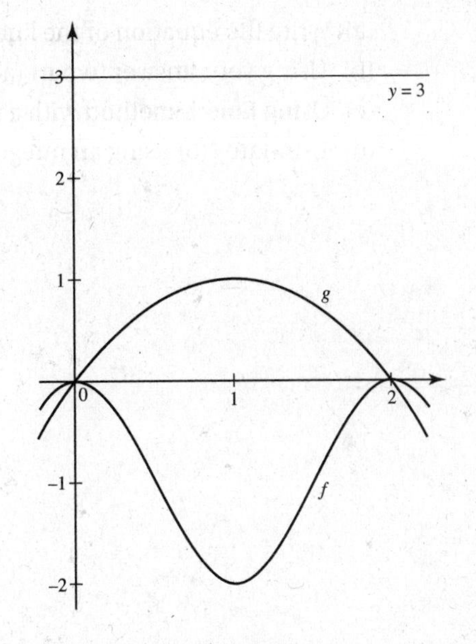

4. Two autos, P and Q, start from the same point and race along a straight road for 10 seconds. The velocity of P is given by $v_p(t) = 6\left(\sqrt{1 + 8t} - 1\right)$ feet per second. The velocity of Q is shown in the graph.

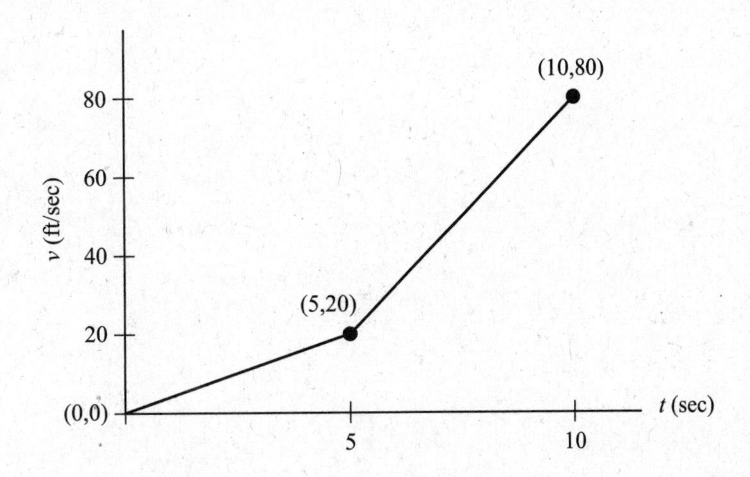

   (a) At what time is P's actual acceleration (in ft/sec^2) equal to its average acceleration for the entire race?
   (b) What is Q's acceleration (in ft/sec^2) then?
   (c) At the end of the race, which auto was ahead? Explain.

GO ON TO THE NEXT PAGE

5. Given that a function f is continuous and differentiable throughout its domain, and that $f(5) = 2$, $f'(5) = -2$, $f''(5) = -1$, and $f'''(5) = 6$.

(a) Write a Taylor polynomial of degree 3 that approximates f around $x = 5$.

(b) Use your answer to estimate $f(5.1)$.

(c) Let $g(x) = f(2x + 5)$. Write a cubic Maclaurin polynomial approximation for g.

6. Let f be the function that contains the point $(-1,8)$ and satisfies the differential equation $\dfrac{dy}{dx} = \dfrac{10}{x^2 + 1}$.

(a) Write the equation of the line tangent to f at $x = -1$.

(b) Using your answer to part (a), estimate $f(0)$.

(c) Using Euler's method with a step size of 0.5, estimate $f(0)$.

(d) Estimate $f(0)$ using an integral.

STOP

END OF TEST

ANSWER KEY
BC Practice Examination 3

Part A

1. **D**	9. **B**	17. **D**	25. **C**
2. **C**	10. **D**	18. **C**	26. **C**
3. **A**	11. **B**	19. **D**	27. **A**
4. **A**	12. **A**	20. **D**	28. **D**
5. **C**	13. **D**	21. **D**	29. **C**
6. **B**	14. **A**	22. **B**	30. **A**
7. **B**	15. **B**	23. **D**	
8. **A**	16. **D**	24. **C**	

Part B

31. **B**	35. **D**	39. **C**	43. **C**
32. **C**	36. **C**	40. **A**	44. **D**
33. **B**	37. **C**	41. **A**	45. **C**
34. **C**	38. **B**	42. **C**	

ANSWERS EXPLAINED

The explanations for questions not given below will be found in the answer section for AB Practice Exam 3 on pages 555–561. Identical questions in Section I of Practice Examinations AB 3 and BC 3 have the same number. For example, explanations of the answers for Questions 1 and 2, not given below, will be found in Section I of Calculus AB Practice Exam 3, Answers 1 and 2, page 555.

Section I Multiple-Choice

Part A

3. **(A)** The series $\frac{1}{4}+\frac{3}{16}+\frac{9}{64}+\cdots$ is geometric with $a=\frac{1}{4}$ and $r=\frac{3}{4}$; it converges to $\frac{\frac{1}{4}}{1-\frac{3}{4}}$.

5. **(C)** $\frac{1}{1+2x}$ is the sum of an infinite geometric series with first term 1 and common ratio $-2x$. The series is $1-2x+4x^2-8x^3+16x^4-\cdots$.

6. **(B)** Assume that $\frac{2x+3}{(x-1)(x-2)}=\frac{A}{x-1}+\frac{B}{x-2}$. Then $2x+3=A(x-2)+B(x-1)$. Because all the choices have a denominator of $x-1$, you are looking for A, so let $x=1$:

$$2(1)+3=A(1-2);\ A=-5$$

8. **(A)** I. $\sum_{n=1}^{\infty} n^2e^{-n}=\sum_{n=1}^{\infty}\frac{n^2}{e^n}$. Using the Ratio Test, we get

$$\lim_{n\to\infty}\left|\frac{a_{k+1}}{a_k}\right|=\lim_{n\to\infty}\left|\frac{(n+1)^2}{e^{n+1}}\cdot\frac{e^n}{n^2}\right|=\lim_{n\to\infty}\left|\left(\frac{n+1}{n}\right)^2\cdot\frac{1}{e}\right|=1^2\cdot\frac{1}{e}<1, \text{ convergent.}$$

II. $\lim_{n\to\infty}3^{1/n}=3^0=1$, so $\lim_{n\to\infty}(-1)^n3^{1/n}$ does not exist and thus the series $\sum_{n=1}^{\infty}(-1)^n3^{1/n}$ diverges because it doesn't pass the nth term test.

III. We would like to compare this series with $\sum_{n=1}^{\infty}\frac{1}{n}$; however, we cannot use the Comparison Test because $\frac{n^2-1}{n^3+1}<\frac{1}{n}$ (see note below). So we must use the Limit Comparison Test: $\lim_{n\to\infty}\frac{\frac{n^2-1}{n^3+1}}{1/n}=\lim_{n\to\infty}\frac{n^3-n}{n^3+1}=1.$

Since $\sum_{n=1}^{\infty}\frac{1}{n}$ diverges, $\sum_{n=1}^{\infty}\frac{n^2-1}{n^3+1}$ also diverges by the Limit Comparison Test.

NOTE: We can rewrite $\frac{n^2-1}{n^3+1}=\frac{1-\frac{1}{n^2}}{n+\frac{1}{n^2}}$; comparing to $\frac{1}{n}$ we see that the numerator $1-\frac{1}{n^2}<1$ and denominator $n+\frac{1}{n^2}>n$; therefore $\frac{n^2-1}{n^3+1}<\frac{1}{n}$.

12. **(A)** $f(x)=\frac{d}{dx}(x\sin\pi x)=\pi x\cos\pi x+\sin\pi x.$

17. **(D)**

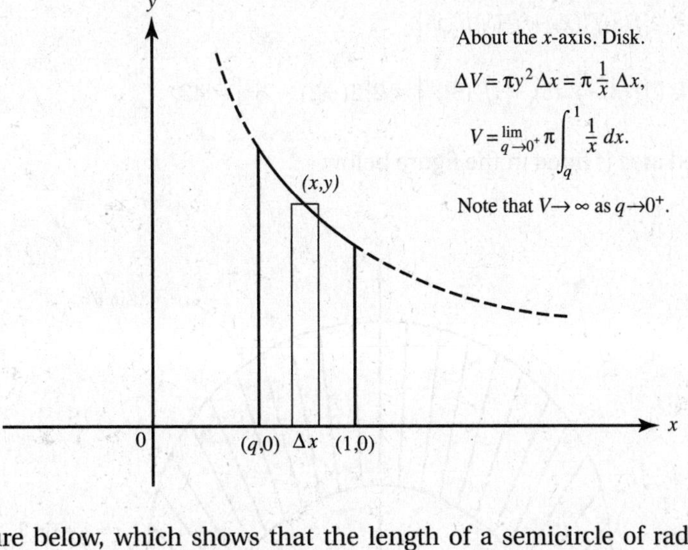

18. **(C)** See the figure below, which shows that the length of a semicircle of radius 2 is needed here. The answer can, of course, be found by using the formula for arc length:

$$s = \int_0^{\pi} \sqrt{\left(\frac{dx}{dt}\right)^2 + \left(\frac{dy}{dt}\right)^2}\, dt = \int_0^{\pi} \sqrt{(-2\cos t)^2 + (-2\sin t)^2\, dt} = 2\int_0^{\pi} dt$$

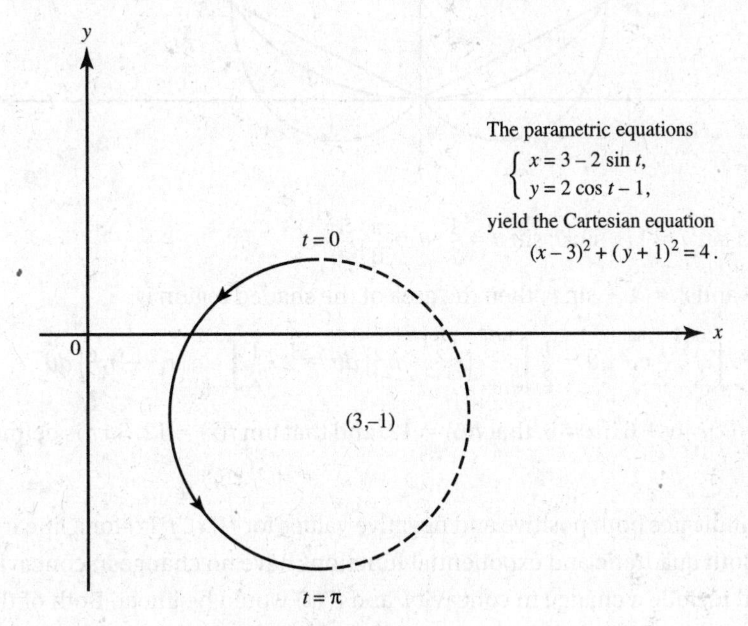

20. **(D)** $\lim_{x\to\infty} f(x) = \infty$; $\lim_{x\to 2^-} f(x) = -\infty$; using long division, $y = x + 2 + \frac{4}{x-2}$.

22. **(B)** Using parts we let $u = x^2$, $dv = e^x dx$; then $du = 2x\,dx$, $v = e^x$, and

$$\int x^2 e^x dx = x^2 e^x - 2\int x e^x dx.$$

We use parts again with $u = x$, $dv = e^x dx$; then $du = dx$, $v = e^x$, and

$$\int x^2 e^x dx = x^2 e^x - 2(x^2 e^x - \int e^x dx) = x^2 e^x - 2x^2 e^x + 2e^x.$$

Now $\int_0^1 x^2 e^x dx = (x^2 e^x - 2x^2 e^x + 2e^x)\Big|_0^1 = (e - 2e + 2e) - (2).$

23. **(D)** $\frac{d}{dx}(f^2(x)) = 2f(x)f'(x),$

$$\frac{d^2}{dx^2}(f^2(x)) = 2\left[f(x)f''(x) + f'(x)f'(x)\right]$$

$$\left.\frac{d^2}{dx^2}\left(f^2(x)\right)\right|_{x=3} = 2\left[f(3)\cdot f''(3) + (f'(3))^2\right] = 2\left[2(-2) + 5^2\right] = 42$$

27. **(A)** The required area is lined in the figure below.

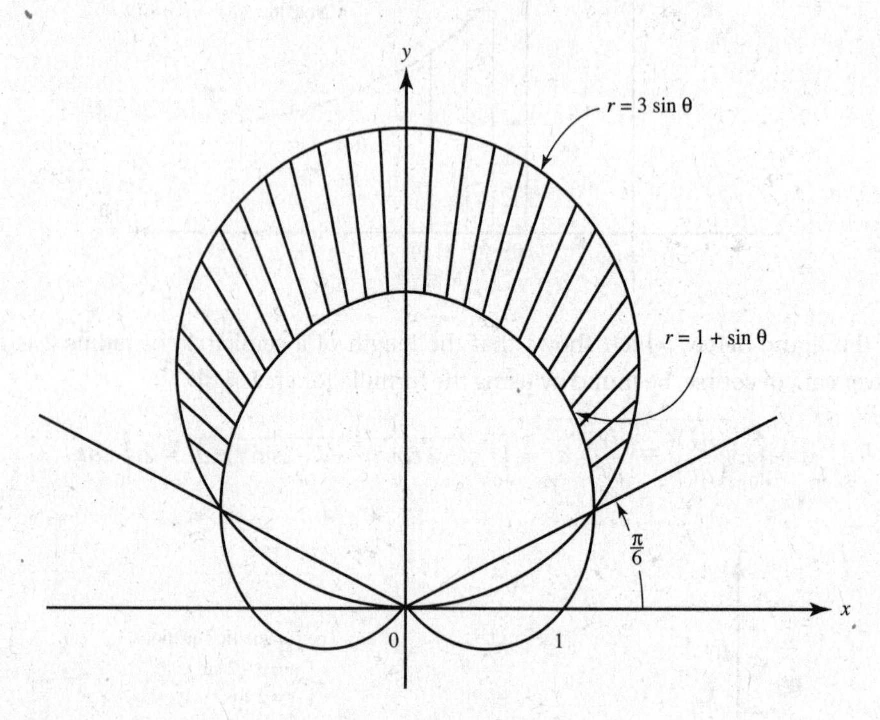

Intersections: $3\sin\theta = 1 + \sin\theta$; $\sin\theta = \frac{1}{2}$; $\theta = \frac{\pi}{6}, \frac{5\pi}{6}$

Let $r_1 = 3\sin\theta$ and $r_2 = 1 + \sin\theta$, then the area of the shaded region is

$$\int_{\pi/6}^{5\pi/6} \frac{1}{2}{r_1}^2\, d\theta - \int_{\pi/6}^{5\pi/6} \frac{1}{2}{r_2}^2\, d\theta = \frac{1}{2}\int_{\pi/6}^{5\pi/6} \left({r_1}^2 - {r_2}^2\right) d\theta = 2\cdot\frac{1}{2}\int_{\pi/6}^{\pi/2} \left({r_1}^2 - {r_2}^2\right) d\theta$$

28. **(D)** Note that $f(x) = x + 6$ if $x \neq 6$, that $f(6) = 12$, and that $\lim_{x\to 6} f(x) = 12$. So f is defined and continuous at $x = 6$.

29. **(C)** The table indicates both positive and negative values for $f''(x)$. $f''(x)$ for a linear function would be zero for all x. Both quadratic and exponential functions have no change in concavity for all x. A cubic function would include a change in concavity, and $f''(x)$ would be linear. Both of these properties are indicated in the table with the given points. Therefore, the function could be a cubic.

30. **(A)** From the integral we get $a = -1, b = 3$, so $\Delta x = \frac{3-(-1)}{n} = \frac{4}{n}$ and $x_k = a + k\cdot\Delta x = -1 + \frac{4k}{n} = \frac{4k}{n} - 1$.
Replace x with x_k and replace dx with Δx in the integrand to get the general term in the summation.

BC PRACTICE EXAMINATION 3

Part B

31. **(B)** Expressed parametrically, $x = \sin 3\theta \cos\theta$, $y = \sin 3\theta \sin\theta$. $\frac{dy}{dx}$ is undefined where $\frac{dx}{d\theta} = -\sin 3\theta$ $\sin\theta + 3\cos 3\theta \cos\theta = 0$.

Use your calculator to solve for θ.

CALCULATOR TIP: Graph in function mode: $y = -\sin(3x) \cdot \sin(x) + 3\cos(3x) \cdot \cos(x)$, and find the roots (zeros) of the function on the interval $\left[0, \frac{\pi}{2}\right]$.

34. **(C)** See the figure below.

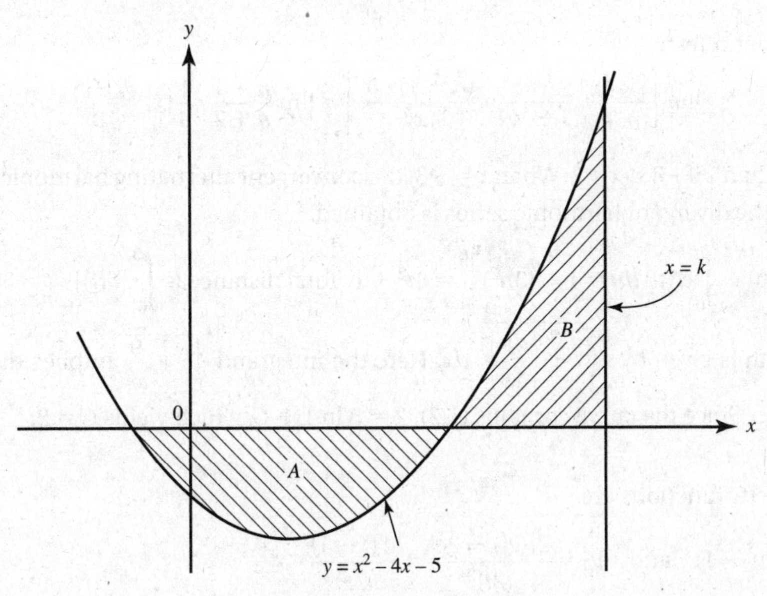

(This figure is not drawn to scale.)

The roots of $f(x) = x^2 - 4x - 5 = (x - 5)(x + 1)$ are $x = -1$ and 5. Since areas A and B are equal, therefore, $\int_{-1}^{k} f(x)d(x) = 0$. Thus,

$$\left(\frac{x^3}{3} - 2x^2 - 5x\right)\Bigg|_{-1}^{k} = \left(\frac{k^3}{3} - 2k^2 - 5k\right) - \left(-\frac{1}{3} - 2 + 5\right)$$

$$= \frac{k^3}{3} - 2k^2 - 5k - \frac{8}{3} = 0.$$

A calculator yields $k = 8$.

37. **(C)** It is given that $\mathbf{a}(t) = \langle \sin t, e^{-t} \rangle$.

$$\mathbf{v}(t) = \left\langle 0 + \int_0^t \sin u\, du,\ 0 + \int_0^t e^{-u} du \right\rangle = \left\langle -\cos u\Big|_0^t,\ -e^{-u}\Big|_0^t \right\rangle = \langle -\cos t + 1, -e^{-t} + 1 \rangle$$

The object's speed is $|\mathbf{v}(t)| = \sqrt{(-\cos t + 1)^2 + (-e^{-t} + 1)^2}$.

Use a calculator to find that the object's maximum speed is 2.217.

BC PRACTICE EXAMINATION 3

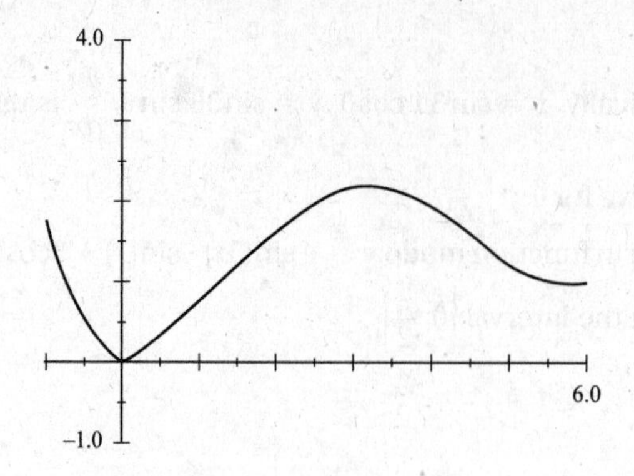

39. **(C)** Use the Ratio Test:

$$\lim_{n\to\infty}\left|\frac{x^{n+1}}{(n+2)\cdot 3^{n+1}}\cdot\frac{(n+1)\cdot 3^n}{x^n}\right| = \lim_{n\to\infty}\frac{n+1}{n+2}\cdot\frac{1}{3}|x| = \frac{|x|}{3},$$

which is less than 1 if $-3 < x < 3$. When $x = -3$, the convergent alternating harmonic series is obtained. When $x = 3$, the divergent harmonic series is obtained.

40. **(A)** $v(t) = v(0) + \int_0^t a(u)\,du = 1 + (3u^2)\Big|_0^t = 3t^2 + 1$. Total distance is $\int_0^3 |v(t)|\,dt = 30$.

41. **(A)** Arc length is given by $\int\sqrt{1+\left(\frac{dy}{dx}\right)^2}\,dx$. Here the integrand $\sqrt{1+\frac{9}{x^2}}$ implies that $\frac{dy}{dx} = \frac{3}{x}$, hence, $y = 3\ln x + C$. Since the curve contains (1,2), $2 = 3\ln 1 + C$, which yields $C = 2$.

42. **(C)** The velocity functions are

$$v_1 = -2t\sin(t^2+1) \quad\text{and}\quad v_2 = \frac{2t(e^t) - 2e^t}{(2t)^2} = \frac{e^t(t-1)}{2t^2}.$$

When these functions are graphed on a calculator, it is clear that they intersect four times during the first 3 sec, as shown below.

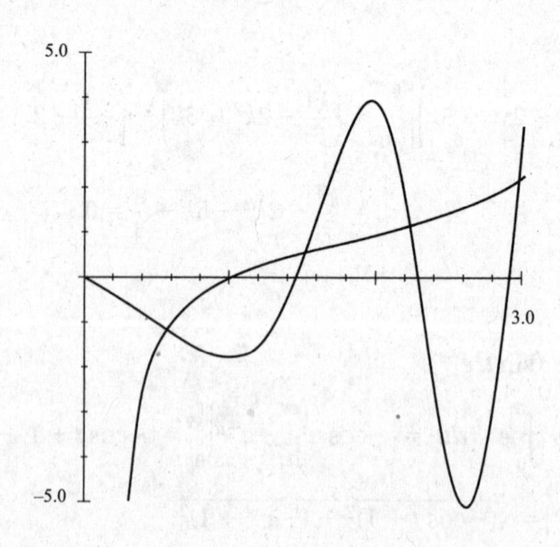

Section II Free-Response

Part A

1.

Green Monster
120 ft/sec
30γ
37′
2′
315′

(a) The following table shows x- and y-components of acceleration, velocity, and position:

	Horizontal	Vertical
acceleration components:	$a_x = 0$	$a_y = -32.172$
velocity $= \int a\,dt$:	$v_x = c_1$	$v_y = -32.172t + c_2$
initial velocities:	$v_x(0) = 60\sqrt{3} = c_1$	$v_y(0) = 60 = c_2$
velocity components:	$v_x(t) = 60\sqrt{3}$	$v_y(t) = -32.172t + 60$
position $= \int v\,dt$:	$x(t) = 60\sqrt{3}t + c_3$	$y(t) = -16.086t^2 + 60t + c_4$
initial position (0,2):	$x(0) = 0 = c_3$	$y(0) = 2 = c_4$
position components:	$x(t) = 60\sqrt{3}t$	$y(t) = -16.086t^2 + 60t + 2$

The last line in the table is the answer to part (a).

(b) To determine how far above the ground the ball is when it hits the wall, find out when $x = 315$, and evaluate y at that time.

$60\sqrt{3}t = 315$ yields $t = \dfrac{315}{60\sqrt{3}} \approx 3.03109$ sec.

$y\left(\dfrac{315}{60\sqrt{3}}\right) \approx 36.075$ ft.

(c) The ball's speed at the moment of impact in part (b) is $|v(t)|$ evaluated at $t = \dfrac{315}{60\sqrt{3}}$.

$|v(t)| = \sqrt{\left(v_x(t)\right)^2 + \left(v_y(t)\right)^2}$

$= \sqrt{(60\sqrt{3}\,)^2 + (-32.172t + 60)^2}$ when $x = 315$,

$\left|v\left(\dfrac{315}{60\sqrt{3}}\right)\right| \approx 110.487$ ft/sec.

2. See solution for AB/BC 2, page 562.

Part B

3. See solution for AB/BC 3, page 563.
4. See solution for AB/BC 4, pages 563–564.
5. (a) The table below is constructed from the information given in Question 5 on page 624.

n	$f^{(n)}(5)$	$a_n = \dfrac{f^{(n)}(5)}{n!}$
0	2	2
1	−2	−2
2	−1	$-\dfrac{1}{2}$
3	6	1

$$f(x) \approx 2 - 2(x-5) - \frac{1}{2}(x-5)^2 + (x-5)^3.$$

(b) $f(5.1) \approx 2 - 2(5.1-5) - \frac{1}{2}(5.1-5)^2 + (5.1-5)^3$

$\approx 2 - 0.2 - 0.005 + 0.001 = 1796.$

(c) Use Taylor's Theorem around $x = 0$.

n	$g^{(n)}(x)$	$g^{(n)}(0)$	$a_n = \dfrac{g^{(n)}(0)}{n!}$
0	$f(2x+5)$	$f(5) = 2$	2
1	$2f'(2x+5)$	$2f'(5) = 2(-2) = -4$	−4
2	$4f''(2x+5)$	$4f''(5) = 4(-1) = -4$	−2
3	$8f'''(2x+5)$	$8f'''(5) = 8(6) = 48$	8

$$g(x) \approx 2 - 4x - 2x^2 + 8x^3.$$

6. (a) At $(-1,8)$, $\dfrac{dy}{dx} = \dfrac{10}{x^2+1} = \dfrac{10}{2} = 5$, so the tangent line is $y - 8 = 5(x - (-1))$.

Therefore $f(x) \approx 8 + 5(x+1)$.

(b) $f(3) \approx 8 + 5(0+1) = 13.$

(c) At $(-1,8)$, $\dfrac{dy}{dx} = 5 \approx \dfrac{\Delta y}{\Delta x}$. For $\Delta x = 0.5$, $\Delta y = 0.5(5) = 2.5$, so move to

$(-1 + 0.5, 8 + 2.5) = (-0.5, 10.5)$.

At $(-0.5,10.5)$, $\dfrac{dy}{dx} = \dfrac{10}{\left(-\frac{1}{2}\right)^2 + 1} = \dfrac{10}{\frac{5}{4}} = 8 \approx \dfrac{\Delta y}{\Delta x}$. For $\Delta x = 0.5$, $\Delta y = 0.5(8) = 4$, so move to

$(-0.5 + 0.5, 10.5 + 4)$.

Thus $f(0) \approx 14.5$.

(d) $\displaystyle\int_{-1}^{0} \frac{10}{x^2+1}\,dx = f(0) - f(-1)$, so $f(0) = 8 + \displaystyle\int_{-1}^{0} \frac{10}{x^2+1}\,dx$

$= 8 + \arctan(x)\big|_{-1}^{0}$

$= 8 + 10(\arctan(0) - \arctan(-1))$

$= 8 + \dfrac{5\pi}{2}.$

Appendix: Formulas and Theorems for Reference

ALGEBRA

1. Quadratic Formula. The roots of the quadratic equation

$$ax^2 + bx + c = 0 \ (a \neq 0)$$

are given by

$$x = \frac{-b \pm \sqrt{b^2 - 4ac}}{2a}.$$

2. Binomial Theorem. If n is a positive integer, then

$$(a+b)^n = a^n + na^{n-1}b + \frac{n(n-1)}{1 \cdot 2}a^{n-2}b^2 + \frac{n(n-1)(n-2)}{1 \cdot 2 \cdot 3}a^{n-3}b^3$$

$$+ \cdots + nab^{n-1} + b^n.$$

3. Remainder Theorem. If the polynomial $Q(x)$ is divided by $(x - a)$ until a constant remainder R is obtained, then $R = Q(a)$. In particular, if a is a root of $Q(x) = 0$, then $Q(a) = 0$.

GEOMETRY

The sum of the angles of a triangle is equal to a straight angle (180°).

Pythagorean Theorem

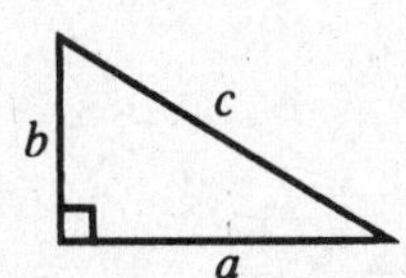

In a right triangle,

$$c^2 = a^2 + b^2.$$

In the following formulas,

A	is	area	B	is	area of base
S		surface area	r		radius
V		volume	C		circumference
b		base	l		arc length
h		height or altitude	θ		central angle (in radians)
s		slant height			

4. Triangle: $A = \frac{1}{2}bh.$

5. Trapezoid: $A = \left(\frac{b_1 + b_2}{2}\right)h.$

6. Parallelogram: $A = bh.$

7. Circle: $C = 2\pi r; A = \pi r^2$.

8. Circular sector: $A = \frac{1}{2}r^2\theta$.

9. Circular arc: $l = r\theta$.

10. Cylinder:

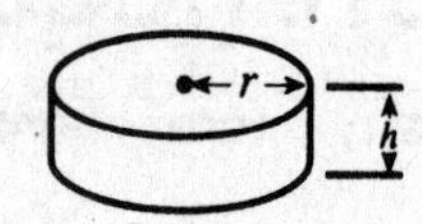

$$V = \pi r^2 h = Bh.$$
$$S \text{ (lateral)} = 2\pi rh.$$
$$\text{Total surface area} = 2\pi r^2 + 2\pi rh.$$

11. Cone:

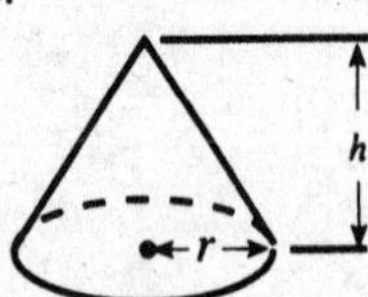

$$V = \frac{1}{3}\pi r^2 h = \frac{1}{3}Bh.$$
$$S \text{ (lateral)} = \pi r\sqrt{r^2 + h^2}.$$
$$\text{Total surface area} = \pi r^2 + \pi r\sqrt{r^2 + h^2}.$$

12. Sphere: $V = \frac{4}{3}\pi r^3$.

$S = 4\pi r^2$.

TRIGONOMETRY

BASIC IDENTITIES

13. $\sin^2\theta + \cos^2\theta = 1$.
14. $1 + \tan^2\theta = \sec^2\theta$.
15. $1 + \cot^2\theta = \csc^2\theta$.

SUM AND DIFFERENCE FORMULAS

16. $\sin(\alpha \pm \beta) = \sin\alpha\cos\beta \pm \cos\alpha\sin\beta$.
17. $\cos(\alpha \pm \beta) = \cos\alpha\cos\beta \mp \sin\alpha\sin\beta$.
18. $\tan(\alpha \pm \beta) = \dfrac{\tan\alpha \pm \tan\beta}{1 \mp \tan\alpha\tan\beta}$.

DOUBLE-ANGLE FORMULAS

19. $\sin 2\alpha = 2\sin\alpha\cos\alpha$.
20. $\cos 2\alpha = \cos^2\alpha - \sin^2\alpha = 2\cos^2\alpha - 1 = 1 - 2\sin^2\alpha$.
21. $\tan 2\alpha = \dfrac{2\tan\alpha}{1 - \tan^2\alpha}$.

HALF-ANGLE FORMULAS

22. $\sin\dfrac{\alpha}{2} = \pm\sqrt{\dfrac{1 - \cos\alpha}{2}}$; $\sin^2\alpha = \dfrac{1}{2} - \dfrac{1}{2}\cos 2\alpha$.

23. $\cos\dfrac{\alpha}{2} = \pm\sqrt{\dfrac{1 + \cos\alpha}{2}}$; $\cos^2\alpha = \dfrac{1}{2} + \dfrac{1}{2}\cos 2\alpha$.

REDUCTION FORMULAS

24. $\sin(-\alpha) = -\sin\alpha$; $\cos(-\alpha) = \cos\alpha$.

25. $\sin\left(\frac{\pi}{2} - \alpha\right) = \cos\alpha$; $\cos\left(\frac{\pi}{2} - \alpha\right) = \sin\alpha$.

26. $\sin\left(\frac{\pi}{2} + \alpha\right) = \cos\alpha$; $\cos\left(\frac{\pi}{2} + \alpha\right) = -\sin\alpha$.

27. $\sin(\pi - \alpha) = \sin\alpha$; $\cos(\pi - \alpha) = -\cos\alpha$.

28. $\sin(\pi + \alpha) = -\sin\alpha$; $\cos(\pi + \alpha) = -\cos\alpha$.

If a, b, c are the sides of triangle ABC, and A, B, C are, respectively, the opposite interior angles, then:

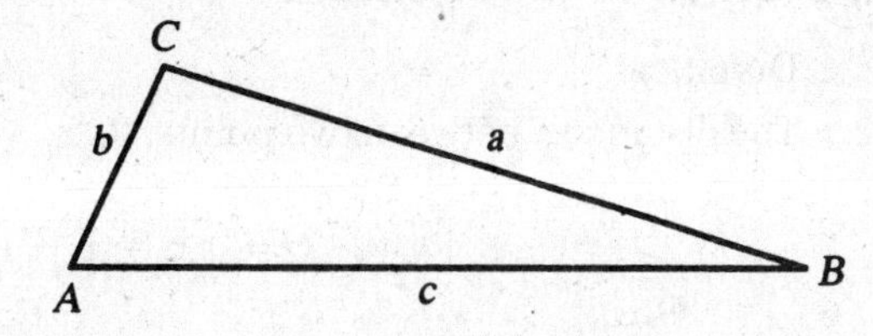

29. LAW OF COSINES. $c^2 = a^2 + b^2 - 2ab\cos C$.

30. LAW OF SINES. $\frac{a}{\sin A} = \frac{b}{\sin B} = \frac{c}{\sin C}$.

31. The area $A = \frac{1}{2}ab\sin C$.

GRAPHS OF TRIGONOMETRIC FUNCTIONS

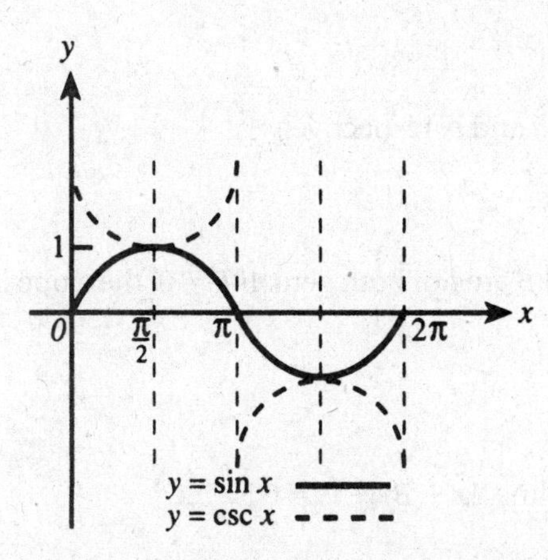

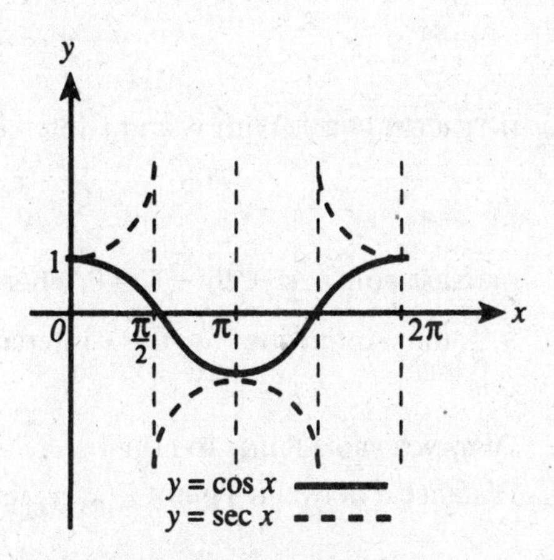

The four functions sketched above, sin, cos, csc, and sec, all have period 2π.

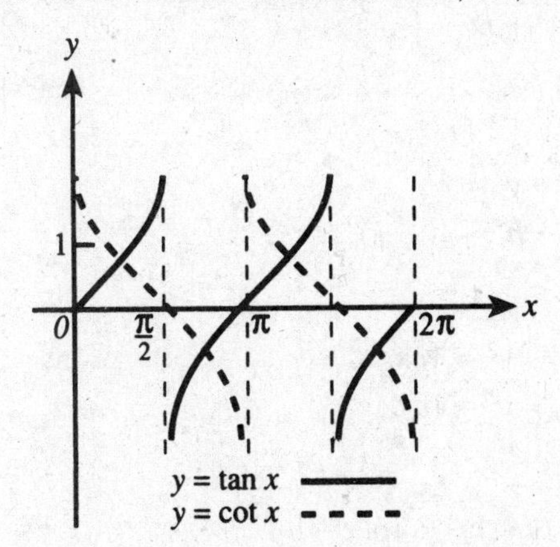

The two functions tan and cot have period π.

INVERSE TRIGONOMETRIC FUNCTIONS

$y = \sin^{-1} x = \arc\sin x$ implies $x = \sin y$, where $-\frac{\pi}{2} \leqq y \leqq \frac{\pi}{2}$.

$y = \cos^{-1} x = \arc\cos x$ implies $x = \cos y$, where $0 \leqq y \leqq \pi$.

$y = \tan^{-1} x = \arc\tan x$ implies $x = \tan y$, where $-\frac{\pi}{2} < y < \frac{\pi}{2}$.

ANALYTIC GEOMETRY

Rectangular Coordinates

DISTANCE

32. The distance d between two points, $P_1(x_1, y_1)$ and $P_2(x_2, y_2)$, is given by

$$d = \sqrt{(x_2 - x_1)^2 + (y_2 - y_1)^2}.$$

EQUATIONS OF THE STRAIGHT LINE

33. POINT-SLOPE FORM. Through $P_1(x_1, y_1)$ and with slope m:

$$y - y_1 = m(x - x_1).$$

34. SLOPE-INTERCEPT FORM. With slope m and y-intercept b:

$$y = mx + b.$$

35. TWO-POINT FORM. Through $P_1(x_1, y_1)$ and $P_2(x_2, y_2)$:

$$y - y_1 = \frac{y_2 - y_1}{x_2 - x_1}(x - x_1).$$

36. INTERCEPT FORM. With x- and y-intercepts of a and b, respectively:

$$\frac{x}{a} + \frac{y}{b} = 1.$$

37. GENERAL FORM. $Ax + By + C = 0$, where A and B are not both zero. If $B \neq 0$, the slope is $-\frac{A}{B}$; the y-intercept, $-\frac{C}{B}$; the x-intercept, $-\frac{C}{A}$.

DISTANCE FROM POINT TO LINE

38. Distance d between a point $P(x_1, y_1)$ and the line $Ax + By + C = 0$ is

$$d = \left|\frac{Ax_1 + By_1 + C}{\sqrt{A^2 + B^2}}\right|.$$

EQUATIONS OF THE CONICS

CIRCLE

39. With center at (0, 0) and radius r: $x^2 + y^2 = r^2$.
40. With center at (h, k) and radius r: $(x - h)^2 + (y - k)^2 = r^2$.

PARABOLA

41. With vertex at (0, 0) and focus at $(p, 0)$: $y^2 = 4px$.
42. With vertex at (0, 0) and focus at $(0, p)$: $x^2 = 4py$.

With vertex at (h, k) and axis

43. parallel to x-axis, focus at $(h + p, k)$: $(y - k)^2 = 4p(x - h)$.
44. parallel to y-axis, focus at $(h, k + p)$: $(x - h)^2 = 4p(y - k)$.

ELLIPSE

With major axis of length $2a$, minor axis of length $2b$, and distance between foci of $2c$:

45. Center at (0, 0), foci at ($\pm c$, 0), and vertices at ($\pm a$, 0):

$$\frac{x^2}{a^2} + \frac{y^2}{b^2} = 1.$$

46. Center at (0, 0), foci at (0, $\pm c$), and vertices at (0, $\pm a$):

$$\frac{y^2}{a^2} + \frac{x^2}{b^2} = 1.$$

47. Center at (h, k), major axis horizontal, and vertices at ($h \pm a$, k):

$$\frac{(x-h)^2}{a^2} + \frac{(y-k)^2}{b^2} = 1.$$

48. Center at (h, k), major axis vertical, and vertices at (h, $k \pm a$):

$$\frac{(y-k)^2}{a^2} + \frac{(x-h)^2}{b^2} = 1.$$

For the ellipse, $a^2 = b^2 + c^2$, and the eccentricity $e = \frac{c}{a}$, which is *less* than 1.

HYPERBOLA

With real (transverse) axis of length $2a$, imaginary (conjugate) axis of length $2b$, and distance between foci of $2c$:

49. Center at (0, 0), foci at ($\pm c$, 0), and vertices at ($\pm a$, 0):

$$\frac{x^2}{a^2} - \frac{y^2}{b^2} = 1.$$

50. Center at (0, 0), foci at (0, $\pm c$), and vertices at (0, $\pm a$):

$$\frac{y^2}{a^2} - \frac{x^2}{b^2} = 1.$$

51. Center at (h, k), real axis horizontal, vertices at ($h \pm a$, k):

$$\frac{(x-h)^2}{a^2} - \frac{(y-k)^2}{b^2} = 1.$$

52. Center at (h, k), real axis vertical, vertices at (h, $k \pm a$):

$$\frac{(y-k)^2}{a^2} - \frac{(x-h)^2}{b^2} = 1.$$

For the hyperbola, $c^2 = a^2 + b^2$, and eccentricity $e = \frac{c}{a}$, which is *greater* than 1.

POLAR COORDINATES

RELATIONS WITH RECTANGULAR COORDINATES

53. $x = r\cos\theta$;
$y = r\sin\theta$;
$r^2 = x^2 + y^2$;
$\tan\theta = \frac{y}{x}$.

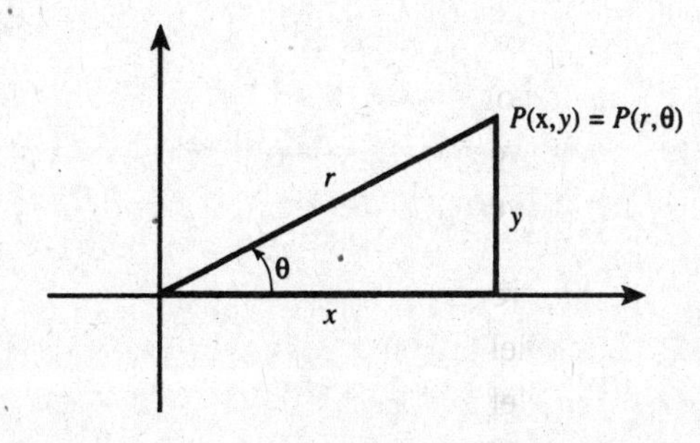

SOME POLAR EQUATIONS

54. $r = a$ circle, center at pole, radius a.

55. $r = 2a \cos \theta$ circle, center at $(a, 0)$, radius a.

56. $r = 2a \sin \theta$ circle, center at $(0, a)$, radius a.

57. $\left.\begin{array}{l} r = a \sec \theta \\ \text{or } r \cos \theta = a \end{array}\right\}$ line, $x = a$.

58. $\left.\begin{array}{l} r = b \csc \theta \\ \text{or } r \sin \theta = b \end{array}\right\}$ line, $y = b$.

roses (four leaves)

59. $r = \cos 2\theta$.

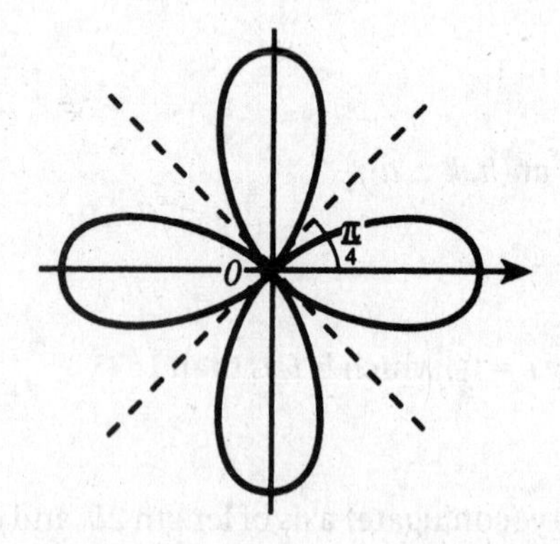

60. $r = \sin 2\theta$.

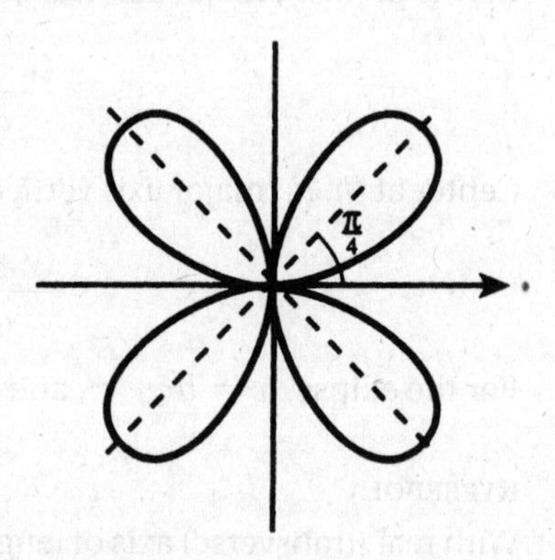

cardioids (specific examples below)

61. $r = a\,(1 \pm \cos \theta)$.

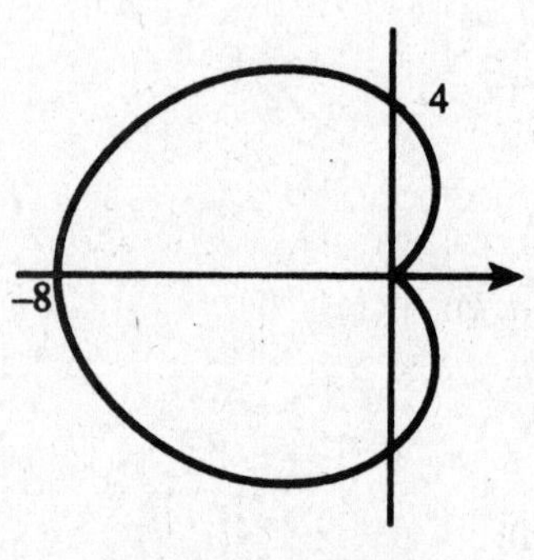

$r = 4\,(1 - \cos \theta)$

62. $r = a\,(1 \pm \sin \theta)$.

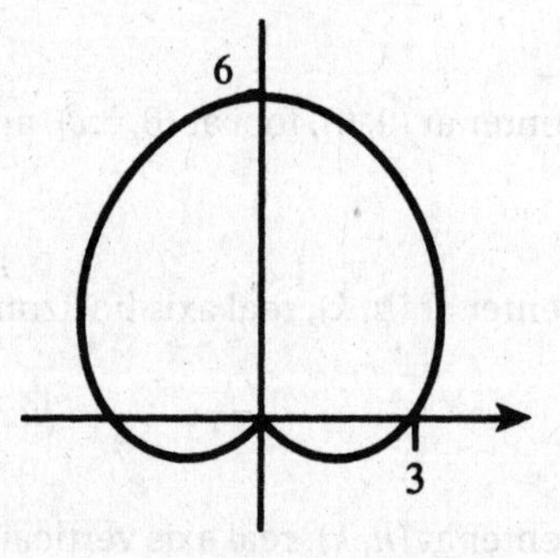

$r = 3\,(1 + \sin \theta)$

63. $r^2 = \cos 2\theta$, lemniscate, symmetric to the x-axis.

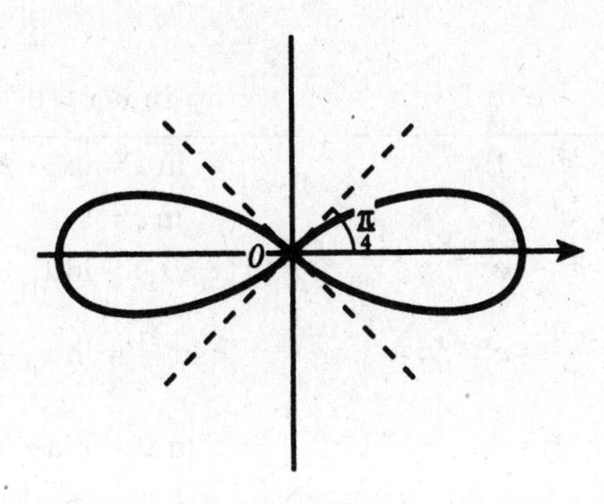

($r^2 = \sin 2\theta$ is a lemniscate symmetric to $y = x$.)

64. $r = \theta$, (double) spiral of Archimedes

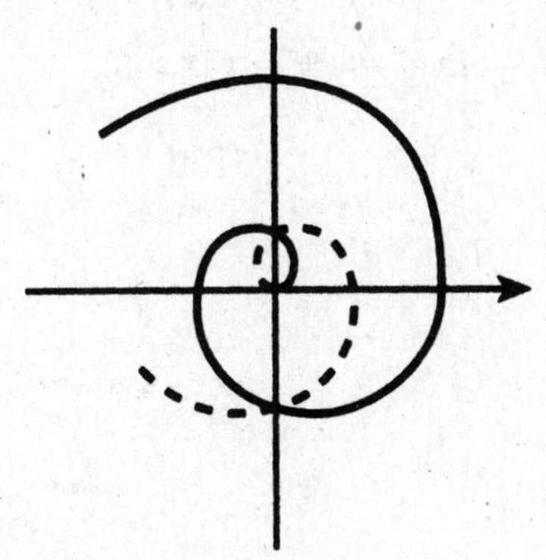

For $\theta > 0$, the curve consists only of the solid spiral.

65. $r\theta = a$ $(\theta > 0)$, hyperbolic (or reciprocal) spiral

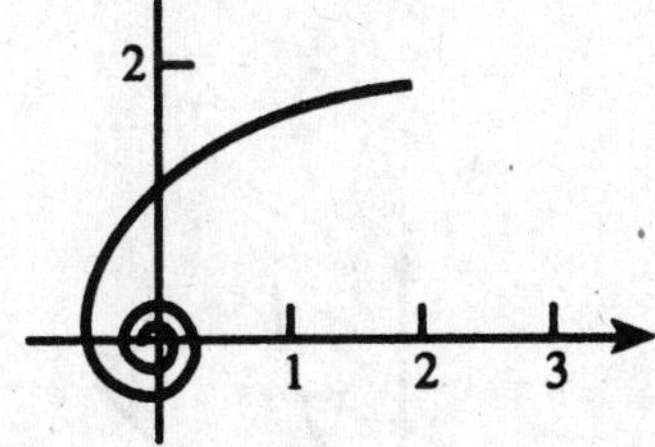

This curve is for $r\theta = 2$.
Note that $y = 2$ is an asymptote.

EXPONENTIAL AND LOGARITHMIC FUNCTIONS

PROPERTIES

e^x	$\ln x\ (x > 0)$
$e^0 = 1;$	$\ln 1 = 0;$
$e^1 = e;$	$\ln e = 1;$
$e^{x_1} \cdot e^{x_2} = e^{x_1 + x_2};$	$\ln (x_1 \cdot x_2) = \ln x_1 + \ln x_2;$
$\frac{e^{x_1}}{e^{x_2}} = e^{x_1 - x_2};$	$\ln \frac{x_1}{x_2} = \ln x_1 - \ln x_2;$
$e^{-x} = \frac{1}{e^x}.$	$\ln x^r = r \ln x\ (r \text{ real}).$

INVERSE PROPERTIES

$f(x) = e^x$ and $f^{-1}(x) = \ln x$ are inverses of each other:

$$f^{-1}\big(f(x)\big) = f\big(f^{-1}(x)\big) = x;$$
$$\ln e^x = e^{\ln x} = x\ (x > 0).$$

GRAPHS

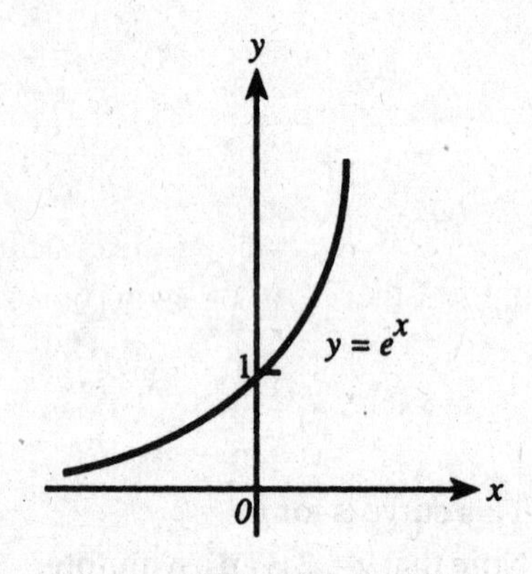

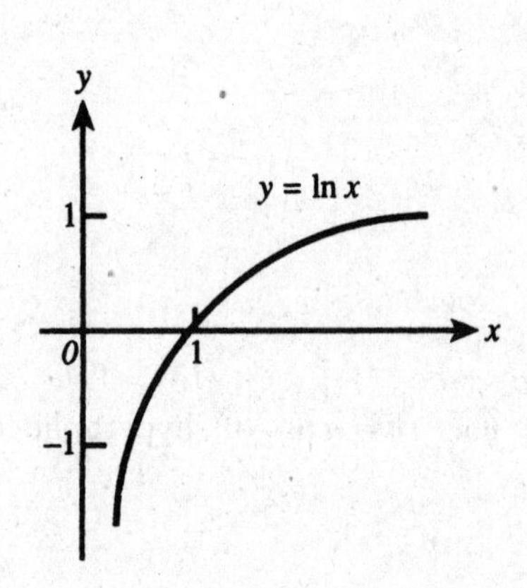

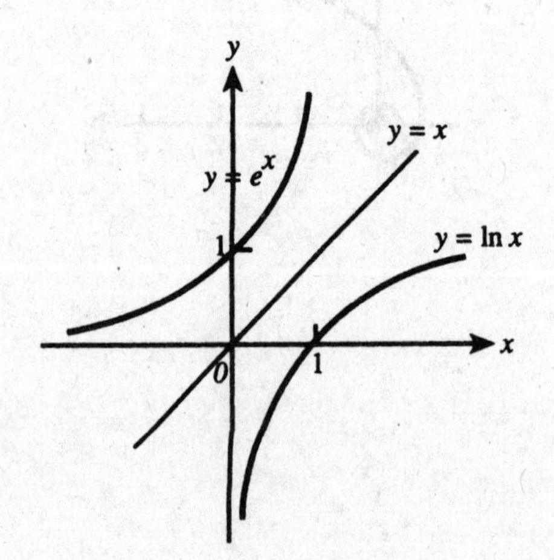

Index

E

F

G